T0142342

# Natural Computing Series

Series Editors: Thomas Bäck  Lila Kari

Natural Computing is one of the most exciting developments in computer science, and there is a growing consensus that it will become a major field in this century. This series includes monographs, textbooks, and state-of-the-art collections covering the whole spectrum of Natural Computing and ranging from theory to applications.

More information about this series at http://www.springer.com/series/4190

Alan J. Lockett

# General-Purpose Optimization Through Information Maximization

Alan J. Lockett
CS Disco Inc.
Austin, TX, USA

ISSN 1619-7127
Natural Computing Series
ISBN 978-3-662-62009-0 ISBN 978-3-662-62007-6 (eBook)
https://doi.org/10.1007/978-3-662-62007-6



This Springer imprint is published by the registered company Springer-Verlag GmbH, DE part of Springer Nature.
The registered company address is: Heidelberger Platz 3, 14197 Berlin, Germany

*Dedicated to my wife Alliene, who patiently and devotedly endured her husband's late night dalliance with abstract mathematics.*

# Preface

This book began out of observations upon reviewing the literature around Estimation of Distributions Algorithms (EDAs) for optimization and recognizing that there was nothing particularly special or different about these algorithms from other methods such as Genetic Algorithms, Evolution Strategies, and Particle Swarm Optimization. It seemed obvious to me that each of these could be formulated as a kind of conditional probability distribution in the same manner as EDAs, perhaps with some degeneracies incorporated. Furthermore, these conditional probabilities were themselves mathematical objects that could be compared and operated on as such, leading to the conclusion that many methods in evolutionary computation could be placed into a shared vector space that could be analyzed using techniques of functional analysis.

The core ideas of this book expanded from that concept, eventually coming to incorporate all iterative stochastic search methods, including gradient-based methods and many others. In the process of exploring these ideas, I found inspiration in the work of Michael Vose and others on Randomized Search Heuristics, which proceeds in roughly the same fashion as this book with a focus on evolutionary methods. The decision to include all iterative optimization methods and not just evolutionary methods was influenced by Wolpert and Macready's work on No Free Lunch (NFL) Theorems. Although these days No Free Lunch is not considered an exciting topic in machine learning and optimization and is typically acknowledged in an offhanded and trite manner as something that *everybody knows*, it was with NFL that the concept of analyzing all methods simultaneously was first developed, and key ideas that should be of broad, ongoing interest to the field, such as problem-solution alignment, were pioneered in the original NFL work. To the NFL literature I have added my own contributions in this book and elsewhere, but the NFL concept itself – that there are problem settings in which no algorithm can outperform others – is viewed as a useful stepping stone into a broader field of analysis that comes from developing a shared mathematical space for optimization algorithms.

The analysis of iterative stochastic optimization algorithms up to the chapter on NFL combined with the chapters on evolutionary annealing and neuroannealing constituted my Ph.D. thesis, advised by Risto Miikkulainen at the University of Texas and completed in 2012. The thesis included speculation that a given problem class formulated as a probability distribution would be solved optimally by a method that maximized an exploration-exploitation tradeoff that I have called the *Information-Maximization Principle*, which provided the title for this book. I subsequently realized and proved that this principle was an application of the Bellman equations to the problem of controlling an iterative optimization method, connecting optimization to control theory. In some ways, this realization reduced the value of the original speculation, because Bellman equations have been understood for some time already. But this is the nature of mathematical and scientific inquiry. The more you examine a problem, the more you appreciate and benefit from the insights of those who have worked on similar problems before, and the more you realize that many disparate fields are connected to each other in disparate and subtle ways.

This book is an expansion, correction, and clarification of the thesis that has taken several years of effort to accomplish. The intuitions of the thesis have held up under further analysis, but refining these intuitions required drawing from several branches of mathematics such as topology, probability theory, stochastic processes, and others in order to state and prove the theorems in this book. Since these fields are generally less familiar to the expected readers, I have taken pains to incorporate substantial background material in order to make the work as self-contained as possible; hopefully those already familiar with the mathematical background will not find this inclusion overly tedious.

The final insight included in this book is the realization that the basic mathematical objects developed to account for stochastic optimization, the trajectors and generators defined in this book, have applications far beyond optimization. The most general way I have found to think about these objects thus far is as stimulus-response systems, a concept which I have briefly included in the first chapter of the book and referenced sporadically throughout. The key intuition for this perspective comes from the Optimization Game, which has to do with far more than optimization and has been given its own chapter in this work. This chapter almost stands on its own and introduces a setting for analysis in its own right that is distinct from the chapters that precede it.

This book attempts to be thorough on the subjects it engages, but it is by no means complete. The parting chapter presents several points of departure that will hopefully be taken up by the readers of this book. Other directions for future work could be added still. For example, if I had this book to write again, I would have included an analysis based on the Wasserstein or Earth Mover distance, which has more desirably continuity characteristics than the total variation norm and would likely resolve some of the more difficult properties of the present analysis, especially concerning deterministic generators and trajectors. However, the scope of the work is already sufficiently broad, and this book has already been written, so an analysis based on the Wasserstein distance will have to be left to the future, perhaps by a reader of this book.

For my own part, it is my perception that each time I examine a piece of this work, I find another opportunity to make the analysis deeper and more general. I hope that the reader may also have this experience and that these ideas can inspire yet further ideas that will lead us to a better understanding of search, optimization, control, and game theory, along with other application domains of which I am as yet unaware.

This book has been a massive labor to produce, due mainly to the fertile theoretical ground it covers. As with any work of this scale, I have by no means accomplished the work alone, but am indebted to many fellow researchers who have reviewed and commented on the ideas at various stages of development. I mention specifically Jonathan Rowe, Per Kristian Lehre, and Kenneth De Jong, who provided specific feedback and useful discussion at critical times.

Above all, I wish to thank my wife, Alliene, who suffered many long evenings and weekends caring for five beautiful but demanding children by herself, and my thesis advisor Risto Miikkulainen, who deserves high praise for his careful encouragement, unflagging optimism, valuable criticisms, and unending enthusiasm.

It is my hope that this book will provide insights and inspiration to researchers and theorists in global optimization, machine learning, evolutionary computation, and control theory, especially to those whose leaning is of a mathematical or theoretical nature. And with this in mind, I submit this voluminous yet incomplete work for the reader's benefit. I hope that you will find enjoyment as you tread the many unexplored paths whose outlines are merely suggested in this book, and look forward to the many discoveries yet to be made out of these pages.

*Alan Lockett*
Georgetown, Texas, USA
January 1, 2020

# Contents

# Chapter 1
# Introduction

Computer programs lie at the center of modern applied mathematics. It is almost inconceivable to think of solving a serious practical problem in the modern age without the aid of a digital machine. One fascinating aspect of this observation is that these discrete programs are commonly used to simulate phenomena whose mathematical formulation is based on continuous spaces. Is the mismatch between *discrete programs* and *continuous mathematics* absolute? Can we imagine continuous spaces of programs? What would be the structure of such spaces, and how would they be constituted?

This text addresses these questions, proposing for the first time a functional analysis of program spaces focused through the lens of iterative optimization. This focus on optimization is appropriate because optimization is a fundamental problem that recurs across scientific disciplines and is pervasive in informatics research, from statistical machine learning to probabilistic models to reinforcement learning and beyond.

We will construct a variety of continuous spaces of optimization problems and solutions based on probabilistic selection over possible solution candidates, and we will examine the relationships between these spaces. Within this formalism, we will be able to shed new light on many existing results. An analysis of performance in optimization methods will lead to equivalence conditions for the No Free Lunch theorems, demonstrating that this classic result arises strictly from the assumption that, in a sense to be made precise, the class of problems to be solved possesses no predictable structure. When these conditions are not satisfied for a given class of problems, the question of how to best solve an optimization problem results in an infinite-dimensional variant of the standard Bellman equations whose solution will be called the Information Maximization Principle. From this principle, a new class of optimization methods called martingale optimization will be derived. Finally, an abstract setting will be derived in which problems of optimization and control reduce fundamentally to game theory, with one or more programs are pitted against each other.

But first, the end goal for this exploration needs to be made clear. We begin with a thought exploration of a system of paired *stimulus-response* processes that generate



A. J. Lockett, *General-Purpose Optimization Through Information Maximization*,
Natural Computing Series, https://doi.org/10.1007/978-3-662-62007-6_1

a type of game. Such systems can be used to model the behavior of algorithms in a range of applications including optimization, control, and game theory.

The language of stimulus and response is not used beyond this chapter, but is introduced here to provide an analogical framework within which to intuitively position what follows. Subsequent material will mostly limit itself to the context of optimization, but the exact formalisms to be developed for optimization apply verbatim to the stimulus-response systems outlined next, and this language accurately reflects the breadth of the approach in this book.

## 1.1 Stimulus-Response Systems

At the most abstract level, we consider a system of interacting stochastic processes that drive each other over time on trajectories through two separate spaces. The first process, the *stimulator*, selects a *stimulus* from a space $X$, the *stimulus space*, conditioned on a history of prior responses. The second process, the *responsor*, selects a *response* from a space $Y$, conditioned on a history of prior stimuli. The interaction of these processes over time generates a *Stimulus-Response system* or SR system. The following definition provides a preview of the kind of analysis developed over the course of this book.

**Definition 1.1.** Suppose $(X, \tau_X)$ and $(Y, \tau_Y)$ are Hausdorff topological spaces called the *stimulus* and *response* spaces, respectively. Let $\mathcal{T}$ be a strictly ordered space corresponding to time, and let $\mathcal{B}a\left[X^{\mathcal{T}} \times Y^{\mathcal{T}}\right]$ be the Baire $\sigma$-algebra over the joint product space of stimulus-response histories. Then a *Stimulus-Response* system, or SR system, is a joint probability distribution on the joint history space $X^{\mathcal{T}} \times Y^{\mathcal{T}}$, defined with respect to the Baire $\sigma$-algebra above.

This definition should be compared with Definition 17.1, which defines the Optimization Game, which is in fact an SR system for which the stimulator is an *optimization player* choosing search points to evaluate and the responsor is an *adversarial fitness* that evaluates those points.

The core feature of this definition is that an SR system is a joint probability distribution over the history of interactions between stimulus and response. One may think of the SR system as a generating a pair of trajectories, one in the stimulus space and one in the response space, representing the history of the stimulator and the responsor, respectively. The stimulator and responsor can be recovered from the SR system as conditional probability distributions over trajectories, each conditioned on the trajectory in the opposite space. Furthermore, these conditional trajectories can be decomposed via marginalization into step-by-step generators that correspond to a single step of an iterative algorithm.

It is also possible to proceed in the opposite direction, starting with an iterative algorithm and formulating generators. Subject to reasonable conditions, for any combination of generators for a stimulator and responsor, there is a unique SR system corresponding to them. Thus one can start with well-known iterative programs such as gradient descent, evolutionary algorithms, reinforcement learning, or turn-based games and construct an SR system for each application of these methods. This SR system is unique *almost everywhere*, that is, it is unique on all trajectories reachable within the application.

Due to this last caveat, it is more elegant from a mathematical point of view to start with the SR system and derive the generators, since otherwise one must worry about equivalence almost everywhere. From a practical point of view, however, it is easier to introduce generators first as abstract reflexes of computer programs, and then assemble them to construct SR systems. This text follows the more practical route. Chapters 3 to 8 construct and analyze the generators for specific search and optimization processes in order to make the conceptual framework clear. Chapters 9 to 11 extend these generators into trajectory-producing stimulators named *trajectors*. Chapters 12 and 13 introduce limited categories of responsors based on random objective functions. Only in Chapter 17 is the full generality of the SR system presented in the form of the Optimization Game, at which point it is then possible to present a top-down derivation of stimulators, responsors, and their generators.

Analyzing the SR system as a whole or as a composition of a stimulator and responsor has theoretical benefits. For example, as discussed in Chapters 12 and 17, the No Free Lunch theorem can be reduced to the claim that there is a responsor for which this joint distribution always factors, meaning that the response is independent of the stimulus. Indeed, SR systems can exhibit varying degrees of dependence between stimulators and responsors. The degree of dependence may be thought of as the *controllability* of the system, with No Free Lunch representing the uncontrollable extreme.

In the remainder of this book, we are primarily concerned with the SR system generated by the interaction of an optimization method and an optimization problem. As the book proceeds, the formalisms developed for optimization will be found useful for analyzing control, game theory, and potentially even general computational systems as well. The term SR system will only be used sporadically, but it should not be forgotten that it is the concept of stimulus and response within a stochastic environment that defines the full application of the methods to be developed in this book, for which search and optimization are just one reflex.

## 1.2 The Optimization Task

Nearly every problem to which human ingenuity is applied either consists of or contains an optimization task. When constructing a building, one wishes to use the least amount of materials and labor to achieve certain standards of size, quality, and content. When planning a trip, one seeks a route of transportation that minimizes distance and maximizes speed. When designing the layout for an integrated circuit, one desires an arrangement of components that minimizes both fabrication errors and surface area. In machine learning research, optimization is ubiquitous, with nearly every learning task being formalized as an optimization problem in which errors are to be minimized and performance is to be maximized. In this setting, learning and optimization are closely intertwined, and a general-purpose optimization method is required in order to implement a general-purpose learner. For this reason, optimization is a valuable place to start when searching for a continuous theory of programs.

An optimization task consists of at least three parts: a configurable system, a set of quantifiable objectives for the system, and potentially a set of observable environmental factors. A solution to an optimization problem prescribes a configuration (possibly as a function of the observable environment) such that the objectives attain their maximal or minimal values. The method of achieving these solutions is often limited by a set of available actions.

The set of admissible configurations is the *search space* or the *search domain*. The objectives for the system are considered as functions taking system configurations as input and producing the objective values as output. As such, in academic settings, optimization is almost always studied in terms of finding the minimal or maximal values of a computable function. The function is referred to as the *objective function* or the *fitness function* depending on the context. In the presence of multiple objectives, the objectives may be combined into a single metric, or each objective can be treated independently, termed *multi-objective optimization*.

Formally, let $X$ be the search domain, and suppose the objective function $f$ is some real-valued function over $X$, i.e. $f \in \mathbb{R}^X$. Optimization is formally considered to mean minimization, since a function $f$ can be maximized by minimizing $-f$. It is then further assumed that $f(x)$ is bounded below so that a minimum exists. Then the goal of the optimization task is to find one or more $x^* \in X$ such that $f$ takes its minimal value on $x^*$, i.e.

$$f(x^*) = \inf_{x \in X} f(x). \tag{1.1}$$

Many times, a set of constraints is provided that restricts the search domain to a subset $C \subseteq X$, and a minimal value for $f$ is sought from among the elements of the set $C$. This setting is termed *constrained optimization*. In this text, the constraints are assumed to be built into the space, so that $X = C$ from the discussion above, so that although problems are generally described in an unconstrained manner, it is possible with minor modifications to incorporate constrained optimization as well.

A simple example is in order. Consider the task of driving a car from a given location to a second fixed location through an urban setting. The car is the system, and its relevant configurable parts consist of the steering wheel, the gas pedal, and the brake. The car can be controlled by specifying at each point in time the rotational force on the steering wheel and the downward pressure on the gas pedal and the brake. In this case, the proper controls for the car depend on the state of the environment. The task of driving a car has several objectives. First, the car must arrive at the correct destination as quickly as possible. Secondly, the car must obey traffic laws, remaining within the appropriate lanes, stopping at red lights, maintaining appropriate speed, and signaling turns in advance. Thirdly, the car must avoid collisions with other vehicles, pedestrians, animals, and objects. A solution to the driving task specifies the force on the steering wheel, brakes, and accelerator as a function of the observed environmental state. The objective function in this case is typically a simulation environment or a real-world test in which the car is controlled by a proposed solution and its performance is measured in terms of the three main criteria above. The search domain is a space of functions that map the observations of the environment to driving decisions. A good solution safely drives the car from its starting point to its destination along an efficient route while following the traffic laws and avoiding collisions.

The example above can be mapped into a formal optimization problem by identifying the car's sensors as an array of $m$ real numbers in $\mathbb{R}^m$ and the three controls as an element of $\mathbb{R}^3$. The search domain $X$ consists of all functions from $\mathbb{R}^m$ to $\mathbb{R}^3$. For any controller $x \in X$, one may define three objectives $f_1$, $f_2$, and $f_3$, such that $f_1(x)$ records the distance from the desired destination at the end of the simulation, $f_2(x)$ counts the number of traffic law violations, and $f_3(x)$ indicates the risk of a collision or other catastrophic mistake over the course of a simulation run. Then a suitable objective function would be $f = \sum_{i=1}^{3} \alpha_i f_i$ with $\alpha_i > 0$, where the $\alpha_i$ balance the importance of each objective. A solution to this optimization task would output a controller $x^*$ for the vehicle such $f(x^*)$ is minimal, that is, such that the vehicle reaches its destination while obeying traffic laws and avoiding collisions.

An iterative optimization method or *optimizer* proposes a sequence of potential solutions to an optimization task, $x_1, x_2, x_3, \dots \subseteq X$. The quality of the solutions should increase as the method proceeds, e.g. $f(x_{100}) < f(x_1)$. Many optimization methods have been proposed and their effectiveness has been demonstrated in a variety of contexts. Some optimization problems can be solved analytically using derivatives or other means. Other problems can be effectively solved by iteratively following the gradient of the objective using methods such as Newton-Raphson [132] or conjugate-gradient descent [73]. When derivatives are not available, they can sometimes be estimated. On many practical problems, particularly those that involve complex simulations, precise estimates of the gradient are highly variable or unpredictable, or they might be too expensive to obtain. In addition, derivative-based methods are *local optimizers* that find a *local optimum* rather than the true *global optimum*; that is, derivatives can only be used to optimize in the neighborhood of a starting point. If an objective function is particularly bumpy or *multimodal*, as in Figure 1.1, then a derivative-based method must be restarted many times with different starting points, or a derivative-free method may be attempted.

Derivative-free methods use trial and error to locate the optimum. Direct search methods test every possible direction from current best solution and then move iteratively in the general direction of the unknown gradient [92]. Genetic algorithms mimic Darwinian evolution by maintaining a population of solutions that are combined and varied iteratively in a manner that prefers to keep solutions with a higher score on the fitness function [60]. Monte Carlo methods such as simulated annealing sample a special Markov chain that theoretically converges on the global optimum, though with many practical caveats [90]. More recent evolutionary algorithms are based on various natural analogies, from the flocking of geese [45] to the foraging of ants [41] or the functioning of the human immune system [48]. A more thorough review of existing optimization methods is provided in Chapter 2, but this brief sum-

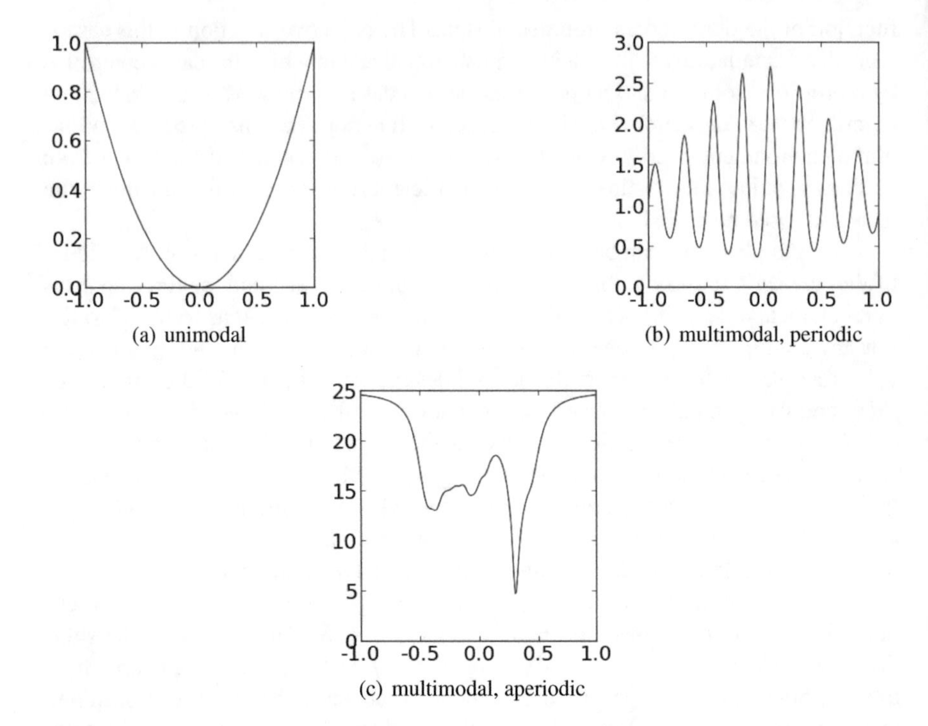


(a) unimodal

(b) multimodal, periodic

(c) multimodal, aperiodic

**Fig. 1.1** Three example objective functions. Figure 1.1(a) is *unimodal* and possesses a unique minimum, easily located analytically or by gradient methods. Figure 1.1(b) is *multimodal* but *periodic*. Gradient methods will fail, but the periodicity can be used to locate the optimum. Figure 1.1(c) is multimodal and irregularly structured. Such problems can be difficult to solve, particularly in high dimension.

mary gives the reader a sense of the overwhelming number of different approaches to the optimization problem. We will use the stimulus-response concept to organize all these methods with an analytic approach.

## 1.3 Spaces of Optimizers

In light of the variety of optimization methods, one may wonder whether a framework that unifies them will be so abstract that it has little practical meaning. A cursory study of these methods gives the first impression that the set of all optimizers for a particular search space is fundamentally discrete and unstructured, and that there is no apparent relationship between any two arbitrary optimizers. This impression will be dispelled by presenting a mathematical analysis that reveals to the contrary that the set of optimizers for a fixed search space is highly structured in mathematical terms. It can, in fact, be represented as a closed, convex set within a normed vector space with well-formed notions of distance, continuity, and limits.

In a finite search space with finitely many output values, the structure of the space is easy enough to understand. A search space is finite if there are only finitely many configurations or inputs to the system. Each of these inputs may be assigned a natural number, so that the inputs are number from 1 to $N$ for some $N < \infty$. The optimization task can be solved by testing each of these inputs in turn; once all inputs have been tested, the optimal input must be known.

An optimizer selects the order in which each possible input is tested. The optimizer may determine the order for later inputs based on the output values for earlier inputs. For instance, if the optimizer has proposed input number 3 as the first input to test, then it may choose to examine input number 4 as the second input to test if the objective value for 3 was negative but might instead choose input number 5 if the objective value for 3 was positive. The optimizer may randomize its choices, but for this example, assume that an optimizer makes only deterministic choices. Additionally, assume that the optimizer does not repeat itself. Since there are only finitely many inputs and outputs, there are only finitely many ways in which an optimizer may order the inputs for testing. Therefore, there are only finitely many optimizers on this space.

Simplifying even further, consider the subset of optimizers that do not depend on the output values at all. These optimizers merely specify at the outset an order in which the inputs will be tested. Suppose that there are only 10 system configurations. Then there are exactly $3,628,800$ such optimizers (i.e. deterministic optimizers that do not vary with output and do not repeat points). Each of these optimizers can be represented as a sequence of 10 numbers, such as:

$$1,2,5,7,8,9,6,3,4,10$$
$$1,2,7,5,8,9,6,3,4,10$$
$$5,7,1,3,4,9,8,10,2,6$$

It is plain to see that the first and second of these optimizers are more similar to each other than to the third optimizer. In fact, one can define a distance metric between any two optimizers in this subset by counting the minimal number of entries that must be swapped in order to convert one optimizer into another. The distance between the first and second optimizer above under this metric is one. The distance between the first and third optimizers is seven. The maximum distance between any two of these optimizers is nine. Far from having no structure at all, the set of output-independent, non-repeating optimizers on a finite space is at least a metric space.

Suppose that an optimizer is allowed to depend on the outcome of objective evaluations, but is still deterministic. Wolpert and Macready introduced a formalism for this setting in their paper on the No Free Lunch theorems for optimization [171]. An optimizer may be specified as a function that takes as input a finite sequence of pairs containing the input value and the corresponding objective evaluation and produces as output the identity of the next input to test. Since the outputs were specified to be finite, they may be numbered as well. There are only finitely many sequences of such pairs no longer than $N$, and thus an optimizer is defined by a table of input-output associations. For example, if there are $M$ outputs, numbered 1 to $M$, then an individual optimizer might look like Table 1.1.

**Table 1.1** Example of tabular representation of a deterministic optimizer

| evaluation history | next choice |
|---|---|
| $\emptyset$ | 1 |
| (1,1) | 3 |
| (1,2) | 6 |
| ... | ... |
| (1,M) | 7 |
| (1,1),(3,1) | 6 |
| (1,1),(3,2) | 2 |
| ... | ... |
| (1,M),(7,1) | 2 |
| ... | ... |
| (1,1),(3,2),(2,6),(5,4),...,(N,5) | 1 |
| ... | ... |

Notice that not all sequences need to be considered for a deterministic optimizer, but only those sequences that the optimizer will produce on some objective function. So only those sequences that begin with input 1 are valid for the optimizer above, since optimizer always tests input 1 first when presented with the empty sequence, $\emptyset$.

The number of entries in such tables is bounded above by $T = N!NM^N$ since the inputs cannot repeat but the outputs can. The extra factor of $N$ reflects the fact that input sequences can have length 1 to $N$; a tighter bound is possible but is unnecessary here. Thus even when optimizers are allowed to consider objective outcomes, the number of deterministic optimizers is finite and bounded above by $T^N$, allowing each entry to take on all $N$ possible outputs.

Most importantly, as Wolpert and Macready noticed, all non-repeating deterministic optimizers on any finite search space can be represented in this table format, regardless of the rationale that led to its formulation. It does not matter whether the algorithm is described by a search heuristic or a biological analogy. Once it is reduced to a table like the one above, it is just another optimizer in the space. Furthermore, one can characterize the distance between any two optimizers as the number of edits that must be made to the table for the first optimizer to convert it into the second. It is also reasonable to speculate that two optimizers with similar tables will perform similarly on the same objective. Further, if the search domain has a known topology, then optimizers may be compared even more meaningfully by incorporating the topological structure over the outputs into the distance metric over optimizers.

The analysis in this section provides an example of the kind of structure that may be observed in optimizer spaces. Fundamentally, an optimizer is a means of selecting which input points will be evaluated next given the inputs evaluated so far and the outputs that resulted. This selection mechanism is precisely what is abstracted as a generator in the stimulus-response framework. By analyzing the outcome of such selection mechanisms independent of the descriptions and procedures used to obtain them, it is possible to compare any two optimizers on practically any search domain and objective through their generators. The study of optimizers in finite spaces is continued in Chapter 3.

It should be evident that even with substantial restrictions, such as a finite search space and deterministic, non-repeating optimizers, a general formalization of optimization methods is an ambitious project. Including randomized optimizers is not difficult. Every run of a stochastic optimizer produces a single input-output sequence, and thus a stochastic optimizer may be regarded as a distribution over the deterministic tables described above, that is, as a (very long) probability vector. However, allowing repetition or infinite search domains requires more powerful mathematical tools. Non-repeating optimizers may produce infinite sequences of inputs without observing the objective value of all inputs, and thus the tables above may require infinitely many entries to represent them. And infinite spaces can certainly not be studied by reasoning about lists and tables.

It is important to justify for why infinite spaces deserve to be studied at all. One might argue that only finite representations are computable, and so the table representations above should suffice for formal analysis. While it is true that digital computers can only represent and manipulate finite objects, many optimizers are designed to search mathematical spaces that are formally infinite. It does not make sense to limit the formal analysis of these optimizers to their finite, computable encodings. Ultimately there are two reasons to study infinite spaces directly. The first reason is that by considering the native topology of the problem, one avoids distortions that may be introduced by projecting the topology into a finite approximation. Secondly, an analysis that accounts for infinite spaces is in some ways simpler than a finite analysis because of the availability of analytic tools developed by the mathematical disciplines such as topology, measure theory, and abstract algebra. The relevant mathematical tools for reasoning about infinite spaces are reviewed in Chapters 4 and 5 along with initial applications to optimization.

An analysis of optimization methods on topological spaces is undertaken in Chapter 6, presenting an abstract treatment of optimization that allows for the simultaneous analysis of general optimizer spaces independent of procedural descriptions and without substantial simplifying assumptions (e.g. finiteness, lack of repetition, etc.). It applies the concepts above to infinite spaces, defines the terms that will be used, and lays the groundwork for subsequent analysis of optimization as an SR system. Chapter 7 applies this framework to population-based optimizers such as genetic algorithms, and Chapter 8 begins to study the continuity properties of these generators. What will be accomplished with this formalization is discussed next.

## 1.4 Optimizer Performance and No Free Lunch

Ordinarily, one is not interested in the similarity of optimizers in terms of how they select inputs. Rather, one seeks an optimization procedure that prioritizes input points with high quality. Whenever an optimizer proposes an input configuration to evaluate, the optimizer makes an error that can be quantified as the difference between the optimal output value and the output value for the proposed input. A good optimizer performs well on a problem if it minimizes its errors very quickly. A performance criterion specifies what kinds of errors are salient and on what time scale the errors are to be considered. Formal definitions of performance criteria are given in Chapter 10. These definitions are accompanied by experimental results for a variety of the optimization methods introduced in Chapters 2 and 7.

The obvious next question is whether there is some optimizer that outperforms all the others, not just experimentally, but theoretically. No Free Lunch theorems have already been mentioned [171, 144, 136, 80]. These theorems demonstrate various optimization settings in which all optimizers have the same average performance. Most of these theorems actually state something stronger still; they demonstrate ways of averaging over objective functions such that the observed fitness values do not differ at all between optimization methods [136]. It has long been known that No Free Lunch is a kind of high-dimensional symmetry [79].

These symmetry considerations limit the power of No Free Lunch. Many believe that No Free Lunch means that there cannot be any sort of universal problem solver. More specifically, it is often stated that for any algorithm, there is a problem on which it performs poorly and conversely, for any problem there is an algorithm that performs well on it. Neither of these statements capture the meaning of No Free Lunch results accurately. No Free Lunch is really about how one chooses problems to solve, and, as will be discussed in Chapter 12, there is always a way to choose problems such that all optimization methods obtain the same results in a probabilistic sense. Within the stimulus-response framework, the "way of choosing problems" is to sample them from a probability distribution, and the conditions for No Free Lunch imply that trajectories of objective values sampled from such a distribution are not self-correlated, that is, they are *path independent* as proven in Theorem 12.7. Such objective functions are inherently unpredictable, and No Free Lunch follows from this fact.

Consider what it means for a random optimization problem to be path independent. When optimizing such an objective, prior evaluations are useless for guessing the outcome of future evaluations. Thus the order in which input points are evaluated is irrelevant. It is impossible for an optimizer to learn anything about such a problem, because the problem reveals nothing about itself. In order to accomplish this feat, the random procedure must scramble the relationship between inputs and outputs to the point that the relationship is fundamentally incompressible. Thus there can be no rule to represent the selected test objective that is smaller than an enumeration of all input-output pairs. The world of No Free Lunch is preeminently unstructured and unlearnable; it is the fuzz between the channels on an old television set.

In small, finite search spaces, the assumption of path independence may make sense. There is no obvious way to compare categorical values with each other, and so one may as well presume that they are arbitrarily interchangeable. However, as soon as the inputs or outputs take on some kind of local structure, the assumption of path independence falls apart. In real problems, concepts such as locality, periodicity, and regularity are important. As soon as such conditions hold, No Free Lunch fails. A simple example of a random test procedure that violates No Free Lunch is the standard Brownian Motion, commonly used to model physical processes involving the diffusion of particles within a substrate, such as the expansion of coffee grounds in water. The position of a single particle at each time step forms an objective that has unbounded variation but is locally predictable with high probability. Another example of such a random test procedure is Solomonoff's universal prior, which prefers functions that are easily computable over functions that are difficult to compute [149]. A random test procedure can be quite general without being subject to No Free Lunch. These facts are detailed in Chapter 13.

If one views the universe as a random test procedure generating a variety of objective functions, then one does not expect to encounter problems in which the outcome in one situation is utterly unpredictable on the basis of previous outcomes in similar situations. This expectation of regularity is not merely utilitarian. It is not sufficient to object that humans expect the world to behave predictably because they have no choice but to do so. To make such an objection is to suggest that every decision made by humans that succeeds is purely serendipitous. While it is true that many of the more complex aspects of human life are subject to severe variability and unpredictability, it is nonetheless the case that many of the everyday aspects of life are highly predictable. When a man takes a step, his foot does not fall through the ground in front of him as it does through the air. The sun proceeds regularly through the sky, and when it sets at night, the time of its rising may be predicted precisely. Apple trees do not produce peaches, and a peach seed will not grow into an apple tree. In these and in many other situations, prior experience is a strong predictor of future outcomes. The very experience of humans as learning machines disproves the claim that the world is unlearnable, so we may reasonably assume that No Free Lunch does not apply to practical problems.

Thus even as this thesis reaffirms No Free Lunch in a wider sphere, it rejects the popular interpretation of No Free Lunch that claims that any particular optimizer is just as good as any other. This claim is categorically false in the context of real-world problems. It is also false on computable problems on infinite spaces, not just real-world problems, since such problems must have finite representations to be computable. The existence of a finite representation is a form of regularity that invalidates No Free Lunch. In sum, some optimizers are better than others when averaged over all possible problems of interest. General-purpose learners exist.

Although it is possible for some optimizer to have better average performance on practical problems, such an optimizer is by no means best on any specific problem. The concept of alignment between problems and methods propounded Wolpert and Macready [171] is a theoretical and experimental fact. As the random procedure for generating test objectives changes, the optimizer with the best average performance changes as well. This notion of alignment places the space of optimizers into a structured relationship with the space of random test procedures as discussed further in Chapter 17.

Despite the claims above, specific solutions to specific problems will almost always perform better than general solutions that work on many problems. This fact has often been raised as an objection to the search for general-purpose optimizers. This objection ignores the effort that human researchers put into finding such specific solutions. In practice, specific solutions are usually identified as the result of a general problem-solving methodology that relies on human learning capabilities. Although specific problem-solving strategies are to be preferred for specific problem classes, general-purpose learners are still necessary to discover such strategies.

## 1.5 Information-Maximizing Optimization

One may regard the optimization process as a zero-sum game in the sense of von Neumann's Game Theory [117]. One player selects an optimizer, and his adversary selects a random procedure for generating objectives. As is proven in Chapter 13, this game is biased in favor of the second player, who can always choose a selection procedure subject to No Free Lunch, since such a selection procedure always exists. If the strategy of the second player is fixed, however, then the first player must select the best optimizer for a fixed strategy. The best choice the first player can make is to play a strategy that minimizes its error. One way to minimize error is to utilize the conditional expectation of the objective function given the outcome of previous objective evaluations. The conditional expectation estimates the true objective function with minimal variability, and variability is directly correlated with optimizer errors. The conditional expectation may or may not be computable in any particular case. If it is not computable, then an approximation may be used. This sort of approach can be viewed as an information-maximizing approach, where points are to be selected in a manner that minimizes the variability of optimizer errors. There is reason to speculate that the optimal optimizer pursues a strategy that maximizes its use of available information and structures its search to improve its access to useful information.

Following this line of thought, this book proposes a new method named *evolutionary annealing* in Chapter 14. Evolutionary annealing is an efficiently computable method for generating strategies that are roughly based on the conditional expectation of the objective function given prior evaluations. Like simulated annealing, evolutionary annealing approximates samples from an increasingly sharp Boltzmann distribution, asymptotically focusing on the global optima. Procedurally, evolutionary annealing resembles an evolutionary algorithm, since it proceeds in phases of selection and variation. Evolutionary annealing selects previously observed points probabilistically in proportion to their fitness in a way that asymptotically samples from the Boltzmann distribution. Then, the selected point is randomly altered to produce a new evaluation point. Evolutionary annealing is provably convergent to the global optimum under certain conditions. The proof is based on a martingale analysis that shows that the global optima become increasingly likely as the information about the objective functions is refined. Experimentally, evolutionary annealing compares favorably with other common optimization methods in a Euclidean search space (Chapter 15), based on the performance criteria presented in Chapter 10.

Evolutionary annealing is a template for new optimizers, and can be applied to arbitrary measure spaces; one needs only to specify the mechanism for generating new points from previously observed ones (in evolutionary terms, the mutation process). To demonstrate this versatility, evolutionary annealing is used to develop a novel procedure for learning artificial neural networks in Chapter 16. Artificial neural networks are parameterized functions representing a network of artificial neurons [71]. The artificial neurons are connected to each other by artificial synapses that are represented by a single real number termed the *weight* of the connection. An artificial neuron computes a function by taking a weighted sum of its input values and passing the sum through a nonlinear squashing function. The network as a whole computes a function by treating the output of a subset of neurons as the output of the function. Arbitrarily complex functions can be represented by wiring sufficiently many neurons together in different ways [38, 146]. Because a neural network computes a function, the space of neural networks can be used as a proxy to search for dynamic control functions, such as those needed to solve the driving task presented in Section 1.2 above.

In Chapter 16, neuroannealing applies evolutionary annealing to the task of learning a neural network to solve control problems. Neuroannealing compares favorably with other methods for training neural networks, and solves some tasks that require complex networks more effectively than previous methods.

In the final analysis, however, choosing a good optimization method requires an understanding of the particular optimization problem to be solved. Evolutionary annealing is an interesting new optimization method based on thorough use of available information. But it is still not the best choice for every optimization problem. Thus the broader contribution of this book is to provide tools that can be used to assess which methods are the proper methods to use for a particular problem, intuitively, theoretically, and experimentally.

## 1.6 Guide to the Reader

A general mathematical analysis of stochastic optimization methods as undertaken in this book requires mathematical tools that may be unfamiliar within the artificial intelligence and machine learning communities. Chapters 4 and 5 to provide substantial mathematical background for the formulae and proofs that follow, but due to the breadth of the relevant fields, these reviews cannot be complete. Basic familiarity with limits, probabilities, and set theory is assumed. Most importantly, the formalization of optimizers that is undertaken here is built on top of measure theory [66, 33] and functional analysis [21, 7], with some elements of topology [114], as well as probability theory, martingales, and stochastic processes [26, 103, 32, 86]. Knowledge in these areas will without doubt be useful to the reader, though significant effort has been made limit the need for prior study in these fields.

In order to prevent this text from becoming a sequence of impenetrable formulae and abstruse theoretical pontification, the definitions, theorems, propositions, and proofs have been infused with connective narrative that should clarify the intent and significance of the more mathematical portions of the text. To a large degree, it should be possible to obtain a workable sense of what has been presented by reading the narrative sections while skipping the proofs and much of the mathematical detail.

Those familiar with the subject matter who wish to skip directly the most significant contributions of this book should read the following sections. Chapter 3 provides a direct intuition of how generators and trajectors operate in finite spaces; this intuition should aid comprehension of the formalism in more complex spaces. Chapter 6 intorduces the topological version of these objects and characterizes them as vectors in a Banach space, while also demonstrating how some concrete optimization methods are represented in this space. Performance criteria are defined at the beginning of Chapter 10, and Chapter 11 describes the experimental performance of a variety of popular optimization methods. The formal proofs of No Free Lunch and the characterization of function priors subject to No Free Lunch are found Chapters 12 and 17, specifically as Theorems 12.7 and 17.8. The implications of these theorems are expounded in Chapter 13, which also introduces the information-maximization principle as a means of identifying the optimal optimizer for a particular function prior. The validity of information-maximization is established based on Bellman equations in Theorem 13.3, which is reprised in the context of the optimization game as Theorem 17.3. The basic evolutionary annealing algorithm is presented in Chapter 14, and Chapter 15 presents an experimental analysis of its performance in Euclidean space. These chapters form the core material of the book.

The other chapters contain material that, while significant, may be of less interest to particular readers. Chapter 7 discusses how particular evolutionary algorithms fit into the formal framework. Chapter 8 provides tools to aid in determining when optimizers are continuous, demonstrated by proving the conditions for continuity in existing optimization methods. This material is crucial to the proof in Chapter 10 that optimizer performance is continuous as the objective function changes, but is not otherwise used later in the text. Chapter 9 discusses the relationship between the sequence of points generated by an optimizer and the decisions made by the optimizer at each time step. This chapter also reviews aspects of stochastic processes that are needed for the proofs of No Free Lunch in Chapter 12 and formulates certain equations that are referenced repeatedly in later chapters. Chapter 16 shows how evolutionary annealing can be applied to train neural networks.

With these guidelines in mind, the reader will hopefully discover in this book a new way of thinking about optimization methods that has the potential to bridge the divide between advocates of different optimization methods and to enable a proper assessment of the value of each method.

## 1.7 Conclusion

Let us conclude with a brief review of the salient points of this book. One might think of the space of optimizers as a large, unexplored territory with pockets of civilization representing well-known and deeply studied forms of optimization, such as gradient, Monte Carlo, or evolutionary methods. However, the space of optimizers is at least as large as the space of objective functions, and many of the unknown and unstudied optimizers may prove to have practical uses. Although the No Free Lunch theorems place some bounds on the degree to which different optimizers can be compared with each other, these bounds are weaker than has been supposed. It is thus possible to develop a rich theory of optimizer performance.

The currency of this unexplored land is information – prior information about the function being optimized and information obtained from evaluating the objective. The best computable optimizer for an unknown objective is one that fully utilizes all sources of information to exclude incorrect objectives. Function evaluations provide a source of increasing information, evoking the idea of a martingale, a stochastic process of constant mean with a resolution that increases with the available information. This book proposes evolutionary annealing, a martingale-driven stochastic optimizer, as an example of such a method. Evolutionary annealing is established theoretically and demonstrated to work well on several optimization problems in real vectors and neural networks. More importantly, evolutionary annealing is a new type of optimization method that is typologically different from existing optimization methods but that arises organically from a functional analysis of the space of optimizers.

As the number of optimization methods proliferates, it is increasingly important to provide a theoretical structure within which these methods can be organized and meaningfully compared. The optimization game provides such a structure, and this structure can be characterized as a stimulus-response system within which the optimizer decisions serve as a stimulus evoking a response in the form of objective evaluations. This conceptualization of optimization as an instance of a stimulus-response system has broader applications beyond search and optimization, and it is anticipated that analyses of this type will become increasingly important and useful in the years to come.

Finally, the theoretical work in this book establishes a case for believing that real-world problems can be approached by some form of general-purpose problem-solving such as that employed by humans in scientific and engineering endeavors; the information-maximizing approach points the way at least towards the intuitive structure of such a problem-solving method.

This book is a modest attempt at the following three goals: (1) to provide a general framework and terminology for analyzing the class of iterative stochastic optimization algorithms as an instance of a stimulus-response system; (2) to analyze optimization performance in a wide variety of scenarios in the pursuit of understanding how to solve general problems; and (3) to use this framework in order to propose improved optimization methods on a principled basis. The discussion will remain primarily at the theoretical level throughout, although three chapters are centered on providing direct experimental results to demonstrate performance criteria and to establish the efficacy of evolutionary annealing and neuroannealing. These experiments notwithstanding, the focus will be on the elegance with which a wide range of optimization methods can be compared as stimulus-response systems and on the surprising relationships that exist between them.

With this summary in mind, after a historical interlude in Chapter 2, the following chapters develop the basic theory of generators and trajectors suggested in Section 1.1. It is hoped that the definitions and formalisms herein will aid the reader in identifying the similarities and differences between the wide variety of optimization methods that now exist. It is further expected that the constructs that follow will be useful for directing future research in new and profitable directions.

# Chapter 2
# Review of Optimization Methods

Modern optimization methods of optimization originated in the seventeenth century with the discovery of the calculus. Until the advent of the digital computer, however, analytic solutions and fast-converging iterative methods were the only practical means of performing optimization. The introduction and proliferation of computing technologies widened both the scope and the number of optimization problems. Nearly all of the optimization methods that will be described in this chapter were developed after 1950, when large research laboratories first acquired programmable computers. The vast majority of common optimization methods were developed after the advent of the personal computer around 1980. Thus a history of optimization methods is necessarily a history of computational methods in optimization, since most of the problems and solutions described in this chapter could not be seriously posed or tested without the computing machinery that is widely available today.

## 2.1 Gradients and Beyond

This chapter presents a brief survey of the primary areas of research in optimization. The best-known methods are based on following the derivatives of an objective. These *gradient methods* or *Newton methods* converge quickly and accurately on convex, unimodal objectives. In practical terms, however, gradient methods have at least two important failure modes when used directly. The first occurs when the gradient must be estimated from noisy data, for example, when the objective function is computed by a stochastic or chaotic simulation. In this case, the noise can hide the gradient signal, preventing convergence. The second failure mode occurs when the objective function is non-convex, or, more generally, when the objective function contains multiple local optima. In either case, gradient methods may converge to a suboptimal point, which can defeat the purpose of the optimization if the objective contains many undesirable local optima.

Many if not most important problems are non-convex. In this case, gradient methods alone may be inadequate to solve an optimization problem. One could run gra-



A. J. Lockett, *General-Purpose Optimization Through Information Maximization*,
Natural Computing Series, https://doi.org/10.1007/978-3-662-62007-6_2

dient methods multiple times from random starting points in hopes finding a quality local optimum, but then the question arises of how random points should be chosen. Depending on the problem, one may not have to run the gradient method many times in order to succeed. Yet in other problems, the desirable optima may be rare or have highly constrained properties. In these cases, the success of many randomized trials of gradient method will depend on the quality of the method for selecting starting points.

In the final analysis, there is a heuristic quality to gradient methods. This point cannot be emphasized enough. In the chapters that follow, we will encounter many heuristic optimization methods that seek the global optimum. These methods are often disparaged by the proponents of gradient methods as inefficient, blind search. And yet gradient methods are subject to heuristic considerations as well; one must select a starting point and a step size. For many problems, there are reasonable ways to choose these quantities, but for many other problems these choices are arbitrary. Even then, gradient methods assume the objective is smoothly differentiable, which it often is not.

The earliest set of computational methods to move beyond analytic gradients have acquired the name of *direct search*. Direct search methods are organized like gradient methods in that they first choose a starting point and then search all possible directions to find the direction of greatest improvement. Such methods can provide convergence guarantees on differentiable objectives, for example, by searching locally on an increasingly refined grid over the search space. These methods ultimately follow the gradient without estimating it, but at an exponential cost in terms of the number of objective evaluations.

In the 1970's and 1980's, researchers increasingly developed stochastic exploration strategies for optimization that avoid gradients entirely. For instance, simulated annealing embeds the optimization problem inside of a sequence of probability distributions that are constructed to asymptotically favor the optima. Simulated annealing still explores the search space one point at a time, but unlike gradient methods and direct search, the point is allowed to explore regions with lower objective values, permitting this method to cross hills and valleys in the objective function in search of the true global optimum. Monte Carlo optimization methods have developed further since that time, many under the guise of evolutionary computation, as discussed below.

Many of the earlier descriptions implicitly assumed that the space being searched was a continuous space. A large number of problems are in fact discrete, and many of these can be represented as a search over a graph structure. This field is known as *combinatorial optimization*. Many important optimization methods are applied specifically to these problems, such as greedy hill-climbing, simulated annealing, and genetic algorithms. Importantly, such problems are not amenable to gradient methods but can be solved by Monte Carlo optimization methods or evolutionary methods.

One of the most celebrated approaches develops a parameterized model that solves a problem, and then uses gradient-based methods to optimize the model parameters. This methodology makes it possible to optimize an objective through a continuous, smoothly differentiable proxy. Methods in this category include back-propagation in neural networks and expectation maximization in graphical models; each of these can be applied to both continuous and discrete problems with an appropriate model.

Evolutionary algorithms represent a stochastic exploration approach to optimization that seeks inspiration from biological processes and analogies. Darwinian evolution was the primary motivation for early research in this field; later work branched into a variety of biological analogies under the moniker *natural computation*. In the past two decades, rigorous mathematical explanations of the core evolutionary algorithms have been developed that make it possible to assess the capabilities and limits of evolutionary methods. These analyses have also pushed the evolutionary computation community into two distinct camps: those who focus on simulated and artificial biology as a testbed for computational innovation and creativity, and those primarily interested in function optimization. The research of the latter group has produced mathematically explicit, quasi-evolutionary methods that quintessentially represent a transition towards more mathematical representations: estimation of distribution algorithms, natural evolution strategies, and differential evolution. Stripped of their pseudo-biological analogies, many of these methods are in fact merely instances of Monte Carlo optimization. The present text analyzes these methods as dynamical systems with a goal of explaining evolutionary computation in terms of their mathematical properties, while recognizing that the study of computational creativity is a separate topic that is interesting in its own right.

Many other optimization scenarios and approaches are possible that are not explicitly considered in this text, including componential search domains such as boolean clause satisfaction (SAT). In these problems, the structure of the domain favors approaches that break the problem down into parts that can be independently analyzed. Such decompositions are problem-dependent. The analytic techniques in this text do apply in such problems, but we do not address them further to avoid getting bogged down in the description of specific problems.

Instead, the methods that are described here are chosen to represent what has been termed *black-box optimization*, in which little or no problem substructure is available to the optimizer up front. On closer inspection, one finds that this distinction is somewhat artificial, since information about the objective can be embedded into most successful black-box methods. Certainly, that approach can be used to shoehorn specific problems into the analytic framework developed hereafter. However, to save space and to promote clarity, this text will focus on the tradition of black-box methods.

## 2.2 Gradient Methods

Gradient-based optimization methods have a long history and continue to be widely used due to their fast convergence to accurate local optima on continuous problems.

This section reviews the origins of gradient-based optimization, leading up to a discussion of its modern variants.

### 2.2.1 Early Gradient Methods

Early optimization methods focused on real-valued functions of one or more real variables. The first published work addressing this problem is Pierre de Fermat's *Methodus ad disquirendam maximam et minima*, written in 1638, in which Fermat presented an analytic method for locating the minima and maxima of a function that corresponds to what is now termed the *first derivative test* [49]. Given a real function over an interval of the real line, the extrema must either lie at the boundary, at the non-differentiable points, or at the points where the function's derivative is zero, i.e. $f'(x) = 0$. If these points are few in number, then the function can be evaluated at each of them to identify the true maximum or minimum. The *second derivative test* provides a means of determining whether internal points are minimal or maximal. Using these tests, the global extrema of a function can be determined analytically for many important objectives. However, the method assumes that the non-differential points are identifiable and few in number, and it requires that the objective function be stated in closed form as an equation. Most importantly, one must be able to locate the zeros of the derivative, a difficult task even for many equations easily stated in closed form.

Fermat's method can also be extended to functions of several variables, but the restriction to intervals limits its applicability. The introduction of Lagrangian multipliers in the eighteenth century provided a means for enforcing more complex constraints [97]. The modernized refinement of Lagrange's method, based on the Karush-Kuhn-Tucker conditions, remains an important technique in constrained optimization [87, 95].

### 2.2.2 Newton's Method

By 1669, Isaac Newton had discovered an iterative method for locating the zeros of a real function, now known as *Newton's method* or the *Newton-Raphson method*. Given a continuous real function $f(x)$ and a starting point $x_0$, the sequence $(x_n)$ defined recursively by

$$x_{n+1} = x_n - \frac{f(x_n)}{f'(x_n)}$$

converges to a root of $f$, i.e. $f(\lim x_n) = 0$. Building on the derivative test methods of Fermat, this result implies that the sequence

$$x_{n+1} = x_n - \frac{f'(x_n)}{f''(x_n)} \tag{2.1}$$

converges to an extremal point of $f$ [163]. Equation 2.1 is not actually due to Newton or Raphson. Newton devised an equivalent derivation as a sequence of polynomials, and it was Raphson who presented the scheme as an iterative search along the real line [132]. Surprisingly, neither Newton nor Raphson recognized the relationship of the method to the calculus, but instead used the method only on polynomials; the generalization to arbitrary functions in Equation 2.1 was not recognized until later in the eighteenth century.

When it applies, the Newton-Raphson method converges quickly. The rate of convergence is quadratic, that is, the distance from the iterate $x_n$ to the local optimum $x$ is inversely proportional to $n^2$. Unfortunately, pure Newton-Raphson has limited applicability. The objective function must have a continuous first derivative and a finite, nonzero second derivative. The starting point $x_0$ must be sufficiently close to the extremum $x$, and if there are multiple local optima close together, then convergence will be slower. In some cases, the iterates may enter a limit cycle. Finally, only a local optimum close to the starting point $x_0$ will be found, and there is no way to choose the starting point $x_0$ without analyzing the objective function or sampling from it. If the derivatives are not available in closed form, they may be estimated by sampling points near $x_n$ and applying the *finite difference* method. The approximation of the derivative with finite differences is termed the *secant method*; a version of the secant method, the Rule of the Double False Position, dates back to ancient India and China [119]. The secant method has a linear convergence rate rather than quadratic.

Newton-Raphson can be generalized to real functions of more than one variable. In this case, the gradient $\nabla f = \left(\frac{\partial f}{\partial x_1}, \ldots, \frac{\partial f}{\partial x_n}\right)$ and the Hessian matrix $\nabla^2 f = \left[\frac{\partial^2 f}{\partial x_i \partial x_j}\right]$ must be computed. Then the iteration is given by

$$x_{n+1} = x_n - \eta_n \left[\nabla^2 f(x_n)\right]^{-1} \nabla f(x_n) \tag{2.2}$$

where $\eta_n > 0$ has been introduced as a time-varying *step size* or *learning rate* to aid convergence as described in the discussion of line search below. In the multi-dimensional version, the gradient functions must each be Lipschitz continuous, and the Hessian must be invertible. If the conditions are satisfied, there is a neighborhood of each local optimum such that the generalized Newton's method converges to that local optimum for all starting points contained within the neighborhood.

Newton's method is one of the most effective known optimization methods, but it applies very narrowly due to the strict conditions of continuous differentiability and an invertible Hessian. Various approximations relax some of these requirements and replace the Hessian and/or the gradient with approximations and simplifications. These approximations were primarily developed during the mid-twentieth century and constitute the most popular and widely known optimization methods today. Of these, the most significant are *gradient descent* and *line search*.

### 2.2.3 Gradient Descent

Equation 2.2 is difficult to compute because of the Hessian. However, an effective method can be developed by omitting the Hessian. Given a point $x_n$, the Taylor expansion of $f$ around $x_n$ is given loosely by

$$f(x_n + \eta_n) = f(x_n) + \eta_n \nabla f(x_n) + o(\eta_n)$$

where, as usual, $o(\eta_n)$ indicates a term that is asymptotically obliterated by $\eta_n$ (i.e. $o(\eta_n)/\eta_n \to 0$). If the step size $\eta_n$ is small, then $\eta_n \nabla f(x)$ is much larger than $o(\eta_n) \approx \eta_n^2 ||\nabla^2 f(x_n)||^2$. In this case the final term can be ignored with minimal error, and the iterates can be computed by

$$x_{n+1} = x_n - \eta_n \nabla f(x),$$

where the step size becomes smaller as $n$ increases. The sequence $x_n$ defined by this iteration typically converges to a local minimum of $f$; a local maximum can be found by reversing the sign of the gradient term. Because this method follows only the gradient and ignores the curvature of the objective, it is known as *gradient descent* or *steepest descent*. As a result of ignoring second-order information, the step size $\eta_n$ must be small in order for gradient descent to succeed, and its convergence rate is consequently slower. Nevertheless, gradient descent is typically easy to implement and avoids the calculation of the Hessian.

Gradient descent has three common failure modes. First, the step size can be too large, causing divergent oscillation away from the optimum. This error can be avoided by choosing the step size $\eta_n$ to respect the Armijo (or Armijo-Goldstein) condition,

$$f(x_n + \eta_n \nabla x) \leq f(x_n) + c_1 \eta_n ||\nabla f(x_n)||^2,$$

where $||z||^2 = \sum_i z_i^2$ is the square of the standard Euclidean norm [8]. If the step size decreases too quickly, then the iterates can converge before reaching a local minimum. This situation can be prevented by observing the Wolfe condition,

$$||\nabla f(x_n + \eta_n \nabla x)||^2 \geq c_2 ||\nabla f(x_n)||^2,$$

with $0 < c_1 < c_2 < 1$ [170].

As a different type of failure, gradient descent may converge to a shallow local optimum rather than a deeper one nearby. A common approach to avoid this problem is to introduce a momentum factor $\mu$ and set

$$x_{n+1} = x_n - \eta_n \nabla f(x) + \mu \Delta_n$$

where $\Delta_n \equiv x_n - x_{n-1}$ [126]. The value $\mu = 0.9$ is standard. This method is often justified by the analogy of rolling a ball over the function surface. The momentum of the ball allows it to roll over small depressions, so that it only gets trapped in a large hole. In certain situations, however, the momentum factor may backfire, causing the iterates to pass over the attraction basin of the global minimum and into that of a local minimum.

### 2.2.4 Line Search

Rather than eliminate the Hessian entirely from Equation 2.2, one can replace it with a more manageable matrix. This generalization of Newton's method is known as *line search*, and it is defined by the equation

$$x_{n+1} = x_n - \eta_n B_n^{-1} \nabla f(x_n)$$

where $B_n$ is a positive definite matrix. In this case, the quantity

$$d_n = -B_n^{-1} \nabla f(x_n)$$

is termed a *descent direction*. The sequence $(x_n)$ will still converge to a local optimum of $x$ provided that the inequality

$$-\nabla f(x_n)^T d_n > 0$$

holds. This inequality guarantees that $x_{n+1}$ moves towards the optimum in general (i.e., the descent direction is not orthogonal to the gradient and does not move against it). Generalized versions of the Armijo and Wolfe conditions can be used to select the step size [92]. Importantly, the step sizes do not necessarily decrease and may increase so long as $f(x_{n+1}) < f(x_n)$. A *backtracking line search* may attempt several values of $\eta_n$ until this monotonicity condition is satisfied.

Line search is a generalization of both Newton's method and gradient descent. In Newton's method, the matrix $B_n$ is given by the Hessian, $\nabla^2 f(x_n)$, whereas in gradient descent $B_n$ is the identity matrix. The key intuition is that the matrix $B_n$ linearly transforms the search space in a manner dependent on $x_n$. In the case of gradient descent, no transformation is performed. The Hessian transforms the space to optimally reflect the curvature of the objective function at the current iterate. Between these extremes there exist transformations that still tailor the iteration to the shape of the objective function at point $x_n$ but are simpler to compute than the Hessian. One popular method that can be characterized in this way is *conjugate gradient descent* [73]; another is to use the second derivative along only a single axis.

### 2.2.5 Gradient Descent in Parametric Models

The description of gradient descent in Section 2.2.3 assumes that the gradient $\nabla f$ must be computed for the objective function $f$. In practice, gradient methods can be applied in many situations where $\nabla f$ is not available through the use of a parametric model. For example, in the car driving task of Section 1.2, the objective function is a simulation and the search domain is a functional space. Clearly, a simulation cannot be differentiated.

Proposed solutions in this sort of task are often formulated as parameterized functions, so that a solution is a function of the form $u(x;\theta)$ where $x$ is the environmental state and $\theta$ is a set of parameters drawn from a finite-dimensional parameter space $\Theta$. For example, in an artificial neural network the parameters are the weights between connected artificial neurons. The objective function $f$ can be written as $f(\theta) = \mathbb{E}_{x\sim\mathbb{P}}\left[C(u(x;\theta))\right]$, where $C$ is an appropriate cost function and $\mathbb{E}_{x\sim\mathbb{P}}$ indicates an average result approximated by stochastically simulating over environmental states $x$. In many cases, the parameterized form $\nabla f(\theta_n)$ depends only on some statistics $(t_1^n,\dots,t_m^n)$ gathered during the evaluation of the costs; that is, $\nabla f(\theta_n) = h(t_1^n,\dots,t_m^n)$. This situation occurs commonly when the objective function is to minimize interpolation error on a given set of input-output pairs, where the statistics gathered are the interpolation errors for each input. Thus in order to compute $x_n$ using gradient descent, one does not need $\nabla_\theta \mathbb{E}_{x\sim\mathbb{P}}\left[C(u(x;\theta_n))\right]$ but only $t_1^n,\dots,t_m^n$, and the particular statistics needed depend in part on the objective function and in part on the parameterization. For example, in an artificial neural network, the gradients often depend entirely or partially on the correlation in activation state between connected neurons.

Thus parameterized gradient descent and direct gradient descent differ in the type of information they extract from the objective evaluation in order to update the next solution. This distinction is important for reasons that will become more clear during the discussion of how optimizers rely on information about the objective in Chapter 6.

## 2.3 Problems with Gradient Methods

The previous section gave a cursory review of gradient-based methods. Since most of the text that follows focuses on gradient-free methods, the amount of material allocated to gradient methods in this chapter requires some justification. First of all, if gradient methods work at all, they work accurately and quickly. Their failure modes are also well understood and can often be identified a priori or during operation.

There are also several reasons why many gradient-free methods exist and continue to be invented. Perhaps the most obvious one is that many optimization tasks are performed on non-continuous search spaces. A prime example is the Traveling Salesman Problem (TSP), in which a salesman is tasked with finding a route that visits each city on a map exactly once using existing roads. Solutions to this problem can be represented in multiple ways, but the representations typically do not induce a continuous search space. Non-continuous tasks require searching on various spaces, including trees, permutations, graphs, subgraphs, and binary codes.

Objective functions on these spaces do not have derivatives and cannot be searched with gradient-based methods.

There are also reasons why gradient-free methods can be preferable even in continuous spaces with differentiable objectives. The main one is that gradient-based methods are local optimizers. In the optimization literature, when it is said that an optimizer converges to an optimum, what is typically meant is that the optimizer produces a sequence of points that converges to some local optimum in the limit. Given an objective function $f$, a point $x$ is a *local minimum* (or a *local maximum*) if there is some open neighborhood $N$ containing $x$ such that $f(x) \leq f(z)$ for all $z \in N$ (for a maximum, $f(x) \geq f(z)$). If this inequality is satisfied when $N$ is the entire search space, then the point is a *global minimum* (or a *global maximum*). A gradient-based method will converge to a local optimum that is determined by the starting point and the step size. A local optimum may be a high-quality solution to the optimization task, or it can be of low quality. Thus a local optimum is sometimes a sufficient solution, and sometimes not.

The local optima of a function divide the search space into a disjoint set of *attraction basins*. An attraction basin of a dynamic system is the set of initial conditions for which the system converges to a particular attractor. Gradient-based optimization approximates a dynamical system given by

$$dz(t) = \nabla f(z(t))\, dt$$

with initial condition $z(0) = z_0$ and $t \in [0, \infty)$. In this system, the local optima are attractors, since the gradient is zero there. For a given local optimum $x$, the attraction basin of $x$ is the set $\{z_0 \mid x = \lim_{t \to \infty} z(t)\}$. Generally, if the starting point $z_0$ is in the attraction basin of $x$ under this system, then the iteration of a gradient method will converge to $x$ unless the step size is large enough that one of the iterates steps across the entire attraction basin. Most gradient methods aim to find the local optimum of the attraction basin that contains the starting point. To study the quality of a gradient-based method on an objective, one should study the attraction basins of the objective.

Even among differentiable functions, the attraction basins of a function may have almost any shape, size, or arrangement within the search space. Figure 2.1(a) shows contours and attraction basins for the sum of ten two-dimensional Gaussian kernels with varying centers and size, given by

$$f(x) = \sum_{i=1}^{10} \frac{1}{2\pi\sigma_i} \exp\left(-\frac{1}{2\sigma_i^2} ||x - \mu_i||^2\right).$$

This function has 10 maxima with one at each $\mu_i$ whose girth is proportional to $\sigma_i$, and the basins displayed for each in Figure 2.1(a) were estimated by running conjugate gradient descent on a $300 \times 300$ grid of starting points arranged from $-1$ to 1 on each axis. Importantly, the narrower kernels have higher objective values at the center, but correspondingly smaller attraction basins. Assume that the starting point is chosen uniformly at random with each component in the interval $[-1, 1]$.

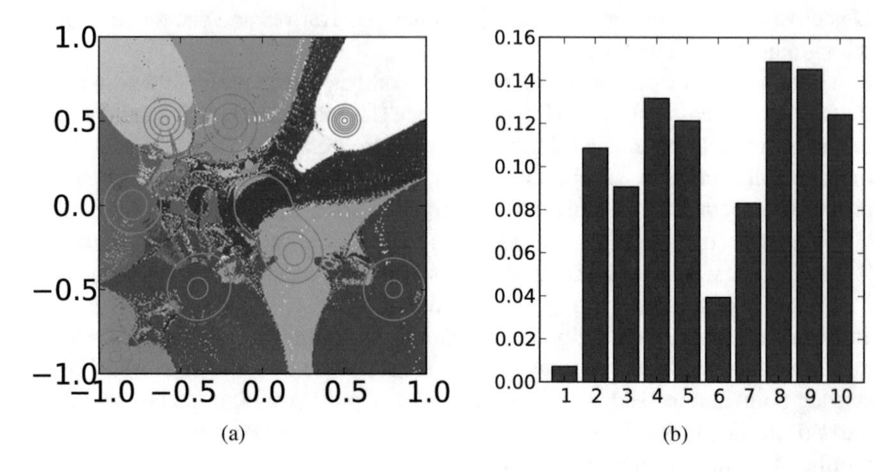


**Fig. 2.1** (a)Attraction basins for a sum of ten Gaussian kernels under conjugate gradient descent on a square region. Plots were generated by sampling a $300 \times 300$ grid to discover which local maximum results from each grid point using conjugate gradient descent. Different colors indicate regions that result in different local maxima. Contour lines are included in gray to indicate the location of the Gaussians. The global maximum is the red region in the upper left. Its area among possible starting points is small; only 639 of the 90,000 sample points (0.7%) converge to the global maximum. (b) A bar chart showing the probability of arriving at the best local maxima, ordered by quality from left to right. Gradient descent is unlikely to find the best solution.

Figure 2.1(b) shows the probability of achieving each of the 10 maxima, ordered by rank with the global maximum at the left. The true optimum is located in the smallest basin and attracts only 639 out of the 90,000 sample points. That is, if the starting point is chosen randomly, then the true optimum is found on less than one out of every 100 runs. If this experiment were repeated in five dimensions with a similar function, the chances of finding the true optimum would be much less than one in 10,000. Therefore, the fact that a gradient-based method will converge quickly and accurately to a solution provides little comfort if the solutions achieved in this manner are severely suboptimal with high probability.

For many tasks, there is a rational means for choosing a starting point that will yield high quality. For example, if the objective is given in closed form as above, it may be analyzed. Yet for many other optimization tasks, there is no way to know at the outset which starting points will yield good results. One must choose the starting point heuristically.

In essence, the use of a gradient method converts a continuous optimization task into a discrete task. An objective with $K$ local optima yields $K$ possible outcomes under Newton's method. It may be that each one of these local optima are acceptable solutions, in which case the gradient search always succeeds. But if only a proper subset of the outcomes are acceptable, then the gradient search is successful only some percentage of the time. This percentage is a property of the objective and may be small.

Several methods have been proposed to improve the quality of the optima obtained by a gradient search. Each of these methods succeeds in some situations and fails in others. The use of a momentum factor was mentioned during the discussion of gradient descent. It can backfire by causing the optimizer to skip over the true optimum in certain situations. Another approach is to add random noise to the computation of the gradient. This approach is called *stochastic gradient descent*, and the random noise is usually embedded into the objective function itself as part of the task. The addition of noise can allow the gradient iteration to escape randomly from shallow local optima. The method succeeds when poor local optima are more shallow than good local optima, since the variance can be tuned just enough so that the iteration escapes poor optima with high probability but is trapped by good optima. But there is no way to know ahead of time whether the poor and good optima fit this profile. An objective can be constructed in which poor optima are relatively deep and good optima are reached by a long series of shallow steps, each of which can be escaped. It is impossible to know how the variance should be set without some experimentation.

If the momentum factor and random noise fail, then another approach is to run a Newton method many times with different starting points. These methods typically require at most several dozen objective evaluations before convergence, making this approach practical. In this case, one is essentially randomizing the starting point and then sampling from a histogram like that of Figure 2.1(b). If the good optima are unlikely, then it is possible that even this method will fail.

Gradient-free methods provide alternatives that answer many of the problems brought up in this section. A full review of these methods is presented over the next few sections.

## 2.4 Direct Search

Direct search is a catch-all term for several gradient-free optimization methods frequently employed within the applied mathematics community over the course of the last century. According to the definition of the term, any method that relies strictly on objective evaluations in order to determine the next point to search is a direct search method [92, 128, 172, 76]. However, the term "direct search" is not used outside of applied mathematics, and it will not be applied more widely here. Some of the earliest variants of direct search were already in use in the Manhattan Project in the early 1940's [92]. The most important categories in this group of approaches are the simplicial methods, represented by the Nelder-Mead algorithm, and pattern search, to which the name *generating set search* is applied following Kolda et al. [92].

### 2.4.1 Simplicial Methods

Simplicial optimization methods search through a $d$-dimensional continuous space manipulating a simplex with its $d+1$ vertices. Because of the way the simplex moves through the search domain, this method is often referred to as the *amoeba method*. The initial simplex is built around a starting point, typically by taking the starting point as a vertex and setting the other $d$ vertices by adding the $d$ coordinate vectors to the starting point. The endpoints are then evaluated under the objective function. At each iteration, the simplex is be transformed in various ways.

The first simplicial method was proposed in 1962 by Spendley et al. [151]. It included two possible transformations to the simplex. The worst vertex could be reflected around the centroid of the opposite face, or the entire simplex could be shrunk towards the best vertex. Nelder and Mead introduced additional transformations [116]. In addition to being reflected, the worst vertex could be moved towards the centroid of the opposite face (contraction), or projected through it twice as far as the reflection would have accomplished (expansion). The transformation to be used is determined by a set of rules that depend on the objective values of the vertices.

The Nelder-Mead algorithm is popular because it works quickly and reasonably well. An implementation of this optimizer is included with several software packages and is widely available. However, Nelder-Mead is not guaranteed to converge, and its widespread use seems to be primarily a matter of convenience.

### 2.4.2 Generating Set Search

In contrast to simplicial methods, generating set searches are designed to ensure convergence to a local optimum. Generating set search maintains an accepted solution that is updated only if the objective value can be improved. The name was coined by Kolda et al., and subsumes the earlier term *pattern search* of Hooke and Jeeves [76, 92]. Recalling the definition of a descent direction in line search from Section 2.2.4, suppose that one wishes to follow the gradient in order to improve the accepted solution, but no gradient information is available. In a $d$-dimensional continuous space, this can be accomplished by testing the objective at $d+1$ points around the current accepted solution. If these points are chosen carefully, then at least one of them will be a descent direction for the objective.

The correct points can be generated from a *positive spanning set*. A set $B = (b_1, \dots, b_k)$ of $d$-dimensional vectors is a positive spanning set if for any vector $z$ in the space, there exists vector $\alpha$ with non-negative components such that $z = \sum_i \alpha_i b_i$. If $\tilde{B}$ is a basis for the space, then a positive spanning set of size $d+1$ can be generated by appending to $\tilde{B}$ a vector $\tilde{b} = -\frac{1}{d} \sum_{x \in \tilde{B}} x$. Or, a positive spanning set of size $2d$ can be generated by extending $\tilde{B}$ with the negative of each vector in $\tilde{B}$.

The simplest generating set search starts at a given point $x_0$ and requires a step-size parameter $\Delta_0$. The method generates a positive spanning set $B$ and then polls each of the directions $x_0 + \Delta_0 b$ for $b \in B$. If $f(x_0 + \Delta_0 b) < f(x_0)$ for at least one direction $b$, then the current solution is updated to $x_1 = x_0 + \Delta_0 b$ and $\Delta_1 = \Delta_0$. The search can either choose the best descent direction or the first discovered. The process continues until a point $x_n$ is found such that none of the directions in $B$ yields an improvement. In this case, $x_{n+1} = x_n$, and the step size $\Delta_n$ is reduced by some factor $\tau < 1$ so that $\Delta_{n+1} = \tau\Delta_n$. The process continues again until $\Delta_n$ falls below a tolerance, in which case the search is complete.

This algorithm converges to a local optimum because the step size $\Delta_n$ can be proven to decrease to zero asymptotically [35, 92]. It can be modified in a number of ways while remaining convergent. A search heuristic can be inserted before each polling step that evaluates any finite number of points on the grid $\{x_n + m\Delta_n b \mid b \in B, m \in \mathbb{Z}, m \neq 0\}$. When the search heuristic is successful, the step size $\Delta_n$ can be increased instead of decreased. The generating set $B$ is allowed to depend on $x_n$ as long as the magnitude of its vectors does not increase.

Several optimization methods that fit this template have been developed and continue to be proposed [10, 35, 158]. These methods work well in practice, but are designed to converge to a local optimum. Once convergence has occurred within a specified tolerance, the search is complete, and new local optima can only be discovered by restarting from a different point. thus the comments about gradient methods above apply equally to convergent direct search methods. A systematically different approach is to use a Monte Carlo search algorithm, as will be discussed next.

## 2.5 Stochastic Optimization

Most of the optimization methods examined so far have shared two properties. First, they have been deterministic. The resulting solution is a function of the starting point. Secondly, they converge to a single local optimum and then either terminate or become asymptotically stable. Once these methods have converged within an acceptable tolerance, they no longer explore new regions of the search space. In contrast, stochastic optimizers search the domain by randomly sampling points based on the objective value of one or more previously evaluated points; this technique is also known as *Monte Carlo optimization*. Because they move randomly, stochastic optimizers can escape local optima with some probability. As a result, they may not always converge to a final solution, or they may explore multiple local optima prior to convergence. The most widely used method of this type is probably simulated annealing, which is reviewed next.

### *2.5.1 Simulated Annealing*

Simulated annealing was developed by Kirkpatrick et al. in the early 1980's [90, 24]. It employs properties of statistical mechanics to locate the minima of a given fitness function. The usual analogy is that of crafting a metallic artifact by repeatedly shaping it at different temperatures. At high temperatures, the metal is malleable and easy to shape, but as such the metal does not easily remain in detailed configurations. As the temperature is gradually lowered, more refined and delicate shapes become possible, but the overall shape is increasingly fixed.

At the core of the simulated annealing algorithm is the Boltzmann distribution. At time $n$, simulated annealing samples from a distribution whose differential probability density given by

$$\mathcal{A}_n^f(dx) = \frac{1}{Z_n} \exp\left(-\frac{f(x)}{T_n}\right) dx, \tag{2.3}$$

where $f$ is the fitness function, $Z_n$ is a normalizing factor known as the *partition function*, and $T_n$ is a sequence of temperatures with $T_n \to 0$. The sequence $T_n$ is known as the *cooling schedule*. The distribution $\mathcal{A}_n^f$ will be referred to as an *annealing distribution* in this book. Simulated annealing samples from $\mathcal{A}_n^f$ repeatedly using the Metropolis algorithm [107, 70]. The process begins with a proposed solution $x$. At each time step, a proposal distribution $\mathbb{Q}$ is used to sample $x_n$. For instance, $\mathbb{Q}$ might be a Gaussian distribution centered at $x$. The proposed solution $x$ is replaced with $x_n$ with probability $\exp\left(-\max\left\{0, f(x) - f(x_n)\right\}/T_n\right)$. For each fixed temperature $T_n$ the algorithm will converge to a sample from $\mathcal{A}_n^f$. As $n \to \infty$, $\mathcal{A}_n^f$ converges in probability to a distribution that samples directly from the optimal points of $f$ [90].

Subject to conditions on the cooling schedule, simulated annealing can be shown to converge asymptotically to the global optima of the fitness function [65, 174]. For combinatorial problems, Hajek [65] showed that simulated annealing converges if the cooling schedule is set according to $T_n \propto 1/\log n$. In practice, simulated annealing has been used effectively in several science and engineering problems. However, it is highly sensitive to the proposal distribution and the cooling schedule.

Whereas simulated annealing lowers the temperature $T_n$ to zero in order to sharpen the Boltzmann distribution, *stochastic tunneling* raises the temperature to higher values in order to soften the function and lower the barriers separating the attraction basins of different local optima. Raising the temperature allows for complete exploration of the local minima of the function and may make it possible to locate the global minima. However, more thorough exploration comes at the cost of much slower convergence.

### 2.5.2 Justification for Stochastic Methods

Stochastic optimizers have some advantages over deterministic methods. Stochastic optimizers do not become trapped by local optima as easily as deterministic optimizers, although eventually most popular stochastic methods do converge around a single point, potentially a local optimum. However, this flexibility comes at a high price. Stochastic methods inevitably converge more slowly than deterministic gradient-based methods because they can explore in the opposite direction of the gradient. This slowdown may be exponential if the stochastic method is particularly thorough. Direct search methods suffer from the same reduction in speed, but many of them can still work faster because they provide a guarantee that a descent direction is eventually followed. Thus, before using a stochastic method, especially a non-gradient-based stochastic method, some justification is required.

Stochastic methods are valuable because they reflect the underlying uncertainty in the optimization task. As will be seen in Chapters 12 and 13, there is substantial reason to believe that the best optimizer under any fixed random test procedure is deterministic. However, if the exact nature of the test procedure is unknown, an optimizer can be made robust against the uncertainty of the test procedure by randomizing. The best deterministic optimizer on one test procedure could have poor performance on a slightly different test procedure. The primary justification for a stochastic method is the underlying uncertainty about the true nature of the problem.

If one desires to study a specific optimization task, then one might learn a great deal about the problem through exhaustive exploration. It would then always be advantageous to design an optimization method that accounts for what one has learned. However, it is important to recognize this procedure for what it is: a long, tedious, manual optimization process in which the researcher has adopted the role of the optimization method and has himself made a series of evaluations in order to remove uncertainty and increase available information. In other words, this approach reflects confidence in humans as superior optimizers. However, the skills, knowledge, and ability required by such a researcher are rare in relation to the number of optimization problems, and the "running time" of the human optimizer is often much longer than that of a general-purpose optimizer. If resources and time permit, then the human optimizer is almost always preferable to automated optimization methods. But if resources and time are a constraint, then a good stochastic optimizer is a viable option. Finally, if the goal of research is to develop human-level problem solving abilities, then comparing the success of human researchers to the success of a good black-box algorithm is useful for assessing the progress of artificial intelligence.

In the end, stochastic methods are useful and can be shown to have good performance on a wide array of metrics. Deterministic or quasi-deterministic optimizers can always perform better on static problems with a sufficient amount of information. The success of an algorithm depends on its alignment with the test procedure used to evaluate it [171].

Note that the statements above do not necessarily hold for dynamic and stochastic objectives. Such objectives are addressed in later chapters, but for now the objective is assumed to be static. That is, if an objective $f$ is evaluated at a point $x$, then the value $f(x)$ is fixed, i.e. subsequent evaluations must return the same value. If subsequent evaluations of $f(x)$ can change, then either the objective function is dynamic (varying over time) or stochastic. The existence of stochastic or dynamic optimization contexts has served as one of the primary justifications for stochastic optimization.

## 2.6 Evolutionary Methods

Evolutionary computation is a major category of stochastic optimization method with origins in the computational simulation of evolutionary processes. The general concept of evolutionary computation has been invented independently numerous times by different researchers [54, 55, 57, 27, 53, 133, 14, 135]. When the biological terminology is removed, we will see that evolutionary computation is actually a just a form of Monte Carlo search, and its application to optimization problems can thus be subsumed into the study of stochastic optimization. This section reviews the most common elements of these methods in order to support this claim.

### *2.6.1 Overview*

The basic structure of a traditional evolutionary algorithm consists of a sequence of subsets of the search space, termed *populations*, with each population in the sequence called a *generation*. Populations consist of *individuals*, generally represented as an array of parameters. The population for each generation is built from the prior generation through processes of competitive selection and random variation. The prior generation is ranked according to fitness, and the most fit individuals are chosen to create the next population either with minor variations called *mutations* or by combining parameters from two or more members in an operation called *crossover*. Many other variants exist, and these variants will be discussed as needed. A short historical summary of evolutionary computation follows.

The earliest research on computational simulation of evolutionary processes was published in 1954 by Barricelli in the context of a cellular automaton [19]. Subsequently, Friedberg and Fogel independently studied *genetic programming* [57, 53]. Separately, Rechenberg and Schwefel developed a technique called *evolution strategies* as an approach to search through Euclidean space with selective competition and local variation [133, 145]. But *genetic algorithms* became the most dominant branch of evolutionary algorithms, having been developed independently by Bremerman [27] and by Holland and his students [74, 14, 135]. Holland championed the cause of genetic algorithms and developed a result known as the schema theorem to explain their success [75]; his work was followed by that of Goldberg, who likewise had a substantial influence in popularizing genetic algorithms [60]. Early work comparing the effects of varying algorithm hyperparameters and population transition mechanisms empirically was performed by De Jong [84]; these experiments were furthered by Brindle [28].

Holland preferred a binary representation of the search space, where each bit of the representation acted analogously to a gene in biology. His schema theorem asserts that genetic algorithms with binary representations probabilistically select and refine subcomponents of a high-dimensional search space by making it more likely that adjacent subcomponents contributing high fitness would survive into later generations. This claim was used as an argument to assert that binary representations were more efficient because binary subcomponents would be more likely to be preserved in population transitions [75, 60]. This argument assumes that problems are structured into small, uncorrelated subcomponents, when in fact real-world problems can be structured in ways that exhibit intricate long-distance dependencies [51]. In these cases, genetic algorithms are likely to struggle to find the correct problem structure. Furthermore, genetic algorithms are highly sensitive to representation in a binary setting. For example, De Jong produced an experiment comparing two different types of genetic algorithms in which a change of representation reversed their order in terms of performance [85]. This research revealed additional problems with early genetic algorithms, such as *premature convergence*, in which bit representations tend to lose their representational diversity so that certain solutions become practically unreachable, with the result that the genetic algorithm might stagnate without finding even a local optimum.

The possibility of premature convergence along with experiments like those of De Jong on the sensitivity of a genetic algorithm to its implementation details created an atmosphere of ambivalence about the prospects of tuning a genetic algorithm to solve yet more complex problems. Further, the No Free Lunch Theorems of Wolpert and Macready [171] demonstrated that, averaged over all problems, any genetic algorithm would perform equivalently to any other genetic algorithm. From that point, research turned to generalizations and abstractions of genetic algorithms, for which it was hoped that novel techniques would outperform genetic algorithms on particular problem domains where genetic algorithms fared poorly.

Many of these new algorithms were proposed as biological analogies. Collectively, they are referred to as *natural computation*. Dorigo presented Ant Colony Optimization (ACO) [41] in 1992 in analogy with the food gathering behavior of ants. For ACO, problems are transformed into a search for a path through a graph. In each generation, a population of *artifical ants* explores the graph stochastically, laying down *pheromones* that bias future searches whenever they locate interesting structure. Particle Swarm Optimization (PSO) was introduced in 1995 by Eberhart and Kennedy [45, 88] to imitate the behavior of flocks of migrating birds in choosing a direction in which to fly. In PSO, rather than randomly mutating population members, candidate solutions in the population travel through the search space with a directional momentum that is incrementally shifted towards the current best solution in the population. Particle Swarm Optimization will be analyzed as an example of natural computation in Chapter 7. Finally, Artificial Immune Systems (AIS) were proposed in stages by several researches to employ an analogy with the structure of the human immune system in order to address issues such as network intrusion detection using a population of *antibodies* [48, 23].

In addition to the biological variants, a substantial field of research developed to learn a probability distribution governing the transition between populations in a genetic algorithm [16, 17, 96, 69]. This research highlights the connection between evolutionary computation and Monte Carlo optimization. It culminated in algorithms that greedily learn Bayesian networks over the best members of the population; these networks are then sampled to produce the next generation [112, 125, 123, 124]. These algorithms are alternately known as Estimation of Distributions Algorithms (EDAs) or Probabilistic Model-Building Genetic Algorithms (PMBGAs) [122]. EDAs are relevant to this text as an example of the application of probability to genetic algorithms and will be reviewed in further detail in Section 2.7.2.

With this historical background in mind, it will be useful to discuss the issues arising in theory and analysis of classical genetic algorithms before returning to newer, probabilistic variants.

### 2.6.2 Genetic Operators

Genetic operators are used to describe the substructure of population transitions in genetic algorithms. A transition from one population to the next is accomplished by applying an ordered sequence of operators to the current population. Typically, these operators include a selection operator, a crossover operator, and a mutation operator. Other genetic operators have been proposed such as diploidy, dominance, and inversion [60]. The most common genetic operators can be subsumed by a two stage process of selection and variation, where selection is a weighted choice over previously observed individuals and variation is an arbitrary stochastic perturbation of a selected individual. A formal account of selection and variation is provided in Chapter 7. The relevant background is summarized in the following paragraphs.

The fundamental logic of evolutionary search is based on Darwinian concept of *competition* forcing the preferential selection of diversely varying features of a population. When applied to optimization, selection is intended to enforce a preference for solutions with better objective values, and variation is intended to guarantee exploration around these preferred solutions. Both features are critical for success. Without selection, exploration is random and hence slow. Without variation, selection cannot reach beyond the properties of the initial population.

The role of a selection operator, then, is to select the parents of the next generation from among the individuals whose fitness has already been measured. Selection can be treated as a probability distribution over candidate solutions observed thus far. The most common types of selection are *proportional selection* [75, 60, 161, 122], *ranking selection* [15, 161], *tournament selection* [28, 161], and *elitist selection* [84]. Proportional selection (also known as *roulette wheel selection*) is an individual-level selection operator that selects members of the previous generation in proportion to their fitness [75, 60]. Formally, for a fitness function $f$, individuals $x, y$ in the prior generation, and the number of occurrences of $x$ in the prior generation $n_x$,

$$\mathbb{P}^t_{\text{prop}}(x) = \frac{n_x f(x)}{\sum_y n_y f(y)}. \tag{2.4}$$

Proportional selection plays a key role in the simple genetic algorithm and in many EDAs. Note that proportional selection assumes that the fitness function is positive. Any bounded function can be shifted so as to be positive, or other monotone increasing transformations may be applied to obtain a positive function (such as exponentiation). These transformations may distort the shape of relative weight of different solutions, but they cannot alter the order in which solutions are preferred.

Ranking selection can operate either at the level of populations or individuals. At the individual level, parents for the next population can sampled randomly from the ranked population according to any monotone decreasing weighting scheme. Ranking selection differs from proportional selection in that the relative weight of population members can be altered by any method that preserves order.

Tournament selection also ranks the population, and then chooses the best individual with probability $p$ and the $k^{\text{th}}$ individual with probability $p(1-p)^{k-1}$ [28, 161].

Elitist selection guarantees that the best solution from all prior generations survives into the next population. When elitist selection is used along with a mutation scheme that eventually visits every point in the space, then asymptotic convergence is obtained with probability one [137].

Selection operators introduce competition into the population. By contrast, crossover and mutation operators provide variation to widen the scope of the search in order to discover highly fit individuals not previously in the population. A crossover operator recombines two candidate solutions to construct a third candidate solution, thus implementing a non-local exploration across a subset of the search space determined by the parent population. In order to perform crossover, two or more parents must be selected according to some probabilistic rule. In addition, a crossover methodology must be supplied to determine how recombination is to occur. It is the form of recombination that defines the subset of the search space being explored given the parent population.

The most basic crossover strategy is *one point crossover*, in which an index inside of a chromosome is selected as a split point, and each parent provides the parameters on one side of the index. The split point may be selected deterministically, or chosen according to some probability. In contrast, *uniform crossover* performs a Bernoulli trial for each parameter, and copies the parameter of the parent chosen by that trial [157]. Many other techniques exist, such as averaging parent parameters or preferentially choosing parameters from the more fit parent. These methods are discussed further in Chapter 7.

Mutation operators are applied at the individual level to slightly vary an individual's parameters. Crossover is limited to the rearrangement of parameter values already present in the parent population; mutation operators introduce new parameter values into the population. Mutation therefore plays the role of local exploration, facilitating detailed exploration of regions with high fitness. In binary spaces, this concept is formalized by the *mutation rate*, that is, the expected percentage of bits in the gene that will be flipped by mutation. In metric spaces, the analogous feature is characterized by the variance of the operator. If an individual can be transformed to any other individual in the space with some nonzero probability after application of the mutation operator, then an evolutionary algorithm equipped with this operator will asymptotically converge to the optima of the fitness function [50, 137]. In some cases, guarantees can even be provided regarding the rate of convergence.

Both crossover and mutation can be construed as a family of probability distributions indexed by candidate solutions. Given an individual, crossover places nonzero probability on solutions that can result from crossing that individual with other members of the population using the crossover methodology. Mutation places nonzero probability on solutions reachable from a given individual. These intuitions are made more explicit in the discussion of Random Heuristic Search that follows.

### 2.6.3 Random Heuristic Search

Some theoretical issues pertaining to the simple genetic algorithm were explored by Vose [161] within a framework he introduced under the name of Random Heuristic Search. The analysis performed by Vose prefigures certain aspects of this text and will therefore be reviewed in this subsection. Vose describes the simple genetic algorithm (SGA) as a search in the space of probability vectors. A random heuristic search [162] consists of two repeated steps. The first step applies a deterministic *heuristic* $\mathcal{G}$ to map the current population $p_n$ to a candidate population $\hat{p}_n = \mathcal{G}(p_n)$, and the second step resamples a new population $p_{n+1}$ from $\hat{p}_n$. The *transition rule* $\tau$ is defined as the composition of the two steps.

For the simple genetic algorithm, the heuristic $\mathcal{G}$ can be broken down further into three steps: selection, mutation, and crossover. The *selection scheme* $\mathcal{F}$ maps a population $p$ to a distribution over bit strings that produces an individual $x$ with probability $\mathcal{F}(p,x)$. Significantly, the selection scheme operates at the level of individuals rather than populations. The *mutation function* $\mu_u^x$ gives the probability of mutating an individual $u$ to an individual $x$. A *crossover function* $\chi_z^{x,y}$ *recombines* individual parents $x$ and $y$ into a third individual $z$ in accordance with a crossover rule.

These three operations can be combined to give an explicit form for the action of the heuristic $\mathcal{G}$ on a population $p$ using SGA. Specifically, for each $z \in \{0,1\}^n$, the probability of obtaining an individual $z$ for the next population is given by

$$\mathcal{G}(p,z) = \sum_{x,y,u,v} \mathcal{F}(p,x)\,\mathcal{F}(p,y)\,\mu_u^x\,\mu_v^y\,\chi_z^{u,v}. \tag{2.5}$$

With some rearrangement, it can be seen that Equation 2.5 is actually a mixture distribution, since it can be rewritten loosely as

$$\mathbb{P}(z \in \tau(p)) = \sum_{x \in A} q(x)\,\nu^x(z), \tag{2.6}$$

where $A$ is the set of individuals $i$ such that $\mathcal{F}(p,i) \neq 0$, $q(x) \equiv \mathcal{F}(p,x)$ and

$$\nu^x(z) = \sum_{y,u,v} \mathcal{F}(p,y)\,\mu_u^x\,\mu_v^y\,\chi_z^{u,v}. \tag{2.7}$$

Here $\mathbb{P}(x \in \tau(p))$ indicates the probability that an individual $z$ occurs in the successor $\tau(p)$ to the population $p$. This equation makes explicit the claim that mutation and crossover can be represented as a family of probability distributions operating on individuals, and that selection could be viewed as a weighted choice over previously observed individuals. This claim will be used implicitly in Chapter 7 to characterize evolutionary algorithms theoretically.

Here, the heuristic $\mathcal{G}$ is defined on binary search spaces. However, this characterization can be generalized to arbitrary Hausdorff topological spaces, which will be done in Chapter 6.

## 2.7 Quasi-Evolutionary Methods

As discussed in Sections 2.6.1 and 2.1, recent trends in evolutionary computation have focused on developing optimization methods with more detailed mathematical justification. These methods bear strong similarities to evolutionary optimization methods in both terminology and intuitive origins but are not adequately described as artificial simulations of biological evolution. They are more aptly described as parallelized stochastic optimization techniques with historical ties to evolutionary computation. For these methods, this term *quasi-evolutionary methods* is adopted in this text. In formal terms, it is not possible to distinguish evolutionary and quasi-evolutionary methods completely. Thus this distinction is primarily rhetorical, and will be reflected in the formalisms of Chapters 6 and 7 only to a limited extent.

This section discusses three kinds of quasi-evolutionary methods: differential evolution, estimation of distribution algorithms, and natural evolution strategies. Evolutionary annealing, introduced in Chapter 14 of this book, is a new type of quasi-evolutionary method.

### 2.7.1 *Differential Evolution*

Differential evolution was introduced by Storn and Price [155] in 1995 as a means of optimizing the coefficients of Chebyshev polynomials. It has since proven itself as a fast and effective optimizer for finite-dimensional real vector spaces.

Differential evolution maintains a population of candidate solutions that may be thought of as an array of slots with one slot per individual in the population. Each slot obeys an independent acceptance-rejection scheme similar to those observed in direct search or simulated annealing. At each generation, a new population is created with one candidate solution generated for each slot. These solutions are evaluated against the objective. Each slot is then filled with either the new candidate from that slot or the prior occupant of the slot, depending on which performs better. Thus the objective value of each slot improves monotonically, and differential evolution with population size $K$ consists of $K$ parallel, monotonically improving searches.

The procedure for generating new candidate solutions involves altering one or more components of the real vector that constitutes the solution. Let $x$ be the member of the population in the $i^{th}$ slot. To generate a new solution $\tilde{x}$ from $x$, the following steps are used. First, three other members of the population are selected, say, $a$, $b$, and $c$. Then for each component $j$ of the vector space being searched, set $\tilde{x}_j = a_j + F(b_j - c_j)$ with probability $CR$ and $\tilde{x}_j = x_j$ otherwise. If no component of $\tilde{x}$ is changed from $x$, then randomly select a component $\tilde{j}$ and apply the change above to that component. The value $F$ is a parameter of the algorithm termed the *weighting factor* that regulates the step size through the search space, and $CR$ is the *crossover rate*, a probability governing the rate at which components are mixed across the population. Typically, $F \in [0,1]$ with $F = .2$ and $F = .9$ being common values. The crossover rate is also often set to $CR = .2$ or $CR = .9$.

Ghosh et al. [59] showed that differential evolution in the form above converges to the global optimum on continuous real-valued functions that possess a unique global optimum. Such a *global* convergence guarantee is quite remarkable and contrasts with the local convergence guarantees of gradient methods and direct search. A number of variants have been developed as well. Some of them move components in the direction of the best member of the population rather than in a random direction (i.e. they replace the vector $a$ above with the best solution in the population). Other methods use crossover operations to further mix the members of the population. Still other methods relax the acceptance criterion to operate more like simulated annealing [39]. Many of these methods have performed well on benchmarks and in optimization competitions, making differential evolution one of the best performing gradient-free stochastic optimizers available today [59].

### 2.7.2 Estimation of Distribution Algorithms

Evolutionary algorithms can be thought of as building a sequence of probability distributions used to sample each generation. The process begins with an initial distribution that is used to sample the first population. Then, the population is scored, and a new population is created stochastically. The new population is just a sample from some abstract probability distribution. This point of view begs the question: can the optimization algorithm be improved by making this abstract distribution explicit?

Estimation of Distribution Algorithms (EDAs) were the first quasi-evolutionary algorithms to seek to learn a distribution explicitly governing the transition between populations [111, 122]. EDAs rank the prior population according to fitness. The worst members of the population are discarded, and the remaining solutions are used as a dataset to estimate parameters for a probabilistic model, usually some sort of graphical model [111, 99, 120]. Because EDAs arose out of the genetic algorithms research, they are typically applied to objectives with binary encodings, and so multinomial Bayesian networks are a suitable probabilistic model.

Baluja et al. introduced the first EDA, Population-Based Incremental Learning (PBIL), in 1994 [16, 17]. PBIL treats each bit in the encoding as an independently sampled Bernoulli random variable. Despite its simplicity, PBIL can outperform traditional genetic algorithms on several problems with significantly fewer evaluations. Hill Climbing with Learning (HWcL) and the Compact Genetic Algorithm (cGA) implement the same concept with distinct update rules for the probability vector [96, 69].

Mühlenbein generalized PBIL by considering a variety of approaches that attempt to implement proportional selection statistically [112]. That is, given proportional selection $\mathbb{P}^t_{prop}$ as in Equation 2.4, the algorithm estimates proportional selection by some distribution $\mathbb{Q}^t$ at each generation so that

$$\mathbb{Q}^t(x) \approx \mathbb{E}\left[\mathbb{P}^t_{prop}(x)\right], \tag{2.8}$$

where the expectation is taken over populations, that is, over the variable $n_x$ in Equation 2.4. The rationale here is that if the initial population is uniformly distributed, then at time $t$ the pointwise expected value of the proportional selection rule yields

$$\mathbb{E}\left[\mathbb{P}^t_{prop}(x)\right] \propto f(x)^t, \tag{2.9}$$

which is a sharpened version of the fitness function. Because of normalization, the result is that as $t \to \infty$, $\mathbb{E}\left[\mathbb{P}^t_{prop}(x)\right]$ goes to one at the maxima of $f$ and zero elsewhere. Proportional selection in genetic algorithms fails to achieve this effect because a specific sample path is followed rather than the pointwise average. That is, in EDAs, proportional selection is performed over all possible populations, at least in theory. By estimating the underlying distribution $\mathbb{P}^t_{prop}$ rather than relying on a single sample path, EDAs seek to benefit from this theoretical convergence to the optimum. The challenge, of course, is that expectation in Equation 2.9 cannot be computed directly, and the estimation may result in a substantial degradation of the theoretical effects. A similar sharpening effect to that observed in Equation 2.9 will be proposed in Chapter 14 without the accompanying context of proportional selection.

A series of subsequent algorithms provided increasingly accurate approximations of Equation 2.9 [112, 111, 125]. The Factorized Distribution Algorithm (FDA) extends this analysis to arbitrary dependencies between variables by estimating $\mathbb{E}\left[\mathbb{P}^t_{prop}\right]$ with a graphical model [111, 99, 120], but FDA does not incorporate a structure learning algorithm. Pelikan introduced the Bayesian Optimization Algorithm (BOA) to provide structure learning details for Bayesian networks and Hierarchical BOA (hBOA) to extend this framework to a hierarchical graphical model [123, 124]. The Real Bayesian Optimization Algorithm (rBOA) translates BOA into an optimization method for real vector spaces [2]. MARLEDA applies similar techniques using a Markov random field rather than a Bayesian network [5].

The class of EDAs thus effectively converts the basic evolutionary algorithm into a probabilistic population-based algorithm that proceeds by matching the structure of a probability distribution to the distribution governing increasingly fit solutions as captured in the mean proportional selection rule of Equation 2.9. It is important to note that the techniques developed for EDAs primarily work for fixed-length binary strings. Evolutionary annealing, introduced in Chapter 14, is similar in some respects to EDAs, but employs mixture distributions that are considerably simpler than Bayesian networks. But evolutionary annealing can be applied naturally to complex domains, and it will be seen that these mixture models approximate more complex models in the limit.

### *2.7.3 Natural Evolution Strategies*

Evolution strategies was mentioned above as a major branch of evolutionary algorithms for real vector spaces; it is reviewed more thoroughly in Chapter 7. Its most common version maintains a Gaussian distribution centered at the current best solution. The method is characterized by global adaptation of the Gaussian covariance parameters embedded into each candidate solution [133, 145, 25]. Evolution strategies employ a naming scheme using parameters $\mu$ for the number of parents, $\lambda$ for the number of children, and sometimes $\rho$ for the number of parents involved in crossover. Thus a $(\mu,\lambda)$-ES has a population of size $\lambda$, from which $\mu < \lambda$ are selected as parents based on their fitness in order to generate $\lambda$ children for the new population by sampling from a Gaussian distribution whose covariance is updated according to specific rules. A $(\mu/\rho+\lambda)$-ES uses populations of size $\mu+\lambda$ in which $\mu$ parents compete for survival with $\lambda$ children by applying Gaussian mutation to a point obtained by crossing $\rho < \mu$ parents. The 'plus' indicates that the parents are retained in the population if they are more fit, whereas the 'comma' indicates that the parents are replaced by their children in each generation.

In 1996, Hansen and Ostermeier [67] introduced a scheme for adapting the mutation parameters to use a population of elliptical Gaussians with arbitrary rotations and named it Correlated Matrix Adaptation Evolution Strategies (CMA-ES). Over time, the algorithm changed substantially so that rather than storing Gaussian parameters on each candidate solutions, a single global set of Gaussian parameters were used to sample each new generation [68]. These changes occurred around the same time as estimation of distribution algorithms were being developed to search binary spaces, and they share important characteristics. The naming of CMA-ES retains the standard $(\mu/\rho \overset{+}{,} \lambda)$ notation of traditional evolution strategies, but the algorithm itself has few similarities with earlier evolution strategies and little if anything to do with artificial evolution.

In the current version of a $(\mu,\lambda)$-CMA-ES, a single $d$-dimensional Gaussian distribution $(\mu_n, \Sigma_n)$ is updated with each generation. The initial population of $\lambda$ solutions is generated randomly. The population is evaluated, and then the mean and covariance is calculated for the best $\mu$ solutions ($\mu < \lambda$). Then the global Gaussian parameters are updated to incorporate this new information in a manner that smoothly integrates the results of subsequent generations. The details of these parameter updates are complex and can be found in the literature [68]. Because CMA-ES uses a single multivariate Gaussian to generate its population, the search cannot adapt itself to a multimodal objective landscape. In practice, CMA-ES converges to a local optimum relatively quickly, and may be restarted in order obtain good results [11].

Wierstra et al. introduced Natural Evolution Strategies (NES) in 2008, and Akimoto et al. subsequently demonstrated that NES is a generalization of the standard CMA-ES algorithm [169, 3]. NES replaces the Gaussian with a generic parameterized probability distribution $\pi(x \mid \theta)$ where the parameters $\theta$ are drawn from a real parameter space and each population samples individuals independently from $\pi(x \mid \theta)$. This representation can be used to generate a meta-optimization problem of choosing the parameters $\theta$ to optimize the expected value of the objective function under $\pi$,

$$J(\theta) = \mathbb{E}_\theta [f(x)].$$

Gradient descent (or ascent) may be applied to this function using

$$\nabla_\theta J(\theta) = \mathbb{E}_\theta [f(x) \nabla_\theta \log \pi(x \mid \theta)],$$

which may be estimated by Monte Carlo integration. A final improvement, termed *natural gradient descent* [169], applies a quasi-Newton method replacing the Hessian with the Fischer information matrix, capturing second-order information about how the distribution $\pi$ changes with the parameters $\theta$.

Natural Evolution Strategies has a firm theoretical foundation and good performance on test problems. Existing versions are limited by the choice of parameterized distribution, which may not align well with the objective being searched. However, this field is relatively new and can be expected to make several useful contributions to the stochastic optimization literature in the future.

## 2.8 Conclusion

This chapter has reviewed the primary branches in deterministic and stochastic optimization research, including Newton and quasi-Newton methods, line search, simplicial methods, generating set search, simulated annealing, evolutionary algorithms, natural computation, and quasi-evolutionary methods. It is hoped that the reader has acquired an appreciation for the diversity and scope of these methods. Each of these approaches exists for a distinct purpose because it performs (or used to perform) reliably well on certain kinds of objectives relative to other concurrent methods.

Given the variety of descriptions and terminology among these various optimizers, it may seem challenging to organize all of them within a single formalism. The subsequent chapters seek to accomplish exactly this goal. The key observation is that every method discussed so far produces a sequence of proposed solutions generated predictably from the prior solutions. The formal study of iterative optimization is built on the analysis of this sequence, beginning with the next chapter.

# Chapter 3
# Optimization in Finite Spaces

Search and optimization are generally considered as tasks to be performed on a digital computer. Computers are finite objects, and even when the spaces being searched are nominally infinite, such as the real numbers, the actual search is limited to finite representations. Thus it is reasonable for a mathematical analysis of search and optimization to begin in finite spaces, where the connection to computational methods is most readily apparent. The formalism to be presented over the next several chapters will address optimization within arbitrary Hausdorff topological spaces. Such spaces are perhaps unfamiliar to many students of heuristic-driven optimization, and hence this chapter provides a concrete grounding of the ideas motivating the more abstract formulation to follow in Chapters 6, 8 and 9. The goal of this chapter is to develop intuitions and notations in the simpler setting of finite spaces. A formalism for optimizers is introduced, and it is shown how local search and genetic algorithms fit into this framework as examples. The final sections point out the issues that begin to arise as the search domain, the value space, or the duration of the search become countably infinite, leading into a general framework in the next few chapters that is applicable to both finite and infinite spaces.

## 3.1 Search Methods, Problems, and Goals

Chapter 1 introduced the distinction between search methods, problems, and goals. To review, one begins with a space to be searched, a set of values that can be assigned, and a goal that can be used to assess the performance of a particular search method. Throughout what follows, the *search space* is denoted by $X$ and the *value space* by $Y$. The nature of the search goal will only be discussed intuitively until Chapter 9 when the performance of an optimization method is studied. For the remainder of this chapter, $X$ and $Y$ are both assumed to be finite collections of points. This section clarifies the types of problems that will be encountered in this book.



A. J. Lockett, *General-Purpose Optimization Through Information Maximization*, Natural Computing Series, https://doi.org/10.1007/978-3-662-62007-6_3

### 3.1.1 Iterative Search and the Trace

This book pertains to iterative search, in which a *search method* tries to solve a *search problem* in accordance with some *search goal*. A search problem can be stated as an attempt to find a point in a search space $X$ based on an associated value drawn from a space $Y$. The search domain $X$ and the value space $Y$ can consist of any finitely representable objects: bit encodings, positive integers up to some limit, finite graphs, or other data structures. Such data structures could represent real-world objects in various forms. For example, the value space could consist of database entries, one for each animal on a large farm. The search domain could be a set of positive integer ids, one for each animal. The search problem then matches each positive integer id to the corresponding database entry, and one search goal might be to find the entry for a pig named Wilbur. In this case, a search method could either scan the ids in order to find the correct entry or conclude that no such entry exists. Or, more efficiently, it could leverage a precomputed index to locate the correct entry. All three elements are necessary in order to form a complete *search context*: method, problem, and goal.

An iterative search method begins by selecting some point in the search domain. The search problem responds by assigning or reporting a value for that point, associating the search point with a point in the value space. This value is used to decide on a new search point, which is also assigned a value. The process iterates until the search goal is satisfied or until some budget of search steps is exhausted. In this chapter, this budget is presumed to be a finite number $N$. Over $N$ steps, the search process generates a *trace*, as follows:

$$\begin{array}{ll} \text{search method} & x_1\ x_2\ x_3\ \dots\ x_{N-1}\ x_N \\ \text{search problem} & y_1\ y_2\ y_3\ \dots\ y_{N-1}\ y_N \end{array} \tag{3.1}$$

Thus the trace is a sequence of pairs $(x_1,y_1),(x_2,y_2),(x_3,y_3),\dots,(x_N,y_N)$ that tells which search point was selected by the search method and which value was assigned to that point by the search problem. Formally, the space of traces of length $N$ is denoted as $(X \times Y)^N$, where $X \times Y$ represents the space of paired search points and values, and the superscript $N$ indicates the number of iterations.

The trace is the key artifact produced by a search, since it records the history of observations during search. It depends on the search method and problem, but not on the search goal. The goal may be assessed either iteratively in order to stop the search early or afterwards to evaluate performance. Future sections use the variable $h$ (for *history*) to indicate a trace when possible, specifically as $h \in (X \times Y)^N$. The notation $t$ is reserved to indicate time, so that $h_t$ is the $t^{th}$ step of the trace.

The analysis below will often require extracting the search points or values from a trace. For a trace $h = ((x_1,y_1),\dots,(x_N,y_N))$, the sequence of search points is denoted by $h_X = (x_1,\dots,x_N)$ and the sequence of values by $h_Y = (y_1,\dots,y_N)$. This notation is used generally; thus if $g$ is a trace, $g_X$ is its sequence of search points and $g_Y$ its sequence of search values.

During a search, the trace is initially empty and subsequently grows based on the output of the search method and the search problem. The notation $\emptyset$ is used to indicate the empty trace. Formally, it is the sole element of the set $(X \times Y)^0$. If we set $h_0 = \emptyset$, then on the first iteration, the search method observes the empty history $h_0$ and produces a search point $x_1 \in X$. The search problem responds by producing a value $y_1 \in Y$ corresponding to $x_1$. The pair $z_1 = (x_1, y_1)$ is then appended to $h_0$ to produce the next history $h_1 \equiv ((x_1, y_1)) = h_0 \| z_1$. Here the symbol $\|$ is used with sequences to indicate concatenation. In the same way, given a trace $h_t$ representing the first $t$ steps of the history, the next trace $h_{t+1}$ is created by appending $z_{t+1} = (x_{t+1}, y_{t+1})$ to $h_t$, i.e., $h_{t+1} \equiv h_t \| z_{t+1}$. Thus the search unfolds as a sequence of traces $h_1, h_2, \ldots, h_N$, such that each trace in the sequence is a prefix of all traces that follow it. This prefix property is denoted by $\sqsubset$, i.e., $h_t \sqsubset h_{t+1} \sqsubset h_{t+2} \ldots$, with possible equality indicated by $\sqsubseteq$, i.e., $h_t \sqsubseteq h_t$. Each trace $h_t$ for $0 \leq t \leq N$ is an element of the set $(X \times Y)^t$.

In summary, an iterative search produces a trace $h \in (X \times Y)^N$ of length $N$ that records the history of the search. This history is the primary object studied in this book. It is generated by the interaction of a search method and a search problem. The next subsection describes how randomness can be incorporated into the trace, followed by a discussion of how this interaction together with a search goal yields many of the search contexts that are commonly studied in search and optimization.

### *3.1.2 Randomized Search*

In many cases, search methods and search problems produce a trace based on some computer program. In this case, the outcome is deterministic: Given a method and a problem, the trace can be known exactly as a function of the two. This trace is obtained by running a program that calls the program for the search method and search problem in succession for a fixed finite number of steps, producing the trace as output.

This book studies what happens if the search method and problem both have access to a source of randomness. Search methods of this type are called *randomized search heuristics* [162]. In this case, even when the search domain and value space are finite, it is not sufficient to describe methods and problems as simple programs. Rather, they can be realized as *randomized programs*; that is, they are programs with a special instruction that, when executed, performs a Bernoulli trial analogous to a single flip of a fair coin. Successive calls to this instruction generate a binary number that can be programmatically cast to an arbitrary finite representation, whether as an integer, a real number, or some structured object. This randomization instruction can be approximated practically on a digital computer by pseudo-random number generators.

In computing probability distributions, a search method or problem can choose to assign zero probability to all search points or values save one, resulting in a deterministic choice in spite of the representation as a probability. In essence, such methods or problems can be computed by a program that never invokes the randomized instruction. Thus the sets of randomized search methods and problems are respectively larger than the sets of deterministic methods and problems. But does this larger set of programs imply that there are problems that can be solved more efficiently by a randomized program that depends on the use of randomness? It is an unanswered question of the theory of computational complexity whether the class RP of problems decidable in polynomial time by a randomized program is strictly larger than the class P of problems decidable in polynomial time by a deterministic program. In later chapters, there will be some evidence that the two classes coincide in some general situations, at least for the purpose of search. Certainly, it is true that in finite spaces, there is always a deterministic search method that performs at least as well as every stochastic search method. In Chapter 18 (Section 18.1.2), a proof concept will be suggested which, if correct, implies that for certain spaces, for every optimally-performing randomized method there is a corresponding deterministic method that performs at least as well.

In the randomized setting, the mechanism of randomness can be separated from the search method and the search problem by requiring that these objects not compute points and values directly. Rather, they can be construed as yielding a probability distribution over search points and values, respectively. Randomized search heuristics (and randomized search problems) can be "derandomized" by separating the generation of a probability distribution from the sampling of that distribution. If a randomized search method is viewed as a program that takes the search history so far and converts it to a probability distribution, it is clear that such a program can be made completely deterministic. Schematically,

$$\text{search history} \xrightarrow{\text{search method}} \text{probability distribution} \xrightarrow{\text{sampler}} \text{search point.}$$

A similar derandomization is possible for the search problem as well. Using the schematic above in iterative succession, the history trace can be generated as one possible interaction between a search method and a problem.

In the following chapters, we will not be precise about how to represent a probability distribution from a computational point of view, since computability is tangential to our focus on the probability distribution as an abstract mathematical object.[1] However, since the current chapter is focused on finite sets, it will be useful to impose finiteness on the acceptable probability distributions for the time being. The most general notion would be to consider probability distributions over search points as finite sampling programs that take no input (*i.e.*, monads) and generate a point in the search space within finite time. Yet this notion is inconvenient for our purposes, firstly because portions of the search method can be snuck inside of the sampling routine so that the schematic delineation is unclear, and secondly because there are an infinite number of such sampling programs. In order to impose finiteness and provide a clear boundary between the search method and the sampler, a probability distribution $\mathbb{P}$ over a finite space will be called *finitely representable*

[1] That is, as a function from sets to real numbers satisfying the probability axioms, discussed further in Chapter 6.

with respect to a finite set $\mathfrak{f} \subseteq \mathbb{R}$ if the probability density $\mathbb{P}(x)$ lies in $\mathfrak{f}$ for any particular point $x$ in the space. In particular, the set $\mathfrak{f}$ might be the set of 64-bit floating point numbers with their standard representation. When the search space $X$ is finite, a finitely representable probability distribution over the search space can be represented as an array of $|X|$ floating point numbers.

To review, the search space and the value space are finite, and the search method and search problem are functions that map history prefixes to finitely representable probability distributions over search points and values, respectively. Although we have not thus far required it, these stringent assumptions of finiteness entail that all search methods and problems that are expressible in this setting are also computable, as will be detailed below.

In what follows, both search methods and problems are assumed to produce probability distributions in the manner described above. The next subsection exhibits a range of common optimization contexts that can be naturally described in the setting described above.

### *3.1.3 Optimization*

The general setting of a search space $X$ and a value space $Y$ together with search methods, problems and goals is abstract. It is important to understand how these abstractions apply to optimization in practice and why this level of abstraction was chosen. What will be seen is that most settings differ from each other in how the value $y_t$ is assigned to search point $x_t$ at time $t$. The value $y_t$ can be chosen deterministically or stochastically. It can be determined by a static function or by a function that depends on the time $t$, or it may even depend on exogenous factors. In each case, the choice of the value space $Y$ bounds what the search method can observe, and the search problem establishes the mechanism by which these observations are made. The value space $Y$ is allowed to be flexible in order to accommodate various ways of passing information to the search method. Some common scenarios are discussed in this subsection, including static, stochastic, and dynamic optimization.

#### 3.1.3.1 Static Black-box Optimization

Although we have required $X$ and $Y$ to be finite in this chapter, many of the examples below pertain to real numbers. As above, finiteness can be maintained by defining a finite set $\mathfrak{f} \subseteq \mathbb{R}$ to stand in for the real numbers. For instance, this set $\mathfrak{f}$ could represent the standard floating point number for some fixed number of bits, which also includes special objects to indicate overflow and invalid operations (i.e., $\infty$, $-\infty$, and NaN, that is, Not a Number). For simplicity, the set $\mathfrak{f}$ will be called the *floating point numbers*. The number of bits will not be relevant so long as it is finite, and thus this detail is omitted from the description. We begin with a trivial example of optimization.

*Example 3.1.* Let the search domain be the two point set $X = \{-1, 1\}$, and let the value space $Y$ be the set of floating point numbers, $Y = \mathfrak{f}$. Consider linear functions of the form $u(x) = ax + b$ for $a, b \in \mathfrak{f}$ and $x \in X$. Then $u(x)$ is a floating point number. Suppose one wishes to find the search point with the maximum value on $u$. One has a program that computes $u$ but does not know the value of $a$ and $b$. The solution is to compute $u(1)$ and $u(-1)$, and then return 1 if $u(-1) < u(1)$ and $-1$ otherwise.[2]

In Example 3.1, the search domain has just two points. The search problem is determined by applying a function $u : X \to Y$ to the search domain. Except for quirks involving the special object NaN, the value space $Y$ is totally ordered, so that it is possible to accomplish the search goal, which is to find the maximum value. This example is an instance of *static optimization*. The function $u$ is variously called the *fitness function*, the *cost function*, the *objective function* or the *target function*. The terminology varies by academic community, although it should be noted that a fitness function is typically maximized and a cost function is minimized. The term *objective* is used for the most part in this book, and $u(x)$ is the *objective value* of the search point $x$ (or, sometimes, the *fitness* or *cost* of $x$). The word *optimization* applies to this setting because the search goal is to find the search point whose objective value is least or greatest. In order for the words "least" and "greatest" to make sense, the set $Y$ needs to be ordered.[3] This optimization is *static* because values are assigned to search points from an unchanging fitness function. As a result, there are only two traces that do not contain repetition once $u$ is fixed:

$$\begin{aligned} &\text{Trace}\,1 : (1, u(1)), \quad (-1, u(-1)) \\ &\text{Trace}\,2 : (-1, u(-1)), \quad (1, u(1)). \end{aligned} \tag{3.2}$$

When the objective function is static, there is no need to repeat search points, since $u(x)$ does not change. Thus only two reasonable optimization methods exist for this optimization problem, one that chooses $x_1 = 1, x_2 = -1$, and one that chooses $x_1 = -1, x_2 = 1$. If $x = 1$ maximizes $u(x) = ax + b$, then $x = -1$ maximizes $u(x) = -ax - b$, and both are valid objective functions. Thus if all values of $a$ and $b$ are equally likely, neither available method consistently finds the optimum any faster than the other. This problem is thus a simple example of No Free Lunch (NFL), discussed in detail in Chapter 12, in which the observed values are independent of non-repeating search trajectories.

Example 3.1 is also an example of *black-box optimization*, in which there is an objective function to which the search method has *oracle access*, i.e., it can only observe the objective values of the search points. The objective function is a *black-box*; its internal composition is hidden from the search. This text is not only about black-box optimization. Note that the black-box restriction in Example 3.1 was imposed by how the search problem was expressed, specifically, $y_t = u(x_t)$.

---

[2] Of course, if $u(1) = \infty$ or $u(1) =$NaN, one could just return 1 without checking $u(-1)$.

[3] Note that the floating point numbers are not totally ordered, due to the fact that NaN is not ordered. However, if NaN is ignored, $\mathfrak{f}$ is totally ordered elsewhere, and this quirk is generally ignored in this text.

#### 3.1.3.2 Static Optimization Without the Black Box

The next example shows how disconnecting the value space $Y$ from the objective function allows gradient methods to be analyzed using the same framework.

*Example 3.2.* Set the search domain $X$ to be the set of floating point numbers, $\mathfrak{f}$, and again let $Y = \mathfrak{f}$ as well. Consider an objective function $u(x) = ax^2 + bx + c$ and its derivative $u'(x) = 2ax + b$, both defined for $x \in X$ and unknown $a, b, c \in \mathfrak{f}$ such that $a$ is positive. Suppose that the search problem assigns $y_t = u'(x_t)$ with the search goal of finding the *minimum* value of $u$ over $X$, which occurs when $x = -b/2a$. Let the search method be gradient descent starting at the origin. That is, it starts with $x_1 = 0$ and then sets $x_{t+1} = x_t - \eta y_t = x_t - \eta u'(x_t)$ where $\eta$ is a small constant learning rate. If $\eta$ is small but not too small, then $x_t$ will quickly approach the search point with the minimum value for $u$.[4]

Example 3.2, like Example 3.1, describes static optimization, but instead of oracle access to the objective function, the search method has oracle access to its derivative. Extending the example, one could alter the search problem to provide $y_t = u'(x_t)/u''(x_t)$ and obtain the Newton-Raphson method, essentially by giving oracle access to the ratio of the first and second derivatives. In either case, the choice of $y_t$ determines the information that the search method can use in order to approach the optimum. At the extreme, one could set $y_1 = -b/2a$, in which case in which case the problem would be solved at the first step. Obviously, this extreme example has little practical value other than to point out the flexibility of the framework.

#### 3.1.3.3 Optimization Subject to Noise

So far the subject of randomness has been avoided. In fact, all components, search method, problem, and goal, can be randomized. As mentioned above, when the search method chooses $x_t$ stochastically, it is called a *randomized search heuristic* [162]; such methods are studied in depth below. The search problem might vary stochastically for many reasons. One common reason is due to measurement noise, as described in the next example.

*Example 3.3.* Suppose we are presented with five equally sized metal bars composed of alloys of unknown origin. We are asked to determine which of these alloys can best withstand a heavy blow. In order to test this property, each alloy is fixed at the edges and hit with a large known force in controlled conditions. The displacement of the alloy at the point of impact is measured, and the alloy with the lowest displacement is selected.

The search domain in this case is the set of alloys. The value space is the set of floating point numbers. The search method is exhaustive; each alloy is tested in any

[4] Whether the learning rate $\eta$ is "small but not too small" is determined by the value of $a$, $b$, and $c$. It can be tested explicitly using a pair of conditions, the Armijo-Goldstein condition and the Wolfe condition.

order. The search problem provides the digitized displacement of the alloy after the experiment. It is determined by two factors: the physical experiment *and* the subsequent measurement of the displacement along with its conversion to a floating point number. Both factors are subject to noise. Unless the instrumentation is perfectly precise, repetitions of the experiments may not produce the same displacement. Formally, one can imagine an objective function $u$ that represents the displacement measured by the experiment. When the alloy $x_t$ is tested in the $t^{th}$ experiment, the search problem provides the observation $y_t = u(x_t) + \varepsilon_t$, where $\varepsilon_t$ is noise, commonly assumed to be a standard normal, independently distributed for each time $t$. If the same alloy is tested at times $t$ and $t+n$ so that $x_t = x_{t+n}$, it does not follow that $y_t = y_{t+n}$. Given the measurement noise, the experiment should be repeated for each alloy (supposing we have multiple copies!) until a certain confidence level is reached that the chosen alloy is likely to be the strongest.

As stressed in Example 3.3, noise of various forms is a standard property of real physical systems that must be addressed when optimization pertains to properties or configurations of such systems. Noise, however, is only one example of *stochastic optimization*, in which the search problem provides feedback that varies according to some probability distribution. In the case of noise, it is assumed that there is an underlying static function that simply cannot be measured accurately. In the general stochastic case, there need be no such function.

#### 3.1.3.4 Optimization in Dynamic Environments

Another class of optimization problems occurs when the system being optimized varies during the course of the search, known as *dynamic optimization*.

*Example 3.4.* Let the search domain be the two point space $X = \{-1, 1\}$, and the value space the floating point numbers $Y = \mathfrak{f}$. Suppose there is a time-varying objective function $u(x,t) = x \sin t$, and the search problem yields $y_t = u(x_t, t)$. If the search goal is to maintain the highest possible value of $y_t$ at all times $t$, then the search method should output $x_t = 1$ when $\sin t$ is positive and $x_t = -1$ when it is negative.

Example 3.4 illustrates how dynamic optimization is closely related to control problems. In fact, most control problems are both dynamic and stochastic, and the identity of the objective $u$ is often unknown. As will be seen in Chapter 13, methods from control theory can also be used to derive the best search method for a given search goal.

#### 3.1.3.5 Optimization and Game Theory

*Example 3.5.* Suppose that the search domain is a set of moves for the first player of a two-player zero-sum game. The value space is the set of moves for the opponent.

The search method plays the game by choosing a move $x_t$, and the search problem responds by choosing a countermove, $y_t$. The search goal is to maximize the value of the game for the first player.

Although Example 3.5 seems disconnected from the previous examples of optimization, in actuality it generalizes them all. As it turns out, all of the above classes of optimization can be expressed as instances of just such a two-player zero-sum game in which the opponent, representing the search problem, is restricted in a variety of ways, either determined by a single function (*static optimization*), limited to the addition of statistical noise (*stochastic optimization*), or forced to vary over time according to strict rules (*dynamic optimization*). This topic is ultimately the centerpiece of this book. It will be taken up again in Chapter 13 with a full discussion in Chapter 17. Note that this game perspective is equivalent to the stimulus-response formulation of Section 1.1.

This subsection has demonstrated many ways in which different limitations on the search problem produce different kinds of optimization problems, from static black-box optimization to dynamic and stochastic problems. This framework is also broad enough to include multi-objective problems as well, although no example of multi-objective optimization has been given. The formal separation of search methods, problems, and goals, along with the arbitrary nature of the search domain and value space, was chosen specifically to treat this wide range of settings within a single framework, with concrete conclusions drawn by narrowing the value space, search problem, and goal as needed. With this background, the next section formalizes the concept of a search method, while focusing for the time being on randomized, static, black-box optimization in finite spaces.

## 3.2 Step-by-Step Generators of Search Methods

The very phrase *iterative search* suggests a procedure that is repeatedly applied. Hence our analysis begins by formalizing the mechanism that iteratively generates search points, naming this mechanism a *generator*. These generators can be more or less similar to each other, establishing a notion of distance between generators that is well defined *and computable* for any two search methods, including those in common use. What initially appears to be a long list of unrelated optimization methods in fact turns out to be a set of points on a high-dimensional map.

### *3.2.1 The Generator*

As discussed in the previous section, we require a search method to produce a probability distribution over search points, from which the *next* search point will be sampled. In the most general form, we allow this probability distribution to depend on the entire history of search points and values preceding the current time step, but on none of the subsequent points. On this basis, the search method can be represented as a map from history prefixes to probability distributions. Thus we define a *generator* of a search method as a function

$$\mathcal{G} : \mathcal{H}_{X,Y} \to \mathcal{P}_X, \tag{3.3}$$

where $\mathcal{H}_{X,Y}$ is the space of history prefixes on the (finite) search space $X$ and the (finite) value space $Y$, and $\mathcal{P}_X$ is the space of finitely representable probability distributions over $X$, where finitely representable means that for each search point $x \in X$ and each probability distribution $\mathbb{P} \in \mathcal{P}_X$, the probability is a floating point number, i.e., $\mathbb{P}(x) \in \mathfrak{f}$, and the sum of all such $\mathbb{P}(x)$ is the floating point representation of the number one.[5]

Consider the space of history prefixes, $\mathcal{H}_{X,Y}$. This space is composed as a union over all histories of fixed length less than the maximum time, $N$. The history of the search up to time $t$ consists of sequences formulated as

$$(x_1, y_1), (x_2, y_2), \ldots, (x_t, y_t)$$

and thus histories prefixes of length $t$ are elements in the space $(X \times Y)^t$, which is finitely representable as a list data structure over pairs of elements from the finite spaces $X$ and $Y$. Thus the mathematical space of history prefixes

$$\mathcal{H}_{X,Y} = \bigcup_{t=0}^{N-1} (X \times Y)^t \tag{3.4}$$

is representable in computational terms as a list of such pairs of length at most $N-1$. In this text, the empty list is represented by $\emptyset$, and longer lists are represented as tuples, i.e., $((x_1, y_1), \ldots (x_t, y_t))$.

Likewise, a distribution $\mathbb{P}$ in the space of probability distributions $\mathcal{P}_X$ can be represented as a fixed-length array of floating point numbers whose length is the size of $X$, that is, $|X|$. The requirement that these numbers sum exactly to floating point one is perhaps excessively restrictive, but we will ignore the complications of rounding errors for now.

The broader point is that each of these distributions has a finite bit representation of bounded size, as does every history prefix. Furthermore, there are a finite number of distributions and prefixes. Consequently, every generator $\mathcal{G}$ has a finite representation as a table with one slot for each history prefix containing a representation of a probability distribution. This is possible because histories have a maximum length and contain only elements of finite sets. Table 1.1 in Chapter 1 gave an example of such a table for deterministic optimizers, in which the next search point is chosen explicitly and not sampled. The representation of a generator follows the same

[5] There is a problem of existence here due to the fact that floating point representations of a probability vector with rational components might not sum exactly to one. The purpose of this chapter is illustrative, to prepare intuitions for the more rigorous formalism of Chapter 6. Consequently, such details are ignored for now.

pattern, except that the entries in the table specify a probability distribution over search points rather than a single search point. Since it is additionally the case that every probability distribution of the form above can sampled exactly on a digital computer equipped with a randomization instruction, it then follows that sampling the next search point from $\mathcal{G}$ for a given history prefix is computable.

The probability distribution that $\mathcal{G}$ assigns to a history $h$ is denoted with square brackets as $\mathcal{G}[h]$. Aside from emphasizing the tabular structure of $\mathcal{G}$, these square brackets allow us to write $\mathcal{G}[h](x)$ for the probability that $\mathcal{G}$ assigns to search point $x$ given history $h$ while avoiding repeated parentheses. The square brackets are merely a notational convenience, since $\mathcal{G}[h]$ is to be interpreted as a function mapping. At a later point, these objects will be viewed as conditional probability distributions, in which case the notation $\mathcal{G}(x \mid h) = \mathcal{G}[h](x)$ could be used as well.

Now suppose that the value space is totally ordered and that the search goal is to minimize a computable function $u : X \to Y$ within $N$ time steps. Then the value $y_t = u(x_t)$ is computable, and hence the entire optimization scenario is computable using the pseudocode in Algorithm 1. Notice that nothing has been said about *how* the optimizer generates search distributions. No matter how it does, the resulting minimization procedure is computable solely due to the finiteness of the search space and the value space, the computability of the function $u$, and the fact that the search distributions are finitely representable.

**Algorithm 1** Pseudocode to Minimize a Function $u$ Using Search Generator $\mathcal{G}$

$h_0 \leftarrow \emptyset$ {Initialize the history}
**for** $t \leftarrow 1 \ldots N$ **do**
  $\mathbb{P}_t \leftarrow \mathcal{G}[h_t]$ {Look up distribution from history}
  $x_t \sim \mathbb{P}_t$ {Sample a search point}
  $y_t \leftarrow u(x_t)$ {Compute the value}
  $h_t \leftarrow h_{t-1} \| (x_t, y_t)$ {Append to the history}
**end for**
**return** $x_t$ such that $u(x_t) = \min_s y_s$

To stress further, the search method could have been a genetic algorithm, the Nelder-Mead method, simulated annealing, or any other stochastic, iterative, black-box optimization method, because all of these are implemented as computable programs. Of course, Algorithm 1 is not the most computationally efficient algorithm implementing these optimization methods, since it relies on enumerating the entries of $\mathcal{G}$ in table form. Nonetheless, from a mathematical perspective, it makes no difference. Each search method can be converted to an equivalent representation as a generator based on how it samples each successive search point. The details of this conversion are provided for common methods in Chapters 6 and 7.

### 3.2.2 Distance and Similarity Among Generators

In the previous section, each search method was represented as a large table with one row per history prefixes containing a representation of a finitely representable probability distribution. Now suppose that the search space $X$ contains at least two elements $x$ and $w$. Then for some small floating point number $\varepsilon$ and some history $h'$, one can take any generator $\mathcal{G}$ that assigns probability at least $\varepsilon$ to $x$ on $h'$ and form a new generator $\mathcal{G}_{\varepsilon,h'}$ by moving an $\varepsilon$ of probability from $x$ to $w$ on $h$. Formally,

$$\mathcal{G}_{\varepsilon,h'}[h](z) = \begin{cases} \mathcal{G}[h'](x) - \varepsilon & \text{if } h = h' \text{ and } z = x, \\ \mathcal{G}[h'](w) + \varepsilon & \text{if } h = h' \text{ and } z = w, \\ \mathcal{G}[h](z) & \text{otherwise.} \end{cases}$$

How different is $\mathcal{G}_{\varepsilon,h'}$ from $\mathcal{G}$? It stands to reason that the two generators differ by an amount proportional to $\varepsilon$. It should also be clear that for *any* other generator $\mathcal{G}'$, it is possible to formulate a sequence of steps $(\varepsilon_1,h_1),(\varepsilon_2,h_2),\ldots,(\varepsilon_M,h_M)$ such that $\mathcal{G}_{\varepsilon_M,h_M} = \mathcal{G}'$. That is, there is a path through the space of generators from $\mathcal{G}$ to $\mathcal{G}'$. Now $\mathcal{G}$ and $\mathcal{G}'$ are arbitrary here; such a path exists for any pair of optimizers defined with respect to the search space $X$ and the value space $Y$. By moving along such paths, one can transition from, say, the simple genetic algorithm to differential evolution to generating set search.

Furthermore, for a small floating point number $\varepsilon > 0$, the steps $\varepsilon_i$ can chosen so that $\varepsilon < \varepsilon_i < 2\varepsilon$. That is, the steps along the path from $\mathcal{G}$ to $\mathcal{G}'$ can be made as fine as the floating point numbers allow.[6] In this sense, the transitions between search methods are therefore smooth and not just connected. The space of generators now appears to be a discretized version of a continuous space.

The distance between two generators $\mathcal{G}$ and $\mathcal{G}'$ can be quantified in many ways. One could take the average difference in the probabilities that they assign across all histories, or the maximum difference. Or, one could start with a notion of distance between probability distributions and then take the average or maximum difference across all histories. To permit generalization to infinite spaces, we choose a notion of distance that begins with the maximum difference between probability distributions and then takes the maximum difference over all histories, defined as

$$\rho(\mathcal{G},\mathcal{G}') = \max_{h\in\mathcal{H}_{X,Y}} \max_{x\in X} \left| \mathcal{G}[h](x) - \mathcal{G}'[h](x) \right|. \tag{3.5}$$

Note that because probabilities are between zero and one, it holds that for all generators $\mathcal{G}$ and $\mathcal{G}'$ that $\rho(\mathcal{G},\mathcal{G}') \leq 2$. Properly speaking, distance metrics are defined by four properties:

- **Nonnegativity**: $\rho(\mathcal{G},\mathcal{G}') \geq 0$.
- **Identity at Zero**: $\rho(\mathcal{G},\mathcal{G}') = 0$ if and only if $\mathcal{G} = \mathcal{G}'$.

[6] Of course, the floating point numbers are a discrete space and this discussion ignores the issue of underflow. The purpose of the discussion, however, is to lead into a continuous treatment of generators, for which underflow is not an issue.

- **Symmetry**: $\rho(\mathcal{G},\mathcal{G}') = \rho(\mathcal{G}',\mathcal{G})$.
- **Triangle Inequality**: $\rho(\mathcal{G},\mathcal{G}') \leq \rho(\mathcal{G},\mathcal{G}'') + \rho(\mathcal{G}'',\mathcal{G})$.

Given that the maximum respects the triangle inequality, it is obvious under inspection that Equation 3.5 satisfies all four properties and hence $\rho$ is a distance metric. It is also a computable function, meaning that for finite spaces, the distance between between two search methods is an observable fact.

Returning to the example of $\mathcal{G}_{\varepsilon,h'}$, which was obtained by perturbing $\mathcal{G}$ by a small epsilon, we find that $\rho(\mathcal{G},\mathcal{G}_{\varepsilon,h'}) = \varepsilon$, just as we would expect. So this distance metric respects our intuitions about how it should behave.

One of the interesting facts about the metric $\rho$ is that although it essentially measures the distance between two programs, it is defined mathematically and not with respect to any particular computation. No matter how the mathematical objects $\mathcal{G}$ and $\mathcal{G}'$ are represented as programs, the distance between them is constant.

### 3.2.3 Performance and Continuity

Once a notion of distance has been established, one can ask more interesting questions about the behavior of a search method. In particular, when do similar search methods behave similarly?

In most practical optimization experiments, performance is determined by examining the history trace. The quantities of interest often answer the questions such as: How many time steps are required to reach the global optimum? Or, how close does the search method come to the optimum in a fixed budget of time steps? Thus it is reasonable to consider performance as being a property of the history trace.

The definition of the distance metric $\rho$ in Equation 3.5 makes it clear that if two generators are close to each other, then the search methods that they generate assign similar probabilities to each history trace. Consequently, any quantity that is computed based on the contents of the history trace should produce similar results for similar search methods. In fact, unless one chooses to compute performance in a way that explicitly amplifies small differences, similar generators will indeed produce similar performance results, simply due to the fact that they yield similar history traces. Mathematically, these questions depend on the *continuity* of the performance function, which forms the subject of Chapter 8.

Suppose, then, that there is a computable function $\phi : (X \times Y)^N \to \mathbb{R}$ that assigns a score to each history trace. Then the overall performance of a search method on a search problem can be computed as a weighted average over history traces

$$V = \sum_{h \in (X\times Y)^N} \phi(h)\, \mathbb{P}(h), \tag{3.6}$$

where $\mathbb{P}(h)$ is the probability that a history trace $h$ will be produced by a particular search method and problem, discussed further below.

If we assume that $\mathbb{P}$ can be chosen arbitrarily, then an *optimal* choice is set $\mathbb{P}(h)$ to place probability one on a history trace $h^*$ that maximizes the performance $\phi$, *i.e.*, such that for all histories $h$, $\phi(h^*) \geq \phi(h)$. If instead the search problem is a fixed function $u : X \to Y$, then the history trace $h$ is determined by the search history $h_X$ and the objective $u$. If we write $h = h(h_X, u)$, then by the same reasoning there is a search history $h_X^*$ that maximizes $\phi$. Any search method that places probability one on the components of $h_X^*$ has maximal performance $V$. Consequently, for any static objective $u$, there is a deterministic search method that equals or outperforms every other search method in terms of $\phi$-performance on $u$, at least in the finite case. In the infinite case, for fixed $u$, this statement remains approximately true: for any $\varepsilon > 0$, there is a deterministic search method come within $\varepsilon$ of the optimal performance.

The broader question is whether or not there is a general search method that outperforms all other search methods on *any* objective function $u$. This question cannot be answered directly. Instead, one must first specify how $u$ is to be chosen. If $u$ is chosen uniformly at random, which is possible since the search space $X$ and the value space $Y$ are both finite, then the No Free Lunch theorem implies that there is no such general search method. But if the choice of $u$ is biased in some way, then a general method to solve a subclass of problems is possible. These issues are discussed in the infinite setting in Chapter 12, where it will be seen that No Free Lunch corresponds roughly to the case where the joint distribution $\mathbb{P}(h) = \mathbb{P}(h_X, h_Y)$ factors.

As seen above, when one begins to ask questions about performance, the probability distribution governing the history trace becomes the object of interest. In Section 3.3, the system of history traces produced by combining a search method and a search problem is used to develop a different perspective on what search methods are and how generators are obtained from them. But first, we make some comments on computational efficiency.

### *3.2.4 Computation and Efficiency*

Algorithm 1 describes a minimization program. For a particular objective function $u$ and a search method generated by $\mathcal{G}$, it is not necessarily the most efficient program applying $\mathcal{G}$ to minimize $u$. Firstly, the algorithm assumed that the entire history prefix would be kept at each time step. Secondly, the algorithm presumed that $\mathcal{G}$ would be stored as a lookup table with one entry per history prefix consisting of an array of floating point numbers, one for each search point. Thirdly, the algorithm samples $\mathcal{G}[h_t]$ using a generic program taking a probability distribution as input. Efficiency can be gained on all three fronts.

With respect to storing the history, the probability distribution $\mathcal{G}[h_t]$ often only depends on the last few time steps. When the generator only depends the previous $n$ time steps, we say that the generator is $n$-steps Markov.[7] For a Markov generator,

[7] In probability theory, a *Markov process* is a stochastic sequence whose values are independent of all but the prior time step.

even the tabular representation of the generator $\mathcal{G}$ can be exponentially compressed, since rather than indexing by the entire history prefix, one can index the table by just the last few time steps. Most common search methods are Markovian. Simulated annealing is 1-step Markov, and a genetic algorithm with population size $K$ is effectively $K$-steps Markov.

Even for non-Markovian methods, the generator representation can usually be compressed substantially from the tabular representation. For example, the search method might only depend on certain statistics from the history, as CMA-ES does when it updates a population mean and covariance based on the history. In order to compress the generator representation, it is useful to know what information the generator requires to sample the next point.

An *information dependency system* is a tuple $(I, \mathcal{I}, U)$ consisting of an *information space* $\mathcal{I}$, an *information function* $I : \mathcal{H}_{X,Y} \to \mathcal{I}$, and an *information update function* $U : \mathcal{I} \times X \times Y \to \mathcal{I}$ such that for any history prefix $h$, search point $x$, and value $y$, the equation $I(h \| (x,y)) = U(I(h), x, y)$ holds, recalling that $\|$ indicates concatenation. The purpose of an information dependency system is to encapsulate the dependence of a generator on the history. The information space provides an intermediate representation from which $\mathcal{G}[h]$ can be computed. The information function processes the history $h$ into this representation, and the information update function provides a further representation of the information function that is expressed with reference only to the (smaller) information space rather the exponentially large space of history prefixes.

A generator $\mathcal{G}$ for a search method is expressible within an information dependency system if $\mathcal{G}$ is decomposable using a proxy function $\tilde{\mathcal{G}} : \mathcal{I} \to \mathcal{P}_X$ with $\mathcal{G} = \tilde{\mathcal{G}} \circ I$, i.e., $\mathcal{G}[h] = \tilde{\mathcal{G}}[I(h)]$. In this case, the information function $I$ mediates the dependence of the search method on the history and represents the only quantities that must be computed in order to sample from $\mathcal{G}[h]$. Since the information function can be updated without requiring the full history, the representation of the generator $\mathcal{G}$ can be compressed at least to a table indexed by the information space rather than the space of history prefixes.

There is some sleight of hand in this construction. After all, a generator may be expressible within many information dependency systems, including at least the trivial system where $\mathcal{I} = \mathcal{H}_{X,Y}$, $I(h) = h$, and $U(I(h), x, y) = h \| (x,y)$. But a generator may be expressible within a system with smaller computational requirements, as with the case of the $n$-step Markov generators, which are expressible with respect to the information space $\mathcal{I} = (X \times Y)^n$.

The most efficient information dependency system for a particular generator will depend on the details of the sampling function as well as tradeoffs in space and time between the information representation and the information updates. Such questions are not addressed here. Nonetheless, more efficient information representations can certainly be used to compute search methods more efficiently. Because this book is not primarily concerned with question of computability, the information required by a search method is not extensively studied here, but absolutely is a quantity of interest for analysis. In particular, it would be useful to understand how placing bounds on the amount of information available for use constrains performance. Such questions are left as future work and are not addressed substantially in this book.

The sampling methodology can also have an impact on computational efficiency. The generic sampling methodology of Algorithm 1 for finitely representable probability distributions increases in complexity roughly with the length of the longest fixed point bit representation for any probability value in the distribution. To sample a fair coin flip (probability of heads = probability of tails = $1/2$) requires only one call to the randomizing instruction. To sample a weighted coin for which the probability of heads is $1/1024$ requires $\log_2 1024 = 10$ calls to the randomizing instruction. Finer probability distinctions are more computationally complex.

Take the example of sampling a discretized Gaussian distribution. For a large finite search space $X$, some members of $X$ would occur with exponentially small probability, and hence the generic sampling method is computationally inefficient on this distribution. In contrast, the Box-Muller transform can sample from a Gaussian distribution by sampling a two uniform variables and combining them. These uniform variables can each be generated using a fixed number of bits and hence a fixed number of calls to the randomizing instruction. Thus the complexity of sampling a Gaussian using a specialized method is far lower than using the generic method. For this reason, it is impossible to draw many conclusions about the computational complexity of search methods based on their generators.

The important issue is whether similar search methods (in terms of the distance metric introduced in the previous subsections) have similar computational complexity. This question cannot be answered firmly. For example, the Box-Muller transform makes it efficient to sample from a Gaussian distribution due to the fact that the probability density of the Gaussian is expressible in terms of elementary functions. Slight perturbations to the Gaussian function may result in a distribution that is no longer expressible in elementary terms and can no longer be efficiently sampled.

Even if computation is forced to used fixed information dependency systems and the generic sampling methodology, there is no guarantee that computational complexity will vary smoothly among similar search methods. In the example of a coin flip, the difference in cost between sampling a fair coin with $1/2$ probability of heads versus a weighted coin with $1/(2+\varepsilon)$ yields an increase of roughly $\log_2 \varepsilon$ calls to the random instruction.

With that said, there are many cases where two computationally efficient search methods can be combined in such a way that the efficiency is preserved. One example of this is when two search methods are linearly combined, so that each search method is used to choose the next search point approximately half of the time. These convex combinations are discussed in more detail in Chapter 11.

The purpose of this excursion into computational complexity has been to introduce the concept of information dependence and to illustrate that the analysis of optimization methods pursued in this book is not attuned to the study of computational complexity, and that this complexity depends on myriad obscure factors underlying the mathematical objects that express the search method. With these things said, we return to the topic of mathematical analysis by looking at the entire history trace.

## 3.3 Search Methods as Conditional Probabilities

The concept of a generator matches up with how one programs an iterative optimization method. A search point is chosen, a value is assigned, another search point is chosen, and so on, *ad infinitum*. But there is another way to characterize search methods based on their behavior over all time steps, and this perspective is important for analyzing features such as performance that depend on the entire history trace.

The contrast between the long-term and one-step behavior mirrors the distinction between the *normal form* and the *extensive form* in game theory, a connection which is explored further in Chapter 13. Although the characterization over all steps at once is a less intuitive description of optimization from a programmatic point of view, it can be more convenient from a mathematical perspective. This material presages the content of Chapter 9, which demonstrates the same relationship for topological spaces.

### 3.3.1 From Generators to Trajectors

The history trace is generated by applying the generator $\mathcal{G}$ of a search method to itself based on the output of an objective function $u$. This self-application is implicit in Algorithm 1. The algorithmic presentation can also be modeled mathematically to produce a probability distribution over history traces.

The search process begins by sampling the first search point $x_1$ from the generator applied to the empty history (denoted $x_1 \sim \mathcal{G}[\emptyset]$), and then the second search point $x_2$ is sampled from the generator applied to the pair $(x_1, u(x_1))$. Overall, the process is repeated for $N$ steps and looks like

$$\begin{aligned}
x_1 &\sim \mathcal{G}[\emptyset] \\
x_2 &\sim \mathcal{G}[(x_1, u(x_1))] \\
x_3 &\sim \mathcal{G}[(x_1, u(x_1)), (x_2, u(x_2))] \\
x_4 &\sim \mathcal{G}[(x_1, u(x_1)), (x_2, u(x_2)), (x_3, u(x_3))] \\
x_5 &\sim \mathcal{G}[(x_1, u(x_1)), (x_2, u(x_2)), (x_3, u(x_3)), (x_4, u(x_4))] \\
&\dots \\
x_N &\sim \mathcal{G}[((x_1, u(x_1)), \dots, (x_{N-1}, u(x_{N-1})))].
\end{aligned}$$

So the history trace is generated iteratively by $\mathcal{G}$.

For a fixed objective $u$, suppose we wish to sample a sequence of search points all at once rather than iteratively. A probability distribution for this purpose can be constructed by induction on the number of time steps; we will call it $\mathcal{A}_u^t$. This probability distribution samples $t$ search points at a time, and so it is a distribution over the product space $X^t$. When $t = 1$, $\mathcal{A}_u^1$ is a distribution over $X$ that corresponds to choosing the first search point with no history. That is,

$$\mathcal{A}_u^1 = \mathcal{G}[\emptyset]$$

If we know $\mathcal{A}_u^{t-1}$, then we can define $\mathcal{A}_u^t$ recursively from $\mathcal{G}$. Specifically, the probability of a sequence $(x_1, \ldots x_t) \in X^t$ is defined as

$$\mathcal{A}_u^t(x_1, \ldots, x_t) = \mathcal{A}_u^t(x_1, \ldots, x_{t-1}) \times \mathcal{G}\left[(x_1, u(x_1)), \ldots, (x_{t-1}, u(x_{t-1}))\right](x_t) \quad (3.7)$$

And hence the probability of a complete search history is a probability distribution $\mathcal{A}_u = \mathcal{A}_u^N$ on the product space $X^N$.

Allowing the objective $u$ to vary, we have a map $u \mapsto \mathcal{A}_u$ that takes an objective function and produces a probability distribution over search histories. This map has a function signature $Y^X \to \mathcal{P}_{X^N}$, where $Y^X$ is the space of objective functions from the search space $X$ to the value space $Y$ and $\mathcal{P}_{X^N}$ is the space of probability distributions over the space of search histories $X^N$.

The domain of this map is unnecessarily large, however, since $\mathcal{A}_u$ only depends the objective values of $u$ observed over the course of the search history. Thus the probability distributions $\mathcal{A}_u$ can be reshuffled and reindexed using value histories drawn from the space $Y^N$. For each $(y_1, \ldots, y_N) \in Y^N$ and $(x_1, \ldots, x_N) \in X^N$, define

$$\mathcal{A}_y(x_1, \ldots x_N) = \mathcal{A}_u(x_1, \ldots, x_N) \qquad \text{for all } u \text{ s.t. for } 1 \leq t < N,\ y_t = u(x_t).$$

It remains to show that $\mathcal{A}_y$ is uniquely defined. Notice that Equation 3.7 can be recursively expanded so that

$$\mathcal{A}_u(x_1, \ldots, x_N) = \prod_{t=1}^{N} \mathcal{G}\left[(x_1, u(x_1)), \ldots, (x_{t-1}, u(x_{t-1}))\right](x_t).$$

Using this expansion, it is apparent that if $u$ and $u'$ are two objectives with $y_t = u(x_t) = u'(x_t)$ for all $1 \leq t < N$, then $\mathcal{A}_u(x_1, \ldots, x_N) = \mathcal{A}_{u'}(x_1, \ldots, x_N)$, which justifies the definition of $\mathcal{A}_y$.

The map $y \mapsto \mathcal{A}_y$ associates each value history with a distribution over search histories. It will be called a *trajector* because it associates trajectories between the search and value spaces. It will be denoted generally by $\mathcal{A}$ to indicate its connection to a search algorithm, or, as will be seen, as an algorithm within a broader computing context. A trajector applied to a history $y$ will be written as either $\mathcal{A}_y$ or $\mathcal{A}[y]$. The map $u \mapsto \mathcal{A}_u$ has a similar structure, and such objects will be called *static trajectors* reflecting the fact that they are defined with respect to a static objective, whereas $\mathcal{A}_y$ can represent a search method for many types of search problems. When a distinction is needed, $\mathcal{A}_y$ will be called a *standard* trajector.

There are two main reasons to focus on the map $y \mapsto \mathcal{A}_y$ over value histories rather than the map $u \mapsto \mathcal{A}_u$ over objective functions. The first is that the domain of search histories $Y^N$ is what actually determines the behavior of a search method, while large portions of $Y^X$ may remain unobserved. The second reason is more abstract and has to do with a mathematical duality between search methods and problems that is only expressible when using value histories as opposed to objective functions. The basic form of this duality is not introduced until Chapter 17, but it is the primary factor driving this representational choice. To give some early intuition as to how the duality will function, imagine swapping the search space $X$ with the value space $Y$ to obtain a trajector $\mathcal{F}$ with signature $X^N \to \mathcal{P}_{Y^N}$, and notice that just as the trajector $\mathcal{A}$ above determines the behavior of the search method, $\mathcal{F}$ can be viewed as determining the behavior of the search problem.

As a final detail, notice that if two value histories $y$ and $y'$ differ only in their final value, then it will hold that $\mathcal{A}_y = \mathcal{A}_{y'}$ in the equations above, because the search method never uses the final value. Based on the argument above for indexing $\mathcal{A}$ by value histories instead of objective functions, it would seem to follow that $\mathcal{A}$ should actually be indexed by $Y^{N-1}$ and not $Y^N$. However, there are two reasons that we do not do this. Firstly, if the search method is run for infinitely many time steps, this detail disappears, and in all subsequent chapters it will generally be assumed that $N \to \infty$. Secondly, the aforementioned duality requires $Y^N$, not $Y^{N-1}$. The fact that search methods cannot see the final value is an artifact of the asymmetry between search methods and problems, namely, that the search method must propose a search point without knowing the value that will be assigned to it, but the search problem gets to assign values with knowledge of the search point to which the value is being assigned. This structure is inherent in the dichotomy between stimulus and response.

In this subsection, the trajector of a search method has been defined as arising from its generator. In the next subsection, the trajector will be derived a priori from a distribution over history traces, and it will be argued that it provides a more fundamental description of the search method than the generator.

### *3.3.2 Extracting Search Methods from History Traces*

The history trace is a result of an interaction between a search method and a search problem. If the method and problem are fixed, then each possible history trace occurs with some fixed probability. Consider the probability distribution $\mathbb{P}$ that governs all possible history traces given a search method and a search problem. Sampling this distribution yields a complete history $h \in (X \times Y)^N$. Thus, since $X$, $Y$, and $N$ are finite, this distribution can be represented as a list of $|X|^N |Y|^N$ numbers; let us presume for now that these are floating point numbers and that the distribution $\mathbb{P}$ is finitely representable.

If we do not know the search method or the search problem but are given the distribution $\mathbb{P}$ over histories for several different search problems, is possible to disentangle the search method from the search problem? In fact, within certain limits, it is mathematically trivial to do so using conditional probabilities and ranging over all possible search problems.

The first observation is that the space of full histories $(X \times Y)^N$, whose elements are sequences of pairs $((x_1,y_1),\dots,(x_N,y_N))$, is equivalent to the space $X^N \times Y^N$, whose elements are pairs of sequences $((x_1,\dots,x_N),(y_1,\dots,y_N))$. The probability distribution $\mathbb{P}$ on $(X \times Y)^N$ thus has an equivalent distribution $\tilde{\mathbb{P}}$ on $X^N \times Y^N$ defined by

$$\tilde{\mathbb{P}}\Big(((x_1,\ldots,x_N),(y_1,\ldots,y_N))\Big) = \mathbb{P}\Big(((x_1,y_1),\ldots,(x_N,y_N))\Big).$$

Now $\tilde{\mathbb{P}}$ is a joint distribution. Recalling that for a history trace $h \in (X \times Y)^N$, we denote $h_X$ as the history of search points and $h_Y$ as the history of values, we find that $\mathbb{P}(h) = \tilde{\mathbb{P}}(h_X, h_Y)$. The dependence of the search points on the history of values can be isolated as a conditional probability,

$$\tilde{\mathbb{P}}(h_X \mid h_Y) = \frac{\tilde{\mathbb{P}}(h_X, h_Y)}{\tilde{\mathbb{P}}(h_Y)} = \frac{\tilde{\mathbb{P}}(h_X, h_Y)}{\sum_{h'_X} \tilde{\mathbb{P}}(h'_X, h_Y)}, \tag{3.8}$$

which is read as "the probability of $h_X$ given $h_Y$". As it happens, this definition is incomplete, because it is possible that the marginal probability $\tilde{\mathbb{P}}(h_Y) = \sum_{h'_X} \tilde{\mathbb{P}}(h'_X, h_Y)$ is zero. To be concrete, if the search problem involves minimizing a constant function $u(x) = 1$, then one can never know what the search method would have done had it observed $u(x) = 0$, because the value zero will never be observed.

If the conditional probability $\tilde{\mathbb{P}}(h_X \mid h_Y)$ is fully defined (i.e., $\tilde{\mathbb{P}}(h_Y) > 0$ for all $h_Y$), it is a function that maps the iterated value space $Y^N$ to some probability distribution over the iterated search space $X^N$.[8] Motivated by the results of the previous subsection, we note that this object is the trajector of a search method,

$$\mathcal{A} : Y^N \to \mathcal{P}_{X^N},$$

and it maps value histories onto probability distributions over search histories. Again, since $X$ and $Y$ are finite spaces, $\mathcal{A}$ is a finite-dimensional object, and we may require $\mathcal{A}$ to be finitely-representable as well, in which case it is also computable.

Incidentally, the same process can be performed for the search problem using the conditional probability $\tilde{\mathbb{P}}(h_Y \mid h_X)$ in order to define the trajector of a search problem as a function

$$\mathcal{F} : X^N \to \mathcal{P}_{Y^N}.$$

The full flexibility of this definition is not explored until Chapter 17. For now, we limit ourselves to the case where the search problem $\mathcal{F}$ is defined by an objective function $u : X \to Y$, in which case

$$\mathcal{F}((y_1,\ldots,y_N)) = (u(x_1),\ldots,u(x_N)),$$

which will be denoted as $\mathcal{F}_u$.

The two functions $\mathcal{A}$ and $\mathcal{F}$ interact in order to generate a probability distribution $\mathbb{P}$ over history traces, which will be written as $\mathbb{P}_{\mathcal{A},\mathcal{F}}$ for now and as $\mathcal{A} \times \mathcal{F}$ in Chapter 17. Furthermore, for each objective function $u$,

$$\mathcal{A}[h_Y](h_X) = \tilde{\mathbb{P}}_{\mathcal{A},\mathcal{F}_u}(h_X \mid h_Y)$$

---

[8] When defined, a conditional probability distribution is always a probability distribution, meaning that its entries are positive and sum to one.

wherever the latter is defined. For this reason, the notation $\mathcal{A}(h_X \mid h_Y) = \mathcal{A}[h_Y](h_X)$ is introduced for trajectors to emphasize that they are essentially conditional probability distributions defined with respect to all possible joint distributions in which they might participate.

Ranging over all possible objective functions distinguishes the trajectors of any pair of search methods. Suppose that two trajectors $\mathcal{A}$ and $\mathcal{A}'$ are distinct. For any given objective $u$, it is possible that the joint distributions have $\mathbb{P}_{\mathcal{A},\mathcal{F}_u} = \mathbb{P}_{\mathcal{A}',\mathcal{F}_u}$ if $\mathcal{F}_u$ places zero probability on the distinctions between $\mathcal{A}$ and $\mathcal{A}'$, but then there is always some other objective $v$ that does emphasize the difference, so that $\mathbb{P}_{\mathcal{A},\mathcal{F}_v} \neq \mathbb{P}_{\mathcal{A}',\mathcal{F}_v}$.

Thus a trajector for a search method can be extracted from the probability distributions over history traces that it produces when run against various objectives. The generator can also be extracted from the trajector, with certain anomalous exceptions that will be discussed next.

### *3.3.3 From Trajectors to Generators*

Trajectors describe how the history trace is produced for a given objective function, so they must be related to the generators introduced previously, which likewise determine the history trace. The relationship can be derived by running Equation 3.7 in reverse using the sequence of marginals $\mathcal{A}_y^t$ defined so that $\mathcal{A}_y^N = \mathcal{A}_y$ and for $1 \leq t < N$

$$\mathcal{A}_y^t(x_1,\dots x_t) = \sum_{w \in X} \mathcal{A}_y^{t+1}(x_1,\dots x_t, w). \tag{3.9}$$

From these marginals, a generator $\mathcal{G}$ can be found by solving for $\mathcal{G}$ in Equation 3.7,

$$\mathcal{G}\left[(x_1,y_1),\dots,(x_{t-1},y_{t-1})\right](x_t) = \frac{\mathcal{A}_y^t(x_1,\dots,x_t)}{\mathcal{A}_y^{t-1}(x_1,\dots,x_{t-1})}. \tag{3.10}$$

The form of this definition – a joint distribution divided by one of its marginals – is transparently a conditional probability. We might even have written that

$$\mathcal{G}\left[(x_1,y_1),\dots,(x_{t-1},y_{t-1})\right](x_t) = \mathcal{A}_y^t(x_t \mid x_1,\dots,x_N). \tag{3.11}$$

But conditional probabilities are only defined when the marginal distribution in the denominator is non-zero. It is easily seen from Equation 3.10 that the definition of the generator has gaps when $\mathcal{A}_y^{t-1}(x_1,\dots,x_{t-1}) = 0$. The trajector cannot tell the generator what to do on such histories, because neither the trajector nor the generator will produce them.

This problem may seem subtle, but it demonstrates that the previous definition of a generator is in some sense wasteful. The gaps that remain illustrate situations for which the generator does not need to assign a probability distribution, because these situations never occur. The search method has zero probability of generating these histories. This fact also illustrates that it is the trajector $\mathcal{A}$ and not the generator $\mathcal{G}$ that is primary. A trajector $\mathcal{A}$ is sufficient to completely determine how history traces are generated on a given search problem, but the generator $\mathcal{G}$ contains extra information about history traces that cannot occur.

### 3.3.4 Similarity and Performance of Trajectors

Briefly, we will demonstrate one of the effects of the definition of trajectors by showing that performance varies smoothly among similar trajectors. This connection is explored for continuous spaces in Chapter 8, but it is worth bringing up now because it hints at how this mathematical formalization of search methods can be applied to questions of performance.

A distance metric can be defined for trajectors just as was done for generators with

$$\rho(\mathcal{A},\mathcal{A}') = \max_{h_Y \in Y^N} \max_{h_X \in X^N} \left| \mathcal{A}(h_X \mid h_Y) - \mathcal{A}'(h_X \mid h_Y) \right|, \tag{3.12}$$

where the symbol $\rho$ has been overloaded to apply to trajectors. Again, for any search space with at least two histories $h_X$ and $h'_X$ we can define a "neighbor" $\mathcal{A}_{\varepsilon,h_Y}$ for $\mathcal{A}$ by shifting an $\varepsilon$ of probability from $\mathcal{A}[h_Y](h_X)$ to $\mathcal{A}[h_Y](h'_X)$, and then $\rho(\mathcal{A},\mathcal{A}_{\varepsilon,h_Y}) = \varepsilon$.

In Section 3.2.3, the performance of a search method was defined weighted average of the performance on each history traces. Given the discussion above, we can now rewrite the performance for a trajector $\mathcal{A}$ as

$$V(\mathcal{A}) = \sum_{h \in (X \times Y)^N} \phi(h)\, \mathcal{A}(h_X \mid h_Y)\, \tilde{\mathbb{P}}(h_Y) \tag{3.13}$$

using the perspective on $\mathcal{A}$ as a conditional probability, where $\tilde{\mathbb{P}}(h_Y)$ is the prior probability of the value history, since $\mathbb{P}(h) = \mathcal{A}(h_X \mid h_Y)\, \tilde{\mathbb{P}}(h_Y)$.

Now suppose that $\mathcal{A}'$ is any trajector such that $\rho(\mathcal{A},\mathcal{A}') \leq \varepsilon$ for some $\varepsilon > 0$, as is the case for $\mathcal{A}_{\varepsilon,h_Y}$ above. In that case, Equation 3.12 implies that

$$\left|V(\mathcal{A}) - V(\mathcal{A}')\right| \leq \sum_{h \in (X \times Y)^N} |\phi(h)| \left|\mathcal{A}(h_X \mid h_Y) - \mathcal{A}'(h_X \mid h_Y)\right| \tilde{\mathbb{P}}(h_Y) \leq C\varepsilon \tag{3.14}$$

where $C = \max_h |\phi(h)| < \infty$. If $\varepsilon$ is made arbitrarily small (assuming no underflow), then this inequality means that the difference between the performance of $\mathcal{A}$ and $\mathcal{A}'$ is negligible.

## 3.4 Conclusion

This chapter has introduced the twin formalisms of *generators* and *trajectors* as mathematical abstractions of search methods. These formalisms have been shown to apply generally to optimization in finite spaces, and some of the interesting features of this theory have been presented, such as the existence of distance metrics among search methods and the possibility of studying optimizer performance analytically.

In subsequent chapters, generators and trajectors will be constructed in arbitrary Hausdorff topological spaces, and the results presented above will be formally explored in greater detail. It is hoped that by introducing these objects in more familiar finite spaces, their properties might seem more familiar and accessible. Nonetheless, many optimization settings in real problems in science and engineering are not typically expressed using finite methods, and forcing such expression is to some degree unnatural.

The main argument for confining analysis of optimization methods to finite spaces is the practical fact that such methods are most often implemented on digital computers. This limitation brings along with it the need to consider questions of computability and efficiency. These questions are certainly important, but they are not truly inherent to the optimization problem, which can exist equally in digital or non-digital settings. Consequently, we leave such questions behind for the moment in order to examine what kinds of optimization problems might exist and how they might be solved in general terms.

# Chapter 4
# Stochastic Generators

The previous chapter formalized iterative stochastic optimization methods using generators and trajectors on the assumption that the search and value spaces were finite and that the stochasticity of these methods was finitely representable. From an analytic perspective, this latter requirement in particular is somewhat artificial and unnatural in that it imposed gaps between neighboring optimizers in terms of the distance metric. These gaps occurred as a consequence of underflow in floating point operations, and they can be removed by lifting the analysis into a continuous space. Once continuous analysis is admitted in order to study arbitrarily fine gradations of probability, the link to computability is irretrievably broken. Quite plainly, arbitrary probability distributions, even over finite spaces, cannot be sampled exactly. Even with a randomized Turing Machine, it is impossible to precisely simulate flipping a coin whose probability of turning up heads is $\pi/4$ by finitely many calls to an instruction that simulates a fair coin.

Rather than bending over backwards to incorporate computational complexity, we choose instead to focus on analysis of optimization using techniques from probability theory and functional analysis. Since there is no obvious way to account for computational complexity in the context of continuous probabilities, there is no compelling reason for us to view search and optimization methods through the lens of finite representations suitable for a digital computer. Consequently, it will be more natural to choose the search space and the value space suitably for study using probability theory.

This chapter reintroduces generators in the setting of probability theory. It first addresses the basic definitions and notation of set-theoretic probability theory used in the remainder of the book, and then defines stochastic generators in these terms. Probability theory is a large topic that cannot be completely covered in these pages, and so the focus here is on laying a foundation of intuitions for what these spaces are and why we use them here. Generators are then defined as abstract objects over abstract spaces. For the purposes of this chapter, the particular kinds of spaces are restricted to familiar examples such as the natural numbers or the real line, while a wider variety of spaces is discussed in Chapter 5 based on general topology.



A. J. Lockett, *General-Purpose Optimization Through Information Maximization*,
Natural Computing Series, https://doi.org/10.1007/978-3-662-62007-6_4

## 4.1 General Notation and Set Theory

This section reviews the notation and definitions of set theory underlying all of the concepts in this book, including sets, function, orders, sequences, upper and lower bounds, and Cartesian products. The notation has been chosen to balance clarity and conventional familiarity. This section also serves as a reference for notation used in this and subsequent chapters.

### *4.1.1 Sets and Functions*

Modern mathematics is built on the foundation of set theory and logic, and basic familiarity with sets, intersections, unions, and simple logic formulae is assumed. To review, a set is a collection of objects drawn from a universe, which usually remains unspecified. Sets are denoted by capital letters, *e.g.*, $A$, $B$, $C$. The membership function $\in$ indicates that an object is a member of a set, so that $x \in A$ means that $x$ is contained in a set $A$; its negation is $x \notin A$. Sets may contain other sets, but they may not contain themselves. Given a logic formula $P(x)$, the notation $\{x \mid P(x)\}$ indicates the set of all objects that satisfy the formula $P(x)$, which is used to define the basic relationships for sets $A$ and $B$:

Subset: $A \subseteq B$ implies that for all $x \in A$, it holds that $x \in B$.
Union: $A \cup B = \{x \mid x \in A \,\text{or}\, x \in B\}$
Intersection: $A \cap B = \{x \mid x \in A \,\text{and}\, x \in B\}$
Difference: $A \setminus B = \{x \mid x \in A \,\text{and}\, x \notin B\}$
Complement: $A^c = \{x \mid x \notin A\}$

If $A \subseteq B$ and $A \neq B$, then $A$ is a *proper* subset of $B$. For any set $A$, we also use the abbreviation $\{x \in A \mid P(x)\} = A \cap \{x \mid P(x)\}$. In addition, the abbreviation $\{P(x)\}$ is commonly used in probability theory for $\{x \mid P(x)\}$.

The existence of certain sets, such as infinite sets and the power set, is guaranteed by axioms. The standard formulation is called ZFC, which stands for the Zermelo-Frankel axioms with the additional Axiom of Choice, which allows one to create a set by choosing one element from each member of an arbitrary collection. The details of ZFC are omitted, but it is worthy of mention that the Axiom of Choice in particular is a topic of considerable debate with several competing, weaker axioms proposed to replace it. Nonetheless, most authors not engaged in foundational explorations of mathematical theory tend to assume the Axiom of Choice, and we follow this convention here. Many of the key results in the subsequent text (such as the No Free Lunch Theorems in Chapter 12) rely on choice, and we will make no consistent effort to delineate where choice underlies results in this text.

A *relation* is a set of pairs, and a function $f$ is a relation such that for any objects $x$, $y$, and $z$, if it holds that $(x,y) \in f$ and $(x,z) \in f$, then $y = z$. The domain of a function is the set

$$\mathrm{Dom}\, f = \{x \mid \text{there exists } y \text{ such that } (x,y) \in f\},$$

and the range is

$$\mathrm{Rng}\, f = \{y \mid \text{there exists } x \text{ such that } (x,y) \in f\}.$$

A function $f$ is typically defined as a map from a set $X$ to a set $Y$, written $f : X \to Y$, for which $X = \mathrm{Dom}\, f$ and $\mathrm{Rng}\, f \subseteq Y$. The set $Y$ is the *co-domain* of the function. In some cases the set $X$ may be larger than $\mathrm{Dom}\, f$, in which case $f$ is said to be a *partial function* for the set $X$. The value of the function at a point $x \in X$ is then written as $f(x)$, which is defined to be the unique $y$ such that $(x,y) \in f$. Several other notations are also frequently used with equivalent meaning, including in particular $f[x]$ and $f_x$. When one wishes to consider a function $f$ as a relation instead of a mapping, then one speaks of the *graph* of the function, which is just the set of pairs making up the function and is hence identical to the function itself as defined above.

The composition of two functions $f$ and $g$ is written $f \circ g$ with the meaning $f \circ g(x) = f(g(x))$. It is defined only if the range of $g$ is a subset of the domain of $f$. The restriction of a function $f : X \to Y$ to a subset $A$ of $X$ is the function $f|_A : A \to Y$ such that for all $x \in A$, $f|_A(x) = f(x)$. If $Z$ is a superset of $X$ (*i.e.*, $X \subseteq Z$), then a function $g$ is an extension of $f$ to $Z$ if the restriction $g|_X = f$.

The *inverse* of a function is the relation $f^{-1} = \{(y,x) \mid f(x) = y\}$. As a relation, this inverse always exists, though it may not be a function. Assuming $f : X \to Y$, its inverse is a function if and only if $f$ is *one-to-one* and *onto*. A function is one-to-one (or *injective*) if for all $x, z \in X$, $f(x) = y$ and $f(z) = y$ together imply $x = z$, and it is onto (or *surjective*) if for all $y \in Y$ there is some $x \in X$ with $f(x) = y$. Such a function is said to be *invertible* and is also known as a *bijection*. A bijection can be used to identify two sets and to transfer properties of one set to another.

In probability theory and analysis in general, the most common usage of the inverse converts it into a set function, so that if $f : X \to Y$ and $A \subseteq Y$,

$$f^{-1}(A) = \{x \mid \text{there exists } y \in A \text{ with } f(x) = y\} \subseteq X,$$

which is a a polymorphic use of notation that is nonetheless distinct from the case of the inverse as a function since sets are written with capital letters. In this case, the set $f^{-1}(A)$ is called the *pre-image* of $A$. Furthermore, the same conversion to a set function may be done with $f$ itself, so that

$$f(B) = \{y \in Y \mid \text{there exists } x \in B \text{ such that } f(x) = y\}.$$

Whereas a bijection relates separate sets to each other, an *equivalence relation* relates elements within the same set to each other. An equivalence relation $\sim$ is a relation that is reflexive (for all $x$, $x \sim x$), symmetric (for all $x, y$, $x \sim y$ implies $y \sim x$), and transitive (for all $x, y, z$, $x \sim y$ and $y \sim z$ together imply $x \sim z$). An equivalence class for an object $x$ is the set of all objects that are equivalent to $x$, i.e., $\{y \mid x \sim y\}$. Equivalence classes partition a set, and the set of such classes is often considered on its own, in which case equivalent objects are said to be *identified*, that is, recast as a single object.

### 4.1.2 Orders and Sequences

An *order* is a relation $\leq$ that is reflexive (for all $x$, $x \leq x$), antisymmetric (for all $x, y$, $x \leq y$ and $y \leq x$ implies $x = y$), and transitive (for all $x$, $y$, $z$, if $x \leq y$ and $y \leq z$, then $x \leq z$). An order is *total* with respect to a set $A$ if for all $x$ and $y$, either $x \leq y$ or $y \leq x$. In this case, the set $A$ is said to be totally ordered. An order that is not total for a set $A$ is *partial* with respect to $A$, and the set $A$ is then partially ordered.

The *supremum* of a set $A$, written $\sup A$, is the least upper bound according to some order. That is, for all $x \in A$, $x \leq \sup A$, and $\sup A \leq y$ for all $y$ that are greater than or equal to each element of $A$. The supremum is more commonly written with the notation

$$\sup_{x \in A} f(x) = \sup\{f(x) \mid x \in A\},$$

which is defined by the right hand side of the equation. Similarly, the *infimum* is the greatest lower bound, written $\inf A$. That is, for all $x \in A$, $\inf A \leq x$, and $y \leq \inf A$ for all $y$ that are less than any element of $A$. The infimum may also be written as $\inf_{x \in A} f(x)$ with the same meaning as for the supremum.

The set of natural numbers $\mathbb{N}$ consists of the counting numbers $0, 1, 2, 3, \ldots$, which can be defined by positing the existence of the number 0 and defining a successor function. The natural numbers are totally ordered in the usual way, which is to say that they are ordered by the smallest total order over the natural numbers that contains the graph of the successor function. The set of all numbers up to a natural number $n$ is written as $[n] = \{x \in \mathbb{N} \mid x \leq n\}$. The integers $\mathbb{Z}$, the rational numbers $\mathbb{Q}$, and the real line $\mathbb{R}$ can all be constructed from the natural numbers, and these constructions will not be reviewed here.

Intervals of the real line will be used frequently. These are defined with respect to the standard order relation on the real line: the open interval $(a, b) = \{x \in \mathbb{R} \mid a < x < b\}$, the closed intervals $[a, b] = \{x \in \mathbb{R} \mid a \leq x \leq b\}$, and the half open intervals $(a, b] = (a, b) \cup \{b\}$ and $[a, b) = (a, b) \cup \{a\}$. The same notation is used with similar meaning for any totally ordered set.

Frequently, one needs to understand how many elements are in a set. The *cardinality* of a set $A$ is denoted $|A|$. Two sets are said to have the same cardinality if there is a bijection between them. If a set $A$ has the same cardinality as the set $[n]$ for some natural number $n \in \mathbb{N}$, then we write $|A| = n$ and say that $A$ has *finite* cardinality. If $A$ has the same cardinality as the natural numbers $\mathbb{N}$, then it is *countably infinite* or just *countable*. The symbol $\infty$ generally represents the cardinality of the natural numbers, *i.e.* $|\mathbb{N}| = \infty$. Often, the term *countable* is used in a lazy fashion to mean *at most* countable; such usage should be clear from context. If a set $A$ has the same cardinality as the real numbers, then it is said to be *uncountably infinite* or just *uncountable*. Standard diagonalization arguments prove that $|\mathbb{N}| \neq |\mathbb{R}|$. We will not discuss larger cardinalities than that of $\mathbb{R}$ in this book.

A sequence is a function whose domain is an ordered set. The domain is called the *index set*. A sequence $a$ over a set $X$ with index set $\mathcal{I}$ is a function $a : \mathcal{I} \to X$ and may be written using parentheses and subscripts as $a = (x_i)_{i\in\mathcal{I}}$. A sequence is said to have the cardinality of its index set. Thus a sequence indexed by $[N]$ for $N \in \mathbb{N}$ is finite, and a sequence indexed $\mathbb{N}$ is countable. Most commonly in this book, the domain of a sequence will be the natural numbers or the set $[N]$ for some $N \in \mathbb{N}$.

A sequence $(x_i)_{i\in\mathcal{I}}$ whose range is also an ordered set is said to be *monotonically increasing* if $i < j$ implies $x_i \leq x_j$. It is strictly monotonically increasing if $i < j$ implies $x_i < x_j$. The sequence is *monotonically decreasing* if $i < j$ implies $x_i \geq x_j$ and strictly monotonically decreasing if $i < j$ implies $x_i > x_j$. In either case, the sequence may be said to be (strictly) monotonic without specifying whether it is increasing or decreasing.

Often, a sequence is used to enumerate a set, and in this case the sequence itself can be treated as a subset of its range. In this case, the notation $(x_i)_{i\in\mathcal{I}} \subseteq A$ indicates that each $x_i$ is an element in the set $A \subseteq X$.

Sequences may be reordered by remapping the sequence indices. Specifically, a function $\pi : \mathcal{I} \to \mathcal{I}$ permutes a sequence $x = (x_i)_{i\in\mathcal{I}}$ to another sequence $x_\pi = (x_{\pi(i)})_{i\in\mathcal{I}}$. In this case, $x_\pi$ is said to be a *permutation* of the sequence $x$.

Subsequences are needed in many cases. If $\mathcal{I}$ is an index set and $\mathcal{K} \subseteq \mathcal{I}$, then for a sequence $x = (x_i)_{i\in\mathcal{I}}$, the subsequence indexed by $\mathcal{K}$ is denoted $x_\mathcal{K} = (x_i)_{i\in\mathcal{K}}$. Usually, subsequences are most useful when they change the cardinality of the index set. The cardinality of a sequence is the cardinality of its index set, so that a finite sequence has a finite index set, a countable sequence has a countable index set, and so on. Finite or countable subsequences are often used in order to constrain or quantify the behavior of uncountable sequences.

Perhaps the most important topic when dealing with sequences is the question of convergence, but since convergence is primarily a topological notion, discussion of the topic is deferred to Section 5.1.

### *4.1.3 Cartesian Products*

Another important concept is the *Cartesian product*. If $A$ and $B$ are sets, then their Cartesian product is the set $A \times B = \{(a,b) \mid a \in A \,\text{and}\, b \in B\}$. More generally, if $X$ is a point set and $(A_i)_{i\in\mathcal{I}}$ is a sequence indexed by an arbitrary (not necessarily ordered) set $\mathcal{I}$ such that each $A_i \subseteq X$, then the Cartesian product of this sequence is

$$\prod_{i\in\mathcal{I}} A_i = \{x \mid x = (x_i)_{i\in\mathcal{I}} \text{ is a sequence over } X \text{ and for all } i \in \mathcal{I}, x_i \in A_i\}.$$

In practical terms, the nature of the set $X$ need not be specified beforehand if the $A_i$ are given, since it suffices to assume $X = \bigcup_{i\in\mathcal{I}} A_i$. Theoretically, however, the space $X$ must exist before the $A_i$ can be formally defined. The most used Cartesian product is one in which the sets $A_i$ are all the same. In this case, one writes $A^{\mathcal{I}}$ for some index set $\mathcal{I}$ with the intent that $A^{\mathcal{I}} = \prod_{i\in\mathcal{I}} A$. A further abbreviation is used when $\mathcal{I} = [n]$, the set of natural numbers up to $n$. In this case one typically writes $A^n$ instead of $A^{[n]}$ for simplicity. In this book, however, such simplicity could cause confusion due to the common usage of products over subsets of indices, such as $A^{\mathcal{I}}$ above. For this reason, the notation $A^{[n]}$ is preferred here, typically in the form of $X^{[t]}$, where $X$ is the search space and $[t]$ is the set of time indices up to some time $t$.

Elements in a Cartesian product are thus defined as sequences. For the Cartesian product of the last paragraph, a mapping $p_i : \prod_{i\in\mathcal{I}} A_i \to A_i$ is called a *canonical projection* if $p_i(x) = x_i$ for all elements $x = (x_i)_{i\in\mathcal{I}}$ in the product set $\prod_{i\in\mathcal{I}} A_i$. There is clearly one canonical projection for each element of the index set. One can also consider a canonical projection to a set of subindices. That is, for $\mathcal{K} \subseteq \mathcal{I}$, the canonical projection to $\mathcal{K}$ is $p_{\mathcal{K}}(x) = (x_i)_{i\in\mathcal{K}}$.

With these basics in mind, the next section motivates the use of probability measures and topology for optimization theory.

## 4.2 Elements of Probability Theory and Integration

This book conceives of the search space and the value space for a search process as probability spaces. Probability theory was developed most notably by A. Kolmogorov [93] in the early part of the twentieth century. It is more complicated than than the simpler, finite theory of probability that preceded it due to definitional complications that arise within probability theory. In this section, these complications are described and the mathematical basis of probability theory and integration are introduced.

### *4.2.1 Foundational Issues in Probability Theory*

The goal of a probability measure is to assign consistent probability values to each possible set of outcomes. But in many spaces, including spaces as basic as the real numbers, no consistent choice is possible. Thus in defining a probability measure, one must also specify the sets of outcomes over which it is consistent. A topological space provides a system of coherent neighborhoods to which consistent probability values can be assigned.

In a finite space, a probability distribution is a relatively simple object. There are a finite number of possible outcomes, and a probability distribution can be fully described by assigning one positive real number to each element of the space such that all of these probabilities sum to one. In infinite spaces, however, complexities arise that are absent in the finite case. The sum of the probabilities becomes an infinite sum with attendant questions of convergence or divergence. In a continuous space, such as the real numbers, the situation is even worse, since the sum is uncountable.

Consequently, rather than assigning probabilities to individual points, one assigns probabilities to regions of space, which are realized as sets of points. So a probability measure $\mathbb{P}$ over a space $X$ should assign a probability to each subset $A$ of $X$, for which we write $\mathbb{P}(A)$, which recognizes that $\mathbb{P}$ is a *set function*. The obvious domain for $\mathbb{P}$ is the power set of $X$, which we will denote $2^X$, and its probability values range from zero to one, so that we can write $\mathbb{P} : 2^X \to [0,1]$.[1] In addition, a probability function should also have certain properties. Primarily, if $A$ and $B$ are two distinct sets, then it ought to be true that $\mathbb{P}(A \cup B) = \mathbb{P}(A) + \mathbb{P}(B)$. That is, probabilities should be disjointly additive; the probability of an outcome in either one of two distinct sets should be equal to the probability of one plus the probability of the other. This statement is simply one of the core aspects of what *probability* means, and it is also the source of a serious problem.

The problem has to do with the nature of continuous spaces. Intuitively, points in these spaces are inextricably entangled with each other. Arbitrary subsets of these spaces can choose points such that even pointwise disjoint sets are still entangled with each other, and this fact can be used to force $\mathbb{P}(A \cup B) > \mathbb{P}(A) + \mathbb{P}(B)$, even when $A$ and $B$ are disjoint. In Euclidean space, such sets are defined more precisely as those sets on which the *exterior measure*, that is, the volume of the set as determined from outside the boundary of the set, differs from the *interior measure*, that is, the volume as determined from inside the boundary of the set. As these terms suggest, the essential issue is that for some sets, the boundary of the set turns out to have volume on its own. Such sets are called *unmeasurable*. One cannot define a notion of volume on unmeasurable sets, nor can one define a probability measure, since probability mass is essentially a normalized measure of volume.

One of the most well-known examples comes from the Banach-Tarski paradox, of which a corollary states that the unit ball in three dimensions can be partitioned into a finite number of subsets that can be reassembled into two complete balls of the same radius by rotation and translation alone. In essence, this paradox implies that volume is not preserved by continuous transformations; where there was one ball with volume $\frac{4}{3}\pi$, there are now two balls with volume $\frac{8}{3}\pi$. As one might expect, the requisite partition of the ball in the Banach-Tarski paradox produces unmeasurable sets whose boundaries have positive volume and are so interlocked that after reassembly there is in fact no gap remaining at all.

These boundary issues can be avoided by excluding unmeasurable sets, which is elegantly accomplished by the introduction of a $\sigma$-algebra.

**Definition 4.1 ($\sigma$-algebra).** A $\sigma$-algebra $\Sigma_X$ over a space $X$ is a set of subsets of $X$ that is:

1. Nonempty: $\emptyset, X \in \Sigma_X$
2. Closed under complements: If $A \in \Sigma_X$, $X \setminus A = \{x \in X \mid x \notin A\} \in \Sigma_X$
3. Closed under countable unions: For $(A_i)_{i=1}^{\infty} \subseteq \Sigma_X$, $\bigcup_{i=0}^{\infty} A_i \in \Sigma_X$
4. Closed under countable intersections: For $(A_i)_{i=1}^{\infty} \subseteq \Sigma_X$, $\bigcup_{i=0}^{\infty} A_i \in \Sigma_X$

[1] The notation $2^X$ comes from set theory. The power set is often denoted $\mathcal{P}[X]$ in many texts, but we do not use the power set much and so reserve this latter to refer to the space of probability measures over $X$.

A $\sigma$-algebra is a subset of the power set, $\Sigma \subseteq 2^X$. Its function in probability theory is to exclude undesirable sets of outcomes (such as those of the Banach-Tarski paradox) from consideration while still allowing probability values to be assigned to the sets of outcomes that are of interest. If a probability measure $\mathbb{P}$ is defined as a function whose domain is a $\sigma$-algebra $\Sigma_X$, then for any set $A \in \Sigma_X$, the probability of an outcome inside of $A$ is given by $\mathbb{P}(A)$. Due to closure under complements, the opposite probability $\mathbb{P}(X \setminus A)$ is also known. If $(A_i)_{i=0}^{\infty}$ is any countable sequence of sets of $\Sigma_X$, then its union $A = \bigcup_{i=0}^{\infty} A_i$ is also a member of $\Sigma_X$, meaning that if multiple groups of outcomes are assigned a probability by the measure $\mathbb{P}$, then the ensemble group of all these outcomes is also assigned a probability. Finally, its intersection $A = \bigcap_{i=0}^{\infty} A_i$ is also a member of $\Sigma_X$, meaning that if the measure $\mathbb{P}$ assigns probability to multiple groups of outcomes, it also assigns a probability to the event that an outcome is simultaneously contained in every group.[2]

Within $n$-dimensional Euclidean space, the most commonly used $\sigma$-algebra is the Lebesgue $\sigma$-algebra, which includes all of the sets whose inner and outer volumes are the same. The analysis in this book ranges over a broader class of spaces than just Euclidean space, and so we will use the Borel $\sigma$-algebra, which is the smallest $\sigma$-algebra that includes all open sets. The idea of an *open set* is a topological idea, which will be discussed below. For the real line (or any totally ordered set), the Borel $\sigma$-algebra, written $\mathcal{B}[\mathbb{R}]$, is the smallest $\sigma$-algebra that contains all open intervals $(a,b)$; this fact implies that it also includes all closed and half-open intervals as well as all countable unions and intersections of closed, open, and half-open intervals. The smallest $\sigma$-algebra on any space is the *trivial* $\sigma$-algebra, which contains only the empty set and the whole space.

The $\sigma$-algebra provides a domain for which it is possible to avoid paradoxical situations, but a further definition is needed in order to enforce the consistency of how a probability measure assigns probabilities to regions of a space. This is done by defining a *probability measure* on a *probability space*.

**Definition 4.2 (Probability Space).** A probability space is a tuple $(\mathbb{P}, \Sigma_X, X)$, where $X$ is any set of points, $\Sigma_X$ is a $\sigma$-algebra over $X$, and $\mathbb{P} : \Sigma_X \to \mathbb{R}$ is a probability measure, meaning that it satisfies:

1. Nonnegativity: For all $A \in \Sigma_X$, $\mathbb{P}(A) \geq 0$.
2. Total Unity: $\mathbb{P}(X) = 1$.
3. Countable Additivity: For disjoint $(A_i)_{i=0}^{\infty} \subseteq \Sigma_X$, $\mathbb{P}\left(\bigcup_{i=0}^{\infty} A_i\right) = \sum_{i=0}^{\infty} \mathbb{P}(A_i)$.

For a probability space $(\mathbb{P}, \Sigma_X, X)$, the sets in $\Sigma_X$ are called *measurable sets* or *events*. Any subset of $X$ not contained in $\Sigma_X$ is called *unmeasurable*.

**Proposition 4.1.** *The following properties result immediately from Definition 4.2:*

*1.* $\mathbb{P}(\emptyset) = 0$.
*2.* $\mathbb{P}(X \setminus A) = \mathbb{P}(X) - \mathbb{P}(A)$.

[2] The fact that a $\sigma$-algebra is closed under countable intersections actually follows from the fact that it is closed under complements and countable unions, and so a $\sigma$-algebra can be defined by just the first three conditions.

*3. For any sequence* $(A_i)_{i=0}^{\infty}$ *with* $A_i \in \Sigma_X$, $\mathbb{P}(\bigcup_{i=0}^{\infty} A_i) \leq \sum_{i=0}^{\infty} \mathbb{P}(A)$.

Probability spaces form the foundation of probability theory. The $\sigma$-algebra $\Sigma_X$ establishes which events can be measured, and the probability measure $\mathbb{P}$ assigns probability mass to each event. Technically, a probability measure is a specialization of another mathematical object called a *measure*. One of the main things we will need to do with probability measures is to integrate over them, and to understand integration in this context, some discussion of measure theory is needed.

### *4.2.2 Measure Spaces and Integration*

A *measure* represents an abstract notion of volume. A measure is essentially a probability measure without the requirement of total unity. In consequence, measures are also defined with reference to a $\sigma$-algebra that identifies the measurable sets.

**Definition 4.3 (Measurable Space).** A measurable space is a pair $(\Sigma_X, X)$, where $X$ is any set of points and $\Sigma_X$ is a $\sigma$-algebra over $X$. A measure space is a tuple $(\mu, \Sigma_X, X)$ where $(\Sigma_X, X)$ is a measurable space and $\mu : \Sigma_X \to \mathbb{R} \cup \{\pm\infty\}$ is a measure, meaning that it satisfies:

1. Nonnegativity: For all $A \in \Sigma_X$, $\mu(A) \geq 0$.
2. Countable Additivity: For disjoint $(A_i)_{i=0}^{\infty} \subseteq \Sigma_X$, $\mu\left(\bigcup_{i=0}^{\infty} A_i\right) = \sum_{i=0}^{\infty} \mu(A_i)$.

A similar set-function $\nu : \Sigma_X \to \mathbb{R} \cup \{\pm\infty\}$ that satisfies countable additivity but not nonnegativity is called a *signed measure*.

The property of *countable additivity* is also known as $\sigma$-additivity in some texts, so that a signed measure is a $\sigma$-additive set function over a $\sigma$-algebra, and a measure is a nonnegative signed measure.

Notice that measures and signed measures may assign infinite volume to some region of space. This fact is easiest to understand in the context of Euclidean space, where it is obvious that the volume of the entire infinite space is infinite. A (signed) measure that assigns finite probability to every measurable set is called a *finite (signed) measure*. It is obvious that every measure is a signed measure, and that every probability measure is a finite signed measure.

Of particular interest are the sets to which a measure assigns zero volume. Such sets are said to have *zero measure*. In measure theory and probability theory, predicates are often stated in terms of sets with non-zero measure, since sets with zero measure are negligible in some sense. Thus it is said that a fact is true *almost everywhere* (in measure theory) or *almost surely* (in probability theory) if it is true except on a set with zero measure. The measure in question is often made explicit, so that a fact can be true $\mu$-almost everywhere, $\mu$-a.e., $\mu$-almost surely, or $\mu$-a.s. This book alternates between the two usages depending on the context.

A function from one measurable space to another is called *measurable* if its inverse preserves measurability (*i.e.*, the pre-image of every measurable set is measurable). That is, if $f : X \to Y$ sends elements of $(X, \Sigma_X)$ to $(Y, \Sigma_Y)$, then $f$ is measurable if for all $A \in \Sigma_Y$, it follows that $f^{-1}(A) \in \Sigma_X$. Notice that even for measurable functions, there may be sets $B \in \Sigma_X$ such that $f(B) \notin \Sigma_Y$. If the function $f$ is real-valued, *i.e.*, $f : X \to \mathbb{R}$, then $\Sigma_Y$ is typically assumed to be the Borel $\sigma$-algebra in this book, and then it suffices for the pre-image of each half-interval $(-\infty, t]$ to be measurable, $f^{-1}((-\infty, t]) \in \Sigma_X$. Measurability is preserved by addition, subtraction, multiplication, division, and limits. Measurable functions have the useful property that their values can be measured over regions. That is to say, measurable functions can be integrated.

It is often the case that multiple measurable spaces are under discussion at once. In this case, a function is said to be $\Sigma_X$-measurable to emphasize that it is measurable with respect to $\Sigma_X$ as opposed to some other $\sigma$-algebra. The $\sigma$-algebra in this case is the one that pertains to the domain of the function. The smallest $\sigma$-algebra on the domain with respect to which a function $f$ remains measurable is called the *$\sigma$-algebra generated by $f$*, about which more will be said in the discussion of conditional probabilities.

Integration can be defined with respect to arbitrary measures; an integral so defined is called a *Lebesgue integral*. The integral familiar from basic calculus courses is the *Riemann integral*, and whenever the Riemann integral exists and is defined, it agrees with the Lebesgue integral. The Lebesgue integral, however, is defined in abstract measure spaces, and even in Euclidean spaces, the Lebesgue integral may exist where the Riemannian integral diverges.[3]

The standard definition of the Lebesgue integral begins with *indicator functions*. Given a subset $A \subseteq X$, an indicator function $\mathbb{1}_A : X \to \{0, 1\}$ is defined so that for $x \in X$,

$$\mathbb{1}_A(x) = \begin{cases} 0 \text{ if } x \in A \\ 1 \text{ otherwise.} \end{cases}$$

The abstract integral over a measure $\mu$ is defined for indicator functions as

$$\int \mathbb{1}_A \, d\mu = \int \mathbb{1}_A(x) \, \mu(dx) = \mu(A),$$

where the integrand $x$ is implied in the leftmost side, and notation $\mu(dx)$ is dummy notation suggesting the measure of a small region around each point $x \in X$. The next step is to define the integral of step functions $u(x) = \sum_{i=1}^{N} a_i \mathbb{1}_{A_i}(x)$ where $N$ is finite, the $a_i$ are finite, positive real numbers, and the $A_i$ are all pairwise disjoint. In that case,

$$\int u \, d\mu = \int_X u(x) \, \mu(dx) = \sum_{i=1}^{N} a_i \, \mu(A_i).$$

This definition is consistent in the sense that two equal step functions obtain the same integral even if they are defined with respect to different measurable sets due to the countable additivity of the measure $\mu$. Now for any measurable, positive, real-valued function $f : X \to \mathbb{R}$, the integral can be defined as the supremum over all step

[3] The canonical example is the indicator function on the rational numbers, which is undefined in the Riemannian sense of converging upper and lower estimates but has Lebesgue integral zero.

functions that it dominates, *i.e.*,

$$\int f\,d\mu = \int_X f(x)\,\mu(dx) = \sup\left\{ \int s\,d\mu \,\middle|\, s \text{ is a simple function and } 0 \leq s \leq f \right\}.$$

And for any measurable, real-valued function $g : X \to \mathbb{R}$, there are measurable, positive, real-valued functions $g_+, g_-$ such that $g(x) = g_+(x) - g_-(x)$, and

$$\int g\,d\mu = \int_X g(x)\,\mu(dx) = \int g_+\,d\mu - \int g_-\,d\mu.$$

A function for which the Lebesgue integral is defined in this way is said to be $\mu$-*integrable* or just *integrable*.

The final step is to define a set function $I_g : \Sigma_X \to \mathbb{R} \cup \{+\infty, -\infty\}$ against the $\sigma$-algebra for $\mu$ by

$$I_g(A) = \int_A g\,d\mu = \int_A g(x)\,\mu(dx) = \int \mathbb{1}_A(x)\,g(x)\,\mu(dx),$$

where the integrand is multiplied by the indicator function $\mathbb{1}_A$ to enforce the restriction to a particular set.

The properties of the integral are such that countable additivity is guaranteed, and so it should not escape notice that the integral $I_g$ is itself a signed measure defined with respect to another measure $\mu$. The integral can also be defined for signed measures. It is easily seen that if $\nu$ is a signed measure, then it can be decomposed into two measures $\nu_+$ and $\nu_-$ such that for all measurable sets $A$, $\nu(A) = \nu_+(A) - \nu_-(A)$. Then one defines

$$\int f\,d\nu = \int f\,d\nu_+ - \int f\,d\nu_-.$$

The fact that this decomposition is unique is known as the *Jordan decomposition theorem*. This decomposition can also be used to create the absolute measure $|\nu| = \nu_+ + \nu_-$, which is by definition nonnegative.

As the Jordan decomposition suggests, signed measures can be added and subtracted pointwise. For signed measures $\mu$ and $\nu$, the signed measure $\mu + \nu$ is defined so that for any measurable set $A$, $(\mu + \nu)(A) = \mu(A) + \nu(A)$. If both $\mu$ and $\nu$ are nonnegative, then obviously $\mu + \nu$ is non-negative as well. They can also be multiplied by scalars, with $(\alpha\mu)(A) = \alpha\mu(A)$ for any real coefficient $\alpha$. These facts together with the standard properties of real arithmetic mean that the set of signed measures is a vector space, the significance of which is discussed below.

Furthermore, it is said that a measure $\nu$ is *absolutely continuous* with respect to another measure $\mu$ if for all measurable $A$, $\mu(A)$ implies $\nu(A)$, written $\nu << \mu$. In this case, the *Radon-Nikodym Theorem* states that for any finite nonnegative measure $\mu$ and any finite signed measure $\nu$ with $\nu << \mu$ there exists a measurable, real-valued function $f : X \to \mathbb{R}$ such that for all measurable $A$,

$$\nu(A) = \int f\,d\mu, \quad \text{and} \quad f = \frac{d\nu}{d\mu}. \tag{4.1}$$

This function is called the *Radon-Nikodym* derivative of $\nu$ with respect to $\mu$. In probability theory, conditional expectations are defined using Radon-Nikodym derivatives. They reveal that integrals are a natural source of transformations among measures.

The integral distributes over finite addition both in terms of the integrand and the measure, *i.e.*,

$$\int \left( \sum_{i=1}^{N} f_i \right) d\mu = \sum_{i=1}^{N} \int f_i \, d\mu \quad \text{and} \quad \int f \, d\left( \sum_{i=1}^{N} \mu_i \right) = \sum_{i=1}^{N} \int f \, d\mu_i .$$

In many cases, infinite sums distribute as well. More generally,

$$\lim_{i \to \infty} \int f_i \, d\mu = \int \left( \lim_{i \to \infty} f_i \right) d\mu$$

if all the terms in the sequence are bounded in absolute value by some nonnegative integrable function $g$, i.e., for all $i$ and $x \in X$, $|f(x_i)| \leq g(x)$; this is the *Dominated Convergence Theorem.* Another important fact used to prove convergence of integrals is *Jensen's inequality* for convex functions. A function $\phi : \mathbb{R} \to \mathbb{R}$ is convex if for all real $a, b, x$ with $a \leq x \leq b$ and all $t \in [0,1]$, it holds that $\phi(x) \leq t\phi(a) + (1-t)\phi(b)$. The function $\phi$ is concave if the $\leq$ can be replaced with $\geq$ in this last inequality and linear if it can be replaced by $=$. Jensen's inequality states that for any convex $\phi$,

$$\phi \circ \int f \, d\mu \leq \int \phi \circ f \, d\mu ,$$

where $\leq$ can be replaced with $\geq$ if $\phi$ is concave and $=$ if it is linear. The two most common instantiations of $\phi$ are given by

$$\left( \int f \, d\mu \right)^2 \leq \int f^2 \, d\mu \quad \text{and} \quad \left| \int f \, d\mu \right| \leq \int |f| \, d\mu .$$

In each case, if $\mu$ is a signed measure rather than a nonnegative measure, then the inequality still holds if $\mu$ on the right hand side is replaced with $|\mu| = \mu_+ + \mu_-$.

It should be noted that two functions that are equal $\mu$-almost everywhere have equal integrals, since the integral is defined based on the measure.

With these basic facts about measure theory and integration in hand, we now proceed to discuss their relevance to probability theory and stochastic analysis.

### 4.2.3 Expectations and Conditional Probabilities

One of the main applications of probability theory is to measure the average value of some function weighted according the likelihood of its inputs. In the context of stochastic optimization, Chapter 3 already introduced the concept of performance as a function of the history trace. In this case, the expected performance of an optimization method is the average performance on history traces, weighted according to how often those traces are chosen by the method. Such values are the results of abstract Lebesgue integrals.

Probability theory uses the analytical tools of measure theory, but often substitutes its own names and notations. Thus a *random variable* is another term for a *measurable function*, with the distinction that random variable names are often written with capital rather than lower case letters. Random variables can serve as integrands, and computing the integral is known as *taking the expectation*. The familiar notation for the expectation of a random variable $H : X \to Y$ over a probability measure $\mathbb{P}$ is

$$\mathbb{E}_{\mathbb{P}}[H] = \int H \, d\mathbb{P} = \int_X H(x) \, \mathbb{P}(dx),$$

using the fact that $\mathbb{P}$ is a nonnegative measure. If the probability measure $\mathbb{P}$ is clear from the context, the abbreviated form $\mathbb{E}[H] = \mathbb{E}_{\mathbb{P}}[H]$ is used with $\mathbb{P}$ omitted. As a non-standard extension of this notation, this book occasionally uses the notation $\mathbb{E}_\mu[H]$ for the integration of $H$ with respect to a (signed) measure $\mu$.

Further abbreviations are used to indicate the expectation of a random variable within a restricted subset of the probability space. So for a measurable event $A$,

$$\mathbb{E}[H;A] = \mathbb{E}[\mathbb{1}_A H] = \int_X \mathbb{1}_A(x) \, H(x) \, \mathbb{P}(dx) = \int_A H \, d\mathbb{P}$$

is the *expectation of H on A*.

Frequently, one wishes to examine two random but dependent events in order to discover the likelihood of one event conditional on the occurrence of the other. This *conditional probability* has already been observed in finite spaces in Chapter 3, but the measure-theoretic version is defined a bit differently.

To begin, suppose that $(X, \Sigma_X, \mathbb{P})$ is a probability space. In this case $\Sigma_X$ is a $\sigma$-algebra, meaning that it specifies events that can be measured, where an *event* is understood to be a region of the space. In this sense, the $\sigma$-algebra can be thought of as an information source, since only events inside the $\sigma$-algebra can be measured.

Now suppose $U : X \to \mathbb{R}$ is a random variable, which is to say that $U$ observes the outcome of a randomized trial and transforms it to a real number. By assumption, $U$ is measurable from $\Sigma_X$ to the Borel $\sigma$-algebra over the reals, $\mathcal{B}[\mathbb{R}]$. If the underlying space contains more information than the real line, for example, say, if $X = \mathbb{R}^3$ is three-dimensional Euclidean space and $\Sigma_X$ contains all open balls in three space, then $U$ cannot possibly convey all of the information about the underlying object from which it is computed. In simple terms, the underlying space has three degrees of freedom where measurement of $U$ only reveals one degree of freedom. Thus observation of $U$ restricts the events in the $\sigma$-algebra $\Sigma_X$. This restriction can be quantified as a $\sigma$-algebra on $X$

$$\sigma(U) = \{U^{-1}(A) \mid A \in \mathcal{B}[\mathbb{R}]\},$$

which is the subset of $\Sigma_X$ that contains all of the pre-images of all Borel-measurable sets in $\mathbb{R}$. In this way, $\sigma(U)$ captures the information about $X$ that is available upon observing the variable $U$. In the case of three-dimensional Euclidean space above, $\sigma(U)$ is necessarily a proper subset of $\Sigma_X$. The $\sigma$-algebra $\sigma(U)$ is unique and is called the *$\sigma$-algebra generated by $U$*. To observe $U$ in a probabilistic sense means to know that an experimental sample lies in some region $A \in \sigma(U)$ of the original space.

The $\sigma$-algebra generated by a random variable is a particular instance of a sub-$\sigma$-algebra, which is defined as a subset of a $\sigma$-algebra that is itself a $\sigma$-algebra. Now suppose that $\mathcal{U}$ is any sub-$\sigma$-algebra of $\Sigma_X$, not necessarily generated by $U$. The probability measure $\mathbb{P} : \Sigma_X \to [0,1]$ has a natural restriction to $\mathcal{U}$ denoted as $\mathbb{P}|_{\mathcal{U}} : \mathcal{U} \to [0,1]$, and this restriction is clearly a probability measure as well.

Suppose further that $V : X \to \mathbb{R}$ is a second random variable, and we wish to know how restricting observations to events $A \in \mathcal{U}$ influences the the observation of $V$ given the shared underlying probability space $X$. Before considering $\mathcal{U}$, note that the probability measure $\mathbb{P}$ can be pushed forward using $V$ to create a signed measure over the real line through the mapping

$$A \mapsto \int_A V \, d\mathbb{P}.$$

This mapping is termed the *pushforward measure* of $V$ and will be denoted $\mathbb{P} \circ V$. As a condition to this definition, it is presumed that $V$ is $\mathbb{P}$-integrable. Then the fact that

$$\text{for all } A \in \Sigma_X, \quad \mathbb{P} \circ V(A) = \int_A V \, d\mathbb{P} \quad \text{implies} \quad V = \frac{d\mathbb{P} \circ V}{d\mathbb{P}}$$

where the later is the Radon-Nikodym derivative of $\mathbb{P} \circ V$ with respect to $\mathbb{P}$. But if both measures are restricted to the $\sigma$-algebra to $\mathcal{U} \subseteq \Sigma_X$, this derivative may be different, yielding the following definition.

**Definition 4.4 (Conditional Expectation).** A *conditional expectation* of a variable $V$ with respect to $\mathcal{U}$ is any $\mathcal{U}$-measurable function $\mathbb{E}[V \mid \mathcal{U}] : X \to \mathbb{R}$ for which

$$\mathbb{E}[V \mid \mathcal{U}] = \frac{d\mathbb{P} \circ V|_{\mathcal{U}}}{d\mathbb{P}|_{\mathcal{U}}} \text{ or, equivalently, } \int_A \mathbb{E}[V \mid \mathcal{U}] \, d\mathbb{P} = \int_A V \, d\mathbb{P} \quad \text{for all } A \in \mathcal{U}.$$

In the case where $\mathcal{U} = \sigma(U)$, the abbreviation $\mathbb{E}[V \mid U]$ is used with the same meaning, described as the *conditional expectation of a variable V with respect to a variable U*.

The conditional expectation is unique $\mathbb{P}|_{\mathcal{U}}$-almost surely, but not $\mathbb{P}$-almost surely. Thus there may be a function $V' : X \to \mathbb{R}$ that is a conditional expectation of $V$ with respect to $\mathcal{U}$ for which

$$\int_B V \, d\mathbb{P} \neq \int_B V' \, d\mathbb{P}$$

provided that $B$ is an event in $\Sigma_X$ but not $\mathcal{U}$. In general, a random variable is its own conditional expectation, but it is not the only one. The main point of the conditional expectation is that sets in the $\sigma$-algebra $\mathcal{U}$ cannot distinguish $V$ from any other $\mathbb{E}[V \mid \mathcal{U}]$. When a conditional expectation is used in an equality, it usually indicates equality $\mathbb{P}|_{\mathcal{U}}$-almost surely rather than absolutely.

If the random variable $V$ is thought of as a generic observer of the space $X$, the conditional expectation can thus be thought of as the limited observer that results when information is restricted to $\mathcal{U}$ or, equivalently for $\sigma(U)$, when the variable $U$ has been observed first.

Now when $\sigma(V) = \mathcal{U}$, that is, when $\mathcal{U}$ contains exactly the information conferred by $V$, then $\mathcal{U}$ does not limit $V$ at all, that is, $V = \mathbb{E}[V \mid \mathcal{U}]$, by which it is meant that $V$ is the *only* conditional expectation of $V$ with respect to $\mathcal{U}$. If, on the other hand, $V$ is independent of $\mathcal{U}$, then $\mathcal{U}$ confers no information about $V$, in which case the conditional expectation is constant, $\mathbb{E}[V \mid \mathcal{U}] = \mathbb{E}[V]$, which means that $\mathbb{E}[V]$ is *one* conditional expectation of $V$.[4] These and other facts are summarized in the following proposition.

**Proposition 4.2.** *The conditional expectation has the following properties:*

1. $\mathbb{E}[\mathbb{E}[V \mid \mathcal{U}]] = \mathbb{E}[V]$.
2. *If $V$ is independent of $\mathcal{U}$,* $\mathbb{E}[V \mid \mathcal{U}] = \mathbb{E}[V]$.
3. *If $V$ is $\mathcal{U}$-measurable,* $\mathbb{E}[V \mid \mathcal{U}] = V$.
4. *If $\mathcal{U} \subseteq \mathcal{W}$, then* $\mathbb{E}\left[\mathbb{E}[V \mid \mathcal{U}] \middle| \mathcal{W}\right] = \mathbb{E}[V \mid \mathcal{U}]$.
5. *If $\mathcal{W} \subseteq \mathcal{U}$, then* $\mathbb{E}\left[\mathbb{E}[V \mid \mathcal{U}] \middle| \mathcal{W}\right] = \mathbb{E}[V \mid \mathcal{W}]$.
6. Linearity*: For $\alpha, \beta \in \mathbb{R}$,* $\mathbb{E}[\alpha V + \beta W \mid \mathcal{U}] = \alpha\,\mathbb{E}[V \mid \mathcal{U}] + \beta\,\mathbb{E}[W \mid \mathcal{U}]$.

*These properties follow from the definition of the integral.*

In this discussion, the probability measure $\mathbb{P}$ was presupposed. To make it explicit, the conditional expectation can be written as $\mathbb{E}_{\mathbb{P}}[V \mid \mathcal{U}]$. It is worthwhile to reiterate that this expectation is only defined when the random variable $V$ is $\mathbb{P}$-integrable, which means that $\mathbb{E}|V|$ exists and is finite.

Conditional probabilities are defined based on the conditional expectation of an indicator function. An indicator function $\mathbb{1}_A$ is integrable on a probability space $(X, \Sigma_X, \mathbb{P})$ whenever $A$ is $\Sigma_X$-measurable, since the pre-image of $\mathbb{1}_A$ is either $A$ or its complement, so $\mathbb{P}(A \mid \mathcal{U}) = \mathbb{E}[\mathbb{1}_A \mid \mathcal{U}]$ is well defined. This object, however, is not a probability measure on its own, since it has two sources of variation: the event $A$ and the $\sigma$-algebra $\mathcal{U}$. The definition below follows the example of generators in Chapter 3 by treating conditional probabilities as functions taking on values from the set of all probabilities $\mathcal{P}[\Sigma_X]$ on the measurable space $(X, \Sigma_X)$.

[4] Technically, two sets $A, B$ are independent if $\mathbb{P}(A \cap B) = \mathbb{P}(A)\mathbb{P}(B)$. Two finite collections of sets are independent if every pair of sets is independent with the pair formed by taking one set from each collection is independent. Two $\sigma$-algebras are independent if every pair of finite collections is independent with the pair formed by taking one finite subset from each $\sigma$-algebra. Two variables are independent if the $\sigma$-algebras they generate are independent. Independence between variables and $\sigma$-algebras is determined by passing from variables to their generated $\sigma$-algebras.

**Definition 4.5 (Conditional Probability).** A *conditional probability* with respect to a $\sigma$-algebra $\mathcal{U}$ in a probability space $(X, \Sigma_X, \mathbb{P})$ with $\mathcal{U} \subseteq \Sigma_X$ is a function $\mathcal{G} : X \to \mathcal{P}[\Sigma_X]$ such that for all $A \in \mathcal{U}$ and for any conditional expectation $\mathbb{E}_{\mathbb{P}}[\mathbb{1}_A \mid \mathcal{U}]$ of the indicator function $\mathbb{1}_A$ with respect to $\mathcal{U}$, the equation

$$\mathcal{G}(x)(A) = \mathbb{E}[\mathbb{1}_A \mid \mathcal{U}](x).$$

holds $\mathbb{P}|_{\mathcal{U}}$-almost surely for $x \in X$.

The notation $\mathbb{P}(A \mid \mathcal{U})$ for $A$ in $\mathcal{U}$ stands for a $\mathcal{U}$-measurable random variable corresponding to the mapping $x \mapsto \mathcal{G}(x)(A)$, and the notation $\mathbb{P}(\cdot \mid x)$ stands for the probability measure resulting from the mapping $x \mapsto \mathcal{G}(x)(\cdot)$. The value $\mathbb{P}(A \mid x) = \mathcal{G}(x)(A)$ is informally called the conditional probability of event $A$ given $x$, and is uniquely defined only for $x$ in a $\mathcal{U}$-measurable set with $\mathbb{P}$-probability one.

The indefiniteness of the conditional expectation may seem a bit confusing at first, but uncertainty is inherent to the study of probability. Intuitively, the idea that two functions may be equal *almost surely* is equivalent to saying that the functions are equal insofar as they can be observed. Clearly, no mathematical system can be expected to reveal unobservable differences.

Finally, conditional probabilities are often taken not with respect to $\sigma$-algebras but with respect to variables. Just as with conditional expectations, if one has a random variable $U$, then the conditional probability $\mathbb{P}(A \mid U)$ is interpreted as $\mathbb{P}(A \mid \sigma(U))$, where $\sigma(U)$ is the $\sigma$-algebra generated by $U$.

Conditional expectations and conditional probabilities, together with independence of sets and variables, distinguish probability theory from measure theory. There is no obvious way to define conditional expectations with respect to signed measures without losing or substantially modifying the properties enumerated in Proposition 4.2, which from the core properties associated with the term *conditional expectation*.

The next section uses the basic concepts of probability theory to extend the definition of generators and trajectors from Chapter 3 for infinite spaces.

## 4.3 Search Generators in Probability Spaces

Chapter 3 defined the generator of a search method for finite spaces. In this section, generators will be extended to arbitrary probability spaces. Some examples are provided for Euclidean space using familiar probability densities.

### *4.3.1 Definition*

The finite version of search generators in Section 3.2.1 defines the generator as a function from search history prefixes to probability distributions. Rather than assuming that search histories are finite, we introduce a totally ordered set $\mathcal{T}$ to play the role of time, called the *temporal index set*. We will focus primarily on the case where $\mathcal{T}$ is either the (infinite) set of natural numbers or the set of numbers up to some fixed maximum time $N$, as in Chapter 3.

The space of history prefixes can be generalized with respect to the temporal index set by

$$\mathcal{H}_{X,Y} = \emptyset \cup \bigcup_{n \in \mathcal{T}} (X \times Y)^{[n]}, \tag{4.2}$$

where $[n] = \{m \in \mathcal{T} \mid m \leq n\}$ is an index prefix[5] and the search space $X$ is a probability space and the value space $Y$ is an arbitrary set. The value $\emptyset \in \mathcal{H}_{X,Y}$ represents the empty history. The space $\mathcal{H}_{X,Y}$ is well-defined as a union of Cartesian products.

The space of probability distributions can be replaced with a space of probability measures. Thus for a measurable space $(X,\Sigma)$, we denote the set of probability measures on $(X,\Sigma)$ variously as $\mathcal{P}_X$, $\mathcal{P}_\Sigma$, $\mathcal{P}[X]$, $\mathcal{P}[\Sigma]$, or $\mathcal{P}[X,\Sigma]$, depending on the distinctions we wish to emphasize. We actually define generators with a larger co-domain, the set of finite signed measures on $(X,\Sigma)$, denoted $\mathcal{M}_X$, $\mathcal{M}_\Sigma$, $\mathcal{M}[X]$, $\mathcal{M}[\Sigma]$, or $\mathcal{M}[X,\Sigma]$. Obviously, $\mathcal{P}_X \subseteq \mathcal{M}_X$, since every probability measure is a finite, signed measure. But $\mathcal{M}_X$ is a complete normed vector space (see Chapter 5 for a definition), whereas $\mathcal{P}_X$ is not, and for this reason the larger co-domain is preferred.

**Definition 4.6 (Search Generator).** Given a search space $X$ and a value space $Y$ such that $(X,\Sigma)$ is a measurable space, a *search generator* is a function $\mathcal{G} : \mathcal{H}_{X,Y} \to \mathcal{M}[X,\Sigma]$ that maps history prefixes to finite signed measures.

As before, the square brackets notation $\mathcal{G}[h]$ is preferred to represent the value of the search generator $\mathcal{G}$ on the history $h$. Then $\mathcal{G}[h]$ is a measure and hence a set function over $\Sigma$. For any $\Sigma$-measurable set $A$ (*i.e.*, $A \in \Sigma$), the notation $\mathcal{G}[h](A)$ represents the measure assigned to the set $A$ by the search generator given history $h$. As measures, search generators can be integrated, and in fact

$$\mathcal{G}[h](A) = \int_A \mathcal{G}[h](dx) = \mathbb{E}_{\mathcal{G}[h]}[\mathbb{1}_A]$$

in general, where the integral is the Lebesgue integral introduced in Section 4.2.2. In finite spaces, this new definition of a search generator corresponds to that of Chapter 3 in that the probability mass of a point $x \in X$ is the measure of the singleton set $\{x\}$, $\mathcal{G}[h](\{x\})$.

The co-domain for search generators has been expanded to include finite, signed generators so that the resulting set of generators is a vector space, but we are primarily interested in what happens when the output of a search generator is always a probability measure.

**Definition 4.7 (Proper Search Generator).** A search generator $\mathcal{G}$ defined on a measurable space $(X,\Sigma)$ is called *proper* if for all histories $h \in \mathcal{H}_{X,Y}$, the measure $\mathcal{G}[h]$ is a probability measure, *i.e.*, $\mathcal{G}[h] \in \mathcal{P}[X,\Sigma]$.

---

[5] The use of $(X \times Y)^{[n]}$ in this way as opposed to simply $(X \times Y)^n$ accounts for the possibility that $\mathcal{T}$ is not countable.

Proper search generators correspond to well-defined search methods. Next, we explore how a search generator creates a search history when sampled sequentially.

### 4.3.2 Generating the Search Process

A proper search generator generates a search history when applied to an optimization problem. Suppose that the optimization problem consists of an objective function $u : X \to Y$, and that $\mathcal{G}[h]$ is proper and may be sampled[6] for any history $h$. Let $h_0 = \emptyset$ be the initial history, and sample $x_1 \sim \mathcal{G}[h_0]$. Then define $h_1 = (x_1, u(x_1)) \in (X \times Y)^1$, and sample $x_2 \sim \mathcal{G}[h_1]$. More generally, for each successive $n \in \mathcal{T}$,

$$\text{sample} \quad x_n \sim \mathcal{G}[h_{n-1}] \quad \text{and define} \quad h_n = (\,(x_i, u(x_i))\,)_{i \leq n}$$

so that $h_n \in (X \times Y)^{[n]} \subseteq \mathcal{H}_{X,Y}$ for all $n$.

Now the sequence of history prefixes $h_1, h_2, \ldots, h_n, \ldots$ is successively generated according to the probability measures $\mathcal{G}[h_0], \mathcal{G}[h_1], \ldots, \mathcal{G}[h_{n-1}], \ldots$, and so one can formalize the probability measure governing $h_n$ given the objective function $u$.

In order to do so, $(X \times Y)^{[n]}$ will need to be cast as a measurable space. Given the objective $u$ and $\Sigma$ as a $\sigma$-algebra for $X$, we can define a $\sigma$-algebra for $X \times Y$ by first defining set expansions

$$A \times u = \{(x, u(x)) \mid x \in A\} \quad \text{and} \quad \Sigma_{X,Y} = \sigma\left(\{A \times u \mid A \in \Sigma\}\right),$$

where the notation $\sigma(\mathcal{B})$ for a collection $\mathcal{B}$ refers to the smallest $\sigma$-algebra containing $\mathcal{B}$. Note that $\sigma(\mathcal{B})$ always exists, although it may be equal to the power set, which serves as an upper bound. This construction projects $\Sigma$ into the product $X \times Y$ via $u$. In the same way, for temporal indices $n \in \mathcal{T}$ define a $\sigma$-algebra over $(X \times Y)^{[n]}$ by

$$\Sigma^n_{X,Y} = \sigma\left(\left\{ \prod_{i \leq n} A_i \,\middle|\, \text{for all } i \leq n, A_i \in \Sigma_{X,Y} \right\}\right).$$

Then $\left((X \times Y)^{[n]}, \Sigma^n_{X,Y}\right)$ is a measurable space.

Let $\mathcal{A}^n_u$ be the probability measure over $h_n$ as generated by $\mathcal{G}$ on $u$. The existence of this measure is guaranteed by Carathéodory's extension theorem, which is discussed in Chapter 9 where a more detailed treatment of the history process is undertaken. As will be discussed at that point, it will be found that for all history prefixes $h$ of length $n$ and all $\Sigma$-measurable sets $A$,

$$\mathcal{G}[h](A) = \mathbb{E}_{\mathcal{A}^{n+1}_u}\left[\mathbb{1}_A \mid \mathcal{H}^+_n\right](h), \tag{4.3}$$

---

[6] A probability measure can be sampled in the most general sense whenever the singleton sets are measurable, as they are, for example, when the Borel $\sigma$-algebra is used.

where

$$\mathcal{H}_n^+ = \{A \times \emptyset \mid A \in \Sigma_{X,Y}^n\} \cup \{A \times (X \times Y) \mid A \in \Sigma_{X,Y}^n\} \subseteq \Sigma_{X,Y}^{n+1}$$

projects $\Sigma_{X,Y}^n$ into $\Sigma_{X,Y}^{n+1}$ by expanding it with the trivial $\sigma$-algebra over $X \times Y$. Notice that with respect to the $(n+1)^{th}$ time step, $\mathcal{H}_n^+$ provides no information, discriminating only the two sets $\emptyset$ and $X \times Y$.

The conditional expectation extracts the information in $\mathcal{A}_u^{n+1}$ that is unknown to $\mathcal{A}_u^n$, which has access only to information contained in $\mathcal{H}_n^+$. This difference in information between time steps is precisely what the search generator provides. A less opaque formulation is given by

$$\mathcal{A}_u^{n+1}(A \times (B \times u)) = \int_A \mathcal{G}[h](B)\ \mathcal{A}_u^n(dh) \tag{4.4}$$

for sets $A \in \Sigma_{X,Y}^n$ and $B \in \Sigma$, where $B \times u = \{(x, u(x)) \mid x \in B\}$ as before.

Equations 4.3 and 4.4 reveal that the search generator $\mathcal{G}$ serves as a conditional expectation of the next search point given the current history or, equivalently, as a Radon-Nikodym derivative of the next search history with respect to the current search history. Again, a more detailed derivation is pursued in Chapter 9, after introducing more tools for extending probability measures to sequences.

### *4.3.3 Extended Example: Simulated Annealing*

To counterbalance the abstract presentation of search generators thus far, it will be worthwhile to consider two practical examples of stochastic optimization. In order to avoid introducing complex probability measures or spaces, the search space will be limited to $n$-dimensional Euclidean space, $X = \mathbb{R}^n$ with the standard Lebesgue $\sigma$-algebra, which includes all sets of practical interest. The value space will be the real line, $Y = \mathbb{R}$. Our first example is simulated annealing, introduced in Section 2.5.1.

Simulated annealing is commonly applied to optimization in Euclidean space. At time $n$ and given an objective function $u : \mathbb{R}^n \to \mathbb{R}$, simulated annealing uses a Monte Carlo technique to generate approximate samples from a Boltzmann distribution whose differential probability density given by

$$\mathcal{B}_n^u(dx) = \frac{1}{Z_n} \exp\left(-\frac{u(x)}{T_n}\right) dx, \tag{4.5}$$

where $Z_n$ is a normalizing factor known as the *partition function*, and $T_n$ is a sequence of temperatures with $T_n \to 0$. The sequence $T_n$ is known as the *cooling schedule*. The distribution $\mathcal{B}_n^u$ will be referred to as an *annealing distribution* in this book.

Simulated annealing samples from $\mathcal{B}_n$ repeatedly using the Metropolis algorithm [107, 70]. The process begins with a proposed solution $x$. At each time step, a proposal distribution $\mathbb{Q}$ is used to sample $x_n$. For instance, $\mathbb{Q}$ might be a Gaussian distribution centered at $x$. The proposed solution $x$ is replaced with $x_n$ with probability

$$p_n(x, x_n) = \exp\left(-\max\{0, u(x) - u(x_n)\}/T_n\right). \tag{4.6}$$

For each fixed temperature $T_n$, successively sampling and replacing points according to Equation 4.6 converges to a sample from $\mathcal{B}^u_n$. As $n \to \infty$, $\mathcal{B}^u_n$ converges in probability to a distribution that samples directly from the optimal points of $u$ [90].

The next step for simulated annealing is chosen according to Equation 4.6, and so this equation should form the basis of the generator for simulated annealing. In fact, if $h^X_{-1}$ represents the last search point in a history $h$, then for all history prefixes $h$ and for Lebesgue $A \subseteq \mathbb{R}^n$,

$$\mathcal{G}[h](A) = \int_A \left[p_n\left(h^X_{-1}, z\right) + \left(1 - p_n\left(h^X_{-1}, z\right)\right) \mathbb{1}_A\left(h^X_{-1}\right)\right] \mathbb{Q}(dz),$$

suffices to define a search generator for simulated annealing and if the proposal distribution $\mathbb{Q}$ is a unit deviation Gaussian, $\mathcal{G}[h](A) =$

$$\left(\frac{1}{2\pi}\right)^{\frac{n}{2}} \int_A \left[p_n\left(h^X_{-1}, z\right) + \left(1 - p_n\left(h^X_{-1}, z\right)\right) \mathbb{1}_A\left(h^X_{-1}\right)\right] \exp\left\{-\frac{1}{2}\left|h^X_{-1} - z\right|^2\right\} dz. \tag{4.7}$$

Although Equation 4.7 is plainly complicated, it is nonetheless a familiar Riemannian integral whenever $A$ is a reasonably simple subset of Euclidean space, as when $A$ is a box defined by $n$ sets of intervals on the real line.

To see how Equation 4.7 implements simulated annealing, note that the integral represents the probability that the chosen point resides in the set $A$. By "chosen point" we mean the outcome of accepting or rejecting a proposed point. There are two ways that the chosen point can land in $A$. The first is that the proposed point is in $A$, and it is accepted. The second is that the original point was in $A$, and the proposed point was rejected. In either case, the outcome depends on the proposed point. The probability that the proposed point is in $A$ is just $\mathbb{Q}(A)$, but the probability that it is accepted differs for each point depending on the objective $u$. Thus to obtain the probability that a proposed point in $A$ is accepted, one has to integrate the acceptance probability with respect to $\mathbb{Q}$ over $A$, that is,

$$\mathbb{P}(\text{The proposed point is in } A \text{ and is accepted} \mid h^X_{-1}) = \int_A p_n(h^X_{-1}, z)\, \mathbb{Q}(dz).$$

The probability that the proposed point is rejected uses the complementary integrand $(1 - p_n(h^X_{-1}, z))$, but this probability should only be included if the prior point $h^X_{-1}$ is in $A$. Thus the component

$$\left(1 - p_n\left(h^X_{-1}, z\right)\right) \mathbb{1}_A\left(h^X_{-1}\right)$$

represents the probability that the proposed point is rejected with a switch to indicate whether the prior point was already in the set $A$. If the prior point was not in $A$, then the switch evaluates to zero and this component is eliminated. Equation 4.7 results from adding the probability for the two ways that the chosen point can land in $A$ and substituting in a unit-deviation Gaussian density

$$\mathbb{Q}(dz) = \left(\frac{1}{2\pi}\right)^{\frac{n}{2}} \exp\left\{-\frac{1}{2}\left|h^X_{-1} - z\right|^2\right\} dz.$$

As demonstrated here, the search generator can be derived by examining the probabilistic choices made by an optimization method in order to determine the next search point.

### *4.3.4 Extended Example: Stochastic Gradient Descent*

The next example is stochastic gradient descent. Gradient descent was introduced in Section 2.2.3; it is widely used and highly effective for smooth optimization problems, particularly when accompanied by second-order methods. But if the error surface is "bumpy", meaning that it contains small local variations, then it is possible for gradient descent to become trapped in low quality local optima. To avoid these traps, one can add noise to the gradient measurement, which will have the effect over several iterations of damping out the bumps, potentially allowing the method to escape the attraction basins of local optima if these basins are small.

Formally, for an objective function $u$, stochastic gradient descent adjusts the update equation for gradient descent to

$$x_{n+1} = x_n - \eta_n \left[\nabla u(x_n) + \varepsilon_{u,x_n}\right]$$

where $\varepsilon_{u,x_n}$ is a random variable potentially depending on both the objective function and the search point $x_n$. In the simplest case, $\varepsilon$ is a standard Gaussian independent of $u$ and $x_n$. More commonly, $x$ represents the parameters of a model, such as an artificial neural network, and the objective has the form

$$u(x) = \sum_{i=1}^{N} L(z_i \mid x)$$

for some loss function $L(z \mid x)$ where the sequence $(z_i)_{i=1}^N$ represents a dataset of points to be modeled.

In this latter case, the optimization proceeds by randomly selecting a subsequence $(i_k)_{k=1}^M$ for $M < N$ that computes the loss for $M$ elements of the dataset. Then the noise is defined as

$$\varepsilon_{u,x} = \sum_{k=1}^{M} \nabla_x L(z_{i_k} \mid x) - \nabla u(x),$$

which in most cases will vary close to zero, especially if $M$ is large. The update equation becomes

$$x_{n+1} = x_n - \eta_n \sum_{k=1}^{M} \nabla_x L(z_{i_k} \mid x).$$

This update is stochastic in that the subsequence $(i_k)_{k=1}^{M}$ is updated for each iteration.

For convenience, we will suppose that the noise is distributed as a multivariate Gaussian whose mean and covariance depend upon the search point. Specifically, suppose the noise has distribution

$$\varepsilon_{u,x} \sim \mathcal{N}(\mu(x), \Gamma(x)),$$

where $\mathcal{N}(\mu, \Gamma)$ represents a multivariate Gaussian with mean vector $\mu$ and covariance matrix $\Gamma$.

To define a generator for stochastic gradient descent, we assume the search space is $X = \mathbb{R}^n$ and the value space is also $Y = \mathbb{R}^n$, since the values used to drive gradient descent are the entries of the gradient vector. For any history $h$, we denote the prior point by $h^X_{-1}$ and suppose that the gradient of this point is $h^Y_{-1} = \nabla h^X_{-1}$. For each history $h$, a search generator for stochastic gradient descent can be defined so that

$$\mathcal{G}[h] = \mathcal{N}\left(h^X_{-1} - \eta_n h^Y_{-1} + \eta_n \mu(h^X_{-1}),\, \Gamma(h^X_{-1})\right) = \mathcal{N}\left(m_n(h^X_{-1}, h^Y_{-1}),\, \Gamma(h^X_{-1})\right),$$

or, equivalently, for all subsets $A$ of $\mathbb{R}^n$, $\mathcal{G}[h](A) =$

$$\frac{1}{\sqrt{|2\pi\Gamma(h^X_{-1})|}} \int_A \exp\left\{-\frac{1}{2}\left(z - m_n(h^X_{-1}, h^Y_{-1})\right)^T \Gamma(h^X_{-1})^{-1}\left(z - m_n(h^X_{-1}, h^Y_{-1})\right)\right\} dz. \tag{4.8}$$

Thus the search generator for stochastic gradient descent is determined by the structure of the noise applied to the gradient.

In the case where the noise results from a random selection of training points, the resulting generator is degenerate as a probability distribution, meaning that it assigns measure zero to a set with full Lebesgue measure. Since there are only $N^M$ ways to select a subset of the training data, there are only finitely many points in Euclidean space that can be generated from a given search point depending on which training examples are selected. Let $K(h)$ represent these $N^M$ points given the history $h$. Then the search generator is given by

$$\mathcal{G}[h](A) = \frac{|A \cup K(h)|}{|K(h)|},$$

which is the fraction of the eligible $N^M$ points contained inside $A$, with each point weighted equally.

The purpose of this example is to show that with appropriate selection of the search and value spaces, gradient methods can be represented by search generators. The concept of search generators is therefore not specific to randomized search heuristics.

## 4.4 Conclusion

In this chapter, the rudiments of probability theory and measure theory have been introduced, and the concept of a search generator has been generalized to measurable search domains. Two examples, simulated annealing and stochastic gradient descent, were presented in order to show how the generator framework captures commonly used optimization methods.

However, the repertory of optimization settings is still limited by the fact that for each search domain, a probability space must be constructed. In order to establish a reasonable source for probability spaces the next chapter introduces general topology, which systematizes concepts of space, neighbors, and continuity from a set theoretic point of view. These constructive tools will make it possible to define search generators in a wide variety of practical spaces.

# Chapter 5
# Search and Optimization in Topological Spaces

Probability theory is built on measurable spaces, typically finite-dimensional Euclidean space. But in order to analyze search and optimization processes, probability measures over sequences, functions, and other complex objects are needed. In many cases these spaces are not just infinite but infinite-dimensional; they can vary infinitely in infinitely many aspects. Finite analysis, and even real analysis, does not suffice to study such spaces. Functional analysis is required, which is presently built on the foundation of general topology to supply meaning for terms such as continuity and convergence.

General topology provides basic set-theoretic concepts of structure and shape in space. Before topology was invented, any study of the structure of space was intricately bound up with the study of numbers. Topology recast the study of space on a set-theoretic basis focused on *neighborhoods* rather than numbers. The use of neighborhoods make it possible to separate notions of shape from notions of size. Consequently, topology allows for much simpler statements and proofs in many fields of mathematics, and for this same reason it has come to play a critical role throughout analysis in the last century.

Since an understanding of basic topology is necessary for our analysis of search and optimization, an additional benefit can be obtained by treating the search domains and value spaces as topological spaces. The Borel $\sigma$-algebra, introduced briefly in Chapter 4 for Euclidean spaces, provides a direct way of transforming topological spaces into measurable spaces. Furthermore, topological spaces are often easier to construct than measurable spaces. In any case such constructions are more frequently pursued in topology than in measure theory, since topology is precisely the study of shape and space.

In this chapter, general topology is introduced along with methods for constructing complex spaces. Examples are then provided to demonstrate how complex but practically useful search domains can be constructed by combining topological spaces. A range of probability measures are then built on these spaces to demonstrate the flexibility that characterizes the topological setting employed in this book.



A. J. Lockett, *General-Purpose Optimization Through Information Maximization*,
Natural Computing Series, https://doi.org/10.1007/978-3-662-62007-6_5

## 5.1 General Topology

In this section, the term *topology* is defined as a collection of sets declared to be *open*. The basic definitions of continuity, convergence, neighborhoods, and topological bases are given, and important properties such as the Hausdorff principle and compactness are reviewed.

### *5.1.1 Motivation and Definition*

An intuition for the origins of topology can be understood by considering the epsilon-delta definition of continuity, by which a function $f:\mathbb{R}\to\mathbb{R}$ is said to be continuous if for every real $x\in\mathbb{R}$ and every $\varepsilon>0$, there is a $\delta>0$ depending on $x$ such that $|f(x)-f(y)|<\varepsilon$ whenever $|x-y|<\delta$. Leaving aside how complicated this definition is, consider what it means. First, one chooses the point $x$ in the domain, which is equivalent to choosing the point $f(x)$ in the range. Second, one chooses an amount of deviation $\varepsilon$ around $f(x)$. Finally, the function $f$ forces an amount of deviation $\delta$ around $x$ that will preserve deviation of at most $\varepsilon$ around $f(x)$. Considered carefully, the choice of $x$ here doesn't really matter, and the definition could be stated without it. Somewhat loosely, the interval $I=(f(x)-\varepsilon, f(x)+\varepsilon)$ is an an open interval on the real line whose midpoint lies in the range of $f$, and its pre-image $f^{-1}(I)$ includes at least those points in $(x-\delta, x+\delta)$, which is again an open interval. The intuitive meaning of continuity is therefore tied to open intervals.

Now consider a function $f:\mathbb{R}^n\to\mathbb{R}^n$ mapping $n$-dimensional Euclidean space to itself. Once again, an epsilon-delta definition of continuity is possible, but now rather than intervals, the definition is based around open balls in space with radii of $\varepsilon$ and $\delta$. So in $n$-dimensional Euclidean space, the central role of open intervals has been replaced with open balls. What if the space becomes more complex, such as the space of all integrable functions or the space of all Turing Machines? Can continuity be defined in such spaces? If so, what should play the role of open intervals or open balls? In topology, the choice of sets that fill this role is subject to a set of axioms, but is otherwise arbitrary.

**Definition 5.1 (Topology).** A *topological space* is a pair $(X,\tau)$ where $X$ is any set and $\tau$ is a *topology*, meaning that it is a collection of sets satisfying:

1. The empty set and the entire space $X$ are elements of $\tau$.
2. For any sequence of sets $(A_i)_{i\in\mathcal{I}}\subseteq\tau$, the union $\bigcup_{i\in\mathcal{I}} A_i\in\tau$.
3. For any finite seqeunce of sets $(A_i)_{i=1}^N\subseteq\tau$, the intersection $\bigcap_{i\in\mathcal{I}} A_i\in\tau$.

The elements of $\tau$ are called *open sets*. If $A\in\tau$ is an open set, its complement $A^c=X\setminus A$ is a *closed set.*

More succinctly, a topology $\tau$ is a superset of $\{\emptyset, X\}$ that is closed under arbitrary unions and finite intersections. In essence, a topology declares certain sets to be open. In fact, any collection $\mathcal{A}$ of subsets of a space $X$ can be turned into a topology on $X$ by augmenting $\mathcal{A}$ with arbitrary unions and finite intersections of its elements, which can be phrased as defining $\tau(\mathcal{A})$ to be the smallest topology containing $\mathcal{A}$ as a subset. If a given topology $\tau$ is the smallest topology containing a collection $\mathcal{A}$, then the collection is called a *topological base* for $\tau$. For example, the open intervals form a topological base for the real line, and the open balls form a topological base for $n$-dimensional Euclidean space. As this last example suggests, topology uses open sets to define continuity.

### *5.1.2 Continuity and Convergence*

Continuity in topological spaces interprets the smoothness of a function as the preservation of open sets by the function inverse. Intuitively, if the pre-image of an open set $G$ under some function is not open, then it contains one or more points from its boundary. The region of space around this boundary point is thus "broken" by the function in the domain without breaking anything in the co-domain. In the domain, the boundary point cannot be separated from the pre-image, but when mapped to the co-domain, the point is no longer on the boundary. It is completely surrounded by points not in $G$ and can therefore be separated from the pre-image of $G$ by the function. In that sense, the function is discontinuous.

**Definition 5.2 (Continuity).** A function $f : X \to Y$ between topological spaces $(X, \tau_X)$ and $(Y, \tau_Y)$ is continuous if the pre-image of every open set is open, *i.e.*, for all $A \in \tau_Y$, it holds that $f^{-1}(A) \in \tau_X$.

**Proposition 5.1.** *A function $f : X \to Y$ between topological spaces $(X, \tau_X)$ and $(Y, \tau_Y)$ is continuous if and only if the pre-image of every closed set is closed.*

Thus continuity of a function is defined as a relationship between the open sets of the domain and the co-domain (or, equivalently, between the closed sets). Although the definition requires every open set to have an open-pre-image, it suffices to prove continuity for some topological base.

**Proposition 5.2.** *A function $f : X \to Y$ between topological spaces $(X, \tau_X)$ and $(Y, \tau_Y)$ is continuous if $\mathcal{A}$ is a topological base for $\tau_Y$ and for all $A \in \mathcal{A}$, $f^{-1}(A) \in \tau_X$.*

The epsilon-delta definition of continuity is a corollary of this proposition, since the open intervals or open balls form a topological base for the real line and $n$-dimensional Euclidean space, respectively.

Often, one wishes to examine the structure of a space around a point. For a point $x$ in a topological space $(X, \tau)$, a *neighborhood* of $x$ is any open set in $\tau$ that contains $x$. The set of all such neighborhoods is called the *neighborhood system* $\mathcal{N}_x$ of $x$, and a *neighborhood base* for $x$ is a subset $\mathcal{B}_x \subseteq \mathcal{N}_x$ such that for any neighborhood $N$ of $x$, there is some neighborhood $B \in \mathcal{B}_x$ such that $B \subseteq N$. That is, the neighborhood base $\mathcal{B}_x$ contains ever smaller sets that shrink around the point $x$. It should not surprise that this concept is closely related to the idea of limits and convergence in a topological setting.

**Definition 5.3 (Convergence).** A sequence of points $(x_i)_{i\in\mathcal{I}}$ in a topological space $(X,\tau)$ converges to a point $x$ in the space if for each neighborhood $N$ of $x$, the sequence is eventually contained within the neighborhood, *i.e.*, there exists $j \in \mathcal{I}$ such that for all $i > j$, $x_j \in N$. In this case, the point $x$ is a *limit* of the sequence.

Notice that this definition of convergence does not require the sequence to have any particular cardinality, and it may indeed be uncountable.

As a more subtle point, this definition does not imply that limits are unique. It only implies that limits are indistinguishable by neighborhoods. It is possible that two distinct points $x$ and $y$ could share neighborhood systems so that $\mathcal{N}_x = \mathcal{N}_y$. In that case, the topology cannot tell the points $x$ and $y$ apart, and the two are said to be *topologically indistinguishable*. In most cases, it is desirable to avoid such a situation, and hence we have the following definition.

**Definition 5.4 (Hausdorff).** A topological space $(X,\tau)$ is *Hausdorff* if for every pair of distinct points $x,y \in X$, $x \neq y$, there are neighborhoods $U$ of $x$ and $V$ of $y$ that are disjoint, $U \cap V = \emptyset$.

**Proposition 5.3.** *In a Hausdorff topological space, limits are unique.*

In this book, only Hausdorff topological spaces will be considered.[1] Non-Hausdorff spaces do occur. For example, spaces of computer programs can easily be constructed to have indistinguishable points by building a topology based on the program's functional behavior. Since two distinct programs may compute the same function, the ability to distinguish these programs is lost. When this happens, it is possible to construct a Hausdorff space from a non-Hausdorff space by defining an equivalence relation among the indistinguishable points (*e.g.*, treating programs as the same if they compute the same thing) and redefining the space based on the equivalence classes. Such a space is called a *quotient space* and will be introduced shortly.

### *5.1.3 Homeomorphisms and Equivalent Topologies*

Given two topological spaces $(X,\tau_X)$ and $(Y,\tau_Y)$, one would like to know whether they are topologically distinct. The key feature of a topology consists of the neighborhoods it admits, and thus two topological spaces are only really distinct if they differ in their neighborhood structure. This concept is captured by the definition of a *homeomorphism*.

**Definition 5.5 (Isomorphism).** Any bijection between two spaces is an *isomorphism*, and two spaces are *isomorphic* if there exists a bijection between them.

[1] There is in fact a weaker notion that avoids indistinguishability. A $T_0$ space only requires that for each pair of points $x,y$ in the space, there is at least one open set that contains one and not the other. This is weaker than the Hausdorff requirement in that there may be points whose neighborhood systems are wholly contained within the neighborhood system of another point. There are still disjoint sets that separate these points, but such sets may not be open.

**Definition 5.6 (Homeomorphism).** A *homeomorphism* between topological spaces $(X, \tau_X)$ and $(Y, \tau_Y)$ is a bijection $h : X \to Y$ such that both $h$ and its inverse $h^{-1}$ are continuous. Two spaces are *homeomorphic* if there exists a homeomorphism between them, denoted $X \simeq Y$, or $X \simeq_h Y$ when the homeomorphism is explicit.

By imposing a continuity requirement on a bijection and its inverse, a homeomorphism forces a one-to-one correspondence between the neighborhoods of $X$ and $Y$ so that these neighborhoods can be considered identical to each other. In topological terms, one thinks of homeomorphic spaces as being distinct representations of the same underlying space. Topological properties of one space are easily pushed through to homeomorphic spaces using the continuity of the homeomorphism that connects them.

### *5.1.4 Properties of Topological Spaces*

Several other properties of points and their neighborhoods are also relevant. A point $x$ is a *limit point* of a set $A$ if every neighborhood of $x$ contains a point in $A$ other than $x$. The name comes from the fact that if $x$ is a limit point of $A$, there is always a sequence in $A$ that converges to $x$. By contrast, an *isolated point* of $A$ is a point $x$ in $A$ that has a neighborhood that has empty intersection with the set $A \setminus \{x\}$. A point $x$ is a *boundary point* of a set $A$ if every neighborhood of $x$ has a nonempty intersection with $A$ and its complement, and $x$ is an *interior point* of a set $A$ if it has a neighborhood $N$ that is completely contained in $A$, *i.e.*, $N \subseteq A$.

These types of points lead to general characterizations of sets. Open and closed sets have already been introduced. The boundary of a set $A$ is

$$\partial A = \{x \mid x \text{ is a boundary point of } A\}.$$

Points in the boundary are not necessarily contained in $A$, nor are they necessarily outside of it. The *closure* of a set $A$, written $\bar{A}$, is the union of a set with its boundary, $\bar{A} = A \cup \partial A$. As the name suggests, the closure of a set is a closed set, and furthermore, $\bar{A} = A$ if and only if $A$ is closed. The *interior* of $A$, written $\mathring{A}$, is the collection of interior points of $A$. It is an open set, and again $\mathring{A} = A$ if and only if $A$ is open. It is always true in addition that $\mathring{A} = A \setminus \partial A$. It is important to note that a set can be both open and closed, or *clopen*, if its boundary is empty. The empty set and the entire space are always clopen. If any other set is clopen, then the space is *disconnected*, which means that it subdivides into separate components that have no overlap.

*Compactness* is an important property of sets that describes when the cardinality of a collection of sets or a sequence can be reduced. Since there are different ways to reduce cardinality, there are multiple compactness properties. By the standard definition, a set $A$ is *compact* if every open cover of $A$ has a finite subcover. An *cover* of a set $A$ is a collection $\mathfrak{U}$ of sets such that $A \subseteq \bigcup_{U \in \mathfrak{U}} U$. A cover is open if all of its sets are open, and a subcover is a subset of a cover. As an example of how compactness is used, suppose that $A$ is a closed and bounded set of points in

$n$-dimensional Euclidean space, which implies that $A$ is a compact set. Let $\mathcal{U}$ be the cover of $A$ formed by all open balls that are centered at a point contained in the closure $\bar{A}$. This collection is uncountable, but by compactness, there is a subcover $\mathcal{U}' \subseteq \mathcal{U}$ with finitely many elements. By finiteness, the collection $\mathcal{U}'$ consists of open balls with radius greater than $r$ for some $r > 0$. Therefore, any closed and bounded subset of Euclidean space can be covered by a finite set of open balls with positive radius bounded away from zero. This example demonstrates that compactness is a powerful property with unexpected and far-reaching consequences.

Every compact set is closed in a Hausdorff space. Every closed subset of a compact set is compact. One useful source of compact sets is the fact that $f : X \to Y$ is a continuous function between topological spaces and $A \subseteq X$ is compact, then the set $f(A) \subseteq Y$ is compact. In metric spaces (introduced below), the Heine-Borel Theorem states that a set is compact if and only if it is closed and bounded, so one strategy for obtaining compact sets is to define a continuous function from a metric space and examine the image of closed and bounded sets. Several other facts about compactness will be introduced as they become relevant in the text.

General topology defines a wide terminology that can be used to build theorems based on the properties of a space rather than its detailed structure. This text cannot cover them all, but the following properties will be referred to in what follows and are thus defined here.

The Hausdorff property is a separation axiom; there are several other such axioms. A topological space is *regular* if for every closed set $F$ and every point $x \notin F$, there are open sets $U$ and $V$ such that $F \subseteq U$ and $x \in V$ with $U \cap V = \emptyset$. That is, in a regular space, there are gaps between closed sets and individual points that can be separated without overlap. Regularity is stronger than the Hausdorff property.

As another separation axiom, a topological space is *normal* if every pair of disjoint closed sets are individually contained inside of disjoint open sets. That is, if $E$ and $F$ are two closed sets with $E \cap F = \emptyset$, then there exist open $U, V$ such that $E \subseteq U$ and $F \subseteq V$ with $U \cap V = \emptyset$. Informally, in a normal space, there is a gap between any two disjoint closed sets, and the neighborhood structure is sufficiently rich to divide this gap without overlap. Normality is stronger than regularity.

Characterizations of a topological space in terms of countable sets are particularly useful, and so there are many tools for this purpose. A space is *first countable* if each point in the space has a countable neighborhood base; most spaces one encounters naturally are first countable, such as Euclidean space or the space of continuous functions. A space is *second countable* if its topology has a countable topological base. In Euclidean space, such a base is given by the open balls centered at rational points with rational radii. Second countability is a stronger requirement than first countability.

A subset $A$ of a topological space $(X, \tau)$ is *dense* if for every open set $G$ in the space the intersection $A \cap G$ is nonempty. Equivalently, $A$ is dense if every point in $X$ is a limit of a sequence in $A$. A space is *separable* if it has a countable, dense subset. Intuitively, spaces that are not separable are in a certain sense so large that they cannot be almost filled by any countable collection of points.

The word "locally" is often applied to a property of a topological space with the usual meaning that every point has some neighborhood within which the property applies. Thus a space is *locally compact* if every point is contained in the interior of a compact set (or, equivalently, if every point has a neighborhood whose closure is a compact set). A collection of sets is *locally finite* if every point has a neighborhood that intersects at most a finite number of sets in the collection.

The prefix $\sigma$ is often added to many properties of a space as well, with the meaning that the space is composed of a countable union of sets for which the property holds. Thus a space is $\sigma$-compact if can be written as a countable union of compact sets, or $\sigma$-bounded if it can be written as a countable union of bounded sets (*e.g.* Euclidean space, which is the union of countably many open balls with positive integral radii). One even has $\sigma$-locally compact spaces, which are composed of a countable union of locally compact spaces. Of course, the $\sigma$-algebra has already been introduced, which is formed from countable unions and intersections over a collection of basic sets.

As a final definition, a space is *paracompact* if every open cover has an open refinement that is locally finite. A *refinement* of a cover is not a subcover but rather a separate cover such that each set in the initial cover has a subset in the refinement. A refinement is open if each of its sets is open. The space is *countably paracompact* if instead every countable open cover has a locally finite open refinement. Countable paracompactness is weaker than paracompactness.

The numerous properties considered in topological characterizations of space can seem overwhelming at first, and some may seem to be duplicative. But these properties are useful in that they become general characteristics upon which one can define theorems with exact pre-conditions that delimit tightly when certain results can be obtained.

Now that the basic terminology of topology has been introduced, we now consider how common spaces can be defined and how more complex spaces can be constructed from them.

## 5.2 Constructive Topology

Topology would not be nearly as useful for our purposes were it not for the various tools available for building new topological spaces by combining old ones. Precisely such constructions are important for understanding the analysis of optimization undertaken in this book, because they illustrate how optimization in various settings can explained or abstracted. This section demonstrates a few basic topologies that are most likely already familiar to the user, and then introduces constructive processes such as product spaces and quotient spaces that make it possible to build complex spaces from simpler ones.

### 5.2.1 Common Topological Spaces

The most basic topology over a set $X$ is the *trivial topology* $\tau = \{\emptyset, X\}$. One important way to describe a topology is delineate the functions that it makes continuous. If $Y$ is another topological space and $f : X \to Y$ is continuous, then for any set $A$ open in $Y$, the set $f^{-1}(A)$ is open, meaning either $f^{-1}(A) = \emptyset$ or $f^{-1}(A) = X$. Stated otherwise, $f$ is a constant function. With respect to the trivial topology, only the constant functions are continuous.

The *discrete topology* over $X$ is the power set of $X$, $\tau = 2^X$. In contrast to the trivial topology, the discrete topology makes *every* function continuous. Any function $f : X \to Y$ has the property that for all $A$ open in $Y$, the pre-image $f^{-1}(A)$ is open in $X$, because every subset of $X$ is open. For practical purposes, the discrete topology is most commonly used when $X$ is a finite set. The search domain and value space from Chapter 3 implicitly used the discrete topology. Although the discrete topology can be used for countable or uncountable topologies, it is not practical for these purposes because it yields an alien notion of continuity. For example, in the discrete topology, the function $\mathbb{1}_{\mathbb{Q}}$ that is one on the rationals and zero on the irrationals is a continuous function. On large spaces, the discrete topology simply cannot support a notion of smooth transitions.

A *metric topology* generalizes familiar concepts from Euclidean space such as open intervals or open balls. The discussion in Section 3.2.2 introduced the concept of a metric, the definition for which is given here in general form.

**Definition 5.7 (Metric).** A function $\rho : X \times X \to \mathbb{R}$ is a *metric* if has the following four properties:

1. **Nonnegativity**: $\rho(x,y) \geq 0$.
2. **Identity at Zero**: $\rho(x,y) = 0$ if and only if $x = y$.
3. **Symmetry**: $\rho(x,y) = \rho(y,x)$.
4. **Triangle Inequality**: $\rho(x,y) \leq \rho(x,z) + \rho(z,y)$.

The metric topology for a metric $\rho$ on a space $X$ is the smallest topology on $X$ that contains all open balls, which are defined as

$$B_\varepsilon(x) = \{y \mid \rho(x,y) < \varepsilon\},$$

where $\varepsilon > 0$ is the *radius* and $x \in X$ is the center. Thus the metric topology has the set of all open balls as a topological base. The pair $(X, \rho)$ is called a *metric space*. In the metric topology, continuity reduces to the familiar epsilon-delta definition mentioned above. A function between two metric spaces is continuous if and only if the pre-image of every open ball is an open ball. Metric spaces are a convenient source of topologies for two reasons. Firstly, they are easily obtained by defining a metric, and secondly, they result in an interpretation of continuity that corresponds to common notions of (locally) smooth change in space.

A topological space $(X, \tau)$ is *metrizable* if there exists a metric $\rho_\tau$ such that the metric topology from $\rho_\tau$ is equivalent to the given topology $\tau$. According to the Urysohn Metrization Theorem, a topological spaces is separable and metrizable if and only if it is regular, Hausdorff, and second countable. Exact conditions for metrizability without separability are given by the Nagata-Smirnov Metrization Theorem, but are not repeated here, except to say that a non-normal space can never be metrizable. This distinction is important, because it means that topological descriptions of space are more general than those obtained by analyzing metrics. A common example of a non-normal and unmetrizable space is the set of functions from the real line to itself with the topology of pointwise convergence. Another canonical example is the *long line*, which will not be described here.

Topologies can also be defined with respect to a total order. The *order topology* on a totally ordered set $(X,<)$ is the smallest topology for which the sets $\{x \mid x < a\}$ and $\{x \mid a < x\}$ are open. By this definition, each interval open $(a,b)$ for $a,b \in X$ is likewise open, since it is the intersection of two of the former sets. Every topological space with the order topology is necessarily a completely normal Hausdorff space. On the real line, the order topology coincides with the metric topology.

Finite Cartesian products over totally ordered sets can be given a topology using the *lexicographical* or *alphabetic ordering*, which orders the Cartesian product like a dictionary orders words, moving from left to right among the components of a point with comparison based on the total order within each component space.

As a final basic topology, a topology can be defined based on a set of functions that one wishes to declare continuous. In this scenario, there is a set $X$, a topological space $Y$, and a set of function $\mathcal{F}$ from $X$ to $Y$. In this case, the basic open sets for $X$ are of the form $f^{-1}(A)$ where $A$ is open in $Y$ and $f \in \mathcal{F}$. A topology is then defined for $X$ as the smallest topology containing these basic open sets. Such topologies are useful because they make it possible to use theorems that apply to continuous functions by controlling which functions are continuous.

### *5.2.2 Tools for Constructing Complex Spaces*

The trivial, discrete, metric, and order topologies provide ways of defining basic topologies for familiar spaces, including all finite, ordered, and metric spaces such as finite sets, the natural numbers, the real line, and Euclidean space. Often, these spaces are sufficient for the analysis of optimization methods, since most optimization problems are embedding inside of one of these four sets. There are cases where more complicated spaces are useful, including in particular spaces of artificial neural networks or configurations for a robot with many joints. Furthermore, the subsequent analysis in this book requires examining topologies on spaces of functions or measures. These more complex spaces are typically constructed from simpler topological spaces using the constructive tools discussed below.

The *subspace* topology for a subset $A$ of a topological space $(X,\tau)$ is the topology

$$\tau_A = \{A \cap B \mid B \in \tau\}$$

consisting of open sets in $X$ intersected with $A$. It is the smallest topology such that any continuous function with $X$ as a domain remains continuous when restricted to $A$. In other words, the subspace topology is designed so that if $f : X \to Y$ is continuous, then $f|_A : A \to Y$ is also continuous.

The *product topology* is the smallest topology on a Cartesian product set that makes all of the associated projections continuous. Supposing $X = \prod_{i\in\mathcal{I}} X_i$ is a Cartesian product and each $X_i$ is a topological space with a given topology $\tau_i$, a topological base for the product topology is given by the collection

$$\mathcal{B} = \left\{ \prod_{i=1}^{n} A_i \;\middle|\; n \in \mathbb{N},\, n < |\mathcal{I}|, \text{ and for all } 1 \le i \le n,\, A_i \in \tau_i \right\},$$

which consists of all finite Cartesian products of open sets from one or more components of $X$. Finiteness is critical here. There is another topology on the Cartesian product called the *box topology* in which *any* Cartesian product of open sets is included in the base, with undesirable consequences. In the product topology, a function $f : Y \to X$ is continuous if $p_i \circ f : Y \to X_i$ is continuous for each component projection $p_i$, but in the box topology this is not generally true. Likewise, for the product topology, Tychonoff's theorem states that the product space $X$ is compact if each $X_i$ is compact,[2] but a similar theorem does not hold in the box topology.

The product topology is one of the most commonly used topologies in this book. The reason for this is that functions and sequences are naturally Cartesian products. Each function from $X$ to $Y$ can be thought of assigning an element of $Y$ to an element of $X$, that is, a function indexes $X$ by $Y$. So the set of functions from $X$ to $Y$ is just the product space $Y^X$. When this function space is given the product topology, it is also called the *topology of pointwise convergence*, because it is the topology for which a sequence of functions $(f_i)_{i\in\mathbb{N}} \subseteq Y^X$ converges to a limit $f$ if and only if for all $x \in X$, $f(x) = \lim_{i\to\infty} f_i(x)$. In addition to its other two names, this topology is also called the *weak topology* by comparison with other, stronger convergence criteria. It was mentioned above that topologies can be defined by the functions that they make continuous. This example also shows that topologies can be defined by the sequences that they cause to converge.

A *disjoint union* is a combination of two potentially overlapping sets that maintains the distinction between the two sets. Formally, the distinction is accomplished by tagging each element of the disjoint union with a tag indicating its source. For two sets $A$ and $B$, the binary disjoint union is defined as

$$A \amalg B = \{(\iota, x) \mid \text{either } \iota = A \text{ and } x \in A \text{ or } \iota = B \text{ and } x \in B\}.$$

Informally, one speaks of $x \in A \amalg B$ for any $x \in A$ or $x \in B$, and if $x$ is in both sets, then one speaks of the "copy" of $x$ in $A$ or $B$, respectively. In general, for an indexed group of sets $(X_i)_{i\in\mathcal{I}}$, the disjoint union is

[2] It is worth stating that Tychonoff's theorem is equivalent to the Axiom of Choice for those familiar with this concept.

$$X = \coprod_{i \in \mathfrak{I}} X_i = \{(\iota, x) \mid \iota \in \mathfrak{I} \text{ and } x \in X_\iota\}.$$

The *inclusion function* is a function $\iota_{X_i} : X_i \to X$ that attaches labels to each element of $X_i$ so that $\iota_{X_i}(x) = (i, X)$. If each $X_i$ is a topological space $(X_i, \tau_i)$, the disjoint union topology is the smallest topology that makes each inclusion function continuous with respect to the given topology $\tau_i$. That is, a set $A \subseteq X$ is open only if the set $\iota_{X_i}^{-1}(A)$ is open for each $i \in \mathfrak{I}$.

The disjoint union is primarily useful as a precursor to the *quotient topology*, which merges the overlaps together smoothly. A *quotient function* $q$ is a surjective (*i.e.*, onto) function from a topological space $X$ to a set $Y$. Since $q$ is surjective, it covers all of $Y$, but it may send multiple points from $X$ to the same value in $Y$. Consequently, one thinks of the quotient as *collapsing* members of $X$ into members of $Y$. In this sense, quotients are associated with partitions or equivalence relations, in which elements are collapsed down to their equivalence classes. In fact, one can always construct an equivalence class corresponding to a quotient function so that each element of $Y$ is identified with an equivalence class. Elements within an equivalence class are said to be *identified* with each other.

The quotient topology for a quotient $q$ is the smallest topology that makes $q$ continuous. A set $A$ is therefore open in $Y$ if and only if its pre-image $q^{-1}(A)$ is open in $X$ with its given topology. One common usage of the quotient topology is an adjunction space. In this case one has two topological spaces $X$ and $Y$ and a function $f : A \to Y$ that maps a subset of $X$ into a subset of $Y$. Then one defines an equivalence on $X \coprod Y$ such that

$$a \sim b \quad \text{if either } a = b, \text{ or } a \in A \text{ and } b \in f(A), \text{ or vice versa.}$$

If $Z$ is the set of equivalence classes generated in this way, a quotient function $q : X \coprod Y \to Z$ is given by mapping each element in the disjoint union to its equivalence class. The adjunction space with respect to $f$ is written $X \cup_f Y$ and is endowed with the quotient topology for $q$. The special case in which $A$ contains a single point is called the *wedge sum*.

By combining basic topological spaces using subspaces, products, disjoint unions, and quotients, one can construct a wide array of topological spaces that are suited for the purposes of this book.

### *5.2.3 From Topological Vector Spaces to Banach Spaces*

Vector spaces are a central topic of this book, since it turns out that spaces of optimizers are in fact easily interpreted as vector spaces. A vector space $V$ over a field $\mathbb{F}$ is defined with respect to two operations: vector addition, which maps $V \times V \to V$, and scalar multiplication, which maps $\mathbb{F} \times V \to V$. Since for our purposes $\mathbb{F}$ is the real line $\mathbb{R}$, we will not bother with explaining what a field is and will treat vector spaces as though all vector spaces were vector spaces over the reals.

**Definition 5.8 (Vector Space).** A *vector space* over the reals is a tuple $(V,+,\cdot)$ where $V$ is a set whose elements are called *vectors*, $+$ is vector addition with signature $V \times V \to V$, $\cdot$ is scalar multiplication with signature $\mathbb{R} \times V \to V$, and the following properties hold:

1. Additive Identity: There is a vector 0 such that for all $x \in V$, $x+0=x$.
2. Additive Commutativity: For all $x,y \in V$, $x+y=y+x$.
3. Additive Associativity: For all $x,y,z \in V$, $(x+y)+z=x+(y+z)$.
4. Additive Inverses: For all $x \in V$, there is a $y \in V$ such that $x+y=0$.
5. Multiplicative Identity: For all $x \in V$, $1 \cdot x = x$.
6. Multiplicative Consistency: For all $a,b \in \mathbb{R}$ and $x \in V$, $a(bx)=(ab)x$.
7. Distributivity of $\times$: For all $a \in \mathbb{R}$ and $x,y \in V$, $a(x+y)=ax+ay$.
8. Distributivity of $+$: For all $a,b \in \mathbb{R}$ and $x \in V$, $(a+b)x=ax+bx$.

A *topological vector space*, or TVS, is a vector space for which $V$ is a topological space $(V,\tau)$ in which addition and multiplication are both continuous with respect to the relevant product topologies over $V \times V$ and $\mathbb{R} \times V$.

Among the main topological consequence of the vector structure is that the neighborhood system at every point is identical to the neighborhood system at the vector 0, called the *origin*. This result follows from the fact that every neighborhood $N$ of zero can be translated to a neighborhood of any other point $x \in V$ as the set $x+N=\{x+y \mid y \in N\}$, and the same can be done in reverse. Thus, in a vector space, one can prove many properties about the topology of the entire space just by proving those properties in the neighborhood of zero. In particular, if a TVS has a countable neighborhood base at the origin, then it is immediately first countable since it has a countable neighborhood base at every point. In this vein, it is often said that a TVS is topologically translation invariant, since the local structure of the space is the same at every point.

In a TVS, two new types of sets are relevant. First, for any set $A$ and real number $t>0$, the set $tA=\{tx \mid x \in A\}$. A set $B$ is *bounded* if for every neighborhood $N$ of the origin, there is some $t_0$ such that $B \subseteq tN$ if $t>t_0$. In a metric TVS, this definition is equivalent to saying that $B$ is contained within a ball of finite radius centered at the origin, but the more general definition works for every TVS. A set $B$ if *balanced* if $tB \subseteq B$ for every $t \leq 1$; in a balanced set, there is a straight line from every point in the set to the origin that remains in the set. Finally, a set $C$ is *convex* if for any $x,y \in C$ and $t \in [0,1]$, the point $tx+(1-t)y \in C$, or, equivalently, the set $C$ contains every possible straight line between any two points of the set.

A TVS is not automatically metrizable, but by the Birkhoff-Kakutani Theorem it is metrizable if and only if the singleton $\{0\}$ is a closed set and the origin has a countable neighborhood base (equivalently, if the origin is closed and the space is first countable). A metrizable vector space is characterized by smooth local transitions in space. Most vector spaces encountered in this book are not only metrizable, but normable.

**Definition 5.9 (Norm).** A function $p: V \to \mathbb{R}$ over a vector space $V$ is a *norm* if it satisfies:

1. Absolute Homogeneity: For all $\alpha \in \mathbb{R}$ and $x \in V$, $p(\alpha x) = |\alpha| p(x)$.
2. Triangle Inequality: For all $x, y \in V$, $p(x+y) \leq p(x) + p(y)$.
3. Nonnegativity: For all $x \in V$, $p(x) \geq 0$.
4. Zero Identity: $p(0) = 0$.

A norm $p(x)$ is typically denoted by $\|x\|$ or $\|x\|_p$. If all properties except the Zero Identity hold, then $p$ is a *seminorm*.

A norm induces a metric $\rho : V \times V \to \mathbb{R}$ such that $\rho(x,y) = \|x - y\|_p$. If a TVS has the topology generated by the metric $\rho$, then it is a normed vector space under the norm $p$. If a TVS has a topology that is equivalent to that generated by some norm, then it is normable. A TVS is normable if and only if it is Hausdorff and has a convex bounded neighborhood of zero. The key difference between a normed vector space and a metric vector space comes from absolute homogeneity, which guarantees that the space spreads out evenly in each direction.

Euclidean space is a primary example of a normed vector space. In $n$ dimensions, the space $\mathbb{R}^n$ is normable with the $p$-norms for Euclidean space,

$$\|x\|_p = \left( \sum_{i=1}^{n} x_i^p \right)^{1/p}.$$

When clear from context, the subscript $p$ may be dropped. The case where $p = 1$ is called the "L1-norm", and when $p = 2$, the term "L2-norm" is used.

The next concept is *completeness*, which is based on Cauchy sequences. A sequence $(x_n)_{n=1}^{\infty}$ in a TVS is *Cauchy* if for every neighborhood $N$ of the origin there is a natural number $M$ such that for all $m, n > M$, $x_m - x_n \in N$. In a normed vector space, this definition is equivalent to the statement that for every $\varepsilon > 0$, there is an $M$ such that for all $m, n > M$, $\|x_m - x_n\| < \varepsilon$. The sequence elements get arbitrarily close to each other, but there is no *a priori* guarantee that the sequence converges within the space. In particular, they may converge to an element outside of the space. A TVS is complete if every Cauchy sequence converges within the space. Without completeness, the space has "holes", that is, there are sequences that converge to a point not in the space.

A complete normed vector space is called a *Banach space*. Banach spaces are one of the most commonly encountered structures in functional analysis, and the characterization of optimizers in this book will place them inside of a Banach space. There are many common examples of Banach spaces, many of which will be used in this book. The familiar Euclidean spaces $\mathbb{R}^n$ are Banach spaces, as is the set $C[X]$ of continuous, real-valued functions on $X$, where continuity is with respect to a given topology on $X$.

Another set of examples are the $\ell^p$ spaces,

$$\ell^p = \left\{ (x_i)_{i=1}^{\infty} \subseteq \mathbb{R} \;\middle|\; \sum_{i=1}^{\infty} |x_i|^p < \infty \right\},$$

which consist of sequences over the real line whose sums converge absolutely when raised to a power $0 < p < \infty$. The sequences can be added pointwise so that for sequences $(x_i)_{i=1}^{\infty}$ and $(y_i)_{i=1}^{\infty}$, $(x+y)_i = x_i + y_i$, and the norm of a sequence is

$$\|x\|_p = \left( \sum_{i=1}^{\infty} |x_i|^p \right)^{1/p}.$$

As $p \to \infty$, one defines $\ell^\infty$ as the set of bounded sequences, *i.e.*, $\sup_i |x_i| < \infty$.

A more general example are the $L^p$ spaces of $p$-integrable functions. Note that functions can be multiplied and added pointwise so that $(f+g)(x) = f(x) + g(x)$ and for $\alpha > 0$, $f(\alpha x) = \alpha f(x)$. The $L^p$ spaces are defined with respect to a measure $\mu$ over real-valued functions on a measurable space $X$. Because two distinct functions can still be equal $\mu$-almost everywhere (that is, equal everywhere except on a set of $\mu$-measure zero), $L^p$ spaces are built not on functions directly, but on equivalence classes of functions that are equal $\mu$-almost everywhere. We will denote this set of equivalence classes as $\mathbb{R}^X_\mu$. Then

$$L^p_\mu = \left\{ f \in \mathbb{R}^X_\mu \;\middle|\; \int |f|^p \, d\mu < \infty \right\},$$

for $0 < p < \infty$, and a norm on $L^p$ is given by

$$\|f\|_p = \left( \int |f|^p \, d\mu \right)^{1/p}.$$

As with sequences, the space $L^\infty_\mu$ is defined as the set of $\mu$-integrable functions that are essentially bounded, *i.e.*, that are bounded except on a set of $\mu$-measure zero, with norm

$$\|f\|_\infty = \operatorname{ess\,sup}_\mu f = \inf_{\{A \,|\, \mu(A)=0\}} \sup_{x \in X \setminus A} f(x),$$

where ess sup represents *essential supremum*, the minimum upper bound of $f$ over all sets of full measure.

There is clearly a tight relationship between the sequence spaces $\ell^p$ and the function spaces $L^p_\mu$. Using the *counting measure*, defined so that $\mu(A) = |A \cap \mathbb{N}|$ counts the number of natural numbers in a set, we find that $\ell^p = L^p_\mu$.

The most relevant Banach spaces for this book are spaces of measures. Consider the space of finite signed measures over a measurable space $(X, \Sigma_X)$,

$$\mathcal{M}[X, \Sigma_X] = \{ \nu \mid \nu \text{ is a finite signed measure, } \textit{i.e.}, |\nu|(X) < \infty \}.$$

Finite signed measures may be added and multiplied pointwise so that for $\Sigma_X$-measurable sets $A$ and measures $\mu, \nu$, $(\mu + \nu)(A) = \mu(A) + \nu(A)$ and for $\alpha \in \mathbb{R}$, $(\alpha\mu)(A) = \alpha(\mu(A))$. Both operations yield finite signed measures contained within the space, so $\mathcal{M}[X, \Sigma_X]$ is a vector space. Further, recalling that $\nu = \nu_+ - \nu_-$ is the unique Jordan decomposition of a signed measure and $|\nu| = \nu_+ + \nu_-$, there is a norm over finite signed measures given by

$$\|\nu\| = \sup_{A \in \Sigma_X} |\nu|(A),$$

called the *total variation*, which is the upper bound of the absolute measure over all measurable sets. Completeness of the space follows from the fact that every Cauchy sequence of finite signed measures converges to a finite signed measure, and the absolute measure of every measurable set must also converge to a finite value. Thus $\mathcal{M}[X, \Sigma_X]$ is a Banach space.

This section has introduced common topologies including discrete, metric, union, product, quotient, and vector spaces. These spaces are used extensively in the remainder of this book, in which search and optimization are studied in Hausdorff topological spaces. In order to apply stochastic optimization to these spaces, however, it is necessary to understand how topology can interact with probability theory in order to generate measurable spaces for a given topology.

## 5.3 Probability Measures in Topological Spaces

In order to study stochastic optimization in topological search domains, there must be a way to convert topological spaces into measurable spaces and to construct probability measures over these spaces. Borel measures satisfy this requirement, and the Carathéodory Extension Theorem provides a way to extend set functions on a simplified domain into measures. These tools are introduced in this section, and it is then shown how to generate a range of probability measures using the constructive tools of topology introduced in the last section.

### *5.3.1 Borel Measures and Carathéodory Extension*

The Borel $\sigma$-algebra for a topological space $(X, \tau)$ is the smallest $\sigma$-algebra such that the $\tau$-open sets are measurable. This $\sigma$-algebra will be denoted variously as $\mathcal{B}_\tau$, $\mathcal{B}_X$, $\mathcal{B}[X]$, or $\mathcal{B}[X, \tau]$, with different notations chosen to emphasize relevant aspects depending on context. The elements of the $\sigma$-algebra are called *Borel sets*. The Borel $\sigma$-algebra aligns a measurable space with the neighborhoods of a topological space so that the structural features of the space define the probabilistic outcomes that can be assessed. The Borel $\sigma$-algebra generally makes enough sets measurable to support robust analysis, and it can be applied to any topological space. A measure defined against a Borel $\sigma$-algebra is called a *Borel measure*.

The Borel $\sigma$-algebra is the $\sigma$-algebra generated by the open sets; that is, $\mathcal{B}_\tau = \sigma(\tau)$. Although one might then expect that if $\mathcal{U}$ is a topological base for $\tau$ then $\mathcal{B}_\tau = \sigma(\mathcal{U})$, this is not true in general because arbitrary unions of open sets are contained in $\tau$, including uncountable unions, whereas $\sigma(\mathcal{U})$ contains only unions that are countably generated from open sets. The following proposition clarifies when $\mathcal{B}_\tau = \sigma(\mathcal{U})$.

**Proposition 5.4.** *If $\mathcal{U}$ is a topological base for a Hausdorff topological space $(X,\tau)$, then $\sigma(\mathcal{U}) = \mathcal{B}_\tau$ whenever one of the following holds:*

1. *Every $\tau$-open set $G = \bigcup_{i=1}^{\infty} U_i$ for some countable sequence $(U_i)_{i=1}^{\infty} \subseteq \mathcal{U}$.*
2. *The base $\mathcal{U}$ is countable.*
3. *The space is second countable, i.e., $\tau$ has a countable topological base.*
4. *The space is Lindelöf, i.e., every open cover has a countable subcover.*
5. *The space is metrizable and separable, i.e., it has a countable dense set.*
6. *The space is separable and paracompact, i.e., open covers have locally finite open refinements.*

*Note that for metrizable spaces, the Lindelöf, separability, and second-countability properties coincide.*

Although a measure is a set function, one does not have to specify the behavior of the measure on every set in order to obtain a complete measure. Rather, a set function called a *pre-measure* can be defined on a limited subset of a $\sigma$-algebra called a *ring*, and this pre-measure can then be extended to a unique measure.

**Definition 5.10 (Ring).** A *ring* $R$ over a set $X$ is a set of subsets of $X$ with:

1. $\emptyset \in R$.
2. For all $A,B \in R$, the intersection $A \cap B \in R$.
3. For all $A,B \in R$, the set difference $A \setminus B \in R$.

**Definition 5.11 (Algebra).** An *algebra* $\mathcal{A}$ over a set $X$ is a ring over $X$ such that for all $A,B \in \mathcal{A}$, the union $A \cup B \in \mathcal{A}$.

Note that since pairwise intersections are contained in a ring, all $n$-ary intersections are contained by implication so long as $n$ is finite. Thus a ring is a collection of sets that contains at least the empty set, all finite intersections, and all finite set differences, and an algebra contains all these items plus all finite unions. The key here is finiteness, which makes it possible to enumerate the value of a measure over finite operations without worrying about limits. A $\sigma$-ring contains countable intersections, and, as we have seen, a $\sigma$-algebra adds countable unions.

Every $\sigma$-algebra is both an algebra and a ring, but these latter two can have less sets. For the purpose of generating measures, a ring can ignore large unions while providing enough sets to cover the space. A pre-measure defined on a ring thus provides a notion of measure over the entire space that is incomplete with respect to large unions. The measure of the remaining sets can be filled in later by invoking disjoint additivity; this is the Carathéodory Extension Theorem. Since every algebra is a ring, the theorem applies to algebras as well as rings.

**Definition 5.12 (Pre-Measure).** A *pre-measure* $\mu_0 : R \to \mathbb{R} \cup \{\pm\infty\}$ is a set function defined on a ring $R$ that satisfies:

- $\mu_0(\emptyset) = 0$.
- For all $A \in R$, $\mu_0(R) \geq 0$.
- For all collections $(A_i)_{i=1}^{\infty} \subseteq \mathcal{U}$ of disjoint sets, $\mu_0\left(\bigcup_{i=1}^{\infty} A_i\right) = \sum_{i=1}^{\infty} \mu_0(A_i)$.

**Theorem 5.1 (Carathéodory Extension Theorem).** *Every pre-measure $\mu_0$ defined on a ring $R$ has an extension to a measure $\mu$ defined on $\sigma(R)$, the smallest $\sigma$-algebra containing $R$, and the extension is unique if $\mu$ is $\sigma$-finite.*

Recall that the prefix $\sigma$- indicates that a space is a countable union of sets for which prefixed property holds, so a measure $\mu$ is $\sigma$-finite if there is a countable union of sets $(A_i)_{i=1}^{\infty}$ that covers the entire space such that $\mu(A_i) < \infty$. In this book, we are generally concerned with finite measures, which are trivially $\sigma$-finite.

The significance of Carathéodory Extension for the purpose of generating Borel measures is that one only needs to define a pre-measure over a ring containing a topological base for the space, as the following corollary states.

**Corollary 5.1.** *Suppose $\mathcal{U}$ is a topological base for a topological space $(X, \tau)$. If any of the conditions in Proposition 5.4 are satisfied, then the $\sigma$-algebra generated by $\mathcal{U}$ is the Borel $\sigma$-algebra on $(X, \tau)$, i.e., $\sigma(\mathcal{U}) = \mathcal{B}_{\tau}$. Consequently, if $R(\mathcal{U})$ is the smallest ring containing $\mathcal{U}$ as a subset and $\mu_0$ is a pre-measure defined on $R(\mathcal{U})$, then there is a measure $\mu$ on $(X, \mathcal{B}_{\tau})$ that extends $\mu_0$, and $\mu$ is unique if it is $\sigma$-finite.*

The Carathéodory Extension Theorem is typically applied to generate measures beneath the surface of other technical machinery. Such tools are now developed to provide basic measures in topological spaces.

### 5.3.2 Constructing Borel Measures

The more practical question is how to obtain a reasonable selection of measures for a given topological space. The concept of a Borel measure provides the framework, and from there we can proceed as in Section 5.2 by establishing classes of basic measures defined from probability mass functions and probability density functions. These basic measures can be projected by restrictions and push-forward measures to subspace, quotient, disjoint union, and product spaces.

If a Hausdorff space $(X, \tau)$ has the discrete topology, then the singletons (*i.e.*, the sets containing a single point) form a topological base for $\tau$, and by Corollary 5.1, all that is needed is a *probability mass function* $p : X \to \mathbb{R}$ that assigns a probability mass to each point. The collection of singletons can be made into a ring just by adding the empty set. So $\mu_0$ defined by $\mu_0(\emptyset) = 0$ and $\mu_0(\{x\}) = p(x)$ is a pre-measure on this ring. Therefore there is a Borel measure $\mu$ that agrees with $\mu_0$ on the singletons. Notice that by the Carathédory Extension Theorem, $\mu$ is $\sigma$-finite and therefore unique if $X$ is countable or if $p$ assigns positive probability mass to at most countably many points of $X$. Similarly, $\mu$ is only finite if $X$ is finite or $p$ assigns positive probability mass to at most finitely many points of $X$. It should be obvious that in finite spaces such as those of Chapter 3, the probability mass function can be computationally represented as an array of real values.

If $(X, \tau)$ is a topological space and $Y$ is an Borel subset of $X$ with the subspace topology, then a Borel measure $\mu$ on $X$ can be converted to a Borel measure $\mu|_Y$ on $Y$ by standard function restriction, so that $\mu|_Y(A) = \mu(A)$ for all Borel $A$ in $Y$, and more generally $\mu|_Y(B) = \mu(B \cap Y)$ for $B$ Borel in $X$. This restricted measure was previously introduced in Section 4.2.2.

For the Euclidean spaces $\mathbb{R}^n$, a general class of finite Borel measures can be obtained by integration with respect to the Lebesgue measure, which yields the common notion of volume in $\mathbb{R}^n$. Thus if $p : \mathbb{R}^n \to \mathbb{R}$ is a nonnegative integrable function, then there is a finite Borel measure $\nu$ defined by

$$\nu(A) = \int_A p \, d\lambda = \int_A p(x) \, dx$$

for all Borel sets $A$ in $\mathbb{R}^n$, where $\lambda$ is the Lebesgue measure, which is defined so that the rightmost equality holds with respect to the Riemann integral when defined. In the case where $\nu(X) = 1$, $p$ is known as a *probability density function*, and the most familiar such density is the multivariate Gaussian density

$$p(x) = \frac{1}{\sqrt{(2\pi)^n |\Sigma|}} \exp\left\{ -\frac{1}{2}(x-\mu)^T \Sigma^{-1} (x-\mu) \right\},$$

where $\mu \in \mathbb{R}^n$ is the mean vector, $\Sigma \in \mathbb{R}^{n \times n}$ is the (symmetric) covariance matrix, $|\Sigma|$ is its determinant, and $\Sigma^{-1}$ is its inverse.

In non-Euclidean metric spaces $(X, \rho)$, there is not an obvious way to generate arbitrary measures, but if there is a Borel-measurable transformation $T : \mathbb{R}^n \to X$ that projects Euclidean space into $X$, then any Borel measure on $\mu$ can be pushed forward through $T$ to create a Borel measure on $X$. Specifically, as mentioned in Section 4.2.3, the *pushforward* measure $\mu \circ T$ is defined so that

$$\mu \circ T(A) = \mu(T^{-1}(A))$$

for all Borel sets $A \subseteq X$. This definition is valid, since Borel-measurability guarantees that $T^{-1}(A)$ is Borel in $\mathbb{R}^n$. The push-forward measure exists generally, whether or not $X$ is metric. Furthermore, if $\mu$ is a finite measure, then so is $\mu \circ T$.

The push-forward measure can also be used to define measures based on quotient topologies. Suppose $q : X \to Y$ is a quotient function that induces the quotient topology on $Y$. In this case, $q$ is continuous by construction and hence Borel-measurable. Therefore, for any Borel measure $\mu$ on $X$, the pushforward measure $\mu \circ q$ is a Borel measure on $Y$.

On a disjoint union space, a collection of pushforward measures can together yield a measure over the space. Suppose $X = \coprod_{i \in \mathfrak{I}} X_i$ has the disjoint union topology, $\alpha$ is a positive measure over $\mathfrak{I}$ with the discrete topology, and $M = \{\mu_i \mid i \in \mathfrak{I}\}$ is a collection of measures. Ideally, one could define

$$\mu(A) = \sum_{i \in \mathcal{I}} \alpha_i \left[\mu_i \circ \iota_{X_i}(A)\right],$$

where $\alpha_i = \alpha(\{i\})$ and $\iota_{X_i}$ is the inclusion function for $X_i$, but this sum is only guaranteed to exist if $\mathcal{I}$ is finite. If the sum does exist, however, then $\mu$ is a valid measure on $X$. In particular, if $\mathcal{I}$ is countably infinite and $\mu$ is finite, the existence of $\mu$ implies that there is an ordering of the summands that is monotonically decreasing. So long as the collection $(X_i)_{i\in\mathcal{I}}$ is finite or countable, the measure $\mu$ is Borel; we will encounter primarily finite disjoint unions in this book.

The product topology presents similar difficulties when the indexing space is infinite. Since this topology is central to much of this book, the resolution of this difficulty is discussed in greater detail.

### 5.3.3 *Baire vs. Borel Measures in Sequence Spaces*

If $X$ is a topological space and $\mathcal{I}$ is an arbitrary set, then the product space $X^{\mathcal{I}}$ is a topological space under the product topology. In particular, the history traces of a search process have this form, which makes this topology particularly salient to the study of search processes. Function spaces can also be viewed as product spaces of this form when endowed with the topology of pointwise convergence, which is equivalent to the product topology.

The Borel $\sigma$-algebra over $X^{\mathcal{I}}$ is well defined, and so $(X^{\mathcal{I}}, \mathcal{B}_{X^{\mathcal{I}}})$ is a measurable space. But if $\mathcal{I}$ is uncountably infinite, the main tool for defining sequence measures, the Kolmogorov Extension Theorem, results in measures that are not Borel when $\mathcal{I}$ is not countable. Instead, these measures are *Baire*.

The Baire $\sigma$-algebra on a topological space $(X, \tau)$ is the smallest $\sigma$-algebra such that contains all compact $G_\delta$ sets, written $\mathcal{B}a[X]$, $\mathcal{B}a[\tau]$, $\mathcal{B}a_X$, or $\mathcal{B}a_\tau$. A set is $G_\delta$ if it can be written as a countable intersection of open sets. The elements of this $\sigma$-algebra are called *Baire sets*, and measures defined over it are called *Baire measures*. Since every open set is Borel, the countable intersection property of $\sigma$-algebras guarantees that every Baire set is also Borel, but not every Borel set is Baire.

One example of a Borel set that is not Baire is particularly relevant to the topic of this book. As mentioned above, a singleton is a set with a single element. In a Hausdorff space, every singleton set is closed and compact, since every open cover can be reduced to a cover by a single set. Consequently, every singleton is Borel, and in countably generated spaces the singletons are Baire as well. But in uncountably generated spaces, not every singleton is Baire.

To begin with a case where the singletons are Baire sets, consider the product $X^{\mathcal{I}}$ where $\mathcal{I}$ is countable and $X$ is Hausdorff and first countable (every point of $X$ has a countable neighborhood base). Each point $x \in X^{\mathcal{I}}$ is thus representable as a function with a countable domain. Recall that under the product topology, every finite Cartesian product of open sets in $X$ is open in $X^{\mathcal{I}}$. Since $X$ is first countable, each point in $X$ has a countable neighborhood base. By the Hausdorff property, for each component $x_i$ of the product $x$, the singleton $\{x_i\}$ is equal to the countable intersection of its neighborhood base in $X$, *i.e.*, $\{x_i\} = \bigcap_{j=1}^{\infty} A_j^i$. The set $\{x\} \subseteq X^{\mathfrak{I}}$ can be written as a countable intersection of such sets by restricting each component one at a time:

$$\begin{aligned}
B_1 &= \phi_{[1]}^{-1}\left(A_1^1\right) \\
B_2 &= \phi_{[2]}^{-1}\left(A_2^1 \times A_1^2\right) \\
B_3 &= \phi_{[3]}^{-1}\left(A_3^1 \times A_2^2 \times A_1^3\right) \\
&\dots \\
B_n &= \phi_{[n]}^{-1}\left(\prod_{k=1}^{n} A_{n-k+1}^k\right).
\end{aligned}$$

Here $\phi_{[n]} : X^{\mathfrak{I}} \to X^{[n]}$ is just the canonical coordinate projection on the first $n$ coordinates according to some fixed enumeration of $\mathfrak{I}$. For any sequence $y \neq x$, there is some $n$ such that $y \notin B_n$, so $\bigcap_{n=1}^{\infty} B_n = \{x\}$. Because this intersection is countable, $\{x\}$ is a compact $G_\delta$ set and therefore Baire.

But now the problem is obvious: What if $\mathfrak{I}$ is *not* countable? In that case the sequence $B_n$ can only restrict countably many coordinates in the sequence space, so this procedure will fail to make $\{x\}$ Baire when $\mathfrak{I}$ is uncountable. As it turns out, at least in the case when $X$ is compact and $\mathfrak{I}$ is uncountable, the singleton $\{x\}$ is not a $G_\delta$ set and is not Baire.

The profound consequence of this fact is that in a Baire-measurable space, the singletons are not necessarily measurable, and hence there are Baire measures that cannot be sampled, not just in a computational sense, but in a fundamentally mathematical sense. With a Borel measure, it is theoretically knowable whether a specific outcome occurs, whereas with some Baire measures, particularly in functional or sequence spaces, the occurrence of an outcome can only be known within some region. Nonetheless, in many cases, it happens that a Baire measure can be extended to an equivalent Borel measure. Such spaces are called *Mařík*.

**Definition 5.13 (Mařík Space).** A topological space $(X, \tau)$ is said to be *Mařík* if every Baire measure on the measurable space $(X, \mathcal{B}a[X])$ can be extended to a unique, regular Borel measure on $(X, \mathcal{B}[X])$.

The condition of regularity need not be understood here, since it comes for free in Mařík spaces. Nonetheless, a measure is said to be *regular* if every measurable set can be approximated from within by compact sets and from without by open sets, *i.e.*,

$$\begin{aligned}
\mu(A) &= \sup\{\mu(K) \mid K \text{ is a compact set}\}, \text{ and} \\
\mu(A) &= \inf\{\mu(G) \mid G \text{ is an open set}\}.
\end{aligned}$$

According to a theorem of Mařík [106], every normal and countably paracompact space is Mařík.

**Theorem 5.2 (Mařík).** *Every normal and countably paracompact space is Mařík and every Mařík space is normal. In particular, the product space $X^{\mathcal{I}}$ is Mařík in each of the following cases:*

- *$X$ is finite.*
- *$\mathcal{I}$ is finite and $X$ is at most countable.*
- *$\mathcal{I}$ is finite and $X$ is locally compact and $\sigma$-compact, e.g., $X = \mathbb{R}^n$ or $\mathbb{N}^n$.*
- *$X$ is a compact space, e.g., $X = \mathbb{R} \cup \{\pm\infty\}$ or $X = [-M, M]$ for $M < \infty$.*
- *$X^{\mathcal{I}}$ is a metric space, e.g., $X$ is metric and $\mathcal{I}$ is countable.*

*If $\mathcal{I}$ has uncountable cardinality, then $X^{\mathcal{I}}$ is not normal and hence not Mařík unless $X$ is compact. Thus, for instance, if $X = \mathcal{I} = \mathbb{R}$, then $X^{\mathcal{I}} = \mathbb{R}^{\mathbb{R}}$ is not normal and thus not Mařík.*

As an important aside, the construction above illustrates a key characterization of the Baire $\sigma$-algebra: It is the smallest $\sigma$-algebra that turns all finite and countable projections into measurable functions. If $\mathcal{K}$ is a subset of $\mathcal{I}$, then define the projection $\phi_{\mathcal{K}} : X^{\mathcal{I}} \to X^{\mathcal{K}}$ so that $\phi\left((x_i)_{i\in\mathcal{I}}\right) = \left((x_i)_{i\in\mathcal{K}}\right)$. If the set $X^{\mathcal{K}}$ has the Borel $\sigma$-algebra for the product topology, then for all Borel sets $A \subseteq X^{\mathcal{K}}$, the pre-image $\phi_{\mathcal{K}}^{-1}(A)$ is a Baire-measurable set if and only if $\mathcal{K}$ is finite or countable. Otherwise, there are Borel sets whose pre-images would have to be restricted at more than countably many coordinates, and the Baire sets cannot accomplish this.

This background is crucial for understanding some subtleties of the results later in this book, such as the No Free Lunch Theorems of Chapter 12. It will reappear starting in Chapter 9, where the Kolmogorov Extension Theorem will be needed to characterize the behavior of search methods over time.

### *5.3.4 The Kolmogorov Extension Theorem*

The Kolmogorov Extension Theorem [93] is a key tool for building Baire measures on product spaces. This extension theorem is usually applied specifically to probability measures, but it can be used to extend a set of finite measures to a product space, provided that this set of measures is bounded above. That is, if $M = \{\mu_i \mid i \in \mathcal{I}\}$ is a collection of measures on $X$, then $M$ is bounded above if there is some number $N < \infty$ such that $\|\mu_i\| < N$ for all $i \in \mathcal{I}$. The result will be to convert a bounded collection of measures $M$ on $X$ into a measure on the product space $X^{\mathcal{I}}$. The version introduced here is based on that of Aliprantis [7] and uses tightness of measures.

**Definition 5.14 (Tightness).** A signed measure $\mu$ on a Borel-measurable space $(X, \mathcal{B}_X)$ is tight if for all $E \in \mathcal{B}_X$,

$$|\mu|(E) = \sup\{|\mu|(K) \mid K \text{ is compact in } X\}. \tag{5.1}$$

That is, $|\mu|(E)$ is the upper bound on the measure of all compact subsets.

Every measure is tight in complete separable metric spaces such as $\mathbb{N}$, $\mathbb{R}^n$, $L^2(\mathbb{R}^n)$, and other common spaces.

The next definition regards the consistency of a bounded collection of measures. Let $\mathcal{I}$ be any set. Although we will call it an index set, it need not be ordered. Suppose $\{(X_i, \mathcal{B}_{X_i}) \mid i \in \mathcal{I}\}$ is a collection of Borel-measurable spaces. An index subset $\mathcal{K}$ of $\mathcal{I}$ can be used to define subsets of this collection. A *family of finite dimensional distributions* over $\mathcal{I}$ is a bounded collection of finite measures indexed by all finite subsets of $\mathcal{I}$, e.g., $M = \{\mu_{\mathcal{K}} \mid \mathcal{K} \subseteq \mathcal{I}, \mathcal{K} \text{ finite}\}$. Each $\mu_{\mathcal{K}}$ must be defined on the Borel product $\sigma$-algebra $\mathcal{B}_{X^{\mathcal{K}}}$. Consistency requires overlapping subsequences to agree on probabilities. [3]

**Definition 5.15 (Kolmogorov Consistency).** Suppose $\{\mu_{\mathcal{K}} \mid \mathcal{K} \subseteq \mathcal{I}, \mathcal{K} \text{ finite}\}$ is a family of finite-dimensional distributions for a set $\mathcal{I}$. This family is consistent if for any other finite subset $\mathcal{H}$ of $\mathcal{I}$ and any $A \in \mathcal{B}_{X^{\mathcal{K}}}$,

$$\mu_{\mathcal{K}\cup\mathcal{H}}(\{a \in X^{\mathcal{K}\cup\mathcal{H}} \mid a_{\mathcal{K}} \in A\}) = \mu_{\mathcal{K}}(A), \tag{5.2}$$

which essentially says that members of the family defined on smaller index sets can be obtained by integrating out all extra indices from members defined on larger index sets.

**Theorem 5.3 (Kolmogorov Extension Theorem).** *Let $\{X_i \mid i \in \mathcal{I}\}$ be a family of Hausdorff topological spaces over a set $\mathcal{I}$, each equipped with their Borel $\sigma$-algebras. Let $\{\mu_{\mathcal{K}} \mid \mathcal{K} \subseteq \mathcal{I}, \mathcal{K} \text{ finite}\}$ be a consistent family of finite-dimensional distributions for $\mathcal{I}$. Assume that each $\mu_{\mathcal{K}}$ is tight. Then there is a unique Baire measure $\mu$ on $X^{\mathcal{I}}$ that extends each $\mu_{\mathcal{K}}$.*

As mentioned above, in the case of a Mařík space (Theorem 5.2), this measure can be extended further to a regular Borel measure. The only requirements are consistency and tightness. As one might imagine,Theorem 5.3 is proved primarily by an application of the Carathéodory Extension Theorem to the ring formed by finite products of Borel sets.

Thus in most cases, a Baire measure over sequences can be constructed for a product space $X^{\mathcal{I}}$ given all of its finite marginal distributions. If one is willing to build a measure for which the components are independent, it suffices to have a bounded collection of tight, positive measures $M = \{\mu_i \mid i \in \mathcal{I}\}$, from which a family of finite dimensional distributions can be constructed as follows.

Let $\mathcal{K}$ be an arbitrary finite subset of $\mathcal{I}$, and let $(A_i)_{i\in\mathcal{K}}$ range over finite $\mathcal{K}$-indexed collection of Borel sets in $X$. Then

$$\mu_{\mathcal{K},0}\left(\prod_{i\in\mathcal{K}} A_i\right) = \prod_{i\in\mathcal{K}} \mu_i(A_i)$$

[3] Many texts give a definition of consistency that uses permutations of multi-indices (see *e.g.* [86]). Permutations are not needed in Definition 5.15 because it is based on unordered sets of subindices. The notation here is based on an unpublished manuscript of K. Border.

defines a pre-measure over $X^{\mathcal{K}}$, since the Cartesian products $\prod_{i\in\mathcal{K}} A_i$ form an algebra. By the Carathéodory Extension Theorem, there is a unique measure $\mu_{\mathcal{K}}$ over $X^{\mathcal{K}}$ that agrees with $\mu_{\mathcal{K},0}$, since each $\mu_i$ is finite.

In this case, the collection $\{\mu_{\mathcal{K}} \mid \mathcal{K} \subseteq \mathcal{I}, \mathcal{K}\,\text{finite}\}$ is a consistent finite family of distributions over $\mathcal{I}$, and the Kolmogorov Extension Theorem yields a unique Baire measure $\mu$ over $X^{\mathcal{I}}$. The only caveat is that although $\mu_{\mathcal{K}}$ is bounded, the collection $\{\mu_{\mathcal{K}} \mid \mathcal{K}\,\text{finite}\}$ may not be bounded as $\mathcal{K}$ grows. Such boundedness does hold, however, at least when each $\mu_i$ has $|\mu_i|(X) \leq 1$, and any bounded collection may be normalized so that this is true. If $X$ is also Mařík, then the measure $\mu$ is Borel. Thus we have proven the following proposition.

**Proposition 5.5.** *Suppose $M = \{\mu_i \mid i \in \mathcal{I}\}$ is a collection of tight, positive measures on $(X, \mathcal{B}_X)$ bounded above so that for all i, $|\mu_i|(X) \leq 1$. There there exists a unique finite Baire measure $\mu$ on $X^{\mathcal{I}}$ that agrees with each $\mu_i$ on the $i^{th}$ component. If $X$ is Mařík, then $\mu$ can be extended to a Borel measure as well.*

This construction is used in Chapter 12 to construct optimization problems satisfying No Free Lunch criteria.

The Kolmogorov Extension Theorem completes this review of how to generate Borel measures in Hausdorff topological spaces.

## 5.4 Extended Example: Artificial Neural Networks

By combining probability mass functions, probability density functions, subspace restrictions, and push-forward measures along with Carathéodory and Kolmorogov extensions, it is possible to build complex measures on a wide variety of topological spaces. This section provides an example in a space of artificial neural networks. Due to its practicality, examples such as this one are, in fact, one of the primary reasons for considering general Hausdorff topological spaces rather than limiting the discussion to Euclidean or finite spaces.

The term *artificial neural networks*, or ANNs, refers to several classes of parametric models historically based on an analogy with the function of biological neurons in the brain. Since that time, these models have attained widespread usage starting in the 1980's and expanding substantially in the 2010's. The similarities between ANNs and biological brains are limited, and the field has continued to develop primarily as an exercise in mathematical modeling of functions.

There are many different ANN architectures. The common features they share is a directed network of components with some components designated to receive inputs, some to produce outputs, and the rest to extract and manipulate internal representations ultimately used to generate outputs. For simplicity, this example will address a feed-forward, fully-connected neural network architecture with a varying number of hidden layers of varying sizes.

Mathematically a fully-connected layer in a neural network computes a function

$$f(x) = h(Wx + b)$$

where $W \in \mathbb{R}^{m\times n}$ is a weight matrix, $b \in \mathbb{R}^m$ is a bias vector, and $h : \mathbb{R} \to \mathbb{R}$ is a vectorized activation function. By "vectorized", it is meant that for $z \in \mathbb{R}^m$, the value of $y = h(z)$ is a vector in $\mathbb{R}^m$ such that for each component $i$, $y_i = h(z_i)$. Common choices for $h$ include:

1. Linear: $h(x) = x$,
2. Hyperbolic Tangent: $h(x) = \tanh x$,
3. Logistic: $h(x) = \frac{1}{1+e^{-x}}$, and
4. Rectified Linear Unit (ReLU): $h(x) = \max\{x, 0\}$,

among many others. The activation function $h$ serves to introduce non-linearities at certain points so that neural networks with several layers can model arbitrary functions, a result proven by Cybenko for three-layer networks with logistic activation [38].

A feed-forward fully-connected neural network computes a function determined by composing the results of several layers. If $f_1, \ldots, f_\ell$ are each neural network layers, then a neural network $N$ computes a function $N(x) = f_1 \circ \ldots \circ f_\ell(x)$. In general, the dimension of the input, the dimension of the output, and the activation function at the output are fixed by the problem being solved, but the dimension of the internal layers can vary.

The space of such networks can be constructed as a disjoint union where each member of the union reflects a choice configuration options including the activation function, the number of layers, and the size of each layer. We will stratify the disjoint union into these three choices, so that a neural network can be sampled by making these three decisions in turn.

Firstly, given the activation $h$, the number of layers $\ell$, the input dimension $n$, and the output dimension $m$, the weight matrix $W$ is a matrix in $\mathbb{R}^{m\times n}$ and the bias $b$ is a vector in $\mathbb{R}^m$. Therefore once the characteristics of the layer are known, the layer function can be realized as an element of the product space

$$\mathcal{L}_{h,\ell,m,n} = \mathbb{R}^{m\times n} \times \mathbb{R}^m.$$

Then, for a finite set of activations $H$, a maximum number of hidden layers $L$, input dimension $i$, output dimension $j$, and for layers of any positive integral size, a space of neural networks can be defined by

$$\mathcal{N}_{L,H} = \coprod_{h\in H} \coprod_{\ell=0}^{L} \coprod_{s\in \mathcal{S}} \prod_{k=0}^{\ell} \mathcal{L}_{h,\ell,s_k,s_{k+1}}$$

where elements of $\mathcal{S}_\ell = \{i\} \times \mathbb{N}^\ell \times \{j\}$ are tuples of length $\ell + 2$ containing the size of each layer, starting with the input of size $i$, going through each hidden layer, which is an arbitrary natural number, and ending with the output of size $j$. The innermost product $\prod_{k=0}^{\ell} \mathcal{L}_{h,\ell,s_k,s_{k+1}}$ is the traditional parameter space of the network, consisting of all the weight matrices and bias vectors for each layer. Here $s_k$ indicates the input size for layer $k$, and $s_{k+1}$ is the output size for the layer. The three disjoint unions each specify different hyperparameters of the networks. Moving from left to right, these range over the activation function, the number of layers, and the sizes for the hidden layers. Essentially, $\mathcal{N}_{L,H}$ has one copy of parameters corresponding to each possible setting of layer size, layer depth, and layer activation function; that is the meaning of the disjoint union.

In order to place a probability measure on the network space $\mathcal{N}_{L,H}$, it suffices to put a probability measure on each $\mathcal{L}_{h,\ell,m,n}$ and then to assign a measure to each of the sets $H$, $[L]$, and $\mathcal{S}_\ell$ that index the three disjoint unions. The overall measure can then be defined using the techniques of Section 5.3.2 for disjoint unions.

Sampling the networks parameters from $\mathcal{L}_{h,\ell,m,n}$ can be accomplished in many ways and is commonly done in order to initialize a neural network. The simplest way is to recognize that $\mathcal{L}_{h,\ell,m,n}$ is isomorphic to $\mathbb{R}^{m(n+1)}$, and then to derive a measure $\gamma_{h,\ell,m,n}$ from a probability density over this Euclidean space, such as a multivariate Gaussian with independent standard normal components,

$$\gamma_{h,\ell,m,n}(dz) = \left(\frac{1}{\sqrt{2\pi}}\right)^{m(n+1)} \exp\left(-\frac{1}{2}\|z\|^2\right) dx = \prod_{i=1}^{m(n+1)} \frac{dz_i}{\sqrt{2\pi}} \exp\left(-\frac{1}{2}z_i^2\right),$$

where $\|z\|$ is the standard Euclidean 2-norm. The measure $\gamma_{h,\ell,m,n}$ is actually defined on the Lebesgue $\sigma$-algebra, but since the every Borel set is also a Lebesgue set, it can be considered as a Borel measure by restricting to the Borel $\sigma$-algebra.

In order to sample the space $\mathcal{S}_\ell$, one observes that $\mathcal{S}_\ell$ with the discrete topology is countable and homeomorphic to $\mathbb{N}^\ell$, again with the discrete topology. Any standard measure for $\mathbb{N}$ can therefore be expanded over the product to obtain a measure $\xi_\ell$ over $\mathcal{S}_\ell$ with independent components. Specifically, we consider a Poisson measure centered at the size of the input $n$ with probability mass

$$\xi_\ell(s) = \prod_{i=1}^{\ell} \frac{n^{s_i}}{s_i!} e^{-n},$$

which uniquely defines a Borel measure over $\mathcal{S}_\ell$ with the discrete topology.

The sets $H$ and $[L] = \{\ell \mid 1 \leq \ell \leq L\}$ are both finite, and when the discrete topology is applied to them, a probability measure can be created simply by specifying a probability mass function. Let $\eta : H \to [0,1]$ and $\lambda : [L] \to [0,1]$ be these probability mass functions, and then putting it all together, a measure over $\mathcal{N}_{H,L}$ is given for any Borel set $A \subseteq \mathcal{N}_{H,L}$ by

$$\mu(A) = \sum_{h\in H} \sum_{\ell=1}^{L} \sum_{s\in\mathcal{S}_\ell} \int_{A\cap\mathcal{L}_{h,\ell,m,n}} \eta(h) \times \lambda(\ell) \times \xi_\ell(s) \times \gamma_{h,\ell,m,n}(dz), \tag{5.3}$$

where $A \cap \mathcal{L}_{h,\ell,m,n}$ is shorthand for the pre-image of $A$ under the inclusion for $\mathcal{L}_{h,\ell,m,n}$, that is, $\iota^{-1}_{\mathcal{L}_{h,\ell,m,n}}(A)$. Essentially, one samples the measure $\mu$ by first choosing and activation function $h$, then a layer size $\ell$, and finally layer sizes $s_\ell$. Then the parameters are initialized by sampling a Gaussian.

This example demonstrates how a measure over a complex space can be built up from basic components using constructive topology as well as the methods of Section 5.3.2. This neural network space has a relatively simple topology. An example of building a measure over a broader space of recurrent neural networks can be found in Chapter 16, which presents a stochastic optimization method for such networks based on measure theory.

## 5.5 Conclusion

Topological spaces provide a general characterization of spaces by their neighborhood structure, and Borel measures based on topologies can serve as a robust source of probability measures over a many different spaces including not only basic spaces such as the natural number, the real line, or Euclidean space, but also in composite spaces such as the space of artificial neural networks examined in Section 5.4.

From here on, this book considers search processes as operating on Hausdorff topological spaces. Thus topological terminology, including topological definitions of continuity and convergence, are critical topics for the remaining chapters. Furthermore, the collection of search generators can be analyzed themselves as topological spaces and even, in fact, as Banach spaces. The study of these spaces is undertaken next.

# Chapter 6
# Algebras of Search Generators

When iterative search processes are viewed as a sequence of stochastic decisions, it is intuitively clear that small changes to the probability law governing those decisions will result in correspondingly small changes to the iterates of the process. The search generators introduced in Definition 4.6 represent how an iterative search process chooses the next point stochastically based on the prior search history. The concept that there can be arbitrarily small changes in the probability of each search point suggests that the collection of search generators should form a continuous space. In fact, with some bounds imposed, the space of search generators is a Banach space, that is, a complete normed vector space. The vector structure implies that the space has a translation-invariant topology; completeness means that the space has no gaps or holes. The norm provides a coherent notion of distance that spreads evenly from each point in the space. All of these features provide a rigorous set of tools on which to base an analysis of search processes.

The vector structure for search generators does not depend the history process *per se*, and hence in this chapter a more general class of objects is introduced, called *measure-valued functionals*, which encompasses not only search generators but the trajectors defined for finite spaces in Chapter 3 as well. At their fullest extent, measure-valued functionals are tightly related to conditional probabilities, as will become evident in Chapter 17.

In addition to the vector operations on the space, other algebraic operations can be defined over search generators. Of particular usefulness is the *convolution* operator, which can be used to construct new search methods or to generate the iterates of the search process. This chapter introduces operators and properties that can be used to study search generators or construct new search generators from simple components, developing definitions and notation that help elucidate the connections between commonly used randomized search heuristics.



A. J. Lockett, *General-Purpose Optimization Through Information Maximization*, Natural Computing Series, https://doi.org/10.1007/978-3-662-62007-6_6

## 6.1 Banach Space of Measure-Valued Functionals

To be a Banach space, the collection of search generators must be a complete, normed vector space. In this section, measure-valued functionals are introduced as a generalization of search generators and trajectors, and the vector operations for these objects are introduced along with a norm on this space. Finally, the space with these operations and norm is shown to be complete and therefore Banach.

### 6.1.1 Measure-Valued Functionals

Search generators were defined in Definition 4.6 for search domains that are Hausdorff topological spaces. Repeating that definition, a search generator is a function from history prefixes to finite signed measures. To define a measure-valued functional, the role of history prefixes is replaced with an arbitrary input space $Z$.

**Definition 6.1 (Measure-Valued Functional).** Given an input space $Z$ and a measurable target domain $(X,\Sigma)$, a *measure-valued functional* is a function $\mathcal{G} : Z \to \mathcal{M}[X]$ that maps each element of the input space to a finite signed measure over the target domain. The space of all such functionals is denoted $\mathcal{MF}^{\infty}_{Z,X}$.

The $\infty$ superscript in $\mathcal{MF}^{\infty}_{X,Z}$ indicates that all measure-valued functionals are included, including those for which the norm defined below does not exist.

**Definition 6.2 (Probability-Valued Functional).** A measure-valued functional $\mathcal{G}$ over an input space $Z$ and a target domain $(X,\Sigma)$ is called a *probability-valued functional* if for all $z \in Z$, the measure $\mathcal{G}[z]$ is a probability measure. The set of all such functionals is denoted $\mathcal{PF}_{Z,X}$.

As with search generators, square brackets, *e.g.*, $\mathcal{G}[z]$, are used to denote the application of a measure-valued functional to an input $z$, so that the usage $\mathcal{G}[z](A)$ may be used to denote the measure assigned to a measurable set $A$. The set of search generators is a particular instance of a space of measure-valued functionals in which the input space consists of history prefixes.

**Proposition 6.1.** *Given a measurable search domain $(X,\Sigma)$ and a value space $Y$, the set of search generators is equal to the set $\mathcal{MF}^{\infty}_{Z,X}$ of measure-valued functionals where $Z = \mathcal{H}^{\mathcal{T}}_{X\times Y}$ is the set of history prefixes over $X$ and $Y$ using the temporal index set $\mathcal{T}$.*

The space of search generators and its subspaces will be denoted with German fonts, with the entire set denoted as $\mathfrak{G}^{\infty} = \mathfrak{G}^{\infty}[X,Y] = \mathfrak{G}^{\infty}_{X,Y}$, where $\mathfrak{G}^{\infty}_{X,Y} = \mathcal{MF}^{\infty}_{\mathcal{H}^{\mathcal{T}}_{X,Y},X}$. The *proper* search generators of Definition 4.7 are precisely the reflexes of the probability-valued functionals when the input space consists of history prefixes.

As developed in Chapter 5, we now focus attention on the case where the search domain is a Hausdorff topological space $(X, \tau_X)$. For symmetry, we shall also generally assume that the value space is a Hausdorff topological space $(Y, \tau_Y)$, which will become more pertinent in Chapter 13 when search generators are used to generate search problems as well. The canonical measurable space for the search domain is then $(X, \mathcal{B}_X)$ where $\mathcal{B}_X$ is the Borel $\sigma$-algebra with respect to $\tau_X$. Unless otherwise stated, we henceforth assume that not only search generators but also measure-valued functionals more generally are defined with this measurable space as a target domain.

### *6.1.2 Vector Operations*

The necessary vector operations for a vector space are vector addition and scalar multiplication. These operations are now defined for measure-valued functionals based on the corresponding operations for finite signed measures.

**Definition 6.3 (Vector Addition of Measure-Valued Functionals).** For a given target domain $X$ and input space $Z$, addition of two measure-valued functionals $\mathcal{G}_1$ and $\mathcal{G}_2$ on these spaces is defined pointwise so that for each input $z \in Z$ and each measurable set $A$,

$$(\mathcal{G}_1 + \mathcal{G}_2)[z](A) = \mathcal{G}_1[z](A) + \mathcal{G}_2[z](A). \tag{6.1}$$

**Proposition 6.2.** *Given two measure-valued functionals $\mathcal{G}_1$ and $\mathcal{G}_2$ on the same target domain $X$ and input space $Z$, $\mathcal{G} = \mathcal{G}_1 + \mathcal{G}_2$ is also measure-valued-functional on $X$ and $Y$.*

*Proof.* It is obvious from Equation 6.1 that $\mathcal{G}$ is a function from inputs to signed measures defined on $X$. All that remains is to show $\mathcal{G}$ yields a finite signed measure in each case, as it must, since by definition both $\mathcal{G}_1[z](A)$ and $\mathcal{G}_2[z](A)$ are finite for all inputs $z$ and measurable sets $A$.

**Definition 6.4 (Scalar Multiplication of Measure-Valued Functionals).** For a given target domain $X$ and input space $Z$, multiplication of a measure-valued functional $\mathcal{G}$ on these spaces by a scalar $\alpha \in \mathbb{R}$ is defined pointwise so that for each input $z \in Z$ and each measurable set $A$,

$$(\alpha\mathcal{G})[z](A) = \alpha\,(\mathcal{G}[z](A)) \tag{6.2}$$

**Proposition 6.3.** *Given a measure-valued functional $\mathcal{G}$ on a target domain $X$ and an input space $Z$ along with a scalar $\alpha \in \mathcal{R}$, it follows that $\mathcal{G}' = \alpha\mathcal{G}$ is a measure-valued functional on $X$ and $Z$.*

*Proof.* Again, it is obvious that $\mathcal{G}'$ is a function from inputs to signed measures, and the resulting measure must be finite because $\alpha$ is finite and $\mathcal{G}[z](A)$ is also finite for all inputs.

**Proposition 6.4.** *With vector addition and scalar multiplication defined as above, the set of measure-valued functionals $\mathcal{MF}^{\infty}_{Z,X}$ for a given target domain X and an input space Z is a vector space over the real numbers.*

*Proof.* According to Definition 5.8, there are eight items which must be demonstrated. These are:

1. Additive Identity: Let $\mathbf{0}$ be the measure-valued functional such that $\mathbf{0}[z](A) = 0$ for all inputs $z$ and measurable sets $A$. Then for any measure-valued functional $\mathcal{G}$, $\mathcal{G} + 0 = \mathcal{G}$.
2. Additive Commutativity: Since addition in Equation 6.1 is commutative addition of real numbers, it holds for all measure-valued functionals $\mathcal{G}_1$ and $\mathcal{G}_2$ that $\mathcal{G}_1 + \mathcal{G}_2 = \mathcal{G}_2 + \mathcal{G}_1$.
3. Additive Associativity: Again, by the associativity of addition over real numbers, it holds that for any measure-valued functionals $\mathcal{G}_1, \mathcal{G}_2, \mathcal{G}_3$, we have $(\mathcal{G}_1 + \mathcal{G}_2) + \mathcal{G}_3 = \mathcal{G}_1 + (\mathcal{G}_2 + \mathcal{G}_3)$.
4. Additive Inverses: For any measure-valued functional $\mathcal{G}$, define $-\mathcal{G}$ so that for all inputs $z$ and measurable sets $A$, $(-\mathcal{G})[z](A) = -(\mathcal{G}[z](A))$. Then from Equation 6.1, $\mathcal{G} + (-\mathcal{G}) = \mathcal{G} - \mathcal{G} = \mathbf{0}$.
5. Multiplicative Identity: By Equation 6.2, for any measure-valued functional $\mathcal{G}$, $(1 \cdot \mathcal{G})[h](A) = \mathcal{G}[h](A)$ for all $h$ and $A$, so $1 \cdot \mathcal{G} = \mathcal{G}$.
6. Multiplicative Consistency: Combining the definitions, it is obvious that for all $\alpha, \beta$ and any measure-valued functional $\mathcal{G}$, $\alpha(\beta\mathcal{G}) = (\alpha\beta)\mathcal{G}$.
7. Distributivity of $\times$: Equations 6.1 and 6.2 imply distributivity of multiplication over addition since all terms in the definition are scalar.
8. Distributivity of $+$: Again, Equations 6.1 and 6.2 imply distributivity of addition over multiplication since all terms in the definition are scalar.

Since all the criteria have been demonstrated, it follows that the space of measure-valued functionals is a vector space.

Proposition 6.4 shows that the collection of measure-valued functionals is indeed a vector space when the target domain $X$ and the input space $Z$ are held fixed.

### *6.1.3 A Generator Norm*

The purpose of a norm in vector space is to capture the magnitude of a vector, especially the magnitude of vector differences, which provides a way of assessing distance between vectors and convergence of vector sequences.

Just as the vector operations were defined pointwise, a norm over search generators can be defined as the pointwise supremum over all the finite signed measures produced on each history prefix. To do this, recall that the space of finite signed measures $\mathcal{M}[X, \mathcal{B}_X]$ is a Banach space with the total variation norm

$$\|\nu\| = \sup_{A \in \mathcal{B}_X} |\nu|(A) \tag{6.3}$$

for each $\nu \in \mathcal{M}[X, \mathcal{B}_X]$, noting that $|\nu| = \nu_+ + \nu_-$. The total variation norm is just the largest absolute volume assigned to any measurable collection of points.

For probability measures $\mathbb{P}$, it is clear that $||\mathbb{P}|| = 1$, since the foregoing statement can be recast to say that the total variation norm of a probability measure is the largest probability assigned to any measurable collection of outcomes. This becomes more interesting when taking the difference for two probability measures $\mathbb{P}$ and $\mathbb{Q}$ as $||\mathbb{P} - \mathbb{Q}|| \in [0,2]$, which provides one way of assessing the closeness of measures. In fact, if $(\mathbb{P}_n)_{n=1}^{\infty}$ is a sequence of measures, then one says that $\mathbb{P}_n$ *converges in norm* to $\mathbb{P}$ if for every $\varepsilon > 0$ there is some $N$ such that for all $m > N$, $||\mathbb{P}_m - \mathbb{P}|| < \varepsilon$. This is one of the strongest forms of convergence for probability measures that is commonly considered, and it reflects convergence in the norm topology for the space of finite signed measures. Weaker forms of convergence of probability measures are often considered as well, and each one generally reflects distinct topologies on a superset of the probability measures.

The total variation norm can be extended to apply to measure-valued functionals by taking the upper bound over inputs, with the caveat that this upper bound does not universally exist.

**Definition 6.5 (Supremum Norm for Measure-Valued Functionals).** The supremum norm for a measure-valued functional $\mathcal{G}$ on a target domain $X$ and an input space $Z$ is the upper bound over all inputs $z \in Z$ of the total variation norm,

$$||\mathcal{G}|| = \sup_{z \in Z} ||\mathcal{G}[z]||. \tag{6.4}$$

If no such upper bound exists for $\mathcal{G}$, then we write $||\mathcal{G}|| = \infty$.

The supremum norm may also be written as $||\mathcal{G}||_{\mathcal{MF}}$ for clarity. The fact that $||\mathcal{G}||$ does not exist for all measure-valued functionals is a minor inconvenience, because we are primarily interested in probability-valued functions such as the proper search generators of Definition 4.7, which yield probability measures on all inputs.

**Proposition 6.5.** *If $\mathcal{G}$ is a probability-valued functional, then $||\mathcal{G}|| = 1$.*

*Proof.* As discussed in the text, the total variation norm of a probability measure is 1, and for all inputs $z$, $\mathcal{G}[z]$ is a probability measure. Consequently, the upper bound in Equation 6.4 is 1.

Given that all probability-valued functionals have a finite supremum norm, we now consider the set of measure-valued functionals which have finite supremum norm,

$$\mathcal{MF}_{Z,X} = \{\mathcal{G} \in \mathcal{MF}_{Z,X}^{\infty} \mid ||\mathcal{G}|| < \infty\}, \tag{6.5}$$

and the corresponding set of search generators

$$\mathfrak{G}_{X,Y} = \{\mathcal{G} \in \mathfrak{G}_{X,Y}^{\infty} \mid ||\mathcal{G}|| < \infty\} = \mathcal{MF}_{\mathcal{H}_{X,Y}^{\mathcal{T}},X}. \tag{6.6}$$

These sets are indeed normed vector spaces.

**Proposition 6.6.** *The collection $\mathcal{MF}_{Z,X}$ of measure-valued functionals with finite supremum norm is a normed vector space containing all probability-valued functionals.*

*Proof.* The proof has two parts. First, it will be shown that $\mathcal{MF}_{Z,X}$ is a vector subspace of $\mathcal{MF}^{\infty}_{Z,X}$, and second that the supremum norm is indeed a norm on this space. Containment of the probability-valued functionals has already been discussed in the text.

Proposition 6.4 established that $\mathcal{MF}^{\infty}_{Z,X}$ is a vector space. So suppose that $\mathcal{G}_1$ and $\mathcal{G}_2$ are both measure-valued functionals with finite supremum norm. Then it holds by the definition of an upper bound that

$$||\mathcal{G}_1 + \mathcal{G}_2|| = \sup_z ||\mathcal{G}_1[z] + \mathcal{G}_2[z]|| \leq \sup_z ||\mathcal{G}_1[z]|| + \sup_z ||\mathcal{G}_2[z]|| = ||\mathcal{G}_1|| + ||\mathcal{G}_2|| < \infty,$$

so that $\mathcal{G} = \mathcal{G}_1 + \mathcal{G}_2$ has finite supremum norm. Furthermore, for any $\alpha \in \mathbb{R}$,

$$||\alpha\mathcal{G}_1|| = \sup_z ||\alpha\mathcal{G}_1[z]|| = |\alpha| \sup_h ||\mathcal{G}_1[z]|| = |\alpha|||\mathcal{G}_1||,$$

whence $\alpha\mathcal{G}_1$ has finite supremum norm as well. Since the property of having a finite supremum norm is closed under vector addition and scalar multiplication, the collection $\mathcal{MF}_{Z,X}$ is a vector subspace of $\mathcal{MF}^{\infty}_{Z,X}$ and hence a vector space in its own right.

According to Definition 5.9, to prove that the supremum norm is a norm, four properties must be satisfied:

1. Absolute Homogeneity: For $\alpha \in \mathbb{R}$, it was shown that $||\alpha\mathcal{G}_1|| = |\alpha|||\mathcal{G}_1||$.
2. Triangle Inequality: It was shown above that $||\mathcal{G}_1 + \mathcal{G}_2|| \leq ||\mathcal{G}_1|| + ||\mathcal{G}_2||$.
3. Nonnegativity: Nonnegativity of the total variation guarantees $||\mathcal{G}|| \geq 0$.
4. Zero Identity: $||\mathbf{0}|| = \sup_z ||\mathbf{0}[z]|| = \sup_h 0 = 0$.

Consequently, the supremum norm is a norm for the vector space $\mathcal{MF}_{X,Y}$.

By restricting the measure-valued functionals to those whose total variation does not increase without bound, a normed vector space of search generators containing the probability-valued functionals has been obtained using the supremum norm. This space will now be shown to be complete and hence Banach.

### *6.1.4 Completeness Under the Supremum Norm*

Section 5.2.3 discussed the concept of completeness, which essentially asserts that a space does not have unfilled gaps between arbitrarily close elements. Such gaps are detected as the limits of Cauchy sequences. A sequence of search generators $(\mathcal{G}_n)_{n=1}^{\infty}$ each with finite supremum norm is Cauchy if for every $\varepsilon > 0$ there is some $N$ such that for all $m, n > N$, $||\mathcal{G}_m - \mathcal{G}_n|| < \varepsilon$. The space of measure-valued functionals with finite supremum norm is complete, although this fact is somewhat more complicated to prove than the other facts in this section.

**Proposition 6.7.** *Every Cauchy sequence in the vector space $\mathcal{MF}_{Z,X}$ converges to a point in the space. Equivalently, the space $\mathcal{MF}_{Z,X}$ is complete.*

*Proof.* Let $(\mathcal{G}_n)_{n=1}^{\infty}$ be a Cauchy sequence in $\mathcal{MF}_{Z,X}$. Let $z$ be an arbitrary input and $A$ be any measurable set in $X$. The sequence $(\mathcal{G}_n[z](A))_{n=1}^{\infty}$ is Cauchy in $\mathbb{R}$, since for any $\varepsilon$ there is some $M$ such that for $m,n > M$,

$$\left|\mathcal{G}_m[z](A) - \mathcal{G}_n[z](A)\right| \quad \leq \quad \|\mathcal{G}_m - \mathcal{G}_n\| \; < \; \varepsilon.$$

But the space $\mathbb{R}$ is complete, and so the sequence converges to a real number, and a potential search generator $\mathcal{G}$ can be defined so that

$$\mathcal{G}[z](A) = \lim_{n\to\infty} \mathcal{G}_n[z](A). \tag{6.7}$$

Now it must be shown that $\mathcal{G}[z]$ so defined is a finite signed measure. The form of Equation 6.7 guarantees that it is a signed measure, since the limit preserves disjoint additivity, but finiteness must be proven. Again because $(\mathcal{G}_n)$ is Cauchy, for any $\varepsilon > 0$ there is a nondecreasing sequence $(M_j)_{j=1}^{\infty}$ such that

$$\left|\mathcal{G}_{M_{j+1}}[z](A) - \mathcal{G}_{M_j}[z](A)\right| \quad \leq \quad \|\mathcal{G}_{M_{j+1}} - \mathcal{G}_{M_j}\| \; < \; \varepsilon 2^{-j}. \tag{6.8}$$

Critically, this $\varepsilon$ as well as the sequence $(M_j)$ is independent of the input $z$ and the measurable set $A$, since it is determined with respect to the Cauchy sequence $(\mathcal{G}_n)$. Also, the value $\mathcal{G}[z](A)$ can be rewritten as the telescoping sum

$$\mathcal{G}[z](A) = \mathcal{G}_{M_1}[z](A) + \sum_{j=1}^{\infty} \mathcal{G}_{M_{j+1}}[z](A) - \mathcal{G}_{M_j}[z](A), \tag{6.9}$$

to which Equation 6.8 can be applied to yield

$$|\mathcal{G}[z](A)| \leq |\mathcal{G}_{M_1}[z](A)| + \sum_{j=1}^{\infty} \left|\mathcal{G}_{M_{j+1}}[z](A) - \mathcal{G}_{M_j}[z](A)\right| < |\mathcal{G}_{M_1}[z](A)| + \varepsilon,$$

where the fact that $\sum_{j=1}^{\infty} 2^{-j} = 1$ has been used. Since the $M_j$ are chosen independent of $h$ and $A$, it follows that for all $z$

$$\|\mathcal{G}[z]\| \; < \; \|\mathcal{G}_{M_1}[z]\| + \varepsilon \; < \; \infty,$$

so that $\mathcal{G}[z]$ is finite. Taking the supremum, one finds that

$$\|\mathcal{G}\| \; < \; \sup_{z'} \|\mathcal{G}_{M_1}\| + \varepsilon \; < \; \infty,$$

which is to say that $\mathcal{G} \in \mathcal{MF}_{Z,X}$. The last step is to show that $\mathcal{G} = \lim_{n\to\infty} \mathcal{G}_n$ in the norm topology on $\mathcal{MF}_{Z,X}$. But Equation 6.9 above can be employed to prove this fact, since it can be rearranged to show

$$\left|\mathcal{G}[z](A) - \mathcal{G}_{M_1}[z](A)\right| \leq \sum_{j=1}^{\infty} \left|\mathcal{G}_{M_{j+1}}[z](A) - \mathcal{G}_{M_j}[z](A)\right| < \varepsilon$$

and $M_1$ was chosen to bound superior differences below $\varepsilon$, so for $n > M_1$,

$$\begin{aligned}\left|\mathcal{G}[z](A) - \mathcal{G}_n[z](A)\right| &\leq \left|\mathcal{G}[z](A) - \mathcal{G}_{M_1}[z](A)\right| + \left|\mathcal{G}_{M_1}[z](A) - \mathcal{G}_n[z](A)\right| \\ &< 2\varepsilon.\end{aligned}$$

For a given $\varepsilon$, then, it holds for all $n > M_1 = M_1(\varepsilon)$ that $\|\mathcal{G} - \mathcal{G}_n\| < 2\varepsilon$. Therefore $\mathcal{G}_n$ converges in norm to $\mathcal{G}$, and $\mathcal{MF}_{Z,X}$ is complete.

Putting together the facts from the prior subsections, we have proven the following.

**Theorem 6.1.** *The space $\mathcal{MF}_{Z,X}$ consisting of measure-valued functionals on the target domain X and input space Z with finite supremum norm is a Banach space.*

*Proof.* This claim follows from the fact that a Banach space is defined as a complete, normed vector space. By Proposition 6.6 $\mathcal{MF}_{Z,X}$ is a normed vector space, and by Proposition 6.7 it is complete.

The Banach space $\mathfrak{G}_{X,Y} = \mathcal{MF}_{\mathcal{H}^{\mathcal{T}}_{X,Y},X}$ has been obtained by using continuously varying stochasticity to fill in the gaps between search methods. Commonly used stochastic search and optimization methods are usually non-degenerate, meaning that they produce a range of values when sampled as opposed to just one singular value. So from an analytic standpoint, each proper search generator is an entirely natural approach to search.

The caveat, of course, is that practical search methods are limited by the toolbox of digital computation. Commonly used search methods are efficiently computable, whereas small perturbations of these methods are no longer efficient. Nonetheless, once it is realized that there is a smooth space of possible search methods, it becomes possible to generate new search methods through constructive and computable processes. The next two sections explore the nature of the proper search generators through the lens of the probability-valued functionals and introduce the convolution operation that is fundamental to the analysis of iterative search and optimization.

## 6.2 Geometry of Probability-Valued Functionals

As discussed above, the proper search generators are the actual objects of interest to this analysis, since they can be used to solve search problems. This section explores the status of the proper search generators within the space $\mathfrak{G}_{X,Y}$ in terms of the more general set of probability-valued functionals within $\mathcal{MF}_{Z,X}$. In particular, these objects form a closed, convex subset that is analogous to a simplex. Within the convex set of probability-valued functionals, a subset of deterministic-valued functionals provides the extreme points that generate the set. Finally, some remarks are made regarding the *smallest* normed space of measure-valued functionals that contains the probability-valued functionals.

### 6.2.1 Convexity of Probability-Valued Functionals

The best analogy for the probability-valued functionals within the vector space of measure-valued functionals is the simplex in Euclidean space. The simplex $\mathcal{S}^n$ in $\mathbb{R}^n$ is the set of $n$-tuples whose components are all nonnegative and sum to one,

$$\mathcal{S}^n = \left\{ x \in \mathbb{R}^n \;\middle|\; \text{For all}\, i,\, x_i \geq 0, \text{ and } \sum_i x_i = 1 \right\}.$$

On the two-dimensional plane, this is a closed line segment from $(0,1)$ to $(1,0)$. In three dimensions, it is the closed triangle with vertices at $(1,0,0)$, $(0,1,0)$, and $(0,0,1)$.

Two important features of the simplex are that it is *closed* and *convex*. Closed sets were discussed in Chapter 5. To see that $\mathcal{S}^n$ is closed, take any point $y \notin \mathcal{S}^n$ and notice that $y$ is at least $\varepsilon > 0$ away from the closest point in $\mathcal{S}^n$. Therefore the ball with radius $\varepsilon$ centered at $y$ is an open set containing $y$ that does not intersect $\mathcal{S}^n$. Thus the complement of $\mathcal{S}^n$ is open, whence $\mathcal{S}^n$ is itself closed.

Convexity was introduced in Section 5.2.3. To see that the simplex is convex, simply take any two points $x, z \in \mathcal{S}^n$ and notice that for any $t \in [0,1]$, $tx + (1-t)z$ necessarily has nonnegative components and sums to one.

To find the analogue of the simplex in the space of search generators, first consider its analogue in the Banach space of finite signed measures, which consists of the probability measures. A probability measure has nonnegative components, and its total measure is one. In fact, for finite measure spaces, the analogy is an isomorphism in which probability densities in finite spaces are isomorphic to elements of $\mathcal{S}^n$. The probability measures are a generalization of the simplex to infinite spaces; this generalization is called the *Choquet simplex*. The set of probability measures is both closed and convex in the Banach space of finite signed measures, for essentially the same reasons.

In the vector space of measure-valued functionals, this simplex-like role is played by the probability-valued functionals. As with the probability measures, the analogy is exact for the most trivial cases, for example, when the input space is finite-dimensional. The probability-valued functionals form a closed, convex set, which is proven now.

**Proposition 6.8.** *The set of probability-valued functionals $\mathcal{PF}_{Z,X}$ is a closed and convex subset of the vector space $\mathcal{MF}_{Z,X}$.*

*Proof.* First, as a minor technical point, note that if $\mathcal{G}$ is a probability-valued functional, then $\|\mathcal{G}\| = 1$ by Proposition 6.5, so $\mathcal{G}$ is indeed an element of $\mathcal{MF}_{Z,X}$.

The fact that the set of probability-valued functionals is closed will be proven by showing that every sequence of probability-valued functionals converges in norm to a probability-valued functional, which is sufficient since $\mathcal{MF}_{Z,X}$ is a normed space. Thus suppose that $(\mathcal{G}_n)_{n=1}^{\infty}$ is a convergent sequence such that each $\mathcal{G}_n$ is a probability-valued functional, and let $\mathcal{G}$ be such that $\|\mathcal{G}_n - \mathcal{G}\| \to 0$. Now for all $n$ and all inputs $z \in Z$, $\mathcal{G}_n[z](X) = 1$, so $\mathcal{G}[z](X) = 1$, or else the sequence would not converge. Secondly, for all $n$, all inputs $z$, and all measurable sets $A$, $\mathcal{G}_n[z](A) \geq 0$, so $\mathcal{G}[z](A) \geq 0$, or again the sequence would not converge. Combining these two facts, we see that $\mathcal{G}$ is a probability-valued functional, and hence the set of probability-valued functionals is closed.

To establish convexity, let $\mathcal{G}_1$ and $\mathcal{G}_2$ be probability-valued functionals and take any $t \in [0,1]$. Set $\mathcal{G} = t\mathcal{G}_1 + (1-t)\mathcal{G}_2$. Then for all inputs $z$ and measurable sets $A$,

$$\mathcal{G}[z](X) = t\mathcal{G}_1[z](X) + (1-t)\mathcal{G}_2[z](X) = 1, \text{ and}$$
$$\mathcal{G}[z](A) = t\mathcal{G}_1[z](A) + (1-t)\mathcal{G}_2[z](A) \geq 0,$$

and so $\mathcal{G}$ is a probability-valued functional. Therefore $\mathcal{PF}_{Z,X}$ is convex.

The closedness and convexity of the probability-valued functionals is a critical fact that can be used to prove, for example, that in the case of static optimization of a single objective, stochastic searching cannot outperform deterministic searching.

### *6.2.2 Deterministic-Valued Functionals as Extreme Points*

In Euclidean spaces, simplexes are convexly generated by their vertices:

$$\mathcal{S}_n = \left\{ \sum_i \alpha_i v_i \,\middle|\, v_i \text{ is a vertex point and for all } i,\ \alpha_i \geq 0 \text{ and } \sum_i \alpha_i = 1 \right\}.$$

This fact is a specialization of the foundational Krein-Milman Theorem, which depends on the following two definitions.

**Definition 6.6.** An *extreme point* of a convex set $C$ is a member of the set that cannot be written as a convex combination of any other two points in the set.

**Definition 6.7.** The *convex hull* of a subset $K$ of a Banach space is the smallest set containing all convex combinations of points in $K$.

**Theorem 6.2 (Krein-Milman).** *Every compact, convex subset of a Banach space is the convex hull of its extreme points.*

The Krein-Milman Theorem is as much an axiom as it is a theorem; it can be proven by assuming the Axiom of Choice, but Krein-Milman together with the Boolean Prime Ideal Theorem suffices to prove the Axiom of Choice. Nonetheless it is generally assumed, and it will add some clarity to the nature of the probability-valued functions, whose extreme points are the degenerate probability-valued functionals.

**Definition 6.8.** A signed measure $\mu$ on a measurable space $X$ is *degenerate* if there exists a unique point $x \in X$ such that for all measurable subsets $A$ of $X$, $x \notin A$ implies that $\mu(A) = 0$.

**Proposition 6.9.** *If $\mathbb{P}$ is a degenerate probability measure such that singletons are measurable and there exists $x \in X$ with $\mathbb{P}(\{x\}) > 0$, $\mathbb{P}(A) = \mathbb{1}_A(x)$.*

*Proof.* Since $\mathbb{P}$ is degenerate, there is a unique point $w \in X$ such that for measurable sets $A$, $\mathbb{P}(A) = 0$ if and only if $w \notin A$. Now $\mathbb{P}(\{x\}) > 0$, so $w \in \{x\}$, *i.e.*, $w = x$. Furthermore, since $\mathbb{P}$ is a probability measure,

$$\mathbb{P}(X) = \mathbb{P}(X \setminus \{x\}) + \mathbb{P}(\{x\}) = 1,$$

and then $x \notin X \setminus \{x\}$ implies that $\mathbb{P}(\{x\}) = 1$. From this fact the conclusion is immediate.

**Definition 6.9.** A measure-valued functional $\mathcal{D}$ is *deterministic* if for all inputs $z$, the probability measure $\mathcal{D}[z]$ is degenerate.

**Theorem 6.3.** *Every deterministic probability-valued function is an extreme point of the set of probability-valued functionals.*

*Proof.* Proposition 6.8 showed that the probability-valued functions are convex as a subset of $\mathcal{MF}_{Z,X}$. Let $\mathcal{D}$ be a deterministic probability-valued function. For all $z \in Z$ and for all measurable $A$, $\mathcal{D}[z](A)$ is either 0 or 1. Suppose $\overline{\mathcal{D}}, \underline{\mathcal{D}}$ are probability-valued functions distinct from each other and from $\mathcal{D}$ and that there is some $\alpha \in (0,1)$ with $\mathcal{D} = (1-\alpha)\underline{\mathcal{D}} + \alpha\overline{\mathcal{D}}$. Then because all three functionals are distinct, there must be some $z \in Z$ and some measurable set $A$ such that one of the four cases holds:

1. $\mathcal{D}[z](A) = 1$ and $\overline{\mathcal{D}}[z](A) > 1$.
2. $\mathcal{D}[z](A) = 1$ and $\underline{\mathcal{D}}[z](A) > 1$.
3. $\mathcal{D}[z](A) = 0$ and $\overline{\mathcal{D}}[z](A) < 0$.
4. $\mathcal{D}[z](A) = 0$ and $\underline{\mathcal{D}}[z](A) < 0$.

In all cases, one of $\underline{\mathcal{D}}$ and $\overline{\mathcal{D}}$ is not a probability-valued functional, which contradicts the assumptions and completes the proof.

**Theorem 6.4.** *Every extreme point of the set of probability-valued functionals is deterministic.*

*Proof.* Suppose that $\mathcal{G}$ is a non-deterministic extreme point of $\mathcal{PF}_{Z,X}$. Then there is some $z \in Z$ and measurable set $A$ such that $\mathcal{G}[z](A) = p$ with $p \in (0,1)$. Let $\overline{A}$ be the complement of $A$ in $X$, i.e. $\overline{A} = X \setminus A$ .

We will define $\underline{\mathcal{G}}$ and $\overline{\mathcal{G}}$ on either side of $\mathcal{G}$ to show that the non-deterministic $\mathcal{G}$ is not extreme. For $w \in Z$ with $w \neq z$, let $\underline{\mathcal{G}}[w] = \overline{\mathcal{G}}[w] = \mathcal{G}[w]$. Pick $\varepsilon > 0$ so that

$$0 < (1-\varepsilon)p < p < (1+\varepsilon)p < 1.$$

For all measurable $F \subseteq A$, let

$$\underline{\mathcal{G}}[z](F) = (1-\varepsilon)\mathcal{G}[z](F) \text{ and } \overline{\mathcal{G}}[z](F) = (1+\varepsilon)\mathcal{G}[z](F).$$

For all measurable $D \subseteq \overline{A}$, let

$$\begin{aligned}\underline{\mathcal{G}}[z](D) &= \frac{1-p+p\varepsilon}{1-p}\mathcal{G}[z](D) \text{ and} \\ \overline{\mathcal{G}}[z](D) &= \frac{1-p-p\varepsilon}{1-p}\mathcal{G}[z](D).\end{aligned}$$

For all other measurable subsets $C$, define

$$\underline{\mathcal{G}}[z](C) = \underline{\mathcal{G}}[z](C \cap A) + \underline{\mathcal{G}}[z](C \cap \overline{A}),$$

and do similarly for $\overline{A}$.

Now $\underline{\mathcal{G}}[z]$ has weight $p\varepsilon$ taken away from $A$ and given to $\overline{A}$, and vice versa for $\overline{\mathcal{G}}[z]$. Both are clearly probability-valued functionals since $\varepsilon$ is sufficiently small. But now it can be verified that

$$\mathcal{G} = \frac{1}{2}\left(\underline{\mathcal{G}} + \overline{\mathcal{G}}\right). \tag{6.10}$$

Therefore $\mathcal{G}$ is not an extreme point.

Given that the deterministic functionals are the extreme points among the probability-valued functionals, one might wonder whether Krein-Milman could be applied, that is, whether $\mathcal{PF}_{Z,X}$ is compact. This book will not prove compactness for all $\mathcal{PF}_{Z,X}$. Intuitively, if in a certain sense the spaces $Z$ and $X$ are too large, there may be points inside this convex set that cannot be approximated by any finite sum over extreme points; this is the essence of non-compactness for a convex set. Nonetheless, in Section 18.1.2, a Hilbert space will be presented that contains a large subset of $\mathcal{PF}_{Z,X}$. Hilbert spaces have more structure than Banach spaces, and in particular, they have the property that their unit ball is compact in the weak topology. Since the probability-valued functionals form a (weakly) closed subset of the unit ball, Krein-Milman can be applied to the probability-valued functionals in this Hilbert space.

One of the benefits of Krein-Milman derives from the Bauer Minimum Principle, which states that every convex function attains its minima and maxima over a convex set on the extreme points of the set. In subsequent chapters, the performance of a search method will be encoded as a convex function, and thus if Krein-Milman holds for some subset of the probability-valued functionals, it also hold that the best performing search method is necessarily deterministic, regardless of the search domain or the value space. This issue is explored to a limited degree in Section 18.1.2.

### *6.2.3 Subspace Generated by Proper Search Generators*

In the previous section, the vector space of all measure-valued functionals, $\mathcal{MF}^{\infty}_{Z,X}$, was reduced in size to the normed space $\mathcal{MF}_{Z,X}$ with the justification that no probability-valued functionals (and hence no search generators) were removed in doing so. It is natural to wonder whether this vector space can in turn be whittled down further to the smallest subspace of $\mathcal{MF}_{Z,X}$ that still contains all of the probability-valued functionals.

This question is surprisingly difficult to answer. It is surprising, because in contrast it is relatively easy to show that $\mathcal{M}_X$ is indeed the smallest vector space containing the Choquet simplex. For any finite signed measure $\mu$, the Hahn-Jordan decomposition yields $\mu = \mu_+ - \mu_-$ where $\mu_+$ and $\mu_-$ are unique, nonnegative measures. Therefore for $\alpha = \mu_+(X)$, $\beta = \mu_-(X)$, $\mathbb{P} = \alpha^{-1}\mu_+$, and $\mathbb{Q} = \beta^{-1}\mu_-$, we find that $\mu = \alpha\mathbb{P} - \beta\mathbb{Q}$, which proves that every finite signed measure is a linear combination of a pair of probability measures. [1]

A similar decomposition can be performed for $\mathcal{MF}_{Z,X}$, so that for any measure-valued functional $\mathcal{G}$ there exist nonnegative measure-valued functionals $\mathcal{G}_+$ and $\mathcal{G}_-$ with $\mathcal{G} = \mathcal{G}_+ - \mathcal{G}_-$, but the linear combinations fail to generate the entire space because $\mathcal{MF}_{Z,X}$ is indexed by the space $Z$. Such decompositions are nonetheless useful, and so some terminology is introduced here.

**Definition 6.10 (Positive and Negative Measure-Valued Functionals).** A measure-valued functional $\mathcal{G} \in \mathcal{MF}^{\infty}_{Z,X}$ is called *positive* if for all $z \in Z$, $\mathcal{G}[z]$ is a positive measure. It is called *negative* if for all $z \in Z$, $-\mathcal{G}[z]$ is a positive measure. Similarly, $\mathcal{G}$ is called *nonnegative* if $\mathcal{G}[z]$ is always a nonnegative measure and *nonpositive* if $\mathcal{G}[z]$ always a nonpositive measure.

**Proposition 6.10 (Decomposition of a Measure-Valued Functional).** *Every measure-valued functional $\mathcal{G} \in \mathcal{MF}^{\infty}_{Z,X}$ has a unique decomposition into two nonnegative measure-valued functionals $\mathcal{G}_+$ and $\mathcal{G}_-$ such that $\mathcal{G} = \mathcal{G}_+ - \mathcal{G}_-$.*

*Proof.* For each $z$ in the input space, define $\mathcal{G}_+[z] = (\mathcal{G}[z])_+$ and $\mathcal{G}_-[z] = (\mathcal{G}[z])_-$ where $\mathcal{G}[z] = (\mathcal{G}[z])_+ - (\mathcal{G}[z])_-$ is the Hahn-Jordan decomposition of the finite signed measure $\mathcal{G}[z]$. The resulting measure-valued functionals $\mathcal{G}_+$ and $\mathcal{G}_-$ are plainly nonnegative, and uniqueness follows from the uniqueness of the Hahn-Jordan decomposition.

[1] To cover two minor details, if exactly one of $\alpha$ or $\beta$ is zero, then $\mathbb{P}$ or $\mathbb{Q}$, respectively, can be omitted from the decomposition. If both are zero, then $\mu = 0 \cdot \mathbb{P}$ for an arbitrary probability measure $\mathbb{P}$.

This decomposition breaks the analogy between the probability-valued functions and the Choquet simplex, because for any measure-valued functional $\mathcal{G}$, it is not obvious how to reduce $\mathcal{G}_+$ and $\mathcal{G}_-$ to probability-valued functionals. If we divide to obtain $\hat{\mathcal{G}} = \mathcal{G}/||\mathcal{G}||$, then we do indeed find that $\hat{\mathcal{G}}$ is nonnegative and has norm 1, but it is not necessarily a probability-valued function. The norm $||\hat{\mathcal{G}}||$ is by definition the upper bound over all $||\hat{\mathcal{G}}[z]||$, but it need not also be the lower bound. Consequently, for some $z \in Z$, it might be true that $||\hat{\mathcal{G}}|| < 1$, and in fact it is trivial to construct such an object.

In order to replicate the simple linear decomposition of the finite signed measures into probability measures, each component of the decomposition of a measure-valued functional would need to be multiplied by a separate scalar for each measure $\mathcal{G}_+[z]$ and $\mathcal{G}_-[z]$. Therefore the space of measure-valued functionals has no corresponding two-step decomposition into probability-valued functionals.

This observation suggests that $\mathcal{MF}_{Z,X}$ may have a smaller vector subspace still containing all of $\mathcal{PF}_{Z,X}$. The identity of this subspace is not pursued further in this text, but the problematic vectors can be described. Consider the map from the measure-valued functionals to the set of bounded nonnegative sequences given by

$$\mathcal{G} \mapsto (||\mathcal{G}[z]||)_{z \in Z}.$$

This map is surjective. The image of $\mathcal{PF}_{Z,X}$ under this map is the sequence consisting of all ones, $(1,1,1,1,1,\ldots)$, and $\mathcal{PF}_{Z,X}$ spans the pre-image of this sequence. But there are many other sequences bounded above by 1, whose pre-images consist of unit vectors in $\mathcal{MF}_{Z,X}$ that may not be finitely constructible from $\mathcal{PF}_{Z,X}$. In order to determine the smallest vector subspace of $\mathcal{MF}_{Z,X}$ containing all of $\mathcal{PF}_{Z,X}$, it must be shown which of these unit vectors can be reached by finite linear combination over the probability-valued functionals.

## 6.3 Convolution and a Banach Algebra

The vector operations introduced earlier in this chapter have a formal quality in that they are not directly related to search methods. This formalism was valuable because it revealed the essentially continuous topology of stochastic search methods taken as a group. But there are more directly applicable operations for constructing search methods as well, of which the convolution operator is preeminently important. This section introduces the convolution operator and demonstrates how when taken together with the vector operation, convolution serves as a multiplication operation that makes a subspace of Markovian search generators into a Banach algebra.

### 6.3.1 Static Black-Box Search Generators

In order to define the convolution operation, a different point of view on search generators will be required, one that was partially developed in Chapter 3. This point of view assumes that there is a static objective function $u : X \to Y$ that determines the points in the value space given the search iterates. The static search generator depends on this objective function. Whereas the search generators above extend a history trace, the static search generator only needs the history of proposed search points. For this purpose, let

$$\mathcal{H}_X^{\mathcal{T}} = \bigcup_{t \in \mathcal{T}} X^{[t]}$$

be the set of search domain histories indexed by the temporal index set $\mathcal{T}$.

**Definition 6.11 (Static Search Generator).** Given a measurable search domain $X$, a value space $Y$, and a temporal index set $\mathcal{T}$, a static search generator is a function $\mathcal{G} : \mathcal{H}_X^{\mathcal{T}} \times Y^X \to \mathcal{M}_X$.

**Proposition 6.11.** *The set of static search generators for search domain $X$, value space $Y$, and temporal index set $\mathcal{T}$ is the vector space $\mathcal{MF}_{Z,X}$ with $Z = \mathcal{H}_X^{\mathcal{T}} \times Y^X$.*

*Proof.* The fact is immediate from the definitions.

**Definition 6.12 (Proper Static Search Generator).** A static search generator $\mathcal{G}$ is called *proper* if $\mathcal{G}$ is a probability-valued functional.

Static search generators are related to search generators by the equation

$$\mathcal{G}[h,u](A) = \mathcal{G}'\left[(x,u(x))_{x \in h}\right](A), \tag{6.11}$$

where $\mathcal{G}$ is a static search generator, $\mathcal{G}'$ is a search generator, $h$ is a search domain history, $u \in Y^X$ is an objective, $A$ is a measurable set in the search domain, and indexing by $x \in h$ indicates iteration of the search domain history.

Consider the set $\mathbf{G}$ of pairs $(\mathcal{G}', \mathcal{G})$ for which Equation 6.11 holds. It is a relation, and it is the graph of a function, which will be denoted as $\mathbf{G}(\cdot)$. This function is linear, but it is neither surjective nor injective.

**Proposition 6.12.** *The set $\mathbf{G}$ is the graph of a linear function that is neither surjective nor injective.*

*Proof.* To be the graph of a function, two things must be true. Firstly, for all search generators $\mathcal{G}'$, there must be some static search generator $\mathcal{G}$ such that $(\mathcal{G}', \mathcal{G}) \in \mathbf{G}$. This requirement is satisfied because Equation 6.11 can be treated as a definition for $\mathcal{G}$. Secondly, if $(\mathcal{G}', \mathcal{G}) \in \mathbf{G}$ and $(\mathcal{G}', \mathcal{K}) \in \mathbf{G}$, then $\mathcal{K} = \mathcal{G}$. But this is also the case, since for all search domain histories $h$, objectives $u$, and measurable sets $A$,

$$\mathcal{G}[h,u](A) = \mathcal{G}'\left[(x,u(x))_{x \in h}\right](A) = \mathcal{K}[h,u](A).$$

So $\mathbf{G}$ is the map of a function, which we will denote $\mathbf{G}(\cdot)$.

The function $\mathbf{G}(\cdot)$ is linear, since Equation 6.11 is preserved under vector addition and scalar multiplication.

Regarding surjectivity, let us say that a proper static search generator $\mathcal{G}$ is *omniscient* if for all objectives $u$ that attain their lower bound,

$$\mathcal{G}[h,u](A) > 0 \text{ implies } A \cup \{x \mid u(x) = \inf_{z \in X} u(z)\} \text{ is nonempty.}$$

That is, $\mathcal{G}$ places all probability mass on exactly on the global minima of each objective $u$, when such global minima exist. As the generator of an optimization method, the omniscient generator always produces the "correct" value. But it is relatively easy to prove that no search generator accomplishes the same, because knowing the minima of each objective requires observing the entire objective, which the standard search generator cannot do. Thus the set $\mathbf{G}$ has no pair with an omniscient static search generator on the right hand side, and its function is not surjective.

With respect to injectivity, the static search generator depends on values generated from a static objective, but the standard search generators can model value processes that assign different values to the same search point at different times, that is, values arising from dynamic or stochastic processes. Suppose $h$ is a full history trace in $(X \times Y)^2$. More concretely, fix $x \in X$ and $y, w \in Y$ with $y \neq w$. Let $z = ((x,y),(x,w))$. Let $\mathcal{G}$ be any standard search generator, and let $\hat{\mathcal{G}}$ be another search generator that agrees with $\mathcal{G}$ everywhere except on $z$. Now then it must be that $\mathbf{G}(\mathcal{G}) = \mathbf{G}(\hat{\mathcal{G}})$, because the static search generator cannot "see" the history $z$, which cannot agree with any static objective. Therefore $\mathbf{G}(\cdot)$ is not injective.

Because $\mathbf{G}(\cdot)$ is linear, the standard search generators $\mathfrak{G}_{X,Y}$ project into a vector subspace of $\mathcal{MF}_{\mathcal{H}_X^{\mathcal{T}} \times Y^X, X}$. This vector subspace is a kind of quotient space[2] that identifies search generators that differ only on histories in which at least one search point is assigned distinct values. More pertinently, this vector subspace corresponds to a class of static search generators that cannot see the future. For this reason, this subspace will be called the *static black-box generators* after the optimization tradition which calls such objects *black-box optimizers*, formalized in the subsequent definition and propositions.

**Definition 6.13 (Static Black-Box Generator).** A static search generator $\mathcal{G}$ is said to have the *black-box property* and is called a *static black-box search generator* if for all search domain histories $h$ and for all objectives $u, v$,

$$u(x) = v(x) \text{ for all } x \in h \quad \text{implies} \quad \mathcal{G}[h,u] = \mathcal{G}[h,v].$$

The set of generators with finite supremum norm having the black-box property is denoted by $\mathcal{BB}_{X,Y}$.

[2] The particular quotient is an algebraic quotient, not a topological quotient. The function $\mathbf{G}(\cdot)$ is surjective but not continuous, since search generators may disagree to an arbitrary degree on history traces that are do not agree with any static objective function.

**Proposition 6.13.** *A static search generator has the black-box property if and only if it is the image of some standard search generator under the function* $\mathbf{G}(\cdot)$ *defined in the text.*

*Proof.* Let $\mathcal{G}$ be a static black-box search generator. Consider history traces $z \in \mathcal{H}^{\mathcal{T}}_{X,Y} = \bigcup_{t\in\mathcal{T}}(X \times Y)^{[t]}$. A history trace $z$ is said to be *objectifiable* if any search point $x \in X$ repeated in $z$ is assigned the same value, that is, $(x,y) \in z$ and $(x,w) \in z$ together imply that $y = w$. An objective function $u$ is said to agree with $z$ if $u(x) = y$ for all $(x,y) \in z$. Note that a history trace $z$ has at least one objective function agreeing with it if and only if $z$ is objectifiable.

Recall that $z_X$ is the sequence of search domain points in the history trace $z$. Define a standard search generator $\mathcal{G}'$ so that for all history traces $z$,

$$\mathcal{G}'[z] = \begin{cases} \mathcal{G}[z_X, u] & \text{if } z \text{ is objectifiable and } u \text{ agrees with } z \\ \mathbf{0} & \text{otherwise,} \end{cases}$$

where $u$ is arbitrary within the constraints given. Since these constraints can always be satisfied for $z$ objectifiable, at least one such $u$ exists. This definition of $\mathcal{G}'$ is unique because if $z$ is objectifiable and $u, v$ are two objectives that both agree with $z$, then $\mathcal{G}[z_X, u] = \mathcal{G}[z_X, v]$ by the definition of the black-box property. The pair $(\mathcal{G}', \mathcal{G})$ satisfies Equation 6.11, and hence $\mathcal{G}$ is the image under $\mathbf{G}(\cdot)$ of a standard search generator.

Conversely, suppose $\mathcal{G}$ is the image under $\mathbf{G}(\cdot)$ of some search generator. Fix a search domain history $h$ and let $u, v$ be objectives such that $u(x) = v(x)$ for all $x \in h$. Then for some standard search generator $\mathcal{G}'$,

$$\mathcal{G}[h,u] = \mathcal{G}'[(x, u(x))_{x\in h}] = \mathcal{G}'[(x, v(x))_{x\in h}] = \mathcal{G}[h,v]$$

by Equation 6.11, which establishes that $\mathcal{G}$ has the black-box property, completing the proof.

**Proposition 6.14.** $\mathcal{BB}_{X,Y}$ *is a vector subspace of the Banach space* $\mathcal{MF}_{\mathcal{H}^{\mathcal{T}}_X \times Y^X, X}$.

*Proof.* The black-box property for a static search generator is equivalent to the existence of a pair in $\mathbf{G}$. Since membership in $\mathbf{G}$ is closed under vector operations as discussed in the text, Proposition 6.13 implies that the black-box property is closed under vector operations, making $\mathcal{BB}_{X,Y}$ a vector subspace.

The static black-box generators differ from the space of search generators by incorporating within themselves a mechanism for generating values from trajectories in the search domain. This property is exploited in the next section to define the convolution operator. A more sophisticated method for including value-generating processes that allows for dynamic and stochastic processes as well is considered in Chapter 13; until then, the black-box search generators provide a more direct tool for this purpose.

### *6.3.2 Definition and Role of Convolution*

Convolution traditionally refers to a process for combining functions through an integral that essentially rolls one function over the other. In this text, convolution for probability distributions performs the intuitive function of applying two probability distributions in sequence, and is so named because of operational similarities with function convolution, with one probability distribution essentially being rolled over the other.

First, a point is sampled from some distribution $\mathbb{P}_1$, and then a point is sampled from some conditional probability distribution $\mathbb{P}_2$ given the outcome of the first sample. The connection to search generators arises from the discussion in Chapter 3 that search generators can be reanalyzed as conditional probabilities, where the choice of the next search iterate is conditioned on the previous search iterates and the resulting search values.

**Definition 6.14.** Let $\mathcal{G}_1$ and $\mathcal{G}_2$ be static search generators. Then their convolution $\mathcal{G}_1 \star \mathcal{G}_2$ is the static search generator defined for each history prefix $h$ and each measurable set $A$ by

$$(\mathcal{G}_1 \star \mathcal{G}_2)[h,u](A) = \int_X \mathcal{G}_2[h\|x,u](A)\, \mathcal{G}_1[h,u](dx), \tag{6.12}$$

where $\|$ indicates concatenation as usual.

Convolution is useful for two major reasons. The first reason is that it provides a constructive operation that can combine two simpler search generators to create a more complex one. This usage will be explored at length in Chapter 7, where convolution will be used to model evolutionary algorithms for optimization by representing an overall evolutionary search as a convolution of three search generators individually representing selection, recombination, and mutation.

The second reason to introduce convolution is the action of convolution reflects the operation of iterative searching. Suppose $\mathcal{G}$ is a proper static search generator for some search method, and $u$ is an objective function to be minimized. Now suppose that $Z_1$ is a sample from the probability measure $\mathcal{G}[\emptyset,u]$; that is, $Z_1$ is a random point in the search domain that serves as the first iterate of the search. Now then let $Z_n$ be sampled from $\mathcal{G}[(Z_1,\ldots,Z_{n-1}),u]$ and consider the stochastic process $(Z_n)_{n\in\mathcal{T}}$. Knowing that $(Z_n)$ is the process generated by $\mathcal{G}$, it follows that the skipping process $(Z_{2n})$ is the process generated by $\mathcal{G}\star\mathcal{G}$. More generally, we find that for all $n$, $Z_n \sim \left(\star_{m=1}^n \mathcal{G}\right)[\emptyset,u]$, where $\star_{m=1}^n \mathcal{G}$ represents $n$ successive self-applications of convolution. Thus when the search generator is convolved with itself, it generates a probability distribution from which each search iterate can be sampled.

As an algebraic operation, this convolution is not commutative, *i.e.*, it is possible that $\mathcal{G}_1 \star \mathcal{G}_2 \neq \mathcal{G}_2 \star \mathcal{G}_1$. One particular case occurs when one of the two convolved generators assigns zero probability mass to some region of the search domain that the other generator assigns positive mass, which results in an asymmetry depending on the order of application. Thus $\mathcal{G}_1 \star \mathcal{G}_2$ may assign positive probability to a measurable set $A$ while $\mathcal{G}_2 \star \mathcal{G}_1$ assigns zero probability to $A$. When three or more convolution operators are used, convolution is assumed to be left associative, *e.g.* $\mathcal{G}_1 \star \mathcal{G}_2 \star \mathcal{G}_3 = (\mathcal{G}_1 \star \mathcal{G}_2) \star \mathcal{G}_3$. This statement is needed, because this convolution operator is not generally associative, due to the fact that the integral in Equation 6.12 hides a search point.

**Proposition 6.15.** *Given three static search generators $\mathcal{A}, \mathcal{B}, \mathcal{C}$, the equation*

$$\mathcal{A} \star (\mathcal{B} \star \mathcal{C}) = (\mathcal{A} \star \mathcal{B}) \star \mathcal{C}$$

*holds if and only if for all objectives $u$ and search histories $h$,*

$$\mathcal{C}[h\|x\|y, u] = \mathcal{C}[h\|y, u] \tag{6.13}$$

*for $\mathcal{A}[h,u]$-almost every $x$ and $\mathcal{A} \star \mathcal{B}[h,u]$-almost every $y$.*

*Proof.* Let $A$ be any measurable set. Expanding the right-associative version gives

$$\begin{aligned}
\mathcal{A} \star (\mathcal{B} \star \mathcal{C})[h,u](A) &= \int_X (\mathcal{B} \star \mathcal{C})[h\|x, u](A)\, \mathcal{A}[h,u](dx) \qquad (6.14)\\
&= \int_X \int_X \mathcal{C}[h\|x\|y, u](A)\, \mathcal{B}[h\|x, u](dy)\, \mathcal{A}[h,u](dx),
\end{aligned}$$

whereas the left-associative version gives

$$\begin{aligned}
(\mathcal{A} \star \mathcal{B}) \star \mathcal{C}[h,u](A) &= \int_X \mathcal{C}[h\|y, u](A)\, (\mathcal{A} \star \mathcal{B})[h,u](dy) \qquad (6.15)\\
&= \int_X \int_X \mathcal{C}[h\|y, u](A)\, \mathcal{B}[h\|x, u](dy)\, \mathcal{A}[h,u](dx).
\end{aligned}$$

Examining these two equations reveals that Equation 6.13 does indeed make them equal and that when Equation 6.13 fails, there is some measurable set $A$ for which they are unequal.

For the two purposes introduced above – componential construction of generators and self-application as search iteration – left associative application of convolution is generally sufficient. There is, however, a vector subspace of static search generators for which convolution is associative. This topic is addressed next.

### *6.3.3 A Banach Algebra of Markov Generators*

Section 6.1 introduced a vector space of measure-valued functionals that subsumes the two different perspectives of search generators and static search generators. Banach spaces are structured around the operations of vector addition and scalar multiplication. When a well-behaved vector multiplication can also be defined, the result is a Banach algebra.

**Definition 6.15 (Banach Algebra).** A Banach algebra $B$ is a Banach space together with an operation $\star : V \times V \to V$ called vector multiplication that satisfies the following properties:

1. Multiplicative Associativity: For all $x,y,z \in B$, $(x \star y) \star z = x \star (y \star z)$.
2. Distributivity of $\star$ over $+$: For all $x,y,z \in B$, $x \star (y+z) = x \star y + x \star z$.
3. Distributivity of $+$ over $\star$: For all $x,y,z \in B$, $(x+y) \star z = x \star z + y \star z$.
4. Multiplicative Triangle Inequality: For all $x,y \in B$, $\|x \star y\| \leq \|x\| \, \|y\|$.

In particular, vector multiplication need not be commutative. If there is also a multiplicative identity $\mathbf{1} \in B$ such that $\|\mathbf{1}\| = 1$ and for all $x \in B$, $\mathbf{1} \star x = x \star \mathbf{1} = x$, then the algebra $B$ is said to be *unital*.

**Definition 6.16 (Banach Subalgebra).** Given a Banach algebra $B$, a subset $B' \subseteq B$ is a *subalgebra* of $B$ if $B'$ is closed under vector addition, scalar multiplication, and vector multiplication.

Convolution of static generators satisfies all of these criteria except for associativity. But there is a vector subspace of the static generators on which convolution is associative, and that subspace corresponds to the most commonly used search methods as well.

**Definition 6.17 (Markov Static Search Generators).** A static search generator $\mathcal{G}$ is said to have the *Markov property* of order $n$ if for all search domain histories $h$, objectives $u$, and $n$-length histories $w \in X^n$,

$$\mathcal{G}[h \| w, u] = \mathcal{G}[w, u]. \tag{6.16}$$

In this case, the static search generator may be described as $n$-Markov, meaning that it depends only on the $n$ prior search points. If $n = 1$, the it may be described simply as Markov.

**Proposition 6.16.** *The n-Markov property is closed under vector addition, scalar multiplication, and convolution.*

*Proof.* The cases of vector addition and scalar multiplication are obvious from Equation 6.16 together with the fact that these operations are defined pointwise.

For convolution, note that if $\mathcal{A}$ and $\mathcal{B}$ are $n$-Markov, then for search histories $h$, objectives $u$, $y \in X$, $w \in X^{n-1}$, and measurable sets $A$,

$$\begin{aligned}
\mathcal{A} \star \mathcal{B}[h \| y \| w, u](A) &= \int_X \mathcal{B}[h \| y \| w \| x, u](A) \; \mathcal{A}[h \| y \| w, u](dx) \\
&= \int_X \mathcal{B}[w \| x, u](A) \; \mathcal{A}[y \| w, u](dx) \\
&= \mathcal{A} \star \mathcal{B}[y \| w, u](A)
\end{aligned} \tag{6.17}$$

by repeated application of Equation 6.16.

The set of $n$-Markov static search generators will be denoted as $\text{Markov}^n_{X,Y}$, and the subset of $\text{Markov}^1_{X,Y}$ for which the black-box property also holds will be denoted as $\mathcal{BB}^{\text{Markov}}_{X,Y} = \text{Markov}^1_{X,Y} \cap \mathcal{BB}_{X,Y}$.

**Proposition 6.17.** *The set of* 1*-Markov static search generators,* $\text{Markov}^1_{X,Y}$*, is a vector subspace of* $\mathcal{MF}_{\mathcal{H}^{\mathcal{T}}_X \times Y^X, X}$ *on which the convolution operator is associative. Additionally, the set* $\mathcal{BB}^{\text{Markov}}_{X,Y}$ *is a vector subspace of both* $\text{Markov}^1_{X,Y}$ *and* $\mathcal{BB}_{X,Y}$.

*Proof.* By Proposition 6.16 this set is closed under vector operations, which proves that it forms a vector subspace. Furthermore, the Markov property guarantees that all triples $\mathcal{A}, \mathcal{B}, \mathcal{C}$ in this set satisfy Equation 6.13, and hence Proposition 6.15 implies that the convolution operator is associative on this space. The final statement follows from the fact that both the black-box property and the Markov property are preserved by vector operations.

It will now be shown that $\text{Markov}^1_{X,Y}$ is a Banach algebra. The only reason to restrict to Markov search generators is associativity, since the other properties, distributivity and the triangle inequality, hold generally for convolution. Because of this, the following two propositions establish distributivity and the triangle inequality separately, and Theorem 6.5 combines these statements.

**Proposition 6.18.** *The convolution operation distributes over addition, and addition distributes over convolution.*

*Proof.* Let $\mathcal{A}, \mathcal{B}, \mathcal{C}$ be arbitrary static search generators. For distributivity of $\star$ over $+$, choose any search domain histories $h$, objectives $u$, and measurable sets $A$, and note that $\mathcal{A} \star (\mathcal{B} + \mathcal{C})\,[h,u](A) =$

$$
\begin{aligned}
&\int_X (\mathcal{B}[h\|x,u](A) + \mathcal{C}[h\|x,u](A))\ \mathcal{A}[h,u](dx) \\
&= \int_X \mathcal{B}[h\|x,u](A)\ \mathcal{A}[h,u](dx)\ +\ \int_X \mathcal{C}[h\|x,u](A)\ \mathcal{A}[h,u](dx) \\
&= (\mathcal{A} \star \mathcal{B} + \mathcal{A} \star \mathcal{C})[h,u](A),
\end{aligned}
$$

where the linearity of the integral is the key justification. Likewise, $(\mathcal{A} + \mathcal{B}) \star \mathcal{C}[h,u](A)$

$$
\begin{aligned}
&= \int_X \mathcal{C}[h\|x,u](A)\,(\mathcal{A}[h,u](dx) + \mathcal{B}[h,u](dx)) \\
&= \int_X \mathcal{C}[h\|x,u](A)\ \mathcal{A}[h,u](dx)\ +\ \int_X \mathcal{C}[h\|x,u](A)\ \mathcal{B}[h,u](dx), \\
&= (\mathcal{A} \star \mathcal{C} + \mathcal{B} \star \mathcal{C})\,[h,u](A)
\end{aligned}
$$

which demonstrates distributivity of $+$ over $\star$.

**Proposition 6.19.** *The convolution operation satisfies the multiplicative triangle inequality.*

*Proof.* Let $\mathcal{A}$ and $\mathcal{B}$ be arbitrary static search generators. Then

$$\begin{aligned}
\|\mathcal{A} \star \mathcal{B}\| &= \sup_{h,u} \sup_{A} \left| \int_X \mathcal{B}[h\|x,u](A)\, \mathcal{A}[h,u](dx) \right| \\
&\leq \sup_{h,u} \sup_{A} \int_X |\mathcal{B}[h\|x,u](A)|\; |\mathcal{A}[h,u](dx)| \\
&\leq \sup_{h,u} \sup_{A} \int_X \|\mathcal{B}\|\; |\mathcal{A}[h,u](dx)| \\
&= \|\mathcal{B}\| \sup_{h,u} |\mathcal{A}[h,u]|\,(X) = \|\mathcal{A}\|\;\|\mathcal{B}\|,
\end{aligned}$$

where in the second step we have used the fact that $|\mathcal{B}[h\|x,u](A)| \leq \|\mathcal{B}\|$ generally, and in the final step we note that $\|\mathcal{A}[h,u]\|_{\mathcal{M}_X} = |\mathcal{A}[h,u]|(X)$.

**Theorem 6.5.** *The set* $\text{Markov}^1_{X,Y}$ *is a Banach algebra with convolution as vector multiplication.*

*Proof.* Proposition 6.17 has already established that the 1-Markov static search generators form a Banach space that is closed under convolution and for which convolution is associative. The multiplicative triangle inequality was proven for convolution in Proposition 6.19, and distributivity was proven as Proposition 6.18. Therefore $\text{Markov}^1_{X,Y}$ is a Banach algebra as desired.

The Banach algebra $\text{Markov}^1_{X,Y}$ is a vector subspace of the static search generators, but in a certain sense it is also a subspace of the search generators as well. To demonstrate this fact, a definition of the Markov property is now given for standard search generators, and it is then proven that this Markov property coincides with the one for static search generators. This proof can only apply to the static black-box search generators $\mathcal{BB}_{X,Y}$, and hence in relating the two Markov properties, the black-box assumption is made.

**Definition 6.18 (Markov Search Generator).** A search generator $\mathcal{G} \in \mathfrak{G}_{X,Y}$ is said to have the $n$-Markov property for $n \geq 1$ if for all full history traces $z$, all $n$-tuples of search points $(x_1,\ldots,x_n)$, and all $n$-tuples of values $y = (y_1,\ldots y_n)$,

$$\mathcal{G}\left[z\|(x_1,y_1)\|\ldots\|(x_n,y_n)\right] = \mathcal{G}\left[(x_1,y_1)\|\ldots\|(x_n,y_n)\right].$$

Such a search generator is also called an $n$-Markov search generator, and $n$ may be omitted if $n = 1$.

**Proposition 6.20.** *Suppose* $\mathcal{G} \in \text{Markov}^n_{X,Y} \cap \mathcal{BB}_{X,Y}$*, that is,* $\mathcal{G}$ *is an n-Markov static black-box search generator. Then* $\mathcal{G}$ *is the image under* $\mathbf{G}(\cdot)$ *of an n-Markov search generator* $\mathcal{G}' \in \mathfrak{G}_{X,Y}$*. If* $n = 1$*, then* $\mathcal{G}'$ *is unique among the* 1*-Markov search generators.*

*Proof.* As with Proposition 6.13, the proof is by construction and relies on the terms *objectifiable* and *agrees* defined there. However, more care must be taken in defining

$\mathcal{G}'$ on non-objectifiable history traces in order to preserve the $n$-Markov property. Working again with history traces $z \in \mathcal{H}^{\mathcal{T}}_{X,Y}$, we modify the terms *objectfiable* and *agrees* to match the $n$-Markov property. Let the *n-suffix* of a history trace $z$ be the $n$-length sequence $\text{suff}(z,n) \equiv (z_{|z|-n}, z_{|z|-n+1}, z_{|z|-n+2}, \dots z_{|z|})$ consisting of the last $n$ elements of $z$, recalling that $|z|$ means sequence length in this text. If $|z| < n$, then let $\text{suff}(z,n) = z$. A history trace $z$ is said to be $n$-objectifiable if its $n$-suffix is objectifiable, and an objective function $u$ is said to $n$-agree with $z$ if $u$ agrees with the $n$-suffix. Then define

$$\mathcal{G}'[z] = \begin{cases} \mathcal{G}[z_X, u] & \text{if } z \text{ is } n\text{-objectifiable and } u \text{ } n\text{-agrees with } z \\ \mathbf{0} & \text{otherwise,} \end{cases}$$

where $u$ is arbitrary within the constraints given. These constraints on $u$ can always be satisfied, and so the object $\mathcal{G}'$ is defined everywhere. Furthermore, $\mathcal{G}'$ is uniquely defined. If $u$ and $v$ both $n$-agree with $z$, then $\mathcal{G}[z_X, u] = \mathcal{G}[\text{suff}(z,n)_X, u]$ by the $n$-Markov property for static search generators, and the black-box property yields that $\mathcal{G}[\text{suff}(z,n)_X, u] = \mathcal{G}[\text{suff}(z,n)_X, v]$ since $u$ and $v$ $n$-agree. A final application of the $n$-Markov property shows that $\mathcal{G}[\text{suff}(z,n)_X, v] = \mathcal{G}[z_X, v]$, and so $\mathcal{G}[z_X, u] = \mathcal{G}[z_X, v]$. As in Proposition 6.13, Equation 6.11 is satisfied and so $\mathbf{G}(\mathcal{G}') = \mathcal{G}$.

Next, it must be shown that the search generator $\mathcal{G}'$ has the $n$-Markov property. If $z$ is not $n$-objectifiable, then $\mathcal{G}'[z] = 0$ in general, and so for any $z'$ such that $\text{suff}(z,n) = \text{suff}(z',n)$, it follows that $\mathcal{G}'[z] = \mathcal{G}'[z']$, which is the essence of Definition 6.18. If $z$ is $n$-objectifiable and $z'$ is any history trace such that $\text{suff}(z,n) = \text{suff}(z',n)$, then $z'$ is $n$-objectifiable as well, and the equations above show that

$$\mathcal{G}'[z] = \mathcal{G}[\text{suff}(z,n)_X, u] = \mathcal{G}[\text{suff}(z',n)_X, u] = \mathcal{G}'[z']$$

for any objective $u$ that $n$-agrees with $z$. Again, this equation is the essence of the $n$-Markov property for search generators, and so $\mathcal{G}'$ is $n$-Markov.

Finally, if $n = 1$, then every history trace $z$ is $n$-objectifiable and $\mathcal{G}'[z] = \mathcal{G}[z_X, u]$ for some objective $u$. If $\mathcal{G}''$ is a 1-Markov search generator such that $\mathbf{G}(\mathcal{G}'') = \mathbf{G}(\mathcal{G}') = \mathcal{G}$, then for all history traces $z$, $\mathcal{G}''[z] = \mathcal{G}''[\text{suff}(z,1)]$ by the Markov property. Set $x$ and $y$ so that $\text{suff}(z,1) = ((x,y))$, and let $u$ be any objective function such that $u(x) = y$. Then by Equation 6.11 and the definition of $\mathcal{G}'$ above,

$$\mathcal{G}''[z] = \mathcal{G}[(x),(u(x))] = \mathcal{G}'[\text{suff}(z,1)] = \mathcal{G}'[z].$$

That is, $\mathcal{G}'$ is unique among 1-Markov search generators.

The preceding proposition shows that the $n$-Markov search generators project into $\mathcal{BB}^{\text{Markov}}_{X,Y}$ via $\mathbf{G}(\cdot)$, forming an algebraic quotient that identifies search generators that disagree on history traces that are not $n$-objectifiable. For $n = 1$, though, this projection is not a quotient but an identity, as restated in the following corollary.

**Corollary 6.1.** *The Banach space* $\mathcal{BB}^{\text{Markov}}_{X,Y}$ *is isometrically isomorphic to the set of* 1*-Markov search generators.*

*Proof.* The restriction of the function $\mathbf{G}(\cdot)$ to the 1-Markov search generators is surjective on $\mathcal{BB}^{\text{Markov}}_{X,Y}$ by Proposition 6.20, and it is injective as well, as established by the uniqueness clause of the same proposition. Thus the restriction of $\mathbf{G}(\cdot)$ is an isomorphism. Suppose $\mathcal{G} = \mathbf{G}(\mathcal{G}')$, then

$$\begin{aligned} \|\mathcal{G}'\| = \sup_z \|\mathcal{G}'[z]\| &\overset{\text{1-Markov}}{=} \sup_{x,y} \|\mathcal{G}'[((x,y))]\| \\ = \sup_{x,u} \|\mathcal{G}'[((x,u(x)))]\| &\underset{\text{1-Markov}}{\overset{\text{black-box}}{=}} \sup_{h,u} \|\mathcal{G}[h,u]\| \quad = \quad \|\mathcal{G}\|, \end{aligned}$$

proving that the isomorphism is also an isometry.

Therefore $\mathcal{BB}^{\text{Markov}}_{X,Y}$ can be regarded either as a set of static search generators or as a set of search generators, and this label will be used for both sets. Unfortunately, although $\text{Markov}^1_{X,Y}$ is a Banach algebra, $\mathcal{BB}^{\text{Markov}}_{X,Y}$ is not a sub-algebra, because the convolution does not preserve the black-box property. To see why, notice that the integral in Equation 6.12 may rely on the objective value $u(x)$, even if $\mathcal{G}_2$ has the black-box property. So convolution is not closed with respect to the black-box property. Still, there is one case in which the black-box property can be guaranteed after convolution, and this case can be used to develop convolution as an operation that produces standard search generators.

### 6.3.4 The R-Module of Value-Agnostic Search Generators

Convolution fails to preserve the black-box property because it depends on the value of a newly-proposed search point internally. Thus the Banach algebra formed by the $n$-Markov static search generators cannot be restricted to the vector subspace with the black-box property. However, if the static search generator determines the next search point independent of its value, then the black-box property is preserved. Furthermore, there is an isometry between the static search generators and the standard search generators for the subset of generators that ignores values. Thus, so long as the right-hand side of the convolution operator is restricted to be *value-agnostic* in this sense, convolution can be projected back as an operation over standard search generators. By doing so, the constructive processes for building search generators can be applied to standard search generators rather than static search generators.

The value-agnostic search generators, defined below, turn the search generators into a *non-unital* right R-module under the convolution operation, which means that they can be applied as the right-hand side of convolution to any standard search generator to produce a new standard search generator. The definition relies on an *abelian group*, which is a set together with an addition operator that is associative and commutative and has an inverse function and an identity. Every vector space is trivially an abelian group under vector addition. The definition also depends on a *non-unital ring* (also called a *rng*), which is an abelian group that also has a multiplication operation that is associative and distributes over addition. *Non-unital* here refers to the lack of an identity function. Every Banach algebra is a non-unital ring (and every unital Banach algebra is a unital ring), and the value-agnostic Markov search generators also form such a ring.[3]

**Definition 6.19 (R-Module).** Given a non-unital ring $R$ with multiplication operator $\star$ and a multiplicative identity $\mathbf{1}$, an abelian group $M$ together with an operation $\star : M \times R \to M$ is a *non-unital right R-module* if the following conditions hold:

- Distributivity of $\star$: For all $x \in M$ and $y, z \in R$, $x \star (y+z) = x \star y + x \star z$.
- Distributivity of $+$: For all $x, y \in M$ and $z \in R$, $(x+y) \star z = x \star z + y \star z$.
- Cross-Associativity: For all $x, y \in M$ and $z \in R$, $x \star (y \star z) = (x \star y) \star z$.

A *non-unital left R-module* is defined similarly, but with signature $R \times M \to M$ instead.

In order to develop the non-unital right R-module of search generators convolved with value-agnostic search generators, we begin with objective-agnostic static search generators.

**Definition 6.20 (Objective-Agnostic).** A static search generator $\mathcal{G}$ is *objective-agnostic* if for all search domain histories $h$ and for all objective pairs $u, v$, $\mathcal{G}[h,u] = \mathcal{G}[h,v]$.

**Proposition 6.21.** *The set of objective-agnostic search generators is closed under vector addition, scalar multiplication, and convolution. Therefore this set forms a vector subspace of any vector space of static search generators that contains it, and it also forms a Banach sub-algebra of any Banach algebra of static search generators that contains it.*

---

[3] In fact, the value-agnostic Markov search generators *almost* form a unital ring. Consider the static search generator defined for all search domain histories $h$, all search points $x \in X$, all objectives $u$, and measurable sets $A$ so that

$$\mathbf{1}[\emptyset, u](A) = 1 \quad \text{and} \quad \mathbf{1}[h \| x, u](A) = \mathbb{1}_A(x).$$

Then $\|\mathbf{1}\| = \sup_{h,u} \|\mathbf{1}[h,u]\| = 1$, and for all Markov static search generators $\mathcal{G}$, all search domain histories $h$, objectives $u$ and measurable sets $A$,

$$\mathcal{G} \star \mathbf{1}[h,u](A) = \int_X \mathbf{1}[h \| x, u](A) \, \mathcal{G}[h,u](dx) = \mathcal{G}[h,u](A).$$

The opposite direction is the problematic one, but nonetheless if $h$ is nonempty,

$$\mathbf{1} \star \mathcal{G}[h,u](A) = \int_X \mathcal{G}[h \| x, u](A) \, \mathbf{1}[h,u](dx) = \mathcal{G}[h,u](A)$$

due to the Markov property. The truth of this statement can be seen from the fact that $\mathbf{1}[h,u]$ is degenerate and assigns all probability mass to the final point of $h$, so that $x$ is just the last element of $h$ in the equation above. That is, $\mathcal{G}[h \| x, u] = \mathcal{G}[h,u]$ under integration by $\mathbf{1}[h,u]$ in the presence of the Markov property. Unfortunately, if $h$ is empty, there does not seem to be a reasonable way to define $\mathbf{1}[\emptyset, u]$ to satisfy the equation above, which is why the text introduces *non-unital* R-modules as opposed to standard (unital) R-modules.

*Proof.* Each of these three claims is trivially seen to be true by alternating over pairs of objectives $u,v$ and substituting into the definitions.

**Proposition 6.22.** *Every objective-agnostic static search generator has the black-box property.*

*Proof.* The definition of an objective-agnostic static search generator is trivially a strengthening of the black-box property.

Based on the two preceding propositions, the objective-agnostic static search generators form a vector subspace of $\mathcal{BB}_{X,Y}$. Furthermore, the objective-agnostic static search generators contained in $\mathcal{BB}^{\text{Markov}}_{X,Y}$ form a Banach sub-algebra of $\text{Markov}^1_{X,Y}$ due to the following proposition. The notation $\mathcal{OA}_{X,Y}$ will be used to refer to the set of objective-agnostic static search generators, which is a Banach space, and $\mathcal{OA}^{\text{Markov}}_{X,Y}$ will refer to the subset which has the 1-Markov property, which is a Banach algebra.

**Proposition 6.23.** *The convolution of a static black-box search generator and an objective-agnostic static search generator has the black-box property.*

*Proof.* Let $\mathcal{G}_1$ have the black-box property and let $\mathcal{G}_2$ be objective-agnostic. Then for all search domain histories $h$, objectives $u$ and $v$ that agree on $h$, and measurable sets $A$,

$$\begin{aligned} \mathcal{G}_1 \star \mathcal{G}_2[h,u](A) &= \int_X \mathcal{G}_2[h\|x,u](A)\ \mathcal{G}_1[h,u](dx) \\ &= \int_X \mathcal{G}_2[h\|x,v](A)\ \mathcal{G}_1[h,v](dx) \\ &= \mathcal{G}_1 \star \mathcal{G}_2[h,v](A), \end{aligned} \tag{6.18}$$

where the substitution of $v$ in to $\mathcal{G}_1$ is justified by the black-box property and the substitution into $\mathcal{G}_2$ is justified by the objective-agnostic property.

**Theorem 6.6.** *With the Banach algebra $\mathcal{OA}^{\text{Markov}}_{X,Y}$ together with the convolution operator, the Banach space $\mathcal{BB}_{X,Y}$ is a non-unital right R-module.*

*Proof.* Proposition 6.23 shows that the convolution operator restricts to an operation

$$\star : \mathcal{BB}_{X,Y} \times \mathcal{OA}^{\text{Markov}}_{X,Y} \to \mathcal{BB}_{X,Y},$$

which serves as the multiplication operator for $\mathcal{BB}_{X,Y}$ as a non-unital right R-module over $\mathcal{OA}^{\text{Markov}}_{X,Y}$. This operation was already proven distributive in Proposition 6.18, and the Markov property together with Proposition 6.15 guarantees cross-associativity, completing the proof.

The next goal is to transport this R-module back to the standard search generators, so that convolution can be used constructively to generate a wider class of search methods. The concept of being objective-agnostic has an analogue for standard search generators as well in which the standard search generator is independent of the values in the full history trace. This analogue projects through to the objective-agnostic static search generators through the function $\mathbf{G}(\cdot)$ of section 6.3.1.

**Definition 6.21 (Value-Agnostic).** A standard search generator $\mathcal{G}' \in \mathfrak{G}_{X,Y}$ is *value-agnostic* if for all full history traces $z, z'$, $\mathcal{G}'[z] = \mathcal{G}'[z']$ whenever $z_X = z'_X$, recalling that $z_X = (x_n)_{n \leq |z|}$ is the search sequence of the full trace $z = ((x_n, y_n))_{n \leq |z|}$.

**Proposition 6.24.** *Every objective-agnostic static black-box generator is the image of a unique value-agnostic search generator under the function* $\mathbf{G}(\cdot)$ *defined in Section 6.3.1, and every value-agnostic search generator projects via* $\mathbf{G}(\cdot)$ *to an objective-agnostic static search generator.*

*Proof.* Proposition 6.13 already showed that every static black-box generator is the image under $\mathbf{G}(\cdot)$ of some standard search generator. Let $\mathcal{G}$ be an objective-agnostic static black-box generator. Using language of the proof of that proposition, define a standard search generator $\mathcal{G}'$ so that for all history traces $z$,

$$\mathcal{G}'[z] = \mathcal{G}[z_X, u] \quad \text{for any objective } u.$$

This definition is unique because for any pair of objectives $u, v$, $\mathcal{G}[z_X, u] = \mathcal{G}[z_X, v]$. Furthermore, for every pair of full history traces $z, z'$ such that $z_X = z'_X$, we see that $\mathcal{G}'[z] = \mathcal{G}'[z']$, because $z_Y$ does not appear on the right-hand side of the definition of $\mathcal{G}'$. That is, $\mathcal{G}'$ is value-agnostic, and so $\mathbf{G}(\mathcal{G}') = \mathcal{G}$. Uniqueness follows from the fact that if $\mathbf{G}(\mathcal{G}') = \mathbf{G}(\mathcal{G}'') = \mathcal{G}$ for $\mathcal{G}', \mathcal{G}''$ both value-agnostic, then Equation 6.11 does not depend on objective values for these two, whence $\mathcal{G}' = \mathcal{G}''$.

Now suppose conversely that $\mathcal{G}'$ is a value-agnostic standard search generator. Then Equation 6.11 proves that $\mathbf{G}(\mathcal{G}')$ is objective agnostic due to the fact that for all search domain histories $h$ and for any pair of objectives $u$ and $v$,

$$\mathcal{G}'\left((x, u(x))_{x \in h}\right) = \mathcal{G}'\left((x, v(x))_{x \in h}\right),$$

which forces $\mathcal{G}[h, u] = \mathcal{G}[h, v]$ and completes the proof.

The set of value-agnostic search generators will be denoted $\mathcal{VA}_{X,Y}$. This space is isometrically isomorphic to the objective-agnostic static search generators, which is to say that the two are indistinguishable as Banach spaces.

**Theorem 6.7.** *The Banach space* $\mathcal{VA}_{X,Y}$ *of value-agnostic search generators is isometrically isomorphic to the space* $\mathcal{OA}_{X,Y}$ *of objective-agnostic static search generators.*

*Proof.* Proposition 6.24 implies that the operator $\mathbf{G}$ can be restricted to an operator $\mathbf{G}|_{\mathcal{VA}_{X,Y}}$ whose range includes all of $\mathcal{OA}_{X,Y}$. Also by Proposition 6.24, this restricted operator is both injective and surjective, meaning that is an isomorphism.

Now suppose $\mathcal{G}' \in \mathcal{VA}_{X,Y}$ and $\mathcal{G} = \mathbf{G}(\mathcal{G}') \in \mathcal{OA}_{X,Y}$. Then

$$\|\mathcal{G}\| = \sup_{h,u} \|\mathcal{G}[h,u]\| = \sup_{h} \|\mathcal{G}[h,u_0]\|$$

by the objective-agnostic property and

$$\|\mathcal{G}'\| = \sup_{z} \|\mathcal{G}'[z]\| = \sup_{h} \|\mathcal{G}'[(x,u_0(x))_{x\in h}]\|$$

by the value-agnostic property. Consequently, $\|\mathcal{G}\| = \|\mathcal{G}'\|$ by Equation 6.11.

Given the equality $\mathcal{VA}_{X,Y} =_{\mathbf{G}} \mathcal{OA}_{X,Y}$, the vector subspace $\mathcal{VA}^{\text{Markov}}_{X,Y}$ can be defined as the image of $\mathcal{OA}^{\text{Markov}}_{X,Y}$ under $\mathbf{G}|_{\mathcal{VA}_{X,Y}}$. By projecting the convolution operation across this transformation, it is easily seen that $\mathcal{VA}^{\text{Markov}}_{X,Y}$ is a Banach algebra. Rather than following this plan exactly, we will define convolution on a superset of $\mathcal{VA}^{\text{Markov}}_{X,Y}$.

**Definition 6.22 ($n$-Step Value-Agnostic).** A search generator $\mathcal{G} \in \mathfrak{G}_{X,Y}$ is $n$-step value-agnostic for $n \geq 1$ if for all full history traces $z$, all $n$-tuples of search points $(x_1,\ldots,x_n)$, and all $n$-tuples of values $(y_1,\ldots y_n)$ and $(y'_1,\ldots,y'_n)$,

$$\mathcal{G}\left[z\|(x_1,y_1)\|\ldots\|(x_n,y_n)\right] = \mathcal{G}\left[z\|(x_1,y'_1)\|\ldots\|(x_n,y'_n)\right].$$

Let $\mathcal{VA}^n_{X,Y}$ indicate the set of all $n$-step value-agnostic search generators. Then it is obvious that if $m > n$, then $\mathcal{VA}^m_{X,Y} \subseteq \mathcal{VA}^n_{X,Y}$. Furthermore, if $\mathcal{V}$ is *any* $n$-step value-agnostic search generator that has the 1-Markov property, then $\mathcal{V}$ is value-agnostic in general, *i.e.*, $\mathcal{V} \in \mathcal{VA}^{\text{Markov}}_{X,Y}$. A more general form of this statement is presented as a proposition without proof. These properties also are preserved by addition and scalar multiplication, stated as a second proposition, and hence both properties, independently or jointly, form Banach spaces for each value of $n$.

**Proposition 6.25.** *Suppose that $\mathcal{V}$ is an n-Markov search generator that is also m-step value-agnostic for $m \geq n$. Then $\mathcal{V}$ is value-agnostic generally. If $n = 1$, then $\mathcal{V} \in \mathcal{VA}^{\text{Markov}}_{X,Y}$ and therefore $\mathcal{BB}^{\text{Markov}}_{X,Y} \cap \mathcal{VA}^m_{X,Y} = \mathcal{VA}^{\text{Markov}}_{X,Y}$.*

**Proposition 6.26.** *Both the n-step value-agnostic property and the n-Markov properties are closed under vector addition and scalar multiplication on $\mathfrak{G}_{X,Y}$ and hence form Banach subspaces.*

Convolution can be defined on search generators for $\mathcal{VA}^1_{X,Y}$, the set of 1-step value-agnostic search generators as follows.

**Definition 6.23 (Convolution of Search Generators).** The convolution operator between Banach space of search generators $\mathfrak{G}_{X,Y}$ and the Banach space $\mathcal{VA}^1_{X,Y}$ so that for $\mathcal{G} \in \mathfrak{G}_{X,Y}$ and $\mathcal{V} \in \mathcal{VA}^1_{X,Y}$, for all full history traces $z$ and measurable sets $A$,

$$\mathcal{G} \star \mathcal{V}[z](A) = \int_X \mathcal{V}[z\|(x,y_0)](A)\, \mathcal{G}[z](dx), \tag{6.19}$$

where $y_0$ is arbitrary. This definition is unique because $\mathcal{V}$ is at least 1-step value-agnostic.

Although convolution is defined for all of $\mathcal{VA}^1_{X,Y}$, it is only associative on $\mathcal{VA}^{\text{Markov}}_{X,Y}$, since only this subset universally satisfies the conditions of Proposition 6.15. Consequently, $\mathcal{VA}^{\text{Markov}}_{X,Y}$ is the largest Banach algebra in $\mathcal{VA}^1_{X,Y}$ using convolution as its multiplication operator. The following theorem can be proven identically to Theorem 6.6 *mutatis mutandis*.

**Theorem 6.8.** *The Banach space* $\mathfrak{G}_{X,Y}$ *is a non-unital right R-module over* $\mathcal{VA}^{\text{Markov}}_{X,Y}$ *under the convolution operator defined in Equation 6.19.*

Thus convolution makes $\mathfrak{G}_{X,Y}$ into a non-unital R-module over $\mathcal{VA}^{\text{Markov}}_{X,Y}$. When expanded to all of $\mathcal{VA}^1_{X,Y}$, convolution can still be applied meaning that convolution is a *group action* in general. The convolution operator will prove useful in the next chapter for constructing search generators.

## 6.4 Conclusion

In this chapter, the Banach space of measure-valued functionals was introduced, of which the search generators introduced in Chapters 3 and 4 provide one instantiation. From these functionals, a normed space was created, which demonstrates that there is a continuum of stochastic iterative search methods bridging the web of commonly used search methods. This surprising fact suggests that rather than searching for new search methods *ex nihilo*, one should instead understand the mathematical principles underlying this Banach space in order to search for computable search methods among the interesting regions within this space.

Within this space, the geometry of the probability-valued functionals plays a simplex-like role, being topologically closed and convex. Although this set does not generate the entire Banach space of measure-valued functionals, it is itself the set generated by deterministic probability-valued functionals, which turn out to be the extreme points of the probability-valued functionals. These deterministic functionals are important because they reflect all possible deterministic iterative search methods, whether computable or not. Thus, as shall be seen in Chapter 13, they hold the potential to elucidate the connection between deterministic versus stochastic searching.

Finally, in the course of proving the Markov static search generators form a Banach algebra, several important facts have been introduced. Firstly, the static search generators were defined and their relationship with the standard search generators was explained using the black-box property. Secondly, the convolution operation was introduced, along with the $n$-Markov property, which makes it associative. Thirdly, this operation was shown to have distributive properties and to satisfy the multiplicative triangle inequality. Finally, the concept of being objective-agnostic (for static search generators) or value-agnostic (for standard search generators) was used in order to preserve the black-box property under convolution. In the next two chapters, this convolution operation will be used to construct a wide range of search generators for common search methods.

# Chapter 7
# Constructing Search Generators for Population-Based Optimizers

The previous chapter introduced a variety of formal tools for describing spaces of search generators, including the deterministic generators, the black-box property, the Markov property, and the convolution operator. In the next few chapters, the pursuit of a formal theory of these objects is paused in order to demonstrate the applicability of these objects and properties to common search methods, from simulated annealing to differential evolution and more. This chapter focuses on applying convolution and other operators in a constructive way to build successful search methods from simple, elemental generators.

This constructive approach has strong parallels with the history of the analysis of randomized search heuristics, most strongly presented by Vose [162]. In this tradition, evolutionary methods decompose into selection, recombination, and mutation components that can be reused to generate new methods. Consequently, the development of the constructive theme below recapitulates this tripartite structure. One of the overall goals of this chapter and indeed of this book is to produce a unified analytic approach to evolutionary computation that relates this field to general methods of iterative optimization using search generators. Thus the text here willingly dives down a rabbit hole, presenting definitions that supposedly distinguish genetic algorithms, evolutionary algorithms, and quasi-evolutionary algorithms. Yet it will be seen these definitions are inadequate to separate the various branches of optimization. Although they capture primary intuitions, they fail establish rigid boundaries among iterative stochastic search methods. Rather than revising the definitions to reestablish "correct" boundaries, the failure of this definitional approach to dividing the communities of evolutionary versus mathematical optimization will instead be used to argue that in fact there is no substantive, formal difference between evolutionary search methods and general-purpose Monte Carlo search.



A. J. Lockett, *General-Purpose Optimization Through Information Maximization*,
Natural Computing Series, https://doi.org/10.1007/978-3-662-62007-6_7

## 7.1 Population-Based Search and Optimization

Evolutionary methods in particular are designed to evaluate several search points simultaneously in batches that are called *populations*. These methods represent just one class of problems in which evaluation of search points can be parallelized. In general, most direct search methods such as Nelder-Mead (see Section 2.4.1) and Generating Set Search (see Section 2.4.2) can also be parallelized and hence the broader term *population-based optimizers* will be used to reflect the entire class of optimization methods for which evaluation can be performed in successive batches.

Population-based optimizers will be built up from search generators using the convolution operator as the core tool. Although convolution was more easily defined for the static search generators, its usage as a group action on the search generators allows more general definitions of search methods that can account for a variety of settings including optimization of static, dynamic, and stochastic objective functions as well as multi-objective optimization and exploratory search. This section lays out the goals and definitions that will guide the process of constructing population-based optimizers.

### 7.1.1 Motivation and Goals

From a formal perspective, the introduction of populations to be evaluated in parallel changes the nature of a search method in one respect only: a population-based method generates an entire population without depending on evaluations of earlier members of the same population. It is this difference in dependency structure that is the primary formal feature; computational parallelization is an implementation detail that can be applied to any method with appropriate dependencies.

The terminology used here is drawn from evolutionary computation. Each batch will be termed a *population*, and successive populations will be referred to as *generations*. A particular point in the search domain may be referred to as an *individual*, and the objective function may be called a *fitness function* with a value referred to as a *fitness*, all following the lexicon of evolutionary algorithms based on extended Darwinian analogies. The term *parallelizable* could have been used instead, but it does not sufficiently capture the idea of evaluating search points in equally sized batches.

The methods described below can be used to construct a number of global search methods, not only for population-based methods. Methods addressed below include Simulated Annealing, which generates a Markov chain by evaluating one point at a time. Nor are these methods limited to evolutionary computation. Within optimization, methods such as Nelder-Mead or Generating Set Search can be naturally described as population-based, although few would normally describe them as *evolutionary*.

A population-based method will be represented here by a proper search generator, and hence methods will be identified with elements in $\mathfrak{G}_{X,Y} \cap \mathcal{PF}$ as introduced in the last chapter. By beginning with $\mathfrak{G}_{X,Y}$ rather than some explicitly population-based approach, a setting has been established within which direct comparisons are possible across methods with different population sizes or typologically distinct approaches to optimization. It is also possible to study methods with dynamic population sizes, although we will not do so here. In what follows, a population-based method will be assumed to have a fixed population size $K$; that is, it will generate batches of $K$ search points using the evaluation information fixed prior to generating the population.

Many evolutionary algorithms and other optimization heuristics can be built up from modular components using the convolution operator from the last chapter. This process is analogous to traditional analyses in evolutionary computation using genetic operators. These components can be defined individually, and their modular structure can be useful for developing general theorems. Selection, recombination and mutation will be studied as component classes that can be used to abstractly characterize evolutionary algorithms. This chapter proposes that an evolutionary algorithm can be identified with the convolution of selection, recombination, and mutation operators each defined using general properties.

Perhaps surprisingly, methods such as Simulated Annealing, Nelder-Mead, and Generating Set Search can all be represented similarly as forms of selection and mutation (without recombination), so the claim that this tripartite structure is necessarily "evolutionary" is more a matter of terminology than substance.

To solidify the claim, equations will be presented that define the most common genetic algorithms and evolution strategies using this modular approach. Similar analysis will be performed for some quasi-evolutionary methods that will highlight some of the ways in which these methods both conform to and deviate from the standard evolutionary computation model.

### 7.1.2 Formalities

Since the methods described in this chapter are generally applied to optimization problems more often than to the more general setting of search, the term *population-based optimizer* will now be used to describe them. A population-based optimizer $\mathcal{G}$ with population size $K > 0$ can be represented as a sequence of $K$ separate search generators $\mathcal{G}_1, \cdots, \mathcal{G}_K \in \mathfrak{G}_{X,Y}$ (not necessarily distinct), each of which is used to generate one individual per population. Then a history trace can be broken up into populations, with one generator $\mathcal{G}_k$ assigned to each slot in the population.

This choice of representation requires tools and notation to convert between histories of search points and histories of populations. To this end, we introduce a *slot extractor* $k : \mathcal{H}_X^{\mathcal{T}} \to [K]$ such that if $h$ is a history trace, then $k(h)$ represents the currently active population slot,

$$k(h) = 1 + (|h| \mod K),$$

recalling that $|h|$ is the length of the history $h$. The slot extractor $k(h)$ will be used repeatedly below.

Formally, a population is an element in the product space $X^K$ consisting of $K$ copies of the search domain $X$. An *evaluated population* is an element of the product space $(X \times Y)^K$ that associates a value to each individual in the population. If $P \in (X \times Y)^K$ is an evaluated population, $P = ((x_i, y_i))_{i \leq K}$, then $P_X$ is the corresponding population, $P_X = (x_i)_{i \leq K}$, and individuals can be referenced as $P_i^X = x_i$. The notation $P_Y$ is used in the same way to access the values.

A history trace $h \in \mathcal{H}_{X,Y}^{\mathcal{T}}$ can be broken up into a history of evaluated populations $H = (P_1, P_2, P_3, \cdots)$ with $P_i \in (X \times Y)^K$ using the mapping

$$H_{i,k} = h_{(i-1)K+k+1}.$$

Let $H(h)$ be the mapping $h \mapsto H$ from the previous line. Then $H(h)$ is the history of complete populations in the history $h$. If $mK < |h| \leq (m+1)K$ for some nonnegative integer $m$, $H(h)$ ignores any elements in $h$ with index greater than $mK$. Now $H(h)$ is a history over populations, *i.e.* $H(h) \in \mathcal{H}_{X^K,Y^K}^{\mathcal{T}/K}$, where $\mathcal{T}/K$ represents the adjusted temporal index set.[1] In reverse, let $\text{traj}_K(H)$ unroll a population history $H \in \mathcal{H}_{X^K,Y^K}^{\mathcal{T}/K}$ to a history trace in $\mathcal{H}_{X,Y}^{\mathcal{T}}$ via

$$\text{traj}_K(H)_j = H_{\lfloor j/K \rfloor,\, 1+(j \mod K)}.$$

Then $\text{traj}_K(H(h)) = h$ if and only if the length of $h$ is a multiple of $K$. Otherwise, it truncates the end of $h$ at the last population boundary. The notation traj reflects turning a sequence of populations into a single "trajectory" in the search space.

A population-based optimizer is distinguished by the fact that it respects the population boundary, and new populations can only be generated based on information available from prior populations. That is, to be a population-based optimizer, an optimizer must be able to evaluate points in parallel. This restriction can be represented by a dynamic version of the value-agnostic property in which a search generator is agnostic with respect to values after the last population boundary.

**Definition 7.1.** A search generator $\mathcal{G}$ is a *population-based optimizer* of population size $K$ if for all history traces $h$ and $h'$, $\mathcal{G}[h] = \mathcal{G}[h']$ whenever $h_X = h'_X$ and $\text{traj}_K(H(h)) = \text{traj}_K(H(h'))$.

This definition dynamically truncates the dependence on search values back to the population boundary while allowing a dependence on the progressively generated individuals of a population. It might seem excessive at first, since one might imagine it sufficient to require $\mathcal{G}[h] = \mathcal{G}[\text{traj}_K(H(h))]$, *i.e.*, that it is enough to simply truncate dependence to the population boundary. Such a conceptualization is inadequate because populations may be generated from a joint distribution; that is,

[1] This adjusted temporal index set can be formed as a quotient in which each successive group of $K$ consecutive elements in $\mathcal{T}$ are identified. We do not address the case where the cardinality of $\mathcal{T}$ is uncountable.

population members may depend on each other. For example, although a parallelized optimization mehod cannot generate the $k^{th}$ member of the population based on the objective evaluation of the $(k-1)^{th}$ member, it may need to inspect the identity of the $(k-1)^{th}$ member, either to avoid duplication (e.g. tabu search), to promote population diversity, or to alter its probability distribution in some other way.

Let $\mathcal{PBO}^K_{X,Y}$ be the set of population-based optimizers of size $K$. For every population size, the set $\mathcal{PBO}^K_{X,Y}$ is closed under vector operations and hence is a vector subspace of the Banach space $\mathfrak{O}_{X,Y}$ for each $K$. There are obvious relationships among these spaces. For example, $\mathcal{PBO}^K_{X,Y} \subseteq \mathcal{PBO}^{nK}_{X,Y}$ for $n \geq 1$, and each containment indicates a vector subspace relationship.

### *7.1.3 The Population Lift and Slot Generators*

Evolutionary algorithms will be built up through constructive operations and analyzed according to these components. In this section, a population-based optimizer $\mathcal{G} \in \mathcal{PBO}_K$ is associated with $K$ optimizers $\mathcal{G}_1, \ldots, \mathcal{G}_K$, one for each population slot with $\mathcal{G}[h] = \mathcal{G}_{k(h)}[h]$, where $k(h)$ is the population indexing function from the previous section. Such a decomposition will be called a *slot decomposition* and its elements *slot generators*. It is always possible but not always meaningful, since one can trivially define $\mathcal{G}_k = \mathcal{G}$. The purpose here is not to decompose but to construct population-based optimizers from a set of slot generators, and this process is the focus of this subsection.

Each successive population is referred to as a *generation*, and one generation of $\mathcal{G}$ samples each of the $\mathcal{G}_k$ in turn to form a new population that can be evaluated in parallel. Expanding based on the definitions, the probability measure governing a particular population $P_X \in X^K$ conditional on a prior history trace $h$ at the population boundary can be written out as a product

$$\mathbb{P}_{\mathcal{G}}(dP \mid h) = \prod_{k=1}^{K} \mathcal{G}_k \left[h \,\|\, (P^X_1, y_1) \,\|\, \ldots \,\|\, (P^X_{k-1}, y_{k-1})\right](dP_k), \tag{7.1}$$

where the $y_k$ are arbitrary values in light of the value-agnostic properties of $\mathcal{G}$ within a single population. As discussed in Chapters 3 and 4, $\mathcal{G}[h](dx)$ is a conditional probability over history traces and can be written as $\mathcal{G}[h](dx) = \mathbb{P}_{\mathcal{G}}(dx \mid h)$. The growing concatenation over $P_k$ in Equation 7.1 reflects the fact that the population is sampled jointly. It also evokes the idea of a sequence of $K$ convolutions in comparison with Definition 6.23, a fact which will be exploited below. To give some motivation, if Equation 7.1 is rewritten as

$$\mathbb{P}_{\mathcal{G}}(dP \mid h) = \prod_{k=1}^{K} \mathbb{P}_{\mathcal{G}_k}\left(dP_k \,\middle|\, h \,\|\, (P^X_1, y_1) \,\|\, \ldots \,\|\, (P^X_{k-1}, y_{k-1})\right), \tag{7.2}$$

then it is clear that Equation 7.1 is an application of Bayes' rule to the joint probability governing the entire population.

To avoid the abusive notation above, we introduce an operator

$$\mathfrak{P}_K : \mathcal{PBO}^K_{X,Y} \to \mathfrak{G}_{X^K,Y^K}$$

called the *population lift operator* that converts individual search generators into a population-level search generator defined so that for population history traces $H \in \mathcal{H}^{\mathcal{T}/K}_{X^K,Y^K}$ and for measurable sets $A_1, \ldots, A_K \subseteq X$,

$$\begin{aligned} &\mathfrak{P}_K \langle \mathcal{G} \rangle [H] \left( \prod_{k=1}^K A_k \right) = \\ &\quad \int_{A_1} \cdots \int_{A_K} \prod_{k=1}^K \mathcal{G} \left[ \mathrm{traj}_K(H) \parallel (x_1, y_1) \parallel \ldots \parallel (x_{k-1}, y_{k-1}) \right] (dx_k), \end{aligned} \tag{7.3}$$

where the values $y_1, \ldots, y_K$ are arbitrary. This use of angle brackets $\langle \ldots \rangle$ to index classes of search generators will be continued below to represent algorithm hyperparameters. The definition of $\mathfrak{P}_K$ is complete due the value-agnostic properties of $\mathcal{PBO}^K_{X,Y}$ and the Carathéodory extension theorem discussed in Chapter 4.

For a population-based optimizer $\mathcal{G}$, the search generator $\mathfrak{P}_K \langle \mathcal{G} \rangle$ will be called the *K-population lift* of $\mathcal{G}$, shortened to *population lift* when $K$ is obvious from context. The population lift can be used to transfer useful properties. For example, it can be used to define the situation where each population is generated strictly from the prior population.

**Definition 7.2 (Population-Markov).** A population-based optimizer $\mathcal{G} \in \mathcal{PBO}^K_{X,Y}$ is said to be *population-Markov* if its population lift $\mathfrak{P}_K \langle \mathcal{G} \rangle$ has the 1-Markov property.

The population lift is also useful for describing other important properties of population-based optimizers. Quite often, evolutionary algorithms generate each individual of the next population independently of the others. In this case, the joint distribution over individuals in the population factorizes, and such an optimizer is termed *factorial.* In some evolutionary algorithms, population members are not only independent but also identically distributed. Such algorithms are termed *factorially homogeneous*. These properties can be defined with respect to the population lift.

**Definition 7.3 (Factorial).** An optimizer $\mathcal{G} \in \mathcal{PBO}^K_{X,Y}$ is factorial if there exist $K$ population-based optimizers $\mathcal{G}_1, \ldots, \mathcal{G}_K \in \mathcal{PBO}^K_{X,Y}$ such that for all population history traces $H \in \mathcal{H}^{\mathcal{T}}_{X^K,Y^K}$ and measurable sets $A_1, \ldots A_K \subseteq X$,

$$\mathfrak{P}_K \langle \mathcal{G} \rangle [H] \left( \prod_{k=1}^K A_k \right) = \int_{A_1} \cdots \int_{A_K} \prod_{k=1}^K \mathcal{G}_k \left[ \mathrm{traj}_K(H) \right] (dx_k). \tag{7.4}$$

The population-based optimizers $\mathcal{G}_k$ are called the *slot generators* of $\mathcal{G}$.

**Definition 7.4 (Factorially Homogeneous).** An optimizer $\mathcal{G} \in \mathcal{PBO}^K_{X,Y}$ is factorially homogeneous if it is factorial and each of its slot generators has $\mathcal{G}_k = \mathcal{G}$.

A process will now be presented to construct factorial slot generators. Define trajectory truncation by the symbol $\triangleleft$ so that

$$(\triangleleft \mathcal{G})\,[h \| (x,y)] = \mathcal{G}\,[h], \tag{7.5}$$

with the base case $(\triangleleft \mathcal{G})\,[\emptyset] = \mathcal{G}\,[\emptyset]$. The effect of this operation is to rewind the history, hiding the last element of the history. This operator can be applied to the same optimizer more than once. Let $\triangleleft_k \mathcal{G}$ represent the optimizer resulting from $k \geq 0$ applications of trajectory truncation, with $\triangleleft_0 \mathcal{G} \equiv \mathcal{G}$. The result of trajectory truncation is always one-step value-agnostic, the proof of which follows trivially from the definition. A stronger but still obvious version is stated in the next proposition.

**Proposition 7.1.** *For any search generator $\mathcal{G} \in \mathfrak{G}_{X,Y}$, $\triangleleft_k \mathcal{G} \in \mathcal{VA}^k_{X,Y}$ for $k \geq 1$.*

Consequently, $\triangleleft \mathcal{G}$ can always appear on the right-hand side of a convolution. The following proposition characterizes how the trajectory truncation operator interacts with definitions for the factorial and factorially homogeneous population-based optimizers. Its proof is essentially a restatement of the definitions.

**Proposition 7.2.** *An optimizer $\mathcal{G} \in \mathcal{PBO}^K_{X,Y}$ is factorial if and only if there exist $\mathcal{G}_1, \ldots, \mathcal{G}_K \in \mathfrak{G}_{X,Y}$ such that for all history traces h,*

$$\mathcal{G}[h] = \triangleleft_{k(h)-1} \mathcal{G}_{k(h)}[h] = \mathcal{G}_{k(h)}[\mathrm{traj}_K(H(h))]$$

*$\triangleleft_{k(h)} \mathcal{G}[h]$-almost surely. $\mathcal{G}$ is factorially homogeneous if and only if for all h,*

$$\mathcal{G}[h] = \triangleleft_{k(h)-1} \mathcal{G}[h] = \mathcal{G}[\mathrm{traj}_K(H(h))]$$

*again $\triangleleft_{k(h)} \mathcal{G}[h]$-almost surely.*

*Proof.* As a first observation, the two equations $\triangleleft_{k(h)-1} \mathcal{G}[h] = \mathcal{G}[\mathrm{traj}_K(H(h))]$ and $\triangleleft_{k(h)-1} \mathcal{G}_{k(h)}[h] = \mathcal{G}_{k(h)}[\mathrm{traj}_K(H(h))]$ hold in general and follow trivially from the definitions. Furthermore, each map $h \mapsto \triangleleft_{k(h)-1} \mathcal{G}_{k(h)}[h] \in \mathcal{PBO}^K_{X,Y}$.

If $\mathcal{G}$ is factorial, then Equation 7.4 and Equation 7.3 together imply that there exist $\mathcal{G}_1, \ldots \mathcal{G}_K$ such that

$$\mathcal{G}[h] \;=\; \mathcal{G}\left[\mathrm{traj}_K(H(h)) \,\|\, (x_1,y_1) \,\|\, \ldots \,\|\, (x_{k(h)-1}, y_{k(h)-1})\right] \;=\; \mathcal{G}_k[\mathrm{traj}_K(H(h))],$$

substituting $H(h)$ for $H$ and $k(h)$ for $k$, with almost-sure properties that ultimately fall back on $\triangleleft_{k(h)} \mathcal{G}[h] = \mathcal{G}[\mathrm{traj}_K(H(h))]$ due to Equation 7.4. Conversely, if there exist $\mathcal{G}_k$ such that $\mathcal{G}[h] = \mathcal{G}_{k(h)}[\mathrm{traj}_K(H(h))]$ for $\triangleleft_{k(h)} \mathcal{G}[h]$-almost every $x \in X$, then Equation 7.4 is satisfied, and $\mathcal{G}$ is factorial. The factorially homogeneous case is proven by setting $\mathcal{G}_k = \mathcal{G}$ above.

The goal of this exercise is to generate a population-based optimizer from $K$ slot generators. The next proposition demonstrates four such constructions, one each for the non-factorial, homogeneous, factorial, and factorially homogeneous cases.

**Proposition 7.3.** *Suppose $(\mathcal{G}_k)_{k=1}^K$ is a sequence of search generators in $\mathfrak{G}_{X,Y}$. Then there is a population-based optimizer $\mathcal{G} \in \mathcal{PBO}_{X,Y}^K$ such that for all history traces h,*

$$\mathcal{G}[h] = \mathcal{G}_{k(h)}[h]$$

*in at least the following three cases:*

1. Non-Factorial: *Each $\mathcal{G}_k \in \mathcal{VA}_{X,Y}^{k-1}$ where $\mathcal{VA}_{X,Y}^0 \equiv \mathfrak{G}_{X,Y}$.*
2. Homogeneous: *The non-factorial case with $\mathcal{G}_k = \mathcal{G}_j$ for all $k, j$.*
3. Factorial: *Each $\mathcal{G}_k = \triangleleft_{k-1}\mathcal{G}_k'$ for some $\mathcal{G}_k' \in \mathfrak{G}_{X,Y}$.*
4. Factorially Homogeneous: *The factorial case with $\mathcal{G}_k = \mathcal{G}_j$ for all $k, j$.*

*In each case, the sequence $(\mathcal{G}_k)_{k=1}^K$ is called a slot decomposition of the search generator $\mathcal{G}$, and each $\mathcal{G}_k$ is called a slot generator of $\mathcal{G}$.*

*Proof.* **Non-Factorial Case:** Suppose $\mathcal{G}_k \in \mathcal{VA}_{X,Y}^{k-1}$ for all $k$ and let $h$ and $h'$ be any history traces such that $h_X = h_X'$ and $\text{traj}_K(H(h)) = \text{traj}_K(H(h'))$. Then the $k$-step value-agnostic properties of $\mathcal{G}_k$ guarantee that $\mathcal{G}_{k(h)}[h] = \mathcal{G}_{k(h)}[h']$, whence $\mathcal{G}[h] = \mathcal{G}[h']$, and $\mathcal{G}$ is a population-based optimizer.

**Homogeneous Case:** This case is intended as a definition.

**Factorial Case:** If each $\mathcal{G}_k = \triangleleft_{k-1}\mathcal{G}_k'$, then necessarily $\mathcal{G}_k \in \mathcal{VA}_{X,Y}^{k-1}$ for all $k$, so $\mathcal{G}$ is a population-based optimizer. It is also factorial as a consequence of Proposition 7.2.

**Factorially Homogeneous Case:** If the factorial case is augmented by $\mathcal{G}_k = \mathcal{G}_j$ for all $k, j$, then the definition of being factorially homogeneous is trivially satisfied.

In sum, a population-based optimizer of population size $K$ can be constructed by proposing a collection of $K$ slot generators conforming to one of the cases in Proposition 7.3. Several examples of this construction will now be given.

### 7.1.4 Examples: Random Walkers

A simple example of a population-based optimizer should help to make these concepts concrete. Consider the case of a random walker in the search domain $X = \mathbb{Z}$, the integers. Given a starting point $x$, a biased random walker flips a weighted coin to decide either to advance one step to $x+1$ with probability $p$ or to go backwards one step to $x-1$ with probability $1-p$. Thus a single random walker can be represented for nonempty history traces $h$ by

$$\mathcal{RW}\langle p,x\rangle[h](\{z\}) = \begin{cases} p & \text{if } z = h_{-1}^X + 1 \\ 1-p & \text{if } z = h_{-1}^X - 1 \\ 0 & \textit{otherwise}, \end{cases} \tag{7.6}$$

where $h_{-1}^X$ is the last search point in the history $h$. For the empty history, set $\mathcal{RW}\langle p,x\rangle[\emptyset](\{z\}) = \delta_x(z)$ to start at $x$, where $\delta_x$ is the Kronecker delta, for which

$\delta_x(z) = 1$ if $z = x$ and zero otherwise. This definition is complete, because the topology of $\mathbb{Z}$ is discrete, and probability measures on $\mathbb{Z}$ are determined by the probability mass assigned to each element.

As mentioned for the case of the population lift operator above, the variables $\langle p,x\rangle$ indicate that this biased random walker is parameterized by the weight of the coin $p$ and the starting point $x$. Parameterized optimizers will be used extensively from this point. In the standard language of machine learning and evolutionary computation, these are *hyperparameters*.

Since the random walker is easily computable and never looks at the objective value of a search point, it is clear that $\mathcal{RW}\langle p,x\rangle$ is value-agnostic. Observe that $\mathcal{VA}_{X,Y} \subseteq \mathcal{PBO}^K_{X,Y}$ for any population size $K$, and hence $\mathcal{RW}\langle p,x\rangle$ is a population-based optimizer. Now consider several variations on this simple random walker.

**Factorial and Homogeneous Independent Paths.** Choose $K$ random walkers, each with identical bias $p$ and starting at zero. Let each walker ignore every other walker, so that the $K$ walkers follow $K$ independent paths. Formally, define for all history traces $h$

$$\mathcal{PRW}\langle p\rangle[h] = \triangleleft_K \mathcal{RW}\langle p,0\rangle[h]. \tag{7.7}$$

Each walker rewinds the history trace $h$ to its last decision $h^X_{-K}$ using trajectory truncation, and then continues its own path. This optimizer is homogeneous in that each individual in the population is sampled from the same base optimizer, $\triangleleft_K \mathcal{RW}$. However, $\mathcal{PRW}$ is not factorially homogeneous because $\triangleleft_i \mathcal{RW}\langle p,0\rangle \neq \triangleleft_j \mathcal{RW}\langle p,0\rangle$ unless $i = j$.

**Homogeneous Dependent Paths.** Choose $K$ random walkers, each with bias $p$ and starting at zero. Rather than rewinding paths, each walker builds on the path determined by the previous walker. This optimizer is in $\mathcal{PBO}^K_{X,Y}$. It is homogeneous, but it is uninteresting because it is equivalent to $\mathcal{RW}\langle p,0\rangle$.

**Dependent Paths.** Choose $K$ random walkers, each with a different bias $p_k$, each starting at zero. Again, each walker builds on the path determined by the previous walker, but in this case the walker uses separate weighted coins for different members of the population.

$$\mathcal{PRW}\langle p_1,\ldots,p_K\rangle[h] = \mathcal{RW}\left\langle p_{k(h)},0\right\rangle[h]. \tag{7.8}$$

This optimizer maintains a single random path through the search space but uses $K$ different weighted coins in a cycle. This optimizer is neither factorial nor homogeneous.

**Independent Paths.** Choose $K$ random walkers, each with a different bias $p_k$, each starting at zero. Let each walker ignore every other walker, so that the $K$ walkers follow $K$ independent paths. This optimizer is given by

$$\mathcal{PRW}\langle p_1,\ldots,p_K\rangle[h] = \triangleleft_K \mathcal{RW}\left\langle p_{k(h)},0\right\rangle[h]. \tag{7.9}$$

Each walker rewinds the path $h$ to its last decision $h_{-K}$ using trajectory truncation, and then continues its own path. The walkers are independent and nonhomogeneous, since they use different biases. Because the walkers are independent of each other, this optimizer is factorial.

### 7.1.5 Example: Simulated Annealing

The random walkers in the last section illustrate simple construction procedures. However, they are weak as optimizers because they do not take objective evaluations into account. This section further develops simulated annealing as an example that does utilize objective evaluations. This example generalizes the previous characterization in Section 4.3.3 to arbitrary search domains and value spaces; a general description of simulated annealing can be found in Section 2.5.1. This example is informative because it contains meaningful substructure that will be used to demonstrate processes of selection and variation.

Recall that simulated annealing consists of a sequence of objective evaluations for some objective function (*i.e.*, fitness function) $f$. At each time step, there is an accepted solution $x$, and a new solution $y$ is proposed. The objective value $f(y)$ is computed, and $y$ replaces $x$ as the accepted solution with probability

$$A(u,v,T) = \exp\left(\frac{1}{T}[u-v]\right) \wedge 1, \tag{7.10}$$

where $u = f(x)$, $v = f(y)$, and the infix operator $\wedge$ indicates the minimum of its arguments, so that $y$ is always accepted if $f(y) < f(x)$. Simulated annealing can thus be seen as a population-based algorithm of size two, with each population consisting of the accepted and proposed solutions.

In addition, simulated annealing requires a proposal distribution that is used to generate $y$ from the accepted solution $x$. The proposal distribution depends on the search space. For this example, let $X = \mathbb{R}^d$, i.e. a $d$-dimensional Euclidean space, and let the value space $Y = \mathbb{R}$ be the real numbers. Then a suitable proposal distribution is a the multivariate Gaussian distribution, denoted here by $\mathcal{N}\langle\mu,\Sigma\rangle$. Suppose for this example that the covariance matrix is fixed to the identity, although most instances of simulated annealing dynamically alter the covariance matrix to keep the acceptance probability close to 0.23. Let the search domain history $h_X$ track the accepted solution and the proposed solution in alternation, so that each point $h_n^X$ in the trajectory is the accepted solution at the $n^{th}$ time step if $n$ is odd, and the proposed solution if $n$ is even. Then set $\mu = h_{-1}^X$, the last accepted solution in the trajectory. The proposal distribution can be represented as a search generator defined for nonempty history traces $h$ by

$$\mathcal{P}[h] = \mathcal{N}\left\langle h_{-1}^X, I\right\rangle, \tag{7.11}$$

where $I$ is the identity matrix of dimension $d$. The search generator $\mathcal{P}$ is clearly value-agnostic and Markov, *i.e.*, $\mathcal{P} \in \mathcal{VA}_{X,Y}^{\text{Markov}}$. For initialization, set $\mathcal{P}[\emptyset] = \mathcal{N}\langle 0, I\rangle$ to be a standard normal.

Given a proposed solution $y \in X$ and an accepted solution $x \in X$, simulated annealing performs a Bernoulli trial to determine whether to accept $y$ or keep $x$. Let $\mathcal{B}\langle p,y,x\rangle$ be a Bernoulli probability measure that produces $y$ with probability $p$ and $x$ with probability $1-p$. Then the acceptance step for simulated annealing can be defined for non-empty history traces $h$ by

$$\mathcal{A}[h] = \mathcal{B}\langle A(h^Y_{-2},h^Y_{-1},T(|h|/2)),h^X_{-1},h^X_{-2}\rangle, \tag{7.12}$$

recalling that $h^X_{-1}$ contains the proposal and $h^X_{-2}$ the accepted solution, and that further $h^Y_{-1}$ contains the objective value of the proposal $f(h^X_{-1})$ and $h^Y_{-2}$ contains the objective value of the selected solution $f(h^X_{-2})$. The temperature $T(n)$ is assumed to be a function of the length of the trajectory, commonly $T(n) = 1/\log n$. For initialization, suppose again that $\mathcal{A}[\emptyset] = \mathcal{N}\langle 0,I\rangle$ is standard normal.

Simulated annealing can be expressed as a generator in at least two ways. First, it can be viewed as a population-based optimizer $\mathcal{SA}$ of population size 2 constructed according to Proposition 7.3 with slot generators $\mathcal{SA}_1 = \mathcal{A}$ and $\mathcal{SA}_2 = \mathcal{P}$, which is valid because $\mathcal{P} \in \mathcal{VA}^1_{X,Y}$. By this construction, $\mathcal{SA}_1$ and $\mathcal{SA}_2$ are used in alternation to accept and propose solutions.

The optimizer $\mathcal{SA}$ defined this way is neither factorial nor homogeneous, but it is in $\mathcal{PBO}^2_{X,Y}$, since only $\mathcal{SA}_1 = \mathcal{A}$ depends on the objective evaluations and $\mathcal{SA}_2 = \mathcal{P}$ does not. The order of these two steps is important to this construction, because if the proposal step was performed first $\mathcal{SA}$ would violate the definition of $\mathcal{PBO}^2_{X,Y}$ by relying on the objective evaluation of the first member of the population to generate the second before the population was completely constructed. While this distinction seems arbitrary in the case of simulated annealing, it captures the difference between a batch optimizer and a one-step-at-a-time optimizer. The distinction is computationally relevant, because a population-based optimizer meeting the definition of $\mathcal{PBO}^K_{X,Y}$ can compute any necessary function evaluations in $K$ parallel processes, whereas the same is not true for all search generators.

However, this arbitrariness captures a relevant fact about simulated annealing, namely, that it does not match well with our natural intuitions about a population-based algorithm. In fact, any evaluation of the performance of simulated annealing would not change if the optimizer had been defined as $\mathcal{SA}' = \mathcal{A} \star \mathcal{P}$, which is valid since $\mathcal{P} \in \mathcal{VA}^{\text{Markov}}_{X,Y}$. In this case, $\mathcal{SA}' \in \mathcal{PBO}^1_{X,Y} = \mathfrak{G}_{X,Y}$; that is, it has population size 1. The difference between $\mathcal{SA}$ and $\mathcal{SA}'$ is that the history trace for $\mathcal{SA}$ includes all of the proposed and accepted search points, whereas the history trace for $\mathcal{SA}'$ includes only the accepted search points.

The profusion of variables, operators, and symbols in this example may seem unnecessary at first. After all, it is possible to write pseudocode for simulated annealing with less effort than it took to describe $\mathcal{SA}$. However, the formalism makes it possible to compare simulated annealing directly with other optimization routines in a way that pseudocode does not allow. For instance, the classic evolutionary strategy known as the $(1+1)$–ES is the norm-limit of $\mathcal{SA}$ as the temperature goes to zero, as is shown in Theorem 7.1.

In addition, the example of simulated annealing has made use of several components and techniques that will be used in defining evolutionary algorithms. The proposal distribution $\mathcal{P}$ plays the role of a mutation operator in evolutionary methods, randomly altering a previously evaluated point. The acceptance optimizer $\mathcal{A}$ mirrors the role of selection in evolutionary methods. The convolution $\mathcal{A} \star \mathcal{P}$ is analogous to the exact form of an evolutionary algorithm, combining selection and variation in sequence. In fact, by the formal definitions that will be given in the next section, simulated annealing is an evolutionary algorithm. If one is uncomfortable calling simulated annealing *evolutionary*, one might say instead that evolutionary algorithms are stochastic Monte Carlo optimization routines. It makes no difference which category subsumes the other. The fact is that there is no formal difference between Monte Carlo optimization and evolutionary optimization, something that only becomes clear when evolutionary algorithms are formally analyzed. [2] With this goal in mind, the discussion now turns explicitly to a formalization of evolutionary algorithms.

## 7.2 Evolutionary Algorithms

In this section, the most common evolutionary algorithms are represented in the formal framework of the previous section. This process demonstrates that the formalization in this chapter and the preceding one do apply to complex practical algorithms. Also, these definitions will be used repeatedly in Chapters 8 and 10 to prove that the performance of most evolutionary algorithms is continuous as the fitness function changes.

### *7.2.1 Characteristics of an Evolutionary Algorithm*

The core characteristics of an evolutionary algorithm are based on an analogy with Darwinian principles and include competition within a population, preferential selection of competitive individuals, reproduction among selected individuals, and random variation of selected individuals. These four processes can be realized into evaluation, selection, recombination, and mutation phases. Selection and recombination occur at the level of populations. Variation occurs at the level of the individual. In formal terms, an evolutionary algorithm can be identified as a convolution of three components, one each for selection, recombination, and mutation processes. Evaluation of the fitness precedes selection. Recombination may be vacuous (asexual reproduction), in which case the algorithm is represented by a convolution of

[2] In existing literature evolutionary computation is occasionally referred to as a form of Monte Carlo optimization, but this statement is intuitively rather than formally derived. The conclusion follows by formalizing evolutionary algorithms mathematically, because Monte Carlo algorithms have always been described mathematically.

selection and mutation, much as simulated annealing was defined in the last section ($\mathcal{SA} = \mathcal{A} \star \mathcal{P}$).

Each of the phases of an evolutionary algorithm can be described by a search generator, just as the acceptance phase and proposal phase of simulated annealing were separated out into two different components. Thus the first step in formalizing evolutionary algorithms is to define what principles make a search generator work as a selection rule, a recombination operator, or a mutation operator. Viewed independently, the optimizers representing each phase are not effective optimizers by themselves in the general case. A selection rule alone is totally ineffective, since it cannot propose new solutions beyond what has already been evaluated. Mutation operators implement a blind random search. Recombination reconfigures evaluated points. In small, discrete spaces, selection plus recombination can be very effective at exploring the space given a sufficiently diverse initial population, but in large spaces, substantial mutation is required to fully explore the space.

### 7.2.2 Selection and Recombination

An evolutionary algorithm will be defined as the convolution of selection, recombination and mutation. These three phases may be thought of as intermediate steps, each of which creates an individual and hands it off to the next phase. So selection chooses parents from among the previously observed points. Recombination invokes one or more additional selection rules to tack on extra parents and then merges these parents with a crossover rule; this merged child is then handed off to the mutation operator, which alters the individual independent of its fitness. These three stages will now be discussed rigorously one at a time.

Selection in evolutionary algorithms is a filtering task, characterized as follows: Given a set of previously observed individuals, select a parent for the next population. Therefore, the selection process must place zero weight on previously unobserved individuals. Only members of the population history can be selected. Given a history trace $h \in \mathcal{H}^{\mathcal{T}}_{X,Y}$, define the previously observed individuals in $h$ as $P(h) = \left\{x \in X \mid \exists n \text{ s.t.} x = h^X_n\right\}$. Taking populations into account, a *selection rule* is a search generator that places zero probability mass on any proposed population that would expand $P(h)$.

**Definition 7.5.** A population-based optimizer $\mathcal{S} \in \mathcal{PBO}^K_{X,Y}$ is a *selection rule* if for all measurable sets $A$, $\mathcal{S}[h](A) = 0$ whenever $A \cap P(h)$ is empty.

It may seem strange to a practitioner of evolutionary algorithms that the selection rule is allowed to select any member of $P(h)$, that is, any preceding search point, and not just the members of the last population. But there are a number of evolutionary methods that select members of populations prior to the last population, such as elitist selection. Methods that store the locally best individual (as in some versions of evolution strategies) also need the flexibility to select from previous generations. Furthermore, several recently proposed techniques such as novelty search [100], curiosity search [141], and the evolutionary annealing method proposed in Chapter 14 store members from each population in an archive, making them available for selection.

Recombination combines some number of selected individuals as parents to form a hybrid child. Although the traditional recombination methods in genetic algorithms utilize only two parents, other methods use an arbitrary number of parents. In evolution strategies, for example, intermediate crossover averages components across several solutions. A recombination operator first selects the parents for each member of the population and then invokes a crossover rule to combine the parents. The number of selected parents (usually just two) is said to be the *order* of the crossover rule and the recombination operator. Parent selection for an $n^{th}$ order operator stacks $n$ populations on top of the current trajectory. A crossover rule consumes these $n$ populations and leaves a single merged population in their place.

The key feature of a crossover rule is that it should combine only the selected parents. It should therefore be independent of all other components of the input trajectory. It should also ignore the objective value of the selected parents, deferring such judgments to the selection operators. From this perspective, for the $k^{th}$ member of the population the selected parents in a crossover rule of order $n$ are just the $k^{th}$ members of the previous $n$ populations in the history trace. Define the sequence

$$\text{parents}(h,n,k,K) = (H(h)^X_{-i,k})_{1\leq i\leq n},$$

recalling that $H(h)$ is the population history of $h$ (with size $K$), negative indices count backwards from the end of the history, and the double index chooses the $k^{th}$ member of the $-i^{th}$ population. Then $\text{parents}(h,n,k,K)$ is the reverse ordered list of the parents available to the crossover rule at slot $k$.

**Definition 7.6.** A value-agnostic search generator $\mathcal{C} \in \mathcal{VA}_{X,Y}$ is a crossover rule of order $n$ if there exists a slot decomposition $\mathcal{C}_1,\ldots,\mathcal{C}_K \in \mathcal{VA}_{X,Y}$ of $\mathcal{C}$ such that for all $k = 1,\ldots,K$ and for all history traces $h,h' \in \mathcal{H}^{\mathcal{T}}_{X,Y}$, $\mathcal{C}_k[h] = \mathcal{C}_k[h']$ whenever $\text{parents}(h,n,k,K) = \text{parents}(h',n,k,K)$. That is, a crossover rule is independent of all but the selected parents and their order.

Such a crossover rule is factorial by definition, which reflects that the manner in which one offspring is generated depends only on its own selected parents and not at all on the state of the other, preceding offspring. It would be possible to define crossover rules to be non-factorial, so that later crossovers depend on the results of earlier ones, but it does not seem necessary. As it is, this definition of crossover accepts a wide range of instantiations that do not necessarily match the concept of crossover in a traditional genetic algorithm. This intuition will be restored with the introduction of crossover masks in Section 7.2.4.

In order to implement recombination, a crossover rule of order $n$ must be combined with $n$ selection rules. The following definition defines a recombination operator as the convolution of $n-1$ selection rules with a crossover rule based on a purely formal decision that the first selection rule will be provided separately and treated as primary, so that given this primary selection $\mathcal{S}$ and a recombination operator $\mathcal{R}$, the full results of selection and recombination are described by the convolution $\mathcal{S}\star\mathcal{R}$.

**Definition 7.7.** A population-based optimizer $\mathcal{R} \in \mathcal{PBO}^K_{X,Y}$ is a recombination operator of order $n$ if either $n = 1$ and $\mathcal{R}$ is a crossover rule of order 1 or $n \geq 2$ and there exists a sequence of $n-1$ selection rules $\mathcal{S}_1, \ldots, \mathcal{S}_{n-1}$ and a crossover rule $\mathcal{C}$ of order $n$ such that

$$\mathcal{R} = \triangleleft \mathcal{S}_1 \star (\triangleleft_2 \mathcal{S}_2 \star (\cdots \star (\triangleleft_{n-1} \mathcal{S}_{n-1} \star \mathcal{C}))).$$

The right association of the convolutions is critical to making this definition of recombination work, since it means that the crossover rule can "see" all of the selected parents. The following proposition justifies that these convolutions are validly defined, that is, that each item appearing on the right-hand side of a convolution above is indeed a one-step value-agnostic search generator in $\mathcal{VA}^1_{X,Y}$ as required by the definition of convolution.

**Proposition 7.4.** *If $\mathcal{A} \in \mathcal{VA}^k_{X,Y}$ and $\mathcal{B} \in \mathcal{VA}^{k+1}_{X,Y}$, then $\mathcal{A} \star \mathcal{B} \in \mathcal{VA}^k_{X,Y}$.*

*Proof.* Suppose that history traces $h, h' \in \mathcal{H}^{\mathcal{T}}_{X,Y}$ are such that $h_X = h'_X$ and $h_Y$ and $h'_Y$ agree on all elements except for the value of the final $k$ search points. That is, if $h_Y = (y_1, \ldots, y_m)$ and $h'_Y = (y'_1, \ldots, y'_m)$, then $y_i = y'_i$ for $i < m - k$ and there exists some $j \geq m - k$ such that $y_j \neq y'_j$ so that $h_Y \neq h'_Y$. Now for any measurable set $A$,

$$\begin{aligned} \mathcal{A} \star \mathcal{B}[h](A) &= \int_X \mathcal{B}[h \| (x,y)](A)\, \mathcal{A}[h](dx) \\ &= \int_X \mathcal{B}[h' \| (x,y)](A)\, \mathcal{A}[h'](dx) = \mathcal{A} \star \mathcal{B}[h'](A) \end{aligned}$$

by the $k$-step value-agnostic property for $\mathcal{A}$ and the $(k+1)$-step value-agnostic property for $\mathcal{B}$. Consequently, $\mathcal{A} \star \mathcal{B} \in \mathcal{VA}^k_{X,Y}$ is $k$-step value-agnostic.

**Corollary 7.1.** *A recombination operator $\mathcal{R}$ is valid and $\mathcal{R} \in \mathcal{VA}^1_{X,Y}$.*

*Proof.* For a recombination operator $\mathcal{R}$, the crossover rule $\mathcal{C}$ is in $\mathcal{VA}_{X,Y} \subseteq \mathcal{VA}^n_{X,Y}$, and the selection rules $\mathcal{S}_k$ have $\triangleleft_k \mathcal{S}_k \in \mathcal{VA}^k_{X,Y}$ for $k \geq 1$. Applying Proposition 7.4 from right to left, one step of value-agnosticism is lost with each application, and there are $n-1$ applications. Thus after all $n-1$ convolutions, $\mathcal{R} \in \mathcal{VA}^1_{X,Y}$.

Inside of the recombination operator, each of the selection rules $\mathcal{S}_k$ are applied in order, with the previous selection hidden by the trajectory-truncation operator. Finally, the crossover rule is invoked to combine the selected points, including the first point selected by an initial selection rule outside of the recombination operator. Formally, the definition expands to

$$\mathcal{R}[h](A) = \int_X \cdots \int_X \mathcal{C}\left[h \| (x_1, y_0) \| \ldots \| (x_n, y_0)\right]\, \mathcal{S}_n[h](dx_n)\, \ldots\, \mathcal{S}_1[h](dx_1)$$

where $y_0 \in Y$ is an arbitrary (and ignored) value. The convolution is performed with right association so that the results of selection are stacked together and not consumed until the crossover rule is reached. Note that there is only one possible proper recombination operator of order 1, and it vacuously reproduces the selected population, representing asexual reproduction.

**Proposition 7.5.** *Suppose $\mathcal{R}$ is a recombination operator of order* 1 *and $\mathcal{S}$ is any selection rule. Then $\mathcal{S} \star \mathcal{R} = \mathcal{S}$.*

*Proof.* With order 1, the definitions imply that $\mathcal{R} = \mathcal{C}$ for some crossover rule $\mathcal{C}$ of order 1. For any history trace $h$, any search point $x$ and any value $y$,

$$\text{parents}(h \| (x,y), 1, k, K) = (x).$$

Consequently, for all pairs of history traces $h$, $\mathcal{C}[h \| (x,y)] = \mathcal{C}[\emptyset \| (x,y)] = \mathcal{C}[((x,y))]$, and so $\mathcal{C}$ is 1-Markov. It is also degenerate with all weight on $x$, *i.e.*, $\mathcal{C}[((x,y))](A) = \mathbb{1}_A(x)$. Then for any selection rule $\mathcal{S}$, any history trace $h$, and any measurable set $A$,

$$\mathcal{S} \star \mathcal{R}[h](A) = \int_X \mathcal{C}[h \| (x,y)](A)\, \mathcal{S}[h](dx) = \int_X \mathbb{1}_A(x)\, \mathcal{S}[h](dx) = \mathcal{S}[h](A),$$

completing the proof.

The results of recombination can then be mutated to provide variation.

### *7.2.3 Mutation and Evolutionary Algorithms*

Mutation in evolutionary algorithms alters a single member of a proposed population. Thus a mutation operator should 1-Markov, altering only the individual proposed to it by selection and recombination. Mutation should also be value-agnostic; it should not be aware of the fitness of the point it is mutating. Conversely, a mutation operator should not simply ignore the individual it is mutating, and so a condition should be included requiring the mutation operator must depend on the object being mutated for at least some history traces. A search generator $\mathcal{G}$ will be called *non-constant* whenever if there exist $h, h' \in \mathcal{H}^{\mathcal{T}}_{X,Y}$ such that $h_X \neq h'_X$ and $\mathcal{G}[h] \neq \mathcal{G}[h']$. Then the term *mutation operator* is defined to mean a non-constant element of $\mathcal{VA}^{\text{Markov}}_{X,Y}$.

**Definition 7.8.** A *mutation operator* is a non-constant 1-Markov value-agnostic search generator.

The condition of being non-constant cannot guarantee more than one case of dependence on the content of the individual, but it does at least prevent the case of degenerate and invariant mutation operators that produce all samples from a single distribution. A stronger condition could be used, for example, by requiring distinct mutation distributions for *every* individual, but this condition seems overly restrictive and is not adopted here.

A quasi-evolutionary algorithm will be defined as the convolution of a selection rule, a recombination operator and a mutation operator. Recall that the recombination operator contains one or more selection rules and a crossover rule. The recombination operator may also be of order one, in which case it simply copies the initial selection rule as stated by Proposition 7.5. A working definition for a strict evolutionary algorithm will be defined based on crossover masks in Section 7.2.4.

**Definition 7.9.** A population-based optimizer $\mathcal{E} \in \mathcal{PBO}^K_{X,Y}$ is a quasi-evolutionary algorithm if it is not value-agnostic and there exist a selection rule $\mathcal{S}$, a recombination operator $\mathcal{R}$ of order 1 or greater, and a mutation operator $\mathcal{V}$ such that $\mathcal{E} = \mathcal{S} \star \mathcal{R} \star \mathcal{V}$.

**Proposition 7.6.** *By implication, $\mathcal{E} \in \mathcal{PBO}^K_{X,Y}$ is also a quasi-evolutionary algorithm if it is not value-agnostic and there is a selection rule $\mathcal{S}$ and a mutation operator $\mathcal{V}$ such that $\mathcal{E} = \mathcal{S} \star \mathcal{V}$, in which case $\mathcal{E}$ has a recombination operator of order* 1.

*Proof.* This proposition is a trivial implication of Proposition 7.5.

Intuitively, a quasi-evolutionary algorithm first samples one or more selection rules to propose a new parent population consisting of the selected individuals, then recombines the parent population to form a new child population, and finally samples a mutation operator to alter the selected individuals. Because crossover and mutation were defined to be value-agnostic, it follows that the selection rules of a quasi-evolutionary algorithm cannot be value-agnostic, and hence that they must rely at least somewhat on the fitness values, which is a reasonable requirement for an optimization method.

The definition of a quasi-evolutionary algorithm and its parts were chosen to exclude algorithms that do not match standard intuitions of how an evolutionary algorithm works. These definitions are at least mildly restrictive, primarily because the crossover rule and the mutation operator must be value-agnostic. Otherwise, any population-Markov population-based optimizer $\mathcal{M}$ that is not value-agnostic would be a quasi-evolutionary algorithm, since $\mathcal{M}$ could be used as a $K^{th}$ order crossover rule, with $K$ selection rules each of which simply pass along one member of the prior population and a vacuous mutation operator that does nothing. The definitions above preclude this possibility among others.

Now that a formal definition of a quasi-evolutionary algorithm and its components has been given, it is possible to state explicit formulae for common genetic algorithms based on the selection, crossover, and mutation methods that they use. Additionally, further definitions will be proposed that will be used to develop a formal definition an evolutionary algorithm as traditionally applied.

### 7.2.4 Genetic Algorithms

Modern genetic algorithms mix and match a variety of selection, crossover, and mutation components to form an optimization routine. This section will review the most common among these components, along with the way in which they are assembled.

With rare exceptions, selection in genetic algorithms is typically restricted to the members of the last population, so that a genetic algorithm unfolds as a sequence of populations, with each population generated strictly from the prior population. Such search generators are *population-Markov*, as defined in Section 7.1.3. To restate, population-Markov search generators depend only on the last population; that is, $\mathcal{G}[h] = \mathcal{G}[h']$ whenever $H(h)_{-1} = H(h')_{-1}$, where $H(h)_{-1}$ is formal notation for the last population as before. The following proposition reveals that in light of the previous definitions for selection, recombination, and mutation, the quality of being population-Markov is entirely determined by the selection rule.

**Proposition 7.7.** *An evolutionary algorithm is population-Markov if and only if its recombination operator and selection rule are, and a recombination operator is population-Markov if and only if each of its subordinate selection rules is.*

*Proof.* The mutation operator is defined to be independent of all history except for the individual generated by selection and recombination. Likewise, the crossover rule for recombination is independent of all history except for the individuals generated by the preceding selection rule or by the its subordinate selection rules. These facts combine to yield the desired result.

Genetic algorithms are population-Markov in general. The most common selection rules historically are proportional selection, tournament selection, and ranking selection. To support these selection rules, assume for the next few paragraphs that the value space $Y$ is totally ordered. Usually, one assumes further that $Y = \mathbb{R}$, though any totally ordered set is sufficient to support the definitions below.

In *proportional selection*, members of the prior population are selected independently proportional to their fitness in the previous population. Ordinarily, the fitness function for proportional selection is assumed to be positive, and the genetic algorithm is maximizing the fitness and so prefers larger fitness values. To adapt proportional selection for minimization as well, we introduce a positive function $g : X \times Y \to [0,\infty)$ so that $g(x,y)$ is positive and increasing as $y$ becomes more optimal. This $g$ will be called the *modulating function* of proportional selection and should be chosen for the optimization task. Given a modulating function $g$ and any measurable set $A$, proportional selection in denormalized form is

$$\mathcal{PS}\langle g\rangle[h](A) \quad \propto \quad \sum_{(x,y)\in H(h)_{-1}} g(x,y)\, \mathbb{1}_A(x), \tag{7.13}$$

and the normalized version is obtained by dividing by the sum of the right-hand side over all search points $x \in H(h)^X_{-1}$ in the last population. Note that $H(h)_{-1}$ is a sequence, and so by $(x,y) \in H(h)_{-1}$, what is intended is to sum over the sequence of pairs in $H(h)_{-1}$, which may contain duplicates. Such duplicates are repeated in the sum according to their multiplicity.

If one desires to maximize a static fitness function $f$ with $f > 0$, then $y = f(x)$ above and the modulating function should be $g(x, f(x)) = f(x)$. The right hand side of Equation 7.13 then adds a proportional probability mass of $f(x)$ once for each time that $x$ appears in the population. This rule will then proportionally select individuals higher values of $f$ weighted according to their dominance in the population. Proportional selection with this choice of modulating function will be termed *standard proportional selection* or *roulette wheel selection*.

Minimization of a static fitness function $f$ with $f > 0$ can be achieved by setting $g(x, f(x)) = -f(x)$. A more neutral choice is $g(x,y) = \exp(-y)$, which does not assume that the fitness is positive. The similarity of this choice with the acceptance probability for simulated annealing should not be missed, and will return in Chapter 14.

Proportional selection is highly sensitive to the magnitude of variation in the fitness function and so can become trapped in steep local minima. *Tournament selection* chooses members of the prior population according to their rank in the population in order to maintain diversity within the population. Like proportional selection, tournament selection is factorial and so chooses each member of the prior population based on the same distribution. This distribution selects the best member of the last population with probability $q$. If the best member is not selected, the second best member is chosen with probability $q$, and then the third, and the fourth, and so on. If the population is exhausted, the selection wraps back around to the best individual. The parameter $q$ is referred to as the *selection pressure* since high values of $q$ force selection to predominately favor the best individuals in the population. Define

$$\text{Rank}\,(x,P) \in \mathbb{N} \cup \{\infty\}$$

to be the rank of the individual $x$ in the evaluated population $P$ according to the order on the value space $Y$, with 0 being the best rank, and $R(x,P) = \infty$ if $x$ does not appear in $P$, so that the probability of such $x$ being selected is zero. In case of ties, assume later members of the population are ranked higher. A version of tournament selection is given in denormalized form by

$$\mathcal{TS}\langle q\rangle [h] (\{x\}) \propto (1-q)^{\text{Rank}(x,H(h)_{-1})}. \tag{7.14}$$

As with proportional selection, $\mathcal{TS}$ is nonzero for at most $K$ points so that the normalization is easily computed.

Like tournament selection, *ranking selection* chooses individuals according to their rank in the prior population, but does so using proportional selection over the rank. Define

$$r_q^h(x) = \begin{cases} 2-q+2(q-1)\,\frac{K-1-\text{Rank}(x,H(h)_{-1})}{K-1} & \text{if}\, x \in H(h)_{-1}^X \\ 0 & \text{otherwise} \end{cases} \tag{7.15}$$

Then *linear ranking selection* is given in denormalized form by

$$\mathcal{RS}\left<q\right>[h]\left(\{x\}\right) \propto r_q^h(x), \tag{7.16}$$

where $q \in [1,2]$ is known as the *selection pressure*. The normalization is computed as before. Notice the similarity to proportional selection. Ranking selection is proportional selection in which the fitness has been replaced with the rank in the population. *Non-linear ranking selection* can be represented in a similar fashion but with more complex detail.

One final feature of selection commonly used in genetic algorithms is *elitism*. Elitism protects the best evaluation point so far from being removed from the population. Elitism can be beneficial to a genetic algorithm because it prevents the algorithm from forgetting the best individual. Given a fitness function $f$, let $\text{best}(P) \in X$ be the search point in the evaluated population $P$ with most optimal fitness on $f$, with ties broken to favor later individuals. Then elitist selection can be represented as altering an entire genetic algorithm $\mathcal{GA} \in \mathcal{PBO}_{X,Y}^K$ by preserving $\text{best}(H(h)_{-1})$ as the first member of the population, so that for all histories $h$ and measurable sets $A$,

$$\mathcal{E}\left<\mathcal{GA}\right>[h](A) = \begin{cases} \mathbb{1}_A(\text{best}(H(h)_{-1})) & \text{if } |H(h)| > 0 \text{ and} k(h) = 1 \\ \mathcal{GA}[h](A) & \text{otherwise,} \end{cases} \tag{7.17}$$

recalling that $\mathbb{1}_A$ is the indicator function for the set $A$. This definition of elitism is essentially a slot decomposition in which the first slot is filled by the best element of the previous population and hence the best for the entire history. The other slots are filled by the genetic algorithm $\mathcal{GA}$. Importantly, elitism is not a selection rule when defined this way, since it not only selects the best individual, but preserves it from alteration as well.

The distinguishing characteristic of a genetic algorithm is undoubtedly recombination with two parents (sexual reproduction). Standard crossover rules of order 2 include one point crossover, multipoint crossover, and uniform crossover. Most often, the same selection rule is often used to select both parents. Sometimes a strongly selective rule is used to choose the "father" while a more uniform selection rule is used to select the "mother". Either way, the "child" is created to combine properties from the father and the mother.

Because crossover rules are specific to the search space, examples will only be given for the case in which the search space $X$ is a $d$-dimensional vector space, $X = R^d$, such as $X = \mathbb{R}^d$ (Euclidean space) or $X = \{0,1\}^d$ (binary space). In this case, many second-order crossover rules can be determined by a random binary vector $M \in \{0,1\}^d$ which will be termed the *crossover mask*. If $M_i = 1$, then the child copies the $i^{th}$ attribute of the father. If $M_i = 0$, then the child copies the $i^{th}$ attribute of the mother. Denote by $\mathbf{1}$ the vector in $\{0,1\}^d$ whose entries are all one, and let $x \otimes y$ be the vector that is the componentwise product of vectors $x$ and $y$. [3] For a history $h$, let $p(h)$ be the selected father and $m(h)$ the selected mother, so that $p(h) = \text{parents}(h,2,k(h),K)_{-1}$ and $m(t) = \text{parents}(h,2,k(h),K)_{-2}$ using the parents

[3] This componentwise product is always well-defined for its use in Equation 7.18, since it is just a conditional selector. Further details will not be developed here.

function in the definition of crossover rules. Define a function $C(h,M)$ by

$$C(h,M) = M \otimes p(h) + (\mathbf{1} - M) \otimes m(h) \tag{7.18}$$

to represent the individual resulting from crossing $p(h)$ with $m(h)$ using the mask $M$. Then given a probability measure $\mathbb{P}_M$ over $M$ that can be used to sample masks, a *masked crossover rule* for histories $h$ and measurable sets $A$ is a crossover rule

$$\mathcal{C}\left<\mathbb{P}_M\right>[h](A) = \sum_{z \in \{0,1\}^d} \mathbb{1}_A(C(h,z))\, \mathbb{P}_M(\{z\}) \tag{7.19}$$

that combines parents using componentwise masks generated by $\mathbb{P}_M$. This $\mathcal{C}$ is clearly a crossover rule of order 2, since it depends strictly on the mother and father as required.

Single point, multipoint, and uniform crossover can be defined by specifying $\mathbb{P}_M$. For uniform crossover, the choice of mask is uniformly random,

$$\mathcal{UC} = \mathcal{C}\left<\text{Uniform}\left(\{0,1\}^d\right)\right>. \tag{7.20}$$

For single point crossover, a random index $i \in \{1,\ldots,d\}$ is chosen, and the mask is set so that $M_j = 1$ for $j \leq i$ and $M_j = 0$ for $j > i$. In multipoint crossover, a fixed number of random indices $i_1,\ldots,i_n$ are chosen and then sorted. $M$ then alternates between series of zeros and a series of ones, starting with ones and with switches occurring at each of the $i_j$. Without stating further details, let $\mathcal{SC}$ denote single-point crossover and let $\mathcal{MC}$ represent multipoint crossover.

Masked crossover best captures the characteristic of a traditional genetic algorithm, and an evolutionary algorithm will be defined as a quasi-evolutionary algorithm with a masked crossover rule. A genetic algorithm will be identified as an evolutionary algorithm that is also population-Markov.

**Definition 7.10.** A population-based optimizer $\mathcal{G} \in \mathcal{PBO}_{X,Y}^K$ is an evolutionary algorithm if it is a quasi-evolutionary algorithm with a crossover rule that either has order 1 or is a masked crossover rule. Additionally, $\mathcal{G}$ is a genetic algorithm if it is also population-Markov.

This definition encompasses most traditional evolutionary algorithms and excludes more recent developments that still conform to the definition of a quasi-evolutionary algorithm as defined above. Once again, a crossover rule of order one may be used, so that every quasi-evolutionary algorithm with a vacuous crossover rule is also an evolutionary algorithm.

In defining a genetic algorithm, the qualities of the mutation operator have so far been ignored because there do not seem to be applicable limitations on it. A mutation operator is even more dependent on the search space and can be almost any probability measure. The most common mutators, however, are Bernoulli mutation in binary spaces and Gaussian mutation in Euclidean space, with Cauchy distributions also used for Euclidean space. In discrete or combinatorial spaces, mutation distributions typically involve random structural operators.

First, consider Gaussian mutation in the search domain $X = \mathbb{R}^d$. The mean of the Gaussian is simply the point being mutated, $h^X_{-1}$, and the covariance is a function of the prior points evaluated, often a constant. Then *Gaussian mutation* with a covariance-generating function $\Sigma : \mathcal{H}^{\mathcal{T}/K}_{X^K,Y^K} \to \mathbb{R}^{d\times d}$ is given by

$$\mathcal{N}\langle\Sigma\rangle[h] = \mathcal{N}\left(h^X_{-1}, \Sigma(H(h))\right), \tag{7.21}$$

where $\mathcal{N}(\mu,\Sigma)$ is the Gaussian distribution with mean $\mu$ and covariance $\Sigma$ and the symbol $\mathcal{N}$ is overloaded to represent Gaussian mutation as well. For genetic algorithms, it is assumed that $\Sigma$ is constant, but for evolution strategies $\Sigma$ often relies on summarized statistics of the full history.

When the search space is binary, $X = \{0,1\}^d$, *Bernoulli mutation* at rate $p$ is given by any value-agnostic search generator $\mathcal{B}$ such that for all histories $h$, search points $x$, and values $y$,

$$\mathcal{B}\langle p\rangle[h\|(x,y)](\{z\}) = \prod_j p^{|z_j - x_{i,j}|}(1-p)^{\left(1-|z_j-x_{i,j}|\right)}, \tag{7.22}$$

which simply says that each bit is flipped with probability $p$.

Genetic algorithms are often used with representational encodings. The concept of an *encoding* has not been introduced heretofore; an encoding function that embeds one space in another. So, for instance, if one wishes to search a real vector space with a binary encoding, then one simply encodes each bit string into a real vector. Bernoulli mutation is the standard mutation for genetic algorithms with binary encodings, whereas Gaussian mutation is the standard mutation for real vector encodings. For the purpose of this text, the encoding is irrelevant. The search domain $X$ should be treated as the encoding space (*i.e.*, $X = \{0,1\}^d$ for binary encodings, but $X = \mathbb{R}^d$ for real encodings). The encoding simply changes the way in which values are selected from the value space $Y$. So if $X$ is the encoding space and $e : X \to Z$ encodes elements of $X$ (*e.g.*, bit strings) into $Z$ (*e.g.*, real vectors), then to optimize a static fitness function $f : Z \to \mathbb{R}$, one simply assigns each search point $x \in X$ a value $y = f \circ e(x)$. The consequence of, for example, using Bernoulli mutation on an encoding to search in a real vector space is then determined automatically by the nature of the encoding $e$ and the interaction of the search generator with the values in the history.

Putting all of these pieces together, a basic genetic algorithm with single-point crossover, proportional selection, and a binary encoding can be written as

$$\mathcal{SGA}\langle p\rangle = (\mathcal{PS} \star ((\triangleleft\mathcal{PS}) \star \mathcal{SC})) \star \mathcal{B}\langle p\rangle, \tag{7.23}$$

which is Goldberg's simple genetic algorithm with a mutation rate of $p$ [60]. A common choice of genetic algorithm for searching in Euclidean space is to use ranking selection with uniform crossover and Gaussian mutation, namely,

$$\mathcal{RGA}\langle q,\sigma\rangle = (\mathcal{RS}\langle q\rangle \star ((\triangleleft\mathcal{RS}\langle q\rangle) \star \mathcal{UC})) \star \mathcal{N}\langle\sigma I\rangle, \tag{7.24}$$

where $q \in [1,2]$ is the selection pressure and $\sigma$ is a small constant rate of mutation. In both cases, the resulting algorithms are formally evolutionary algorithms, since they are composed of a population-Markov selection rule, a recombination operator with masked crossover, and a mutation operator.

Most standard genetic algorithms can be written down by mixing and matching the components described in the section along with domain-specific mutation operators, as was done in Equations 7.23 and 7.24. More esoteric genetic algorithms could also be represented in similar fashion with the definition of additional components.

### *7.2.5 Evolution Strategies*

Evolution strategies differ from genetic algorithms primarily in the choice of selection and mutation operators, and in the fact that crossover is rarely used in evolution strategies or is used with different crossover rules if so. Additionally, traditional evolution strategies also adapt their algorithm parameters dynamically, which implies, for instance, that they cannot be population-Markov and hence do not generally satisfy the formal definition of genetic algorithms given in the last subsection.

Standard evolution strategies are denoted as either $(\mu,\lambda)$–ES or $(\mu+\lambda)$–ES. In this notation, $\mu$ is the number of parents, and $\lambda$ is the number of children. The parents are always the $\mu$ best members of the last population, so if $K = 10$ and $\mu = 3$, then the parents are the top three members of the last population by fitness. Notice that this requires the value space $Y$ to be totally ordered, which will be assumed throughout this subsection.

A $(\mu+\lambda)$–ES has population size $K = \mu+\lambda$, and in each generation, the parents are retained unchanged from the prior generation, and $\lambda$ new solutions are sampled from the parents. A $(\mu,\lambda)$–ES has a population size $K = \lambda > \mu$; it discards the parents and replaces them with the children at each time step. The simplest evolution strategy is the $(1+1)$–ES, which is equivalent to simulated annealing at zero temperature. The most commonly used is probably the $(10,100)$–ES, which tends to find solutions to basic benchmark problems with reasonable speed and accuracy.

Selection in evolution strategies first sorts the prior population by rank and then selects the next population. Ranking can be represented as a selection rule defined so that for histories $h$ and search points $x$

$$\mathcal{R}[h](\{x\}) = \delta_{k(h)}\left(\text{Rank}(x, H(t)_{-1}) + 1\right), \tag{7.25}$$

where $\delta_{k(h)}$ is the Kronecker delta that is 1 if the argument is $k(h)$ and 0 otherwise. Thus $\mathcal{R}$ simply sorts the prior population by fitness rank. The $\lambda$ children are then selected uniformly from among the top-ranked $\mu$ members of the last population using one of value-agnostic search generators:

$$\mathcal{U}\langle\mu\rangle[h](\{x\}) = \frac{1}{\mu}\sum_{i=1}^{\mu}\delta_{H(h)^X_{-1,i}}(x) \tag{7.26}$$

$$\mathcal{U}_+\langle\mu\rangle[h](\{x\}) = \begin{cases} H(h)^X_{-1,k(h)} & \text{if} k(h) \leq \mu \\ \mathcal{U}\langle\mu\rangle[h](\{x\}) & k(h) > \mu \end{cases} \tag{7.27}$$

The rule $\mathcal{U}$ is used for so-called “comma” selection, where the parents are discarded and only the children remain in the population. The alternative version $\mathcal{U}_+$ is for “plus” selection, where both parents and children remain in the new population. Selection in evolution strategies is given by $\mathcal{ESS}\langle\mu\rangle \equiv \mathcal{R}\star\mathcal{U}\langle\mu\rangle$ for “comma” selection, and $\mathcal{ESS}_+\langle\mu\rangle \equiv \mathcal{R}\star\mathcal{U}_+\langle\mu\rangle$ for “plus” selection; both a valid because $\mathcal{U}$ and $\mathcal{U}_+$ are value-agnostic. This two-part decomposition is just one formal representation and is somewhat arbitrary. While $\mathcal{ESS}\langle\mu\rangle$ and $\mathcal{ESS}_+\langle\mu\rangle$ are well-defined, unique elements in $\mathcal{PBO}^K_{X,Y}$ for a given $\mu$ and $\lambda$, the decomposition $\mathcal{R}\star\mathcal{U}$ is just one way of expressing it, just as $1+3$ and $2+2$ are two different ways of expressing 4.

Evolution strategies usually do not employ recombination, but when they do, they often use higher-order crossover rules. The resulting algorithms are termed either as a $(\mu/\rho+\lambda)$–ES or a $(\mu/\rho,\lambda)$–ES, where $\rho$ is the order of the crossover. Two crossover rules are commonly used: intermediate crossover and dominant crossover. Dominant crossover is a higher order generalization of uniform crossover to $\rho$ parents. Intermediate crossover averages the parent components. Like the crossover methods used for genetic algorithms, these two methods assume that the search space has a product space structure. Additionally, intermediate crossover requires that the search space be a vector space with addition and scalar multiplication. Intermediate crossover is easy to express as a point distribution on the average of the parents. With

$$\text{average}(h) = \frac{1}{\rho}\sum_{i=1}^{\rho} H(h)^X_{-i,k(t)}, \tag{7.28}$$

intermediate crossover is given by

$$\mathcal{IC}\langle\rho\rangle[h](\{x\}) = \delta_{\text{average}(h)}(x). \tag{7.29}$$

Dominant crossover can be expressed by generalizing the idea of crossover masks so that the mask ranges from 1 to $\rho$, i.e. $M \in \{1,\ldots,\rho\}^d$. Further, let $p(i,h)$ be the $i^{th}$ parent on the history $h$ with order $\rho$, $p(i,h) = \text{parents}(h,\rho,k(h),K)_{-i}$. Also, let $M \otimes_i p(i,h)$ denote the point that is zero for each component where $M \neq i$ and equal to $p(i,h)$ when $M = i$. Then the crossover random variable can be redefined to

$$C(h,M) = \sum_{i=1}^{\rho} M \otimes_i p(i,h). \tag{7.30}$$

And then if $\mathbb{P}_M = \text{Uniform}\left(\{1,\ldots,\rho\}^d\right)$, dominant crossover is given by

$$\mathcal{DC}\langle\rho\rangle[h](A) = \sum_{z\in\{1,\ldots,\rho\}^d} \mathbb{1}_A(C(h,z))\,\mathbb{P}_M(z), \tag{7.31}$$

which is clearly a generalization of the masked crossover rule for higher orders. These retroactively generalized masked crossover rules will be used to state that the evolution strategies with the above recombination operators are evolutionary algorithms according to the formal definition above.

The final element is mutation. As with genetic algorithms, the type of mutation has to be customized to the search domain. Most often, evolution strategies are employed in $\mathbb{R}^d$, and Gaussian mutation is used. One of the main differences between genetic algorithms and evolution strategies, however, is the level of effort expended to adapt the mutation parameters over the course of optimization. Sometimes, evolution strategies apply adaptation at a global level, so that all elements of the next population are generated from the same mutation operator. Let $\Sigma : \mathcal{H}^{\mathcal{T}/K}_{X^K,Y^K} \to \mathbb{R}^{d\times d}$ be a matrix-valued function that takes a population history and performs history-specific computations to produce a covariance matrix for mutation (see e.g [25, 67]). Then the $(\mu/\rho,\lambda)$–ES with dominant crossover and Gaussian mutation is an element of $\mathcal{PBO}^{\lambda}_{\mathbb{R}^d,\mathbb{R}}$ and can be written as

$$\mathcal{ES}\langle\mu,\rho\rangle = \mathcal{ESS}\langle\mu\rangle \star \left(\triangleleft\mathcal{ESS}\langle\mu\rangle \star \left(\cdots \star \left(\triangleleft_{\rho-1}\mathcal{ESS}\langle\mu\rangle \star \mathcal{DC}\langle\rho\rangle\right)\right)\right) \star \mathcal{N}\langle\Sigma\rangle \tag{7.32}$$

Notice that this equation identifies evolution strategies as a formal evolutionary algorithm, since it is the convolution of a selection rule, a recombination operator with (generalized) masked crossover, and a mutation operator. Notably, this method fails to meet the formal definition of a genetic algorithm. The covariance function includes some information about previous populations, so this evolution strategies method is not population-Markov. Also, if intermediate crossover had been used, then the algorithm would only be a quasi-evolutionary algorithm by the definitions above. As mentioned, crossover is a relatively recent innovation in evolution strategies, and so this violation may be regarded as marking a early trend towards the quasi-evolutionary algorithms analyzed below.

The characterization in this subsection applies to general evolution strategies, but does not apply to CMA-ES. Apart from their reliance on a trajectory-specific covariance matrix $\Sigma(H(h))$, modern versions of CMA-ES have more in common with EDAs than with traditional evolution strategies, and these methods are discussed together in Section 7.3.3.

Rather than having a global adaptation procedure, evolution strategies often adapt mutation parameters with each point. In this case, the mutation parameters are carried along with the selected point. The point itself is mutated using the current mutation parameters, and then the mutation parameters are themselves mutated using a global adaptation scheme. Adaptive mutation of this form cannot be represented formally in $\mathcal{PBO}^K_{X,Y}$ due to the dependence on fitness values to update the mutation parameters. However, if the mutation parameters range over a space $\Theta$, then adaptive mutation can be described on an extended state space as an element of $\mathcal{PBO}^K_{X\times\Theta,Y}$, where $X\times\Theta$ is the Cartesian product. Any fitness function can be extended to this product space by defining $\tilde{f}(x,\theta) = f(x)$. Finally, since $X$ can be embedded in $X \times \Theta$, elements of $\mathcal{PBO}^K_{X,Y}$ can be projected trivially into $\mathcal{PBO}^K_{X\times\Theta,Y}$ using this embedding, and so adaptive evolutionary strategies can be compared directly with non-adaptive ones for theoretical and practical purposes. The issue of space extension will be explored further in Section 7.3.1.

### 7.2.6 The $(1+1)$–ES as the Norm Limit of Simulated Annealing

The standard theoretical example of an evolution strategy is the $(1+1)$–ES in $\mathbb{R}^d$. In terms of the previous subsection, this optimizer is represented by

$$\mathcal{ES}_+\langle 1,1\rangle[h] = \begin{cases} \mathcal{ESS}_+\langle 1\rangle[h] & \text{if}\, k(h)=1 \\ \mathcal{ESS}_+\langle 1\rangle \star \mathcal{N}\langle\sigma(h)^2 I\rangle[h] & \text{if}\, k(h)=2 \end{cases} \tag{7.33}$$

It operates on a population of size two. The first member of the population is always the current best solution, and the second member is a proposed replacement mutated from the current best. The function $\sigma(h)$ is a globally adaptive parameter that controls the standard deviation of mutation. The standard deviation is controlled so that it improves approximately 23% of the proposed solutions.

The description of the $(1+1)$–ES is reminiscent of simulated annealing. There is an accepted solution and a proposed solution. The proposed solution is generated from the acceptance probability using Gaussian variation. The standard deviation of the Gaussian distribution is controlled so that the running best has a 0.23 probability of being replaced. The only notable difference is that the $(1+1)$–ES lacks an explicit acceptance probability. In fact, by gradually reducing the cooling schedule, the $(1+1)$–ES can be shown to be the norm-limit of $\mathcal{SA}$. While this fact has always been obvious to the intuition, the formalisms proposed in this text allow it to be proven mathematically; without the formalism, the result could not be clearly achieved.

To obtain this result, redefine simulated annealing using the terminology of the last several sections, especially that of Section 7.1.5:

$$\mathcal{SA}\langle T\rangle[h] = \begin{cases} \mathcal{B}\langle A(h^Y_{-2}, h^Y_{-1}, T(|h|/2)), h^X_{-1}, h^X_{-2}\rangle & \text{if}\, k(h)=1 \\ \mathcal{N}\langle\sigma(h)I\rangle & \text{if}\, k(h)=2, \end{cases} \tag{7.34}$$

where $T:\mathbb{N}\to\mathbb{R}$ is a cooling schedule for simulated annealing.

**Theorem 7.1.** *Suppose $T_n$ is a cooling schedule such that $T_n \to 0$ as $n\to\infty$. Then $||\mathcal{SA}\langle T_n\rangle - \mathcal{ES}_+\langle 1,1\rangle|| \to 0$. That is, the $(1+1)$–ES is the limit of simulated annealing using the supremum norm for search generators, assuming both use the same initial distribution, i.e., $\mathcal{S}\langle T\rangle[\emptyset] = \mathcal{ES}_+\langle 1,1\rangle[\emptyset]$.*

*Proof.* Assume that the initial population is generated from the same initial distribution. Fix the history $h$ with $|h|>2$ to represent any non-initial history. First, note that if $k(h)=2$, then $\mathcal{SA}\langle T_n\rangle[h] = \mathcal{ES}_+\langle 1,1\rangle[h]$ for all $n$ by the definitions. Thus the interesting case occurs when $k(h) = 1$. Consider the sequence of acceptance probabilities

$$A_n(h) = A(h^Y_{-2}, h^Y_{-1}, T_n(|h|/2)).$$

As $T_n \to 0$, either $A_n(h) \to 0$ if $h^Y_{-2} > h^Y_{-1}$ or $A_n(h) \to 1$ otherwise. Let

$$\tilde{\mathcal{B}}_n[h] = \mathcal{B}\left\langle A_n(h), h^X_{-1}, h^Y_{-2}\right\rangle[h],$$

When $k(h) = 1$, then $\mathcal{SA}\langle T_n\rangle = \tilde{\mathcal{B}}_n$, and for $A$ measurable, ,

$$\mathcal{ES}_+\langle 1\rangle[h](A) = \mathcal{ESS}_+\langle 1\rangle[h](A) = \begin{cases} \mathbb{1}_A(h^X_{-1}) & \text{if} h^Y_{-1} \geq h^Y_{-2} \\ \mathbb{1}_A(h^X_{-2}) & \text{otherwise,} \end{cases}$$

where $\mathbb{1}_A$ is the indicator function for the set $A$.

If neither $h^X_{-1}$ nor $h^X_{-2}$ are in $A$, then

$$\mathcal{SA}\langle T_n\rangle[h](A) = \mathcal{ES}_+\langle 1,1\rangle[h](A) = 0.$$

If both are in $A$, then $\mathcal{SA}\langle T_n\rangle[h](A) = \mathcal{ES}_+\langle 1,1\rangle[h](A) = 1$. Thus the only possible differences are realized when $A$ contains only one of the points. It suffices to let $A$ contain only $h^X_{-1}$ and not $h^X_{-2}$, since convergence on $A$ implies convergence on $X \setminus A$ as a consequence. There are now two cases.

*Case 1*: $h^Y_{-1} \geq h^Y_{-2}$. In this case, $A_n(h) \to 1$, so $\tilde{B}_n \to 1$, since $B$ contains $h^X_{-1}$. So

$$|\mathcal{SA}\langle T_n\rangle[h](A) - \mathcal{ES}_+\langle 1,1\rangle[h](A)| = \left|\tilde{B}_n[h](A) - 1\right| \to 0. \tag{7.35}$$

*Case 2*: $h^Y_{-1}) < h^Y_{-2}$. In this case, $A_n(h) \to 0$, so $\tilde{B}_n \to 0$, since $A$ does not contain $h^X_{-2}$. So

$$|\mathcal{SA}\langle T_n\rangle[h](A) - \mathcal{ES}_+\langle 1,1\rangle[h](A)| = \tilde{B}_n[h](A) \to 0. \tag{7.36}$$

Thus in either case, since $h$ and $A$ were arbitrary, then for $\varepsilon > 0$ there exist $h^*$ and $A^*$ such that for $n$ large,

$$\begin{aligned} ||\mathcal{SA}\langle T_n\rangle - \mathcal{ES}_+\langle 1,1\rangle|| &< |\mathcal{SA}\langle T_n\rangle[h^*](A^*) - \mathcal{ES}_+\langle 1,1\rangle[h^*](A^*)| + \varepsilon \\ &< 2\varepsilon, \end{aligned} \tag{7.37}$$

and so the proof holds.

This simple theorem justifies the approach of unifying population-based optimizers through the formalism of search generators because it allows a rigorous comparison of two typologically distinct optimization methods, one a Monte Carlo method and the other an evolutionary method. It also helps to identify opportunities for new algorithms, as is done with evolutionary annealing in Chapter 14.

## 7.3 Quasi-Evolutionary Algorithms

The previous section explored the relationship of evolutionary algorithms to the proposed formalism. This section examines how some of the natural computation and quasi-evolutionary methods from Chapter 2 can be formalized.

### 7.3.1 Particle Swarm Optimization

Particle Swarm Optimization (PSO) is used to search rectangular regions within $\mathbb{R}^d$ [88, 45]. To generalize the algorithm somewhat, the formalization will assume that the search domain $X$ is a closed rectangular subset of a vector space. Optimization is performed with respect to a fitness function $f : X \to Y$ where $Y$ is totally ordered. Arbitrary bounded subsets of a vector space (say, $\Omega \subseteq X$) can be searched in this way through the use of a feasibility region by restricting the fitness function so that $f(x) = \infty$ for $x \notin \Omega$ (see [92]).

PSO maintains a population of particles metaphorically representing a flock of birds. Each particle has a position and a velocity in the search space, and the particles are ranked by fitness. As in differential evolution, the population may be viewed as a sequence of slots. PSO stores the best overall solution (called the *global best*) as well as the best solution that has occurred at each slot over all prior populations (the *local best*). The position and velocity are often initialized uniformly at random over the search space. At each generation, the velocity of each particle is shifted towards the position of both the global best and the local best. The position of the particle is then updated according to the velocity. For the $k^{th}$ slot of the population, the update rules for the velocity $v^{n+1}$ and the position $x^{n+1}$ after the $n^{th}$ generation are

$$v_k^{n+1} = \omega v_k^n + \phi_p U_{p,k}^n x_{p,k}^n + \phi_g U_{g,k}^n x_g^n \tag{7.38}$$

$$x_k^{n+1} = x_k^n + v_k^{n+1}, \tag{7.39}$$

where $\omega$ decays the current velocity; $x_{p,k}^n$ and $x_g^n$ are the local and global best positions at the $n^{th}$ generation, respectively; $\phi_p$ and $\phi_g$ control the sensitivity to the local best and global best solutions; and $U_{p,k}^n, U_{g,k}^n$ are uniform random variables on $[0,1]$. If the velocity $v_k^{n+1}$ will cause the particle to exit the search space, then the velocity can be clipped so that the position $x_k^{n+1}$ is on the boundary of $X$. Another approach is to allow particles to exit the rectangular boundary, but to draw them back in by using the feasibility modification to the objective function. Feasibility guarantees that $x_g^n$ and $x_{p,k}^n$ are inside of $X$, and as long as $|\omega| < 1$, the particle will eventually be pulled back into the search space.

The three parameters $\omega$, $\phi_p$ and $\phi_g$ control the algorithm and are arbitrary. Studies have shown that if $\phi_p = 0$, then the optimal settings for $\omega$ and $\phi_g$ over an average of benchmarks occur when $\omega \in [-0.6, 0]$ and $\phi_g \in [2, 4]$ (see [121]). The use of negative values for the velocity decay $\omega$ wreaks havoc with the flock analogy but is nonetheless effective.

To formalize PSO, the search space must be extended to include the velocities. Otherwise, there is no way to infer the velocity from the particle positions, since the velocity is initialized randomly. To this end, let the extended search space by $X^2 = X \times X$ with an altered objective $\tilde{f}(x,v) = f(x)$. If the initial velocity is set to zero, as is sometimes done, then the velocities may be inferred from the particle trajectories and this extension is not necessary.

Let $\mathcal{PSO}\left\langle \omega, \phi_p, \phi_g, K \right\rangle$ be an instantiation of *PSO* with the specified parameters and population size $K$. The optimizer $\mathcal{PSO}$ will be defined by constructing a process $Z$ on $Y$ and then setting $\mathcal{PSO}$ to be the distribution of this process so that $Z \sim \mathcal{PSO}\left\langle \omega, \phi_p, \phi_g, K \right\rangle$. Let the vectors $u$ and $\ell$ represent the upper and lower bounds for $X$. Use the subscript $n,k$ to represent the state of the $k^{th}$ slot in the $n^{th}$ population, i.e. $Z_{n,k} = Z_{(n-1)K+k+1}$. Initialize the process so that $Z_{0,k} \sim \text{Uniform}\left([\ell,u]^2\right)$. At every step, the process consists of a position and a velocity component, $Z_{n,k} = \left(X_{n,k}, V_{n,k}\right)$. Let the update rules in Equation 7.38 govern the position variable $X_{n,k}$ and the velocity $V_{n,k}$, with variables $X^n_{p,k}$ and $X^n_g$ to represent the local and global best. Then $\mathcal{PSO}$ can be represented as

$$\mathcal{PSO}\left\langle \omega, \phi_p, \phi_g, K \right\rangle [h]\,(A) = \mathbb{P}\left(Z_{|H(t)|,(|h| \bmod K)} \in A \mid Z_n = h_n \, \forall n < |h|\right). \quad (7.40)$$

$\mathcal{PSO}$ can be reduced to the search space $X$ by marginalizing out the velocities when required in order to compare PSO with other methods.

Formally, PSO is a quasi-evolutionary algorithm by the definition above with a vacuous mutation operator. PSO can be characterized as using three selection rules, one that selects the previous position and velocity, one that selects the global best, and one that selects the local best. The crossover rule then applies Equation 7.38 to these three items to generate the recombined point. Since the recombined point is also the output, the mutation operator must be vacuous. Thus PSO introduces a unique and complex crossover rule, but is formally a quasi-evolutionary algorithm. It is not a genetic algorithm, since there is no masked crossover rule that implements PSO. It is also important to note that PSO is not population-Markov and hence not an evolutionary algorithm as defined formally above, since it depends on the global and local best solutions at each time step.

### *7.3.2 Differential Evolution*

Like PSO, differential evolution is designed to search a rectangular region in $\mathbb{R}^d$. Once again, this method can be easily generalized to any search domain that is a closed rectangular subset of a vector space. Suppose that the search space $X$ is a subset of a finite-dimensional vector space defined by a closed rectangular region with upper and lower boundaries $u$ and $\ell$.

As mentioned in Chapter 2, Ghosh et al. [59] recently proved that differential evolution converges to the true global optimum on functions with two continuous derivatives. The techniques employed by Ghosh et al. are similar to the techniques employed in this book, and thus their result provides further evidence that theorems about stochastic optimizers can be stated and proven in a suitable formalization.

Differential evolution passes through three phases: selection, mutation, and crossover. Notably, crossover in differential evolution crosses a mutated vector with the selected vector, and so differential evolution does not meet the formal definition of an evolutionary algorithm.

Differential evolution uses a form of selection that will be termed *local best selection*. This method selects the best member of the trajectory so far along each component. Let $b_k(h) \in X$ be the best solution found so far at the $k^{th}$ slot along the history $h$,

$$b_k(h) = \{x \mid \text{there exists } y \text{ s.t. } (x,y) = H(h)_{n,k} \text{ and } y < w \text{ for all } (z,w) = H(h)_{m,k}\},$$

where the subscripts indicate the $k^{th}$ member of the $n^{th}$ (or $m^{th}$) population. Local best selection always selects $b_k(t)$ for the $k^{th}$ member of the population, *i.e.*, with population size $K$ it is the point distribution given by

$$\mathcal{LBS}\left\langle K\right\rangle[h](A) = \mathbb{1}_A\left(b_{k(h)}(h)\right). \tag{7.41}$$

Mutation in differential evolution adds the difference between two members of a population to a third member of the population. The vector to which the difference is added is termed the *donor vector*, and $K$ donor vectors are chosen from among the local best vectors in one of three ways. In random mutation, each donor vector may be a chosen uniformly at random from among the local best vectors. In target-to-best mutation, every donor vector may be fixed as the global best vector (denoted as $h^*$ for a history $h$). In best mutation, the $k^{th}$ donor vector may be chosen as a particular point along the line from the $k^{th}$ local best to the global best vector. Once the donor vector is selected, then two other distinct vectors are chosen randomly from among the local best vectors, and the donor vector is moved in direction of their difference, multiplied by a weighting factor denoted $F$. These possibilities are expressed as

$$Y_{\text{rand},k}(h) = b_{R_1}(h) + F\left(b_{R_2}(h) - b_{R_3}(h)\right) \tag{7.42}$$
$$Y_{\text{target},k}(h) = b_k(h) + F\left(h^* - b_k(h)\right) + F\left(b_{R_1}(h) - b_{R_2}(h)\right) \tag{7.43}$$
$$Y_{\text{best},k}(h) = h^* + F\left(b_{R_1}(h) - b_{R_2}(h)\right), \tag{7.44}$$

where $R_1$, $R_2$, and $R_3$ are distinct uniformly random indices between 1 and $d$, inclusive, and $d$ is the dimension of the space. Some versions also add a second difference chosen randomly from among the remaining local best vectors. Let $Y_*$ stand for any one of $Y_{\text{rand}}$, $Y_{\text{target}}$, or $Y_{\text{best}}$. Then mutation in differential evolution can be represented by

$$\mathcal{DM}_*\left\langle F,K\right\rangle[h](A) = \mathbb{1}_A\left(Y_{*,k(h)}(h)\right). \tag{7.45}$$

Differential evolution recombines the local best vectors with the mutated vectors to create the next population. The two crossover strategies are termed *binomial* and *exponential*. Both schemes can be described using crossover masks, and each is parameterized by a *crossover rate* denoted by $CR$. Binomial crossover is so named because each mutated component is selected as a Bernoulli trial with probability $CR$ (*i.e.*, $\mathbb{P}(M_k = 0) = CR$), recalling that $M_k = 0$ implies that the "mother" (the mutated vector) is chosen. However, if $M_k = 1$ for all $k$, the sample is rejected, so that at least one mutated component is included. Exponential crossover copies a subsequence of adjacent components from the mutated vector onto the local best vector. A random index $I$ in $1,\ldots,d$ is chosen along with a random length $L$ also in $1,\ldots,d$. Then $M_k = 0$ if $k \geq I$ and $k < I + L$, applying modular arithmetic as necessary.

Recalling the crossover mask rule $\mathcal{C}\langle \mathbb{P}_M \rangle$ and letting $M[CR]$ be the selected crossover mask, differential evolution is thus given by

$$\mathcal{DE}_* \langle F, CR, K \rangle = \mathcal{LBS}\langle K \rangle \star \left( \mathcal{DM}_* \langle F, K \rangle \star \mathcal{C}\left\langle \mathbb{P}_{M[CR]} \right\rangle \right). \tag{7.46}$$

As with PSO, DE is also a quasi-evolutionary algorithm according to the formal definition in this chapter. The convolution $\mathcal{DM}_* \star \mathcal{C}$ fits the definition of a fourth-order crossover rule, provided that the vectors $t^*$, $b_k$, $b_{R_1}$, $b_{R_2}$, and $b_{R_3}$ are selected by a selection rule. The mutation operator for DE is vacuous, as it is for PSO. DE is not formally an evolutionary algorithm. It does employ a masked crossover rule $\mathcal{C}$, but its entire recombination operator $\mathcal{DM}_* \star \mathcal{C}$ is more complex and does not fit the definition of a masked crossover rule. Also, DE is not population-Markov due to its use of the local best solutions. Indeed, PSO and DE share quite a few structural similarities, and it is interesting that the proposed formalism draws them out.

### 7.3.3 Parameterized Quasi-Evolutionary Methods

Both Estimation of Distribution Algorithms and Natural Evolution Strategies have an explicit representation as a parameterized probabilistic model. These methods thus fit into the formalism naturally. Each of them is described by a distribution $\pi(dx \mid \theta)$ where $\theta$ is drawn from a parameter space. The parameters $\theta$ are reestimated once per generation based on the population history, so that one may write $\theta = \hat{\theta}(H(h))$ to reflect the dependence of the parameters on the prior populations and their evaluations. Then all EDA and NES instances can be represented by

$$\mathcal{G}\langle \pi, \theta \rangle [h](A) = \int_A \pi\left(dx \mid \hat{\theta}(H(h))\right) \tag{7.47}$$

for an appropriate model $\pi$ and a parameter estimation function $\hat{\theta}$.

Also, both EDAs and NES are quasi-evolutionary algorithms according to the formalism above. EDAs use truncation selection to choose the best $M$ members of the prior population. These points are then used to build a probabilistic model. This model depends only on the selected points and not on their objective value, so the construction and sampling of the model are value-agnostic. Therefore, the model-sampling process can also be described formally as a crossover rule of order $M$. Thus the EDA consists of truncation selection, recombination through model-sampling, and vacuous mutation.

NES and CMA-ES use truncation selection in the same way as traditional evolution strategies. Just like EDAs, they then build a probabilistic model from the selected points. However, these methods differ from EDAs in that they maintain a set of global parameters that are adjusted to follow the gradient of certain meta-parameters. Thus NES implementations are evolutionary algorithms for the same reason that EDAs are, but EDAs are population-Markov, unlike NES. Neither NES nor EDAs are strict evolutionary algorithms, since the model-building process cannot be implemented as a masked crossover rule.

### *7.3.4 Non-Quasi-Evolutionary Algorithms*

Under analysis, each of the quasi-evolutionary algorithms studied in this section has formally satisfied the definition of a quasi-evolutionary algorithm proposed in this chapter. This result is not surprising, since all of these methods were ultimately inspired by previous work on evolutionary algorithms. However, it does beg the question of whether there exist search generators that are not formally quasi-evolutionary algorithms for some population size $K$, *e.g.*, $K = 1$. Under further consideration, simulated annealing, generating set search, and even Nelder-Mead[4] are also formally quasi-evolutionary algorithms by the criterion above.

At least one optimization method from Chapter 2 is not a quasi-evolutionary algorithm even by this definition. Gradient-based methods with estimated gradients use the same search domain and value space as the population-based optimizers described in this chapter, but they depend on the raw objective values of more than one point. Since a selection rule can only choose one previously observed point as a parent, it cannot encode the estimated gradient, which depends on at least two points. Since a crossover rule and mutation operator must be value-agnostic, they cannot compute the gradient either. Thus optimizers that are not quasi-evolutionary algorithms do exist under this definition.

It is worthwhile to consider how to distinguish between algorithms that fall traditionally within the ambit of evolutionary computation from those that do not. If the population size $K$ were required to be greater than one, some versions of generating set search would still be included, since they can generate a fixed number of points that can be evaluated in parallel. Further, the $(1+1)$–ES, a traditional evolutionary algorithm, would be excluded under a certain interpretation and included under an-

[4] It is difficult to cast Nelder-Mead into the format of a selection rule, recombination and mutation. To see how it might be done in $d+1$ dimensions, $2d+2$ selection rules can be used, with $d+1$ rules selecting the simplex endpoints in some fixed order and an additional $d+1$ points selecting the same endpoints in rank order. The if-then rules of Nelder-Mead can then be implemented within a value-agnostic crossover rule.

other. A quasi-evolutionary algorithm could be defined as factorial or homogeneous, but then one or more evolutionary algorithms would be excluded.

When defining evolutionary algorithms, only masked crossover rules were allowed, which reflect traditional intuitions about evolutionary computation. As a positive effect, PSO, DE, EDAs, and NES fail to meet the formal definition of an evolutionary algorithm under this assumption. But intermediate crossover is excluded by this definition and along with it some evolution strategies and even certain forms of neuroevolution, such as NEAT. Yet even this definition is more inclusive than some might prefer, since other optimizers not traditionally included within evolutionary computation could be expressed as evolutionary algorithms without crossover, such as simulated annealing.

In the final analysis, the attempt to draw a line between which optimization methods are *evolutionary* and which are not reveals more than anything that this distinction is a false one. There is no formal difference between evolutionary search methods and other stochastic search methods, such as Monte Carlo search. It is not important to distinguish which algorithms are and are not evolutionary or quasi-evolutionary algorithms. A formal analysis such as the one undertaken in this chapter actually serves to undermine such categorical schemes, as the reasoning in the prior paragraphs show. In fact, it is a major benefit that the formal setting removes the ability to distinguish certain methods categorically, because it reveals important similarities among these methods along several dimensions that would not be evident otherwise.

## 7.4 Conclusion

Population-based optimizers were reviewed in this chapter to show how these optimizers fit into the formalism of search generators. This exercise demonstrated that methods as diverse as genetic algorithms, evolution strategies, particle swarm optimization, differential evolution, and estimation of distributions algorithms all fit within the proposed framework for formal analysis. In addition, some of the advantages of this approach were demonstrated by proving that the $(1+1)$–ES method is the limit of simulated annealing with respect to the optimizer norm.

This discussion has proposed a definition for evolutionary and quasi-evolutionary algorithms that requires selection, recombination, and mutation, all explicitly defined. These definitions captured some meaningful distinctions among search methods, but it was not possible to draw a meaningful categorical distinction between evolutionary, quasi-evolutionary algorithms and non-evolutionary methods such as Nelder-Mead and generating set search. Indeed, it is not clear that such a distinction is inherently useful, since the formalism of search generators provides a single setting within which all of these algorithms can be compared. It is a confirmation of this approach that several algorithms proposed with distinct motives and inspirations bear structural similarities to each other that can be made clear using the tools provided by this analysis.

The following chapters will not treat population-based optimizers separately from other optimizers, but the equations and formulae in this chapter demonstrate effectively that the subsequent results apply equally to evolutionary, quasi-evolutionary, and other optimization methods. The next chapter discusses the continuity of various optimizers, subsequent chapters introduce several general-purpose methods defined explicitly using the formal framework of this and preceding chapters. These results begin an analysis that will culminate in extended No Free Lunch theorems for optimization and a characterization of high quality optimization routines as information maximizers in Chapters 12 and 13.

# Chapter 8
# Continuity of Search Generators

The goal of an analysis of search and optimization is ultimately to draw conclusions about the performance of different methods. Up to this point, the spatial characteristics of search generators have been analyzed, and several optimization methods have been formalized as search generators. Starting with this chapter, analytic tools will now be developed to address these larger question of performance. This chapter focuses on continuity, which was defined in Chapter 5 as a topological concept. Continuity considers the question of whether it is reasonable to assume that the outputs of a function will be similar when the inputs are similar. The continuity of search generators can be used to demonstrate continuity of performance in many cases, and as it turns out, most non-deterministic optimizers are continuous on a broad range of history traces.

## 8.1 Background and Motivation

Search generators are defined as functions from history traces to finite signed measures over the search space. In this context, continuity answers a simple question:

- Given similar histories, will an optimizer $\mathcal{G}$ choose similar search points?

A history trace $h$ can be broken into two parts: the search history $h_X$ and the evaluation history $h_Y$. If a search generator $\mathcal{G}$ is continuous when $h_Y$ is held constant and $h_X$ varies, it will be called continuous *in trajectories*, meaning trajectories through the search domain. If it continuous when $h_X$ is held constant and $h_Y$ varies, it will be called continuous *in objectives*, referring to the presumed objective that generates values in the value space. If it is continuous when both vary, then it is continuous with respect to the norm topology on $\mathfrak{G}_{X,Y}$. When a search generator is simply called continuous without qualification, it is this latter concept of continuity with respect to the norm that is intended. These concepts will be made precise below.



A. J. Lockett, *General-Purpose Optimization Through Information Maximization*,
Natural Computing Series, https://doi.org/10.1007/978-3-662-62007-6_8

### 8.1.1 Norm-Continuity of Search Generators

Continuity is a central topological concept. As discussed in Chapter 5, a function is continuous if the pre-image of every open set is open. An open set is nothing more than a set that is declared to be open by the topology. A topology is in fact defined by the sets that it declares to be open, and the open sets are arbitrary within certain consistency constraints. Thus the continuity of a function is always continuity with respect to a particular topology on each of the input and output spaces. For the most part, the topologies in this chapter are metric topologies, for which the open sets are defined by a distance metric as defined in Section 5.2.1.

Each search generator $\mathcal{G}$ with search domain $X$ and value space $Y$ is a function from the space of history traces $\mathcal{H}_{X,Y}^{\mathcal{T}}$ to the Banach space of finite signed measures, $\mathcal{M}_X$. The concept of continuity with respect to trajectories and objectives above will require us to topologize these two spaces. Without a concept of what similar histories are we cannot say what it means for two search generators to produce similar search points on similar history inputs. As a Banach space, $\mathcal{M}_X$ has the norm topology determined by the total variation norm as discussed in Section 5.2.3. A natural topology on the set of history prefixes $\mathcal{H}_{X,Y}^{\mathcal{T}}$ can also be easily defined using the methods of Section 5.2.2, specifically the product topology and the disjoint union topology.

Recall that the search domain $X$ is generally assumed to be a Hausdorff[1] topological space with topology $\tau_X$ that generates a measurable space $(X, \mathcal{B}_{\tau_X})$ through the Borel $\sigma$-algebra on $X$. We will now also assume that $Y$ is also a Hausdorff topological space with topology $\tau_Y$. Recall the totally ordered temporal index set $\mathcal{T}$ that provides the indices of iteration for an iterated search method. The following spaces are all topological spaces under the product topology:

$$X \times Y, \quad X^{\mathcal{T}}, \quad Y^{\mathcal{T}}, \quad (X \times Y)^{\mathcal{T}}.$$

The relevant finite products of open sets in $\tau_X$ and $\tau_Y$ form a topological base for these spaces. Finally, the space of histories was defined as a union of products indexed by temporal indices from $\mathcal{T}$. Rather than repeat this definition, we give a new one. Recognizing that each of the temporal products is disjoint from the others, the following history prefix spaces can be defined as disjoint unions:

$$\mathcal{H}_X^{\mathcal{T}} = \emptyset \amalg \coprod_{t \in \mathcal{T}} X^{[t]}, \quad \mathcal{H}_Y^{\mathcal{T}} = \emptyset \amalg \coprod_{t \in \mathcal{T}} Y^{[t]}, \quad \text{and} \quad \mathcal{H}_{X,Y}^{\mathcal{T}} = \emptyset \amalg \coprod_{t \in \mathcal{T}} (X \times Y)^{[t]},$$

recalling that $[t] = \{s \in \mathcal{T} \mid s \leq t\}$. Here $\mathcal{H}_X^{\mathcal{T}}$ contains search histories, $\mathcal{H}_Y^{\mathcal{T}}$ contains evaluation histories, and $\mathcal{H}_{X,Y}^{\mathcal{T}}$ is the full history trace with both search points and values. These definitions are equivalent to those definitions given before except that they come with an automatic topology ultimately built from $\tau_X$ and $\tau_Y$ using the product topology and the disjoint union topology in succession.

---

[1] Definition 5.4

Consider what it means for histories to be “close”. Let $h \in \mathcal{H}_X^{\mathcal{T}}$ be any search domain history, and let $(h_n)_{n\in\mathbb{N}} \subseteq \mathcal{H}_X^{\mathcal{T}}$ be a sequence converging to $h$, *i.e.*, $h_n \to h$. Then $X^{|h|}$ is an open neighborhood of $h$, whence for all sufficiently large $n$, $h_n \in X^{|h|}$ or, stated otherwise, the histories $h_n$ and $h$ eventually have the same length: $|h_n| = |h|$. This conclusion follows from the structure of the disjoint union topology. Secondly, if the length $|h|$ is finite, then for any open collection of open sets $(A_t)_{t\leq|h|}$ with $A_t \subseteq X$ such that the component $h_t \in A_t$, where $h_t$ is the $t^{th}$ component of $h$, it is also true that $h_{n,t} \in A_t$ for $n$ sufficiently large, where $h_{n,t}$ is the $t^{th}$ component of $h_n$. This observation is the application of the product topology on $X^{|h|}$. Finally, set $A_t$ is a member of the topology $\tau_X$ on the search domain. In this way, the concept of neighborhood for the history spaces is built off of the concept of neighborhoods in $X$ and $Y$.

If $X$ and $Y$ are metric spaces and $\mathcal{T}$ is countable (*i.e.*, $\mathcal{T} = \mathbb{N}$), then the history spaces are all metrizable. If $d_X$ is a metric on $X$ and $d_Y$ is a metric on $Y$, then $d_{X,Y}$ defined by

$$d_{X,Y}((x,y),(x',y')) = d_X(x,x') + d_Y(y,y')$$

is a metric on $X \times Y$. Then for $*$ being any one of $X$, $Y$, or $X,Y$,

$$\rho_*(h,h') = \Big||h| - |h'|\Big| + \sum_{i\leq|h|\wedge|h'|} \frac{d_*(h_i,h'_i)}{1+d_*(h,h')},$$

is a metric on $\mathcal{H}_*^{\mathcal{T}}$ where $\wedge$ is the minimum operator so that $|h| \wedge |h'|$ is the length of the shorter history. The purpose of the fraction $d_* 1 + d_*$ is to squeeze $d_*$ to the interval $[0,1)$. Due to this squeezing, it is true for this metric space as it was in the preceding paragraph that if $h_n \to h$, then $h_n$ eventually has $|h_n| = |h|$. In fact, $d(h_n,h) < 1$ is sufficient to imply that $|h_n| = |h|$.

The goal now is to obtain a definition for what makes a search generator continuous in trajectories and in objectives. To this end, we first note that there is a homeomorphism (a bijection continuous in both directions)

$$\kappa : \mathcal{H}_X^{\mathcal{T}} \times \mathcal{H}_Y^{\mathcal{T}} \to \mathcal{H}_{X,Y}^{\mathcal{T}}$$

that combines search domain histories with evaluation histories to produce full history traces. For all history traces $h$, this homeomorphism is given by $\kappa(h_X,h_Y) = h$. Its inverse is just $\kappa^{-1}(h) = (h_X,h_Y)$, and its continuity is due to the continuity of index projections under the product topology.

For any search generator $\mathcal{G}$, the history can be split into a search domain history and a domain history using $\kappa$, and either argument may be fixed to create two functions

$$\mathcal{G}_{h_Y} : h_X \mapsto \mathcal{G}[\kappa(h_X,h_Y)] \quad \text{and} \quad \mathcal{G}_{h_X} : h_Y \mapsto \mathcal{G}[\kappa(h_X,h_Y)].$$

With this notation the two desired versions of restricted continuity can be defined by holding either the search domain history or the evaluation history fixed.

**Definition 8.1 (Continuity of Trajectories).** A search generator $\mathcal{G} \in \mathfrak{G}_{X,Y}$ is *continuous in trajectories* if the function $\mathcal{G}_{h_Y}$ is continuous for all $h_Y \in \mathcal{H}_Y^{\mathcal{T}}$.

**Definition 8.2 (Continuity of Objectives).** A search generator $\mathcal{G} \in \mathfrak{G}_{X,Y}$ is *continuous in objectives* if the function $\mathcal{G}_{h_X}$ is continuous for all $h_X \in \mathcal{H}_X^{\mathcal{T}}$.

**Proposition 8.1.** *Every value-agnostic search generator is continuous in objectives.*

*Proof.* $\mathcal{G}_{h_X}$ is constant, so the pre-image of any open set is $\mathcal{H}_Y^{\mathcal{T}}$.

**Proposition 8.2.** *A search generator $\mathcal{G} \in \mathfrak{G}_{X,Y}$ is continuous if and only if it is continuous in both trajectories and objectives.*

*Proof.* Suppose $\mathcal{G}$ is continuous. Then for any sequence of search domain histories $(h_n^X)_{n\in\mathbb{N}}$ with $h_n^X \to h_X \in \mathcal{H}_X^{\mathcal{T}}$, it is clear that for any evaluation history $h_Y \in \mathcal{H}_Y^{\mathcal{T}}$ that $\kappa(h_n^X, h_Y) \to \kappa(h_X, h_Y)$. Therefore *a fortiori* $\mathcal{G}_{h_Y}[h_n^X] \to \mathcal{G}_{h_Y}[h_X]$, and $\mathcal{G}_{h_Y}$ is continuous. Similarly, for evaluation histories $(h_n^Y)_{n\in\mathbb{N}}$ with $h_n^Y \to h_Y \in \mathcal{H}_Y^{\mathcal{T}}$, it follows that $\mathcal{G}_{h_X}$ is continuous for all $h_X \in \mathcal{H}_X^{\mathcal{T}}$.

Suppose $\mathcal{G}$ is continuous in both trajectories and objectives. Now suppose that $(h_n)_{n\in\mathbb{N}} \subseteq \mathcal{H}_{X,Y}^{\mathcal{T}}$ with $h_n \to h \in \mathcal{H}_{X,Y}^{\mathcal{T}}$. The proof will follow if $\|\mathcal{G}[h_n] - \mathcal{G}[h]\| \to 0$. This result is accomplished by separating the trajectories from objectives using intermediate fixed points:

$$\begin{aligned}
\|\mathcal{G}[h_n] - \mathcal{G}[h]\| &= \|\mathcal{G}[\kappa(h_n^X, h_n^Y)] - \mathcal{G}[\kappa(h_X, h_Y)]\| \\
&\leq \|\mathcal{G}[\kappa(h_n^X, h_n^Y)] - \mathcal{G}[\kappa(h_n^X, h_Y)]\| \\
&\quad + \quad \|\mathcal{G}[\kappa(h_n^X, h_Y)] - \mathcal{G}[\kappa(h_X, h_Y)\|
\end{aligned}$$

Examining the right-hand side, the first term converges to zero because $h_n^X$ is fixed and $\mathcal{G}$ is continuous in objectives. The second term converges to zero because $h_Y$ is fixed and $\mathcal{G}$ is continuous in trajectories. Therefore the left-hand side converges to zero as well, and the proof is complete.

In the second half of the proof for the preceding proposition, an intermediate value was injected through the triangle inequality for norms. In this case, the intermediate value was $\mathcal{G}[\kappa(h_n^X, h_Y)]$. This injection is a standard proof strategy and will be reused numerous times in what follows.

If a search generator is continuous in objectives, then it can be expected to perform similarly on similar search problems. If a search generator is continuous in trajectories, then it can be expected to make similar decisions on similar trajectories. If a search generator is continuous generally, then it can be expected to make similar decisions on similar histories generally. Because several of the subsequent theorems about optimization performance are formulated to apply only to continuous search generators, it is important to know which optimizers are continuous. For example, in Chapter 10, the continuity of performance criteria will be studied, and it will be seen that performance criteria are continuous on continuous optimizers, meaning that continuous optimizers can be expected to have similar performance on similar problems. The remainder of this chapter develops definitions and theorems that establish the conditions under which search generators for common optimization methods such as evolutionary algorithms, differential evolution, and stochastic gradient descent are continuous. We begin with a study of continuity in deterministic optimizers.

### 8.1.2 Continuity of Deterministic Search Generators

Deterministic probability-valued functionals were introduced in Section 6.2.2. In this section, more tools for working with deterministic search generators will be provided, along with a new definition of continuity that is more convenient for working with deterministic search generators.

Determinism in this book is equated with degeneracy of probability distributions, as seen in Definition 6.9. Unlike many other properties discussed so far, determinism does not induce a vector subspace of measure-valued functionals. It is easy to see why. First of all, the zero measure is not degenerate, since the existence of a unique point conferring probability mass implies that the total measure of the space is non-zero. Secondly, if $\mathbb{P}_x$ and $\mathbb{P}_y$ are degenerate probability distributions such that $\mathbb{P}_x(A) = \mathbb{1}_A(x)$ and $\mathbb{P}_y(A) = \mathbb{1}_A(y)$ for $x, y \in X$ with $x \neq y$, then the sum $\mathbb{P}_x + \mathbb{P}_y$ cannot be degenerate because it has no point satisfying the degeneracy condition. If $y \in A$ but $x \notin A$, then $(\mathbb{P}_x + \mathbb{P}_y)(A) = 1$ and vice versa, so neither $x$ nor $y$ can be the unique point of degeneracy. But if both $x \notin A$ and $y \notin A$, then $(\mathbb{P}_x + \mathbb{P}_y)(A) = 0$. So no other points can be the unique point of degeneracy either. Consequently, degeneracy is not closed under vector addition, so it does not form a vector subspace.

What happens is that the set of probability-valued functionals blends deterministic trajectories by switching probabilistically between them. Because deterministic measure-valued functionals are degenerate at each trajectory, they cannot switch trajectories. Focusing in on deterministic search generators removes the fabric of continuity. Although it will not be explored in this text, there is another topology in which deterministic search generators are continuous. This topology is based on the Wasserstein distance, also known as the Earth Mover distance. This topic is an interesting direction for future work not pursued here.

Perhaps unsurprisingly, most deterministic search generators are discontinuous in the norm topology in general. Given a deterministic proper search generator $\mathcal{D}$ and a history trace $h$, the search point proposed by $\mathcal{D}$ on $h$ is the unique point of degeneracy.[2] Given a particular history trace $h$, denote the unique point of degeneracy by $\xi(h)$. Then $\mathcal{D}[h](A) = \mathbb{1}_A(\xi(h))$ by Proposition 6.9 since $\mathcal{D}$ is proper. Choose a distinct history trace $h' \neq h$, and then

[2] Note that we may assume the measurability of the singletons in this case, because the $\sigma$-algebra is assumed to be Borel over $X$ and the topology of $X$ is Hausdorff. Without this measurability, the unique point would still exist by the definition, but it might not be possible to observe it.

$$\left|\mathcal{D}[h]-\mathcal{D}[h']\right|(X) = \int_X \left|\mathcal{D}[h]-\mathcal{D}[h']\right|(dx)$$
$$= \begin{cases} 0 \text{ if } \xi(h)=\xi(h') \text{ and} \\ 2 \text{ if } \xi(h)\neq\xi(h'). \end{cases}$$

But now if $\xi(h)\neq\xi(h')$, it follows that

$$\|\mathcal{D}[h]-\mathcal{D}[h']\| \geq |\mathcal{D}[h]-\mathcal{D}[h']| = 2.$$

Because $h$ and $h'$ were arbitrary, it follows that $\mathcal{D}$ is only continuous if $\xi(h)=\xi(h')$ for all pairs $h,h'$. We have proven the following.

**Theorem 8.1.** *A deterministic proper search generator is continuous if and only if it is constant.*

By "constant", what is meant is that a search generator is constant as a function; that is, it produces the same measure as output irrespective of the history input. A constant search generator is not just value-agnostic but history-agnostic as well. Such a search generator is clearly useless, and hence all useful deterministic proper search generators are norm-discontinuous. These include Newton and quasi-Newton methods as well as simplicial methods and basic generating set search (without a randomized search heuristic). Therefore another approach is needed to discuss continuity in deterministic search generators.

Now the reasoning above clearly implies the existence of a map

$$\Xi : \mathcal{D} \mapsto \xi$$

that is bijective from the deterministic proper search generators to the function space $X^{\left(\mathcal{H}^{\mathcal{T}}_{X,Y}\right)}$ of functions from history traces to search points. As in the last chapter, we will write $\Xi\langle\mathcal{D}\rangle=\xi$ using angle brackets. The function space forming the range of $\Xi$ is a Cartesian product and hence can be endowed with the product topology. This topology on Range $\Xi$ can be projected back to form a topology on the deterministic proper search generators given by

$$\tau_\Xi = \{\Xi^{-1}(A) \mid A \text{ open in the product topology on Range } \Xi\}.$$

This type of topology can be constructed for any function whose codomain is a topological space, and the topology it induces on the domain of the function is called the *initial topology*. Let $\mathfrak{DG}_{X,Y}$ be the set of deterministic proper search generators, which is a topological space under $\tau_\Xi$ whose structure is determined by the initial topology induced by $\Xi$.

When applied to functions, the product topology is also known as the *topology of pointwise convergence*, because in this topology a sequence of functions $\{f_n\}_{n\in\mathbb{N}}$ converges to a function $f$ if $\lim f_n(x)=f(x)$ for all $x\in X$. The topology of pointwise convergence is sometimes also called *weak* because it admits a metric in only limited circumstances. One topological base for this topology is formed by intervals bounded on either side by functions. For functions $f,g$, define $f<g$ to mean that

there exists at least one $x_0 \in X$ such that $f(x_0) < g(x_0)$ and $f(x) \leq g(x)$ for all $x$ in the domain. An open interval $(f,g)$ is defined by

$$(f,g) = \{h \mid f < h < g\},$$

and the topology of pointwise convergence is the smallest topology containing every such interval.

Now each deterministic proper search generator $\mathcal{D}$ has a corresponding image $\xi = \Xi\langle\mathcal{D}\rangle$, and as a function

$$\xi : \mathcal{H}^{\mathcal{T}}_{X,Y} \to X,$$

$\xi$ is continuous if $\xi^{-1}$ maps every open set in $X$ to an open set in $\mathcal{H}^{\mathcal{T}}_{X,Y}$. Thus we have the following definition and propositions mirroring the situation for norm-continuity of search generators in general.

**Definition 8.3 (Deterministically Continuous).** A deterministic proper search generator $\mathcal{D} \in \mathfrak{D}\mathfrak{G}_{X,Y}$ is *deterministically continuous* if $\Xi\langle\mathcal{D}\rangle$ is continuous. $\mathcal{D}$ is *deterministically continuous in trajectories* if the map

$$h_X \mapsto \Xi\langle\mathcal{D}\rangle(\kappa(h_X,h_Y))$$

is continuous for all fixed $h_Y$, and $\mathcal{D}$ is *deterministically continuous in objectives* if the map

$$h_Y \mapsto \Xi\langle\mathcal{D}\rangle(\kappa(h_X,h_Y))$$

is continuous for all fixed $h_X$.

**Proposition 8.3.** *A deterministic proper search generator $\mathcal{D} \in \mathfrak{D}\mathfrak{G}_{X,Y}$ is deterministically continuous if and only if it is deterministically continuous in both trajectories and objectives.*

*Proof.* The proof is identical to that of Proposition 8.2, *mutatis mutandis*.

The concept of *deterministic continuity* applied to the set $\mathfrak{D}\mathfrak{G}_{X,Y}$ provides a way to assess how deterministic search methods vary as the history input varies. As will be shown, Newton and quasi-Newton methods are deterministically continuous in objectives, and they are also deterministically continuous in trajectories when the value history is drawn from a continuously differentiable objectives.

### *8.1.3 Sequential Continuity and Continuity at a Point*

The notions of continuity above differ in some notable ways from the concept of continuity used in classical analysis. For metric spaces, the epsilon-delta definition of continuity is equivalent to the topological definition of continuity. But there are older definitions of continuity based on sequence limits that differs subtly from both. These distinctions are relevant here because these differences influence which of the traditional proof strategies for continuity are acceptable.

**Definition 8.4.** A function $f : Z \to W$ between two topological spaces $Z$ and $W$ is sequentially continuous for every sequence $(z_n)_{n\in\mathbb{N}} \subseteq Z$ converging to a point $z \in Z$, $f(z_n) \to f(z)$.

**Proposition 8.4.** *Every continuous function is sequentially continuous.*

*Proof.* Suppose $f$ is continuous and $z_n \to z$. Let $A$ be any open neighborhood of $f(z)$. Then $f^{-1}(A)$ is an open neighborhood of $z$. Consequently $z_n$ is eventually contained in $f^{-1}(A)$, and therefore $f(z_n)$ is eventually contained in $A$. Since $A$ was arbitrary, we have that $f(z_n) \to f(z)$ as required.

The converse of the preceding proposition is *not* true, however. In "large" spaces the closure of a set can include points that are not the limits of any countable sequence. Such limit points can only be "reached" by a sequence indexed by a *directed set*, that is, by a set that has a partial or total order such that every pair of elements in the set has an upper bound. Such a sequence called a *net*. Convergence with respect to a net is equivalent to the convergence of all chains in the partial order when considered as sequences.

Topological spaces in which every sequentially continuous function is continuous are called *sequential spaces*. Every first-countable space is a sequential space, including all metric spaces. In addition, every quotient space formed from a first-countable space is a sequential space, and every sequential space can be represented as a quotient of some first-countable space. Thus sequentially continuity is equivalent to topological continuity in the most common spaces. Nonetheless, the convergence of a function on convergent arguments does not automatically imply topological continuity.

From sequential continuity, it is easy to see what *continuity at a point* should mean. That is, a function $f$ is sequentially continuous at a point $z$ in the domain if $f(z_n) \to f(z)$ whenever $z_n \to z$. Then a function is sequentially continuous if it is sequentially continuous at every point in the domain. This type of definition can also be extended to the epsilon-delta setting for metric spaces. A function $f : Z \to W$ between two metric spaces is continuous at a point $z$ if for every $\varepsilon > 0$ there is a $\delta > 0$ such that $d_Z(z, z') < \delta$ implies that $d_W(f(z), f(z')) < \varepsilon$. And again, if a function is continuous at every point in the epsilon-delta sense, then it is continuous everywhere. Since epsilon-delta continuity is equivalent to topological continuity for metric spaces, the concept generalizes to functions on topological spaces generally, as in the next definition.

**Definition 8.5.** A function $f : Z \to W$ between two topological spaces is continuous at a point $z$ in the domain if there exists some open neighborhood $A$ of $f(z)$ such that the restricted function $f|_{f^{-1}(A)}$ is continuous as a function from $f^{-1}(A)$ to $A$, where both sets are endowed with the subspace topology.

**Proposition 8.5.** *A function $f : Z \to W$ between two topological spaces is continuous if and only if it is continuous at every point in its domain.*

*Proof.* If $f$ is continuous, then trivially for all $z \in Z$, the entire space $W$ is a neighborhood of $f(z)$ such that $f|_{f^{-1}(W)}$ is continuous. Suppose on the other hand $f$ is continuous at every point in its domain. Then for every point $z$ in the domain there is an open set $A_z \subseteq W$ that such that $f|_{f^{-1}(A)}$ is continuous. Let $G$ be any open set in $W$. The $G = \bigcup_{z\in Z}(A_z \cap G)$ is the union of the intersection of itself with all of the neighborhoods $A_z$. Consequently,

$$f^{-1}(G) = f^{-1}\left(\bigcup_{z\in Z}(A_z \cap G)\right) = \bigcup_{z\in Z} f|^{-1}_{f^{-1}(A_z)}(A_z \cap G).$$

Now each $A_z \cap G$ is the intersection of two open sets and is therefore open, so by continuity its preimage under $f|_{f^{-1}(A_z)}$ is open. Finally, $f^{-1}(G)$ is open as the union of an arbitrary collection of open sets, whence $f$ is continuous.

This notion of continuity at a point naturally extends to the space of search generators. As a function, each search generator $\mathcal{G}$ is continuous at a history trace $h$ if there is some neighborhood of $\mathcal{G}[h]$ in the space of measures such that the search generator is continuous on histories drawn from the pre-image of this neighborhood under $\mathcal{G}$. Obviously, this statement is rather complicated. Thus it will usually be easier to prove continuity when the space of histories $\mathcal{H}^{\mathcal{T}}_{X,Y}$ is metric, in which case the epsilon-delta approach can be used, or sequential, in which case the convergent sequence approach can be used. Otherwise, since the codomain $\mathcal{M}_X$ is a normed space, a hybrid epsilon-open-set approach can be used, in which $\varepsilon > 0$ is fixed and then used to show that there exists an open set $H$ of history traces containing $h$ such that $\|\mathcal{G}[h] - \mathcal{G}[h']\| < \varepsilon$ for all $h' \in H$. If this statement can be proved, then $\mathcal{G}$ is continuous at $h$. Variants of this latter strategy are used repeatedly below.

Since *deterministic continuity* as well as continuity *in objectives* and *in trajectories* are ultimately forms of topological continuity, the meaning of statements such as "$\mathcal{D}$ is deterministically continuous at history trace $h$" or "$\mathcal{G}$ is continuous in objectives at history trace $h$" should be interpreted as meaning topological continuity at a point with respect to the relevant topology.

Perhaps the main reason why continuity at a point matters in what follows is due to the interaction of topology and measure theory. When the history trace is generated stochastically from a search generator, there are certain histories that are visited with zero probability. Continuity over the entire space of histories can then be replaced by the weaker requirement of continuity on a set of history traces with full measure. Recall that a set of full measure for a signed measure $\mu$ on a measurable space $X$ is a measurable set $F$ such that $|\mu|(F) = |\mu|(X)$; that is, the set $X \setminus F$ has measure zero. In terms of a probability measure $\mathbb{P}$, a set has full measure if $\mathbb{P}(X) = 1$, hence the phrase "with probability one" implicitly invokes a set of full measure. When the inputs of a function are subject to some measure, then the function is said to be continuous *almost everywhere* or continuous *almost surely* if it is continuous at every point in a set that has full measure in the input space. If it is not clear from context, the measure in question may specified by saying that something true $\mu$-almost everywhere or $\mathbb{P}$-almost surely. Consequently, in the context of measure spaces, continuity at a point is a critical concept.

These points are somewhat technical in nature but should help to explain the structure of the continuity proofs in this chapter.

## 8.2 Continuity of Evolutionary Algorithms

As discussed in Chapter 7, an evolutionary algorithm can be represented as a convolution of selection, recombination, and variation processes, $\mathcal{E} = \mathcal{S} \star \mathcal{R} \star \mathcal{V}$. Evolutionary algorithms can be continuous or discontinuous, depending on the details of the genetic operators, and so each of the three components of an evolutionary algorithm must be examined for continuity. Mutation operators are value-agnostic and therefore trivially continuous in objectives. Typically, mutation operators are continuous in trajectories as well, as will be seen for Bernoulli or Gaussian mutation. Crossover rules are likewise value-agnostic and therefore continuous in objectives. Thus in order to determine when evolutionary algorithms as a whole are continuous, it will be necessary to understand when selection rules are continuous and when convolution preserves continuity. Cases where evolutionary algorithms are continuous or discontinuous will be addressed with general theorems in this section. These results can then be used in conjunction with the results of Chapter 10 to conclude when the performance of evolutionary algorithms changes continuously with the objective function. That is, these results will make it possible to determine a particular algorithm will perform similarly on similar objectives.

### 8.2.1 Continuity of Convolution

Evolutionary algorithms were constructed in Chapter 7 using the convolution operator. Therefore, any study of continuity using this construction relies crucially on how convolution preserves continuity. Because the convolution depends not just on the given history trace but on a search point proposed by a search generator, the continuity of convolution depends on how this next point is generated and whether this generation process preserves continuity in some way.

In this subsection, a concept called *extensible continuity* is introduced to address the way that convolution examines an internally proposed search point. It is then shown that a convolution $\mathcal{A} \star \mathcal{B}$ is continuous at a history if $\mathcal{A}$ is continuous at that history and $\mathcal{B}$ is extensibly continuous there. In the next subsection, a separate approach to continuity will be pursued based on whether samples from the first search generator converge.

**Definition 8.6.** Given measurable sets $A \subseteq X^k$ and $B \subseteq Y^k$, a search generator $\mathcal{G} \in \mathfrak{G}_{X,Y}$ is *k-extensibly continuous* to $A$ and $B$ at a history trace $h$ if for all $\varepsilon > 0$ there exists an open neighborhood $H$ of $h$ such that for all $h' \in H$, $x \in A$ and $y \in B$

$$\left\| \mathcal{G}[h' \| \kappa(x,y)] - \mathcal{G}[h \| \kappa(x,y)] \right\| < \varepsilon.$$

The phrases "$k$-extensibly continuous in objectives" and "$k$-extensively continuous in trajectories" are defined similarly with $h_X$ and $h_Y$ held constant, respectively.

If a search generator is 1-extensibly continuous, then it will called *extensibly continuous* without the 1. When $k = 0$, then for $x \in X^0$, $y \in Y^0$, we assume that $\kappa(x,y) = \emptyset$, the empty history. In that case $h \| \emptyset = h$, so if a search generator is 0-extensibly continuous, then the definition implies that is continuous in the standard topological sense.

The definition of extensible continuity is fairly complicated, but some simple cases can quickly be stated for which some variety of extensible continuity holds, including in particular $k$-Markov search generators, value-agnostic search generators, crossover rules, and mutation operators.

**Proposition 8.6.** *Every k-Markov search generator $\mathcal{G} \in \mathfrak{G}_{X,Y}$ is k-extensibly continuous to $X^k$ and $Y^k$ at every history trace in $\mathcal{H}^{\mathcal{T}}_{X,Y}$.*

*Proof.* The $k$-Markov property implies that $\mathcal{G}[h' \| \kappa(x,y)] = \mathcal{G}[h \| \kappa(x,y)]$ for all $h, h'$ with $x \in X^k$ and $y \in Y^k$. Hence the entire set $\mathcal{H}^{\mathcal{T}}_{X,Y}$ is a suitable open neighborhood to carry the proof for any $\varepsilon > 0$ and any history trace $h$.

**Corollary 8.1.** *Every mutation operator is extensibly continuous to $X$ and $Y$ at every history trace.*

*Proof.* By the definition in Chapter 7, mutation operators are 1-Markov.

**Proposition 8.7.** *Every k-step value-agnostic search generator is k-extensibly continuous in objectives to $X^k$ and $Y^k$ at every history trace in $\mathcal{H}^{\mathcal{T}}_{X,Y}$.*

*Proof.* If $h$ and $h'$ are history traces with $h_X = h'_X$, it follows for any $k$-step value-agnostic search generator $\mathcal{G} \in \mathcal{VA}^k_{X,Y}$, $x \in X^k$, and $y \in Y^k$ that $\mathcal{G}[h' \| \kappa(x,y)] = \mathcal{G}[h \| \kappa(x,y)]$. Thus the entire domain $\mathcal{H}^{\mathcal{T}}_{Y}$ is a suitable open neighborhood to carry the proof for any $\varepsilon > 0$ and any history trace $h$.

**Corollary 8.2.** *Every crossover rule and mutation operator is k-extensibly continuous in objectives to $X^k$ and $Y^k$ at every history trace.*

*Proof.* According to the definitions in Chapter 7, all crossover rules and mutation operators are value-agnostic.

**Proposition 8.8.** *For a search generator $\mathcal{G}$, the truncation $\triangleleft_k \mathcal{G}$ is k extensibly continuous to $X^k$ and $Y^k$ at a history trace h if and only if $\mathcal{G}$ is continuous at h. The same statement holds if "continuous" is replaced by either "continuous in objectives" or "continuous in trajectories".*

*Proof.* Since $\triangleleft_k \mathcal{G}[h \| \kappa(x,y)] = \mathcal{G}[h]$, for all $h, h' \in \mathcal{H}^{\mathcal{T}}_{X,Y}$, $x \in X^k$, and $y \in Y^k$,

$$\left\| \triangleleft_k \mathcal{G}[h' \| \kappa(x,y)] - \triangleleft_k \mathcal{G}[h \| \kappa(x,y)] \right\| = \left\| \mathcal{G}[h'] - \mathcal{G}[h] \right\|,$$

which equates $k$-extensible continuity for $\triangleleft_k \mathcal{G}$ with continuity for $\mathcal{G}$.

Now we discuss some preliminaries for proving continuity of convolution. Let $\mathcal{A}$ and $\mathcal{B}$ be search generators such that $\mathcal{A} \star \mathcal{B}$ is defined. Let $h$ and $h'$ be arbitrary history traces. The supremum in the total variation norm implies that for every $\varepsilon > 0$ there exists some measurable set $A \subseteq X$ such that

$$\left\| \mathcal{A} \star \mathcal{B}\left[h'\right] - \mathcal{A} \star \mathcal{B}\left[h\right] \right\| < \frac{\varepsilon}{3} + \left| \mathcal{A} \star \mathcal{B}\left[h'\right] - \mathcal{A} \star \mathcal{B}\left[h\right] \right| (A). \tag{8.1}$$

This equation allows us to exchange the norm for the absolute measure of set. The problem here is that the absolute value prevents us from looking inside the convolution.

To correct this, recall that the Hahn decomposition theorem asserts that for any measurable set $A$ and any signed measure $\mu$, there are two measurable sets $A_+$ and $A_-$ (the positive and negative sets) such that $|\mu|(A) = |\mu(A_+)| + |\mu(A_-)|$. This decomposition can be applied to $\mu = \mathcal{A} \star \mathcal{B}\left[h'\right] - \mathcal{A} \star \mathcal{B}\left[h\right]$ to separate the positive and negative cases. The right hand side of Equation 8.1 can now be expanded to three terms,

$$\frac{\varepsilon}{3} + \left| \left(\mathcal{A} \star \mathcal{B}\left[h'\right] - \mathcal{A} \star \mathcal{B}\left[h\right]\right)(A_+) \right| + \left| \left(\mathcal{A} \star \mathcal{B}\left[h'\right] - \mathcal{A} \star \mathcal{B}\left[h\right]\right)(A_-) \right|,$$

and if the right two terms are each bounded by $\frac{\varepsilon}{3}$, then $\|\mathcal{A} \star \mathcal{B}[h'] - \mathcal{A} \star \mathcal{B}[h]\| < \varepsilon$.

It is now possible to expand the convolution. Letting $A_*$ represent either $A_+$ or $A_-$, apply the definition of convolution for an arbitrary value $y \in Y$ and inject a mixed term to obtain

$$\begin{aligned}
&\left| \left(\mathcal{A} \star \mathcal{B}\left[h'\right] - \mathcal{A} \star \mathcal{B}\left[h\right]\right)(A_*) \right| \\
&\quad = \left| \int_X \mathcal{B}\left[h' \| (x,y)\right](A_*)\, \mathcal{A}\left[h'\right](dx) - \int_X \mathcal{B}\left[h \| (x,y)\right](A_*)\, \mathcal{A}\left[h\right](dx) \right| \\
&\quad \leq \left| \int_X \mathcal{B}\left[h' \| (x,y)\right](A_*)\, \mathcal{A}\left[h'\right](dx) - \int_X \mathcal{B}\left[h' \| (x,y)\right](A_*)\, \mathcal{A}\left[h\right](dx) \right| \\
&\qquad + \left| \int_X \mathcal{B}\left[h' \| (x,y)\right](A_*)\, \mathcal{A}\left[h\right](dx) - \int_X \mathcal{B}\left[h \| (x,y)\right](A_*)\, \mathcal{A}\left[h\right](dx) \right|
\end{aligned}$$

These two terms in the final inequality can now be bounded separately. In each case, like terms can be gathered and the absolute value pushed inside, all while preserving the inequality. Further reduction using the supremum norm and total variation norms respectively as upper bounds where possible yields

$$\int_X \left|\mathcal{B}\left[h' \| (x,y)\right](A_*)\right| \left| \mathcal{A}\left[h'\right] - \mathcal{A}\left[h\right] \right| (dx) \leq \|\mathcal{B}\| \, \|\mathcal{A}[h'] - \mathcal{A}[h]\| \tag{8.2}$$

for the first term and

$$\int_X \left| \mathcal{B}\left[h' \| (x,y)\right](A_*) - \mathcal{B}\left[h \| (x,y)\right](A_*) \right| \; |\mathcal{A}[h]|(dx) \qquad (8.3)$$
$$\leq \int_X \left\| \mathcal{B}\left[h' \| (x,y)\right] - \mathcal{B}\left[h \| (x,y)\right] \right\| \; |\mathcal{A}[h]|(dx)$$

for the second. Notice that both inequalities remove the dependence on $A_*$, and hence these two cases reduce to one.

Equations 8.2 and 8.3 represent the quantities that must be bounded in order to establish in theorem about the continuity of the convolution $\mathcal{A} \star \mathcal{B}$ with respect to the properties of $\mathcal{A}$ and $\mathcal{B}$. The first theorem about such properties can now be stated.

**Theorem 8.2.** *Let $\mathcal{A} \in \mathcal{VA}^k_{X,Y}$ and $\mathcal{B} \in \mathcal{VA}^{k+1}_{X,Y}$. Let $h$ be any history trace. Let $A \subseteq X^k$, $B \subseteq Y^k$, and let $C \subseteq X$ be a set of full measure for $\mathcal{A}[h]$. Suppose $\mathcal{A}$ is $k$-extensibly continuous to $A$ and $B$ at $h$ and that $\mathcal{B}$ is $(k+1)$-extensibly continuous at $h$ to $A \times C$ and $B \times Y$. Then $\mathcal{A} \star \mathcal{B}$ is $k$-extensibly continuous to $A$ and $B$ at $h$. The statement still holds if "continuous" is replaced by either "continuous in objectives" or "continuous in trajectories".*

*Proof.* Fix $\varepsilon > 0$. Let $h$ be the history trace referenced in the theorem statement. Equations 8.2 and 8.3 will be used to prove the conclusion, but $h \| \kappa(x,y)$ and $h' \| \kappa(x,y)$ will be substituted for $h$ and $h'$ respectively to demonstrate $k$-extensible continuity.

Starting with Equation 8.2, fix $\eta > 0$ and by the $k$-extensible continuity of $\mathcal{A}$ at $h$ conclude that there is some open neighborhood $H_1$ of $h$ such that $\|\mathcal{A}[h' \| \kappa(x,y)] - \mathcal{A}[h \| \kappa(x,y)]\| < \eta$ for all $h' \in H$, $x \in A$, and $y \in B$. Therefore, for $h' \in H_1$,

$$\|\mathcal{B}\| \; \|\mathcal{A}[h' \| \kappa(x,y)] - \mathcal{A}[h \| \kappa(x,y)]\| \quad < \quad \eta \|\mathcal{B}\| \qquad (8.4)$$

Returning to Equation 8.3, note that a set of full measure for $\mathcal{A}[h]$ is also a set of full measure for $|\mathcal{A}[h]|$. Using $\eta$ and the $(k+1)$-extensible continuity of $\mathcal{B}$ at $h$, choose an open neighborhood $H_2$ of $h$ such that

$$\left\| \mathcal{B}\left[h' \| \kappa(x,y) \| (x_0,y_0)\right] - \mathcal{B}\left[h \| \kappa(x,y) \| (x_0,y_0)\right] \right\| < \eta$$

for all $h' \in H_2$, $x \in A$, $y \in B$, $x_0 \in C$, and $y_0 \in Y$. But then for $h' \in H_2$,

$$\int_X \left\| \mathcal{B}\left[h' \| \kappa(x,y) \| (x_0,y_0)\right] - \mathcal{B}\left[h \| \kappa(x,y) \| (x_0,y_0)\right] \right\| |\mathcal{A}\left[h \| \kappa(x,y)\right]|(dx) < \eta \|\mathcal{A}\|.$$

Let $H = H_1 \cap H_2$ and note that $H$ is an open neighborhood of $h$. Consequently, the two terms above can be combined to show that for all $h' \in H$,

$$\left| \left(\mathcal{A} \star \mathcal{B}\left[h' \| \kappa(x,y)\right] - \mathcal{A} \star \mathcal{B}\left[h \| \kappa(x,y)\right]\right)(A_*) \right| \quad < \quad \eta(\|\mathcal{A}\| + \|\mathcal{B}\|).$$

By setting $\eta = \frac{\varepsilon}{3\|\mathcal{A}+\mathcal{B}\|}$ as we may, the preceding equation proves that

$$\left\|\mathcal{A}\star\mathcal{B}\left[h'\|\kappa(x,y)\right]-\mathcal{A}\star\mathcal{B}\left[h\|\kappa(x,y)\right]\right\| \quad < \quad \frac{\varepsilon}{3}+\frac{\varepsilon}{3}+\frac{\varepsilon}{3}=\varepsilon$$

for all of $h' \in H$, which is an open neighborhood of $h$. Therefore $\mathcal{A}\star\mathcal{B}$ is $k$-extensibly continuous to $A\times C$ and $B\times Y$ at $h$.

**Corollary 8.3.** *Let $\mathcal{A}\in\mathfrak{G}_{X,Y}$ and $\mathcal{B}\in\mathcal{VA}^1_{X,Y}$. Let $h$ be a history trace, and let $C$ be a set of full measure for $\mathcal{A}[h]$. If $\mathcal{A}$ is continuous at a history trace $h$ and $\mathcal{B}$ is extensibly continuous to $C$ and $Y$ at $h$, then $\mathcal{A}\star\mathcal{B}$ is continuous at $h$. The statement remains true if "continuous" is replaced by "continuous in objectives" or "continuous in trajectories".*

*Proof.* Note that $\mathfrak{G}_{X,Y}=\mathcal{VA}^0_{X,Y}$. As stated in the text, a search generator that is continuous at a history trace $h$ is 0-extensibly continuous to $X^0$ and $Y^0$ at $h$. The corollary is then a transparent simplification of Theorem 8.2.

Using the definitions given in Chapter 7, Theorem 8.2 can be applied to evolutionary algorithms to deduce continuity from the continuity of the selection rules and the extensible continuity of the crossover rule and mutation operator. The caveat, however, is that extensible continuity for crossover rules is only automatic in objectives, not trajectories. In fact, most selection and crossover rules are discontinuous in trajectories. Thus the overall claim is that an evolutionary algorithm is continuous in objectives if its selection rules are continuous in objectives. This claim is now proven, as arguments about continuity in trajectories are postponed until the next subsection.

**Corollary 8.4.** *A recombination operator is extensibly continuous in objectives to $X$ and $Y$ at a history trace $h$ if each of its selection rules are continuous in objectives at $h$.*

*Proof.* Let $\mathcal{R}$ be a recombination operator. Then either $\mathcal{R}$ is a crossover rule, in which case it is extensibly continuous to $X$ and $Y$ in objectives by Corollary 8.2, or $\mathcal{R}=\triangleleft\mathcal{S}_1\star(\cdots\star(\triangleleft_n\mathcal{S}_n\star\mathcal{C}))$ for selection operators $\mathcal{S}_1,\ldots,\mathcal{S}_n$ and a crossover rule $\mathcal{C}$. The crossover rule $\mathcal{C}$ is $(n+1)$-extensibly continuous in objectives to $X^{n+1}$ and $Y^{n+1}$ at $h$ by Corollary 8.2. If the selection rules are continuous at $h$, Proposition 8.8 ensures that each $\triangleleft_k\mathcal{S}_k$ is $k$-extensibly continuous in objectives to $X^k$ and $Y^k$ at $h$. By $n$ recursive applications of Theorem 8.2, $\mathcal{R}$ is extensibly continuous in objectives to $X$ and $Y$ at $h$.

**Corollary 8.5.** *An evolutionary algorithm is continuous in objectives at $h$ if its selection rules are continuous in objectives at $h$.*

*Proof.* Let $\mathcal{E}$ be a evolutionary algorithm. Then $\mathcal{E}=\mathcal{S}\star\mathcal{R}\star\mathcal{V}$ where $\mathcal{R}$ is a recombination operator and $\mathcal{V}$ is a mutation operator. By Corollary 8.1, $\mathcal{V}$ is extensibly continuous in objectives to $X$ and $Y$ at $h$. $\mathcal{S}$ is continuous in objectives at $h$ by assumption. Furthermore, $\mathcal{R}$ is extensibly continuous in objectives to $X$ and $Y$ at $h$ by the previous corollary. Applying Theorem 8.2, $\mathcal{S}\star\mathcal{R}$ is continuous in objectives at $h$, and so by Corollary 8.3, $\mathcal{S}\star\mathcal{R}\star\mathcal{V}$ is continuous in objectives as well.

The previous theorem and its corollaries provide several tools to show that evolutionary algorithms are continuous in objectives when the selection rules are continuous in objectives. A broader conclusion is possible by inspecting the probability distribution governing the history trace, as described next.

### *8.2.2 Sample Convergence and Continuity*

Selection and recombination in evolutionary algorithms typically can only choose search points from a finite set of parents. The parents must come from the prior generation, and there are only finitely many ways that the parents can be recombined. Selection and crossover are often norm-discontinuous for the same reason that deterministic optimizers are norm-discontinuous; convergence of the selected points does not imply convergence in the total variation norm. In evolutionary algorithms, mutation varies the recombined point, spreading it out so that after mutation, any point in the search space can be generated. The mutation process restores continuity under certain conditions even when the selection and crossover rule are not continuous. These concepts are developed formally in this subsection. First, the concept of a finite support set is introduced to encapsulate the role of the parents in selection.

**Definition 8.7 (Finite Support Set).** A finite set $S \subseteq X$ is called a finite support set for a signed measure $\mu$ on $X$ if $S$ has full $\mu$-measure and for all $x \in S$ and measurable $A \subseteq X$, $\mu(A) > 0$ whenever $x \in A$.

The two requirements for a finite support set are that it be finite and minimal. If $\{x\}$ is measurable for all $x \in X$, then the final condition implies that $\mu(\{x\}) > 0$ if and only if $x \in S$. Note that if the search domain $X$ is Hausdorff, then $\{x\}$ is measurable under the Borel $\sigma$-algebra, which is generally assumed in this text.

**Definition 8.8 (Finitely Supported Search Generator).** A search generator $\mathcal{G}$ is finitely supported with size $s < \infty$ on a set of histories $H$ if for all $h \in H$, $\mathcal{G}[h]$ has a finite support set $S_h$ such that $s = \sup_h |S_h|$, *i.e.*, the support sets are of finitely bounded in size by $s$.

The goal for sample convergence will be to identify a finitely supported region and track how the elements in the finite support sets change as the history changes. This purpose will be served by continuous *trajectory tracking functions* $\tau : \mathcal{H}^{\mathcal{T}}_{X,Y} \to X$ that will predict the contents of $S_{h'}$ for each history trace $h'$ within some neighborhood.

**Definition 8.9 (Finitely Trackable).** A search generator $\mathcal{G}$ is *finitely trackable* at a history trace $h$ if $\mathcal{G}$ is finitely supported on a neighborhood $H$ of $h$ with size $s$ and there is collection of trajectory tracking functions $\{\tau_i \mid 1 \leq i \leq s\}$ such that each $\tau_i$ is continuous and for all $h' \in H$, $S_{h'} = \{\tau_i(h') \mid 1 \leq i \leq s\}$.

Suppose $\mathcal{G}$ is finitely trackable at $h$. Then there is a finite collection of tracking functions that fully describe $\mathcal{G}$ within some neighborhood of $h$. The trajectory tracking functions will be associated with probability values that change with the history trace. For the most part, this probability is given at $h$ by $\mathcal{G}[h](\{\tau_i(h)\})$, but there is a special case that must be considered when $\tau_i(h) = \tau_j(h)$, that is, when two trajectories $i$ and $j$ cross. Therefore, we define $\{\mathcal{G}_{\tau_i}\}$ ambiguously as any collection of functions satisfying

$$\mathcal{G}[h'](\{x\}) = \sum_{\{i \mid \tau_i(h')=x\}} \mathcal{G}_{\tau_i}(h') \tag{8.5}$$

for all $h' \in H$ and $x \in S_{h'}$, where $H$ is the neighborhood on which $\mathcal{G}$ is trackable. Equation 8.5 is always satisfiable, but in what follows these functions will be further constrained to be continuous at $h$, which will reduce the ambiguity but not remove it.

Since the $\mathcal{G}_{\tau_i}$ track how the probability of each member of the finite support set changes with the history, they will be called *probability tracking functions*. Because the space of histories and the space of real numbers are both topological spaces (using the Euclidean metric topology for $\mathbb{R}$), continuity is well-defined for $\mathcal{G}_{\tau_i}$ according to these topologies. These objects can now be put together to define sample convergence.

**Definition 8.10 (Sample Convergent).** A search generator $\mathcal{G}$ is sample convergent at a history trace $h$ if it is finitely trackable at $h$ with a set of probability tracking functions that are each continuous at $h$.

The intuition for this definition is that we can separate a search generator into finitely many points that must be tracked. If we can define a mechanism for tracking points that results in smoothly changing probabilities assigned to the tracked points, then the search generator is sample convergent. As with other forms of continuity explored in this chapter, sample convergence can be considered along trajectories or objectives by holding one or the other fixed.

**Definition 8.11 (Sample Convergent in Objectives).** A search generator $\mathcal{G}$ is sample convergent *in objectives* at a history trace $h$ if it is finitely trackable at $h$ and there exists a set of probability tracking functions such that for all $h_X$ the map $h_Y \mapsto \mathcal{G}_{\tau_i}(\kappa(h_X, h_Y))$ is continuous at $h_Y$.

**Definition 8.12 (Sample Convergent in Trajectories).** A search generator $\mathcal{G}$ is sample convergent *in trajectories* at a history trace $h$ if it is finitely trackable at $h$ and there exists a set of probability tracking functions such that for all $h_Y$ the map $h_X \mapsto \mathcal{G}_{\tau_i}(\kappa(h_X, h_Y))$ is continuous at $h_X$.

**Proposition 8.9.** *A search generator $\mathcal{G}$ is sample convergent at a history trace $h$ if and only if it is sample convergent in both objectives and trajectories at $h$.*

*Proof.* This claim follows from the fact that the probability tracking function is a map $(h_X, h_Y) \mapsto \mathcal{G}_{\tau_i}[\kappa(h_X, h_Y)]$, and sample convergence in objectives or in trajectories is just the continuity of the map when restricted in either coordinate. ∎

The name *sample convergent* is chosen to reflect the fact that a sample from a sample convergent optimizer converges along a sequence of trajectories or objectives. Loosely, for a sequential history space, if $Z_h \sim \mathcal{G}[h]$ for all $h$, then $Z_{h_n}$ converges in distribution to $Z_h$ when $h_n \to h$. Sample convergence is not just important for evolutionary algorithms; for example, it can also be used to give conditions for when stochastic gradient descent is continuous.

Now consider how to analyze continuity in the context of a convolution $\mathcal{A} \star \mathcal{B}$ where $\mathcal{A}$ is sample convergent at a history trace $h$. Let $s$ be the maximum size of the finite support set $S_h$ for $\mathcal{A}$ at $h$, and let $\{\tau_i \mid 1 \leq i \leq s\}$ be the trajectory tracking functions for $\mathcal{A}$. Within some open neighborhood $H$ of $h$, sample convergence implies that the search generator $\mathcal{A} \star \mathcal{B}$ can be written for each $h' \in H$ and all measurable $A$ as

$$\mathcal{A} \star \mathcal{B}[h'](A) = \sum_{x \in \{\tau_i(h') \mid 1 \leq i \leq s\}} \mathcal{A}[h'](\{x\}) \; \mathcal{B}\left[h' \| (x,y)\right](A),$$

where $y$ is arbitrary since necessarily $\mathcal{B} \in \mathcal{VA}^1_{X,Y}$. The use of the sum in place of the integral reflects the fact that $\mathcal{A}$ is degenerate at $h'$, with all support lying in the finite set $S_{h'}$. Using the probability tracking function $\mathcal{A}_{\tau_i}$ and defining $\mathcal{B}_{i,h'} = \mathcal{B}\left[h' \| (\tau_i(h'), y)\right]$ this equation simplifies further to

$$\mathcal{A} \star \mathcal{B}[h'](A) = \sum_{i=1}^{s} \mathcal{A}_{\tau_i}(h') \; \mathcal{B}_{i,h'}(A). \tag{8.6}$$

We now proceed as in Equation 8.1 and by the same logic as used there separate into two cases $A_+$ and $A_-$, with $A_*$ referring variably to either case. As there, we find that

$$\begin{aligned}
&\left| \left(\mathcal{A} \star \mathcal{B}[h'] - \mathcal{A} \star \mathcal{B}[h]\right)(A_*) \right| \\
&\quad = \left| \sum_{i=1}^{s} \mathcal{A}_{\tau_i}(h') \; \mathcal{B}_{i,h'}(A_*) - \sum_{i=1}^{s} \mathcal{A}_{\tau_i}(h) \; \mathcal{B}_{i,h}(A_*) \right| \\
&\quad \leq \left| \sum_{i=1}^{s} \mathcal{A}_{\tau_i}(h') \; \mathcal{B}_{i,h'}(A_*) - \sum_{i=1}^{s} \mathcal{A}_{\tau_i}(h) \; \mathcal{B}_{i,h'}(A_*) \right| \\
&\quad\quad + \left| \sum_{i=1}^{s} \mathcal{A}_{\tau_i}(h) \; \mathcal{B}_{i,h'}(A_*) - \sum_{i=1}^{s} \mathcal{A}_{\tau_i}(h) \; \mathcal{B}_{i,h}(A_*) \right|.
\end{aligned}$$

Again, the final equality generates two cases. The first term is analogous to Equation 8.2. Pushing absolute values inside and bounding by norms yields

$$\sum_{i=1}^{s} \left| \mathcal{A}_{\tau_i}(h') - \mathcal{A}_{\tau_i}(h) \right| \left| \mathcal{B}_{i,h'}(A_*) \right| \quad \leq \quad \|\mathcal{B}\| \sum_{i=1}^{s} \left| \mathcal{A}_{\tau_i}(h') - \mathcal{A}_{\tau_i}(h) \right|. \tag{8.7}$$

For the second term, restoring the integrals from the sums we see that

$$\left|\sum_{i=1}^{s} \mathcal{A}_{\tau_i}(h)\, \mathcal{B}_{i,h'}(A_*) - \sum_{i=1}^{s} \mathcal{A}_{\tau_i}(h)\, \mathcal{B}_{i,h}(A_*)\right|$$
$$= \int_X \left| \mathcal{B}\left[h' \| (x,y)\right](A_*) - \mathcal{B}\left[h \| (x,y)\right](A_*)\right| \, |\mathcal{A}[h]|(dx)$$

by which we can apply Equation 8.3 to obtain

$$\left|\sum_{i=1}^{s} \mathcal{A}_{\tau_i}(h)\, \mathcal{B}_{i,h'}(A_*) - \sum_{i=1}^{s} \mathcal{A}_{\tau_i}(h)\, \mathcal{B}_{i,h}(A_*)\right| \tag{8.8}$$
$$\leq \int_X \left\| \mathcal{B}\left[h' \| (x,y)\right] - \mathcal{B}\left[h \| (x,y)\right]\right\| \, |\mathcal{A}[h]|(dx).$$

Equations 8.7 and 8.8 provide the basis for proving continuity under convolution with sample convergence.

**Theorem 8.3.** *Let $\mathcal{A} \in \mathfrak{G}_{X,Y}$ and $\mathcal{B} \in \mathcal{VA}^1_{X,Y}$. Let $h$ be any history trace. Let $A \subseteq X^k$, $B \subseteq Y^k$, and let $C$ be a set of full $\mathcal{A}[h]$-measure. If $\mathcal{A}$ is sample convergent at $h$ and $\mathcal{B}$ is $(k+1)$-extensibly continuous to $A \times C$ and $B \times Y$ at $h$, then $(\lhd_k \mathcal{A}) \star B$ is $k$-extensibly continuous to $A$ and $B$ at $h$. The statement still holds if the phrases "sample convergent" and "extensibly continuous" are respectively replaced by either "sample convergent in objectives" and "extensibly continuous in objectives" or "sample convergent in trajectories" and "extensibly continuous in trajectories".*

*Proof.* Fix $\varepsilon > 0$. The proof uses the same logical structure as that of Theorem 8.2, but with sample convergence replacing extensible continuity for $\mathcal{A}$ by means of Equations 8.7 and 8.8. These equations are now applied with $\lhd_k \mathcal{A}$, $h \| \kappa(x,y)$, and $h' \| \kappa(x,y)$ substituted for $\mathcal{A}$, $h$, and $h'$, respectively, where $x \in A$ and $y \in B$.

Fix $\eta > 0$ and refer to Equation 8.7. To clarify, the term in question is

$$\|\mathcal{B}\| \sum_{i=1}^{s} \left| (\lhd_k \mathcal{A})_{\tau_i}(h' \| \kappa(x,y)) - (\lhd_k \mathcal{A})_{\tau_i}(h \| \kappa(x,y)) \right|$$

under the substitutions, where the $\tau_i$ are the trajectory tracking functions for $\lhd_k \mathcal{A}$ and the $(\lhd_k \mathcal{A})_{\tau_i}$ are the probability tracking functions. Next, observe that

$$\left| (\lhd_k \mathcal{A})_{\tau_i}(h' \| \kappa(x,y)) - (\lhd_k \mathcal{A})_{\tau_i}(h \| \kappa(x,y)) \right| = \left| \mathcal{A}_{\tau_i}(h') - \mathcal{A}_{\tau_i}(h) \right|.$$

Now because each $\mathcal{A}_{\tau_i}$ is a continuous function, we can choose open neighborhoods $H_1^i$ of $h$, one for each $\tau_i$, such that $|\mathcal{A}_{\tau_i}(h') - \mathcal{A}_{\tau_i}(h)| < \frac{\eta}{s}$ for all $h' \in H_1^i$. Since $s$ is finite, the intersection $H_1 = \bigcap_i H_1^i$ is an open neighborhood of $h$ wholly contained in each of these neighborhoods for which

$$\|\mathcal{B}\| \sum_{i=1}^{s} \left| \mathcal{A}_{\tau_i}(h') - \mathcal{A}_{\tau_i}(h) \right| < \eta \|\mathcal{B}\|$$

for all $h' \in H_1$. Moving to the second term and examining Equation 8.8, the proof of Theorem 8.2 with $k = 0$ applies almost verbatim to bound this term by $\eta \| \lhd_k \mathcal{A} \| = \eta \|\mathcal{A}\|$ on some neighborhood $H_2$ of $h$. Concluding, we find that

$$\left| \left( (\lhd_k \mathcal{A}) \star \mathcal{B}[h'] - (\lhd_k \mathcal{A}) \star \mathcal{B}[h] \right) (A_*) \right| \quad < \quad \eta \left( \|\mathcal{A}\| + \|\mathcal{B}\| \right),$$

so that setting $\eta = \frac{\varepsilon}{3(\|\mathcal{A}\|+\|\mathcal{B}\|)}$ suffices to prove that $(\lhd_k \mathcal{A}) \star \mathcal{B}$ is $k$-extensibly continuous to $A$ and $B$ at $h$ using an analogue of Equation 8.1.

**Corollary 8.6.** *Let $\mathcal{A} \in \mathfrak{G}_{X,Y}$ and $\mathcal{B} \in \mathcal{VA}^1_{X,Y}$, and let $C$ be a set of full $\mathcal{A}[h]$-measure. If $\mathcal{A}$ is sample convergent at a history trace $h$ and $\mathcal{B}$ is extensibly continuous to $C$ and $Y$ at $h$, then $\mathcal{A} \star B$ is continuous at $h$. The statement still holds if the phrases "sample convergent" and "extensibly continuous" are respectively replaced by either "sample convergent in objectives" and "extensibly continuous in objectives" or "sample convergent in trajectories" and "extensibly continuous in trajectories".*

*Proof.* This is Theorem 8.3 with $k = 0$.

**Corollary 8.7.** *A recombination operator is extensibly continuous in objectives to $X$ and $Y$ at a history trace $h$ if each of its selection rules are either sample convergent in objectives or continuous in objectives at $h$.*

*Proof.* Let $\mathcal{R}$ be a recombination operator. Then either $\mathcal{R}$ is a crossover rule, in which case it is extensibly continuous to $X$ and $Y$ in objectives by Corollary 8.2, or $\mathcal{R} = \lhd \mathcal{S}_1 \star (\cdots \star (\lhd_n \mathcal{S}_n \star \mathcal{C}))$ for selection operators $\mathcal{S}_1, \ldots, \mathcal{S}_n$ and a crossover rule $\mathcal{C}$. The crossover rule $\mathcal{C}$ is $(n+1)$-extensibly continuous in objectives to $X^{n+1}$ and $Y^{n+1}$ at $h$ by Corollary 8.2. If the selection rule $\mathcal{S}_k$ is continuous at $h$, Proposition 8.8 ensures that each $\lhd_k \mathcal{S}_k$ is $k$-extensibly continuous in objectives to $X^k$ and $Y^k$ at $h$, and Theorem 8.2 can be applied when it occurs on the left side of a convolution. If on the other hand the selection rule is sample convergent at $h$, Theorem 8.3 can be applied to $\lhd_k \mathcal{S}_k$ directly. Thus by $n$ interwoven applications of either Theorem 8.2 or Theorem 8.3, $\mathcal{R}$ is extensibly continuous in objectives to $X$ and $Y$ at $h$.

**Corollary 8.8.** *An evolutionary algorithm is continuous in objectives at $h$ if each of its selection rules are either continuous in objectives at $h$ or sample convergent in objectives at $h$.*

*Proof.* By Corollary 8.2, mutation operators are extensibly continuous to $X$ and $Y$ in objectives everywhere. Suppose that $\mathcal{E}$ is an evolutionary algorithms with standard decomposition $\mathcal{E} = \mathcal{S} \star \mathcal{R} \star \mathcal{V}$. By Corollary 8.7, $\mathcal{R}$ is extensibly continuous to $X$ and $Y$ at $h$. Now $\mathcal{S} \star \mathcal{R}$ is continuous in objectives by either Theorem 8.2 or Corollary 8.6, and an final application of Theorem 8.2 to $(\mathcal{S} \star \mathcal{R}) \star \mathcal{V}$ shows that $\mathcal{E}$ is continuous in objectives at $h$.

The preceding theorem and its corollaries suffice to show establish continuity in objectives for a wide range of search and optimization methods even when using discontinuous selection rules, provided that they exhibit sample convergence. But sample convergence can be used to establish continuity with respect to trajectories as well. Since mutation operators are extensibly continuous in both objectives and trajectories according to Corollary 8.1, the obstacle to obtaining continuity in trajectories is found in the crossover rule.

The next theorem shows that masked crossover rules are sample convergent in general if they have sample convergent selection rules. Since most crossover rules used in evolutionary algorithms are masked crossover rules, this fact implies that the continuity of most evolutionary algorithms – in objectives or trajectories – can be established from the sample convergence of the selection rules.

**Theorem 8.4.** *The convolution of a selection rule and a recombination operator with a masked crossover rule is sample convergent at h if its selection rules are also sample convergent at h. The statement remains true if the sample convergence is in objectives or trajectories.*

*Proof.* The application of a masked crossover rule implies that the search domain $X$ is a $d$-dimensional vector space. Let $\mathcal{S}$ be a selection rule that is sample convergent. Let $\mathcal{R}$ be a recombination operator with a masked crossover rule. Then $\mathcal{S}\star\mathcal{R} = \mathcal{S}\star(\triangleleft\mathcal{S}_1\star(\cdots\star(\triangleleft_{n-1}\mathcal{S}_{n-1}\star\mathcal{C}\langle\mathbb{P}_M\rangle)))$ for sample convergent selection rules $\mathcal{S}_1,\ldots,\mathcal{S}_{n-1}$ and a masked crossover rule $\mathcal{C}\langle\mathbb{P}_M\rangle$ of order $n$. Assume that $n > 1$, since otherwise the crossover rule is trivial and there is nothing to prove. Let $\mathcal{S}_0 = \mathcal{S}$ to simplify the notation.

Each $\mathcal{S}_k$ is finitely trackable for $0 \le k < n$. This fact must be extended to make $\mathcal{R}$ finitely trackable. A generalized crossover mask selects one of the $n$ parents for each of the $d$ components, so there are $n^d$ possible masks. Let $(m_j)_{j=1}^{n^d}$ enumerate these masks, and let $S_h^k$ be the finite support set with size $s_i$ for $\mathcal{S}_i$ at $h$. Then the set

$$S_h = \left\{ \sum_{i=1}^{n} m_k \otimes_i x_i \;\middle|\; 1 \le k \le n^d \quad \text{and} \quad x_i \in S_h^i \text{ for all } 0 \le i < n \right\}$$

is a finite support set for $\mathcal{S}\star\mathcal{R}$ in light of Equations 7.30 and 7.31 together with the definition of convolution.

In order to associate a trajectory tracking function for each element of $S_h$, note that each element of $S_h$ may be specified (perhaps with multiplicity) by a multi-index $\alpha = (k, j_0, \ldots j_{n-1})$ where $1 \le k \le n^d$ and $1 \le j_i \le s_i$ for $0 \le i < n$. The index $k$ selects a mask, and the indices $j_i$ select the $j^{th}$ tracking function for $\mathcal{S}_i$. Then if $x_\alpha \in S_h$, a trajectory tracking function $\tau_\alpha$ can be defined as

$$\tau_\alpha(h') = \sum_{i=1}^{n} m_k \otimes_i \tau_{j_i}(h') \qquad \text{where } \tau_{j_i} \text{ is the } j_i^{th} \text{ tracking function for } \mathcal{S}_i.$$

This $\tau_\alpha$ is defined on an open neighborhood $H$ of $h$ given by $H = \bigcap_{i=0}^{n-1} H_i$ where each $H_i$ is an open neighborhood of $h$ on which $\mathcal{S}_i$ is finitely trackable. Each map $h' \mapsto m_k \otimes_i \tau_{j_i}(h')$ is continuous because it is the composition of three continuous operations: (1) application of $\tau_{j_i}$; (2) projection from $X$ to the $i^{th}$ component of $x$; and (3) injection back into $X$. Therefore each $\tau_\alpha$ is continuous as the sum of continuous functions. It is then apparent that $\mathcal{S} \star \mathcal{R}$ is finitely trackable by the collection $\{\tau_\alpha\}$.

The final step is to present a set of probability tracking functions for $\mathcal{S} \star \mathcal{R}$ that are continuous at $h$. For this purpose, define

$$(\mathcal{S} \star \mathcal{R})_{\tau_\alpha}(h') = \sum_{i=1}^{n} \mathbb{P}_M(m_k) \times (\mathcal{S}_i)_{\tau_{j_i}}(h')$$

for all $h' \in H$, where $(\mathcal{S}_i)_{\tau_{j_i}}$ is the $j_i^{th}$ probability tracking function for $\mathcal{S}_i$. Then $(\mathcal{S} \star \mathcal{R})_{\tau_\alpha}$ satisfies Equation 8.5 by the definition of $\mathcal{C}\langle \mathbb{P}_M \rangle$, and it is continuous at $h$ because each $(\mathcal{S}_i)_{\tau_{j_i}}$ is continuous at $h$. Therefore $\mathcal{S} \star \mathcal{R}$ is sample convergent at $h$.

**Corollary 8.9.** *An evolutionary algorithm with a masked crossover rule is continuous at a history h if its selection rules are sample convergent at h.*

*Proof.* Let $\mathcal{E} = \mathcal{S} \star \mathcal{R} \star \mathcal{V}$ be an evolutionary algorithm with its standard decomposition. Then $\mathcal{S} \star \mathcal{R}$ is sample convergent by Theorem 8.4, and $\mathcal{V}$ is extensibly continuous to $X$ and $Y$ at $h$. Thus $\mathcal{E}$ is continuous at $h$ by Theorem 8.3.

Sample convergence has provided the first tool in this chapter that permits one to conclude that an evolutionary algorithm is continuous not just in objectives but also in trajectories. This tool is now applied to study proportional selection with a modulating function as defined in Equation 7.13. As a reminder, selection rules for single-objective optimization, such as proportional selection, generally assume that the value space $Y$ is totally ordered.

**Theorem 8.5.** *Proportional selection with modulating function g is sample convergent everywhere if g is continuous in both arguments. If g is continuous in the first argument, then proportional selection is sample convergent in trajectories everywhere, and if g is continuous in the second argument, then proportional selection is convergent in objectives everywhere.*

*Proof.* Let $h$ be an arbitrary history trace. The claims will first be shown for unnormalized proportional selection, given for measurable $A$ by

$$\mathcal{UPS}\langle g \rangle [h](A) = \sum_{(x,y) \in H(h)_{-1}} g(x,y)\, \mathbb{1}_A(x), \tag{8.9}$$

and then the results will carry over for normalized proportional selection since the normalization process preserves the continuity of the probability tracking functions.

The obvious finite support set for $\mathcal{UPS}$ regardless of $g$ is

$$S_h = \{x \mid (x,y) \in H(h)_{-1} \text{ for some } y \in Y\},$$

and its size is bounded by the population size $K$. The trajectory tracking function $\tau_i$ is given by

$$\tau_i(h') = H(h')^X_{-1,i}$$

for $1 \leq i \leq K$ and $\tau_i(h')$, which plucks out the $i^{th}$ member of the last population. Note that $\tau_i$ is continuous within a neighborhood of $h$ in which all elements have the same length as $h$, since it is just a static projection from a product space. Setting $y_i(h') = H(h')^Y_{-1,i}$ to represent the value assigned to the $i^{th}$ member of the last population, a valid set of probability tracking functions is transparently given by

$$\mathcal{UPS}_{\tau_i}(h') = g\left(\tau_i(h'), y_i(h')\right), \tag{8.10}$$

which satisfies Equation 8.5. It remains to show that $\mathcal{UPS}_{\tau_i}$ is continuous at $h$ for all $i$. By assumption, $g$ is continuous in both arguments. Furthermore, the functions $\tau_i$ and $y_i$ are continuous everywhere, since there is some open neighborhood $h$ on which these functions are simply static projections from a product space. [3] Therefore $\mathcal{UPS}_{\tau_i}$ is continuous as the composition of continuous functions. All these facts together imply that $\mathcal{UPS}\langle g\rangle$ is finitely trackable at $h$.

Since $\mathcal{UPS}\langle g\rangle$ is finitely trackable at $h$ with a set of probability tracking functions that are continuous at $h$, it follows that $\mathcal{UPS}\langle g\rangle$ is sample convergent at $h$, and hence so it $\mathcal{PS}\langle g\rangle$.

The final claims about sample convergence in trajectories or objectives can be established by observing that each one corresponds to holding one or the other argument of $g$ fixed in Equation 8.10.

**Corollary 8.10.** *The simple genetic algorithm $\mathcal{SGA}$ of Equation 7.23 is continuous everywhere.*

*Proof.* Recall that $\mathcal{SGA}\langle p\rangle = (\mathcal{PS}\langle g\rangle \star ((\lhd\mathcal{PS}\langle g\rangle) \star \mathcal{SC})) \star \mathcal{B}\langle p\rangle$ with $g(x,y) = |y|$. The search space is $\{0,1\}^d$ with the discrete topology (i.e. all sets are open). The function $g$ is continuous in both arguments, and therefore $\mathcal{PS}\langle g\rangle$ is sample convergent everywhere by Theorem 8.5. The conclusion then follows from Corollary 8.9.

The same proof strategy can be used to show that genetic algorithms in any space are continuous everywhere when they use masked crossover and proportional selection with a continuous modulating function. For example, a real-coded genetic algorithm with proportional selection, uniform crossover, and Gaussian mutation is continuous in this way.

Proportional selection (also called roulette wheel selection) is no longer commonly used as a selection rule because of its sensitivity to numeric values, its requirement of a nonnegative fitness function, and its inability to prefer more refined solutions near an optimum. It has been replaced by tournament selection and ranking selection. Whereas roulette wheel selection makes a genetic algorithm continuous, tournament and ranking selection are discontinuous at some points. The following subsections identify these discontinuities, leading up to a full characterization of when exactly the more commonly used selection rules are continuous.

[3] Recall that under the disjoint union topology for history traces, every history has an open neighborhood within which all histories have the same length.

### 8.2.3 Sample Divergence and Discontinuity

In the previous section, sample convergence was used to show that many genetic algorithms are continuous on a large set of objectives. In this section, similar proofs will be used to demonstrate a converse result, that selection rules whose samples diverge are a source of discontinuities in the optimizer. The concept of sample divergence is defined next, followed by the inverse of Theorem 8.3.

**Definition 8.13 (Sample Divergent).** A search generator $\mathcal{G}$ is *sample divergent* at a history trace $h$ if it is finitely trackable at $h$ but not sample convergent at $h$.

Similarly, it will be said that a finitely trackable search generator $\mathcal{G}$ is sample divergent *in objectives* or *in trajectories* if it is not sample convergent in objectives or in trajectories, respectively. Being sample divergent logically requires that the search generator be continuously trackable, but that every possible set of probability tracking functions contain at least one discontinuous function.

Proving sample divergence or discontinuity works similar to how sample convergence or continuity are proven, but with the goal of showing that quantities are large rather than small. In general, in order to show a discontinuity, the total variation norm is bounded below by

$$\left\|\mathcal{G}[h'] - \mathcal{G}[h]\right\| \geq \left|\mathcal{G}[h'] - \mathcal{G}[h]\right|(A)$$

for every pair $h, h'$ and arbitrary measurable sets $A$. Using the Hahn decomposition again, $A$ partitions into two sets $A_+$ and $A_-$ such that the difference $(\mathcal{G}[h'] - \mathcal{G}[h])$ is positive on $A_+$ and negative on $A_-$. In this case, the inequality above becomes

$$\left\|\mathcal{G}[h'] - \mathcal{G}[h]\right\| \geq \left|\mathcal{G}[h'] - \mathcal{G}[h]\right|(A_+) + \left|\mathcal{G}[h'] - \mathcal{G}[h]\right|(A_-).$$

Now if for some $\varepsilon_0 > 0$ it can be shown that for all $\varepsilon < \varepsilon_0$ both terms on the right are greater than $\varepsilon$ no matter how close $h$ and $h'$ are, then $h$ is a point of discontinuity for $\mathcal{G}$. Topologically, every open neighborhood $H$ of $h$ has some $h' \in H$ such that the norm $\|\mathcal{G}[h'] - \mathcal{G}[h]\| > \varepsilon_0$. The discontinuity is proven by finding such an $\varepsilon_0$.

**Theorem 8.6.** *Let $\mathcal{A} \in \mathfrak{G}_{X,Y}$ be sample divergent at a history trace h, and let $\mathcal{B} \in \mathcal{VA}^1_{X,Y}$ be extensibly continuous to C and Y at h, where C is a set of full $\mathcal{A}[h]$-measure. Suppose that $|\mathcal{B}[h\|c]| \neq \mathbf{0}$ for all $c \in C$. Then $\mathcal{A} \star \mathcal{B}$ is discontinuous at h. The statement also holds separately either "in objectives" or "in trajectories".*

*Proof.* Let $h'$ be an arbitrary history trace. Let $\{\tau_i\}_{i\in[s]}$ be any collection of tracking functions for $\mathcal{A}$. Using the notation $\mathcal{A}_{\tau_i}$ and $\mathcal{B}_{i,h}$ from Equation 8.6 and letting $A$ be arbitrary with $A_*$ one of $A_+$ or $A_-$, it holds for any $i$ that

$$\left|\mathcal{A}\star\mathcal{B}[h]-\mathcal{A}\star\mathcal{B}[h']\right|(A_*)\geq\left|\mathcal{A}_{\tau_i}(h)\mathcal{B}_{i,h}(A_*)-\mathcal{A}_{\tau_i}(h')\mathcal{B}_{i,h'}(A_*)\right|$$
$$\geq\left|\left[\mathcal{A}_{\tau_i}(h)-\mathcal{A}_{\tau_i}(h')\right]\mathcal{B}_{i,h}(A_*)\right.$$
$$\left.+\ \mathcal{A}_{\tau_i}(h')\left[\mathcal{B}_{i,h}(A_*)-\mathcal{B}_{i,h'}(A_*)\right]\right|,$$

where the second inequality results from inserting a factor $\mathcal{A}_{\tau_i}(h')\mathcal{B}_{i,h}(A_*)$ in the middle.

By the sample divergence of $\mathcal{A}$, there is some $i$ and some $\eta_0$ such that for any neighborhood $H$ of $h$, there exists $h'\in H$ with $|\mathcal{A}_{\tau_i}(h)-\mathcal{A}_{\tau_i}(h')|>\eta_0$. Using the continuity of $\mathcal{B}$, let $H$ be an open neighborhood of $h$ such that $\|\mathcal{B}_{i,h}-\mathcal{B}_{i,h'}\|<\eta$ for $\eta>0$. In fact, we now have $\|\mathcal{A}\star\mathcal{B}[h]-\mathcal{A}\star\mathcal{B}[h']\|\geq$

$$\left|\mathcal{B}_{i,h}(A_*)\right|\left|\left[\mathcal{A}_{\tau_i}(h)-\mathcal{A}_{\tau_i}(h')\right]+\frac{\mathcal{A}_{\tau_i}(h')}{\mathcal{B}_{i,h}(A_*)}\left[\mathcal{B}_{i,h}(A_*)-\mathcal{B}_{i,h'}(A_*)\right]\right|.$$

Consequently, if $\eta<\frac{\mathcal{B}_{i,h}(A_*)}{2\mathcal{A}_{\tau_i}(h')}\eta_0$, substituting $-\eta<[\mathcal{B}_{i,h}(A_*)-\mathcal{B}_{i,h'}(A_*)]<\eta$,

$$\left\|\mathcal{A}\star\mathcal{B}[h]-\mathcal{A}\star\mathcal{B}[h']\right\| \quad\geq\quad \frac{1}{2}\left|\mathcal{B}_{i,h}(A_*)\right|\eta_0=\varepsilon_0.$$

Note that the last few inequalities do not work for arbitrary $A$; instead, the set $A$ must be constrained to avoid division by zero. First, $\mathcal{B}_{i,h}(A_*)>0$, which can be guaranteed for either $A_+$ or $A_-$ for some $A$ since $|\mathcal{B}[h\|\tau_i(h)]|\neq\mathbf{0}$ by assumption. Secondly, $\mathcal{A}_{\tau_i}(h')>0$ since $\tau_i(h')$ is in the finite support of $\mathcal{A}[h']$. In light of the above inequalities, $\mathcal{A}\star\mathcal{B}$ is discontinuous at $h$.

Theorem 8.4 stated that a masked crossover rule preserves sample convergence from its selection rules. An analogue to this theorem is true; masked crossover also preserves sample divergence. The following Theorem and Corollary can be proven in a similar way to Theorem 8.4 and its corollaries, and so the proofs are omitted.

**Theorem 8.7.** *A recombination operator with a masked crossover rule is sample divergent at h if any one of its selection rules is sample divergent at h. The same is true "in objectives" and "in trajectories".*

**Corollary 8.11.** *An evolutionary algorithm with a masked crossover rule is discontinuous at h if any one of its selection rules is sample divergent at h and its mutation operator satisfies the conditions on $\mathcal{B}$ in Theorem 8.6. The same is true "in objectives" and "in trajectories".*

The theorems above provide mechanisms for locating points of discontinuity for common search methods. This theme is explored next.

## 8.2.4 Discontinuities of Specific Selection Rules

Theorem 8.5 showed that generalized proportional selection is sample convergent where the modulating function is continuous on the image of the objective function. The next result shows the opposite. Proportional selection is sample divergent when the composition of the modulating function and the objective is discontinuous. As above, note that for single-objective optimization, the value space $Y$ is assumed to be totally ordered.

**Theorem 8.8.** *Proportional selection with modulating function $g$ is sample divergent at a history trace $h$ whenever $g$ is discontinuous at $(H(h)^X_{-1,i}, H(h)^Y_{-1,i})$ for some $1 \leq i \leq K$, where $K$ is the population size. The statement holds "in objectives" if $g$ is only discontinuous in the second argument at this point and "in trajectories" if $g$ is only discontinuous in the first argument.*

*Proof.* As in the proof of Theorem 8.5, we use unnormalized proportional selection, $\mathcal{UPS}$. It has already been shown that $\mathcal{UPS}$ is finitely trackable everywhere. Let $\{\tau_i\}_{1 \leq i \leq K}$ be any set of trajectory tracking functions, and note that Equation 8.10 applies, except that now $g$ is discontinuous with the same arguments for some $i$. Therefore $\mathcal{UPS}\langle g \rangle$ is discontinuous as well. Normalization does not remove the discontinuity, and so proportional selection with $g$ is discontinuous.

The proof of Theorem 8.8 can be leveraged to conclude that tournament selection and ranking selection are also sample divergent on the majority of objectives for certain histories. The following definition will make explicit the trajectories on which this discontinuity occurs. Specifically, the discontinuity occurs whenever the history either contains two search points with equal value or assigns two different values to the same search point, both within some fixed time window.

**Definition 8.14 (Ambivalent Value).** A history trace $h \in \mathcal{H}^{\mathcal{T}}_{X,Y}$ is *of ambivalent value* at degree $d$ if there exist $x, x' \in X$ and $y, y' \in Y$ such that for some $i, j$ with $|h| - d < i, j \leq d$ either (1) $x \neq x'$ but $(x,y) = h_i$ and $(x',y) = h_j$ or (2) $y \neq y'$ but $(x,y) = h_i$ and $(x,y') = h_j$. Otherwise, $h$ is of *unambivalent value* at degree $K$. The history trace $h$ is of ambivalent value *at full degree* if $d = |h|$; the degree may be omitted if clear from the context.

**Theorem 8.9.** *Tournament selection (Equation 7.14) with selection pressure $q \in [0,1)$ and ranking selection (Equation 7.16) with selection pressure $q \in [0,2)$ are both sample divergent at every history trace that is of ambivalent value at the degree of the selection rule and sample convergent at every history trace that is of unambivalent value at the same degree.*

*Proof.* Let $\text{Rank}(x,P)$ be the ranking function of Section 7.2.4, which is the rank of an individual $x$ in the evaluated population $P$, ranging from 0 to $K-1$ with 0 as the best rank. In order to discuss sample divergence, note that by the same logic as from unnormalized proportional selection in Theorem 8.5, both tournament selection and ranking selection are finitely trackable, and their probability tracking functions are continuous function of the ranking function, which captures their entire dependence on the history. Whether or not they are sample convergent at a particular history $h$ therefore depends exactly on whether the ranking function is continuous at $(H(h)^X_{-1,i}, H(h)_{-1})$ for all $1 \leq i \leq K$. It will be shown that the ranking function is continuous at this point if and only if $h$ is of unambivalent value at degree $K$.

The last population is $H(h)_{-1} = ((x_1,y_1),\ldots,(x_K,y_K))$. Since ranks are discrete, the question of continuity depends for each $i$ on whether the set

$$R = \left\{ h' \;\middle|\; \text{Rank}(H(h')^X_{-1,i}, H(h')_{-1}) = \text{Rank}(y_i, H(h)_{-1}) \right\}.$$

Suppose $h'$ is "near" $h$. In order for the $i^{th}$ rank in $h'$ to differ from $h$, either some point with lower rank must surpass $y_i$ or a point with higher rank must be surpassed by it.

Suppose $h$ is of unambivalent value at degree $K$. If follows that for every $i$ there is an open clearance around both $x_i$ and $y_i$ within which rank must preserved. Projecting these clearances into the product topology for $\mathcal{H}^{\mathcal{T}}_{X,Y}$ yields the set $R$, which is therefore open. Thus the ranking function is continuous at the required point, and the selection rule is sample convergent.

Suppose $h$ is of ambivalent value at degree $K$. Then because either $x_i = x_j$ with $y_i \neq y_j$ or $y_i = y_j$ with $x_i \neq x_j$ for some $i, j$ with $i \neq j$, there is no such clearance within which rank is preserved. Suppose the former and fix any open neighborhood $N$ of $x_i$. For any $x_j \in N$ with $x_i \neq x_j$, the relative rank of the two points changes. Hence if $N'$ is the upward projection of $N$ into $\mathcal{H}^{\mathcal{T}}_{X,Y}$, $N'$ is open, but $N' \cap R$ is not open. Therefore $R$ cannot be open, and we conclude that the boundary of $R$ is sharp when the point of equality is disrupted. Since the exact same argument is true when $y_i = y_j$, it follows that the ranking function is discontinuous at the required point, and the selection rule is sample divergent.

An analysis of truncation selection will complete this survey of continuity in evolutionary algorithms. Truncation selection, used by evolution strategies and estimation of distribution algorithms, also depends indirectly on the rank. A *truncation selection rule* is a selection rule with population size $K$ that places probability one on the best $T$ members of the last population, with $1 \leq T < K$. In Section 7.2.5, evolution strategy selection was defined as $\mathcal{ESS}$ and $\mathcal{ESS}_+$. Both of these selection rules are truncation selection rules. Like tournament selection and ranking selection, truncation selection rules depend continuously on the ranking function. Therefore they are sample convergent and divergent exactly when the history trace is unambivalent or ambivalent at the degree of the population size. This fact is stated in the next proposition. Its proof is identical to that of Theorem 8.9.

**Proposition 8.10.** *A truncation selection rule is sample divergent on histories of ambivalent value and sample convergent on histories of unambivalent value, both at the degree of the population size.*

As a corollary to this proposition along with Theorems 8.3 and 8.6, evolution strategies are therefore continuous under certain conditions.

**Corollary 8.12.** *Evolution strategies with intermediate or dominant crossover are continuous at a history trace h if and only if h is of unambivalent value.*

The principles from the theorems above are not restricted to evolutionary algorithms. Sample convergence is an important and useful concept that can be used to demonstrate the continuity or discontinuity of quasi-evolutionary algorithms and even stochastic gradient descent, as is done in the next two sections.

## 8.3 Quasi-Evolutionary Algorithms

Quasi-evolutionary algorithms can be described in terms of selection, crossover, and mutation operators, but they typically use much more complex selection or crossover mechanisms. Techniques for determining the continuity of evolutionary algorithms were described in at a general level in Section 8.2 so that the same techniques can be applied to demonstrate the discontinuity of the most popular quasi-evolutionary algorithms with respect to objectives. This section develops this result for the parameterized model-building methods and the class of locally improving optimizers, which includes differential evolution.

### *8.3.1 Parameterized Methods*

In Chapter 2, the class of parameterized quasi-evolutionary methods was introduced. These methods include estimation of distribution algorithms (EDAs) and natural evolution strategies (NES), which subsumes Correlated Matrix Adaption (CMA-ES). The most popular methods in this class are discontinuous at histories of ambivalent value, but are for the most part continuous elsewhere, at least in objectives. When they arise, the discontinuities in objectives are due to the use of truncation selection.

An EDA with truncation selection can be represented as the convolution of a truncation selection rule and a model sampler. The model sampling procedure typically does not depend on the fitness values, but only on the selected members of the population. In this case, the model sampler is value-agnostic and therefore continuous in objectives, and so the EDA can be shown to be continuous or discontinuous in objectives by applying Theorem 8.3 or Theorem 8.6, depending on whether the trajectory in question is of ambivalent value. The same logic also holds for Natural Evolution Strategies, including CMA-ES.

The proofs in Section 8.2 were developed abstractly. A proof that is specific to a known algorithm may be help to make the meaning of these results more concrete. With this goal in mind, this section shows directly that the Bayesian Optimization Algorithm (BOA), a popular EDA, is discontinuous in objectives. To this end, let $\mathcal{BOA}\langle T,K\rangle$ represent BOA with a population size of $K$ and a truncation size of $T$. $\mathcal{BOA}\langle T,K\rangle$ builds a directed graphical model from the best $T$ individuals out of a population of size $K$ by employing a greedy hill-climbing search through graph structures using the K2 metric [123, 124, 122, 72]. For this example, a binary search domain is assumed, $X = \{0,1\}^d$; the value space is $Y = \mathbb{R}$; and the temporal index set $\mathcal{T}$ is finite. Recall that a history $h$ agrees with an objective $u : X \to Y$ if $h = ((x_i, u(x_i)))_{i\leq|h|}$ for some $(x_i)_{i\leq|h|}$; that is, $h_Y$ is determined from $h_X$ by $u$.

**Proposition 8.11.** *$\mathcal{BOA}\langle T,K\rangle$ is discontinuous in objectives on $X = \{0,1\}^d$ when $1 < T < K$.*

*Proof.* Let $f_n(x) = -\frac{1}{n}\delta_1(x_0)$ with $\delta$ as the Kronecker delta and $x_0$ the first bit of $x$, and let $f(x) = 0$. Then $f_n \to f$. Note that for any $x \in X$, $f_n(x) = -\frac{1}{n}$ if $x_0 = 1$ and $f_n(x) = 0$ if $x_0 = 0$. But $f(x) = 0$ for all $x$. Define

$$A = \{x \in X : x_0 = 1\}, \; B = \{x \in X : x_0 = 0\}.$$

Assume without loss of generality that truncation selection prefers elements in $B$ over elements in $A$ when their fitness is equal. Let $h$ be a trajectory of length $K$, and let $h$ have exactly $T$ elements in $A$ and $K-T$ elements in $B$. The $T$ elements in $A$ are more optimal on $f_n$ than the elements in $B$ because BOA builds a model out of the best $T$ elements. Fix $h$ agreeing with $f$ and suppose that $h_n$ agrees with $f_n$ and has $h_n^X = h_X$ Then for some fixed $\varepsilon > 0$ determined by the smoothing procedure used for model estimation in the Bayesian network,

$$\mathcal{BOA}\langle T,K\rangle[h_n](A) = 1-\varepsilon. \tag{8.11}$$

But on $f$ all elements in $h_X$ are equally optimal. Thus the model constructed by BOA for $h$ should produce elements from $B$ with approximate probability $\frac{K-T}{K}$, i.e. $\mathcal{BOA}\langle T,K\rangle[h](A) \approx \frac{K-T}{K} \neq 1-\varepsilon$.

Since $\mathcal{H}_{X,Y}^{\mathcal{T}}$ is here a sequential space, we have $h_n \to h$ but $\mathcal{BOA}\langle T,K\rangle[h_n] \not\to \mathcal{BOA}\langle T,K\rangle[h]$. Therefore $\mathcal{BOA}$ is discontinuous.

This proof shows how discontinuities can appear in algorithms such as BOA. Notice that the discontinuity in the proof above exists because the objective function was chosen to be identically zero, which implies that all trajectories are of ambivalent value. Like most EDAs, BOA is generally continuous in objectives on trajectories of unambivalent value.

### *8.3.2 Differential Evolution and Locally Improving Optimizers*

In Sections 8.2 and 8.3.1, optimizers that depend on the fitness rank of evaluation points have been repeatedly shown to be discontinuous in objectives exactly on trajectories of ambivalent fitness. Differential evolution also depends on the objective rank but in a different way from the previously analyzed optimizers, and thus a different type of continuity proof is required.

To generalize the result, consider population-based optimizers with population size $K$ that depend only on the local best solutions for each member of the population. Given a history $h$, define the best running population by $best\,(h) \in X^K$ so that $h$ is treated as $K$ separate trajectories in the search domain, one per population slot, and $best\,(h)$ stores the best individual along each of these trajectories. Formally, $best(t,f)_k = \max\left\{H(h)^Y_{n,k} \mid n \leq |H(h)|\right\}$, where $(x,y) = H(h)_{n,k}$ is the $k^{th}$ individual in the $n^{th}$ population derived from the trajectory $h$ together with its value. To resolve ambiguities, let $best(h)$ take on the value corresponding to the largest $n$. Such optimizers will be termed *locally improving*.

**Definition 8.15 (Locally Improving).** An optimizer $\mathcal{G} \in \mathcal{PBO}^K_{X,Y}$ is *locally improving* if $\mathcal{G}[h] = \mathcal{G}[h']$ if and only if $best\,(h) = best\,(h')$.

As an aside, the space of locally improving optimizers forms a vector subspace of $\mathfrak{O}_{X,Y}$, because the locally improving property is preserved by vector operations.

In all but the simplest search domains, locally improving optimizers are continuous in objectives on histories satisfying an analogue of the unambivalent value requirement of the previous sections. The following definition extends the definition of ambivalent value to account for the structure of locally improving optimizers.

**Definition 8.16.** A history $h \in \mathcal{H}^{\mathcal{T}}_{X,Y}$ is of *componentwise ambivalent value* at degree $K$ if for some $k$ with $1 \leq k \leq K$, there exist $m,n$ such that (1) $H(h)^X_{m,k} \neq H(h)^X_{n,k}$, (2) $H(h)^Y_{m,k} = H(h)^Y_{n,k}$, and (3) $H(h)^Y_{n,k} \leq H(h)^Y_{i,k}$ for all $i$. Otherwise, $h$ is of *componentwise unambivalent value* at degree $K$.

The main concept is the same as in Theorem 8.9. Any objective function can be modified to add a mode (or even a plateau) of arbitrary size along histories of ambivalent value, and when this is done, the vector $best(h)$ changes discontinuously with the objective function as the added mode becomes arbitrarily small. The proof is given next.

**Theorem 8.10.** *Every locally improving optimizer is continuous in objectives at h if and only if h is of componentwise unambivalent value at the degree of the population size.*

*Proof.* Let $\mathcal{G} \in \mathcal{PBO}^K_{X,Y}$ be locally improving with population size $K$. Simply by the fact that $\mathcal{G}$ is locally improving, it follows that there is a function

$$\hat{\mathcal{G}} : X^K \to \mathcal{M}_X$$

such that $\mathcal{G}[h] = \hat{\mathcal{G}}[best(h)]$. Since $\hat{\mathcal{G}}$ does not depend on $Y$ in any way, it follows that $\mathcal{G}$ is continuous in objectives at $h$ if and only if $best(h)$ is continuous at $h$ when $h_X$ is held fixed. But when $h_X$ is held fixed, the range of $best(h)$ is necessarily constrained to a finite subset of $X^K$. Thus as in Theorem 8.9, it suffices to consider the set

$$R = \left\{ h'_Y \;\middle|\; h'_X = h_X \text{ and } best(h') = best(h) \right\}.$$

The continuity in objectives of $\mathcal{G}$ depends on whether this set is open in $\mathcal{H}^{\mathcal{T}}_{Y}$.

Suppose $h$ is of componentwise unambivalent value. Then $R$ is open as before, since for all $k$ there is an open neighborhood around the value of $best(h)_k$ that can be projected up to form an open neighborhood of $h$ in $\mathcal{H}^{\mathcal{T}}_{X,Y}$. In this case, best is continuous with respect to $h_Y$ with $h_X$ fixed, so $\mathcal{G}$ is continuous in objectives at $h$.

Suppose $h$ is of componentwise ambivalent value. Then as with the ranking function, for all $k$, every neighborhood of $h_Y$ contains a point $h'_Y$ for which it holds that $best(\kappa(h_X, h'_Y)) \neq best(h)$. Therefore the set $R$ is not open, and $\mathcal{G}$ is discontinuous in objectives at $h$.

This theorem makes it clear exactly when a locally improving optimizer is continuous in objectives. These facts will be employed in the next two chapters to conclude that the performance of locally improving optimizers is continuous. That is, these optimizers perform similarly on similar objectives.

## 8.4 Static Search Generators

The discussion in the previous sections has primarily focused on evolutionary and quasi-evolutionary algorithms using the structure of the convolution operator. However, many other search methods can also be characterized as resulting from convolution. Furthermore, in the next chapter, the behavior of a search generator run on a specific search problem will be studied, and hence it will be necessary to consider how the history generated by a search generator changes when the problem changes. To begin to answer such questions, the static search generators introduced in Chapter 6 will be used, since they incorporate a specific search problem as an input. First, the continuity of the static black-box search generators is shown to follow under certain conditions from the continuity of a corresponding standard search generator. Then stochastic gradient descent is presented as an example of a static search generator without the black-box property, but still provably continuous using the methods of this chapter.

### 8.4.1 Continuity Transfer Through the Black-Box Property

Recall that a static search generator has the black-box property as defined in Section 6.3.1 if it is the image of some standard search generator under the function $\mathbf{G}(\cdot)$ defined by the relationship in Equation 6.11. The space of such static search generators is denoted $\mathcal{BB}_{X,Y}$.

Since a static search generator is a function $\mathcal{G} : \mathcal{H}^{\mathcal{T}}_{X,Y} \times Y^X \to \mathcal{M}_X$, it is topologically continuous if the pre-image under $\mathcal{G}$ of every open set in $\mathcal{M}_X$ is open in $\mathcal{H}^{\mathcal{T}}_{X,Y} \times Y^X$. This concept is identical to the continuity of standard search generators, except that the input space is the product of a search history and an objective rather than a full history trace. Thus all of the considerations from the previous section apply, including continuity at a point and the use of $\varepsilon$ arguments to prove continuity.

Continuity of a static search generator $\mathcal{G}[h_X, u]$ means continuity with respect to the pair $(h_X, u)$. If the map $h_X \to \mathcal{G}[h_X, u]$ is continuous for each fixed objective $u$, then $\mathcal{G}$ is continuous *in trajectories*. If the map $u \to \mathcal{G}[h_X, u]$ is continuous for each fixed search history $h_X$, then $\mathcal{G}$ is continuous in objectives. If $\mathcal{G}$ is continuous in both trajectories and objectives, then it is continuous in the topological sense. In order to establish continuity in trajectories at $(h_X, u)$, it suffices for $\mathcal{G}$ to be the image under $\mathbf{G}(\cdot)$ of a standard search generator $\mathcal{G}'$ that is continuous at some history trace $h$ that agrees with $u$; *i.e.*, $h_Y = (u(h_1^X), \ldots u(h_{|h|}^X))$.

**Theorem 8.11.** *Suppose that $\mathcal{G} \in \mathcal{BB}_{X,Y}$ is the image under $\mathbf{G}(\cdot)$ of a standard search generator that is continuous in trajectories at a history trace $h$ that agrees with an objective $u$. Then $\mathcal{G}$ is continuous in trajectories at $(h_X, u)$.*

*Proof.* Fix $\varepsilon > 0$. Let $\mathcal{G}'$ be a standard search generator such that $\mathcal{G} = \mathbf{G}(\mathcal{G}')$. But then Equation 6.11 implies that for any objective $u$ and any histories $h, h'$ that agree with $u$,

$$\left\| \mathcal{G}[h'_X, u] - \mathcal{G}[h_X, u] \right\| = \left\| \mathcal{G}'[h'] - \mathcal{G}'[h] \right\|.$$

If $\mathcal{G}'$ is continuous in trajectories at $h$, then there is an open neighborhood $H_X$ of $h_X$ such that $\|\mathcal{G}'[h'] - \mathcal{G}'[h]\| < \varepsilon$ whenever $h'_Y = h_Y$ and $h'_X \in H_X$. Since $h'$ agrees with $u$, it follows that $\mathcal{G}$ is continuous in trajectories at $(h_X, u)$.

Continuity in objectives is slightly more difficult to prove, since there is a mismatch in the domain, with $Y^X$ as for static search objectives versus $\mathcal{H}_Y^{\mathcal{T}}$. Given any search history $h_X$, define a map $\zeta_{h_X} : Y^X \to Y^{[|h_X|]}$ by

$$\zeta_{h_X}(u') = \left( u'(h_1^X), \ldots, u'(h_{|h_X|}^X) \right).$$

The function $\zeta_{h_X}$ is a projection of an objective $u' \in Y^X$ into $Y^{[|h_X|]}$. If the function $\zeta_{h_X}$ is continuous, then it can be used to transfer open neighborhoods of from $\mathcal{H}_Y^{\mathcal{T}}$ to $Y^X$. Specifically, if $h$ is a history that agrees with $u$, then $\zeta_{h_X}(u) = h_Y$. If $H_Y$ is any open neighborhood of $h_Y$, then, fixing the history length, $h_Y$ has some other open neighborhood $H'_Y \subseteq H_Y$ contained wholly in $Y^{[|h_X|]}$ (in fact, $H'_Y = H_Y \cap Y^{[|h_Y|]}$), and $U = \zeta_{h_X}^{-1}(H'_Y)$ is an open neighborhood of $u$.

So when is $\zeta_{h_X}$ continuous? The set $Y^X$ has the product topology, which is defined so that all finite projections are continuous. Therefore, the projection $\zeta_{h_X}$ is continuous if and only if $|h_X|$ is finite. This finiteness becomes the condition under which continuity in objectives can be transferred from standard search generators to static search generators.

**Theorem 8.12.** *Suppose that $\mathcal{G} \in \mathcal{BB}_{X,Y}$ is the image under $\mathbf{G}(\cdot)$ of a standard search generator that is continuous in objectives at a history trace $h$ that agrees with an objective $u$. If $h$ has finite length, then $\mathcal{G}$ is continuous in objectives at $(h_X, u)$.*

*Proof.* Fix $\varepsilon > 0$ and let $\mathcal{G}'$ be a standard search generator such that $\mathcal{G} = \mathbf{G}(\mathcal{G}')$. Again, Equation 6.11 implies that for any objectives $u, u'$, any search history $h_X$, and any value histories $h_Y, h'_Y$ that agree with $u$ on $h_X$,

$$\left\|\mathcal{G}[h_X, u'] - \mathcal{G}[h_X, u]\right\| = \left\|\mathcal{G}'[\kappa(h_X, h'_Y)] - \mathcal{G}'[\kappa(h_X, h_Y)]\right\|.$$

If $\mathcal{G}'$ is continuous in objectives at $h = \kappa(h_X, h_Y)$, then there is some open neighborhood $H_Y$ of $h_Y$ such that the right-hand side is less than $\varepsilon$. As discussed in the text above, $H'_Y = H_Y \cap Y^{[|h|]}$ is also a neighborhood of $h_Y$ with this same property. If $h$ has finite length, then $\zeta_{h_X}$ is continuous, and therefore $U = \zeta_{h_X}^{-1}(H'_Y)$ is an open neighborhood of $u$ satisfying $\|\mathcal{G}[h_X, u'] - \mathcal{G}[h_X, u]\| < \varepsilon$ for all $u \in U$.

Summarizing, as long as the temporal index set $\mathcal{T}$ is of countable cardinality or less, the operator $\mathbf{G}(\cdot)$ transfers continuity in both trajectories and objectives from standard search generators to static search generators. Even if $\mathcal{T}$ is not countable, the continuity still transfers, but continuity in objectives is limited to histories of finite length. These facts will be used to assess the continuity of performance metrics in the next chapter.

### 8.4.2 Stochastic Gradient Descent

The concept of sample convergence is not only useful for assessing the continuity of evolutionary and quasi-evolutionary methods. Returning briefly to quasi-Newton methods, a sample convergence argument can be used to conclude that stochastic gradient descent is continuous in objectives. Furthermore, a new mapping from static search generators to standard search generators can be introduced that represents an encoding of gradient methods applied to differentiable objectives.

Let the search domain be $X = \mathbb{R}^d$. As a first step, recall that a search generator for gradient descent on a differentiable objective $u : \mathbb{R}^d \to \mathbb{R}$ can be defined by assuming that the value space is $Y = \mathbb{R}^d$ and that the value assigned to each search point $x$ is $y = \nabla u(x)$. In this case, gradient descent is a deterministic proper search generator. Using the map $\Xi$ of Section 8.1.2, basic gradient descent with learning rate $\eta$ is a generator $\mathcal{G}$ such that

$$\Xi \langle \mathcal{G} \rangle (h) = h^X_{-1} - \eta h^Y_{-1} = x_{|h|} - \eta \nabla u(x_{|h|}),$$

where the notation $x_n$ is used for convenience to indicate the $n^{th}$ element of the search history $h_X$.

Stochastic gradient descent in $\mathbb{R}^d$ commonly adds a Gaussian noise factor to the gradient. This noise can be thought of as a mutation centered at the basic gradient descent update with a isotropic noise applied. Thus if $\mathcal{N}\langle\sigma\rangle$ is Gaussian mutation with variance $\sigma^2$ as in Chapter 7, then stochastic gradient descent can be represented as

$$\mathcal{SG}\langle\sigma\rangle = \mathcal{G}\star\mathcal{N}\langle\sigma\rangle\,.$$

Now $\mathcal{G}$ is sample convergent. It clearly finitely trackable with size 1, and because it is deterministic, the single value to track has a continuous probability tracking function with a constant value of 1. Since Gaussian mutation is extensibly continuous, Theorem 8.3 implies that $\mathcal{SG}$ is continuous.

This continuity is somewhat unsatisfying, since it relies on the weak assumption that the values are chosen so that $y = \nabla u(x)$. To improve this conclusion, we can define a static search generator that does implement this choice. Suppose that a pair $(\mathcal{A}, \mathcal{A}')$ consisting of a static search generator $\mathcal{A}$ and a standard search generator $\mathcal{A}'$ satisfies

$$\mathcal{A}[h_X, u] = \mathcal{A}'[((x, \nabla u(x)))_{x \in h_X}] \tag{8.12}$$

for all $h_X \in \mathcal{H}_X^{\mathcal{T}}$ and for all differentiable $u$. This constraint is an analogue of Equation 6.11, but in which the objective is mapped to a different value, specifically the gradient. As with the operator $\mathbf{G}(\cdot)$, Equation 8.12 establishes a linear relationship between static search generators and standard search generators, but with a change of value space (from $Y = \mathbb{R}$ to $Y = \mathbb{R}^d$) that is specific to the search domain.

Let $\mathbf{N}(\cdot)$ represent the operator implementing Equation 8.12 in analogy with $\mathbf{G}(\cdot)$. Then the domain of $\mathbf{N}(\cdot)$ represents the class of first-order Newton methods that depend on the gradient of an objective (but not its Hessian).

The search generator $\mathcal{SG}$ corresponds to a static search generator $\mathcal{SG}'$ in the domain of $\mathbf{N}(\cdot)$. The continuity of $\mathcal{SG}'$ can be established using the same methods as in the previous subsection, but with $\mathbf{N}(\cdot)$ as a transfer function in place of $\mathbf{G}(\cdot)$. There is a caveat in that Equation 8.12 says nothing about what happens when the objective $u$ is not differentiable, and hence continuity can only be transferred on search histories for which the objective is differentiable.

Thus we find that the static search generator $\mathcal{SG}'$ implementing stochastic gradient descent is continuous in objectives at $(h_X, u)$ if $u$ is differentiable at each search point in $h_X$.

## 8.5 Conclusion

This chapter has provided tools to assess the continuity of various search generators with respect to histories, including search histories and value histories. Continuity is an important analytical tool because, as the next two chapters will show, continuous search generators perform similarly on similar problems.

The next chapter addresses what happens when the search generator is run for several steps on a particular search problem. In this context, continuity in objectives is more important than continuity in trajectories, since the search generator controls the trajectory but not the objective. Because the continuity of search generators is important for analyzing optimizer performance, this chapter has reviewed the continuity of a variety of optimization methods discussed in previous chapters.

Deterministic search generators have been shown to be discontinuous everywhere in objectives, although because of their singular nature, these generators can still have continuous performance if they are sample convergent.

Evolutionary and quasi-evolutionary methods are continuous in objectives at least on histories of unambivalent value. In the next chapter, histories of ambivalent value will be shown to have measure zero when optimization is run on a sufficiently large search space, so that these methods are continuous almost everywhere with respect to the histories they generate.

The concept of sample convergence or divergence has proven to be theoretically important for search methods in general, even though it only applies to search generators that are singular. This importance derives from the fact that most popular search and optimization methods can be decomposed into a convolution of a singular search generator and a nonsingular one, as was done for stochastic gradient descent.

A final question about continuity is the following: When does continuity of an search generator imply continuity on the stochastic process that it generates? This question is addressed in the next chapter, along with an analysis of the long-running behavior of search generators.

# Chapter 9
# Generating the Optimization Process

A major goal of the previous chapters has been to enable a rigorous mathematical analysis of optimizer performance within a common space of optimization methods. Performance is not directly determined by how an optimization method chooses a particular point given a history; it arises as a consequence of the totality of decisions made by an optimization method over the course of time. Therefore, the search generators analyzed in the prior chapters must be connected to a method for generating complete history traces over the entire temporal index set. This construction was performed for finite spaces in Section 3.3. In this chapter, it will be repeated more generally for the class of static search generators on Hausdorff topological spaces with countable temporal index sets, paving the way for an analysis of optimizer performance on static objectives in Chapter 10 and an extension of the No Free Lunch theorems to infinite-dimensional spaces in Chapter 12. The most general version of this extension, though, is delayed until Chapter 17, where the concept of search problems is explored in its full generality. By restricting to static objectives in the next few chapters, a more straightforward analysis of optimizer behavior will be possible.

## 9.1 Construction of the Optimization Process

As mentioned above, this chapter analyzes what happens when search generators are used to optimize a single given static objective function. In this case, the static search generators of Section 6.3.1 provide the appropriate tool, since they describe how to generate search points given a search history and an objective. To simplify the analysis, the temporal index set is assumed here to be the natural numbers, *i.e.*, $\mathcal{T} = \mathbb{N}$.

The method of how to generate a history trace from a search generator has already been discussed twice, in Section 3.3 for the finite case. Section 4.3.2 provided some of the first steps of the extension for search generators, but did not go beyond



A. J. Lockett, *General-Purpose Optimization Through Information Maximization*,
Natural Computing Series, https://doi.org/10.1007/978-3-662-62007-6_9

exhibiting in what sense a generator generates search points. What follows is the complete extension process for static search generators.

When a static search generator $\mathcal{G}$ is run on a particular objective $u$, it generates a history trace $h$. In the static case, the value history $h_Y$ is completely determined by $u$. Thus there is no need to consider how $h_Y$ is generated, and so we focus on the search history. If $h_X \in X^{\mathcal{T}}$ is a complete search history generated by $\mathcal{G}$ and $h^X_{<t}$ represents history prefix up to but not including time $t$, then the phrase "generated by $\mathcal{G}$" is supposed to mean that $h^X_t$ is sampled as

$$h^X_t \sim \mathcal{G}[h^X_{<t}, u],$$

under that assumption that $h^X_{<1} = \emptyset$. The question, as in the other two sections referenced, is how to generate a probability distribution governing $h_X$ that is faithful to $\mathcal{G}$ as run on $u$.

In order to define a probability distribution for $h_X$, the set $X^{\mathcal{T}}$ must be a measurable space, meaning that a $\sigma$-algebra is required on $X^{\mathcal{T}}$. The simple choice for this purpose would the Borel $\sigma$-algebra on the product topology, and this simple choice suffices when $\mathcal{T} = \mathbb{N}$ as assumed in this chapter. However, for general (*i.e.*, uncountable) temporal index sets $\mathcal{T}$, this simple choice is inadequate, because the primary tool for constructing probability distributions, the Kolmogorov Extension Theorem, can fail to define complete probability measures on this $\sigma$-algebra, as discussed in Sections 5.3.3 and 5.3.4. Thus the proper choice is the Baire $\sigma$-algebra on $X^{\mathcal{T}}$, which is the smallest $\sigma$-algebra that makes all finite and countable projections of sequences in $X^{\mathcal{T}}$ measurable. In the case where $\mathcal{T} = \mathbb{N}$, this distinction is vacuous; the Baire $\sigma$-algebra is equal to the Borel $\sigma$-algebra on countable product spaces. The distinction becomes relevant, however, when the product space is large, such as $Y^X$, which occurs in Chapter 12 when probability measures are placed on the space of objectives. For this reason, the distinction is emphasized here.

The goal now is to generate from a static search generator $\mathcal{G}$ and an objective $u$ a probability distribution $\mathcal{A}[u]$ over search histories $h_X$ such that $\mathcal{A}[u]$ is measurable with respect to the Baire $\sigma$-algebra; this goal is accomplished using the Kolmogorov Extension Theorem (see Section 5.3.4) applied to iterated samples from the generator $\mathcal{G}$. In order to apply this theorem, four criteria must be met: (1) the generator must satisfy weak measurability criteria; (2) there must be a consistent family of finite-dimensional distributions for $h_X$; (3) each of these measures must be tight; and (4) the set of such measures must be bounded above in magnitude. The measurability criterion for $\mathcal{G}$ is defined first.

**Definition 9.1.** A static search generator $\mathcal{G}$ is *setwise measurable* at $u$ if the map $z \mapsto \mathcal{G}[z,u](B)$ is measurable for all measurable sets $B \subseteq X$.

As in Chapter 3, the family of finite-dimensional distributions is generated iteratively for fixed $u$. Let $\mathcal{A}^{\emptyset}_u = \mathcal{G}[\emptyset, u]$, which is a measure on $X^{[1]}$, and then for $t \in \mathcal{T}$ and Borel sets $A \subseteq X^{[t]}$ and $B \subseteq X$, recursively define $\mathcal{A}^{[t]}_u$ as a measure on $X^{[t+1]}$ by

$$\mathcal{A}^{[t]}_u(A \times B) = \int_A \mathcal{G}[z,u](B)\, \mathcal{A}^{[t-1]}_u(dz), \tag{9.1}$$

which depends on setwise measurability. Recall that $[t] = \{s \in \mathcal{T} \mid s \leq t\}$. This definition of $\mathcal{A}_u^{[t]}$ is complete by the Carathéodory Extension Theorem. Furthermore, for any finite set $\mathcal{K} \subseteq \mathcal{T}$, let $t_{\max} = \max \mathcal{K}$ and define $\mathcal{A}_u^{\mathcal{K}}$ from $\mathcal{A}_u^{[t_{\max}]}$ by "integrating out" the missing indices. More formally, let $(A_i)_{i \in \mathcal{K}}$ be Borel sets in $X$, and let $(B_i)_{i \in [t_{\max}]}$ be defined so that $B_i = A_i$ if $i \in \mathcal{K}$ and $B_i = X$ otherwise. Then

$$\mathcal{A}_u^{\mathcal{K}}\left(\prod_i A_i\right) = \mathcal{A}_u^{[t_{\max}]}\left(\prod_i B_i\right),$$

which provides a unique definition of $\mathcal{A}_u^{\mathcal{K}}$, again by Carathéodory. Noting that each set $[t]$ is a finite subset of $\mathcal{T}$, the set

$$\left\{\mathcal{A}_u^{\mathcal{K}} \mid \mathcal{K} \text{ is a finite subset of } \mathcal{T}\right\}$$

is a family of finite-dimensional distributions. It is Kolmogorov consistent by construction.

This family is not necessarily bounded in magnitude. Using the total variation norm for $\mathcal{A}_u^{[t]}$, all that can be said is that $\left\|\mathcal{A}_u^{[t]}\right\| \leq \|\mathcal{G}\|^{|[t]|}$, which grows without bound whenever $\|\mathcal{G}\| > 1$. Nor is $\mathcal{A}_u^{\mathcal{K}}$ necessarily tight. There are no simpler conditions that will force either boundedness or tightness, and so the next two definitions are introduced.

**Definition 9.2 (Tight Generator).** A static search generator is *tight* on an objective $u$ if each $\mathcal{A}_u^{\mathcal{K}}$ as defined above is tight. If a static search generator is tight on all objectives, it is simply said to be tight.

**Definition 9.3 (Eventually Bounded).** A static search generator is *eventually bounded* on an objective $u$ if the family of finite-dimensional distributions $\{\mathcal{A}_u^{\mathcal{K}} \mid \mathcal{K} \text{ finite}\}$ is bounded, *i.e.*, if there exists $N < \infty$ with $\|\mathcal{A}_u^{\mathcal{K}}\| < N$ for all $\mathcal{K}$. If a static search generator is eventually bounded on all objectives, it is simply said to be eventually bounded.

In order for the desired probability distribution $\mathcal{A}[u]$ over complete search histories to exist, it is necessary but not sufficient that $\|\mathcal{G}[z,u]\| \to 1$ for almost every $z$; this convergence is ensured when $\mathcal{G}$ is eventually bounded on $u$. Eventually, the search generator must approach a proper search generator at a sufficiently fast rate as the length of the history grows. Every proper static search generator is obviously eventually bounded. Given these definitions, the Kolmogorov Extension Theorem directly implies the following proposition.

**Proposition 9.1.** *If a static search generator $\mathcal{G}$ is setwise measurable, tight, and eventually bounded on an objective $u$, there exists a finite signed measure $\mathcal{A}[u]$ on the measurable space $\left(X^{\mathcal{T}}, \mathcal{B}a\left[X^{\mathcal{T}}\right]\right)$ that extends the family $\{\mathcal{A}_u^{\mathcal{K}} \mid \mathcal{K} \subseteq \mathcal{T} \text{ finite}\}$.*

Thus the finite signed measure $\mathcal{A}[u]$ exists under certain conditions. For practical reasons, the most interesting condition is tightness, since measurability typically holds and eventual boundedness is automatic for the proper static search generators, which includes all computable search methods. Section 5.3.4 presented conditions on the search domain that make tightness automatic as well. Reviewing, all measures are tight when the search domain is a complete, separable metric space, such as $\mathbb{N}$, $\mathbb{R}^n$, or any finite set.

Given the measure $\mathcal{A}_u$, it will be useful to have a way to reference its time coordinates. A *stochastic process* is a collection of random variables on a shared measure space. In the case of search generators, let $Z = (Z_t)_{t \in \mathcal{T}}$ be the collection of $X$-valued random variables such that for all $h_X \in X^{\mathcal{T}}$,

$$Z_t(h_X) = h_t^X. \tag{9.2}$$

Then $Z_t$ is a projection from $X^{\mathcal{T}}$ to the $t^{th}$ coordinate. As a projection, each $Z_t$ is a measurable function by the properties of the Baire $\sigma$-algebra, and this fact guarantees that $Z_t$ is a random variable. The stochastic process $Z_t$ will be called the *optimization process*, since it represents the progression of the search history over time.

Thus a measure $\mathcal{A}[u]$ has been derived that describes the probabilistic outcomes of running a static generator $\mathcal{G}$ on an objective $u$. This object $\mathcal{A}[u]$ will be referred to as an *extension* of $\mathcal{G}$ to the temporal index set $\mathcal{T}$. Furthermore, the optimization process applied to the measure $\mathcal{A}[u]$ provides a random variable that inspects the individual search point $Z_t$ at time $t$. The object $\mathcal{A}[u]$ provides a mechanism to generate the entire search history all at once for an objective. As an analogy that will be exploited more in Chapter 17, if optimization is treated as a game between a search method and a problem, then in comparison with game theory, the generator is the *extensive form* and its extension is the *normal form*. The next step is to study objects like $\mathcal{A}[u]$ in their own right.

## 9.2 Static Search Trajectors

In the previous section, it was shown that tight and eventually bounded static search generators with search domain $X$ extend to a measure $\mathcal{A}[u]$ over sequences in $X^{\mathcal{T}}$ for each objective $u$. Consider the map $u \mapsto \mathcal{A}[u]$ that takes $Y^X$ to $\mathcal{M}_{X^{\mathcal{T}}}$, the space of finite signed measures on $X^{\mathcal{T}}$. It represents how the search history changes probabilistically as the objective changes. As was done for finite spaces in Chapter 3, such objects will be called *trajectors*. In this case, the trajector is indexed by an objective, so it will be called a *static search trajector*. This section introduces the static search trajectors and examines their relationship with the static search generators.

### 9.2.1 Trajectors and Their Relationship to Generators

Static search trajectors capture the behavior of an optimization method over its entire running time on a particular objective. As in the chapter on finite domains, there is a difference between indexing by objectives ($\mathcal{A}[u]$) as opposed to value histories ($\mathcal{A}[y]$), but the latter will not be studied again until Chapter 13, since they will not be needed until then. The following definition will be used in the mean time, which focuses on the case of optimization with respect to a single static objective.

**Definition 9.4 (Static Search Trajector).** A Baire static search trajector is a function $\mathcal{A} : Y^X \to \mathcal{M}\left[X^{\mathcal{T}}, \mathcal{B}a_{X^{\mathcal{T}}}\right]$. A Borel static search trajector is a function $\mathcal{A} : Y^X \to \mathcal{M}\left[X^{\mathcal{T}}, \mathcal{B}_{X^{\mathcal{T}}}\right]$. If $\mathcal{A}$ is described as a static search trajector, it is assumed to be Baire. If $\mathcal{T}$ is countable, the two concepts coincide.

As before, square brackets are used to indicate the application of the trajector as a function, so for trajector $\mathcal{A}$ and objective $u$, the notation $\mathcal{A}[u]$ refers to the value of the trajector on $u$. Also, as in the last section, the optimization process is a collection $Z = (Z_t)_{t\in\mathcal{T}}$ of projections to each temporal coordinate, each of which are $\mathcal{A}[u]$-measurable. The definition of a trajector works for arbitrary temporal index sets, but when $\mathcal{T}$ is countable, the distinction between Baire and Borel is irrelevant. However, it is worth stating the following proposition. Recall that Theorem 5.2 lists many common instances in which a product space is Mařík.

**Proposition 9.2.** *Every Baire static search trajector can be extended uniquely to a Borel static search trajector whenever $X^{\mathcal{T}}$ is normal and countably paracompact, or, more generally, when $X^{\mathcal{T}}$ is a Mařík space.*

*Proof.* By Theorem 5.2, every normal and countably paracompact space is Mařik, and every Baire measure can be extended to a unique, regular Borel measure on a Mařik space by definition. The extension for static search trajectors is obtained by applying the extensions pointwise for each objective.

The previous section demonstrated that there is often an extension from a static search generator to a static search trajector. Defining static search trajectors independently of this extension avoids requiring the conditions of tightness and eventually boundedness, but begs the question of when trajectors are related to generators. It is easier to begin with trajectors and derive generators than it is to extend generators, as will be seen momentarily.

Section 4.2.3 introduced the concept of a conditional expectation with respect to a $\sigma$-algebra or a variable. A generator can be defined from a trajector as an appropriate conditional expectation. Consider the $\sigma$-algebra generated by successive evaluation of the optimization process,

$$\mathcal{Z}_t = \sigma(Z_1, \ldots Z_t) = \sigma\left(\left\{\bigcap_{s\leq t} Z_s^{-1}(A_s) \,\middle|\, A_s \in \mathcal{B}_X \text{ for all } s \leq t\right\}\right), \tag{9.3}$$

which represents the information obtained by observing $Z_1,\ldots,Z_t$. A sequence of such $\sigma$-algebras generated by a stochastic process is called a *filtration*, and $\mathcal{Z}_t$ is the filtration of the optimization process. More will be said about filtrations in the next section. As a side note, the far right-hand side of Equation 9.3 assumes that $X$ is being measured with respect to the Borel $\sigma$-algebra ($\mathcal{B}_X$); any other appropriate $\sigma$-algebra can be substituted. Given a trajector $\mathcal{A}$, consider for each objective $u$, Borel set $B \subseteq X$, and search history $h_X \in X^{[t]}$,

$$\mathcal{G}[h_X,u](B) = \mathbb{E}_{\mathcal{A}[u]}\left[\mathbb{1}_B \circ Z_{t+1} \mid \mathcal{Z}_t\right](h'_X), \tag{9.4}$$

where $h'_X \in X^{\mathcal{T}}$ is any complete history such that $h_X \sqsubseteq h'_X$; that is, $h_X$ is a prefix of $h'_X$. Equation 9.4 is a constraint $\mathcal{G}$ rather than a definition, since this conditional probability is a Radon-Nikodym derivative and potentially represents not just a single function but an equivalence class of functions from $X^{\mathcal{T}}$ to $\mathbb{R}$. Any search generator $\mathcal{G}$ satisfies the constraint above is a member of this equivalence class.

In the looser notation of probability theory, Equation 9.4 equates to

$$\mathcal{G}[h_X,u](B) = \mathbb{P}_{Z\sim\mathcal{A}[u]}\left(Z_{t+1} \in B \mid Z_1 = h^X_1,\ldots Z_t = h^X_t\right).$$

This notation expresses that any static generator will satisfy the constraint provided that it agrees with $\mathcal{A}[u]$ on the probability assigned to the next search point given a history $h_X$.

It is important to understand the intuitive role of the equivalence class introduced by Equation 9.4. It consists of functions equal almost everywhere with respect to the measure $\mathcal{A}[u]$. In sufficiently large search domains, $\mathcal{G}[h_X,u]$ can be ambiguously defined by the constraint. However, the set

$$H_X = \{h'_X \mid \text{there exists } h_X \sqsubseteq h'_X \text{ such that } \mathcal{G}[h_X,u] \text{ is ambiguous}\}$$

has $\mathcal{A}[u]$-measure zero; *i.e.*, $\mathcal{A}[u](H_X) = 0$. Informally, $H_X$ is the set of histories that never happen, and hence the ambiguity of $\mathcal{G}$ when defined by Equation 9.4 is restricted to nonsense situations. On any histories that can actually be proposed by $\mathcal{A}[u]$, $\mathcal{G}$ is well defined as a single object by the constraint.

### 9.2.2 The Extension Operator

Proposition 9.1 and Equation 9.4 introduce two distinct procedures relating generators to trajectors. When both are applicable, these two processes express the same relationships, although Equation 9.4 encompasses a wider set of generators and trajectors since it does not require tightness or eventual boundedness. This relationship will now be studied analogously to how the operator $\mathbf{G}(\cdot)$ relating static search generators to standard search generators was studied based on Equation 6.11.

To this end, let $\mathbf{E}$ be the set of pairs $(\mathcal{G},\mathcal{A})$ such that $\mathcal{G}$ is a static search generator, $\mathcal{A}$ is a static search trajector, and the pair satisfies Equation 9.4 for all objectives $u$,

Borel sets $B \subseteq X$, and search history prefixes $h_X \in X^{[t]}$ for any $t \in \mathcal{T}$. It is clear from the existence of the conditional expectation in Equation 9.4 that for every trajector $\mathcal{A}$ there exists at least one generator $\mathcal{G}$ such that $(\mathcal{G}, \mathcal{A}) \in \mathbf{E}$. The trajector $\mathcal{A}$ will be called an *extension* of $\mathcal{G}$, justifying the notation $\mathbf{E}$. Define

$$\mathbf{E}_0 = \{\mathcal{G} \mid \text{there exists } \mathcal{A} \text{ with } (\mathcal{G}, \mathcal{A}) \in \mathbf{E}\},$$

and notice that $\mathbf{E}$ is the graph of a surjective function, as proven in the next proposition.

**Proposition 9.3.** $\mathbf{E}$ *is the graph of a surjective function from* $\mathbf{E}_0$ *to the set of Baire static search trajectors.*

*Proof.* The goal is to show that the set $\mathbf{E}$ defines a function and not a relation, or, stated otherwise, that each generator extends to exactly one trajector. Suppose $(\mathcal{G}, \mathcal{A}) \in \mathbf{E}$ and $(\mathcal{G}, \mathcal{A}') \in \mathbf{E}$. If $\mathcal{A} \neq \mathcal{A}'$, then for some objective $u$, it follows that $\mathcal{A}[u](D) \neq \mathcal{A}'[u](D)$, but for all $t$ and all Borel $B \subseteq X$,

$$\mathcal{G}[h_X, u](B) = \mathbb{E}_{\mathcal{A}[u]}\left[\mathbb{1}_B \circ Z_{t+1} \mid \mathcal{Z}_t\right](h'_X) \;=\; \mathbb{E}_{\mathcal{A}'[u]}\left[\mathbb{1}_B \circ Z_{t+1} \mid \mathcal{Z}_t\right](h'_X) \qquad (9.5)$$

for almost every $h'_X$ with $h_X \in X^{[t]}$ and $h_X \sqsubseteq h'_X$.

Recall that the Baire $\sigma$-algebra is the smallest $\sigma$-algebra that makes finite projections measurable, and hence a Baire measure is in fact defined by its finite projections. Define the push-forward measures $\nu_{t,u} = Z_t \circ \mathcal{A}[u]$ and $\nu'_{t,u} = Z_t \circ \mathcal{A}'[u]$ in the usual way, which is to say that for all Borel $B \subseteq X$,

$$\nu_{t,u}(B) = \mathcal{A}[u]\left(Z_t^{-1}(B)\right) \quad \text{and} \quad \nu'_{t,u}(B) = \mathcal{A}'[u]\left(Z_t^{-1}(B)\right).$$

Now if $\nu_{t,u} = \nu'_{t,u}$ for all times $t$ and all objectives $u$, then it will follow that $\mathcal{A}[u] = \mathcal{A}'[u]$ as desired, since the coordinate projections determine the finite projections, which in turn determine the Baire measures $\mathcal{A}[u]$ and $\mathcal{A}'[u]$.

The remainder of the proof proceeds by induction. If $t = 1$, then

$$\mathbb{E}_{\mathcal{A}[u]}\left[\mathbb{1}_B \circ Z_1 \mid \mathcal{Z}_0\right](h'_X) \;=\; \mathbb{E}_{\mathcal{A}[u]}\left[\mathbb{1}_B \circ Z_1\right] \;=\; \mathcal{A}[u](Z_1^{-1}(B)) \;=\; \nu_{1,u}(B)$$

and similarly for $\nu'_{1,u}$, so $\nu_{1,u} = \nu'_{1,u}$ by Equation 9.5. If $\nu_{s,u} = \nu'_{s,u}$ for all $s \leq t$, then for any $\mathcal{Z}_t$-measurable random variable $X$,

$$\mathbb{E}_{\mathcal{A}[u]}[X] \;=\; \mathbb{E}_{\mathcal{A}'[u]}[X]$$

because $\nu_{s,u}$ and $\nu'_{s,u}$ encode the information generated by $Z_s$ on $\mathcal{A}[u]$ and $\mathcal{A}'[u]$, respectively. Now for fixed $B$, the map $h_X \mapsto \mathcal{G}[h_X, u](B)$ is a $\mathcal{Z}_t$-measurable random variable according to Equation 9.4 and the definition of conditional expectation, so

$$\begin{aligned}
v_{t+1,u}(B) = \mathcal{A}[u](Z_{t+1}^{-1}(B)) &= \mathbb{E}_{\mathcal{A}[u]}\left[\mathbb{1}_B \circ Z_{t+1}\right] \\
&= \mathbb{E}_{\mathcal{A}[u]}\left[\mathbb{E}_{\mathcal{A}[u]}\left[\mathbb{1}_B \circ Z_{t+1} \mid \mathcal{Z}_t\right]\right] \\
&= \mathbb{E}_{\mathcal{A}[u]}\left[\mathcal{G}[h_X, u](B)\right] \\
&= \mathbb{E}_{\mathcal{A}'[u]}\left[\mathcal{G}[h_X, u](B)\right] = \mathcal{A}'[u](Z_{t+1}^{-1}(B)) = v'_{t+1,u}(B),
\end{aligned}$$

which proves that $\mathcal{A}[u] = \mathcal{A}'[u]$, so $\mathbf{E}$ is the graph of a function. Furthermore, because for all static search trajectors $\mathcal{A}$ there is always at least one static search generator $\mathcal{G}$ satisfying Equation 9.4, this function is surjective.

Thus there is an extension operator consistent with $\mathbf{E}$ that will be denoted $\mathbf{E}(\cdot)$ in analogy to the standard-to-static conversion operator $\mathbf{G}(\cdot)$. But the domain of this operator is still unclear. Which static search generators are contained in $\mathbf{E}_0$ and thus have extensions to trajectors? As the next proposition shows, the set $\mathbf{E}_0$ contains at least the tight and eventually bounded static search generators, and the resulting extension is identical to that obtained by Kolmogorov extension for these generators.

**Proposition 9.4.** *Suppose that $\mathcal{G}$ is a static search generator that is tight and eventually bounded for all objectives. Then $\mathcal{G} \in \mathbf{E}_0$, and in particular, if $\mathcal{A}$ is generated by pointwise Kolmogorov extension from $\mathcal{G}$ using Proposition 9.1, then $(\mathcal{G}, \mathcal{A}) \in \mathbf{E}$.*

*Proof.* The tightness and eventual boundedness of $\mathcal{G}$ guarantees by Proposition 9.1 that there exists a trajector $\mathcal{A}$ such that for each objective $u$, $\mathcal{A}[u]$ is formed by Kolmogorov extension from $\mathcal{G}$ using the finite family of distributions $\{\mathcal{A}_u^{\mathcal{K}} \mid \mathcal{K} \text{ finite}\}$ from Section 9.1. The proof will be complete if $\mathcal{G}$ and $\mathcal{A}[u]$ together satisfy Equation 9.4.

Consider what $\mathbb{E}_{\mathcal{A}[u]}\left[\mathbb{1}_B \circ Z_{t+1} \mid \mathcal{Z}_t\right](h'_X)$ in Equation 9.4 means. Let $g_B : X^{\mathcal{T}} \to \mathbb{R}$ be any function such that

$$g_B(h'_X) = \mathbb{E}_{\mathcal{A}[u]}\left[\mathbb{1}_B \circ Z_{t+1} \mid \mathcal{Z}_t\right](h'_X). \tag{9.6}$$

That is, $g_B$ is a conditional expectation, a member of the equivalence class defined by the right-hand side above. By definition, for all $D \in \mathcal{Z}_t$,

$$\int_D g_B(h'_X)\ \mathcal{A}[u](dh'_X) \quad = \quad \int_D \mathbb{1}_B\left(Z_{t+1}\left(h'_X\right)\right)\ \mathcal{A}[u](dh'_X).$$

This equation is exactly the definition of a conditional expectation. For $\mathcal{K} \subseteq \mathcal{T}$, let $\phi_{\mathcal{K}} : X^{\mathcal{T}} \to X^{\mathcal{K}}$ be a projection from complete search histories to the coordinates in $\mathcal{K}$. Then $\phi_{[t]}$ is $\mathcal{Z}_t$-measurable, and there is some Borel set $A \subseteq X^{[t]}$ such that $D = \phi_{[t]}^{-1}(A)$ and additionally $D = \phi_{[t+1]}^{-1}(A \times X)$. Using the latter along with the fact that $\mathcal{A}[u]$ is a Kolmogorov extension of $\mathcal{A}_u^{[t+1]}$,

$$\int_D \mathbb{1}_B\left(Z_{t+1}\left(h'_X\right)\right)\ \mathcal{A}[u](dh'_X) = \int_A \int_X \mathbb{1}_B\left(z'_{t+1}\right)\ \mathcal{A}_u^{[t+1]}(dz')$$
$$= \int_A \int_B \mathcal{A}_u^{[t+1]}(dz')\ =\ \mathcal{A}_u^{[t+1]}(A \times B)$$
$$= \int_A \mathcal{G}[z,u](B)\ \mathcal{A}_u^{[t-1]}(dz),$$

where in the final equality Equation 9.1 has been applied. Consequently, for all times $t \in \mathcal{T}$, for all Borel sets $A \subseteq X^{[t]}$, $B \subseteq X$, and for all objectives $u$,

$$\int_{\phi_{[t]}^{-1}(A)} g_B(h'_X)\, \mathcal{A}[u](dh'_X)\ =\ \int_A \mathcal{G}[z,u](B)\, \mathcal{A}_u^{[t-1]}(dz),$$

which, by varying $A$ and $t$, is equivalent to the statement that

$$\mathcal{G}[h_X,u](B)\ =\ g_B(h'_X)\ =\ \mathbb{E}_{\mathcal{A}[u]}\left[\mathbb{1}_B \circ Z_{t+1} \mid \mathcal{Z}_t\right](h'_X)$$

for $\mathcal{A}[u]$-almost every $h'_X$ with $h_X \sqsubseteq h'_X$, which is equivalent to Equation 9.4.

So the domain of the extension operator $\mathbf{E}(\cdot)$ contains at least the tight and eventually bounded static search generators. This domain may be larger than these two properties, but this issue is not clarified here. It is enough to note that as a surjective function, every static search trajector arises as the extension of some static search generator. In other words, all possible search trajectories are determined by some protocol that generates one search point at a time.

### 9.2.3 Spaces of Trajectors

Two perspectives on an optimization method have now been introduced: generators and trajectors. Static search generators encode how a method chooses the next point given a cumulative history and an objective, whereas static search trajectors capture how a method chooses a trajectory through the search domain given an objective. As mentioned above, the distinction between these two perspectives is similar to the distinction between *extensive form* and *normal form* in game theory, a topic which is discussed at length in Chapter 13.

Static search trajectors are obviously measure-valued functionals as introduced in Chapter 6. Specifically, the space of static search trajectors with finite norm is $\mathcal{MF}_{Y^X, X^{\mathcal{T}}}$, and it is a Banach space. Addition and scalar multiplication are defined pointwise on this space exactly as was done for generators. The structure of this Banach space is distinct from that of the Banach space of static search generators, however.

Suppose that three static search generators $\mathcal{G}$, $\mathcal{G}'$, and $\alpha\mathcal{G}+\beta\mathcal{G}'$ all extend via $\mathbf{E}(\cdot)$ to static search trajectors $\mathcal{A}$, $\mathcal{A}'$, and $\mathcal{A}''$, respectively. One might wish that $\mathcal{A}'' = \alpha\mathcal{A} + \beta\mathcal{A}'$, but that is not generally the case. To see why, consider Equation 9.1, which applies to all three generators by Proposition 9.4. Using $\mathcal{A}_u^{[t]}$ and similar for the distribution induced by projecting to the first $t$ time steps,

$$\left(\alpha\mathcal{A}+\beta\mathcal{A}'\right)^{[t]}_{u}(A\times B) = \alpha\int_{A}\mathcal{G}[z,u](B)\,\mathcal{A}^{t}_{u}(dz) \;+\; \beta\int_{A}\mathcal{G}'[z,u](B)\,\mathcal{A}'^{t}_{u}(dz).$$

Suppose that $\mathcal{A}'' = \alpha\mathcal{A}+\beta\mathcal{A}'$. The right-hand side would have to be equal to

$$\int_{A}\left(\alpha\mathcal{G}+\beta\mathcal{G}'\right)[z,u](B)\,\mathcal{A}''^{[t]}_{u}(dz),$$

which is only true if $\mathcal{A} = \mathcal{A}'$ almost everywhere. This topic is covered in more detail in Chapters 10 and 13, but it suffices for this section to note that addition and scalar multiplication are plainly separate operations for generators versus trajectors.

Given that the linear structure of the trajectors is separate from that of the generators, one wishes to know whether certain linear properties are preserved by extension. The most important such property is the black-box property, which forms the space $\mathcal{BB}^{\mathcal{T}}_{X,Y}$ of static black-box search generators defined in Section 6.3.1. Recall that these are the generators that cannot "see the future"; they cannot distinguish objectives that are equal on the search history. The next definition proposes a version of this property for trajectors, and it is then shown that this property is linear in trajector space and that it extends the black-box property of generator space. First, however, some preliminary notation and facts must be introduced.

Suppose that $u,v : X \to Y$ are measurable objectives. Then the set

$$\{u=v\} = \{x \mid u(x) = v(x)\}$$

on which the two objectives are equal is measurable. To prove this, let $h_{u,v}(x) = (u(x),v(x))$ project $X$ to $Y\times Y$. Then $h_{u,v}$ is a measurable function. Furthermore, the diagonal $D_Y = \{(y,y) \mid y\in Y\}$ is Borel measurable in $Y\times Y$ (and we have generally assumed that both $X$ and $Y$ are Borel). Now then, notice that $\{u=v\} = h^{-1}_{u,v}(D_Y)$, which is measurable because $h_{u,v}$ is.

Next, the equality of objectives needs to be overlaid onto the $\sigma$-algebra $\mathcal{Z}_t$. Given any measurable $A\subseteq X^{\mathcal{T}}$, define

$$A^{t}_{\{u=v\}} = \{z\in A \mid \text{for all } s\leq t,\, u(z_s) = v(z_s)\}$$

to contain the restriction of any set to those elements whose values are equal under $u$ and $v$ up to time $t$. Now given $A\in\mathcal{Z}_t$, it follows that $A^{t}_{\{u=v\}}\in\mathcal{Z}_t$. Suppose that $B_s\subseteq X$ is measurable. Then $B_s\cap\{u=v\}$ is also measurable and $Z^{-1}_{s}(B_s\cap\{u=v\})$ is $\mathcal{Z}_s$-measurable. Consequently, with $\phi_{[t]}$ as the canonical projection from $X^{\mathcal{T}}$ to $X^{[t]}$ and $B=\prod_{s\leq t}B_s$ for some collection $(B_s)_{s\leq t}$,

$$\left[\phi^{-1}_{[t]}(B)\right]^{t}_{\{u=v\}} = \phi^{-1}_{[t]}\left(\prod_{s\leq t}B_s\cap\{u=v\}\right) = \bigcap_{s\leq t}Z^{-1}_{s}\left(B_s\cap\{u=v\}\right)\in\mathcal{Z}_t.$$

Due to the construction of the product topology, this fact is sufficient to conclude that $A^t_{\{u=v\}}$ is $\mathcal{Z}_t$ measurable for any $\mathcal{Z}_t$-measurable $A$. These sets can be used to define the black-box property for trajectors, which states that $\mathcal{A}[u]$ and $\mathcal{A}[v]$ must be equal on such sets.

**Definition 9.5 (Black-Box Property).** A static search trajector $\mathcal{A} \in \mathcal{MF}_{Y^X,X^{\mathcal{T}}}$ has the black-box property if for any time $t \in \mathcal{T}$, any pair of measurable objectives $u$ and $v$, and any $\mathcal{Z}_t$-measurable set $A$,

$$\mathcal{A}[u]\left(A^t_{\{u=v\}}\right) = \mathcal{A}[v]\left(A^t_{\{u=v\}}\right). \tag{9.7}$$

**Proposition 9.5.** *Suppose $\mathcal{A}$ and $\mathcal{A}'$ are static search trajectors with the black-box property. For any $\alpha, \beta \in \mathbb{R}$, $\alpha\mathcal{A} + \beta\mathcal{A}'$ has the black-box property.*

*Proof.* This statement is a trivial consequence of the definition of the black-box property and the pointwise nature of the vector operations on trajectors.

**Proposition 9.6.** *Suppose $\mathcal{G}$ is a static black-box search generator. Then $\mathcal{A} = \mathbf{E}(\mathcal{G})$ has the black-box property if it exists. Conversely, if a static search trajector $\mathcal{A}$ has the black-box property, then there exists at least one $\mathcal{G}$ such that $\mathcal{A} = \mathbf{E}(\mathcal{G})$ and $\mathcal{G}$ has the black-box property.*

*Proof.* The relationship between $\mathcal{A}$ and $\mathcal{G}$ is defined by Equation 9.4. If $\mathcal{G}$ has the black-box property, for any $h_X \in X^{\mathcal{T}}$ with $u(h^X_s) = v(h^X_s)$ for all $s \leq t$,

$$\mathbb{E}_{\mathcal{A}[u]}\left[\mathbb{1}_{B_{s+1}} \circ Z_{s+1} \mid \mathcal{Z}_s\right](h_X) = \mathbb{E}_{\mathcal{A}[v]}\left[\mathbb{1}_{B_{s+1}} \circ Z_{s+1} \mid \mathcal{Z}_s\right](h_X) \tag{9.8}$$

for all $0 \leq s \leq t$, where $(B_s)_{s \leq t}$ is a collection of measurable sets in $X$ as in the preceding text and $B_{t+1} = X$. With $\phi_{[t]}$ as the canonical projection from $X^{\mathcal{T}}$ to $X^t$ and $B = \prod_{s \leq t} B_s$, the above equations restrict all coordinates up to time $t$ to be identical for both $u$ and $v$, which implies that

$$\mathcal{A}[u]\left(\left[\phi^{-1}_{[t]}(B)\right]^t_{\{u=v\}}\right) = \mathcal{A}[v]\left(\left[\phi^{-1}_{[t]}(B)\right]^t_{\{u=v\}}\right),$$

and by the Carathéodory extension theorem,[1] this equation extends from $B$ to all measurable sets, that is to say, to Borel sets on $X^{[t]}$, which determines $\mathcal{Z}_t$. Therefore $\mathcal{A}$ has the black-box property.

The converse statement that $\mathcal{G}$ has the black-box property when $\mathcal{A}$ does is seen to be true by the reverse sequence of statements, with the caveat that the conditional

[1] The collection $\{A^t_{\{u=v\}} \mid A \in \mathcal{Z}_t\}$ forms a $\sigma$-algebra over $(X^{\mathcal{T}})^t_{\{u=v\}}$, and the measure $\mathcal{A}[u]$ passes to a measure $\mathcal{A}[u]^t_{u=v} : A^t_{\{u=v\}} \mapsto \mathcal{A}[u]\left(A^t_{\{u=v\}}\right)$ defined over this collection. The invocation of Carathéodory refers to this measure, which can be generated as the $\sigma$-algebra completion of the collection of sets $\left\{\left[\phi^{-1}_{[t]}(B)\right]^t_{\{u=v\}} \,\middle|\, B \in \mathcal{B}_{X^{[t]}}\right\}$.

expectation is only determined for almost every $h_X$. However, the appropriate black-box $\mathcal{G}$ can be constructed by setting $\mathcal{G}[h^X_{<t}, u] = \mathcal{G}[h^X_{<t}, v]$ to be any desired value for all $h_X$ in a set of measure zero.

The net result of the preceding definition and propositions is that the space $\mathbf{E}\left(\mathcal{BB}^{\mathcal{T}}_{X,Y}\right)$ is a vector space of trajectors, and by intersecting with the space $\mathcal{MF}^{\mathcal{T}}_{Y^X, X^{\mathcal{T}}}$ of measure-valued functionals with finite norm, one obtains a Banach space that extends the static black-box generators into trajector space. A similar projection can be obtained for the $n$-Markov static search generators.

**Definition 9.6.** A static search trajector $\mathcal{A} \in \mathcal{MF}_{Y^X, X^{\mathcal{T}}}$ has the $n$-Markov property if

$$\mathbb{E}_{\mathcal{A}[u]}\left[\mathbb{1}_B \circ Z_{t+1} \mid \mathcal{Z}_t\right] = \mathbb{E}_{\mathcal{A}[u]}\left[\mathbb{1}_B \circ Z_{t+1} \mid \sigma(Z_{t-n+1}, \ldots, Z_t)\right] \tag{9.9}$$

for all objectives $u$, all times $t \geq 0$, and all measurable $B \subseteq X$, with the interpretation $\sigma(Z_{t-n+1}, \ldots, Z_t) = \mathcal{Z}_t$ when $n \geq t$.

Comparing Equation 9.9 to Equation 9.4 reveals that as a statement about the dependence on the search history, this criterion is precisely the same as the $n$-Markov property for generators, which is summarized in the next proposition.

**Proposition 9.7.** *Suppose $\mathcal{G}$ is an $n$-Markov static search generator. Then $\mathcal{A} = \mathbf{E}(\mathcal{G})$ has the $n$-Markov property if it exists. Conversely, if a static search trajector $\mathcal{A}$ has the $n$-Markov property, then there exists at least one $\mathcal{G}$ such that $\mathcal{A} = \mathbf{E}(\mathcal{G})$ and $\mathcal{G}$ has the $n$-Markov property.*

Like the black-box property, the $n$-Markov property is linear in trajector space as well as generator space, even though the linear operations in each space are distinct.

**Proposition 9.8.** *Suppose $\mathcal{A}$ and $\mathcal{A}'$ are both $n$-Markov static search trajectors. Then for all $\alpha, \beta \in \mathbb{R}$, $\alpha\mathcal{A} + \beta\mathcal{A}'$ is an $n$-Markov static search trajector.*

*Proof.* Equation 9.9 means that for any function $g : X^{\mathcal{T}} \to \mathbb{R}$, if the statement

$$\int_A g(h_X)\ \mathcal{A}[u](dh_X) = \int_A \mathbb{1}_B\left(h^X_t\right)\ \mathcal{A}[u](dh_X)$$

holds for all $A \in \sigma(Z_{t-n+1}, \ldots, Z_t)$, then it holds for every $A \in \mathcal{Z}_t$ (the converse is also true but trivial). Clearly, if Equation 9.9 holds for both $\mathcal{A}$ and $\mathcal{A}'$ then it holds for $\alpha\mathcal{A} + \beta\mathcal{A}'$ as well by the linearity of the integral.

Thus recalling the space $\text{Markov}^{n,\mathcal{T}}_{X,Y}$ of $n$-Markov static search generators, we find that $\mathbf{E}\left(\text{Markov}^{n,\mathcal{T}}_{X,Y}\right)$ is a vector space of trajectors, all of which have the $n$-Markov property for trajectors. Again, by intersecting with $\mathcal{MF}_{Y^X, X^{\mathcal{T}}}$, a Banach space of $n$-Markov trajectors can be obtained.

Notice that Equation 9.9 is really a statement about the almost-everywhere behavior of generators according to Equation 9.4. Other properties of generators can be extended across $\mathbf{E}$ similarly, and in at least some cases they produce vector spaces, as with the black-box and $n$-Markov properties. Notice, however, that if a static search trajector $\mathcal{A}$ inherits a property because $\mathcal{A} = \mathbf{E}(\mathcal{G})$ for some generator $\mathcal{G}$ having the property, then if there is some other generator $\mathcal{G}'$ such that $\mathcal{A} = \mathbf{E}(\mathcal{G}')$, it does not necessarily follow that $\mathcal{G}'$ shares the properties of $\mathcal{G}$ as well. It is possible for $\mathcal{G}'$ to violate the property on search histories within a set of $\mathcal{A}[u]$-measure zero. This is merely a technical point, because any computational or physical implementation of a search method need not concern itself with events that never occur.

This section has developed the concept of a static search trajector on a Hausdorff topological space and has shown that these trajectors can be related to static search generators matching the extensions in Section 9.1 by conditioning on values of the optimization process. These trajectors form Banach spaces just as the generators do, and in fact pairs of spaces are related by the extension operator $\mathbf{E}(\cdot)$. The next chapter examines the performance of optimization methods using trajectors, but in order to support this discussion, the next section introduces important properties of stochastic processes in general and the optimization process in particular.

## 9.3 Information, Stopping Times, and Continuity

The last two sections relied on the optimization process $Z = (Z_t)_{t\in\mathcal{T}}$ and the filtration $\mathcal{Z}_t$ in order to describe the generation of the search history from a trajector. These objects are obviously crucial to measuring the performance of an optimization method. The optimization process provides access to the search points, and its filtration determines what information is available at each point in time. In order to assess performance, though, it may be necessary to observe $Z_t$ for cases where the time $t$ depends on the progress of the optimization. For example, one may wish to observe the first time that a search point lies within some radius around an optimum. These dynamic time variables are called a *random times*, and the theoretical machinery needed to analyze them is introduced now.

### 9.3.1 Filtrations and Information

One of the most important intuitions underlying the theory of stochastic processes is the concept of the $\sigma$-algebra as an information source. Information in this sense refers to the events that are measurable, which is wholly determined by the $\sigma$-algebra. Let $\mathcal{F}$ be a $\sigma$-algebra. The distinct sets within the $\mathcal{F}$ represent observable events. Suppose $x$ and $z$ are two points in an $\mathcal{F}$-measurable space that cannot be separated by sets in $\mathcal{F}$. That is, for any set $A \in \mathcal{F}$, it holds that $x \in A \iff z \in A$. Then $x$ and $y$ are indistinguishable by any $\mathcal{F}$-measurable random variable $Y$. No observation of the state of $Y$ can ever distinguish whether the state $x$ or the state $z$ has occurred, or even whether the two represent different states. Specifically, for any event $B$ in the range of $Y$, the definitions imply that $Y(x) \in B \iff Y(z) \in B$. The question of whether $x$ occurred or $z$ occurred cannot be resolved by $Y$; the $\sigma$-algebra $\mathcal{F}$ does not contain that information.

It is easy to see a specific instance of this principle using the optimization process. Suppose that $x, z \in X^{\mathcal{T}}$ are sequences of search points such that $x_s = z_s$ for all $s \leq t$, and consider the $\sigma$-algebra $\mathcal{Z}_t$, which is the smallest $\sigma$-algebra making $Z_1, \ldots Z_t$ measurable. Now $Z_s(x) = x_s = z_s = Z_s(z)$ for all $s \leq t$, and so the random variables $Z_1, \ldots, Z_t$ cannot be used to determine whether $x$ and $z$ are the same or different. The event

$$\{z \mid \text{for all } s \leq t,\, x_s = z_s \text{ but } x \neq z\}$$

is not $\mathcal{Z}_t$-measurable. To observe this event requires a larger $\sigma$-algebra than $\mathcal{Z}_t$. In fact, it requires the Borel $\sigma$-algebra on $X^{\mathcal{T}}$, since this is the "first" $\sigma$-algebra that contains enough sets to separate every pair of sequences.[2] In a broader sense, the Borel $\sigma$-algebra is the smallest collection such that the boundary of every open and closed set is an observable event. That is, a Borel $\sigma$-algebra is the smallest $\sigma$-algebra that has sufficient information to determine when a trajectory within the search space has entered or exited a closed set.

In general, stochastic processes progressively reveal information about the underlying measurable space they are defined on. The amount of information can grow as new random variables expose new facets of the space, but the information cannot shrink, since the observations of previous random variables cannot be taken back. The concept of a *filtration* was introduced to reflect the increasing information made available a stochastic process.

**Definition 9.7 (Filtration).** A *filtration* is an indexed collection of $\sigma$-algebras $(\mathcal{F}_i)_{i\in\mathcal{I}}$ such that for any $i, j \in \mathcal{I}$ with $i < j$, $\mathcal{F}_i \subseteq \mathcal{F}_j$.

Notice that a filtration is defined independently and not as a derivative of any particular stochastic process, so its connection to a particular process must be defined. A stochastic process $(Y_i)_{i\in\mathcal{I}}$ is said to be *adapted* to a filtration $(\mathcal{F}_i)_{i\in\mathcal{I}}$ if for all $i \in \mathcal{I}$, $Y_i$ is $\mathcal{F}_i$-measurable.

If each $Y_i$ is a random variable between measure spaces $(\Omega, \Sigma_\Omega)$ and $(Z, \Sigma_Z)$, then every $Y_i$ is $\Sigma_\Omega$ measurable. If $(Y_i)_{i\in\mathcal{I}}$ is also adapted to $(\mathcal{F}_i)_{i\in\mathcal{I}}$, then it follows that each $\mathcal{F}_i \subseteq \Sigma_\Omega$. If each $\mathcal{F}_i$ is the smallest $\sigma$-algebra such that $Y_j$ is measurable for $j \leq i$, then $\mathcal{F}_i = \sigma(Y_1, \ldots, Y_i)$. This filtration is called the *natural filtration* of the process $(Y_i)_{i\in\mathcal{I}}$. The natural filtration is a function of the stochastic process. The filtration $(\mathcal{Z}_t)_{t\in\mathcal{T}}$ introduced in the last section is the natural filtration for the stochastic process $(Z_t)_{t\in\mathcal{T}}$.

Just as the natural filtration defines a filtration from a stochastic process, a stochastic process can be defined from a filtration. Supposing that $\mathcal{F}_i \subseteq \Sigma$ for all $i$ and that $Y$ is a $\Sigma$-measurable random variable, the *Lévy process* is a stochastic process defined by

$$Y_i = \mathbb{E}\left[Y \mid \mathcal{F}_i\right].$$

[2] An application of the Hausdorff axiom, assumed to hold here for the underlying topological space.

This process is useful for turning a $\Sigma$-measurable random variable into a sequence of $\mathcal{F}_i$ random variables. Given a filtration, the total information content of the filtration is the union

$$\mathcal{F}_\infty = \sigma\left(\bigcup_{i\in\mathcal{I}} \mathcal{F}_i\right),$$

and in the special case where $\mathcal{F}_\infty = \Sigma$, a Lévy process has the limit

$$Y = \lim_{i\to\infty} Y_i,$$

so that the Lévy process progressively recovers the target random variable. In Chapter 14, a Lévy process will be constructed in order to direct the optimization process towards the global optimum in the limit.

For the optimization process with its filtration $(\mathcal{Z}_t)_{t\in\mathcal{T}}$, it holds that $\mathcal{Z}_\infty = \mathcal{B}_{X^\mathcal{T}}$. That is, observing the entire optimization process is equivalent to make observations through the Borel $\sigma$-algebra on $X^\mathcal{T}$, with the usual caveat that the Baire $\sigma$-algebra suffices when $\mathcal{T}$ is countable.

Given these basic facts about filtrations, we can now investigate what happens to the optimization process when the index $t$ varies randomly.

### 9.3.2 Stopping Times

A *random time* for a stochastic process $(Y_i)_{i\in\mathcal{I}}$ on a measurable space $(\Omega,\Sigma)$ is a random variable $T:\Omega\to\mathcal{I}$ where $\mathcal{I}$ is assigned the Borel $\sigma$-algebra for the order topology. The order topology guarantees that for all $i\in\mathcal{I}$, the sets $\{\omega\in\Omega \mid T(\omega)\leq i\}$ and $\{\omega\in\Omega \mid T(\omega)<i\}$ are contained in $\Sigma$. These two sets are often abbreviated $\{T\leq i\}$ and $\{T<i\}$, respectively, with $\{T=i\}$ used similarly; see *e.g.* [86]. This notation will simplify several statements below and will therefore be adopted here as well.

Now $T$ is $\Sigma$-measurable. Given the entire process, the value of the random time can be determined exactly. Given only a prefix of the process, it is not necessarily possible to know whether or not the random time has passed at the current time. This latter fact is crucial, because one typically wishes to use a random time in order to discover when some event has occurred, and it is inconvenient if one has to observe the entire process to its end to discover this, especially if the process runs indefinitely. A *stopping time* resolves this dilemma by requiring that the passage of the timing event can be measured by some filtration, that is to say, that the timing event is measurable at the time it occurs, and not at the end of the process.

**Definition 9.8 (Stopping Time).** A random time $S:\Omega\to\mathcal{I}$ is a stopping time for a filtration $(\mathcal{F}_i)_{i\in\mathcal{I}}$ if $S^{-1}(\{s \mid s\leq i\})\in\mathcal{F}_i$ for all $i\in\mathcal{I}$.[3]

---

[3] This condition is often equivalently stated in abbreviated form as $\{S\leq i\}\in\mathcal{F}_i$.

Notice that if $\{S \leq i\}$ is an $\mathcal{F}_i$-measurable event and $\{S \leq i-1\}$ is an $\mathcal{F}_{i-1}$-measurable event, then it follows from $\mathcal{F}_{i-1} \subseteq \mathcal{F}_i$ that $\{S = i\}$ is a measurable event, so the exact value of the stopping time can be determined, at least in a discrete time setting.

A stopping time need not be finite. In that case, the event $\{S = \infty\}$ is an element of the reference $\sigma$-algebra $\Sigma$ but is not typically measurable in any member $\mathcal{F}_i$ of the filtration. If the stopping time is finite on a particular sequence in the state space, then it is said to *hit* on that sequence. One cannot always determine whether a stopping time will hit on a particular sequence given only a finite number of components. The possibility that $S = \infty$ must always be considered.

In the next chapter, stopping times will be used as criteria for deciding when to measure the performance of an optimizer. They will also show up again in Chapter 12 in the proof of the No Free Lunch Identification Theorem (Theorem 12.7), which depends on filtering the optimization process by a series of stopping times.

For any stopping time $S$ of a process $Y$ on a measurable space $(\Omega, \Sigma)$, there is a random variable $Y_S = Y_{S(\omega)}(\omega)$ representing the value of the process at the stopping time. To see why this function is a random variable, let $A$ be any measurable set in the range of each $Y_i$. Then $Y_i^{-1}(A) = \{\omega \mid Y_i(\omega) \in A\}$ is measurable for each $i$. Likewise $S^{-1}(\{i\}) = \{S = i\}$ is measurable for each $i$, and so the intersections

$$Y_i^{-1}(A) \cap S^{-1}(\{i\}) = \{\omega \mid S(\omega) = i \text{ and } Y_i(\omega) \in A\}$$

are measurable for each $i$. Finally, the event that $Y_S \in A$ (technically, the set $\{\omega \mid Y_{S(\omega)}(\omega) \in A\}$) can be written as the grand union of these intersections,

$$\bigcup_{i \in \mathcal{I}} Y_i^{-1}(A) \cap S^{-1}(\{i\}),$$

which is measurable whenever $\mathcal{I}$ is countable. A slightly more sophisticated argument can be used to show that $Y_S$ remains a random variable for some larger index sets as well, such as the case where $\mathcal{I} \subseteq \mathbb{R}$.

The preceding paragraph established that $Y_S$ is a random variable, but measurability was still with respect to the overall $\sigma$-algebra $\Sigma$. In fact, a $\sigma$-algebra with less information can be used to measure $Y_S$, namely,

$$\mathcal{F}_S = \{B \in \Sigma \mid B \cap \{S \leq i\} \in \mathcal{F}_i \text{ for all } i \in \mathcal{I}\}.$$

This $\sigma$-algebra includes all events in $\Sigma$ that can be measured by the filtration at time $S$ if they event has happened at or before that that time.

If $(S_k)_{k \in \mathcal{K}}$ is an increasing sequences of stopping times, *i.e.*, $S_m(\omega) \leq S_k(\omega)$ whenever $m \leq n$, then the sequence of $\sigma$-algebras $\left(\mathcal{F}_{S_k}\right)_{k \in \mathcal{K}}$ is a filtration, and for any process $(Y_i)_{i \in \mathcal{I}}$, the filtered process $\left(Y_{S_k}\right)_{k \in \mathcal{K}}$ is a stochastic process adapted to this filtration.

Suppose that $S$ is a stopping time of the natural filtration for the optimization process, $(\mathcal{Z}_t)_{t \in \mathcal{T}}$. Then $S$ is a function over sequences, i.e. $S = S(z)$ for $z \in X^{\mathcal{T}}$.

The stopped variable $Z_S$ is a random variable on $\mathcal{Z}_S$ representing the search point identified by the stopping time $S$.

In order measure the performance of an optimizer, it will be necessary to take the integral with respect to $\mathcal{A}[u]$ of functions depending on the sequence $Z_1,\dots,Z_S$. In particular, in the next section, it will be seen that continuity can transfer from generators to trajectors under the integral so long as the integrand depends on at most finitely many time steps. Given that the value of $S$ is potentially unbounded, integrals involving $Z_S$ can depend on an unbounded number of time steps. Nonetheless, by factoring out the stopping time, continuity can be preserved.

The concept of factoring out the stopping time requires us to develop some notation. First, notice that for any measurable set $A \subseteq X^{\mathcal{T}}$, the set $A$ can be partitioned into a union of joint sets based on the value of a stopping time $S$ as

$$A = \bigcup_{t\in\mathcal{T}} A \cap \{S = t\}. \tag{9.10}$$

In this case, so long as the stopping time $S$ is almost surely finite, an integral with respect to a trajector can be reduced to an infinite sum of integrals that depend on finitely many time steps.

The phrase "depend on finitely many time steps" is to be interpreted by defining $\mathcal{A}_u^{[t]}$ using Equation 9.1 in light of Equation 9.4, which is always valid for finite $t$ regardless of tightness or eventual boundedness. In this case one finds by expanding recursively that for measurable $(A_s)_{s\leq t}$ with each $A_s \subseteq X$,

$$\mathcal{A}_u^{[t]}\left(\prod_{s\leq t} A_s\right) = \int_{A_1}\cdots\int_{A_t}\prod_{s\leq t}\mathcal{G}\left[(z_1,\dots,z_{s-1}),u\right](dz_s), \tag{9.11}$$

where $(z_1,\dots,z_0) = \emptyset$. Furthermore, for any $D \in \mathcal{Z}_t$ and any $W_t : X^{[t]} \to \mathbb{R}$,

$$\int_D W_t(h_1^X,\dots,h_t^X)\,\mathcal{A}[u](dh_X) = \int_{\phi_{[t]}(D)} W_t(z_1,\dots,z_t)\,\mathcal{A}_u^{[t]}(dz),$$

where $\phi_{[t]} : X^{\mathcal{T}} \to X^{[t]}$ projects onto the first $t$ coordinates as before. Thus $W$ depends on finitely many steps. A suitable $W_t$ can be obtained from any $\mathcal{Z}_t$-measurable random variable.

**Proposition 9.9.** *Suppose $W : X^{\mathcal{T}} \to \mathbb{R}$ is a $\mathcal{Z}_t$-measurable random variable, $\mathcal{A}$ is a static search trajector and $u$ is an objective. Then there exists $W_t : X^{[t]} \to \mathbb{R}$ such that $W = W_t \circ \phi_{[t]}$ for $\mathcal{A}[u]$-almost every search history.*

*Proof.* Let

$$W_t(z) = \int_{\phi_{[t]}^{-1}(\{z\})} W(h_X)\,\mathcal{A}[u](dh_X),$$

which is feasible because $\{z\}$ is Borel-measurable in $X^{[t]}$. Note that $W = \mathbb{E}_{\mathcal{A}[u]}\left[W \mid \mathcal{Z}_t\right]$ since $W$ is $\mathcal{Z}_t$-measurable. Now for any $D \in \mathcal{Z}_t$,

$$\int_D W_t \circ \phi(h_X)\, \mathcal{A}[u](dh_X) = \int_{\phi_{[t]}(D)} W_t(z)\, \mathcal{A}_u^{[t]}(dz) = \int_D W(h_X)\, \mathcal{A}[u](dh_X),$$

and therefore $W_t \circ \phi = \mathbb{E}_{\mathcal{A}[u]}[W \mid \mathcal{Z}_t]$. Stated otherwise, $W_t \circ \phi = W$ for $\mathcal{A}[u]$-almost every search history.

**Proposition 9.10.** *Given a static search trajector $\mathcal{A}$ and an objective $u$, if $S$ is a stopping time such that $S(h_X) < \infty$ for $\mathcal{A}[u]$-almost every $h_X$, then for any $\mathcal{Z}_S$-measurable random variable $W$ and any measurable $A$,*

$$\mathbb{E}_{\mathcal{A}[u]}[\mathbb{1}_A W] = \sum_{t\in\mathcal{T}} \int_{A\cap\{S=t\}} W_t(z)\, \mathcal{A}_u^{[t]}(dz) = \sum_{t\in\mathcal{T}} \int_{A\cap\{S=t\}} W_t(z) \prod_{s\leq t} \mathcal{G}[(z_1,\ldots,z_{s-1}),u](dz_s).$$

*Proof.* Equation 9.10 together with the disjoint additivity of measures yields that $\mathcal{A}[u](A) = \sum_{t\in\mathcal{T}} \mathcal{A}[u](A\cap\{S=t\})$, and Proposition 9.9 allows us to replace $W$ with $W_t$, since when $S=t$, $W$ is $\mathcal{Z}_t$-measurable. The final equality is just Equation 9.11.

The previous proposition can also be applied to $W = 1$ in order to compute $\mathcal{A}[u](A) = \mathbb{E}_{\mathcal{A}[u]}[\mathbb{1}_A]$. Similarly, for arbitrary $W$, another important subcase occurs when the expectation $\mathbb{E}_{\mathcal{A}[u]}[W] = \mathbb{E}_{\mathcal{A}[u]}[\mathbb{1}_{X^\mathcal{T}} W]$. The result can also be used to demonstrate that continuous generators extend to continuous trajectors when integrating over stopped processes with almost surely finite stopping times, discussed next.

### *9.3.3 Continuity of the Optimization Process*

In order to analyze optimizer performance, it will be important to answer the following question: When does continuity of a generator imply the continuity of the optimization process it generates? Specifically, suppose $u_n(x) \to u(x)$ for all $x \in X$, and let $\mathcal{G} \in \mathcal{PF}$ be a static generator that is continuous in objectives. If $\mathcal{A} = \mathbf{E}(\mathcal{G})$ is the extension of $\mathcal{G}$, does $\mathcal{A}[u_n] \to \mathcal{A}[u]$ in the total variation norm for $\mathcal{M}[X^\mathcal{T}]$? Because the optimization process is infinite, it may be possible for $\mathcal{A}[u_n]$ to diverge from $\mathcal{A}[u]$ even if $\mathcal{G}$ is continuous everywhere. For this reason, it is not possible to extend continuity in objectives for $\mathcal{G}$ to continuity $\mathcal{A}$. It is possible, however, to prove that continuity does extend to the expectations of the stopped optimization process, which is why the theory around stopping times has been introduced in the previous subsections. These proofs are developed now.

Chapter 8 addressed continuity for standard search generators, but this chapter has focused on the static search generators. Recall that Theorems 8.11 and 8.12 proved that continuity can be transferred from standard to static generators through the operator $\mathbf{G}(\cdot)$, without any restrictions when $\mathcal{T}$ is countable. The following discussion relies on these theorems in order to extend the previous continuity results to trajectors. If $\mathcal{G} \in \mathcal{BB}^\mathcal{T}_{X,Y}$ is a static black-box search generator such that $\mathcal{G}'$ is a standard search generator with $\mathcal{G} = \mathbf{G}(\mathcal{G}')$, then $\mathcal{A} = \mathbf{E}\circ\mathbf{G}(\mathcal{G}')$ is a static search trajector. If $\mathcal{G}'$ is continuous in objectives, then $\mathcal{G}$ is continuous in objectives. The goal here is to give conditions under which continuous behavior can be derived for $\mathcal{A}$.

Consider a convergent sequence of objectives $u_n \to u$. Continuity of a trajector $\mathcal{A}$ at $u$ requires that $\mathcal{A}[u_n]$ and $\mathcal{A}[u]$ must eventually be close together. Suppose $\mathcal{A} = \mathbf{E}(\mathcal{G})$ for some generator $\mathcal{G}$ that is continuous in objectives at $u$. Along any particular search history, the optimization processes produced by $u$ and $u_n$ close to $u$ cannot move far apart within finite time. The distance between these two processes can only grow as fast as the difference $\|\mathcal{G}[\cdot,u] - \mathcal{G}[\cdot,u_n]\|$, and this distance is small due to continuity. But as the time grows without bound, these small amounts can add up to be large amounts. Thus we will only be able to prove that continuity is preserved by $\mathbf{E}(\cdot)$ when observations are limited to finite time. This case can be expressed by using a stopping time $S$ that is almost surely finite.

The performance of an optimization method depends on the method and the problem on which it is run. We might regard the overall performance of an optimizer on a particular problem as the weighted average of its performance on every possible run of the optimizer. This average can be found be integrating over $\mathcal{A}[u]$. It is important to know whether average performance changes only slightly when the optimizer or the objective function are altered slightly. The next theorem shows that if an optimizer is almost surely continuous in objectives, then the expected value of $\mathcal{Z}_S$-measurable random variables changes continuously with the objective if $S < \infty$ almost surely. If the performance of an optimizer is assessed within finitely many optimization steps, the next theorem will imply that the average performance should not change much if the objective is not changed much. To support these concepts, a definition of "almost sure" continuity is introduced for a generator that encapsulates the idea that a generator is continuous on all of the histories that it generates.

**Definition 9.9 (Almost Surely Continuous).** Given an objective $u : X \to Y$, a static search generator $\mathcal{G}$ is *almost surely continuous* in objectives or trajectories at $u$ if there exists a static search trajector $\mathcal{A} = \mathbf{E}(\mathcal{G})$ and a set $H_X \subseteq X^{\mathcal{T}}$ with full $\mathcal{A}[u]$-measure such that $\mathcal{G}$ is continuous in objectives or trajectories at $h_X, u$ whenever $h_X \sqsubseteq h'_X$ for some $h'_X \in H_X$. If $\mathcal{G}'$ is a standard search generator with $\mathcal{G} = \mathbf{G}(\mathcal{G}')$, then $\mathcal{G}'$ is said to be almost surely continuous at $u$ whenever $\mathcal{G}$ is.

**Theorem 9.1.** *Let $\mathcal{G}$ be a static search generator with finite norm that is almost surely continuous in objectives at u, and let $\mathcal{A} = \mathbf{E}(\mathcal{G})$ be its extension to a static search trajector. Let $W : X^{\mathcal{T}} \to \mathbb{R}$ be a $\mathcal{Z}_S$-measurable random variable for some stopping time with $S < \infty$ $\mathcal{A}[u']$-almost everywhere for all $u'$ in some neighborhood of u, and suppose W is $\mathcal{A}[u']$-integrable.*[4] *The map*

$$v \mapsto \mathbb{E}_{\mathcal{A}[v]}[W]$$

*is continuous at u, where $v \in Y^X$ is assigned the product topology (i.e., the topology of pointwise convergence) and $\mathbb{R}$ has the usual Euclidean topology.*

[4] i.e., $\mathbb{E}_{\mathcal{A}[u']}|W| < \infty$.

*Proof.* First, suppose that $W$ is $\mathcal{Z}_t$ measurable for finite $t$, so that it can be integrated with a finite product as in Equation 9.11; the proxy $W_t$ from Proposition 9.9 will be used for this purpose. Suppose $J$ and $L$ are two index sets of positive integers less than or equal to $t$. $J$ and $L$ will be termed *complementary* if $J \cap L = \emptyset$ and $J \cup L = \{1,\dots,t\}$. Let $\mathcal{K}$ be the set of all complementary pairs of index sets. There are exactly $2^t$ such pairs. These complementary sets can be used to multiply out differences of products like those in Equation 9.11.

Suppose $u_n(x) \to u(x)$ for all $x \in X$ and fix $\varepsilon > 0$. Pick any $z \in X^{[t]}$ and define an abbreviation $z_{<s} = (z_1,\dots,z_{s-1})$. Require further that $\mathcal{G}$ is continuous in objectives at $z_{<s}, u$ for all $s \le t$. By inserting a term $\mathcal{G}[z_{<s}, u_n]$,

$$\int W_t \prod_{s\le t} \mathcal{G}[z_{<s}, u](dz_s) = \int W_t \prod_{s\le t} [(\mathcal{G}[z_{<s}, u](dz_s) - \mathcal{G}[z_{<s}, u_n](dz_s)) + \mathcal{G}[z_{<s}, u_n](dz_s)] \tag{9.12}$$

$$= \int W_t \sum_{J,L\in\mathcal{K}} \left[ \prod_{j\in J} (\mathcal{G}[z_{<j}, u](dz_j) - \mathcal{G}[z_{<j}, u_n](dz_j)) \times \prod_{\ell\in L} \mathcal{G}[z_{<\ell}, u_n](dz_\ell) \right]. \tag{9.13}$$

Equation 9.13 expands the product in Equation 9.12 by cross multiplying the difference with $u_n$. This sum contains $2^t$ terms, one for each pair of complementary index sets. With the exception of the complementary sets given by $J_0 = \emptyset, L_0 = \{1,\dots,t\}$, every pair of complementary index sets in $\mathcal{K}$ yields a product in Equations 9.13 with at least one factor of the form

$$\mathcal{G}[z_{<j}, u](dz_j) - \mathcal{G}[z_{<j}, u_n](dz_j).$$

Because $t$ is finite, $z$ is fixed, and $\mathcal{G}$ is continuous in objectives at $z_{<j}, u$, it is possible for each $j$ to choose $n = n(j)$ so that

$$\int |W_t| \left| \mathcal{G}[z_{<j}, u] - \mathcal{G}[z_{<j}, u_n] \right| (dz_j) < \frac{\varepsilon}{2^t \|\mathcal{G}\|^m}$$

where $m < t$, implicitly using that $W$ is integrable. Now for $J \ne J_0$,

$$\int |W_t| \prod_{j\in J} |\mathcal{G}[z_{<j}, u](dz_j) - \mathcal{G}[z_{<j}, u_n](dz_j)| \prod_{\ell\in L} |\mathcal{G}[z_{<\ell}, u_n]|(dz_\ell) < \frac{\varepsilon}{2^t},$$

where the $\|\mathcal{G}\|^m$ on bottom is cancelled by bounding all $m < t$ occurrences of $\mathcal{G}[z_{<\ell}, u_n]$ by $\int |\mathcal{G}[z_{<\ell}, u_n]|(dz_\ell) \le \|\mathcal{G}\|$. Further, the term for $J_0, L_0$ reduces to

$$\int W_t \prod_{s\le t} \mathcal{G}[z_{<s}, u_n](dz_s),$$

which can be subtracted from both sides of Equation 9.13 to obtain

$$\left| \int W_t \left[ \prod_{s \leq t} \mathcal{G}\left[z_{<s}, u\right](dz_s) - \prod_{s \leq t} \mathcal{G}\left[z_{<s}, u_n\right] \right] (dz_s) \right|$$

$$\leq \sum_{J,L \in \mathcal{K} \setminus \{J_0, L_0\}} \int |W_t| \prod_{j \in J} \left| \mathcal{G}\left[z_{<j}, u\right](dz_j) - \mathcal{G}\left[z_{<j}, u_n\right] \right| (dz_j)$$

$$\times \prod_{\ell \in L} \left| \mathcal{G}\left[z_{<\ell}, u_n\right] \right| (dz_\ell)$$

$$< 2^t \frac{\varepsilon}{2^t} = \varepsilon. \tag{9.14}$$

Since $t$ is finite, the $n(j)$ above can be replaced by $n_{\max} = \max_j n(j)$ to show

$$\left| \mathbb{E}_{\mathcal{A}[u_n]}[W] - \mathbb{E}_{\mathcal{A}[u]}[W] \right| \to 0. \tag{9.15}$$

Now suppose $W$ is an integrable, $\mathcal{Z}_S$-measurable random variable. From the fact that $\mathcal{A}[u](\{S = \infty\}) = 0$, it follows that $\mathcal{A}[u](\{S > t\}) \downarrow 0$. Also, $\mathbb{E}_{\mathcal{A}[u]}\left[\mathbb{1}_{\{S>t\}} W\right] \downarrow 0$ due to the integrability of $W$. For any $\varepsilon > 0$, there exists a $\tau < \infty$ such that

$$\left| \mathbb{E}_{\mathcal{A}[u]}[W] \right| = \left| \mathbb{E}_{\mathcal{A}[u]}\left[\mathbb{1}_{\{S \leq \tau\}} W\right] + \mathbb{E}_{\mathcal{A}[u]}\left[\mathbb{1}_{\{S > \tau\}} W\right] \right| < \left| \mathbb{E}_{\mathcal{A}[u]}\left[\mathbb{1}_{\{S \leq \tau\}} W\right] \right| + \frac{\varepsilon}{2}.$$

The same is true for all $u'$ in some neighborhood of $u$. Proposition 9.10 yields

$$\mathbb{E}_{\mathcal{A}[u']}\left[\mathbb{1}_{\{S<\tau\}} W\right] = \sum_{t<\tau} \int_{\{S=t\}} W_t(z)\, \mathcal{A}_{u'}^{[t]}(dz) = \sum_{t<\tau} \mathbb{E}_{\mathcal{A}[u']}\left[\mathbb{1}_{\{S=t\}} W\right]. \tag{9.16}$$

Since $\mathbb{1}_{\{S=t\}} W$ is a $\mathcal{Z}_t$-measurable random variable, Equation 9.15 implies

$$\left| \mathbb{E}_{\mathcal{A}[u]}\left[\mathbb{1}_{\{S=t\}} W\right] - \mathbb{E}_{\mathcal{A}[u_n]}\left[\mathbb{1}_{\{S=t\}} W\right] \right| < \frac{\varepsilon}{2\tau}$$

for sufficiently large $n$ and all $t < \tau$, and consequently

$$\left| \mathbb{E}_{\mathcal{A}[u]}\left[\mathbb{1}_{\{S<\tau\}} W\right] - \mathbb{E}_{\mathcal{A}[u_n]}\left[\mathbb{1}_{\{S<\tau\}} W\right] \right| < \frac{\varepsilon}{2} \text{ and } \left| \mathbb{E}_{\mathcal{A}[u]}[W] - \mathbb{E}_{\mathcal{A}[u_n]}[W] \right| < \varepsilon,$$

which completes the proof.

**Corollary 9.1.** *Under the same general assumptions as Theorem 9.1 on $\mathcal{A}$, $\mathcal{G}$, $S$, and $u$. Let $\mathcal{A}[v]|_{\mathcal{Z}_S}$ represent the measure $\mathcal{A}[v]$ restricted to the $\sigma$-algebra $\mathcal{Z}_S$. Then the map $v \mapsto \mathcal{A}[v]|_{\mathcal{Z}_S}$ is continuous at $u$, where $\mathcal{A}[v]|_{\mathcal{Z}_S}$ is regarded as an element of the Banach space $\mathcal{M}[X^{\mathcal{T}}, \mathcal{Z}_S]$ of finite signed measures.*

*Proof.* Suppose first that $\mathcal{A}[u]$ is a positive measure for some neighborhood of $u$. Take any $D \in \mathcal{Z}_S$ and note that $\mathcal{A}[u](D) = \mathbb{E}_{\mathcal{A}[u]}[\mathbb{1}_D]$. It follows from Theorem 9.1 with $W = \mathbb{1}_D$ that the map $v \mapsto \mathcal{A}[v](D)$ is continuous at $u$.

Now in particular $X^{\mathcal{T}} \in \mathcal{Z}_S$, and the map $v \mapsto \mathcal{A}[v](X^{\mathcal{T}})$ dominates all of the maps for other $D \subseteq X^{\mathcal{T}}$ in some neighborhood of $u$, since $\mathcal{A}[v]$ is positive. Consequently, for $u_n \to u$,

$$\left\| \mathcal{A}[u]|_{\mathcal{Z}_S} - \mathcal{A}[u_n]|_{\mathcal{Z}_S} \right\| \to 0.$$

Since this convergence holds for positive measures, it also holds for negative measures. The Hahn-Jordan decomposition then guarantees the final result for arbitrary trajectors.

It may not yet be clear why "almost sure" continuity is needed, as opposed to continuity at every history. For certain evolutionary and quasi-evolutionary algorithms of Chapter 8, continuity in objectives occurs exactly for histories of unambivalent value (see Definition 8.14 and Theorem 8.9 as well as Theorem 8.10). Nonetheless, these algorithms are almost surely continuous, provided that the histories of unambivalent value have full measure. The following theorem presents conditions under which this result is guaranteed, so that Theorem 9.1 can be applied. In essence, the optimizer must place full measure on points that would extend a trajectory ambivalently.

**Theorem 9.2.** *Let $u$ be an objective, and let $\mathcal{G} \in \mathcal{PBO}_K$ be a population-based optimizer that is continuous in objectives or trajectories at all histories $h$ of unambivalent value at degree $K$. Let $A_{h,u}$ be the set of points in $X$ for which $h \| (x, u(x))$ is of ambivalent value on $u$. If $\mathcal{G}[h](A_{h,u}) = 0$ for all histories $h$ of unambivalent value at degree $K$, then $\mathcal{G}$ is almost surely continuous in objectives or trajectories.*

*Proof.* The proof is by induction on the length of the history. Every history of length 1 is of unambivalent value. Suppose that histories of length $n-1$ or less are unambivalent at degree $K$ with full $\mathcal{A}[u]$-measure. That is, there is a set $C \subseteq X^{\mathcal{T}}$ of full $\mathcal{A}[u]$-measure such that $\mathcal{G}$ is continuous at $h$ whenever $|h| < n$ and $h \sqsubseteq h' \in C$, and such $h$ is necessarily of unambivalent value at degree $K$. Let $h$ be such a history with length $n-1$. Let $U = (X \times Y) \setminus A_{h,u}$ so that $\mathcal{G}[h](U) = 1$. That is, extensions of $h$ to length $n$ are of unambivalent value with full $\mathcal{A}[u]$-measure, and histories of length $n$ are in unambivalent at degree $K$ with full $\mathcal{A}[u]$-measure. Therefore histories of arbitrary length generated from $\mathcal{G}$ are of unambivalent value at degree $K$ with full $\mathcal{A}[u]$-measure, which completes the induction.

In combination with Theorem 9.1, Theorem 9.2 is sufficient to prove the continuity of performance for evolutionary algorithms in many cases. For example, if the search domain is $d$-dimensional Euclidean space, $X = \mathbb{R}^d$, then a real-coded genetic algorithm with tournament selection, masked crossover, and Gaussian selection is continuous on almost every generated history for objective functions without fitness plateaus, that is, on all objective functions whose level sets have Lebesgue measure zero.

Theorem 9.1 considered what would happen if the generator is held constant but the objective is changed. If instead the objective is held constant but the generator is altered slightly, a similar theorem holds without continuity assumptions. Integrals over $\mathcal{Z}_S$-measurable random variables for $S < \infty$ change continuously with the optimizer, regardless of whether the generator is continuous. This result will be used to demonstrate that performance criteria are continuous over optimizers.

Before stating the theorem, consider how $\mathbf{E}(\mathcal{G})$ changes with $\mathcal{G}$. In general, it is possible that $\mathcal{G}$ and $\mathcal{G}'$ could be arbitrarily close, and yet $\mathbf{E}(\mathcal{G})$ may exist whereas

$\mathbf{E}(\mathcal{G}')$ does not. Therefore the map $\mathcal{G}'' \mapsto \mathbb{E}_{\mathbf{E}(\mathcal{G}'')}[W]$ may be defined for $\mathcal{G}$ but not $\mathcal{G}'$. Hence the continuity of this map cannot be assessed without clarifying the scope of $\mathcal{G}'$ appropriately. The simplest solution is to require that both $\mathcal{G}$ and $\mathcal{G}'$ be tight and eventually bounded. Since these two properties induce a vector space that contains all of the proper static search generators, this solution is sufficient for the purposes of this book.

**Theorem 9.3.** *Let $\mathcal{G}$ be a tight and eventually bounded static search generator of finite norm with extension $\mathcal{A} = \mathbf{E}(\mathcal{G})$. Let $u : X \to Y$ be an objective. Let $W : X^{\mathcal{T}} \to \mathbb{R}$ be a $\mathcal{Z}_S$-measurable random variable for some stopping time with $S < \infty$ $\mathbf{E}(\mathcal{G}')[u]$-almost everywhere for all $\mathcal{G}'$ in some neighborhood of $\mathcal{G}$, and suppose that $W$ is similarly $\mathbf{E}(\mathcal{G}')[u]$-integrable. Then the map*

$$\mathcal{G}' \to \mathbb{E}_{\mathbf{E}(\mathcal{G}')}[W]$$

*is well defined over all tight and eventually bounded static search generators with finite norm, and it is continuous at $\mathcal{G}$.*

*Proof.* Let $(\mathcal{G}_n)_{n \in \mathcal{I}}$ be any net converging to $\mathcal{G}$ in the norm topology of the tight and eventually bounded static search generators with finite norm. Repeat the proof of Theorem 9.1, replacing $\mathcal{G}[z_{<s}, u_n]$ and similar by the proper analogue of $\mathcal{G}_n[z_{<s}, u]$. The proof then follows *mutatis mutandis*.

Theorem 9.1 and 9.3 are sufficient to prove the continuity of finitely determined performance criteria on continuous optimization methods, which is done in Chapter 10.

## 9.4 Conclusion

This chapter discussed the properties of the infinite optimization process, which is well-defined for every proper static search generator. It also briefly introduced the relevant background in stochastic processes that will be used to establish the theorems of subsequent chapters. Finally, it has been shown that static search trajectors weakly preserve continuity in the sense that the expected value of $\mathcal{Z}_S$-measurable random variables for almost surely finite $S$ converges when applied to similar objectives and optimizers. The next chapter develops the analysis of performance based on the results of this chapter.

# Chapter 10
# Performance Analysis

Given a particular objective function to be optimized, it would be useful to know which optimization method will perform best on that objective. Indeed, the entire purpose of studying the static search generators is to provide tools to answer this very question. To this end, different categories of performance criteria are analyzed theoretically in this chapter. Several performance criteria are shown to be continuous and non-linear with respect to static search generators, implying that similar optimization methods perform similarly and that linearly interpolated generators produce optimization methods may outperform the methods being interpolated. These facts are demonstrated experimentally in Chapter 11. Further, the categories of performance criteria described in this chapter make it possible to identify the conditions under which No Free Lunch theorems hold in infinite-dimensional spaces, to be undertaken in Chapter 12.

## 10.1 Performance Criteria

This section introduces *performance criteria* that formalize common notions of what it means for an optimization method to perform well on an objective. A performance criterion takes a static search generator implementing an optimization method along with an objective function and outputs a real number, providing an objectively determined score for each optimization method on each cost function. As a convention (which will sometimes be violated), this score should be nonnegative with a value of zero considered perfect performance.

### *10.1.1 Defining Performance*

A performance criterion is defined as the expected value of a random variable defined on the optimization process.



A. J. Lockett, *General-Purpose Optimization Through Information Maximization*,
Natural Computing Series, https://doi.org/10.1007/978-3-662-62007-6_10

**Definition 10.1.** Recall that $\mathbf{E}_0$ is the set of all static search generators that extend to a static search trajector. Let $Y_m^X \subseteq Y^X$ be the set of measurable objectives. A *static performance criterion* $\psi : \mathbf{E}_0 \times Y_m^X \to \mathbb{R} \cup \{\infty\}$ has

$$\psi(\mathcal{G}, u) = \begin{cases} \mathbb{E}_{\mathcal{A}[u]}[V_u] & \text{if } V_u \text{ is } \mathcal{A}[u]\text{–integrable} \\ \infty & \text{otherwise.} \end{cases}$$

where $\mathcal{A} = \mathbf{E}(\mathcal{G})$ is the extension of $\mathcal{G}$ and $u \mapsto V_u$ is a map called the *kernel* of $\psi$ that yields random variables on all of $Y_m^X$.

Performance criteria can be used to compare optimization methods to each other, and to analyze how the performance varies as the method or the objective changes. Ultimately, an analysis of performance should reveal how to select a particular optimizer for a particular task. This issue will be approached experimentally in the next chapter and theoretically in Chapters 12 and 13.

Most of the performance criteria considered in this chapter are defined with respect to the error magnitude at each optimizer step. In order to compute an error, the value space $Y$ is assumed for the moment to be a strictly ordered vector space (*e.g.*, $Y = \mathbb{R}$) so that errors have a uniquely directed magnitude, values can be subtracted, and $Y$ has a zero value. In particular, in order to have a meaningful error magnitude, the objective $u : X \to Y$ must have a lower bound inside $Y$, which will be denoted as $u_* = \inf_{x \in X} u(x)$. With $Y = \mathbb{R}$, this statement means $u(x) > -\infty$ for all $x \in X$. On a similar note, the assumption that $\mathcal{T} = \mathbb{N}$ continues in this chapter.

**Definition 10.2.** The *error history* $E(h_X) = (E_t(h_X))_{t \in \mathcal{T}}$ of a history $h_X \in X^{\mathcal{T}}$ on an objective $u : X \to Y$ is the sequence on $Y^{\mathcal{T}}$ given by

$$E_t(h_X) = u(h_t^X) - u_* \tag{10.1}$$

for any $u$ that is bounded below, *e.g.*, $u_* > -\infty$. When the objective function must be stated explicitly, the error history may be written as $E^u(h_X)$ and its components as $E_t^u(h_X)$.

**Proposition 10.1.** *Given an objective $u : X \to Y$ that is bounded below, if $Y^{\mathcal{T}}$ has the Baire $\sigma$-algebra and $X^{\mathcal{T}}$ has either Baire or the Borel $\sigma$-algebra, then the error history $E^u : X^{\mathcal{T}} \to Y^T$ is a random variable if and only if $u$ is measurable (i.e., $u$ is a random variable). If $X^{\mathcal{T}}$ and $Y^{\mathcal{T}}$ both have the Borel $\sigma$-algebra, then $E^u$ is again a random variable if and only if $u$ is measurable.*

*Proof.* Let $\phi_{\mathcal{K}}^Z : S^{\mathcal{T}} \to S^{\mathcal{K}}$ be a projection of the space $S$ for some set $\mathcal{K} \subseteq \mathcal{T}$. For $z \in X^{\mathcal{T}}$, define $u_*^{\mathcal{K}}(z) = (u(z_k) - u_*)_{k \in \mathcal{K}}$ apply $u$ pointwise, which is measurable if and only if $u$ is. Then define $E_{\mathcal{K}}^u = u_*^{\mathcal{K}} \circ \phi_{\mathcal{K}}^X$, which is measurable if and only if both $u$ and $\phi_{\mathcal{K}}^X$ are measurable. Given that $E^u \circ \phi_{\mathcal{K}}^Y = E_{\mathcal{K}}^u$, it follows that $E^u \circ \phi_{\mathcal{K}}^Y$ is measurable if and only if both $u$ and $\phi_{\mathcal{K}}^X$ are measurable.

Suppose $Y^{\mathcal{T}}$ has the Baire $\sigma$-algebra, which is the smallest $\sigma$-algebra that makes all finite projections measurable. Then $E^u$ will be measurable if and only if $E^u \circ \phi_{\mathcal{K}}^Y$ is measurable for all finite $\mathcal{K}$. By the preceding paragraph, this is true if and only if both $\phi^X_{\mathcal{K}}$ and $u$ are measurable. Now $\phi^X_{\mathcal{K}}$ is measurable for finite $\mathcal{K}$ when $X^{\mathcal{T}}$ has either the Borel or the Baire $\sigma$-algebra, so then $E^u$ is measurable if and only if $u$ is measurable.

Suppose instead that both $X^{\mathcal{T}}$ and $Y^{\mathcal{T}}$ has the Borel $\sigma$-algebra. Then in the same way, $E^u$ will be measurable if and only if $E^u \circ \phi^Y_{\mathcal{K}}$ is measurable for all $\mathcal{K}$, including infinite subsets of $\mathcal{T}$, which is true if and only if $u$ is measurable.

The error history of the optimization process, denoted $E(Z) = E^u(Z)$, will thus be termed the *error process*. Proposition 10.1 shows that when the objective $u$ is measurable, this error process is indeed a stochastic process. The sequence of search points along the optimization process that corresponds to the sequence of best values so far will be termed the *running minimum process*, denoted by $Z^*$, with $Z^*_t(h_X) = \text{arginf}_{\{Z_s:s\leq t\}} u(Z_s(h_X))$. That is, $Z^*_t$ is the best known solution discovered by the optimization process at time $t$. The collection $Z^* = (Z^*_t)_{t\in\mathcal{T}}$ is again a stochastic process if and only if $u$ is measurable. Define the *minimum error history* $E^* = E^*(h_X)$ as the running minimum of the error history with

$$E^*_t(h_X) = \inf_{s\leq t} E_s(h_X) = u(Z^*_t(h_X)) - u_*, \tag{10.2}$$

which is likewise measurable if and only if $u$ is. The *minimum error process* is the minimum error history of the optimization process, $E^*(Z) = E(Z^*)$. These definitions will be used to define classes of performance criteria.

The remainder of this section gives examples of possible performance criteria that correspond broadly to the kinds of results reported in the experimental literature on optimizers. These examples are given in four groups: (1) evaluation by average error, (2) hitting times for an error bound, (3) probability of attaining an error bound, and (4) error at a stopping time.

### *10.1.2 Evaluation by Average Error*

A first approach to evaluating optimization methods is to average the magnitude of the errors at each time step. This metric combines the total accuracy along with the speed of convergence, at the risk of disproportionately penalizing optimizers for early errors due to exploration of the objective. Such a metric is not traditionally reported, but could prove useful, since it contains information about the convergence speed of the optimizer.

Define $\phi_w$ as the performance criterion with kernel

$$V_u(h_X) = w \cdot E^*(h_X) = \sum_{t\in\mathcal{T}} w_t E^*_t(h_X)$$

where $w = (w_t)_{t\in\mathcal{T}} \subseteq [0,\infty)$ is a sequence of weights that can be used to discount later values. When $\mathcal{G}$ is an extensible static search generator with $\mathcal{A} = \mathbf{E}(\mathcal{G})$ and $u$ is a measurable objective,

$$\phi_w(\mathcal{G},u) = \mathbb{E}_{\mathcal{A}[u]}\left[w \cdot E^*(h_X)\right] = \sum_{t\in\mathcal{T}} \int_{X^{\mathcal{T}}} w_t E_t^*(z)\, \mathcal{A}[u](dz),$$

where the far right-hand side assumes that the sum converges absolutely so that the integral exists and that the sum and integral may be exchanged.

Three basic choices for $w_n$ are (1) $w_t = 1$, which treats all errors equally but only results in $\phi_w$ finite when the objective is optimized at a fast enough rate, (2) $w_t = 2^{-t}$, which places more weight on earlier errors but yields $\phi_w$ finite whenever the objective is almost surely finite on the initial population, and (3) $w_t = 1$ for $t \leq N$ for some fixed $N < \infty$ and $w_t = 0$ for $t > N$, which considers only a finite number of time steps. Another possible scheme might ignore initial errors up to a finite time, allowing optimizers to explore more broadly in earlier stages without penalty.

The function $\phi_w$ using any of the three methods described above has two primary advantages. First of all, it captures a natural intuition for evaluating an optimizer, namely, the magnitude of errors it makes before finding a good optimum. Secondly, by taking a sum of these errors, $\phi_w$ measures the convergence rate of an optimizer. The disadvantage of $\phi_w$ is that it can be sensitive to the early errors of an optimizer, especially when $w_t = 2^{-t}$. Also, if $w_t$ is set according to either the second or third option above, then later errors will be ignored, and an asymptotically convergent optimizer that converges late will be outscored by a non-convergent optimizer that attains good but suboptimal solutions earlier on (which may or may not be a desirable feature).

One may wish to estimate the value of a performance criterion in order to evaluate various optimizers. If $w_t$ is set according to the first option ($w_t = 1$ for all $t$), then there is no reliable way to approximate the value of $\phi_w$ through sample runs. No matter how many times an optimizer converges to the correct solution, it is always possible that there is a set of sample runs with positive probability on which the algorithm never reaches the global optimum. In this case, the integrand is infinite on a set of positive probability, and thus it is possible to have $\phi_w = \infty$ even if the cumulative error appears small and finite for all observed runs. In fact, many optimizers of interest will have $\phi_w = \infty$ on a large number of problems (e.g. most genetic algorithms). Thus setting $w_t = 1$ for all $t$ is practically undesirable unless one has a proof that an algorithm converges in probability to the global optimum on all objectives of interest.

If $w_t$ is set according to either $w_t = 2^{-t}$ or $w_t = 1$ for $t \leq N$, then the value of $\phi_w(\mathcal{G},u)$ can be estimated using Monte Carlo methods by running several instances of the optimizer $\mathcal{G}$ on $u$ for a fixed number of iterations. In the first case, the number of iterations is chosen to satisfy a tolerance, $2^{-t} < \varepsilon$; in the second, the number of iterations is simply the bound $N$. The minimum error sequence $E^*(Z)$ is non-increasing, and thus $\phi_w$ converges for either choice of $w$ provided that $E_1^*(Z)$ is almost surely finite. In Chapter 11, results will be reported for both $w_t = 2^{-t}$ and $w_t = 1$ for $t < N$.

### 10.1.3 Evaluation by Hitting Time

In existing literature, when evaluating a proposed optimizer, the optimizer is often run on a benchmark set of problems for which the optima are known (see e.g. [29, 9, 64]). A common performance criterion for ranking optimizers is to count the number of points that must be generated before obtaining a solution whose fitness is within a fixed error from the globally optimal fitness.

For a fixed error $\varepsilon > 0$, define the hitting time for $\varepsilon$ as the first time when a search point has global error less than $\varepsilon$, *i.e.*,

$$\tau_\varepsilon(h_X) = \inf\{t \mid E_t(h_X) \leq \varepsilon\}.$$

The hitting time depends only on the error process and is a stopping time since $E_t$ is $\mathcal{Z}_t$-measurable. For an extensible static search generator $\mathcal{G}$ with $\mathcal{A} = \mathbf{E}(\mathcal{G})$ and a measurable objective $u$, define a performance criterion by

$$\psi_\varepsilon(\mathcal{G}, u) = \mathbb{E}_{\mathcal{A}[u]}[\tau_\varepsilon],$$

which is the average hitting time for $\varepsilon$ over all runs of the optimizer with generator $\mathcal{G}$ on the objective $u$.

This formula has a serious flaw for non-convergent optimizers. If $\mathcal{A}[u]$ has a positive probability of failing to attain error less than $\varepsilon$, then $\psi_\varepsilon = \infty$. Additionally, from the standpoint of approximation, only finite computational time is available, and thus cases in which $\tau_\varepsilon$ is large cannot be distinguished computationally from cases in which it is infinite.

One alternative is to place a finite limit on the stopping time; that is, for $N < \infty$,

$$\psi_\varepsilon^N(\mathcal{G}, u) = \mathbb{E}_{\mathcal{A}[u]}[\tau_\varepsilon \wedge N],$$

where the notation $\tau_\varepsilon \wedge N = \min\{\tau_\varepsilon, N\}$, which is also a stopping time. The criterion $\psi_\varepsilon^N(\mathcal{G}, u)$ can be estimated reasonably by running $\mathcal{G}$ on $u$ several times for at most $N$ evaluations. This performance criterion also reflects a natural criterion for comparing optimizers; it measures the average number of steps the optimizer must be run out of some maximum budget before it produces a solution correct within error $\varepsilon$. Unlike $\phi_w$, $\psi_\varepsilon^N$ is generally bounded across optimizers and objectives; optimizers will have $\psi_\varepsilon^N \leq N$ on all objectives. Unfortunately, $\psi_\varepsilon$ and $\psi_\varepsilon^N$ are discontinuous on many objective functions, as will be discussed below.

### 10.1.4 Evaluation by Success Probability

The hitting time tests how long it takes on average to attain an error threshold $\varepsilon$. However, it does not test how often the threshold is attained. Define the sets $T_\varepsilon = \{h_X \mid \tau_\varepsilon(h_X) < \infty\}$ and $T_\varepsilon^N = \{h_X \mid \tau_\varepsilon(h_X) < N\}$ to represent respectively the sequences that asymptotically attain a given error bound and those that attain it within a fixed number of evaluations. These sets are measurable whenever the error sequence is measurable. Then the *success probability* is the probability of attaining a bound asymptotically, and the *finite success probability* is the probability of attaining the bound within a finite time window [165]. Each of these are performance criteria given by

$$\sigma_\varepsilon(\mathcal{G},u) = \mathcal{A}[u](T_\varepsilon), \quad \sigma_\varepsilon^N(\mathcal{G},u) = \mathcal{A}[u](T_\varepsilon^N). \tag{10.3}$$

To see that $\sigma_\varepsilon$ and $\sigma_\varepsilon^N$ are performance criteria, recall that $\mathcal{A}[u](A) = \mathbb{E}_{\mathcal{A}[u]}[\mathbb{1}_A]$ where $\mathbb{1}_A$ is the indicator set of $A$, i.e. $\mathbb{1}_A(z) = 1$ if $z \in A$ and is zero otherwise. The finite success probability is the preferred criterion, since $\sigma_\varepsilon^N$ can be estimated experimentally, whereas $\sigma_\varepsilon$ cannot. Notice that $\sigma_\varepsilon$ does not conform to the convention that lower performance values should be better and zero should be optimal. The convention is ignored here because the success probability has an intuitive meaning in its own right. In situations where the convention is important, the performance criterion $1 - \sigma_\varepsilon$ can be used instead.

Given the finite success probability, it is of interest to know the average hitting time for sequences that attain the error bound. The average hitting time on successful trajectories is a performance criterion, given by

$$\hat{\psi}_\varepsilon^N(\mathcal{G},u) = \mathbb{E}_{\mathcal{A}[u]}\left[\mathbb{1}_{T_\varepsilon} \times (\tau_\varepsilon \wedge N)\right]. \tag{10.4}$$

On its own, this quantity is not useful, since it may be zero when the optimizer fails, i.e. when $\mathcal{A}[u](T_\varepsilon^N) = 0$. However, the pair $\left(\hat{\psi}_\varepsilon^N, \sigma_\varepsilon^N\right)$ disambiguates this situation, and these two values can be reported together for completeness [9].

### *10.1.5 Evaluation by Error at a Stopping Time*

Optimizers are often tested by running the algorithm for a fixed number of evaluations and then reporting the final error. As a generalization of this type of evaluation, suppose that an optimizer is run until some criterion is satisfied, not necessarily connected to the number of evaluations. As one example of why this generalization may be useful, suppose that rather than stopping after a fixed number of evaluations, one wishes to stop an optimizer after it uses up a fixed amount of resources, such as CPU cycles or calendar time. Such a criterion can be modeled as a stopping time, and the error magnitude at this stopping time is a performance criterion.

Let $T$ be a stopping time equal to the generation in which this resource limit is first expended, and define a performance criterion by

$$\zeta_T(\mathcal{G},u) = \mathbb{E}_{\mathcal{A}[u]}\left[E_T^*\right], \tag{10.5}$$

so that $\zeta_T$ is the smallest error attained within the allocated resources, where $E_t^*$ is the running error at time $t$.

One stopping time that will be used extensively is the number of unique points evaluated. In an environment where function evaluation is expensive, the objective value of repeated points can be retrieved from a cache. In this case, it is reasonable to suggest that repeated evaluation points are irrelevant to overall performance. Given a sequence $z \in X^{\mathcal{T}}$, let

$$T_m(z) = \inf\{t \in \mathcal{T} \mid z_1, \ldots, z_t \text{ contains } m \text{ unique points}\}. \tag{10.6}$$

Performance criteria based on $T_m$ are used to derive No Free Lunch theorems in Chapter 12, extending previous results of this type that only applied to optimizers that never repeat any point.

Performance criteria defined on this sequence will be studied almost exclusively from this point, and thus it is worthwhile to define this sequence independently.

**Definition 10.3 (Unique Stopping Sequence).** The sequence of stopping times given by $(T_m)_{m\in\mathcal{T}}$ is termed the *unique stopping sequence*.

As a variation on $\phi_w$ above, one may define the average minimum error after each unique individual by

$$\phi_T(\mathcal{G}, u) = \sum_{m=L}^{U} \zeta_{T_m}(\mathcal{G}, u) = \mathbb{E}_{\mathcal{A}[u]}\left[\sum_{m=L}^{U} E^*_{T_m}\right] \tag{10.7}$$

for some lower bound $L \geq 1$ and upper bound $U < \infty$. The criterion $\phi_T$ is finite whenever $E^*_{T_L} = u\left(Z^*_{T_L}\right) - u_*$ is almost surely finite.

Usually, it is not difficult to estimate $\zeta_{T_m}$ or $\phi_T$. Most optimizers produce unique points with some frequency, so that $T_m < \infty$ almost surely if $m < |X|$. In infinite spaces, it is even common to have $T_m = m$ $\mathcal{A}[u]$-almost surely. If an optimizer does not produce $m$ unique points, or does so slowly, this property of the optimizer will generally be known ahead of time either analytically or constructively. If $T_m = \infty$, then the set of unique points in the optimization process is of size at most $m-1$, so $E^*_{T_m} = E^*_{T_{m-1}}$ and $\zeta_{T_m} = \zeta_{T_{m-1}}$. Thus the infinite case is easy to handle when it can be identified. It is only difficult to approximate $\zeta_{T_m}$ when unique points are generated slowly. In this case, assuming that $T_m = \infty$ will produce an overestimate of the performance criterion. Optimizers that produce unique points slowly are generally undesirable, and thus an overestimate of the performance criterion for these optimizers is not problematic.

A substantial number of performance criteria have now been introduced. The next two sections discuss the mathematical properties of performance criteria, such as nonlinearity, decomposability, and continuity.

## 10.2 Properties of Performance Criteria

It is clear that a wide variety of performance criteria exists. These criteria can be analyzed in general according to their mathematical properties. This section examines three such properties that a performance criterion may possess: (1) nonlinearity, (2) progressive decomposability, and (3) dependence on the error history. The question of continuity in performance criteria is a larger topic and will be addressed separately in the next section.

### *10.2.1 Nonlinearity*

All non-trivial performance criteria are nonlinear in both arguments. A performance criterion is trivial if it does not depend on the generator, i.e. $\psi(\mathcal{G},u) = \psi(u)$, or if it only depends on the first element of the error sequence, i.e. $\psi(\mathcal{G},u) = \mathbb{E}_{\mathcal{A}[u]}[g(E_1(Z))]$ for some $g$.

For a given objective function, the location and nature of the optima are nonlinear qualities. The location of the global optimum for $u+v$ bears no general relationship to the location of the optimum for $u$ or $v$. The error sequence has $E_t^{u+v}(z) \neq E_t^u(z) + E_t^v(z)$ for most non-constant $u,v$. Thus for any useful performance criterion $\psi$, including the ones defined above, one expects that $\psi(\mathcal{G},u+v) \neq \psi(\mathcal{G},u) + \psi(\mathcal{G},v)$ in general. Trivial parameter assignments, such as, for example, $w_n = 0$ for $\phi_w$, are ignored here and elsewhere.

Non-trivial performance criteria are also nonlinear with respect to generators as well. Continuing the use of product expansions as in Equation 9.11, let $\phi_{[t]} : X^{\mathcal{T}} \to X^{[t]}$ be the canonical coordinate projection, and for two static search generators $\mathcal{G}$ and $\mathcal{H}$ and $A \in \mathcal{Z}_t$,

$$\mathbf{E}(\mathcal{G}+\mathcal{H})[u](A) = \int_{\phi_{[t]}(A)} \prod_{s\leq t} \mathcal{G}+\mathcal{H}[z_{<s},u](dz_s)$$

assuming that the extension of $\mathcal{G}+\mathcal{H}$ exists, which it does if both $\mathcal{G}$ and $\mathcal{H}$ are tight and eventually bounded. It is thus clear that $\mathbf{E}(\mathcal{G}+\mathcal{H})[u] \neq \mathbf{E}(\mathcal{G})[u] + \mathbf{E}(\mathcal{H})[u]$ except under special circumstances because of the cross terms under the product. In general, $\psi(\mathcal{G}+\mathcal{H},u) \neq \psi(\mathcal{G},u) + \psi(\mathcal{H},u)$.

The nonlinearity of most performance criteria has an important consequence: It opens the possibility that a convex combination over a bank of one-step optimizers may outperform any of the given optimizers. Chapter 11 will present some experimental evidence supporting this possibility, and the topic will be discussed further in Chapter 13.

### *10.2.2 Progressive Decomposability*

Theorems 9.1 and 9.3 proved that the expected value of a random variable on the optimization process changes continuously with the objective if the value of the random variable is determined by a finite number of optimization steps. A progressively decomposable performance criterion can be broken down into an infinite sum of finitely determined random variables.

**Definition 10.4 (Progressive Decomposability).** A performance criterion $\psi$ is *progressively decomposable* if there exists a sequence of functions $h_t : X^{[t]} \times Y^X \to \mathbb{R}$ such that

$$\psi(\mathcal{G},u) = \sum_{t\in\mathcal{T}} \mathbb{E}_{\mathcal{A}[u]}\left[h_t\left((Z_s)_{s\leq t}, u\right)\right], \tag{10.8}$$

for all objectives $u$ and all $\mathcal{G} \in \mathbf{E}_0$ with $\mathcal{A} = \mathbf{E}(\mathcal{G})$.

Progressive decomposability means that a performance criterion can be analyzed as the sum of infinitely many performance criteria that each depend on the state of the optimizer up to a fixed time step. This fact is used to prove that performance criteria are continuous in certain cases. Perhaps surprisingly, all of the performance criteria presented thus far are progressively decomposable.

**Proposition 10.2.** *The performance criterion $\phi_w$ is progressively decomposable.*

*Proof.* Because all terms are positive, Tonelli's theorem implies that

$$\phi_w(\mathcal{G},u) = \sum_{t\in\mathcal{T}} w_t \mathbb{E}_{\mathcal{A}[u]}\left[E_t^*\right], \tag{10.9}$$

which is progressively decomposable with $h_t(z,u) = w_t E_t^*$.

**Proposition 10.3.** *The performance criteria $\psi_\varepsilon$ and $\psi_\varepsilon^N$ are progressively decomposable.*

*Proof.* Rewriting the expected value,

$$\psi_\varepsilon(\mathcal{G},u) = \sum_{t\in\mathcal{T}} \mathcal{A}[u]\left(\{h_X \mid E_t^*(h_X) \geq \varepsilon\}\right), \tag{10.10}$$

which follows from $\mathcal{A}[u](\{h_X \mid E_t^*(h_X) \geq \varepsilon\}) = \mathcal{A}[u](\{h_X \mid \tau_\varepsilon(h_X) > t\})$. Then

$$\mathcal{A}[u]\left(\{h_X \mid E_t^*(h_X) \geq \varepsilon\}\right) = \mathbb{E}_{\mathcal{A}[u]}\left[\mathbb{1}_{(\varepsilon,\infty)}\left(E_t^*\right)\right], \tag{10.11}$$

which concludes the proof for $\psi_\varepsilon$ with $h_t(z,u) = \mathbb{1}_{(\varepsilon,\infty)}(u(z_t^*) - u_*)$. The result follows for $\psi_\varepsilon^N$ by additionally setting $h_t(z,u) = 0$ for $t > N$.

**Proposition 10.4.** *The performance criterion $\zeta_T$ is progressively decomposable.*

*Proof.* The functional $\zeta_T$ can be rewritten as follows:

$$\begin{aligned} \zeta_T(\mathcal{G},u) &= \mathbb{E}_{\mathcal{A}[u]}\left[E_T^*\right] \\ &= \sum_{t\in\mathcal{T}} \mathbb{E}_{\mathcal{A}[u]}\left[\mathbb{1}_{\{T=t\}}\, E_t^*\right]. \end{aligned} \tag{10.12}$$

The result follows with $h_t(z,u) = \mathbb{1}_{\{T=t\}}(z)\, E_t^*(z)$. Notice that the stopping time $T$ may depend on $u$ without violating this result.

**Proposition 10.5.** *The performance criteria $\sigma_\varepsilon$ and $\sigma_\varepsilon^N$ are progressively decomposable.*

*Proof.* It is possible to rewrite $\sigma_\varepsilon$ as

$$\sigma_\varepsilon(\mathcal{G},u) = \sum_{t\in\mathcal{T}} \mathcal{A}[u](\{\tau_\varepsilon = t\}) = \sum_{t\in\mathcal{T}} \mathbb{E}_{\mathcal{A}[u]}\left[\mathbb{1}_{\{\tau_\varepsilon=t\}}\right]. \tag{10.13}$$

Since $\tau_\varepsilon$ is a stopping time, $\{\tau_\varepsilon = t\}$ is $\mathcal{Z}_t$-measurable. Let $B_\varepsilon^u \subseteq \mathbb{R}^t$ be given by $B_\varepsilon^u = \{x \in \mathbb{R}^t : |u(x_m) - u_*| \leq \varepsilon \text{ and } |u(x_s) - u_*| > \varepsilon \, \forall s < t\}$. Then $h_t(z,u) = \mathbb{1}_{B_\varepsilon^u}(z)$ makes $\sigma_\varepsilon$ progressively decomposable. Letting $h_t = 0$ for $t \geq N$ proves that $\sigma_\varepsilon^N$ is progressively decomposable as well.

In fact, it is simple to prove that every performance criterion is progressively decomposable by conditioning on the natural filtration of the optimization process, $\mathcal{Z}_t$.

**Theorem 10.1.** *Every performance criterion as defined in Definition 10.1 is progressively decomposable.*

*Proof.* Given a kernel $V_u$ for some performance criterion, consider the collection of random variables $\left\{\mathbb{E}_{\mathcal{A}[u]}[V_u \mid \mathcal{Z}_t] \;\middle|\; t \in \mathcal{T}\right\}$. Each member of this collection is $\mathcal{Z}_t$-measurable for the appropriate $t$, so by Proposition 9.9, there exists a collection of functions $\{h_t^u \mid t \in \mathcal{T}, u \in Y_m^X\}$ with $h_t^u : X^{[t]} \to \mathbb{R}$ and

$$\mathbb{E}_{\mathcal{A}[u]}[V_u \mid \mathcal{Z}_t] = h_t^u \circ \phi_{[t]} \qquad \mathcal{A}[u]\text{-almost everywhere}$$

for the canonical projection $\phi_{[t]} : X^{\mathcal{T}} \to X^{[t]}$. Now define

$$h_1(z,u) = h_1^u(z) \quad \text{and for } t > 1, \quad h_t(z,u) = h_t^u - h_{t-1}^u, \tag{10.14}$$

Note that $\mathbb{E}_{\mathcal{A}[u]}[h_1(\phi_{[1]}(Z), f)] = \phi(\mathcal{G},u)$, and for $t > 1$, $\mathbb{E}_{\mathcal{A}[u]}[h_t(\phi_{[t]}(Z),u)] = 0$. As a result, for any performance criterion $\psi$,

$$\psi(\mathcal{G},u) = \mathbb{E}_{\mathcal{A}[u]}[V_u] = \sum_{t\in\mathcal{T}} \mathbb{E}_{\mathcal{A}[u]}[h_t(\phi_{[t]}(Z),u)].$$

That is, $\psi$ is progressively decomposable.

In general, the progressive decomposition of a performance criterion is not unique. A decomposition that is unique wherever the performance criteria is finite can be generated by stipulating the additional condition

$$\forall s < t, \quad \mathbb{E}_{\mathcal{A}[u]}\left[h_t\left(\phi_{[t]}(Z),u\right) \mid \mathcal{Z}_s\right] = 0. \tag{10.15}$$

This condition states that the expectation of $h_t$ given the information available at time $s < t$ is an unbiased estimator of $h_t$. This condition is satisfied by the definitions in Equation 10.14. The progressive decomposition that satisfies Equation 10.15 is called the *principal progressive decomposition*, and it is unique where finite, as stated in Theorem 10.2.

It will be necessary to identify the conditions under which the performance criterion is finite. Suppose $(h_t)_{t\in\mathcal{T}}$ is a progressive decomposition of a performance criterion $\psi$. Let $F_\psi \subseteq X^{\mathcal{T}} \times Y^X$ be the set of pairs $(z,u)$ for which $\sum_{t\in\mathcal{T}} h_t(\phi_{[t]}(z),u) < \infty$. For any other progressive decomposition $(\tilde{h}_t)_{t\in\mathcal{T}}$ of $\psi$, $\sum_{t\in\mathcal{T}} \tilde{h}_t(\phi_{[t]}(z),u) < \infty$ as well, since there is a proper generator $\mathcal{G}_{z,u}$ that produces $z$ with probability one on $u$,[1] and in this case

$$\psi(\mathcal{G}_{z,u},u) = \sum_{t\in\mathcal{T}} h_t(\phi_{[t]}(z),u) = \sum_{t\in\mathcal{T}} \tilde{h}_t(\phi_{[t]}(z),u) < \infty. \tag{10.16}$$

The set $F_\psi$ determines where $\psi$ is finite. Define $F_\psi(u) = \{z : (z,u) \in F_\psi\}$, and note that $\psi(\mathcal{G},u) < \infty$ if and only if $|\mathcal{A}[u]|(F_\psi(u)) = |\mathcal{A}[u]|(X^{\mathcal{T}})$ based on the definition of $F_\psi$. A property of a performance criterion $\psi$ holds *up to finiteness* if it holds for all $(z,u)$ in $F_\psi$, or for all $(\mathcal{G},u)$ such that $\psi(\mathcal{G},u) < \infty$.

**Theorem 10.2.** *The principal progressive decomposition of a performance criterion is unique up to finiteness and corresponds to the decomposition in the proof of Theorem 10.1.*

*Proof.* Suppose a performance criterion $\psi$ has two progressive decompositions $(h_t)_{t\in\mathcal{T}}$ and $(\tilde{h}_t)_{t\in\mathcal{T}}$ that both satisfy Equation 10.15 such that $h_t \neq \tilde{h}_t$ for some $t$. Let $D \subseteq F_\psi$ be the set of pairs $(z,u)$ in $F_\psi$ on which $h_t \circ \phi_{[t]}$ and $\tilde{h}_t \circ \phi_{[t]}$ differ; i.e, $h_t(\phi_{[t]}(z),u) \neq \tilde{h}_t(\phi_{[t]}(z),u)$. For each $(z,u) \in D$, let $t_0(z,u)$ be the least time $t$ such that inequality holds. Pick $(z_0,u_0) \in D$ to minimize $t_0 = t_0(z_0,u_0)$, and let $\mathcal{G}$ be a proper static search generator that extends to a static search trajector $\mathcal{A}$ such that $\mathcal{A}[u_0]$ places probability one on $z_0$. Now then,

$$\begin{aligned}
\psi(\mathcal{G},u_0) - \psi(\mathcal{G},u_0) &= \sum_{t\in\mathcal{T}} \mathbb{E}_{\mathcal{A}[u]}[h_t(\phi_{[t]}(Z),u_0) - \tilde{h}_t(\phi_{[t]}(Z),u_0)] \\
&= \sum_{t\in\mathcal{T}} \mathbb{E}_{\mathcal{A}[u]}\left[\mathbb{E}_{\mathcal{A}[u]}\left[h_t(\phi_{[t]}(Z),u_0) - \tilde{h}_t(\phi_{[t]}(Z),u_0) \mid \mathcal{Z}_{t_0}\right]\right] \\
&= h_{t_0}(\phi_{[t_0]}(z_0),u_0) - \tilde{h}_{t_0}(\phi_{[t_0]}(z_0),u_0) \\
&\quad + \sum_{t<t_0} \mathbb{E}_{\mathcal{A}[u]}[h_t(\phi_{[t]}(Z),u) - \tilde{h}_m(\phi_{[t]}(z),u)] \\
&= h_{t_0}(\phi_{[t_0]}(z_0),u_0) - \tilde{h}_{t_0}(\phi_{[t_0]}(z_0),u_0) \neq 0.
\end{aligned} \tag{10.17}$$

In the second line, conditioning on $\mathcal{Z}_{t_0}$ removes all terms greater than $t_0$ by Equation 10.15. The sum over $t < t_0$ in the next to last line is equal to zero because $t_0$ is minimal over all of $D$, whence $h_t = \tilde{h}_t$ for $t < t_0$. The final line is nonzero, which is a contradiction. The set $D$ has no shortest trajectory and is therefore empty. So for all $t$, $h_t = \tilde{h}_t$ on $\tilde{F}_\phi$, and the principal progressive decomposition is unique up to finiteness.

[1] Again, we are assuming that $\mathcal{T} = \mathbb{N}$ and here $z \in X^{\mathcal{T}}$. So even when $\mathcal{A} = \mathbf{E}(\mathcal{G})$ is considered as a Baire trajector, the singleton $\{z\}$ is measurable, and hence $\mathcal{G}_{z,u}$ exists.

Using $(h_t)_{t\in\mathcal{T}}$ as given in the proof of Theorem 10.1, for any $s < t$,

$$\begin{aligned}\mathbb{E}_{\mathcal{A}[u]}\left[h_t(\phi_{[t]}(Z),u)\,\middle|\,\mathcal{Z}_s\right] &= \mathbb{E}_{\mathcal{A}[u]}\left[\mathbb{E}_{\mathcal{A}[u]}\left[h(Z,u)\,|\,\mathcal{Z}_t\right]\middle|\,\mathcal{Z}_s\right] \\ &\quad - \mathbb{E}_{\mathcal{A}[u]}\left[\mathbb{E}_{\mathcal{A}[u]}\left[h(Z,u)\,|\,\mathcal{Z}_{t-1}\right]\middle|\,\mathcal{Z}_s\right] \\ &= \mathbb{E}_{\mathcal{A}[u]}\left[h(Z,u)\,|\,\mathcal{Z}_s\right] - \mathbb{E}_{\mathcal{A}[u]}\left[h(Z,u)\,|\,\mathcal{Z}_s\right] = 0,\end{aligned}$$

whence $(h_t)_{t\in\mathcal{T}}$ is the principal progressive decomposition of $\psi$.

Theorem 10.2 means that the only unbiased estimator of future performance given past performance is the principal progressive decomposition introduced in Theorem 10.1. This decomposition is distinct from the examples of progressive decompositions for $\phi_w$, $\psi_\varepsilon$, $\sigma_\varepsilon$, and $\zeta_T$ given earlier in this section, all of which are statistically biased.

### *10.2.3 Dependence on the Error Sequence*

The performance criteria specified above all have the property that they depend primarily on the error process. This quality is captured by the following definition.

**Definition 10.5 (Sole Dependence).** A performance criterion $\psi$ is *solely dependent* on the error process $E^u = (E^u_t)_{t\in\mathcal{T}}$ if there is a measurable function $H : Y^{\mathcal{T}} \to \mathbb{R}$ such that $\psi(\mathcal{G},u) = \mathbb{E}_{\mathcal{A}[u]}[H(E^u)]$ where $\mathcal{A} = \mathbf{E}(\mathcal{G})$.

The performance criteria $\phi_w$, $\psi_\varepsilon$, and $\sigma_\varepsilon$ above are solely dependent on the error process, which can be verified by inspecting their definitions. Performance criteria that are based on stopping times, such as $\zeta_T$ and $\phi_T$ are not solely dependent on the error process in general, because the value of the stopping time may change based on factors other than the error, such as the evaluation cost along a particular trajectory.

As mentioned above, sometimes one wants to ignore repeated evaluation points when analyzing optimizer performance. A performance criterion is *uniquely dependent* on the error process if it depends only on the evaluation of unique points. This property can be determined by using the unique stoping sequence $(T_m)_{m\in|X|}$ from Section 10.1.5, which yields the index of the $m^{th}$ unique point of the optimization process. The unique stopping sequence can be used to pick out the errors at unique points.

**Definition 10.6 (Unique Error Process).** Given the error process $E^u = (E^u_t)_{t\in\mathcal{T}}$, the unique error process is the subsequence of $E^u$ determined by the unique stopping sequence $(E^u_{T_m})_{m\in|X|}$.

Given any stochastic process (*i.e.*, sequence of random variables), the stopped sequence is necessarily a stochastic process as well, that is, each member of the sequence is again a random variable. Thus the unique error process is indeed a stochastic process whenever the objective $u$ is a measurable function by Proposition 10.1.

**Definition 10.7 (Unique Dependence).** A performance criterion $\psi$ is *uniquely dependent* on the error process if there is a measurable function $H : Y^{\mathcal{T}} \to \mathbb{R}$ such that

$$\psi(\mathcal{G},u) = \mathbb{E}_{\mathcal{A}[u]}\left[H\left(\left(E^u_{T_m}\right)_{m\in|X|}\right)\right].$$

Of the performance criteria above, only $\sigma_\varepsilon$, $\zeta_{T_m}$, and $\phi_{T_m}$ are uniquely dependent on the error sequence in general. The criterion $\phi_w$ obviously has one term for each point including the repeated points. The criteria $\psi_\varepsilon$ and $\psi_\varepsilon^N$ compute the hitting time without excluding repeated points. The finite success probability $\sigma_\varepsilon^N$ is not uniquely dependent even though $\sigma_\varepsilon$ is because it includes repeated points to determine when $N$ evaluations have been performed.

Each of the criteria that are not uniquely dependent on the error process can be replaced by a similar criterion that is uniquely dependent by making simple alterations. For example, the expected hitting time can be modified to $\tilde{\psi}_\varepsilon = \mathbb{E}_{\mathcal{G}f}[\tau_\varepsilon - R_{\tau_\varepsilon}(Z)]$ where $R_m(z)$ is the number of repeated points in $z$ up to the $m^{th}$ component. The unique average error $\phi_{T_m}$ is a uniquely dependent variant of $\phi_w$, and $\sigma_\varepsilon^{T_N}$ is uniquely dependent as well. However, these modified criteria can only be computed in finite time for generators that eventually generate unique points almost surely, and they are only practical if the size of expected gaps in time between unique points is uniformly bounded.

This section has introduced properties such as nonlinearity, progressive decomposability, and dependence on the error process. The next section addresses the question of continuity.

## 10.3 Continuity of Performance

Continuous performance criteria are of interest because a continuous performance criterion must score an optimization method similarly on similar objective functions. The primary tools to prove the continuity of performance criteria are Theorems 9.1 and 9.3.

A performance criterion can be continuous or discontinuous in either argument. In accordance with the terminology adopted thus far, a performance criterion is continuous in objectives if small changes to the objective result in small changes to the performance, where the space of objectives has the standard product topology, *i.e.*, the topology of pointwise convergence. The criterion is continuous in optimizers if small changes to the generator do not greatly affect the performance. Continuity in objectives is examined first.

### 10.3.1 Continuity in Objectives

The first continuity theorem is a slight extension of Theorem 9.1. This theorem inherits several conditions and adds a requirement that the kernel change continuously in a uniform sense across search histories, and it will be shown below that these conditions are in fact satisfied in many cases for the performance criteria introduced in this chapter.

The primary reason that additional conditions on the kernel are needed is that the product topology on $Y^X$ is insufficient to constrain change in the performance criterion based on change in the objective. Recall that the product topology has a topological base that restricts only finitely many coordinates. Consequently, if $u_n \to u$ in the product topology, it is guaranteed that $u_n(x) \to u(x)$ for every $x$, but the rate of convergence is not bounded. Traditionally, the product topology on functions has been called *weak* for this reason. To bound the rate of convergence, a stronger topology on $Y^X$ is needed. However, it is often not possible to formulate such topologies for all of $Y^X$, and thus in addition to the need for a new topology, there is a need to consider only a subset of functions.

A simple choice that will be used below is to build a subspace around continuous functions. Define $B_m(X,Y)$ to be the set of all bounded measurable functions from $X$ to $Y$ with respect to their given topologies. Clearly, for $u \in B_m(X,Y)$, the lower bound $u_*$ exists, and since there is an upper bound as well, $u(x) - u_*$ is finite for all $x$. Every continuous function is measurable, so $B_m(X,Y)$ contains all bounded continuous functions. Furthermore, when $Y$ is a vector space, $B_m(X,Y)$ is a vector space with pointwise addition and pointwise scalar multiplication, and so it has a supremum norm that is given by $\|u\| = \sup_{x\in X} \|u(x)\|$ that remains finite under these operations. The only deficiency of this space is that it is not closed under Cauchy limits, and so it is a normed vector space that is not Banach. This choice comes with the limitation of assuming that $Y$ is a vector space, but the definition of the error process already introduced this assumption in many cases, and the requirement that $Y$ is totally ordered as well essentially identifies $Y$ as the real line $\mathbb{R}$.

Generalizations of these concepts to non-metric $Y$ are possible but will not be pursued in this text. However, the conditions on the kernels of performance criteria will nonetheless be stated for general subsets $W \subseteq Y^X_m$, with the substitution $W = B_m(X,Y)$ only enforced as necessary. With this in mind, the particular form of uniform continuity that the kernel must possess in order to guarantee continuity of performance is now stated, followed by a continuity theorem for performance criteria with respect to the topology on $W$.

**Definition 10.8 (Uniform Continuity).** Let $(W,\tau)$ be a topological space with $W \subseteq Y^X_m$. The kernel $V : Y^X_m \to \mathbb{R}^{X^{\mathfrak{T}}}$ of a performance criterion is *uniformly continuous with respect to histories* over $W$ at a measurable objective $u$ if for every $\varepsilon > 0$ there is an $\tau$-open neighborhood $U_\varepsilon$ of $u$ such that for all $u' \in U_\varepsilon$,

$$|V_u(z) - V_{u'}(z)| < \varepsilon \qquad \text{for all } z \in X^{\mathfrak{T}}.$$

**Theorem 10.3.** *Suppose $\mathcal{G}$ is a static search generator of finite norm with extension $\mathcal{A} = \mathbf{E}(\mathcal{G})$ that is almost surely continuous in objectives at $u$. Given a performance criterion $\psi$, suppose the kernel $V$ of $\psi$ is uniformly continuous with respect to histories over a topological space $(W, \tau)$ at $u$, and that for all $u'$ in some $\tau$-neighborhood $U$ of $u$, $V_{u'}$ is $\mathcal{Z}_S$-measurable for a stopping time $S$ such that $S < \infty$ $\mathcal{A}[u'']$-almost everywhere for all $u'' \in U$. Then $\psi$ is continuous in objectives at $\mathcal{G}$ and $u$ when considered as a function from $\mathbf{E}_0 \times W$ to $\mathbb{R}$ if $\psi(\mathcal{G}, u)$ is finite.*

*Proof.* If $\psi(\mathcal{G}, u)$ is finite, then $V_u$ is $\mathcal{A}[u]$-integrable. By Theorem 9.1, the map

$$u' \to \mathbb{E}_{\mathcal{A}[u']}[V_u]$$

is continuous at $u$, but in order to complete the proof, the $V_u$ inside the expectation must be replaced with $V_{u'}$. This is possible by the uniform continuity of $V$ since for any $\varepsilon$, there is an open neighborhood of $u$ on which

$$\left|\mathbb{E}_{\mathcal{A}[u']}[V_u] - \mathbb{E}_{\mathcal{A}[u']}[V_{u'}]\right| \leq \mathbb{E}_{\mathcal{A}[u']}\left|V_u(Z) - V_{u'}(Z)\right| < \varepsilon\|\mathcal{A}\|,$$

which implies both that $V_{u'}$ is $\mathcal{A}[u']$-integrable and that $\psi$ is continuous in objectives at $\mathcal{G}$ and $u$. To see why the latter is true,

$$\begin{aligned}\left|\psi(\mathcal{G},u) - \psi(\mathcal{G},u')\right| &= \left|\mathbb{E}_{\mathcal{A}[u]}[V_u(Z)] - \mathbb{E}_{\mathcal{A}[u']}[V_{u'}(Z)]\right| \\ &\leq \left|\mathbb{E}_{\mathcal{A}[u]}[V_u(Z)] - \mathbb{E}_{\mathcal{A}[u']}[V_u(Z)]\right| + \left|\mathbb{E}_{\mathcal{A}[u']}[V_u(Z) - V_{u'}(Z)]\right| \\ &< \varepsilon + \varepsilon\|\mathcal{A}\|.\end{aligned}$$

The facts that $\varepsilon$ is arbitrary and $\|\mathcal{A}\| < \infty$ prove the conclusion.

**Proposition 10.6.** *Each element $E_t^u$ of the error process is uniformly continuous with respect to histories over $B_m(X,Y)$ at every objective $u \in B_m(X,Y)$ where $Y$ is a totally ordered normed vector space (e.g., $\mathbb{R}$).*

*Proof.* Fix $\varepsilon > 0$. The objective of the proof is to bound

$$\left\|E_t^u(z) - E_t^{u'}(z)\right\| \leq \left\|u(z_t) - u'(z_t)\right\| + \left\|u_* - u'_*\right\| < \varepsilon$$

for all search histories $z \in X^{\mathcal{T}}$ and for $u$ and $u'$ sufficiently close in $B_m(X,Y)$. But under the supremum norm for $B_m(X,Y)$, for $u$, $u'$ sufficiently close,

$$\left\|u(z_t) - u'(z_t)\right\| \leq \left\|u - u'\right\| < \frac{\varepsilon}{5},$$

and thus only the term $\|u_* - u'_*\|$ remains to be accounted for. Choose $z_*$ such that $\|u_* - u(z_*)\| < \eta$ and choose $z'_*$ such that $\|u'_* - u'(z'_*)\| < \eta$. Note that

$$\|u(z'_*) - u'(z'_*)\| \leq \|u - u'\| < \frac{\varepsilon}{5}.$$

Consider $\|u(z_*) - u(z'_*)\|$. There exists $y_\eta \in Y$ with $\|y_\eta\| < \eta$ such that

$$u(z_*) - y_\eta \;<\; u_* \leq u(z_*') \quad \text{and so} \quad u(z_*) - u(z_*') < y_\eta.$$

For $y_\varepsilon$ with $\|y_\varepsilon\| < \frac{\varepsilon}{5}$, $\|u - u'\| < \frac{\varepsilon}{5}$ gives $u(z_*') - y_\varepsilon < u'(z_*') < u(z_*') + y_\varepsilon$, and

$$u(z_*') - y_\varepsilon - y_\eta \;<\; u(z_*') - y_\eta \;<\; u_*' \;\leq\; u'(z_*) \;<\; u(z_*) + y_\varepsilon$$

which implies $-2y_\varepsilon - y_\eta < u(z_*) - u(z_*')$ and $\|u(z_*) - u(z_*')\| < \frac{2\varepsilon}{5} + \eta$. Observe

$$\begin{aligned} \|u_* - u_*'\| &\leq \|u_* - u(z_*)\| + \left\|u(z_*) - u(z_*')\right\| + \left\|u(z_*') - u'(z_*')\right\| + \left\|u_*' - u'(z_*')\right\| \\ &< \frac{3\varepsilon}{5} + 3\eta, \end{aligned}$$

wherefore with $\eta = \frac{\varepsilon}{15}$, $\|u_* - u_*'\| < \frac{4\varepsilon}{5}$ and thus $\|E_t^u(z) - E_t^{u'}(z)\| < \varepsilon$, completing the proof.

**Proposition 10.7.** *Each element $E_t^{*,u}$ of the minimum error process is uniformly continuous with respect to histories over $B_m(X,Y)$ at every objective $u \in B_m(X,Y)$ where $Y$ is a totally ordered normed vector space (e.g., $\mathbb{R}$).*

*Proof.* This is a trivial consequence of the preceding proposition. For $\varepsilon > 0$,

$$\left\|E_t^{*,u}(z) - E_t^{*,u'}(z)\right\| \;\leq\; \left\|u(z_t^*) - u'(z_t^*)\right\| + \left\|u_* - u_*'\right\| \;<\; \varepsilon.$$

As in the proof of Proposition 10.6, $u'$ can be chosen so that

$$\|u(z_t^*) - u'(z_t^*)\| \;\leq \|u - u'\| < \frac{\varepsilon}{5},$$

and $\|u_* - u_*'\| < \frac{4\varepsilon}{5}$, which establishes the required bound.

The following corollaries apply the two preceding theorems to the classes of performance criteria defined in Section 10.1, beginning with the average error $\phi_w$.

**Corollary 10.1.** *Suppose $\mathcal{G}$ is a static search generator of finite norm with extension $\mathcal{A} = \mathbf{E}(\mathcal{G})$ that is almost surely continuous in objectives at $u$, and that $w = (w_t)_{t\in\mathcal{T}}$ has $\sum_{t\in\mathcal{T}} |w_t| < \infty$. Then $\phi_w$ is continuous in objectives at $\mathcal{G}$ and $u$ for all $u \in B_m(X,Y)$ considered as a function on $\mathbf{E}_0 \times B_m(X,Y)$.*

*Proof.* Define $\phi_{w,t}(\mathcal{G},u) = \mathbb{E}_{\mathcal{A}[u]}\left[E_t^{*,u}\right]$ and notice that

$$\phi_w = \sum_{t\in\mathcal{T}} w_t \; \phi_{w,t}(\mathcal{G},u).$$

For $B_m(X,Y)$, boundedness implies that $E_t^{*,u} < \infty$ for all $t$, and so $\phi_{w,t}$ is finite everywhere. In addition, $E_t^{*,u}$ is $\mathcal{Z}_S$-measurable for the finite, degenerate stopping time $S = t$. Thus $\phi_{w,t}$ is a performance criterion and by Proposition 10.7, its kernel is uniformly continuous with respect to measures over $B_m(X,Y)$ at all objectives in that set. By Theorem 10.3, $\phi_{w,t}$ is continuous at $\mathcal{G}$ and $u$ in the desired sense.

Now fix $\varepsilon > 0$ and consider the difference

$$\left|\phi_w(\mathcal{G},u) - \phi_w(\mathcal{G},u')\right| \leq \sum_{t \leq t_\varepsilon} |w_t| \left|\phi_{w,t}(\mathcal{G},u) - \phi_{w,t}(\mathcal{G},u')\right| + \frac{\varepsilon}{2},$$

where $t_\varepsilon$ is chosen so that

$$\sum_{t > t_\varepsilon} |w_t| \left|\phi_{w,t}(\mathcal{G},u) - \phi_{w,t}(\mathcal{G},u')\right| < \frac{\varepsilon}{2},$$

which is possible for any pair $u, u'$ because $w$ converges absolutely and the difference is finite and bounded above by $|\phi_{w,1}(\mathcal{G},u) - \phi_{w,1}(\mathcal{G},u')|$. Now then, the continuity of $\phi_{w,t}$ together with the finite sum implies that $u'$ can be chosen so that each term is bounded above by $\varepsilon/2t_\varepsilon$, and thus the original difference is bounded by $\varepsilon$, proving continuity.

The criterion $\zeta_T$ is continuous under the same conditions, provided that the stopping time $T$ stops sufficiently fast and does not introduce discontinuities with respect to objectives. To represent the idea of stopping "sufficiently fast", Big-O notation is used. Recall that a function $h(t) = O(g(t))$ for $t \in \mathcal{T}$ if there exists a $t_0 \in T$ and a constant $c > 0$ such that for all $s > t_0$, $h(s) \leq c\,g(s)$. That is, $h(t) = O(g(t))$ if $h$ is eventually dominated by $g$. Again, the assumption $\mathcal{T} = \mathbb{N}$ is used here.

**Corollary 10.2.** *Suppose $\mathcal{G}$ is a static search generator of finite norm with extension $\mathcal{A} = \mathbf{E}(\mathcal{G})$ that is almost surely continuous in objectives at u, and that $T = T_u$ is a stopping time dependent on the objective that is uniformly continuous with respect to histories over $B_m(X,Y)$ on some neighborhood U. If $\mathcal{A}[u](\{T_u = t\}) = O\left(t^{-(2+\varepsilon)}\right)$ for some $\varepsilon > 0$ and all $u \in U$, then $\zeta_T(\mathcal{G},u)$ is finite and continuous in objectives at $\mathcal{G}$ and u when considered as a function on $\mathbf{E}_0 \times B_m(X,Y)$.*

*Proof.* Since $\zeta_T(\mathcal{G},u) = \mathbb{E}_{\mathcal{A}[u]}[T_u]$, expanding gives

$$\mathbb{E}_{\mathcal{A}[u]}[T_u] = \sum_{t \in \mathcal{T}} t\,\mathcal{A}[u](\{T_u = t\}) \leq \sum_{t \leq t_0} t\,\mathcal{A}[u](\{T_u = t\}) + c \sum_{t > t_0} \frac{1}{t^{1+\varepsilon}}$$

The term at the far right is a subseries of a convergent series, known in this form as either the Dirichlet series or the Riemann zeta function. Therefore, $\zeta_T(\mathcal{G},u)$ is finite. Since the kernel of $\zeta_T$ is $u \mapsto T_u$, which is uniformly continuous with respect to histories and $\mathcal{Z}_{T_u}$-measurable with $T_u < \infty$ almost surely, Theorem 10.3 implies the desired result.

Corollary 10.2 begs the question of when $T$ varies uniformly with the objective $u$. One simple answer is that any stopping time that is independent of the objective function will have this property. The issue of fast-converging stopping probabilities is more difficult, but if there is an upper bound on the stopping time, then eventually the stopping probability will be zero rather than just converging to zero, which would satisfy the requirement.

As another example, for a stopping time that limits the number of CPU cycles used, it seems reasonable to assume that in most cases the required number of cycles would change continuously with the objective function. Additionally, each iteration takes at least one CPU cycle, and thus the limit on CPU cycles is also an upper bound on the number of iterations that could ever be performed. There are, of course, limiting cases. For an example of discontinuity, consider the functions

$$u_n(x) = n^{-1} \exp(-x)\sin(nx)$$

on the interval $(0,1)$. Then $u_n \to 0$, and the zero function is trivial to compute whereas each $u_n$ requires approximately the same time to compute on most computers. One may expect discontinuities at constant functions. In practice, however, most of the variation in computational time is due to the choice of optimization method rather than to small changes in the objective function.

The performance criteria $\psi_\varepsilon$, $\psi_\varepsilon^N$, $\sigma_\varepsilon$, and $\sigma_\varepsilon^N$ require more stringent conditions in order to prove convergence. Each of these metrics stop when the error has $u(z_t) - u_* \leq \varepsilon$, but there exist sequences of objectives $u_n \to u$ such that $u_n(z_t) - u_n^* > \varepsilon$ while $u - u_* = \varepsilon$. In other words, these sequences of objectives hit the error target at the limit.

As a simple example of discontinuity, let $X = (0,2)$ and $Y = \mathbb{R}$ and set

$$u_n(x) = \begin{cases} (\varepsilon + n^{-1})x & x < 1 \\ \varepsilon + n^{-1} & x \geq 1 \end{cases} \quad \text{and} \quad u(x) = \begin{cases} \varepsilon x & x < 1 \\ \varepsilon & x \geq 1, \end{cases}$$

so that $u_n \to u$ in $B_m(X,Y)$. Let $\mathcal{G}$ be a proper static search generator that produces points uniformly over $(0,2)$. To evaluate $\psi_\varepsilon$, note that $\mathcal{G}[\cdot, u]((0,1)) = 1/2$. Each search point is chosen independently so that $\tau_\varepsilon = 1$ with probability $1/2$, $\tau_\varepsilon = 2$ with probability $1/2^2$, and then

$$\psi_\varepsilon(\mathcal{G}, u_n) = \sum_{t \in \mathcal{T}} \frac{t}{2^t} > 1, \quad \text{but} \quad \psi_\varepsilon(\mathcal{G}, u) = 1,$$

since the entire range of $u$ falls within $\varepsilon$ of the optimum. The discontinuity is caused by objectives with plateaus located at a distance of precisely $\varepsilon$ away from the optimum. This problem does not arise if the histories with error $\varepsilon$ have $\mathcal{A}[u]$ measure zero. In this case, the proof is easier if one relies directly on Theorem 9.1 rather than Theorem 10.3.

**Theorem 10.4.** *Suppose $\mathcal{G}$ is a static search generator of finite norm with extension $\mathcal{A} = \mathbf{E}(\mathcal{G})$ that is almost surely continuous in objectives at $u$. Suppose also that the set*

$$Z_\varepsilon = \left\{ z \in X^{\mathcal{T}} \;\middle|\; |u(x_t) - u_*| = \varepsilon \text{ for some } t \right\}$$

*has $\mathcal{A}[u]$-measure zero. Then $\psi_\varepsilon^N$ and $\sigma_\varepsilon^N$ are continuous in objectives at $\mathcal{G}$ and $u$ when considered as functions on $\mathbf{E}_0 \times B_m(X,Y)$.*

*Proof.* On the set $X^{\mathcal{T}} \setminus Z_\varepsilon$, it is not possible to have $u(z_t^*) - u_* = \varepsilon$. Thus for any sequence $u_n \to u$ in $B_m(X,Y)$, $u_n(z_m^*) - u_{n,*}$ must eventually be on the same side of $\varepsilon$ as $u(z_t^*) - u_*$. A progressive decomposition of $\psi_\varepsilon^N$ is $h_t(z,u') = \mathbb{1}_{(\varepsilon,\infty)}(u'(z_t^*) - u'_*)$ for $t \leq N$. On $X^{\mathcal{T}} \setminus Z_\varepsilon$, $h_t(z,u_n) = h_t(z,u)$ for all $n > M$ with $M$ independent of $z$, a fact which results from the topology of $B_m(X,Y)$. A progressive decomposition of $\sigma_\varepsilon^N$ is $h_t(z,u) = \mathbb{1}_{B_\varepsilon^{u'}}(z)$ with

$$B_\varepsilon^{u'} = \left\{ x \in X^{[t]} \;\middle|\; \left|u'(x_t) - u'_*\right| \leq \varepsilon \text{ and } \left|u'(x_s) - u'_*\right| > \varepsilon \; \forall s < t \right\}$$

for $t \leq N$. Once again, $h_t(z,u_n) = h_t(z,u)$ for all $n > M$ on $X^{\mathcal{T}} \setminus Z_\varepsilon$.

The claim of the theorem now reduces to the question of whether

$$\mathbb{E}_{\mathcal{A}[u_n]}[h_t(\phi_{[t]}(Z),u)] \to \mathbb{E}_{\mathcal{A}[u]}[h_t(\phi_{[t]}(Z),u)] \tag{10.18}$$

for all $t \leq N$. Let $W(z) = h_t(\phi_{[t]}(z),u)$, and then $W$ is a $\mathcal{Z}_t$-measurable random variable that is $\mathcal{A}[u']$-integrable for all objectives $u'$. The criteria for Theorem 9.1 are satisfied, and so the convergence of Equation 10.18 holds for the product topology on $Y^X$ and *a fortiori* for the topology of $B_m(X,Y)$ as well.

The last several results have demonstrated cases where performance criteria change continuously as the objective changes, in each case requiring a static search generator that is continuous on all histories that occur with non-zero probability. The next subsection addresses the question of continuity as the generator changes rather than the objective.

### 10.3.2 Continuity in Optimizers

Performance criteria are continuous in optimizers whenever their kernel is determined by an almost surely finite stopping time, without the complications that arose analyzing continuity in objectives. The following theorem is analogous to Theorem 10.3 but with much weaker assumptions. For instance, continuity is determined for all objectives in $Y_m^X$, and the only material requirement is that the kernel be $\mathcal{Z}_S$ measurable for an almost surely finite stopping time $S$. For a discussion of the other assumptions that are required, such as the use of tight and eventually bounded generators, see the text around Theorem 9.3. The next result is a direct consequence of that theorem.

**Theorem 10.5.** *Let $\mathcal{G}$ be a tight and eventually bounded static search generator of finite norm with extension $\mathcal{A} = \mathbf{E}(\mathcal{G})$. Let $u \in Y_m^X$ be an objective. If the kernel $V_u$ of a performance criterion $\psi$ is a $\mathcal{Z}_S$-measurable random variable for some stopping time with $S < \infty$ $\mathbf{E}(\mathcal{G}')[u]$-almost everywhere for all $\mathcal{G}'$ in some neighborhood of $\mathcal{G}$ and if $V_u$ is $\mathbf{E}(\mathcal{G}')[u]$-integrable for $\mathcal{G}'$ in the same neighborhood, then $\psi$ is continuous at $\mathcal{G}$ and $u$.*

*Proof.* Since $\psi(\mathcal{G}',u) = \mathbb{E}_{\mathbf{E}(\mathcal{G})[u]}[V_u]$, the result follows from Theorem 9.3.

Regarding the specific performance criteria introduced above, $\psi_\varepsilon^N$ and $\sigma_\varepsilon^N$ are transparently $\mathcal{Z}_N$-measurable and integrable.[2] Likewise $\zeta_T$ is $\mathcal{Z}_T$-measurable, but the integrability and almost sure finiteness of $T$ must be proven separately. Finally, $\phi_w$ can in many cases be shown to be close to a clipped version in which only finitely many weights are used. Nonetheless, stricter conditions on the error process are generally needed to guarantee continuity in optimizers for $\phi_w$. For example, requiring $w_t = 0$ for $t > N$ is sufficient to obtain that $\phi_w$ is continuous in optimizers for all tight, eventually bounded generators and all bounded, measurable objectives.

Theorem 10.5 proves that for performance criteria that can be measured in finite time, performance always changes smoothly as one moves smoothly from one optimizer to another. That is, similar optimizers perform similarly on the same objective.

### 10.3.3 Sample Convergence and Performance Continuity

The concept of sample convergence was introduced in Chapter 8 to represent the concept of an optimization method depending primarily on one or more deterministic trajectories through the search domain (Definition 8.10). In that chapter, sample convergence was used to determine when the generator are continuous after convolution; in this case, it was the selection and recombination of evolutionary algorithms that were sample convergent.

Sample convergence is relevant beyond evolutionary algorithms. For instance, Newton and quasi-Newton methods are sample convergent on continuously differentiable objectives, and Nelder-Mead is sample convergent on trajectories of unambivalent value (Definition 8.14). Just as sample convergence can be used to prove that compositions under convolution are continuous, it can also be used to demonstrate continuity for certain performance criteria as well.

One technical difficulty is that the concept of sample convergence was defined for standard search generators, and thus far performance criteria have been defined for static search generators. The question, then, is how to transfer sample convergence to static search generators through the conversion operator $\mathbf{G}(\cdot)$, just as continuity was transferred in Section 8.4.1. With respect to performance criteria, only sample convergence in objectives will matter, and so only this case is addressed below.

In analogy to the development of sample convergence for standard search generators, a static search generator $\mathcal{G}$ is finitely supported with size $s$ on a set $H \subseteq \mathcal{H}_X^{\mathcal{T}} \times Y^X$ if for all $(h_X,u) \in H$, there exists sets $S_{h_X,u}$ with $\sup_{h_X,u} \left|S_{h_X,u}\right| = s < \infty$ that form the support of $\mathcal{G}[h_X,u]$; *i.e.*, $|\mathcal{G}[h_X,u]|(X) = |\mathcal{G}[h_X,u]|(S_{h_X,u})$. The generator $\mathcal{G}$ is finitely trackable at a search history $h_X$ and an objective $u$ if there is an open neighborhood $H$ of $(h_X,u)$ on which $\mathcal{G}$ is finitely supported with size $s$ and

[2] Integrability follows from the fact that $\mathcal{A}[u]$ is a finite measure plus the fact that both integrands are bounded.

there exists a collection of $s$ trajectory tracking functions $\tau_i : \mathcal{H}_X^{\mathcal{T}} \times Y^X \to X$ each continuous in the second argument[3] with $S_{h'_X,u'} = \{\tau_i(h'_X,u') \mid 1 \le i \le s\}$ for all $(h'_X,u') \in H$. Given a set of trajectory tracking functions, the probability tracking functions $\mathcal{G}_{\tau_i} : \mathcal{H}_X^{\mathcal{T}} \times Y^X \to \mathbb{R}$ are chosen to satisfy an updated version of Equation 8.5,

$$\mathcal{G}[h'_X,u'](\{x\}) = \sum_{\{i \mid \tau_i(h'_X,u')=x\}} \mathcal{G}_{\tau_i}(h'_X,u'), \tag{10.19}$$

and then finally sample convergence in objectives can be defined.

**Definition 10.9 (Sample Convergence in Objectives).** A static search generator $\mathcal{G}$ is *sample convergent in objectives* at a search history $h_X$ and an objective $u$ if $\mathcal{G}$ is finitely trackable at $(h_X,u)$ with a set of probability tracking functions that are each continuous in objectives at $(h_X,u)$.

**Theorem 10.6.** *Suppose that a static black-box search generator $\mathcal{G} \in \mathcal{BB}_{X,Y}^{\mathcal{T}}$ is the image under $\mathbf{G}(\cdot)$ of a standard search generator that is sample convergent in objectives at a history trace $h \in \mathcal{H}_{X,Y}^{\mathcal{T}}$ that agrees with an objective $u$. If $h$ has finite length, then $\mathcal{G}$ is sample convergent in objectives at $(h_X,u)$.*

*Proof.* Let $\mathcal{G}'$ be a standard search generator such that $\mathcal{G} = \mathbf{G}(\mathcal{G}')$, and let $\mathcal{G}'$ be sample convergent in objectives at $h \in \mathcal{H}_{X,Y}^{\mathcal{T}}$. The key idea is to convert the tracking functions for $\mathcal{G}'$ into tracking functions for $\mathcal{G}$, and the key difficulty is in guaranteeing the required continuity properties. The technique for doing so is essentially identical to that used in the proof of Theorem 8.12, which relies on the continuity of the function $\zeta_{h_X}(u') = (u'(h_1^X),\ldots,u'(h_{|h_X|}^X))$ defined and discussed in the text preceding that theorem.

Since $\mathcal{G}'$ is sample convergent in objectives, it has trajectory tracking functions $(\tau'_i)_{1\le i\le s}$ and $\left(\mathcal{G}'_{\tau'_i}\right)_{1\le i\le s}$ that are continuous in objectives. That is, recalling the trace forming operator $\kappa(h_X,h_Y) = h$, the maps

$$h_Y \mapsto \tau_i(\kappa(h_X,h_Y)) \quad \text{and} \quad h_Y \mapsto \mathcal{G}'_{\tau'_i}(\kappa(h_X,h_Y))$$

are continuous. Now for $u$ agreeing with $h = \kappa(h_X,h_Y)$ with $|h| < \infty$, define

$$\tau_i(h_X,u) = \tau'_i(h) \quad \text{and} \quad \mathcal{G}_{\tau_i}(h_X,u) = \mathcal{G}'_{\tau'_i}(h),$$

which respectively provide trajectory tracking functions and probability tracking functions for $\mathcal{G}$. The fact that $\mathcal{G} = \mathbf{G}(\mathcal{G}')$ guarantees everything except continuity. It remains to show that these functions are all continuous in the second argument as required. By composing with $\zeta_{h_X}$, we see

$$\tau_i(h_X,u) = \tau'_i\left(\kappa\left(h_X,\zeta_{h_X}(u)\right)\right) \quad \text{and} \quad \mathcal{G}_{\tau_i}(h_X,u) = \mathcal{G}'_{\tau'_i}\left(\kappa\left(h_X,\zeta_{h_X}(u)\right)\right).$$

[3] Continuity is only required as the objective function changes because only sample convergence in objectives is needed here.

Since $\zeta_{h_X}$ is continuous when $h_X$ has finite length, these statements prove the desired continuity since the composition of continuous functions is continuous.

The next theorem shows that optimizers that are *almost surely* sample convergent induce convergence of the performance criterion under similar conditions to Theorem 10.3. The concept of *almost sure* sample convergence is defined first; this property describes generators that are sample convergent in objectives across all search histories that have non-zero measure.

**Definition 10.10 (Almost Sure Sample Convergences).** A static search generator $\mathcal{G}$ with extension $\mathcal{A} = \mathbf{E}(\mathcal{G})$ is *almost surely* sample convergent (or divergent) on an objective $u$ if there is a set $A \subseteq X^{\mathcal{T}}$ such that $|\mathcal{A}[u]|(A) = |\mathcal{A}[u]|(X)$ and for all search histories $h_X \in \mathcal{H}_X$, $\mathcal{G}$ is sample convergent (or divergent) in objectives at $h_X$ and $u$ whenever $h_X \sqsubseteq h'_X$ for some $h'_X \in A$, that is, for $h_X$ a prefix of $h'_X$.

**Theorem 10.7.** *Suppose $\psi$ is a performance criterion with a kernel that is $\mathcal{Z}_S$-measurable for some stopping time $S$ and $\mathcal{G}$ is a static search generator with finite norm that is almost surely sample convergent on all objectives $u$ within an open neighborhood $U \subseteq Y^X_m$. If $S < \infty$ $\mathcal{A}[u]$-almost everywhere for all $u \in U$, $\psi(\mathcal{G}, u) < \infty$ for all $u \in U$, and the principal progressive decomposition of $\psi$ yields functions $h_t(z,u)$ that are continuous in the second argument at $z, u$ for all $z \in X^{[t]}$ and $u \in U$, then $\psi$ is continuous in objectives at $\mathcal{G}$ and $u$ for all $u \in U$.*

*Proof.* Let $(\tau_i)_{1\leq i\leq s}$ and $(\mathcal{G}_{\tau_i})_{1\leq i\leq s}$ be the trajectory tracking functions and probability tracking functions for $\mathcal{G}$, respectively. These can be chosen to hold on some neighborhood of almost every point in $\mathcal{H}_X$. To reason about the continuity of $\psi$, fix $\varepsilon > 0$ and apply Proposition 9.10 to yield for all $u \in U$,

$$\begin{aligned}\psi(\mathcal{G},u) &= \sum_{t\in\mathcal{T}} \int_{\{S=t\}} h_t(z,u) \prod_{s\leq t} \mathcal{G}\left[(z_1,\ldots,z_{s-1}),u\right](dz_s) \\ &\approx \sum_{t\leq T_0} \int_{\{S=t\}} h_t(z,u)\, \mathcal{K}_u(dz) \quad \pm \frac{\varepsilon}{6},\end{aligned}$$

where $T_0 = T_0(u)$ is chosen to limit the sum to finitely many terms and $\mathcal{K}_u$ is introduced to abbreviate the product, which is possible since $\psi$ is finite on this region. Choosing $T_0 = \max\{T_0(u), T_0(u')\}$, observe $|\psi(\mathcal{G},u) - \psi(\mathcal{G},u')| \leq$

$$\sum_{t\leq T_0} \int_{\{S=t\}} |h_t(z,u) - h_t(z,u')|\, |\mathcal{K}_u|(dz) + \sum_{t\leq T_0} \int_{\{S=t\}} |h_t(z,u')|\, |\mathcal{K}_u - \mathcal{K}_{u'}|(dz) + \frac{\varepsilon}{3},$$

which is obtained by inserting a term $h_t(z,u')\,\mathcal{K}_u(dz)$ and pushing the absolute value through. Now $|h_t(z,u) - h_t(z,u')|$ is small for $u, u'$ close due to the continuity of $h_t(z,u)$ and $|\mathcal{K}_u|$ is bounded above by $\|\mathcal{G}\|^{T_0}$ for all $u$. Therefore, for $u'$ in a small enough neighborhood of $u$,

$$\sum_{t\leq T_0} \int_{\{S=t\}} |h_t(z,u) - h_t(z,u')|\, |\mathcal{K}_u|(dz) < \frac{\varepsilon}{3}.$$

For the other term, $h_t(z,u')$ is integrable by construction, so $|h_t(z,u')| < \infty$ almost surely.

The final step is to prove that $|\mathcal{K}_u - \mathcal{K}_{u'}|$ is small for $u,u'$ close. It is at this point that sample convergence is used. After $t$ steps (with $t$ finite), there are at most $s^t$ possible search histories that have nonzero probability, since at each step, the support set for $\mathcal{G}$ is finite by sample convergence. Letting $(z^j)_{j=1}^{s^t}$ enumerate these possibilities, with multiplicity to account for crossing paths,[4] and defining $\mathcal{K}_u^j$ and $\mathcal{K}_{u'}^j$ to denote the multiplicative accumulation of probability mass along these trajectories,

$$\int_{\{S=t\}} |h_t(z,u')|\, |\mathcal{K}_u - \mathcal{K}_{u'}|\,(dz) \quad = \quad \sum_j |h_t(z^j,u')| \left|\mathcal{K}_u^j - \mathcal{K}_{u'}^j\right|,$$

where the terms $\mathcal{K}_u^j$ are

$$\mathcal{K}_u^j = \prod_{s\leq t} \mathcal{G}_{\tau_i}((z_1^j,\ldots,z_{s-1}^j),u) = \prod_{s\leq t} \beta_{s,u}^j.$$

Here the sum of Equation 10.19 has been removed for simplicity since $z^j$ accounts for multiplicity and the $\beta_{t,u}^j$ are introduced for notational convenience.

The continuity of the probability tracking functions implies $\left|\beta_{s,u}^j - \beta_{s,u'}^j\right|$ can be made small, so the goal is to introduce these terms into the difference of products. Define

$$\gamma_\rho^j = \prod_{s<\rho} \beta_{s,u'}^j \prod_{\rho\leq s\leq t} \beta_{s,u}^j$$

for $0 \leq \rho \leq t+1$. Adjacent terms by $\rho$ differ in exactly one multiplicand, so

$$\left|\gamma_\rho^j - \gamma_{\rho-1}^j\right| \leq \left|\beta_{\rho,u}^j - \beta_{\rho,u'}^j\right| \, \|\mathcal{G}\|^{t-1}.$$

Furthermore, by injecting $t$ terms,

$$\left|\mathcal{K}_u^j - \mathcal{K}_{u'}^j\right| \; = \; \left|\gamma_0^j - \gamma_{t+1}^j\right| \; \leq \; \sum_{1\leq\rho\leq t+1} \left|\gamma_\rho^j - \gamma_{\rho-1}^j\right|,$$

and so finally,

$$\sum_j |h_t(z^j,u')| \left|\mathcal{K}_u^j - \mathcal{K}_{u'}^j\right| \leq \sum_j |h_t(z^j,u')| \sum_{1\leq\rho\leq t+1} \left|\beta_{\rho,u}^j - \beta_{\rho,u'}^j\right| \, \|\mathcal{G}\|^{t-1} \; < \; \frac{\varepsilon}{3},$$

where the final bound is obtainable because the sums each have finitely many terms. Therefore $\psi(\mathcal{G},u)$ is continuous in objectives at $u$.

Theorem 10.7 is stronger than Theorem 10.3 because it proves that sample convergent performance criteria can be continuous in objectives without requiring

[4] *I.e.*, if $\tau_m(h_X,u) = \tau_n(h_X,u)$ for $m \neq n$, then two identical paths are enumerated among the $z^j$, with each one assigned a probability as though the paths had not crossed, so that the sum of the two paths has the correct probability of Equation 10.19.

the objective to converge uniformly. In particular, each of the corollaries in Section 10.3.1 has an analogue for sample convergent optimizers that is the same in all respects, except that the (subspace) product topology for $Y_m^X$ can be used without needing to restrict the objectives to $B_m(X,Y)$.

Since standard measures of performance can be computed in finite time and the vast majority of optimizers are either sample convergent or almost surely continuous on most objectives, Theorems 10.7 and 10.3 together imply that standard measures of performance are generally continuous on all optimizers and most objectives.

## 10.4 Conclusion

This chapter introduced a flexible framework for analyzing performance criteria for optimization methods. Specific categories of performance criteria were presented, most of which correspond to the experimental quantities that are commonly reported in the literature. All performance criteria were shown to be progressively decomposable into sums of finite expectations, and this fact was leveraged to prove that most performance criteria are continuous subject to certain conditions.

Up to this point, the properties of performance criteria have been discussed in the abstract, but the value of these performance criteria can also be measured experimentally, which is done next in Chapter 11. The experiments in that chapter will demonstrate concretely the features of the performance criteria introduced in this chapter, including continuity and convergence. More specifically, they demonstrate that in practical terms, certain optimization methods appear to perform better than others on problems of interest using the performance criteria defined here.

After that experimental interlude, Chapter 12 will study the important theoretical question of whether some optimization methods can be better than others in general. This is the classic No Free Lunch Theorem for optimization, which states that it is possible to choose optimization problems to confound any method and render all methods equal in performance. This theorem will be generalized optimization problems sampled from measure spaces, and the conditions in which No Free Lunch results hold will be elucidated in terms of how problems are chosen. These proofs will rely heavily on the unique stopping sequence and the concept of unique dependence on the error sequence. In intuitive terms, it will be seen that No Free Lunch only holds perfectly in settings where learning is impossible.

# Chapter 11
# Performance Experiments

Several performance criteria were defined and analyzed in Chapter 10. This chapter reports the results of experiments run to estimate the values of these performance criteria on a bank of standard optimizers and objectives. The complete results are provided for reference in tabular form in Appendix A. They are summarized and discussed below. In addition, the theoretical continuity of performance criteria is illustrated through several examples. The final section of the chapter applies principal components analysis to the experimental performance values in order to visualize the position of the standard optimizer set in the space of long-running optimizers.

## 11.1 Experimental Setup

Experiments were performed in real-vector space on a set of twelve standard benchmarks. The search space was $X = \mathbb{R}^d$ with the topology induced by the standard Euclidean metric, $d(x,y) = \sum_{i=1}^{d} |x_i - y_i|^2$. The benchmarks were optimized within a benchmark-specific hypercube $Q \subseteq \mathbb{R}^d$, and feasibility regions were used to prevent the optimizers from escaping the constraints. That is, for each benchmark $f$, the experiment was performed with an altered objective $\tilde{f}$ given by

$$\tilde{f}(x) = \begin{cases} f(x) & x \in Q_f \\ \infty & \text{otherwise} \end{cases}, \tag{11.1}$$

where $Q_f$ is the hypercube constraining $f$.

### *11.1.1 Benchmarks*

The twelve benchmarks are defined in Table 11.1. These benchmarks are commonly used to test global optimizers. They cover a broad cross-section of possible objec-



A. J. Lockett, *General-Purpose Optimization Through Information Maximization*,
Natural Computing Series, https://doi.org/10.1007/978-3-662-62007-6_11

tive functions, including objectives that are convex, multimodal, periodic, differentiable, nowhere differentiable, deceptive, and irregularly shaped. All of the optimizers tested were continuous and bounded. The definitions and descriptions of these benchmarks can be found in the literature [84, 6, 113, 166, 22, 1]. Notably, two versions of Ackley's function exist; both are included in the comparisons. The less common one is termed *log-ackley* and is due to [1]. The more common version of the benchmark is exponentiated and centered and is simply termed *ackley*.

Each benchmark was tested in five, ten, and 25 dimensions ($d = 5, 10, 25$), except that *shekel* and *langerman* were tested in five and ten dimensions only (since they are not defined in 25 dimensions). The feasible region for each benchmark was a bounded hypercube with the range for each component shown in the table. The minima for these functions are known, as shown in Table 11.1 for five dimensions with precision up to $10^{-4}$. The actual values are known up to machine-level precision ($10^{-16}$), and these more accurate values were used for testing the accuracy of the experiments. Heat maps of the eleven benchmarks with $d = 2$ are shown in Figure 11.1.

### *11.1.2 Algorithms*

The performance of nine algorithms was measured. These algorithms were (1) conjugate gradient descent (CG), (2) Nelder-Mead (NM), (3) a generating set search (GSS), (4) simulated annealing (SA), (5) a real-coded genetic algorithm (rGA), (6) an evolution strategy (CMA-ES), (7) differential evolution (DE), (8) particle swarm optimization (PSO), and (9) the real-coded Bayesian Optimization Algorithm (rBOA). These algorithms, described in Chapter 2, cover a broad spectrum of stochastic optimization algorithms. They are known to be effective on a wide array of fitness functions and most of them perform reasonably well on the selected benchmarks. For all of the algorithms, parameters were set according to the literature where available and hand-tuned otherwise to optimize performance.

**Table 11.1** Benchmarks for Experimental Validation with dimension $d = 5, 10, 25$. Minimum for $d = 5$.

| Name | Definition | Minimum | Domain |
|---|---|---|---|
| sphere | $\sum_{i=1}^{d} x_i^2$ | 0.0000 | (-5.12 , 5.12) |
| ackley | $-20\exp(-\frac{.02}{d}\lVert x\rVert^2) - \exp(\frac{1}{d}\sum_{i=1}^{d}\cos(2\pi x_i)) + 20 + e$ | 0.0000 | (-30 , 30) |
| log-ackley | $\sum_{i=1}^{d-1} e^{-0.2}\sqrt{x_i^2 + x_{i+1}^2} + 3\cos(2x_i) + 3\sin(2x_{i+1})$ | -13.3796 | (-30 , 30) |
| whitley | $\sum_{i=1}^{d}\sum_{j=1}^{d}\frac{w(x_i,x_j)^2}{4000} - \cos(w(x_i,x_j)) + 1$, with $w(y,z) = 100\left(y^2 - z\right)^2 + (1-z)^2$ | 0.0000 | (-30 , 30) |
| shekel | $\sum_{i=1}^{30}\frac{1}{\sum_{j=1}^{d}\left(x_j - a_{ij}\right)^2 - c_i}$ | -10.4056 | (-5 , 15) |
| rosenbrock | $\sum_{i=1}^{d-1} 100(x_i^2 - x_{i+1})^2 + (1 - x_i)^2$ | 0.0000 | (-5.12 , 5.12) |
| rastrigin | $10d + \sum_{i=1}^{d} x_i^2 - 10\cos(2\pi x_i)$ | 0.0000 | (-5.12 , 5.12) |
| salomon | $-\cos(2\pi\lvert x\rvert) + 0.1\lvert x\rvert + 1,\ \lvert x\rvert \equiv \left(\sum_i x_i^2\right)^{1/2}$ | 0.0000 | (-30 , 30) |
| langerman | $-\sum_{i=1}^{5} c_i \exp(-y_i/\pi)\cos(\pi y_i),\ y_i = \sum_{j=1}^{d}(x_j - a_{ij})^2$ | -0.9650 | (-5 , 15) |
| schwefel | $d^{-1}\sum_{i=0}^{d} -x_i \sin\sqrt{\lvert x_i\rvert}$ | -418.9829 | (-512 , 512) |
| griewank | $1 + \sum_{i=1}^{d}\frac{x_i^2}{4000} - \prod_i \cos(x_i/\sqrt{i})$ | 0.0000 | (-600 , 600) |
| weierstrass | $\sum_{i=1}^{d}\sum_{j=1}^{20} 0.5^j \cos\left(2\cdot 3^j\pi(x_i + 0.5)\right) + d\sum_{j=1}^{20} 0.5^j \cos\left(3^j\pi\right)$ | 0.0000 | (-0.5 , 0.5) |

Conjugate gradient descent with estimated gradients was tested using the publicly available fmin_cg implementation from the SciPy package with its defaults. The Nelder-Mead algorithm was described in Section 2.4.1 and was implemented in the standard form. GSS was based on the direct search algorithm described in Section 2.4.2 using the positive spanning set of size $d+1$ and no search heuristic.

Simulated annealing was run as a single chain with a logarithmic cooling schedule (Section 2.5.1). It was restarted randomly with probability 0.001 after each point.

The rGA method was a standard real-coded genetic algorithm using linear ranking selection with pressure 1.8, uniform crossover, and Gaussian mutation (Equation 7.24). The mutation variance for rGA was set to 0.05 for all problems except *schwefel* and *griewank*, where it was set to 10.

CMA-ES is the Correlated Matrix Adaption algorithm of Hansen and Ostermeier (Section 2.7.3) and was tested with four different population sizes: 100, 750, 1250, and 2500 [68]. At each generation, 50% of the population was used to build an updated normal distribution.

DE [155] was trained with four different parameter settings, one each with crossover rates 0.2 and 0.9 and learning rates 0.2 and 0.9 (Section 2.7.1).

PSO [45] was trained with both the global and local adaptation rates set to 2.0 (Section 7.3.1). The velocity decay was tested with two different values, $-0.5$ and 1.0, following results by Pedersen [121] on optimal parameter settings for PSO.

The rBOA method is an Estimation of Distribution Algorithm (EDA), a class of optimizers introduced in Section 2.7.2. It was implemented as described by Ahn et al. in [2].

Many optimizers converge quickly to a local optimum, and restarting an optimizer can be an effective strategy to bootstrap its performance. To demonstrate this idea, CG, NM, GSS, and CMA-ES were restarted on convergence to improve performance. Results of this nature have been reported previously in the literature for CMA-ES [11]. The restarted versions are referred to as CG-R, NM-R, GSS-R, and CMA-ES-R, respectively. Other methods could also benefit from restarting, but these four methods should benefit most, since they converge quickly.

All algorithms were run on all benchmarks 200 times for each tested parameter setting. These 200 runs are sufficient to guarantee statistical significance on the estimated success rates $\sigma_{\varepsilon}^{N}$ for each algorithm at the 95% level within $\pm 0.5\%$ [160]. The variance on other performance criteria was large, but not deleteriously so (see Figures 11.8 and 11.9 for visual examples and the tables in Appendix A for exact numbers). When a single number is shown as the result of an experiment, that number represents the best value achieved on any parameter setting for that algorithm, unless otherwise stated. Experiments with different parameters are shown separately in Appendix A.

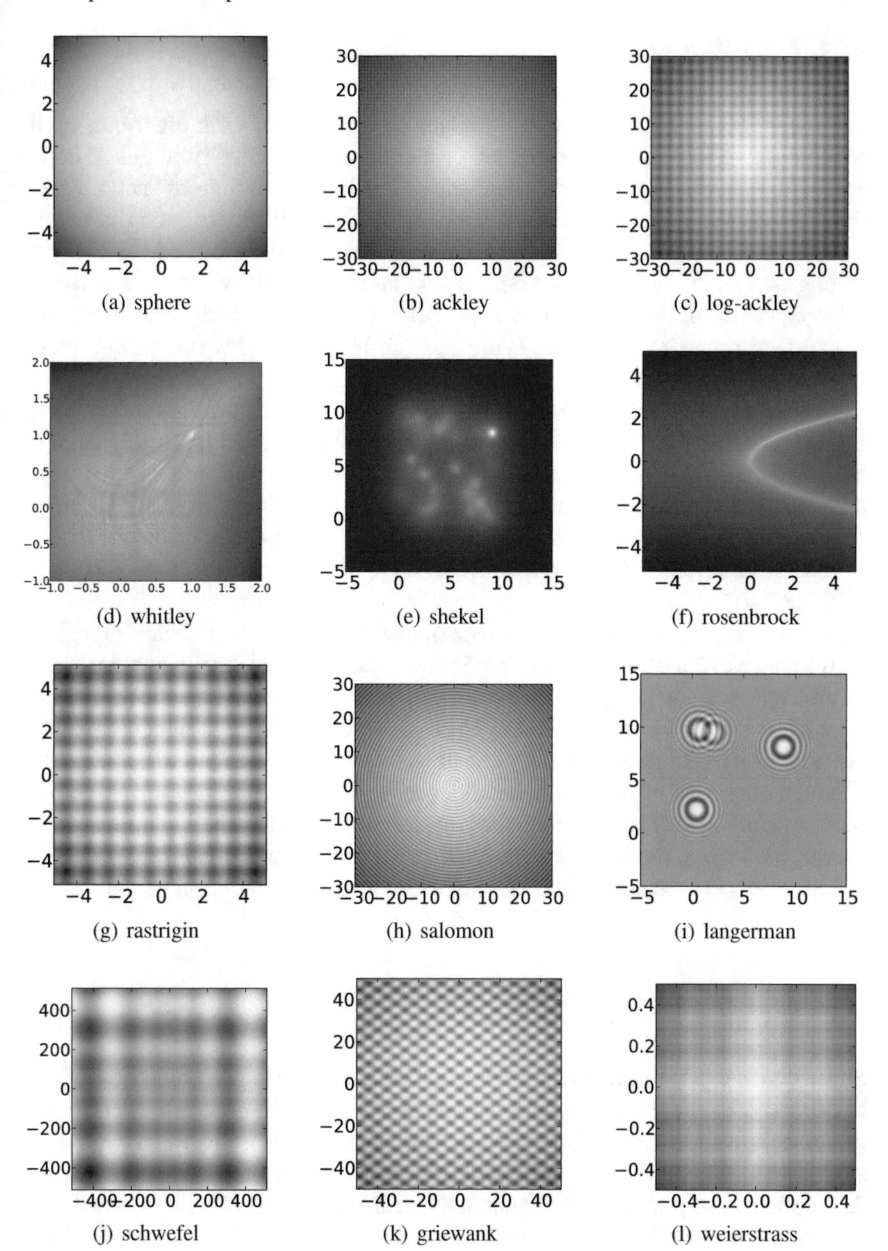


**Fig. 11.1** Heat maps for the twelve benchmark functions in two dimensions ($d = 2$). The benchmarks *whitley* and *griewank* are scaled to show the critical region. These benchmarks include unimodal, multimodal, periodic, irregular, and discontinuous functions, resulting in a broad test of an optimizer's capabilities.

### *11.1.3 Scaling Factors*

The experimental results contain estimates of the error of each algorithm on the benchmarks. The performance criteria $\zeta_T$ and $\phi_w$ are computed from this error. Because the magnitude of the error depends on the internal scaling of each objective function, comparisons across benchmarks are not numerically meaningful without scaling. For example, by multiplying the scaled error values in Table A.15 by the scaling factors in Table A.1, it would seem that rGA has its worst performance on *schwefel* out of all the benchmarks. However, comparing the performance of all optimizers on *schwefel*, rGA does better than all but three other optimizers, two of which benefitted from restarts. Thus scaling is necessary.

Scaling factors were computed by estimating the performance of random search on each benchmark. To this end, $10,000$ points were sampled uniformly from the bounding cube for each benchmark objective, and the minimum error from the optimum was recorded. This procedure was repeated 100 times and the results were averaged. The scaling factors computed in this way were only computed once for each dimension and were reused throughout the book. They are listed for each benchmark in Table A.1.

The scaling factors used in these experiments reveal the ratio of each algorithms performance to the performance of random search. There are other ways that scaling could have been accomplished. For example, the norm of each objective could have been estimated as the scaling factor, but it is not always easy to obtain a practical estimate of $||f||$. More importantly, the objective was not assumed to be integrable, although each one of these benchmarks can be integrated on the search domain. Since random search is a suitable comparison point for analyzing optimizer performance, the scaling factors that were used are meaningful and do make it possible to compare the performance of a single optimizer across several benchmarks.

## 11.2 Experimental Results

Figures 11.2 to 11.6 provide a visualization of the performance criteria introduced in Section 10.1 for the benchmarks in five dimensions; the complete experimental results are given in Appendix A. The performance criteria in the figures group the evaluations into virtual populations of 100 each so that the experiment contains $2,500$ successive populations. Population-based optimizers with larger populations and optimizers that do not use populations are thus compared in the same setting.

Figure 11.2 displays a scaled instance of $\phi_w$ with weights $w_{100n} = \frac{1}{2500}$ for $10 \leq n \leq 2500$ and zero otherwise. That is,

$$\phi_1(\mathcal{G}, u) = \frac{1}{s_u} \frac{1}{2490} \sum_{n=10}^{2500} \mathbb{E}_{\mathbf{E}(\mathcal{G})[u]} \left[ u(Z^*_{100n}) - u_* \right], \tag{11.2}$$

where $s_u$ is the objective-specific scaling factor from Table A.1. The initial factor of $\frac{1}{2490}$ was used to scale the magnitude of the sum, and the sum was started at $n = 10$ in order to ignore the initial error of the first 1,000 evaluations. This bar chart provides a sense of how different optimizers compare to each other on each objective. DE, CMA-ES, CMA-ES-R, GSS-R, and NM-R perform best on this criterion. As expected, CG and even CG-R perform poorly overall on this benchmark set.

Figure 11.3 displays another scaled instance of $\phi_w$, this time with exponential decay,

$$\phi_2(\mathcal{G}, u) = \frac{1}{s_u} \sum_{n=10}^{2500} \frac{1}{2^{n-10}} \mathbb{E}_{\mathbf{E}(\mathcal{G})u}\left[u(Z^*_{100n}) - u_*\right]. \tag{11.3}$$

Once again, the sum was started after 1,000 evaluations to avoid early errors. Unlike $\phi_1$, $\phi_2$ places higher emphasis on early errors. Thus by comparing Figure 11.3 with Figure 11.2 it is possible to obtain a sense of the convergence speed of each optimizer on the benchmarks. The values of $\phi_2$ are larger than $\phi_1$ for most optimizers, reflecting the earlier errors. Importantly, this effect is less pronounced in CMA-ES, GSS, and NM, which converge faster than the other algorithms. The term "converge" here is intended to mean "cease to propose substantially new evaluation points" rather than "converge to an optimum", although it is known that each of the algorithms mentioned do converge to a local optimum. The restarted versions of these three algorithms do perform worse when using $\phi_2$ rather than $\phi_1$ because restarting lowers the average error substantially in later evaluations, which are less important under $\phi_2$.

The scaled values for $\zeta_{T_{250,000}}$, the minimum global error at the $250,000^{th}$ unique evaluation, are shown in Figure 11.4. Whereas $\phi_1$ and $\phi_2$ give the average error under different weightings, $\zeta_{T_{250,000}}$ gives the error at the final evaluation. In $\mathbb{R}^m$, non-unique points have measure zero under the selected optimizers, so $T_{250,000} = 250,000$ on these experiments. The criterion $\zeta_T$ gives little information about the speed of convergence but is useful for comparing the absolute performance of different optimizers on a fixed objective. In Figure 11.4, DE is most reliable algorithm, with CMA-ES not far behind. This conclusion can also be drawn from the results for $\phi_1$, but is less clear. In $\phi_2$, the fast-converging optimizers appear preferable – especially CMA-ES and GSS. Restarting improves performance, with CMA-ES-R, GSS-R, NM-R, and CG-R all performing well on $\zeta_T$. On $\zeta_T$, at least, DE still appears preferable to the restarted optimizers.

Comparing the different criteria reveals a tradeoff between solution quality and convergence speed. DE achieves solution quality by exploring the space more thoroughly. CMA-ES provides slightly worse solution quality in much faster time. The restarted algorithms also converge slower but achieve higher quality. The desired tradeoff can be achieved to some extent by choosing the appropriate algorithm.

The average hitting time $\psi_\varepsilon^N$ represents the convergence time directly. This performance criterion is displayed with $\varepsilon = 0.01$ and $N = 250,000$ in Figure 11.5, scaled to represent the number of virtual populations of size 100 before the hitting time. The values are noisy and generally quite large due to the high variability of this criterion. Overall, it is difficult to draw conclusions from Figure 11.5, and the

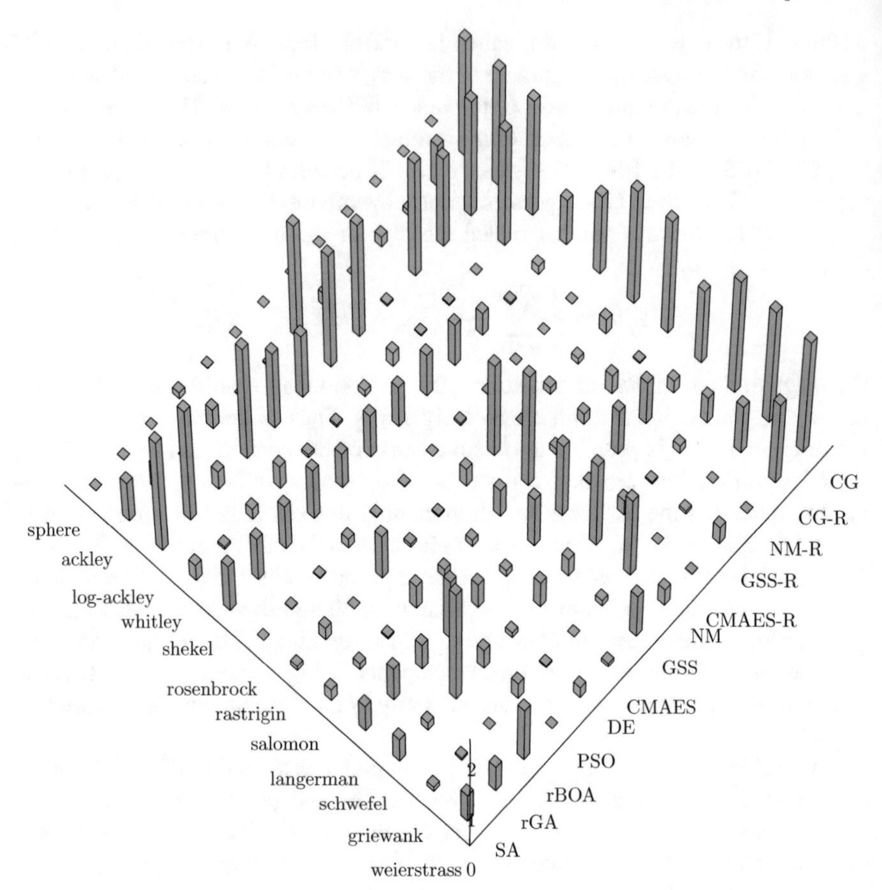


**Fig. 11.2** Performance values for selected optimizers on the twelve benchmarks using the performance criterion $\phi_1$, which averages global error over 250,000 evaluations, starting after 10,000 evaluations. Lower values are better. DE, CMA-ES, CMA-ES-R, GSS-R, and NM-R perform best on this performance criterion.

success-only hitting time $\hat{\psi}_{\varepsilon}^{N}$ is preferable. Since only a percentage of the trial runs hit the error threshold, each average includes a substantial number of copies of the maximum, $N$. Consistently low values are only achieved when the optimizer converges on almost every run, as PSO does on *weierstrass*. The criteria $\psi_{\varepsilon}^{N}$, $\hat{\psi}_{\varepsilon}^{N}$, and $\sigma_{\varepsilon}^{N}$ are somewhat sensitive to the scaling of the objective function, but their values have a much more consistent meaning when comparing the performance of an optimizer on different objectives, so scaling was not applied to the error threshold.

A clearer picture of the convergence speed is given by Figure 11.6, which shows the criterion $\hat{\psi}_{\varepsilon}^{N}$ from Equation 10.4 with the same parameters. This criterion lacks the high variability of $\psi_{\varepsilon}^{N}$. Optimizers that universally failed to attain the error target are shown with values of 2,500 in Figure 11.6. This figure shows that CG, CMA-ES, GSS, NM, and rBOA are the optimizers with the fastest convergence speeds,

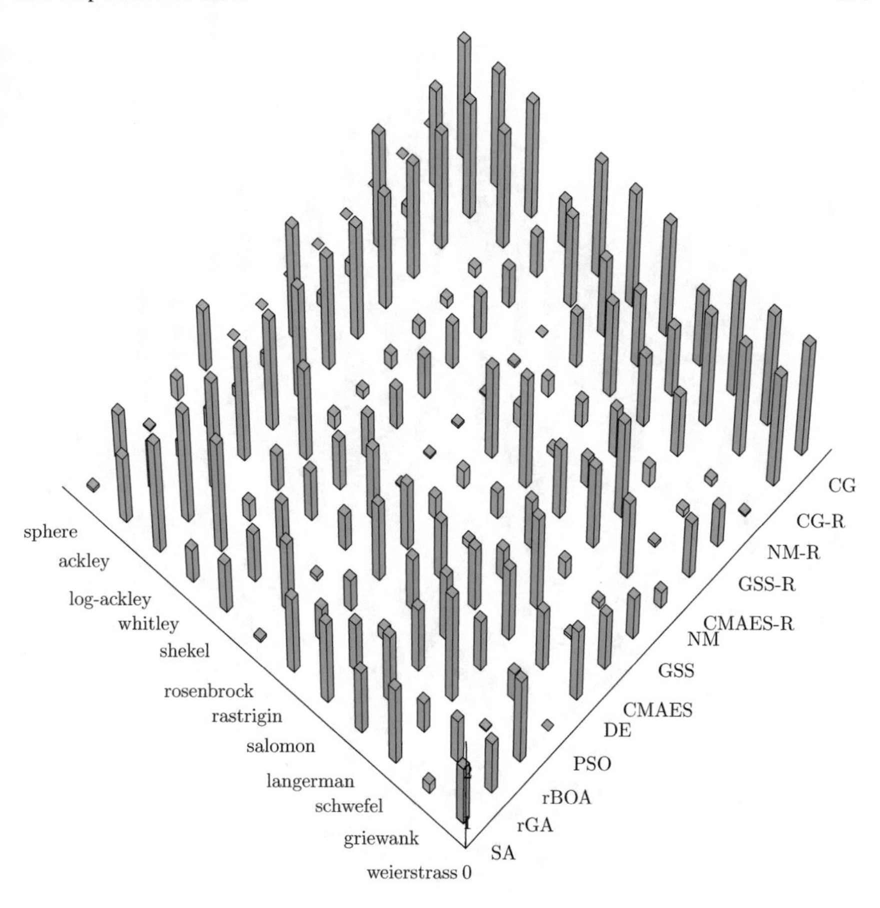


**Fig. 11.3** Performance values for selected optimizers on the twelve benchmarks using the performance criterion $\phi_2$, which sums global error with an exponential decay over 250,000 evaluations, starting after 10,000 evaluations. Lower values are better. Unlike $\phi_1$, $\phi_2$ counts earlier errors more heavily, and thus prefers optimizers that converge faster, such as NM, GSS, and CMA-ES.

although each of them fail on some subset of the benchmarks. Restarting, which improves performance on $\zeta_T$, predictably weakens performance on $\hat{\psi}^N_\varepsilon$.

The success probability $\sigma^N_\varepsilon$ complements the values of $\hat{\psi}^N_\varepsilon$. It is shown in Figure 11.7. Unlike in the other figures, higher values of $\sigma^N_\varepsilon$ indicate higher probability of success and thus larger bars are better. Figure 11.7 shows which optimizers are the most reliable overall. The restarted optimizers have the highest success probabilities, with some lapses. Among the optimizers that do not restart, DE and CMA-ES are the most reliable and consistent at reaching the error target, with CMA-ES appearing preferable in this figure. Comparing with the values for $\zeta_T$ in Figure 11.4, it can be surmised that when DE fails to reach the error target $\varepsilon$, it still attains a local minimum close in value to the true global minimum, whereas CMA-ES makes larger errors when the error target is not attained. Thus, the performance criteria

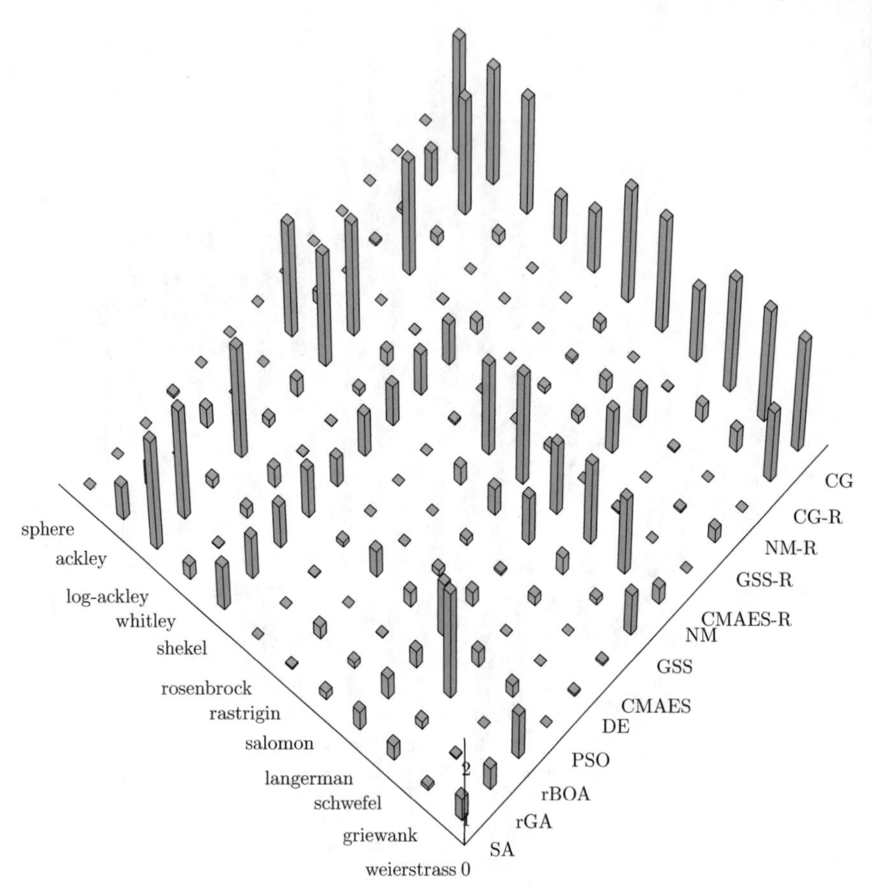


**Fig. 11.4** Performance values for selected optimizers on the twelve benchmarks using the performance criterion $\zeta_{T_{250,000}}$, which reports the minimum global error after $250,000$ unique evaluations. Lower values are better. Because it does not sum over multiple time steps, $\zeta_T$ communicates little information about the convergence rate. DE performs best among the optimizers, with CMA-ES close behind. Restarting improves performance with enough evaluations, and so CMA-ES-R, GSS-R, NM-R, and CG-R each perform well.

that use hitting times ignore catastrophic failures in favor of frequent successes, and CMA-ES appears more reliable than DE in this regard.

Overall, the choice of performance criterion should reflect the preferences of the practitioner, balancing tradeoffs of convergence speed, solution quality, and consistency. If both solution quality and speed are important, then the pair $(\hat{\psi}^N_\varepsilon, \sigma^N_\varepsilon)$ is a good choice. In this case, the three restarted algorithms performed best on the benchmarks. If solution quality and consistency are paramount but speed is less of a concern, then $\zeta_T$ is the best choice. To balance convergence speed as well, $\phi_1$ can be used. In either case, DE appears to be the most reliable non-restarted optimizer. Its performance could be further bootstrapped by restarting as well. The value of

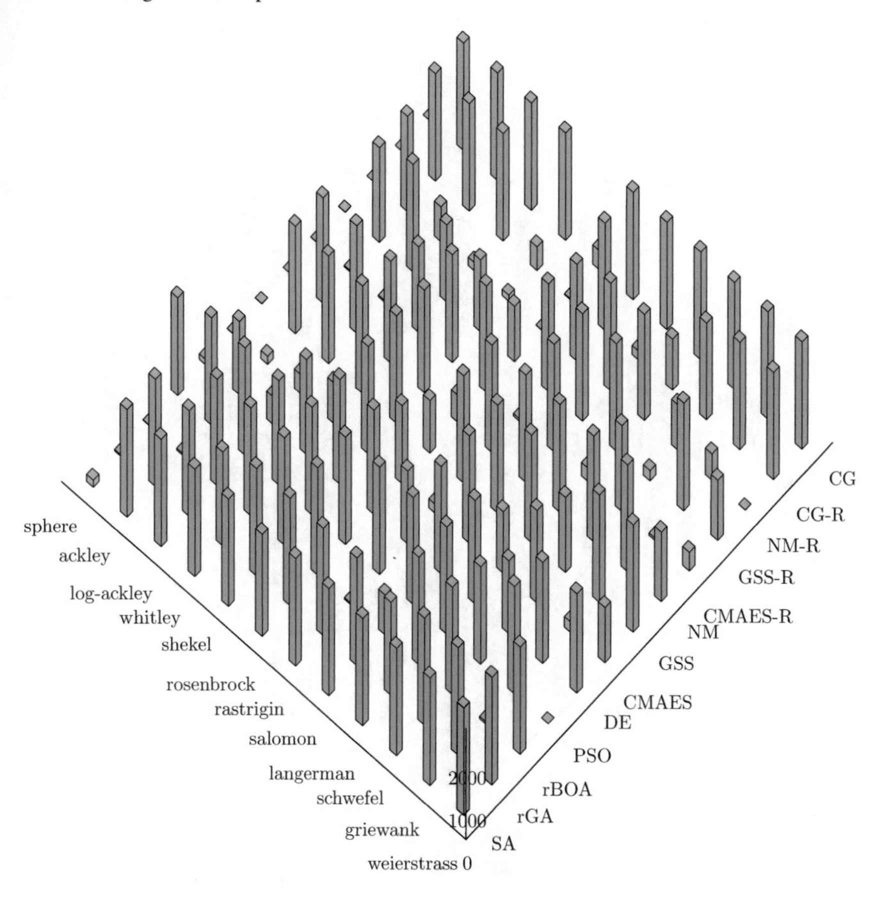


**Fig. 11.5** Performance values for selected optimizers on the twelve benchmarks using the performance criterion $\frac{1}{100}\psi_{\varepsilon}^{N}$ with $\varepsilon = 0.01$ and $N = 250,000$. This criterion records the average number of evaluations before the minimum global error drops below $\varepsilon$, capped at a maximum of $N$. Lower values are better. Scale as shown runs from zero to $2,500$ and represents the number of generations until the hitting time with a notional population size of 100. Because many trial runs fail on these benchmarks, the numbers are typically high, and often near $2,500$. Very low values, as seen for *sphere*, indicate fast convergence. In general, the success-only hitting time $\hat{\psi}_{\varepsilon}^{N}$ is more preferable than $\psi_{\varepsilon}^{N}$ for measuring performance.

restarting appears as a constant theme through these experiments. This topic is discussed more thoroughly in the next section.

## 11.3 Restarting to Bootstrap Performance

The restarted algorithms were undeniably the best performers out of all the optimizers on nearly every performance criterion and benchmark. Given that high number

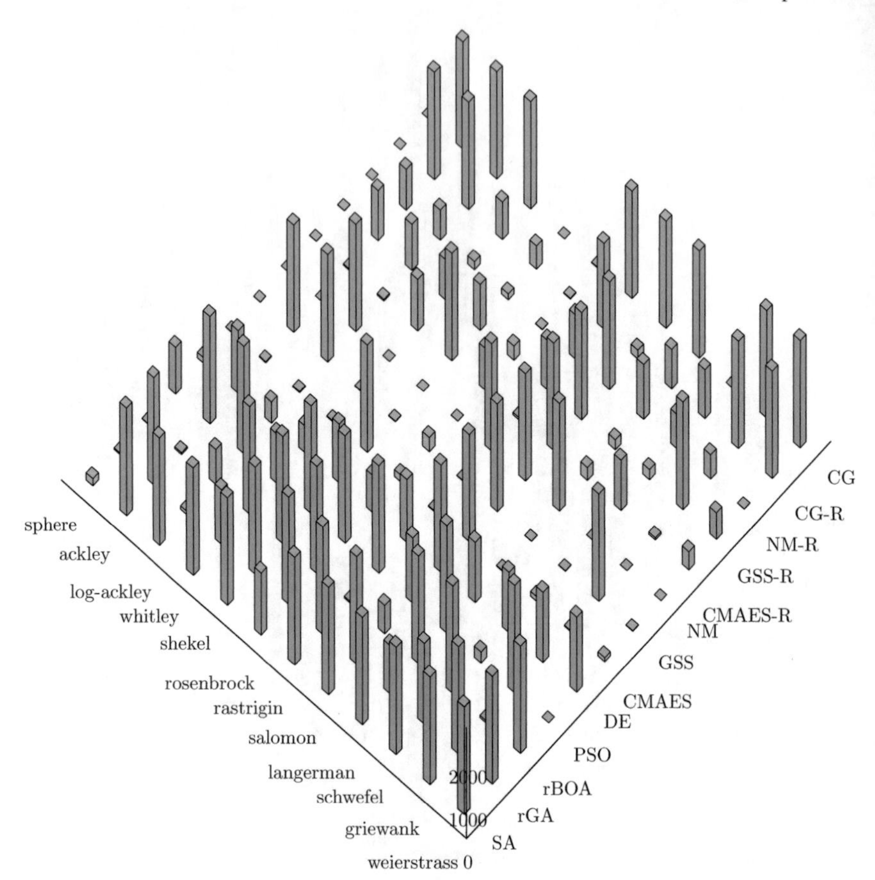


**Fig. 11.6** Performance values for selected optimizers on the twelve benchmarks using the performance criterion $\frac{1}{100}\hat{\psi}_{\varepsilon}^{N}$ with $\varepsilon = 0.01$ and $N = 250,000$ (see Equation 10.4). This criterion records the average number of evaluations before the minimum global error drops below $\varepsilon$ on trial runs where this error is attained. If the error threshold is never attained, the value is set at $N$. Lower values are better. Scale as shown runs from zero to $2,500$ and represents the number of generations until the hitting time with a notional population size of 100. These values give a clear picture of the relative convergence speed of various optimizers. For example, on successful trials, CMA-ES, GSS, and NM converge very quickly, whereas DE converges, but more slowly.

of evaluations performed $(250,000)$, it is perhaps not surprising that restarting would have a beneficial effect. Most of the optimizers tested tend to converge quickly to a small region of the search space, which they then sample in increasing detail. This convergence is faster in lower dimensions, as can be seen from the tables in Appendix A. In 25 dimensions, the restarted optimizers are less distinguishable from their non-restarted versions. The reason is that in higher dimensions, the optimizers converge more slowly, and thus are restarted less often. If the experiments were run for substantially more evaluations, then the beneficial effect of restarting might reappear even in higher dimensions.

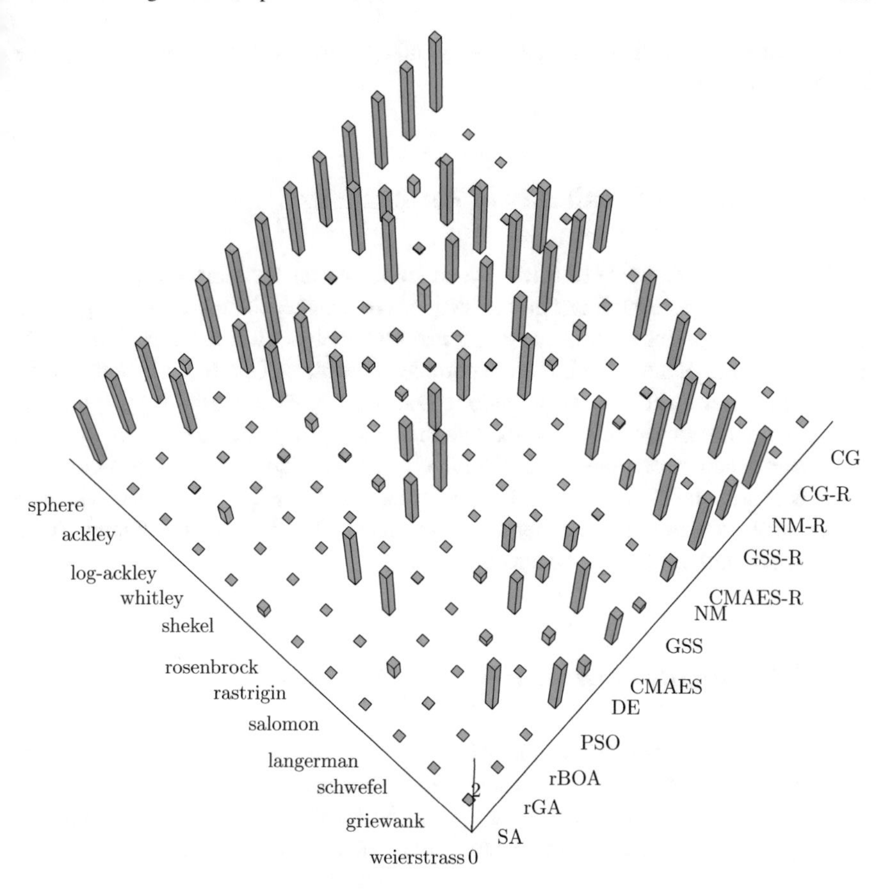


**Fig. 11.7** Performance values for selected optimizers on the twelve benchmarks using the performance criterion $\sigma_\varepsilon^N$ with $\varepsilon = 0.01$ and $N = 250,000$. This criterion computes the probability of attaining global error within $\varepsilon$ of the true optimum. Values run from zero to one; higher values are better. Overall, CMA-ES (especially CMA-ES-R) performs best on this performance criterion. DE, GSS-R, and NM-R also perform well on different objectives.

It is not necessary to wait for convergence to restart an optimizer, and there may be a benefit to restarting after a fixed number of evaluations. To observe the effect mathematically, suppose an optimizer has a success probability of $\sigma_\varepsilon^{25000} = 0.05$ on a particular objective after $25,000$ evaluations. If the optimizer is restarted every $25,000$ evaluations, then after $250,000$, the optimizer will have been run 10 times, and its success probability can be calculated. This value may be computed recursively as the sum of $p_n = p_{n-1} + (1 - p_{n-1}) \times p_1$ for $1 \leq n \leq 10$ with $p_1 = 0.05$. In this case, $\sigma_\varepsilon^{250000} = 0.40$. If the same success probability can be achieved with $10,000$ evaluations, then $\sigma_\varepsilon^{250000} = 0.72$ will be reached. In this way, fast but infrequent convergence can be parlayed into reliable but potentially slow convergence.

There is always a tradeoff between speed and quality, but it is possible to improve both with effective restarts.

## 11.4 Illustrating Continuity of Performance

In Chapter 10, substantial effort was expended to demonstrate the continuity of performance criteria as optimizers and objectives are changed. These results suggested that simplified approximations to an optimizer or an objective can be used to predict the performance of a similar but more complex optimizer. This result is both practical and important, and it is worthwhile to demonstrate such continuity graphically. This section includes results that illustrate three facts proven in the Chapter 10. First, the performance of continuous or sample convergent algorithms changes continuously with the objective. Second, the performance of similar optimizers is similar on the same problems. Third, when the conditions of the theorems in Chapter 10 are not met, discontinuities may be encountered.

### *11.4.1 Continuity in Objectives*

In Section 10.3.1, it was proven that $\phi(\mathcal{G}, u_n) \to \phi(\mathcal{G}, u)$ if $u_n \to u$ subject to certain conditions. This section looks at how the performance changes for a fixed optimizer as the objective changes. For this purpose, the benchmark objectives *shekel* and *langerman* were convexly combined to form a line in objective space given by

$$f_\alpha(x) = \alpha\, langerman(x) + (1-\alpha)\, shekel(x). \tag{11.4}$$

The optimizer NM-R was run for 200 trials on a range of objectives $f_\alpha$ with $\alpha = 0.0, 0.05, 0.10, \ldots, 0.95, 1.0$. The results are shown for different performance criteria in Figure 11.8. In these experiments, it is difficult to know the minimum $f_\alpha^*$ exactly, and so the best observed value on any trial was taken as the minimum. As long as the true success probability is positive for one of the two optimizers, it is reasonable to estimate the minimum in this way.

As Figure 11.8 shows, the performance changes smoothly as $\alpha$ runs from zero to one on four different performance criteria: $\phi_1$, $\zeta_T$, $\hat{\psi}_\varepsilon^N$, and $\sigma_\varepsilon^N$. Lines indicating the first standard deviation are shown, with the performance value in bold. The variance cannot be computed from these experiments for $\sigma_\varepsilon^N$, but should be less than 0.005 with high probability. Referring to Figure 11.7, it can be seen that NM-R succeeds frequently on *shekel* but rarely on *langerman*. The smooth and nonlinear transition in performance values as $\alpha$ runs from zero to one is expected, since NM-R is sample convergent $\mathcal{G}_f$-*a.s.* on objectives without plateaus. In this situation, Theorem 10.7 implies

$$\psi(\text{NM-R}, f_{\alpha_n}) \to \psi(\text{NM-R}, f_\alpha)$$

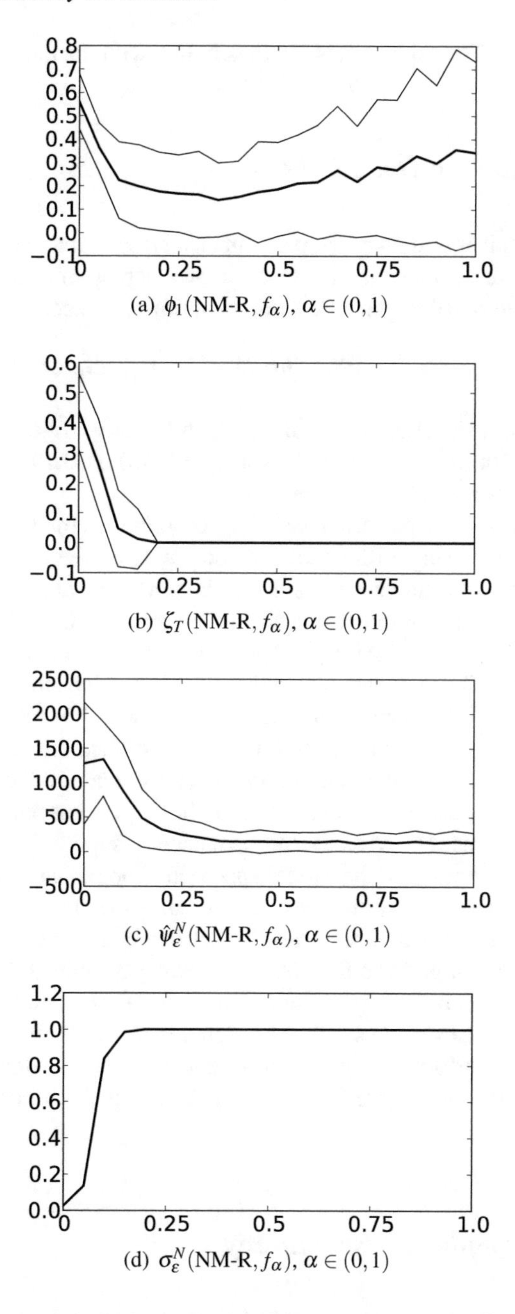


(a) $\phi_1(\text{NM-R}, f_\alpha)$, $\alpha \in (0,1)$



(b) $\zeta_T(\text{NM-R}, f_\alpha)$, $\alpha \in (0,1)$



(c) $\hat{\psi}^N_\varepsilon(\text{NM-R}, f_\alpha)$, $\alpha \in (0,1)$



(d) $\sigma^N_\varepsilon(\text{NM-R}, f_\alpha)$, $\alpha \in (0,1)$

**Fig. 11.8** Change in performance by NM-R as the objective changes smoothly from *langerman* ($\alpha = 0$) to *shekel* ($\alpha = 1$). The x-axis ranges over values of $\alpha$, the y-axis over performance values. The first standard deviation is also plotted on either side of the performance where possible. Panels show the performance criteria $\phi_1$, $\zeta_T$, $\hat{\psi}^N_\varepsilon$, and $\sigma^N_\varepsilon$, respectively. As predicted by the theory, performance on these optimizers changes smoothly and nonlinearly as a function of the objective.

whenever $\alpha_n \to \alpha$. The experiments are thus in line with the theory.

### 11.4.2 Continuity in Optimizers

In Section 10.3.2, it was shown that performance criteria are continuous as the optimizer changes. To demonstrate this fact, the one-step optimizers for DE and PSO were convexly combined to generate a line in optimizer space, given by

$$\mathcal{G}_\alpha[h_X,u] = \alpha\mathcal{PSO}<-.5,2,2,100>[h_X,u] + (1-\alpha)\mathcal{DE}_{\text{rand}}<.2,.2,100>[h_X,u], \tag{11.5}$$

recalling $\mathcal{PSO}$ from Equation 7.40 and $\mathcal{DE}$ from Equation 7.46. The optimizer $\mathcal{G}_\alpha$ was tested with 200 trials on *schwefel* for $\alpha = 0.0, 0.05, 0.10, \ldots, 0.95, 1.0$. PSO outperforms DE on *schwefel* in general.

Figure 11.9 shows the performance of $\mathcal{G}_\alpha$ on *schwefel* for various values of $\alpha$. Once again, the change in performance is smooth but non-linear, as predicted by the theory. The most interesting aspect is that although performance initially worsens for $\alpha$ in $(0, 0.15]$, it then improves consistently until $\alpha = 0.95$. In Figure 11.9(b), it can be seen that at $\alpha = 0.95$, $\mathcal{G}_\alpha$ outperforms both PSO and DE, although the result is statistically insignificant. Given that PSO is significantly better than DE on *schwefel* for most of the performance criteria, it is surprising that the best values of $\alpha$ are closer to DE rather than PSO. Convex combinations of optimizers were proposed as part of the formal analysis in Chapter 6, and the theory developed in Chapter 10 predicted convex combinations might outperform pure algorithms. The result of this experiment provides further evidence to support this claim. This discovery reinforces the value of the formal approach adopted in this book.

As mentioned, the fact that the best performance occurs for $\alpha$ other than zero or one confirms the conjecture in Section 10.2.1 that convex combinations of existing optimizers may outperform the optimizers being combined. Consequently, the problem of *convex control* of optimizers is worthy of further study. The problem of convex control may be stated as follows: given a bank of optimizers $\mathcal{G}_1, \ldots, \mathcal{G}_N$, an objective $u$, and a performance criterion $\psi$, find the convex combination $\hat{\alpha} \in \mathbb{R}^N$ that minimizes $\psi(\mathcal{G}_\alpha, u)$, where $\mathcal{G}_\alpha = \sum_i \alpha_i \mathcal{G}_i$. This topic is discussed again briefly in Chapter 18.

### 11.4.3 An Example of Discontinuity

The proofs of continuity in Section 10.3.1 contained several conditions that must be met to guarantee continuity. It was stated that discontinuities can be expected at functions with substantial plateaus. There are two reasons that support this claim. First, functions with plateaus induce trajectories of ambivalent value that cause several specific optimizers to be discontinuous on sets of positive $\mathcal{A}[u]$-measure. Sec-

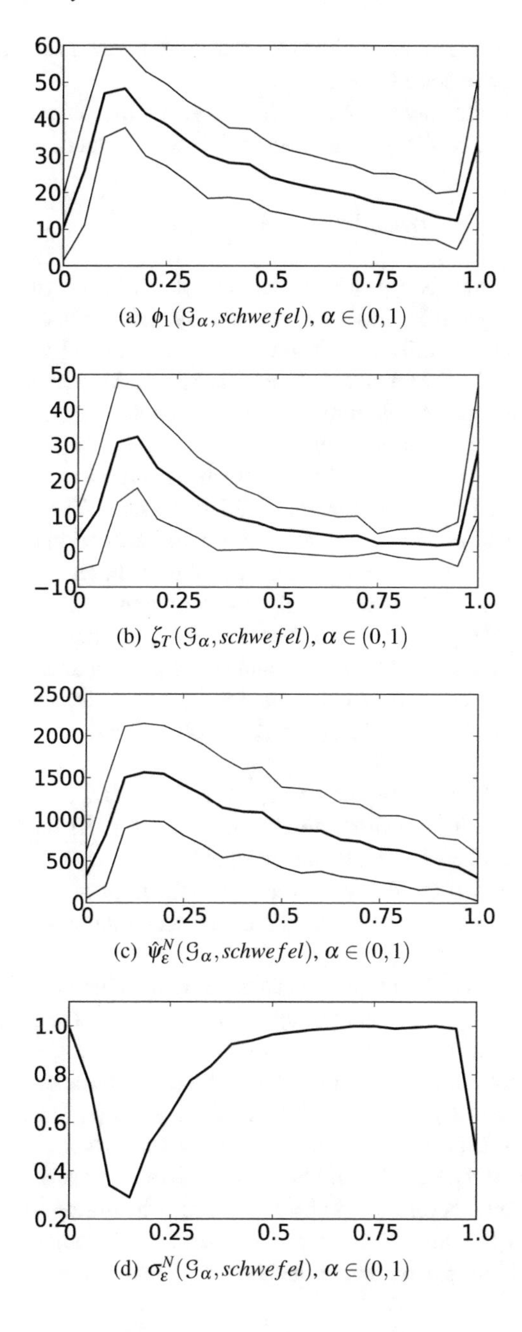


(a) $\phi_1(\mathcal{G}_\alpha, schwefel)$, $\alpha \in (0,1)$

(b) $\zeta_T(\mathcal{G}_\alpha, schwefel)$, $\alpha \in (0,1)$

(c) $\hat{\psi}_\varepsilon^N(\mathcal{G}_\alpha, schwefel)$, $\alpha \in (0,1)$

(d) $\sigma_\varepsilon^N(\mathcal{G}_\alpha, schwefel)$, $\alpha \in (0,1)$

**Fig. 11.9** Change in performance as the optimizer changes smoothly from PSO with $\omega = -.5, \phi_g = \phi_p = 2$ $(\alpha = 0)$ to DE with CR=.2, F=.2 $(\alpha = 1)$. The x-axis ranges over values of $\alpha$, the y-axis over performance values. The first standard deviation is also plotted on either side of the performance where possible. The panels show the performance criteria $\phi_1$, $\zeta_T$, $\hat{\psi}_\varepsilon^N$, and $\sigma_\varepsilon^N$, respectively, with $\varepsilon = 25$ for *schwefel*. As predicted by the theory, performance on these optimizers changes smoothly and nonlinearly as a function of the optimizer. Interestingly, at $\alpha = .95$, $\mathcal{G}_\alpha$ outperforms PSO and DE on $\zeta_T$, although the result is not statistically significant. Convex combinations of algorithms were formally proposed in this book, and Section 10.2.1 suggested that convex combinations may outperform pure algorithms. This example validates this conjecture and confirms the value of the formal approach in this book.

ond, functions with plateaus can introduce discontinuities in the hitting time even for optimizers that are continuous everywhere.

To demonstrate the discontinuities that occur as the objective passes through a continuous function, a new objective on $\mathbb{R}^d$ was created, *triangle*, defined on $(-30,30)^d$ by

$$triangle(x) = \min_i \left[1 - \frac{1}{30}|x_i|\right]. \tag{11.6}$$

This objective function is a $d$-dimensional simplex with height 1 and base width 60. A range of objectives was then defined by $t_\alpha(x) = \alpha triangle(x)$ for $\alpha = -0.1, -0.09, -0.08, \ldots, 0.09, 0.1$. When $\alpha < 0$, the minimal values of $t_\alpha(x)$ occur at 0. When $\alpha > 0$, the minimal values of $t_\alpha$ are around the boundary of the space. When $\alpha = 0$, every point has minimal value. Thus as $\alpha$ passes through zero, the minimal points shift discontinuously.

Figure 11.10 shows the values of the performance criteria $\phi_1$, $\zeta$, $\hat{\psi}_\varepsilon^N$, and $\sigma_\varepsilon^N$ as $\alpha$ runs from $-0.1$ to $0.1$ for two optimizers in 25 dimensions. The two optimizers were (1) DE and (2) a real-coded genetic algorithm with proportional selection, uniform crossover, and gaussian mutation, named rGA-2 to distinguish it from rGA, which used ranking selection. DE is potentially discontinuous when $\alpha = 0$, but rGA-2 is continuous on all of $C[\mathbb{R}^{25}]$, which includes $t_\alpha$ for all values of $\alpha$. The objective $t_0$ fails the requirements of Corollary 10.4, and so the performance of $\hat{\psi}_\varepsilon^N$ and $\sigma_\varepsilon^N$ each have a potential discontinuity at $\alpha = 0$, which is realized for DE in $\hat{\psi}_\varepsilon^N$ and $\sigma_\varepsilon^N$ for DE and is visible in Figures 11.10(f) and 11.10(h). For these graphs, $\varepsilon$ was set at 0.0001.

The objective $t_0$ fails the requirements of Corollary 10.4, and so the performance of $\hat{\psi}_\varepsilon^N$ and $\sigma_\varepsilon^N$ each have a discontinuity at $\alpha = 0$ on DE (right panels) but not on rGA-2 (left panels). The plots are interpolated, but careful inspection of Figure 11.10(g) shows that $\sigma_\varepsilon^N$ descends below 1 to the right of zero (it is still equal to 1 at $\alpha = 0.01$), whereas Figure 11.10(h) jumps discontinuously from 1 to 0 between $\alpha = 0$ and $\alpha = 0.01$.

The hitting time for DE drops to zero as $\alpha$ approaches zero from the left, and the success probability is constant at one. From the right, the hitting time is fixed at $N = 250{,}000$, and the success probability is zero. The graphs for DE appear left continuous, but jump discontinuously to the right of zero. As noted above, the plots are interpolated. At $\alpha = 0.01$, DE immediately has $\hat{\psi}_\varepsilon^N > 100$ versus values of $\hat{\psi}_\varepsilon^N < 50$ on the left. In contrast, the value of $\hat{\psi}_\varepsilon^N$ for rGA-2 also descends to zero from the left, but on the right it still has $\hat{\psi}_\varepsilon^N$ relatively small at $\alpha = 0.01$. Notice that $\phi_1$ and $\zeta$ are continuous for both optimizers despite the discontinuity in $\hat{\psi}_\varepsilon^N$ and $\sigma_\varepsilon^N$ due to Corollaries 10.1 and 10.2. Once again, theoretical inquiry correctly predicted potentially useful information about the performance of the optimizers.

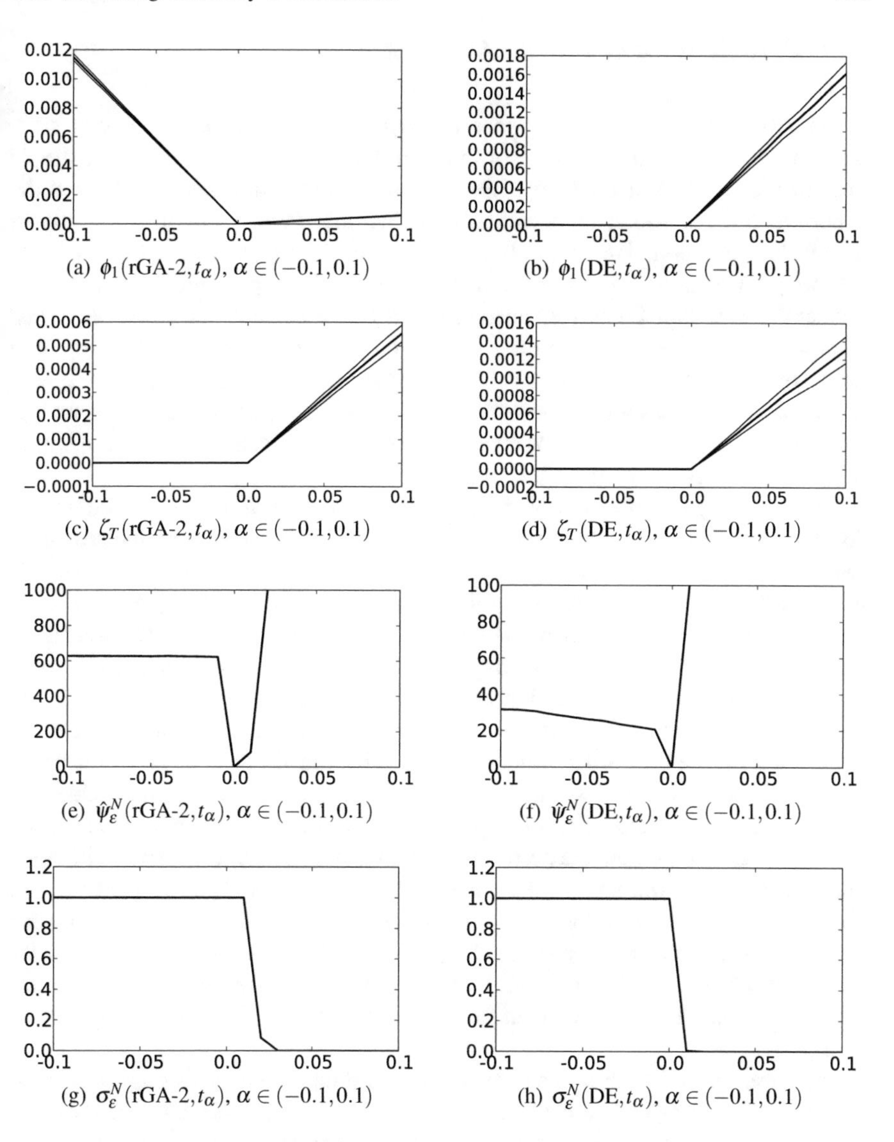


(a) $\phi_1(\text{rGA-2}, t_\alpha)$, $\alpha \in (-0.1, 0.1)$

(b) $\phi_1(\text{DE}, t_\alpha)$, $\alpha \in (-0.1, 0.1)$

(c) $\zeta_T(\text{rGA-2}, t_\alpha)$, $\alpha \in (-0.1, 0.1)$

(d) $\zeta_T(\text{DE}, t_\alpha)$, $\alpha \in (-0.1, 0.1)$

(e) $\hat{\psi}_\varepsilon^N(\text{rGA-2}, t_\alpha)$, $\alpha \in (-0.1, 0.1)$

(f) $\hat{\psi}_\varepsilon^N(\text{DE}, t_\alpha)$, $\alpha \in (-0.1, 0.1)$

(g) $\sigma_\varepsilon^N(\text{rGA-2}, t_\alpha)$, $\alpha \in (-0.1, 0.1)$

(h) $\sigma_\varepsilon^N(\text{DE}, t_\alpha)$, $\alpha \in (-0.1, 0.1)$

**Fig. 11.10** Change in performance of rGA-2 and DE for zero-centered prisms of different heights in 25 dimensions, measured on $t_\alpha$ from Equation 11.6 for 21 values of $\alpha$ evenly spaced in the interval $[-0.1, 0.1]$. The x-axis ranges over values of $\alpha$, the y-axis over performance values. Panels in the left column show the performance of rGA-2, and panels in the right column show the performance of DE. The rows show the performance criteria $\phi_1$, $\zeta_T$, $\hat{\psi}_\varepsilon^N$, and $\sigma_\varepsilon^N$, respectively. DE has a discontinuity at $t_0$.

## 11.5 Principal Components Analysis

Chapter 6 emphasized that the space of optimizers is a vector space with well-defined notions of distance between any two optimizers. In this section, a simple visualization of the space will be given for the optimizers that were included in the experiments. This visualization relies on the fact that there is a performance-based duality between optimizers and objective functions. This duality will be formally explored in Chapter 17. In essence, the results that will be presented in that section indicate that for a given performance criterion $\psi$, the average performance against an objective function under $\psi$ forms a line through the vector space of long-running optimizers introduced in Chapter 9. The set of all possible distributions over objective functions induces an uncountable linear basis over optimizer space dependent on $\psi$. The performance of an optimizer on a particular objective is thus a projection onto one of the components of this basis.

This property of optimizers was studied in the context of genetic algorithms by Ashlock et al. [9]. They developed a set of test optimizers by varying the crossover rules of a genetic algorithm. The resulting optimizers were then run on an array of benchmarks. For a given crossover rule, the tuple $\left(\hat{\psi}_{\varepsilon}^{N}, \sigma_{\varepsilon}^{N}, \mathrm{Var}(\tau_{\varepsilon})\right)$ was recorded for each benchmark. The set of all tuples for each crossover rule was treated as a signature of the crossover rule unique to it. The system of Ashlock et al. works because of the duality described in the prior paragraph, but its effectiveness is not limited to genetic algorithms. It can also be applied to any performance criterion.

To demonstrate the results of such a characterization visually, the performance of each of the optimizers tested was used to create an array of performance values for each of the performance criteria $\phi_1$, $\zeta_T$, $\hat{\psi}_{\varepsilon}$, and $\sigma_{\varepsilon}^{N}$. Principal components analysis (PCA) was used to project the performance into a three-dimensional space. PCA requires a square matrix, and there were thirteen algorithms and twelve benchmarks. To make a square matrix, the values for CG were excluded, so that the values for each performance criterion constituted an $12 \times 12$ matrix. For the performance criteria $\phi_1$, the influence of CG-R on PCA was so strong that it skewed the visualization, and so PCA was performed without for this performance criteria, leaving an $11 \times 11$ matrix.

PCA was applied to this matrix to create a 12-dimensional basis projection such that the earlier components have larger eigenvalues. For $\phi_1$, the first three components found by PCA had an average range of 5.64. The remaining eight components had an average range of 1.39, so that the first three components do capture a substantial amount of the variation; results for other performance criteria were similar. These components were plotted in a three-dimensional scatterplot in Figures 11.11–11.13.

Reviewing the results for $\phi_1$, the first PCA component separates NM ($x = -5$) from PSO ($x = -2.5$) and the rest ($x \in (0,2)$). The second PCA component separates rBOA, GSS, SA, and rGA ($y < 0$) from the restarted optimizers, NM, DE, and CMA-ES ($y > 0$). The third component separates PSO ($z = -3$) from the rest ($z > -1$). More generally, for $\phi_1$, (1) the restarted optimizers other than CG-R group together,

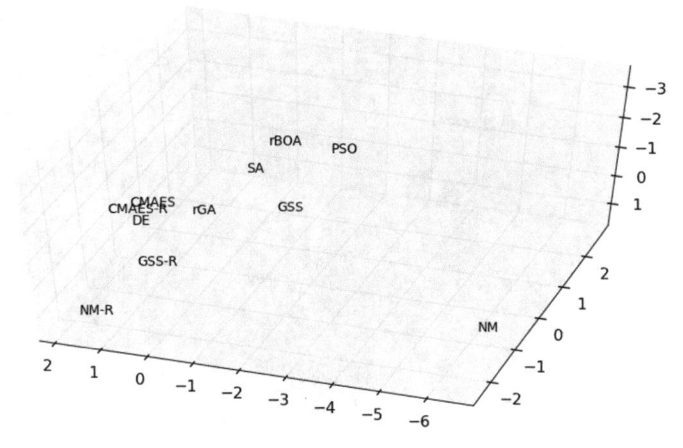


(a) First three PCA components of optimizers in the $\phi_1$ basis, without CG-R

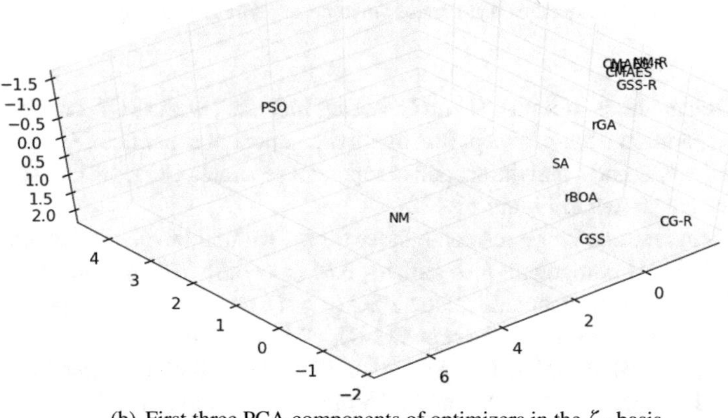


(b) First three PCA components of optimizers in the $\zeta_T$ basis

**Fig. 11.11** PCA plots for the test algorithms on the benchmarks. These plots show the proximity between various optimizers based on their performance on $\phi_1$ and $\zeta_T$. The resulting layout of optimizers reveals interesting new relationships among the algorithms.

(2) CMA-ES and DE are relatively close to each other and are closest to the restarted optimizers, and (3) SA and rGA generally appear together.

When CG-R is included for $\phi_1$, then the first PCA component has a range of 10.75, as opposed to a range of 6.35 without it. Additionally, CG-R is located at −9.27 on this scale, whereas the other 11 algorithms fall between 0 and 1.45. Thus $\phi_1$ strongly separates CG-R from the other algorithms, matching the intuition that gradient-based methods should behave in a noticeably different manner than

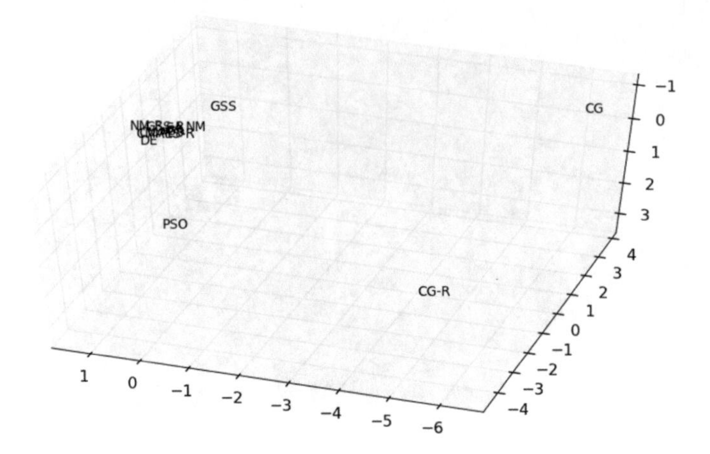


**Fig. 11.12** The first three PCA components in the $\phi_1$ basis, with CG and CG-R included. The first component separates conjugate gradient descent from the other methods, and the second component separates the restarted version from the non-restarted version. This plot shows that CG and CG-R are indeed distinct from the other methods in terms of performance on $\phi_1$.

gradient-free methods. If both CG and CG-R are included, with rBOA omitted, then a similar separation occurs, except that the first component separates CG and CG-R from the other methods, and the second component separates CG from CG-R. These distinctions are shown in Figure 11.12.

The salient features for $\zeta_T$ resemble those for $\phi_1$, without the disruptive influence of CG-R. The first component also isolates NM ($x = 6.5$) from the other methods ($x < 3$). The second component isolates PSO ($y = 4$), and the third component separates the restarted algorithms (except CG-R), CMA-ES, and DE ($z = -1.5$) from the others ($z > 0$). Once again, DE and CMA-ES are close both to each other and to the restarted optimizers. Also, SA and rGA are near to each other. CG-R does not cluster with the other restarted optimizers, in part because it performs worse than the others on $\zeta_T$, particularly on problems like *log-ackley* and *weierstrass*.

As might be expected, $\hat{\psi}_\varepsilon^N$ separates optimizers first based on the convergence speed. NM, GSS, CMA-ES, and all the restarted optimizers have $x > 0$, while DE has $x = 0$ and the other algorithms have $x < 0$. The third component has CG-R at one end ($z = -3$) and quasi-evolutionary methods except rBOA at the other ($z = 1$), with direct search methods clustered together at the center ($z \in (-1,0)$). Interestingly, this projection places the restarted version of optimizers close to the version without restarts in each case. Notably, SA and rGA are still relatively nearby for $\hat{\psi}_\varepsilon^N$.

As for $\sigma_\varepsilon^N$, the first component separates the all of the restarted optimizers off from the rest. The second component places NM-R on the far negative side ($y = -3$) and SA on the other extreme ($y = 2$), with the rest distributed evenly. The third component separates CG-R ($z = 4$) from the others ($z < 2$). SA and rGA are still close together, and DE is as close to CMA-ES as it is to any other algorithm.

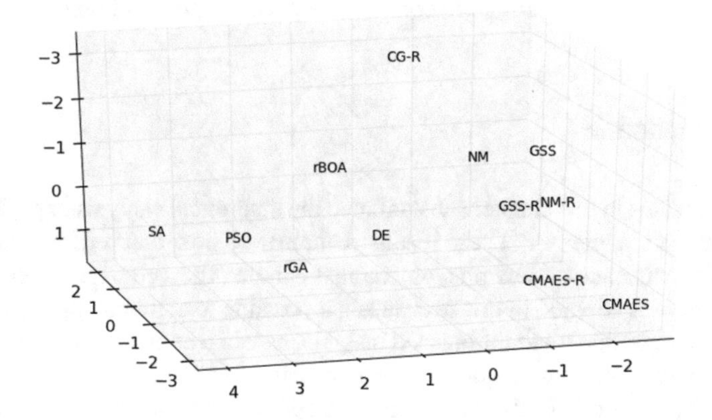


(a) First three PCA components of optimizers in the $\hat{\psi}_{\varepsilon}^{N}$ basis

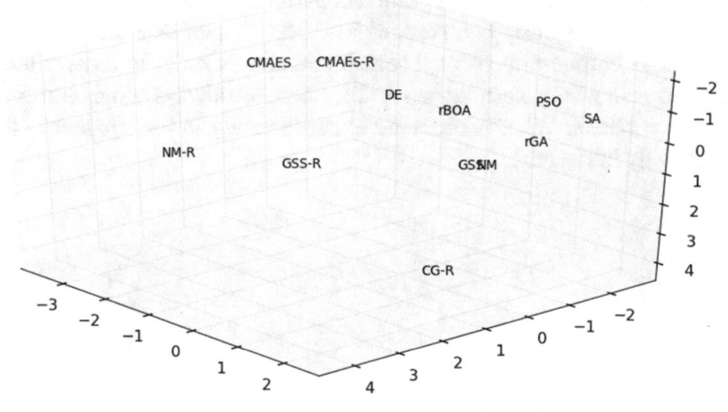


(b) First three PCA components of optimizers in the $\sigma_{\varepsilon}^{N}$ basis

**Fig. 11.13** PCA plots for eleven algorithms on eleven benchmarks. These plots show the proximity between various optimizers based on their performance on $\hat{\psi}_{\varepsilon}^{N}$ and $\sigma_{\varepsilon}^{N}$. The resulting layout of optimizers reveals interesting new relationships among the algorithms.

The graphs in Figures 11.11–11.13 thus demonstrate that the formal analysis of optimizers and their performance can enable new ways of looking at the relationships between optimizers. The picture that emerges provides insights that are not predicted by the origins of these optimizers. For example, simulated annealing and genetic algorithms are close in all of the graphs above. Even though such a result is unintuitive, it is substantiated by two theoretical observations: Theorem 7.1, which states that the $(1+1)$–ES is the norm-limit of simulated annealing, and the discussion of Expected Proportional Selection in Section 14.1.3. Thus theoretical inquiry is a useful tool for uncovering the connections between different optimizers.

## 11.6 Conclusion

The experiments in this chapter substantiated the theoretical analysis of performance undertaken in Chapter 10. These results demonstrate how theoretical analysis can suggest the existence of new phenomena that can be observed experimentally. The formal approach adopted in this text makes it possible to compare algorithms using novel techniques that yield unforeseen insights, as when convex combinations were shown to outperform pure algorithms in some cases.

In earlier sections of this chapter, certain optimizers were shown to outperform others on the benchmarks. For example, the restarted algorithms, DE, and CMA-ES collectively perform much better than PSO, rGA, rBOA, SA, GSS, and NM. To the extent that such claims are restricted to the experiments performed, they cannot be disputed. But how will these optimizers perform on practical objectives on which they have not previously been tested? The next chapter extends the No Free Lunch theorems to infinite-dimensional search domains in order to answer this question: In any domain where learning is possible, there are always some optimizers that are better than others. What is observed experimentally in this chapter is thus proven theoretically in the next.

# Chapter 12
# No Free Lunch Theorems

Every student of machine learning and artificial intelligence has heard of the No Free Lunch (NFL) theorem, often formulated with the broad and somewhat misleading statement that *every algorithm pays for good performance on some problems with bad performance on other problems*. This phrasing better describes the duality theory that was discovered simultaneously with NFL than it does NFL itself as subsequently defined, but the wording nonetheless adequately captures the intuition behind the name. A search or optimization algorithm that solves every problem well would be *too good to be true*, and so finding such an algorithm would be analogous to *getting a free lunch*. The question of whether such *free lunches* exist pertains both to the development of NFL theorems as reviewed and updated in this chapter and to problem-solution duality theory, which is explored further in subsequent chapters.

In this chapter, a probabilistic version of NFL in Hausdorff topological spaces is presented, and this NFL property is shown to be equivalent to a notion of *path independence*, in which the probability distribution governing observed values is independent of all paths through the search space. This description of NFL can be parlayed to a claim that NFL is purely a property of search and optimization problems without reference to algorithms. In the context of duality, it also supports identifying NFL as the dual problem with which blind random search is best aligned. NFL theory is important for the study of search and optimization performance because it addresses the question of what kind of performance is possible in an extreme cases, leading to the recognition of certain limits to search and optimization in general.

## 12.1 Overview of No Free Lunch

The greatest difficulty in approaching NFL is in finding a formal definition that captures the intuition of paying for good performance. In their original paper proving NFL for optimization, Wolpert and Macready represented this intuition as an average over algorithm performance across all problems [171], proving that this average is constant, so long as the algorithm does not repeat search points (with repetition,



A. J. Lockett, *General-Purpose Optimization Through Information Maximization*,
Natural Computing Series, https://doi.org/10.1007/978-3-662-62007-6_12

performance is worse for most interesting metrics). Subsequent authors have expanded these concepts [37, 42, 43, 81], and later authors have tended to replace the concept of average performance with a statement about problem classes that exhibit symmetries that induce constant average performance across algorithms, either due to set-theoretic observations [144, 143, 136] or due to probability theory and stochastic processes [13, 4]. In this chapter, NFL is treated as a property of problem classes, which are represented as unsigned measures. This section reviews the results that led to this transition and sets up the formal setting for NFL used in the remainder of the chapter.

### *12.1.1 What Is No Free Lunch?*

Wolpert and Macready's formulation of NFL considers the average performance of an algorithm across all objective functions. They assume that the search domain $X$ and the value space $Y$ are finite, in which case the space of objective functions $Y^X$ is also finite. Search and optimization methods are presumed to be *non-repeating*, meaning they never propose the same search point more than once, which implies that the time horizon of the search is finite with $|\mathcal{T}| \leq |X|$. There is also an assumption that search methods are deterministic as well, although this restriction is easily removed.

Defining performance criteria as in Definition 10.1, the assumption of finiteness simplifies the definition so that a performance criterion is any function $\psi : \mathcal{BB}^{\mathcal{T}}_{X,Y} \times Y^X \to \mathbb{R}$, since every function is measurable and every kernel is integrable. NFL does not apply to performance criteria that examine the global properties of objectives since these criteria can obviously be rigged to depend on the search path. Thus NFL only admits performance measurements that depend only on the observed values.

**Definition 12.1 (Value-Dependent).** A performance criterion $\psi$ with kernel $V_u : X^{\mathcal{T}} \to \mathbb{R}$ is value-dependent if there exists measurable $\hat{V} : Y^{\mathcal{T}} \to \mathbb{R}$ such that $\hat{V}(u(h_X)) = V_u(h_X)$ where $u(h_X) \in Y^{\mathcal{T}}$ is interpreted as the value sequence that result from applying $u$ to every element of $h_X$ in order so that $u(h_X)_t = u(h^X_t)$. The revised kernel $\hat{V}$ is then called the *value kernel* of $\psi$, and its relationship with $V$ is abbreviated as $V = \hat{V} \circ u$.

The original NFL result can be stated using static search trajectors after reviewing a few definitions. First, recall that a static search trajector $\mathcal{A}$ is *proper* if for all objectives $u$, $\mathcal{A}[u]$ is a probability measure, *i.e.*, $|\mathcal{A}[u]|(X^{\mathcal{T}}) = \mathcal{A}[u](X^{\mathcal{T}}) = 1$. Second, $\mathcal{A}$ is *deterministic* if for every objective $u$ there is a search history $h_{X,u}$ such that $\mathcal{A}[u](\{h_{X,u}\}) = |\mathcal{A}[u]|(X^{\mathcal{T}}) = 1$; that is, $h_{X,u}$ has full measure.[1] Finally, a formal definition of *non-repeating* is required.

[1] This is a repetition of Definition 6.9.

**Definition 12.2 (Non-Repeating).** A static search trajector $\mathcal{A}$ is *non-repeating* if the set

$$\left\{h_X \,\middle|\, \text{there exist } i, j \in \mathcal{T} \text{ with } i \neq j \text{ such that } h_i^X = h_j^X\right\}$$

has $\mathcal{A}[u]$-measure zero for all objectives $u$. A static search generator $\mathcal{G}$ is non-repeating if it extends to a static search trajector $\mathcal{A} = \mathbf{E}(\mathcal{G})$ that is non-repeating.

**Theorem 12.1 (Original No Free Lunch [171]).** *Suppose that the search domain $X$ and the value space $Y$ are finite with temporal index set $|\mathcal{T}| = |X|$. If $\mathcal{G}$ and $\mathcal{G}'$ are any two proper, non-repeating, and deterministic static black-box search generators, then for any value-dependent performance criterion $\psi$,*

$$\sum_{u \in Y^X} \psi(\mathcal{G}, u) = \sum_{u \in Y^X} \psi(\mathcal{G}', u). \tag{12.1}$$

The obvious meaning of this theorem is that average performance over all possible objective functions is a constant independent of the search method. The subtle requirement of value-dependence deserves some discussion. If the performance criterion depends on global properties of the objective, then the equality may not hold. In particular, dependence on the error sequence as introduced in Chapter 10 implies that a performance criterion is not value-dependent. This subtlety actually points the way towards understanding NFL; by summing over all values, the natural symmetries of the function space $Y^X$ generate all possible value histories regardless of the chosen search points, so long as repetition is forbidden. This observation has been generalized as *set-theoretic NFL*, discussed below.

In finite spaces, the original NFL result automatically generalizes to non-deterministic, non-repeating static search trajectors. There are finitely many non-repeating search histories and finitely many objectives, and thus there are finitely many proper, non-repeating, and deterministic trajectors, which can be enumerated as $(\mathcal{A}_k)_{k \in \mathcal{K}}$. For any performance criterion $\psi = \mathbb{E}[V]$ and for every proper non-repeating static black-box search trajector $\mathcal{A} = \mathbf{E}(\mathcal{G})$, there are weights $(w_k)_{k \in \mathcal{K}}$ with $\sum_k w_k = 1$ such that

$$\mathcal{A} = \sum_{k \in \mathcal{K}} w_k \mathcal{A}_k \quad \text{and} \quad \psi(\mathcal{G}, u) = \sum_{k \in \mathcal{K}} w_k \psi(\mathcal{G}_k, u) \quad \text{where } \mathcal{A}_k = \mathbf{E}(\mathcal{G}_k).$$

This fact was observed by Wolpert and Macready [171]. By plugging in this equality and reversing the order of the sums, it follows that Equation 12.1 holds for all pairs of proper non-repeating static black-box generators in finite spaces. Generalizing it beyond finite domains and value spaces is more difficult.

Theorem 12.1 is not the simplest statement of the NFL property; the dependence on performance criteria can be removed. In finite spaces,

$$\psi(\mathcal{G}, u) = \mathbb{E}_{\mathcal{A}[u]}[V_u] = \sum_{h_X \in X^{\mathcal{T}}} \hat{V}(u(h_X))\, \mathcal{A}[u](\{h_X\})$$

where $\mathcal{A}$ is the extension of $\mathcal{G}$, $X^{\mathcal{T}}$ is as usual the set of all search histories, and $V_u : X^{\mathcal{T}} \to \mathbb{R}$ is the value-dependent kernel of $\psi$. If $\mathcal{G}$ is deterministic, then at most one term of this sum is non-zero, but we will allow for the case where $\mathcal{G}$ is non-deterministic as well.

When $|\mathcal{T}| = |X|$, each non-repeating $h_X$ is simply an enumeration of $X$ so that $u(h_X)$ is a complete description of $u$. By the definition of $Y^X$, for each value history $h_Y \in Y^{\mathcal{T}}$ (including those with repeated elements), there is exactly one $u \in Y^X$ such that $u(h_X) = h_Y$. Therefore, letting $u_h$ be the objective determined by the pair $(h_X, h_Y)$, one obtains

$$\sum_{u \in Y^X} \mathbb{E}_{\mathcal{A}[u]}[V_u] = \sum_{h_Y \in Y^{\mathcal{T}}} \hat{V}(h_Y) \sum_{h_X \in N} \mathcal{A}[u_h](\{h_X\}) = \sum_{h_Y \in Y^{\mathcal{T}}} \hat{V}(h_Y), \tag{12.2}$$

where the sum over $\mathcal{A}[u_h]$ equals 1 because $\mathcal{A}$ is proper and non-repeating. In fact, Equation 12.2 essentially proves Theorem 12.1, since the right-hand side has no dependence on the trajector $\mathcal{A}$; it uses only the facts that $\psi$ is value-dependent and $\mathcal{A}$ is non-repeating.

It is important to account for repetitions when discussing NFL, because under the assumption of oracle access to a static fitness function, repeating a search point forces a value to be repeated as well, which will necessarily violate NFL unless the search problem degenerately produces a single, constant value. NFL is a property that prevents a search method from exercising any degree of control over the observed values, and for static objective functions, repetition always results in effective control of the value. Eliminating repetition from consideration is valid, because if we assume that accessing the oracle incurs a cost, repeatedly observing the same search point is obviously wasteful. Repetitions can always be avoided by keeping a list of the previously accessed points and skipping them when proposed, a method known as *taboo search*.

If one takes Equation 12.1 as the meaning of NFL, at least two kinds of extensions can be considered. First, one may ask whether the equation holds when the sum is taken in a different way. Second, one can examine whether some version of this equation also holds when the search domain $X$ and the value space $Y$ are infinite spaces. Two main approaches to both extensions have been followed. The first approach, termed *Set-Theoretic NFL*, removes the probabilistic aspects of the problem formulation and searches for subsets of objectives that yield NFL results. The second, known as *Probabilistic NFL*, takes the probabilistic description as fundamental and treats the sum as an expectation with respect to a probability measure, which then must then satisfy an NFL property. Although this text views the probabilistic formulation as the more natural extension, the set-theoretic perspective provides an elegant intuition for why NFL holds that supports the probabilistic NFL property introduced in the next section. Consequently, both approaches are now reviewed in turn.

### 12.1.2 Set-Theoretic NFL: Closure Under Permutation

The set-theoretic approach focuses on the fact that NFL dissociates value histories from search histories and uses this property in order to abstract away the complications due to probability theory and consider the subject at a more elemental level.

In Equation 12.1, the NFL equality was derived from a uniform prior over objective functions. Wolpert and Macready conjectured that the general equality would hold for many if not most function priors, particularly priors that place positive probability on a large number of functions [171]. In retrospect, this claim was too expansive. The equality holds only for very few function priors that satisfy strict conditions.

The first result in this direction was obtained by Schumacher [144] using permutations on the search space based on observations previously made by Radcliffe and Surrey [131]. Suppose that $\mathcal{D}$ is a deterministic black-box static search generator. Then there is a map $\xi : Y^X \to X^{\mathcal{T}}$ that assigns to each objective $u \in Y^X$ the unique point of degeneracy for $\mathcal{D}[u]$. That is, $\xi(u)$ is the search history selected by $\mathcal{D}$ in response to $u$.[2] The search history $\xi(u)$ generates a value history denoted in the next few paragraphs as $\xi_Y(u) \in Y^{\mathcal{T}}$, which is found by applying $u$ to each element of $\xi(u)$. Schumacher called the map $\xi$ a *search operator* and $\xi_Y(u)$ the *performance vector* of $\mathcal{D}$ on $u$.

Radcliffe and Surrey noticed that if $\xi$ and $\xi'$ are two search operators and $u$ is an objective, then there always exists a second objective $u'$ such that the pair $(\xi, u)$ and $(\xi', u')$ generate the same performance vectors, *i.e.*, $\xi_Y(u) = \xi'_Y(u')$. This observation was in fact crucial to the proof of Theorem 12.1 offered in the text above. Schumacher made these relationships more explicit using permutations.

A permutation $\pi : X \to X$ can be applied to an objective $u : X \to Y$ to obtain a new objective $u \circ \pi$ that shuffles the inputs. A permutation can likewise be applied to a search operator $\xi$ to obtain a new search operator $\xi \circ \pi$ defined so that $\xi \circ \pi(u) = \pi(\xi(u \circ \pi))$, where the outer reference to $\pi$ is presumed to permute each element of $\xi(u \circ \pi)$ individually. These definitions produce the following Lemma.

**Lemma 12.1 (Permutation Duality [143]).** *Suppose $\xi$ is a search operator, $u$ is an objective, and $\pi$ is a permutation. Then $\xi \circ \pi(u) = \pi(\xi(u \circ \pi))$ and $\xi_Y(u \circ \pi) = (\xi \circ \pi)_Y(u)$.*

*Proof.* If $h_X = \xi(u \circ \pi)$ and $h_Y = \xi_Y(u \circ \pi)$, then $h_i^Y = u \circ \pi(h_i^X)$. If $h'_X = \xi \circ \pi(u)$ and $h'_Y = (\xi \circ \pi)_Y(u)$, then $h'^X_i = \pi(\xi(u \circ \pi)_i) = \pi(h_i^X)$ and $h'^Y_i = u(h'^X_i) = u(\pi(h_i^X)) = h_i^Y$.

Schumacher used permutations in order to define a concept of *closure under permutation* that implies an NFL property.

**Definition 12.3 (Closure Under Permutation).** A set of functions $\mathcal{F} \subseteq Y^X$ is *closed under permutation* (c.u.p.) if for any permutation $\pi$, $u \in \mathcal{F}$ implies that $u \circ \pi \in \mathcal{F}$.

[2] Here we assume that this point of degeneracy is measurable as a singleton. Section 8.1.2 discussed this relationship for standard search generators.

**Theorem 12.2 (Sharpened NFL [143]).** *Suppose $\mathcal{F} \subseteq Y^X$ is c.u.p. for $X$ and $Y$ finite with $|\mathcal{T}| = |X|$. Then any pair $\xi, \xi'$ of search operators produce the same performance vectors over $\mathcal{F}$, that is,*

$$\{\xi_Y(u) \mid u \in \mathcal{F}\} = \{\xi'_Y(u) \mid u \in \mathcal{F}\}.$$

*Proof.* As discussed in the text around Equation 12.2, by manipulating the objective function – essentially by permuting the objective values – one can force a particular performance vector to be assigned to an arbitrary search history. Now consider the search trajectory $\xi(u) \in X^{\mathcal{T}}$ for an arbitrary objective $u \in \mathcal{F}$. The proof consists in observing that there exists a permutation $\pi$ such that $\xi'_Y(u \circ \pi) = \xi_Y(u)$, and then the fact that $\mathcal{F}$ is c.u.p. implies that $u \circ \pi \in \mathcal{F}$ so that $\xi_Y(u) \in \{\xi'_Y(u) \mid u \in \mathcal{F}\}$. The theorem then follows by symmetry since the same argument applies verbatim for $\xi'_Y(u)$.

Suppose $\pi$ is the desired permutation. By Lemma 12.1, $\xi'_Y(u \circ \pi) = (\xi' \circ \pi)_Y(u)$, and we want this latter to equal $\xi_Y(u)$, which implies $\xi(u) = \xi' \circ \pi(u) = \pi(\xi'(u \circ \pi))$. Observe that both $\xi(u)$ and $\xi'(u \circ \pi)$ enumerate $X$, and make this enumeration explicit as $\xi(u) = (x_i)_{i \in \mathcal{T}}$ and $\xi'(u \circ \pi) = (x'_i)_{i \in \mathcal{T}}$, where the $x'_i$ are to be defined next. Using $\xi(u)$, we will iteratively construct $\pi$ as follows. First, $x'_1$ is fixed independent of the choice of $\pi$ by the fact that $\xi'$ is a black-box search operator, so we can define $\pi(x'_1) = x_1$. Now, assuming that $\pi$ is defined for $x'_1, \ldots, x'_{i-1}$, then $x'_i$ is fixed independent of $\pi$, and so we can define $\pi(x'_i) = x_i$. Given that $\xi'$ is non-repeating and $(x_i)_{i \in \mathcal{T}}$ enumerates $X$, $\pi$ is one-to-one and onto, whence it is a permutation of the search domain $X$ for which $\xi(u) = \xi' \circ \pi(u) = \pi(\xi'(u \circ \pi))$ as desired, completing the proof.

**Corollary 12.1.** *Suppose $\mathcal{F} \subseteq Y^X$ is c.u.p. for $X$ and $Y$ finite with $|\mathcal{T}| = |X|$. If $\mathcal{G}$ and $\mathcal{G}'$ are any two proper, non-repeating, and deterministic static black-box search generators, then for any value-dependent performance criterion $\psi$,*

$$\sum_{u \in \mathcal{F}} \psi(\mathcal{G}, u) = \sum_{u \in \mathcal{F}} \psi(\mathcal{G}', u). \tag{12.3}$$

*Proof.* Suppose $\xi$ and $\xi'$ are the search operators corresponding to $\mathcal{G}$ and $\mathcal{G}'$, respectively, and let $\hat{V}$ be the value-dependent kernel of $\psi$. By Theorem 12.2, the sums can be reindexed using the set $H_Y = \{\xi_Y(u) \mid u \in \mathcal{G}\}$ to obtain

$$\sum_{u \in \mathcal{F}} \psi(\mathcal{G}, u) = \sum_{h_Y \in H_Y} \hat{V}(h_Y) = \sum_{u \in \mathcal{F}} \psi(\mathcal{G}', u).$$

The original result of Wolpert and Macready can be recovered by setting $\mathcal{F} = Y^X$. As before, the theorem is not limited to deterministic search methods because in finite spaces every stochastic search method can be written as a weighted sum over all deterministic search methods.

Theorem 12.2 provides a sufficient condition for NFL as defined in Equation 12.3. This condition is also necessary.

**Theorem 12.3.** *Suppose there is a subset $\mathcal{F} \subseteq Y^X$ with $X$ and $Y$ finite and $|\mathcal{T}| = |X|$. If Equation 12.3 holds for all proper, non-repeating, deterministic static black-box search generators, then $\mathcal{F}$ is c.u.p.*

*Proof.* Suppose $\tilde{u} \in \mathcal{F}$ and $\xi$ is the search operator for a proper, non-repeating, deterministic static black-box search generator. Fix $h_Y = \xi_Y(\tilde{u})$ and choose $\hat{V}$ such that $\hat{V}(h_Y) = 1$ and $\hat{V}(h''_Y) = 0$ for $h''_Y \neq h_Y$. Let $\pi : X \to X$ be any permutation. Set $h_X = \pi(\xi(\tilde{u}))$ and let $\xi'$ be the search operator that produces $h_X$ regardless of $u$, that is, $\xi'(u) = h_X$ for all $u \in Y^X$. Equation 12.3 implies

$$\sum_{u \in \mathcal{F}} \hat{V}(\xi_Y(u)) = \hat{V}(\xi_Y(\tilde{u})) = 1 = \sum_{u \in \mathcal{F}} \hat{V}(\xi'_Y(u)).$$

But $\xi'_Y(u)$ is just $u(h_X)$, meaning that $u$ is applied to each element of $h_X$ in order. So the fact that the rightmost sum equals one together with the fact that $\xi'_Y(u)$ uniquely identifies $u$ for $\xi'$ non-repeating implies that there is some function $u' \in \mathcal{F}$ such that $u'(h_X) = h_Y = \tilde{u}(\xi(\tilde{u}))$. It then follows that $u' = \tilde{u} \circ \pi \in \mathcal{F}$. Since $\pi$ was arbitrary, this fact implies that $\mathcal{F}$ is c.u.p.

Therefore, Equation 12.3 is satisfied only by sets $\mathcal{F}$ that are closed under permutation. If one takes this equation as the proper generalization of Equation 12.1, then closure under permutation expresses the whole theory of NFL.

Igel and Toussaint [81] asked how many c.u.p. subsets of $Y^X$ exist. They proved that the percentage of subsets of $Y^X$ that are c.u.p. is exactly

$$\frac{2^{\binom{|X|+|Y|-1}{|Y|}} - 1}{2^{\left(|Y|^{|X|}\right)} - 1}. \tag{12.4}$$

This fraction vanishes double exponentially fast as $|X|$ and $|Y|$ increase. The obvious conclusion is that function priors subject to NFL are extraordinarily rare.

In some ways, the fact that NFL is rare does not make it less important. One of the main conclusions of NFL is that an optimizer's success on an optimization task depends on how well the optimizer is aligned with the class of problems likely to appear. This duality has already been emphasized above and is discussed more deeply in the following chapters.

The sets of functions that are not c.u.p. can be quite large and general. The fact that an optimizer can be well-aligned with a very general set of functions weakens the claim that all optimizers perform equivalently, especially since closure under permutation seems to be an unreasonable assumption for any practical class of problems.

The sum over objectives in Equation 12.1 does not generalize nicely to infinite search domains and value spaces, but the concept of closure under permutation does. Rowe et al. developed an NFL theorem for c.u.p. sets of arbitrary cardinality. The language and proof of this theorem is similar in essence to that of Theorems 12.2 and 12.3, except that it applies to c.u.p. sets of arbitrary cardinality [136]. The theory requires the temporal index set $\mathcal{T}$ to have cardinality equal to that of search domain $X$, *i.e.*, $|\mathcal{T}| = |X|$. When $X$ is has cardinality greater than that of the integers, its cardinality said to be *transfinite*. The index set set $\mathcal{T}$ is then a well-ordered, transfinite set, which implies the existence of *limit points* within $\mathcal{T}$ that are approached but not attained by countable subsequences of $\mathcal{T}$. Rowe et al. extended the proof of Theorem 12.2 by defining how search operators behave at limit points and then demonstrating that this limit behavior interacts appropriately with permutations of the search domain. Transfinite sequences are not needed elsewhere in this text, and so the reader is referred to their paper for details on this procedure.

The set-theoretic approach has the benefit of simplicity and elegance. It can be presented within a few pages of text and demonstrates the universal existence of NFL conditions for arbitrary search domains. Based on these attributes, Rowe et al. concluded that probabilistic approaches to NFL were necessarily weaker and less powerful than the set-theoretic approach. They averred that the introduction of measure theory into the discussion was an unnecessary complication irrelevant to the essence of NFL.

This conclusion is somewhat premature, however. When practitioners consider transfinite spaces, there is usually a concept of spatial structure included, as in $n$-dimensional Euclidean space. The idea of a search method that exhaustively explores the entire transfinite space in sequence seems artificial, since spatial concepts such as continuity or density can allow one to skip many search points without compromising the quality of the search. But then, this observation is particularly revelatory, since permutation invariance necessarily disrupts spatial structure. Thus one expects that when any sort of structural or spatial regularity is imposed on a set of search problems, permutation invariance will be violated and NFL cannot hold.

A more serious issue for the set-theoretic approach is that it limits generalization of Equation 12.1 to deterministic sets of objectives. The prior paragraph raised the issue of spatial regularity, suggesting that it conflicts with permutation invariance. But narrow consideration of sets of objectives does not naturally reveal complications due to spatial interactions. If one wishes to know the boundaries of NFL theorems, that is, if one wishes to be able to say for sure when NFL does not apply, one must understand how NFL interacts with space, which brings the question back to measure theory and ultimately topology.

Furthermore, the assumption of measuring performance with respect to uniform probabilities over a set of functions is baked into the c.u.p. approach. Can non-uniform function priors exhibit NFL symmetry properties? Do NFL properties arise strictly from permutation invariance, or is there a still more essential characterization of NFL? These questions cannot be answered by set-theoretic NFL alone. There remains a role for probabilistic NFL to play; this role is explored in the following sections.

### 12.1.3 Probabilistic NFL: Continuous Lunches Are (Not) Free!

Returning to Equation 12.1, one notices that these sums are essentially expectations of the performance with respect to a random source of objectives. Suppose that $U$ is a random variable taking values in $Y^X$ and then the NFL property can be stated as

$$\mathbb{E}\left[\psi(\mathcal{G},U)\right]=\mathbb{E}\left[\psi(\mathcal{G}',U)\right] \tag{12.5}$$

for all pairs $\mathcal{G},\mathcal{G}'$ and for all value-dependent performance criteria $\psi$. In this case, one wishes to know which probability laws cause Equation 12.5 to hold. It is obvious from Corollary 12.1 that Equation 12.5 holds for all probability laws that induce a uniform distribution of $U$ with respect to some c.u.p. set $\mathcal{F}$. As suggested above, the question is whether there are others.

To explore this issue, one supposes that there is a probability space $(\Omega,\mathbb{P},\Sigma)$ and that $U:\Omega\to Y^X$ is a random variable on this space, called a *random objective*. Then $U(\omega)$ is an objective function for $\omega\in\Omega$. Since it is a function, one can write $U(\omega)(x)$ for $x\in X$, and this is usually abbreviated by omitting the $\omega$ and simplifying the notation to $U(x)$. Given a search history $h_X\in X^{\mathcal{T}}$, we also write $U(h_X)$ to mean the element of $Y^{\mathcal{T}}$ that results from applying $U(\omega)$ to each element of $h_X$ in order.

Given that $U$ is a random variable, the expectations $\mathbb{E}\left[\psi(\mathcal{G},U)\right]$ are well defined. However, since $\mathcal{G}$ is also probabilistic (with extension $\mathcal{A}=\mathbf{E}(\mathcal{G})$) and $\psi$ is an integral, this can be expanded to

$$\mathbb{E}\left[\psi(\mathcal{G},U)\right]=\mathbb{E}_{\mathbb{P}}\left[\mathbb{E}_{\mathcal{A}[U]}\left[\hat{V}\circ U\right]\right]=\mathbb{E}_{\mathbb{P}\times\mathcal{A}[U]}\left[\hat{V}\circ U\right], \tag{12.6}$$

where the notation $\mathbb{P}\times\mathcal{A}[U]$ has been introduced to denote the joint probability measure arising by taking both expectations at once, with measurability defined on the space $\left(\Omega\times X^{\mathcal{T}},\Sigma\otimes\mathcal{B}a_{X^{\mathcal{T}}}\right)$. Recall that $\hat{V}\circ U(h_X)=\hat{V}(U(h_X))=V_u(h_X)$, where $V$ is the kernel of $\psi$ and $\hat{V}$ is its value kernel.

As with set-theoretic NFL, one wishes to deal with *performance vectors* rather than performance criteria to simplify Equation 12.5. Define a map

$$\Xi_Y:(\omega,h_X)\mapsto\left(U(\omega)(h_t^X)\right)_{t\in\mathcal{T}},$$

which is measurable and projects $\Omega\times X^{\mathcal{T}}$ into $Y^{\mathcal{T}}$. It then follows that

$$\mathbb{E}_{\mathbb{P}\times\mathcal{A}[U]}\left[\hat{V}\circ U\right]=\mathbb{E}_{\mathbb{P}\times\mathcal{A}[U]}\left[\hat{V}\circ\Xi_Y\right]=\mathbb{E}_{\mathbb{Y}_{\mathcal{A}}}\left[\hat{V}\right]=\int_{Y^{\mathcal{T}}}\hat{V}(h_Y)\,\mathbb{Y}_{\mathcal{A}}(dh_Y)$$

where $\mathbb{Y}_{\mathcal{A}}=(\mathbb{P}\times\mathcal{A}[U])\circ\Xi_Y$ is the pushforward measure induced by $\Xi_Y$.

The last two paragraphs skirted over the question of measurability for $\Xi_Y$. This issue is critical, however. The function $U:\Omega\to Y^X$ was declared to be a random variable, but to be a random variable, there must be a $\sigma$-algebra specified on $Y^X$. If $Y$ is a topological spaces, then the obvious choice is the Baire $\sigma$-algebra, $\mathcal{B}a_{Y^X}$, which arises from the product topology on $Y^X$ as well as from Kolmogorov extension.[3] With this choice, $\Xi_Y$ is Baire-measurable if and only if $\mathcal{T}$ is countable. This statement is true because $\Xi_Y$ is a composition of $U$, which is measurable, with a projection $\phi_{h_X} : Y^X \to Y^{\mathcal{T}}$ that selects projection coordinates according to $h_X$. Such projections are Baire-measurable if and only if $\mathcal{T}$ has countable cardinality. However, if one chooses the Borel $\sigma$-algebra, $\mathcal{B}_{Y^X}$, for $Y^X$, then all projections are measurable, and so $\Xi_Y$ is always Borel-measurable. Likewise, the definition of *value-dependent* specified that $\hat{V}$ is measurable but did not specify the $\sigma$-algebra on $Y^{\mathcal{T}}$; here we assume that it matches that chosen for $Y^X$, either Borel or Baire. Further discussion of these distinctions is postponed until the next section. In any case, the $\sigma$-algebra on which $\Xi_Y$ and hence $\mathbb{Y}_{\mathcal{A}}$ are defined does not depend on $\mathcal{A}$.

Taking $U$ and $\mathbb{P}$ as fixed, we now have a map $\mathcal{A} \mapsto \mathbb{Y}_{\mathcal{A}}$. If this map is constant with respect to proper, non-repeating $\mathcal{A}$, there is a single measure $\mathbb{Y}$ independent of $\mathcal{A}$ such that $\mathbb{Y}_{\mathcal{A}} = \mathbb{Y}$ and then for all proper, non-repeating $\mathcal{G}, \mathcal{G}' \in \mathcal{BB}^{\mathcal{T}}_{X,Y}$ with extensions $\mathcal{A} = \mathbf{E}(\mathcal{G})$ and $\mathcal{A}' = \mathbf{E}(\mathcal{G}')$,

$$\mathbb{E}_{\mathbb{P}}\left[\psi(\mathcal{G},U)\right] = \mathbb{E}_{\mathbb{Y}_{\mathcal{A}}}\left[\hat{V}\right] = \mathbb{E}_{\mathbb{Y}}\left[\hat{V}\right] = \mathbb{E}_{\mathbb{Y}_{\mathcal{A}'}}\left[\hat{V}\right] = \mathbb{E}_{\mathbb{P}}\left[\psi(\mathcal{G}',U)\right]. \tag{12.7}$$

Although the symbols obscure it somewhat, this equation is merely a generalization of Equation 12.2.

The following equivalences can be observed based on Equation 12.5 following Auger and Teytaud [13].[4] Due to equivalence, each of these items can be taken as a definition of probabilistic NFL.

**Definition 12.4.** For any measurable $A \subseteq Y^{\mathcal{T}}$, the *performance indicator* of $A$ is the value-dependent performance criterion $\psi_A$ whose value kernel is the indicator function for $A$, *i.e.*, $\hat{V} = \mathbb{1}_A$.

**Theorem 12.4.** *Suppose that either $Y^X$ has the Borel $\sigma$-algebra and the cardinality of $\mathcal{T}$ is arbitrary, or that $Y^X$ has the Baire $\sigma$-algebra and $\mathcal{T}$ has at most countable cardinality. Then the following statements are equivalent:*

1. *Equation 12.5 holds for all proper, non-repeating $\mathcal{G}, \mathcal{G}' \in \mathcal{BB}^{\mathcal{T}}_{X,Y}$ and for all value-dependent $\psi$.*
2. *Equation 12.5 holds for all proper, non-repeating $\mathcal{G}, \mathcal{G}' \in \mathcal{BB}^{\mathcal{T}}_{X,Y}$ and for all performance indicators $\psi_A$, where $A$ is any measurable subset of $Y^{\mathcal{T}}$.*
3. *The map $\mathcal{A} \to \mathbb{Y}_{\mathcal{A}}$ is constant for proper, non-repeating $\mathcal{A}$.*

*Proof.* Statement (1) implies (2), and Equation 12.7 demonstrates that (3) implies (1), so proof is only needed to show that (2) implies (3). With $\mathcal{A} = \mathbf{E}(\mathcal{G})$ and $\mathcal{A}' = \mathbf{E}(\mathcal{G}')$, statement (2) implies that for all measurable $A \subseteq Y^{\mathcal{T}}$,

$$\mathbb{Y}_{\mathcal{A}}(A) = \mathbb{E}_{\mathbb{Y}_{\mathcal{A}}}\left[\mathbb{1}_A\right] = \mathbb{E}_{\mathbb{Y}_{\mathcal{A}'}}\left[\mathbb{1}_A\right] = \mathbb{Y}_{\mathcal{A}'}(A),$$

[3] For those familiar with the theory of stochastic processes, the Baire $\sigma$-algebra is the one generated by cylinder sets. See *e.g.* [86].

[4] Lemma 2.3 of [13]

which means that $\mathbb{Y}_{\mathcal{A}} = \mathbb{Y}_{\mathcal{A}'}$ for all $\mathcal{A}, \mathcal{A}'$ that are proper and non-repeating, which completes the proof.

As a side note, the cardinality of $\mathcal{T}$ enters the proof due to the assumptions on the measurability of $\mathbb{Y}_{\mathcal{A}}$ and $\hat{V}$ above. Furthermore, the proof is vacuously satisfied if $|\mathcal{T}| > |X|$, because there are no non-repeating search generators in this case.

Theorem 12.4 is the probabilistic analog of Theorem 12.2. Whereas in set-theoretic NFL, all algorithms produce the same set of performance vectors, in probabilistic NFL, all algorithms produce the same distribution over performance vectors. This fact connects probabilistic and set-theoretic NFL. It should be noted, however, that while c.u.p. sets clearly exist, the existence of random objectives that render $\mathcal{A} \mapsto \mathbb{Y}_{\mathcal{A}}$ constant is far from obvious. A general existence proof is offered below in Theorem 12.10.

The probabilistic approach to NFL in continuous domains was first expounded by Auger and Teytaud [12, 13] in a form equivalent to statement (3) of Theorem 12.4. They were dissatisfied with the wildness inherent in allowing the measure $\mathbb{P}$ on the space $\Omega$ to be arbitrary and sought to place reasonable limits on the distribution of $U$. As a first step, they fixed the search domain to be the unit interval, $X = [0,1]$, with value space $Y = \mathbb{R}$ and required the process $(U(x))_{x\in[0,1]}$ to be a *measurable process*. By itself, this requirement is not particularly strong; it is satisfied by both Borel and Baire measurability on $Y^X$ as described above.

**Definition 12.5.** The stochastic process $(U(x))_{x\in[0,1]}$ generated by a random objective $U$ is a *measurable process* if the map

$$(\omega, x) \mapsto U(\omega)(x)$$

is measurable as a function from $\left(\Omega \times [0,1], \Sigma \otimes \mathcal{B}([0,1])\right)$ to $\mathcal{B}(\mathbb{R})$.

**Proposition 12.1.** *The process $(U(x))_{x\in[0,1]}$ is a measurable process if and only if the $\sigma$-algebra used to measure $Y^X$ has the Baire $\sigma$-algebra as a subset.*

*Proof.* Notice that the function $\Xi_Y$ is essentially the same as the process map $(\omega, x) \mapsto U(\omega)(x)$ when $|\mathcal{T}| = 1$. Since $\Xi_Y$ is Baire-measurable, the process is measurable as well. Conversely, a countable number of copies of the measurable processes can be stacked to define $\Xi_Y$ for $\mathcal{T}$ countable, so that the Baire-measurability of $\Xi_Y$ implies that at least the Baire sets are measurable.

The purpose of Proposition 12.1 is to show that the restriction to measurable processes is identical to the assumption that $\mathbb{P}$ induces a Baire measure on $Y^X$, so that the results below applying to Baire measures likewise apply to measurable processes.

Auger and Teytaud sought to show that NFL could not exist in continuous domains under reasonable assumptions on the measure [13]. As Theorem 12.4 and Proposition 12.1 show, measurability as a process is not enough to exclude NFL, and so stronger hypotheses are required.

Their choice of hypothesis was the requirement that the process have a *proper median*. A *proper median* is a real number such that half of the objective values are strictly above this number and half strictly below it, with the proper median itself having zero measure.[5] This median is held in common among all objectives sampled from a fitness measure. Thus it is not enough that each objective function individually have a median value with zero measure; every objective function must also have the same median value.

Auger and Teytaud offered a proof that if $\mathbb{P}$ produces a measurable process with a proper median, then NFL results such as those enumerated in Theorem 12.4 could not hold. The less specific phrasing that "Continuous Lunches are Free!" was selected as the title of their paper.

It is not clear why the authors believed that this proper median requirement was a legitimate requirement to impose on randomized objective function, but it was critical to their proof. As will be seen below, removing the proper median assumptions makes it possible to demonstrate NFL; the most trivial example is an almost surely constant fitness function, which is ruled out by the requirement that the proper median have zero measure. Additionally, these assumptions should be removed, because they require global information about the objective and do not naturally arise from the optimization setting in which oracle access to the objective function is assumed. Even with the proper median constraint, there are serious questions about the correctness of their proof that proper medians prevent NFL.

Alabert et al. studied the same setting, but exchanged the proper median requirement for a requirement that a random fitness be square integrable [4]. They demonstrated that all square integrable processes that satisfy the NFL property are constant almost everywhere on $[0,1]$ with respect to the Lebesgue measure.

This line of research reveals that NFL does not seem to coexist easily when the values assigned to neighboring search points become entangled. The proper median requirement is merely one form of such entanglement. Square integrability is a more natural concept than a proper median, but it plays the same essential role: It ties together the values in one part of the search domain with values in another part of the domain. In this case, square integrability places a limit on how fast the process variance can grow. A more explicit form of entanglement, namely, Lipschitz continuity, has likewise been shown to eliminate NFL [83]. What all these researchers have correctly identified is that NFL reflects a lack of interdependence among the values assigned to search points. Thus, to violate NFL, one need only introduce such interdependences by some means. This concept is formalized as *path independence* later in this chapter, and its implications are explored further at that time.

In any case, it is worthwhile to consider under what conditions NFL can exist rather than to simply declare that it does not exist. The non-existence of NFL does not automatically imply the existence of a master algorithm for search or optimization. In fact, NFL is only one part of a broader duality theory explored in Chapter 17 that reaffirms the NFL-based conclusion that certain methods are aligned with cer-

[5] *I.e.*, the set of points on which the objective function takes the value of the proper median must have Lebesgue measure zero, and the set of points on which the objective function is greater than the proper median must have Lebesgue measure $\frac{1}{2}$.

tain problems. If one wishes to solve a particular class of problems, no such master algorithm can generally outperform a tailored, problem-specific solution.

The next section continues the discussion of probabilistic NFL, developing the concept of path independence. But first, some discussion of the intuitive nature of the probabilistic setting is needed in light of both set-theoretic and probabilistic NFL results.

### 12.1.4 Search Problems

The detailed discussion of prior NFL results risks losing the lede. For Wolpert and Macready, the intuitive goal was to show that algorithms *pay for good performance* in some sense. They formalized this concept as an average over performance with respect to all objective functions, but one should not confuse the formalism with the goal. In fact, subsequent formalisms have narrowed rather than expanded the content, diffusing the clarity of the original NFL claims.

By pushing NFL from a statement about average performance across problems to a statement about producing a constant set of performance vectors (set-theoretic NFL) or a constant distribution over performance vectors (probabilistic NFL), NFL becomes a special case, whereas the narrative thrust of Wolpert and Macready's work was focused more around the need for aligning algorithms with problem classes, with the core NFL theorem merely proving that there was no effective way to align one algorithm with all problems.

Despite this loss of focus, the theory surrounding NFL has certainly revealed the absurdity of trying to address all problems with one solution. A class of permutation-invariant objectives leaves no room for drawing any conclusions based on the search history as it progresses, and random objectives that are conditionally independent of the search trajectory are likewise inscrutable. NFL is perhaps best understood as merely the problem class to which blind random search is best aligned. This concept will be made explicit in due course.

Furthermore, formalizing NFL required a development of formal interpretations of what the terms *algorithm* and *problem* mean. Chapter 3 introduced what this text means by the terms *search method* and *search problem*, but the previous chapters have focused more on defining methods than problems. In order to provide a more complete exploration of the performance of different search methods, search problems need to be rigorously explored as well.

In the last several chapters, the search problem was represented as a fixed static objective function $u : X \to Y$, and the interaction of a search method and a search problem was modeled as by a static search generator. In this chapter, we have already seen more general forms of search problems, such as the random objective $U$ in the last section.

Moving forward, we suppose as a first step that the static objective is not fixed, but is chosen stochastically. This situation has an intuitive interpretation. Suppose that one is faced with a problem that one wishes to solve by search or optimization, where the objective value of a proposed solution can only be obtained by evaluation. Suppose further that the evaluations are reliable; evaluating a single proposed solution always returns the same result. In this case there is some objective function that governs the observed values, but the objective is unknown. One then wishes to know how the objective is to be chosen (*e.g.*, uniformly at random for Wolpert and Macready) and more importantly what can be known about the objective within the oracle context.

If the search domain is larger than the number of evaluations performed, the objective function is unknowable. There are questions about the objective functions that simply cannot be answered by evaluating proposed solutions. Is the objective continuous, or, will solutions similar to those already evaluated perform similarly? Has the true optimum been found, or is there a more optimal solution still to be evaluated? If there is a better solution, is it similar to solutions previously evaluated? To answer each of these question, information about the objective is required that cannot be observed by evaluating less than every point in the search domain. Each question requires prior knowledge about the objective.

Mathematically, these limitations correspond precisely to the restrictions on observations arising from the Baire $\sigma$-algebra over $Y^X$. As discussed previously, the Baire $\sigma$-algebra is the smallest $\sigma$-algebra within which all finite projections are measurable. Equivalently, the Baire $\sigma$-algebra permits one to measure the results of countably many evaluations about an objective $u \in Y^X$ but does not allow global information about uncountable search domains, such as whether $u$ is continuous, or whether the infimum $u_*$ exists and what it might be. For these reasons, this chapter models search problems as an unsigned measure defined on the Baire $\sigma$-algebra for $Y^X$. Naturally, other conceptions of fitness measures can be considered as well. In order to represent the case where some global information is available, we also introduce a concept of a fitness measure defined on the Borel $\sigma$-algebra, within which individual objective functions are measurable.

**Definition 12.6 (Fitness Measure).** A *fitness measure* is an unsigned measure over the space $Y^X$ of objective functions. If the measure is defined on the Baire $\sigma$-algebra, it is a *Baire fitness measure*. If it is defined on the Borel $\sigma$-algebra, it is a *Borel fitness measure*.

Every Borel fitness measure restricts trivially to an equivalent Baire fitness measure, but if either the search domain $X$ or the value space $Y$ are uncountable, Baire fitness measures are not automatically equivalent to Borel fitness measures. As discussed in Chapter 4, every Baire fitness measure can be extended to a unique regular Borel fitness measure if and only if $Y^X$ is a Mařík space.

In probabilistic NFL, the random objective $U : \Omega \to Y^X$ was defined on a probability space $(\Omega, \Sigma, \mathbb{P})$. The introduction of the surrogate space $\Omega$ adds a layer of indirection that seems based more in tradition than necessity. Consequently, the remainder of this chapter discards $\Omega$ in favor of defining fitness measures directly. The relationship between the two settings is simple: The pushforward measure $\mathbb{P} \circ U$ defines a fitness measure. Or, as an alternative, one may take $\Omega = Y^X$ and then set $U : Y^X \to Y^X$ to be the identity function so that $\mathbb{P}$ itself is a fitness measure.

Some remarks should be made about the nature of the value space $Y$. Whereas the preceding chapters on optimization and performance assumed that $Y$ was either totally ordered or a vector space, this chapter does not make such assumptions and requires only that $Y$ be a Hausdorff topological space. Thus $Y$ could be a discrete space for search, a partially ordered vector space for multi-objective optimization, or even the a space containing gradient information for gradient-based optimization. Because this flexibility is possible, NFL then appears to be a property that applies to a broader category than just search methods or optimization methods. It is, in truth, a symmetry property on the space of fitness measures defined above, as will be proven below.

Returning to the discussion above, the Baire fitness measure corresponds to the case in which one has strict oracle access to the objective function, perceiving the attributes of the function only through evaluations. The Borel fitness measure makes it possible to observe the entire objective function as a whole, so that questions such as continuity or minima may become answerable. However, even in the Borel case, such questions cannot be answered by evaluation alone but require a different type of measurement.

In retrospect, it is easy to see why NFL holds in so many settings. The search method observes objective values by presenting points in the search domain for evaluation. In effect, when the objective function is chosen uniformly at random, the value at any search point is likewise chosen uniformly at random as well. Consequently, the observed value cannot give any information about what values will be observed at other search points, because it is statistically independent of these values. Any search method is rendered impotent, because the observed information is hidden behind a screen of randomness.

The next section defines NFL as a property of fitness measures building on the results described above. The specific goal is to make precise what a "screen of randomness" means, which will be accomplished by the definition of path independence, a form of conditional independence in which the fitness measure necessarily decouples from the search trajectory.

## 12.2 NFL as Path Independence

Using measure theory, a necessary and sufficient criterion for NFL can be developed that identifies NFL as a property of fitness measures, specifically, the property that the observed value at any search point is statistically independent of all other observations at any other point. This property will be called *path independence*, and it can be defined without referring to search algorithms at all. The NFL property itself will be defined based on the equivalences noted in Theorem 12.4, but with the concept of *non-repetition* replaced by filtering on the unique stopping sequence as defined in Definition 10.3. These definitions will be stated for the static search trajectors as used in the last several chapters, but with the twist that the temporal index set $\mathcal{T}$ will be generalized to any totally ordered set. The path independence result, termed the NFL Identification Theorem, will result from an examination of the stochastic process of fitness values arising from these trajectors.

### 12.2.1 Baire and Borel Settings

The first step towards NFL is to define the domain of objects to which the theorem will apply. The concept of a static search trajector has been extensively explored in Chapter 9. In this chapter, a setwise measurability criterion is added that reflects that performance will now be integrated over objectives.

**Definition 12.7 (Setwise Measurable).** For Hausdorff $X$ and $Y$, a Baire static search trajector $\mathcal{A} : Y^X \to \mathcal{M}[\mathcal{B}a_{X^{\mathcal{T}}}]$ is *setwise measurable* if the map $u \mapsto \mathcal{A}[u](E)$ is Baire-measurable as a function from $Y^X$ to $\mathbb{R}$ for all Baire sets $E$. A Borel static search trajector is setwise measurable if the same map is Borel-measurable for all Borel sets $E$.

A setwise measurable trajector can be integrated with respect to objectives. In particular, if $\mathcal{A}$ is a Baire (or Borel) static search trajector and $\mathbb{F}$ is a Baire (or Borel) fitness measure[6], then a Baire (or Borel) product measure $\mathcal{A} \times \mathbb{F}$ can be defined on $X^{\mathcal{T}} \times Y^X$. For any Baire (or Borel) sets $E \subseteq X^{\mathcal{T}}$ and $G \subseteq Y^X$, this measure is defined so that

$$\mathcal{A} \times \mathbb{F}(E \times G) = \int_G \mathcal{A}[u](E)\, \mathbb{F}(du), \tag{12.8}$$

which extends uniquely to a complete Baire (or Borel) measure by the Carathéodory Extension Theorem. This definition depends on setwise measurability, without which $\mathcal{A}[u](E)$ cannot serve as an integrand. A simpler definition that works for broader problem classes is possible using a different definition of a trajector, to be presented in Chapter 17, but this measure suffices to prove the NFL results of this chapter.

In order to obviate the repeated statements of whether the trajector and the fitness measure are Baire or Borel, a setting can be defined to encapsulate these ideas. In doing so, it will also be convenient to incorporate some constraints on the temporal index set $\mathcal{T}$ as well.

**Definition 12.8 (Baire Setting).** A triple $(\mathcal{A}, \mathbb{F}, \mathcal{T})$ consisting of a static search trajector, a fitness measure, and a temporal index set is called a *Baire setting* if $\mathcal{A}$ is a Baire static search trajector, $\mathbb{F}$ is a Baire fitness measure, and $\mathcal{T}$ is countable.

The association of a Baire setting with a countable index set, *i.e.*, $\mathcal{T} = \mathbb{N}$, reflects the fact that only countable projections of the search history $X^{\mathcal{T}}$ are Baire measurable and only countably many objective values can be observed from the Baire fitness measure, so that there is no need or use for a larger index set in this setting.

[6] Definition 12.6

**Definition 12.9 (Borel Setting).** A triple $(\mathcal{A}, \mathbb{F}, \mathcal{T})$ consisting of a static search trajector, a fitness measure, and a temporal index set is a called a *Borel setting* if $\mathcal{A}$ is a Borel static search trajector, $\mathbb{F}$ is a Borel fitness measure, and $\mathcal{T}$ is first countable.[7]

In the Borel setting, the index set may be uncountable, reflecting the fact that entire objective functions may be observed. However, in order to prove theorems, there will still be a need to reduce the index set to a countable subset by sampling. Forcing $\mathcal{T}$ to be first countable guarantees that countable approximations can be developed at any point in time. In particular, we have in mind uncountable index sets as $\mathcal{T} = [1, \infty)$, in which the optimization process unrolls in continuous time.

Two final points of notation are required for the expanded cardinality of $\mathcal{T}$. First, $\mathcal{T}$ is required to be well-founded and have a least element, which will be universally denoted by 1. Also, because $\mathcal{T}$ may be smaller or larger than the search domain $X$ and the value space $Y$, the notation $\mathcal{T}_X$ and $\mathcal{T}_Y$ is introduced to represent index sets that have the same cardinality as $X$ and $Y$, respectively.

### 12.2.2 The Fitness Process

The concept of a distribution $\mathbb{Y}_{\mathcal{A}}$ over performance vectors was introduced in Section 12.1.3. In this section, the same concept is expanded and adapted for fitness measures, deriving a stochastic process called the *fitness process* corresponding to the sequence of objective values observed at unique search points. The derivation is more complicated than before because the concept of a non-repeating search method is generalized to cover *eventually non-repeating* search methods that may repeat themselves for a finite period of time. As with probabilistic NFL, the label NFL will be applied when the fitness process is independent of the search method $\mathcal{A}$.

Suppose that $z \in X^{\mathcal{T}}$ is a sequence of search points. For each $t \in \mathcal{T}$, define $W_t^z : Y^X \to Y$ so that $W_t^z(u) = u(z_t)$ is the projection of $u$ onto the $t^{th}$ coordinate in $z$. The function $W_t^z$ is Borel and Baire measurable for all $t \in \mathcal{T}$. A stochastic process $W^z$ can now be defined for each $z \in X^{\mathcal{T}}$ by

$$W^z = (W_t^z)_{t \in \mathcal{T}}, \tag{12.9}$$

where each element of the process is a random variable from $(Y^X, \mathcal{B}a_{Y^X})$ to $(Y, \mathcal{B}_Y)$. Recall that a stochastic process is just a collection of indexed random variables (i.e., measurable functions).

When $z$ is generated by some search method, there are two sources of randomness, a trajector $\mathcal{A}$ and a fitness measure $\mathbb{F}$. Thus we consider Baire probability measures on the space $X^{\mathcal{T}} \times Y^X$. With this setting, for all $t \in \mathcal{T}$, define $W_t : X^{\mathcal{T}} \times Y^X \to Y$ so that $W_t(z, u) = u(z_t)$ as before. Once again, $W_t$ is both Borel and Baire measurable. To see why, note that $W_t$ is a composition of three maps

[7] That is, $\mathcal{T}$ has the property that every interval $(s,t) \subseteq \mathcal{T}$ contains a countable increasing sequence $(k_n)_{n \in \mathbb{N}}$ such that $\lim_n k_n = t$.

$(z,u) \mapsto (z_t,u) \mapsto (z_t,u(z_t)) \mapsto u(z_t)$, each of which is a finite projection and thus Baire-to-Baire or Borel-to-Borel measurable. Their composition is therefore both Borel and Baire measurable. Then there is a collection of Baire measurable random variables $W$ given by

$$W = (W_t)_{t\in\mathcal{T}}, \tag{12.10}$$

and $W$ is a stochastic process with values in $Y$. It will be called the *history process* since it contains the entire history trace of a search or optimization method run on a problem. Notice that the definition of $W$ does not depend on any trajector or fitness measure.

Given either a Baire or Borel setting $(\mathcal{A},\mathbb{F},\mathcal{T})$, each random variable $W_t$ in the history process induces a Borel measure on $Y$. This measure determines the fitness value of the $t^{th}$ search point stochastically. This process needs to be filtered for unique search points, which will replace the concept of a non-repeating search method.

Stopping times and filtrations were introduced in Section 9.3, including the natural filtration of a process, which is a sequence of minimal $\sigma$-algebras against which each successive variable of a stochastic process is measurable. This filtration is understood to contain the information generated by the process for reasons that were discussed in Section 9.3. Thus for the process $(W_t)_{t\in\mathcal{T}}$, the natural filtration $(\mathcal{W}_t)_{t\in\mathcal{T}}$ is defined such that each $\mathcal{W}_t$ is the smallest $\sigma$-algebra that makes $W_s$ a measurable function for each $s \leq t$. Intuitively, $\mathcal{W}_t$ represents the information that can be observed from the history process up to time $t$.

Reviewing the definitions, a stopping time of the history process is a function $T : X^{\mathcal{T}} \times Y^X \to \mathcal{T} \cup \{\top\}$ such that for all $t \in \mathcal{T}$, the set

$$\{(z,u) \in X^T \times Y^X \mid T(z,u) \leq t\}$$

is a measurable event in the $\sigma$-algebra $\mathcal{W}_t$. That is, it may not be possible to identify in advance when a stopping time will stop, but it is always possible to determine whether it has already stopped. The element $\top$ is a special element disjoint from $\mathcal{T}$ and greater than any element in $\mathcal{T}$ included to account for events that never happen. The set $\{T < \top\}$ replaces the notation $\{T < \infty\}$ from Section 9.3 because with the expanded cardinality of $\mathcal{T}$ is possible to have $\infty \in T$; the semantics of the $\{T < \top\}$ are otherwise the same.

Next, the unique coordinates are identified by the unique stopping sequence from Definition 10.3. Let $\# : X^{\mathcal{T}} \times \mathcal{T} \to \mathcal{T}_X$ be a function mapping a history $z$ and an index $s$ to the number of unique elements of $X$ in $z_{[1,s]}$. The set $\mathcal{T}_X$ is used because the number of unique elements is limited by the size of $X$. If $\#(z,s) = t$, then there are $t$ unique elements in $z$ in the first $s$ indices of $z$ (including index $s$). The unique stopping sequence $(T_t)_{t\in\mathcal{T}_X}$ such that $T_t(z,u) = \inf\{s \in \mathcal{T} \mid \#(z,s) = t\}$. If the set under the infimum is empty, then $T_t(z,u) = \top$ by convention. In either the Borel or the Baire setting, each $T_t$ is a stopping time since $\#(z,s) = t$ can be determined by examining the first $s$ coordinates of $z$ (Theorem 12.5). Furthermore, the unique stopping times are ordered. If $j < t$, then $T_j < T_t$. Every stopping time induces a $\sigma$-algebra that includes events prior to the stopping time as discussed in Section 9.3.

For the stopping time $T_t$, let $\mathcal{S}^j_{T_t} = \{(z,u) \in X^{\mathcal{T}} \times Y^X \mid T_t(z,u) \leq j\}$ be the set on which $t$ unique search points have been observed by time $j$. The stopped $\sigma$-algebra for the history process after $t$ unique points is given by

$$\mathcal{W}_{T_t} = \left\{ A \in \mathcal{B}a_{X^{\mathcal{T}}} \otimes \mathcal{B}a_{Y^X} \,\middle|\, A \cap \mathcal{S}^j_{T_t} \in \mathcal{W}_j \,\text{for all}\, j \in \mathcal{T} \right\} \tag{12.11}$$

for the Baire setting. In a Borel setting $\mathcal{B}a_{X^{\mathcal{T}}} \otimes \mathcal{B}a_{Y^X}$ is replaced by $\mathcal{B}_{X^{\mathcal{T}}} \otimes \mathcal{B}_{Y^X}$.

Next, it needs to be shown when there is also a random variable

$$W_{T_t} : X^{\mathcal{T}} \times Y^X \to Y \tag{12.12}$$

that maps the history process $W$ to the first coordinate $W_s(z,u)$ such that $\#(z,s) = t$. The variable $W_{T_t}$ represents the $t^{th}$ unique search point in the search history. Its existence indicates that the trajectory of unique objective values can be measured, which is required to define NFL. If $W_{T_t}$ exists, it would be $\mathcal{W}_{T_t}$-measurable. However, there are two potential problems. First, the set $\mathcal{S}^j_{T_t}$ must be proven to be $\mathcal{W}_j$-measurable (that is, $T_t$ must be Borel or Baire measurable). Secondly, $W_{T_t}$ must be proven to be measurable. Finally, the event that $T_t = \top$ needs to be considered. Each of these issues is dealt with in turn.

**Theorem 12.5.** *For all $t \in \mathcal{T}_X$, the stopping time $T_t$ is Borel measurable in the Borel setting and Baire measurable in the Baire setting with $\mathcal{W}_t$ as the target $\sigma$-algebra for measurability in either case.*

*Proof.* To prove that $T_t$ is Borel (or Baire) measurable, it suffices to show that for all $s$ the sets $\{(z,u) \mid T_t(z,u) < j\}$ and $\{(z,u) \mid T_i(z,u) = j\}$ are Borel (or Baire) measurable, since these sets generate the Borel $\sigma$-algebra for the order topology on $\mathcal{T}_X$. The set $\{(z,u) \mid T(z,u) = \top\}$ is another special case, but since this set is the complement of $T^{-1}(\mathcal{T}_X)$, its measurability follows automatically when the above sets are proven measurable.

For the first type of set, note that $\{(z,u) \mid T_t(z,u) \leq 1\}$ is either the entire space (if $t = 1$) or the empty set (if $t > 1$) and is therefore Baire and Borel measurable. Assume for transfinite induction that $\{(z,u) \mid T_k(z,u) \leq j\}$ is Borel (or Baire) measurable for all $k < t$. Note the identities

$$\{(z,u) \mid T_t(z,u) < j\} = \bigcup_{k<t} \{(z,u) \mid T_k(z,u) \leq j\} = \bigcup_{k_n \uparrow t} \{(z,u) \mid T_{k_n}(z,u) \leq j\}. \tag{12.13}$$

In the Borel setting the existence of a countable increasing sequence $k_n \to t$ was used to reduce a potentially uncountable union to a countable one on the far right in Equation 12.13 based on the assumption that $\mathcal{T}$ is first countable. In the Baire setting $(k_n)_{n\in\mathbb{N}}$ simply enumerates the interval $[1, j)$. Because the rightmost union above is a countable union of measurable sets, the set on the far left of Equation 12.13 is Borel (or Baire) measurable. To complete the induction, it remains to show that $T^{-1}(\{j\}) = \{(z,u) \mid T_t(z,u) = j\}$ is measurable.

Define $g_j$ and $\#_j$ so that $X^{\mathcal{T}} \times Y^X \xrightarrow{g_j} X^{[1,j]} \xrightarrow{\#_j} [1,j]$ where $g_j(z,u) = z_{[1,j]}$ and $\#_j(z)$ counts unique elements for $z \in X^{[1,j]}$. Applying these maps in sequence yields $T_t^{-1}(\{j\}) = (g_j \circ \#_j)^{-1}(\{i\})$ since both indicate the set of sequences indexed by $[1,j]$ that contain $t$ unique elements. The measurability of $T_t^{-1}(\{j\})$ will follow if for all $j \leq t$ the $g_j$ are Borel (or Baire) measurable and the set $(g_j \circ \#_j)^{-1}(\{t\})$ is Borel measurable. The functions $g_j$ are projections with the same cardinality as $[1,j]$. In the Baire setting, $[1,j]$ is at most countable, and so the $g_j$ are Baire measurable. In the Borel setting, all projections are Borel measurable. Finally, $\#_j$ is a continuous function from $X^{[1,j]}$ to $\mathcal{T}$ with the order topology. It is therefore Borel measurable, and since $\{t\}$ is a closed (and thus Borel) set, so is $\#_j^{-1}(\{t\})$.

Given that $T_t$ is measurable, Proposition 12.2 below demonstrates the measurability of $W_{T_t}$ whenever all $T_t$ take on values inside $\mathcal{T}_X$ with probability one.

**Definition 12.10 (Non-Repeating Setting).** A Borel or Baire setting $(\mathcal{A}, \mathbb{F}, \mathcal{T})$ is *non-repeating* if $T_t = t$ with $\mathcal{A} \times \mathbb{F}$-probability one for all $t \in \mathcal{T}_X$. The setting is *eventually non-repeating* if $T_t < \top$ with $\mathcal{A} \times \mathbb{F}$-probability one for all $t \in \mathcal{T}_X$.

**Proposition 12.2.** *If a Borel or Baire setting $(\mathcal{A}, \mathbb{F}, \mathcal{T})$ is eventually non-repeating, each element of the sequence $U = (W_{T_t})_{t \in \mathcal{T}_X}$ is a $Y$-valued random variable, and $U$ is a stochastic process with natural filtration $(\mathcal{W}_{T_t})_{t \in \mathcal{T}_X}$.*

*Proof.* The claim is that $W_{T_t}$ is $\mathcal{W}_{T_t}$-measurable for all $t$ with $|[1,t]| \leq |X|$. For every $E \in \mathcal{B}_Y$ and every $k \in \mathcal{T}_X$, define sets

$$Z(E) = \left\{(z,u) \in X^{\mathcal{T}} \times Y^X \mid u(z_{T_t(z,u)}) \in E\right\}$$
$$Z(k) = \left\{(z,u) \in X^{\mathcal{T}} \times Y^X \mid T_t(z,u) \leq k\right\}. \tag{12.14}$$

From Equation 12.11, the claim will hold if $Z(E) \cap Z(k)$ is in $\mathcal{W}_k$. Define the capped stopping time $S_t^k(z,u) = \min\{T_t(z,u), k\}$, which is a stopping time and is measurable because $T_t$ is measurable by Theorem 12.5. But whereas $T_t$ is measurable with respect to the whole process $W$, $S_t^k$ is $\mathcal{W}_k$ measurable because it only depends on the first $k$ indices of $W$.

Because $T_t$ lies in $\mathcal{T}_X$ with probability one for all $t \in \mathcal{T}_X$, it suffices to show that the stopped projection $(z,u) \mapsto u(z_{S_t^k(z,u)})$ is Borel or Baire measurable (depending on the setting) as a map to $\mathcal{W}_k$ for any $k \in \mathcal{T}_X$. First, the map $(t,z,u) \mapsto u(z_t)$ is a composition of measurable projections and is therefore measurable to $\mathcal{W}_k$ for every $k \in \mathcal{T}$ (where the first argument has $t \leq k$ and measurability in the first argument is Borel with respect to the order topology). Second, the map $(z,u) \mapsto (S_t^k(z,u), z, u)$ is measurable due to the measurability of $S_i^k$. The composition of these two maps is $(z,u) \mapsto u(z_{S_t^k(z,u)})$, which must be measurable as the composition of measurable maps.

The index $t \in \mathcal{T}$ was arbitrary, so the sequence $U = (\mathcal{W}_{T_t})_{t \in \mathcal{T}_X}$ is a collection of random variables, as desired.

The stochastic process $U$ will be called the *fitness process* of an optimizer $\mathcal{A}$ on a fitness measure $\mathbb{F}$. Let $U_t = W_{T_t}$. The fitness process induces a measure $\mathbb{Y}_{\mathbb{F}}^{\mathcal{A}}$ on $Y^{\mathcal{T}_X}$ derived from $\mathcal{A} \times \mathbb{F}$. This measure is Baire or Borel depending on the setting. First define $\Phi_U : X^{\mathcal{T}} \times Y^X \to Y^{\mathcal{T}_X}$ so that $(\Phi_U(z,u))_t = U_t(z,u)$, and then for each measurable set $E \subseteq Y^{\mathcal{T}_X}$

$$\mathbb{Y}_{\mathbb{F}}^{\mathcal{A}}(E) = \mathcal{A} \times \mathbb{F}\left(\Phi_U^{-1}(E)\right), \quad \text{that is,} \quad \mathbb{Y}_{\mathbb{F}}^{\mathcal{A}} = (\mathcal{A} \times \mathbb{F}) \circ \Phi_U, \tag{12.15}$$

which is the pushforward measure of $\Phi_U$ with respect to $\mathcal{A} \times \mathbb{F}$. This construction is a standard way of obtaining a measure on a product space from a stochastic process. The function $\Phi_U$ is measurable whenever $U$ is; such constructions will be used again below. NFL is formulated in the next section as a statement about the fitness process and its induced measure.

### 12.2.3 Path Independence

Using the fitness process, NFL for fitness measures can now be defined analogously to Theorem 12.4 as the case in which the induced measure of the fitness process $\mathbb{Y}_{\mathbb{F}}^{\mathcal{A}}$ is constant. This NFL property is proven equivalent to path independence, a more general criterion requiring the probability over fitness trajectories to be constant across search paths.

**Definition 12.11 (Probabilistic NFL).** A Borel (or Baire) fitness measure $\mathbb{F}$ is *of class P-NFL* on an index set $\mathcal{T}$ if the map $\mathcal{A} \to \mathbb{Y}_{\mathbb{F}}^{\mathcal{A}}$ is constant for all proper static search trajectors $\mathcal{A}$ such that the setting $(\mathcal{A}, \mathbb{F}, \mathcal{T})$ is an eventually non-repeating Borel (or Baire) setting.

That is, a Borel (or Baire) fitness measure $\mathbb{F}$ is of class P-NFL if its fitness process has the same distribution for every eventually non-repeating Borel (or Baire) optimizer, with limitations on the temporal index set determined by the choice of Borel or Baire measures. To demonstrate that this definition of NFL is a reasonable generalization, let $\hat{V}$ be the value kernel of some value-dependent performance criterion and let $\mathbb{F}$ be a fitness measure. Let $\mathcal{A}$ and $\mathcal{A}'$ be any pair of static search trajectors such that $(\mathcal{A}, \mathbb{F}, \mathcal{T})$ and $(\mathcal{A}', \mathbb{F}, \mathcal{T})$ are both non-repeating settings. Then consider the equation

$$\int_{X^{\mathcal{T}} \times Y^X} \hat{V}(u(h_X))\, \mathcal{A} \times \mathbb{F}(dh_X, du) = \int_{X^{\mathcal{T}} \times Y^X} \hat{V}(u(h_X))\, \mathcal{A}' \times \mathbb{F}(dh_X, du) \tag{12.16}$$

where $u(h_X)$ applies $u$ to each element of $h_X$ in order. This equation is a generalization of Equation 12.5, most easily seen through comparison with Equation 12.6. As with probabilistic NFL, the integrator in Equation 12.16 can be replaced with the pushforward measure $\mathbb{Y}_{\mathbb{F}}^{\mathcal{A}}$ to yield the simplified version

$$\int_{Y^{\mathcal{T}}} \hat{V}(h_Y)\, \mathbb{Y}_{\mathbb{F}}^{\mathcal{A}}(dh_Y) = \mathbb{E}_{\mathbb{Y}_{\mathbb{F}}^{\mathcal{A}}}\left[\hat{V}\right] = \mathbb{E}_{\mathbb{Y}_{\mathbb{F}}^{\mathcal{A}'}}\left[\hat{V}\right] = \int_{Y^{\mathcal{T}}} \hat{V}(h_Y)\, \mathbb{Y}_{\mathbb{F}}^{\mathcal{A}'}(dh_Y) \tag{12.17}$$

With these equations, Theorem 12.4 can be restated for the generalized setting, proving that class P-NFL does indeed implement an NFL result.

**Theorem 12.6.** *Suppose $\mathbb{F}$ is a Borel (or Baire) fitness measure. The following statements are equivalent:*

1. *Equation 12.16 holds for all value kernels $\hat{\mathbb{V}}$ and all pairs of setwise-measurable Borel (or Baire) static search trajectors $\mathcal{A},\mathcal{A}'$ such that $(\mathcal{A},\mathbb{F},\mathcal{T})$ and $(\mathcal{A}',\mathbb{F},\mathcal{T})$ are both eventually non-repeating Borel (or Baire) settings.*
2. *Equation 12.16 holds for all value kernels of the form $\mathbb{1}_A$ for $A \subset Y^{\mathcal{T}}$ measurable and all pairs of setwise-measurable Borel (or Baire) static search trajectors $\mathcal{A},\mathcal{A}'$ such that $(\mathcal{A},\mathbb{F},\mathcal{T})$ and $(\mathcal{A}',\mathbb{F},\mathcal{T})$ are both eventually non-repeating Borel (or Baire) settings.*
3. *$\mathbb{F}$ is of class P-NFL on $\mathcal{T}$.*

*Proof.* Statement (1) obviously implies (2). Equation 12.17 shows that

$$\mathbb{E}_{\mathbb{Y}^{\mathcal{A}}_{\mathbb{F}}}\left[\mathbb{1}_A\right] = \mathbb{Y}^{\mathcal{A}}_{\mathbb{F}}(A) = \mathbb{Y}^{\mathcal{A}'}_{\mathbb{F}}(A) = \mathbb{E}_{\mathbb{Y}^{\mathcal{A}'}_{\mathbb{F}}}\left[\mathbb{1}_A\right],$$

for all measurable $A$, which proves that (2) implies (3). Finally, if (3) holds, then Equation 12.17 follows, which then implies (1).

NFL implies that information extracted from any series of fitness evaluations is insufficient to suggest search points where more desirable fitness values may be found. Thus NFL is not really about algorithms as much as it is about search trajectories. Path independence captures this intuition based on non-repeating search histories. In order to handle cases where the cardinality of $\mathcal{T}$ is greater than that of $X$, a search history will be called non-repeating if it does not repeat elements of $X$ until all elements of $X$ have been exhausted.

**Definition 12.12 (Non-Repeating Search History).** A search history $z \in X^{\mathcal{T}}$ is non-repeating if for any pair of indices $i,j \in T$ with $i \leq j$, $i \neq j$ and $z_i = z_j$ together imply that $|[j]| > |X|$.

In Section 12.2.2, the stochastic process $W^z$ of projections $W^z_t(u) = u(z_t)$ was defined. As in Equation 12.15, for each $z$ and each Borel (or Baire) fitness measure, this process induces a Borel (or Baire) measure $\mathbb{W}^z_{\mathbb{F}}$ on $Y^{\mathcal{T}_X}$. To obtain this measure, define $\Phi_{W^z} : Y^X \to Y^{\mathcal{T}_X}$ so that $(\Phi_{W^z}(u))_t = W^z_t(u)$. Then let

$$\mathbb{W}^z_{\mathbb{F}}(E) = \mathbb{F}\left(\Phi^{-1}_{W^z}(E)\right) \tag{12.18}$$

for each event $E$ of $Y^{\mathcal{T}_X}$. A fitness measure is of class P-NFL if and only if this measure is independent of non-repeating $z$, called *path independence*.

**Definition 12.13 (Path Independence).** A Borel (or Baire) fitness measure $\mathbb{F}$ is path independent over an index set $\mathcal{T}$ if the map $z \mapsto \mathbb{W}^z_{\mathbb{F}}$ is constant for all non-repeating $z$. If $\mathbb{F}$ is not path independent, then it is path dependent.

The term "path independence" captures the intuition that there are no paths through the search space that provide more information about the fitness value of the next unique point. Only the number of fitness values requested matters.

Path independence is obviously related to class P-NFL, since it implies that whatever points an optimizer chooses to evaluate, the distribution over fitness values is the same. In fact, the two are equivalent. To prove this claim, path independence is first shown to be necessary for class P-NFL, then sufficient. These results are combined in the NFL Identification Theorem below.

To demonstrate necessity, deterministic trajectors are used, especially deterministic trajectors that are *objective-agnostic*, that is, independent of the fitness values. Both of these properties have already been defined in various contexts.[8] Section 12.1.2 introduced the notation $\xi(u)$ for the search path followed by a deterministic trajector and called it a *search operator*. In this section, this term will be replaced with the more general term *deterministic core*. A definition of *objective-agnostic* is also given for static search trajectors. It should be noted that if $\mathcal{G}$ is an objective-agnostic static search generator that extends to a static search trajector $\mathcal{A}$ then $\mathcal{A}$ is objective agnostic. The converse is not absolutely true, but it is true almost surely.

**Definition 12.14 (Deterministic Core).** Given a Borel (or Baire) static search trajector $\mathcal{A}$, a function $\xi : Y^X \to X^{\mathcal{T}}$ is the *deterministic core* of $\mathcal{A}$ if $\mathcal{A}[u](E) = \mathbb{1}_E(\xi(u))$ for each Borel (or Baire) set in $X^{\mathcal{T}}$.

**Definition 12.15 (Objective-Agnostic Trajector).** A Borel (or Baire) search trajector $\mathcal{A}$ is *objective-agnostic* if the map $u \mapsto \mathcal{A}[u]$ is constant.

**Definition 12.16 (Invariant Trajectory).** If a Borel (or Baire) static search trajector $\mathcal{A}$ is deterministic and objective-agnostic, then its deterministic core $\xi$ is constant and equal to some sequence $z \in X^{\mathcal{T}}$, called the *invariant trajectory* of $\mathcal{A}$.

**Proposition 12.3.** *If the index set $\mathcal{T}$ satisfies the conditions for a Borel (or Baire) setting, then for any $z \in X^{\mathcal{T}}$ there is a Borel (or Baire) static search trajector $\mathcal{A}_z$ that is deterministic and objective-agnostic and has $z$ as its invariant trajectory. Furthermore, for any Borel (or Baire) fitness measure $\mathbb{F}$, the triple $(\mathcal{A}_z, \mathbb{F}, \mathcal{T})$ is a Borel (or Baire) setting.*

**Proposition 12.4.** *Suppose $(\mathcal{A}, \mathbb{F}, \mathcal{T})$ is a Borel (or Baire) setting, and that $\mathcal{A}$ is deterministic and objective-agnostic with non-repeating invariant trajectory $z_0$. Then the fitness process has $\mathbb{Y}^{\mathcal{A}}_{\mathbb{F}} = \mathbb{W}^{z_0}_{\mathbb{F}}$.*

*Proof.* The definitions imply that $(\mathcal{A}, \mathbb{F}, \mathcal{T})$ is a non-repeating setting so that the fitness process $U$ and the history process $W$ coincide. For every Borel (or Baire) event $E \subseteq Y^{\mathcal{T}_X}$, the following equalities hold with $\mathcal{A} \times \mathbb{F}$-probability one:

[8] See Definition 6.20 for objective-agnostic static search generators and Definition 6.9 deterministic probability-valued functionals, which incorporates both generators and trajectors.

$$\begin{aligned}
\Phi_U^{-1}(E) &\overset{\mathcal{A}\times\mathbb{F}-a.s.}{=} \{(z_0,u)\,|\,\exists v\in E\, s.t.\, \forall t\in \mathcal{T}_X\; v_t = U_t(z_0,u)\}\\
&\overset{\mathcal{A}\times\mathbb{F}-a.s.}{=} \{(z_0,u)\,|\,\exists v\in E\, s.t.\, \forall t\in \mathcal{T}_X\; v_t = W_t(z_0,u)\}\\
&= \{z_0\}\times\left\{u\,\middle|\,\exists v\in E\, s.t.\, \forall t\in \mathcal{T}_X\; v_t = W_t^{z_0}(u)\right\},
\end{aligned}$$

where two sets are equal "$\mathcal{A}\times\mathbb{F}$-*a.s.*" if they have equal measure under $\mathcal{A}\times\mathbb{F}$. The last equality makes it possible to apply Equation 12.8 to $\{z_0\}\times\Phi_{W^{z_0}}^{-1}(E)$ since

$$\Phi_{W^{z_0}}^{-1}(E) = \left\{u\,\middle|\,\exists v\in E\, s.t.\, \forall t\in \mathcal{T}_X \quad v_t = W_t^{z_0}(u)\right\}.$$

This equation implies

$$\mathbb{Y}_{\mathbb{F}}^{\mathcal{A}}(E) = \mathcal{A}\times\mathbb{F}\left(\Phi_U^{-1}(E)\right) = \int_{\Phi_{W^{z_0}}^{-1}(E)} \mathcal{A}(u)(\{z_0\})\, d\mathbb{F}(u) = \mathbb{F}\left(\Phi_{W^{z_0}}^{-1}(E)\right) = \mathbb{W}_{\mathbb{F}}^{z_0}(E).$$

That is, $\mathbb{Y}_{\mathbb{F}}^{\mathcal{A}} = \mathbb{W}_{\mathbb{F}}^{z_0}$.

**Lemma 12.2.** *If a Borel (or Baire) fitness measure $\mathbb{F}$ is of class P-NFL on an index set $\mathcal{T}$, it is path independent over $\mathcal{T}$.*

*Proof.* Assume $\mathbb{F}$ is of class P-NFL on $\mathcal{T}$ and path dependent. Take $z_1, z_2 \in X^{\mathcal{T}}$ to be distinct non-repeating search histories such $\mathbb{W}_{\mathbb{F}}^{z_1} \neq \mathbb{W}_{\mathbb{F}}^{z_2}$. Such trajectories must exist since $\mathbb{F}$ is path dependent. Let $\mathcal{A}_1$ and $\mathcal{A}_2$ be fitness agnostic optimizers with invariant trajectory $z_1$ and $z_2$ respectively. Then Proposition 12.4 implies that $\mathbb{Y}_F^{\mathcal{A}_1} \neq \mathbb{Y}_F^{\mathcal{A}_2}$, which contradicts the claim that $\mathbb{F}$ is of class P-NFL.

**Lemma 12.3.** *If a Borel (or Baire) fitness measure $\mathbb{F}$ is path independent over an index set $\mathcal{T}$, then it is of class P-NFL.*

*Proof.* Suppose $E$ is any Borel (or Baire) subset of $Y^{\mathcal{T}_X}$. From Equation 12.15,

$$\mathbb{Y}_{\mathbb{F}}^{\mathcal{A}}(E) = \mathcal{A}\times\mathbb{F}\left(\Phi_U^{-1}(E)\right). \tag{12.19}$$

The proof centers on obtaining a Borel (or Baire) event $E_Y \subseteq Y^X$ such that $\Phi_U^{-1}(E) = X^{\mathcal{T}}\times E_Y$, for which $\mathcal{A}$ may be averaged out of Equation 12.19 using Equation 12.8. To this end, note that

$$\begin{aligned}
\Phi_U^{-1}(E) &= \{(z,u)\,|\,\exists v\in E\, s.t.\, \forall t\in \mathcal{T}_X\; v_t = U_t(z,u)\}\\
&= \left\{(z,u)\,\middle|\,\exists v\in E\, s.t.\, \forall t\in \mathcal{T}_X\; v_t = W_{T_t(z,u)}(z,u)\right\}\\
&= \left\{(z,u)\,\middle|\,\exists v\in E\, s.t.\, \forall t\in \mathcal{T}_X\; v_t = W_t^{(z_{T_t})_{t\in\mathcal{T}_X}}(z,u)\right\}.
\end{aligned}$$

In this last equation, $\tilde{z} = (z_{T_t})_{t\in\mathcal{T}_X}$ represents the sequence $z$ filtered for uniqueness. Now $\tilde{z}$ is non-repeating, so path independence implies that for any non-repeating sequence $a\in X^{\mathcal{T}_X}$, with $\mathbb{F}$- and therefore $\mathcal{A}\times\mathbb{F}$-probability one,

$$\Phi_U^{-1}(E) \overset{\mathcal{A}\times\mathbb{F}-a.s}{=} \{(z,u) \mid \exists v \in E\, s.t. \forall t \in \mathcal{T}_X \quad v_t = W_t^a(u)\}$$
$$= \quad X^{\mathcal{T}} \times \{u \mid \exists v \in E\, s.t. \forall t \in \mathcal{T}_X \quad v_t = W_t^a(u)\},$$

where in the second line the $z$ has been pulled outside the set since it does not appear on the right. Renaming this final set as

$$E_Y = \{u \mid \exists v \in E\, s.t. \forall t \in \mathcal{T}_X \quad v_t = W_t^a(u)\},$$

Equation 12.8 can be applied to obtain

$$\mathbb{Y}_{\mathbb{F}}^{\mathcal{A}}(E) = \int_{E_Y} \mathcal{A}(u)\left(X^{\mathcal{T}}\right) d\mathbb{F}(u) = \mathbb{F}(E_Y),$$

where the right-hand side is independent of $\mathcal{A}$. Therefore the map $\mathcal{A} \mapsto \mathbb{Y}_{\mathbb{F}}^{\mathcal{A}}$ is constant, and $\mathbb{F}$ is of class P-NFL on $\mathcal{T}$.

**Theorem 12.7 (NFL Identification Theorem).** *A Borel (or Baire) fitness measure $\mathbb{F}$ is of class P-NFL on an index set $\mathcal{T}$ if and only if it is path independent over $\mathcal{T}$.*

*Proof.* This result combines Lemma 12.2 and Lemma 12.3.

The NFL Identification Theorem characterizes probabilistic NFL as a statement about fitness measures. Such a characterization yields a concept equivalent to class P-NFL that does not depend at all on search methods or optimizers. If each path is identified with its histogram as done by Igel and Toussaint [81], then the necessity of path independence is either previously proven or at least strongly suggested for finite spaces by Igel and Toussaint [81] and for infinite spaces by Rowe et al. [136]. The proof above expresses path independence as a probabilistic concept, proving it is both a necessary and sufficient condition for NFL.

## 12.2.4 Comments of the Scope of the Class P-NFL

The NFL Identification Theorem is broader than previous NFL theorems, so it will prove worthwhile to consider what it asserts in greater detail. Firstly, the reduction to path independence demonstrates that one need only prove that an NFL result holds for blind search, and the result automatically generalizes to all methods. Secondly, this NFL result applies to all *eventually non-repeating* proper static search trajectors; which objects are non-repeating depends on the fitness measure. Thirdly, this result is not limited to *black-box* search methods, nor is a separate result needed for optimization as distinct from search, because with NFL being a property of a fitness measure as applied to individual search paths, there is no way to circumvent NFL by any algorithmic choice.

What path independence means more than anything else is that the values generated by the fitness measure cannot be controlled through the search path. This can be expressed in terms of statistical independence of the value trajectory from the search history. The measure $\mathcal{A} \times \mathbb{F}$ derived at the beginning of this section is a joint measure over search histories and objective functions, that is, it measures the space $X^{\mathcal{T}} \times Y^X$, but it can be used to induce a measure over the space $X^{\mathcal{T}} \times Y^{\mathcal{T}}$ of history traces. Specifically, suppose for simplicity that $\mathcal{A}$ is non-repeating and consider the map $H : (h_X, u) \mapsto (h_X, u(h_X))$, where once again $u(h_X) \in Y^{\mathcal{T}}$ is the value history resulting from applying $u$ to each element of $h_X$ in order. The map $H$ is measurable in the appropriate settings, and so the pushforward measure $(\mathcal{A} \times \mathbb{F}) \times H$ is well-defined over history traces.

Using the looser notation of statistics, let $\mathbb{P}_H = \mathbb{P}_H(h_X, h_Y)$ represent this joint probability measure. Notice that the measure $\mathbb{W}^z_{\mathbb{F}}$ above is nothing other than the conditional probability $\mathbb{P}_H(h_Y \mid h_X = z)$. If $\mathbb{F}$ is path independent, then the map $z \to \mathbb{W}^z_{\mathbb{F}}$ is constant, and so this conditional probability does not depend on $z$. Therefore

$$\mathbb{P}_H(h_Y \mid h_X) = \mathbb{P}(h_Y), \qquad \text{that is,} \qquad \mathbb{P}_H(h_Y, h_X) = \mathbb{P}_H(h_Y)\,\mathbb{P}_H(h_X).$$

Thus path independence is nothing other than statistical independence.

Another way of stating this independence is that since the value history is independent of the search history, blind search obtains equivalent results to more complicated algorithms on an NFL fitness measure. One can, in fact, consider $\mathbb{W}^z_{\mathbb{F}}$ to be the result of blindly choosing a search history without consulting any objective values. It then does not matter which history $z$ is chosen. In terms of duality, the class of P-NFL fitness measures forms the dual problem set for the class of blind search methods.

Generalizing NFL from non-repeating search methods to eventually non-repeating search methods required substantial technical machinery. It should be clarified what has been gained by this knowledge. Although the claim is often made – and was even made above – that a search method can cache previously guessed search points in order to avoid repetition, such caching is only practical in finite spaces. Even in large finite spaces, the size of such a cache is limited and will eventually be exceeded. Thus one really should consider what happens when there are at least some repetitions.

Consider the simple genetic algorithm as a simple case in which even the weaker requirement of eventual non-repetition can fail due to premature convergence (see Section 2.6.1). In premature convergence, the population collapses to a single individual, and a recombination-only approach will then repeat itself *ad infinitum*. In this case, fixes are available. One can detect population collapse and restart the algorithm, or else one can add mutation operators to make new points eventually reachable. Both of these fixes make the underlying search method eventually non-repeating rather than non-repeating. Thus the category of eventually non-repeating search methods is not vacuous, and it does contain objects that are not included in the class of non-repeating search methods.

The scope of this result is broader in other senses as well. At no point was a search objective or performance criterion invoked. It therefore does not matter whether the purpose is to find the minimum of an objective or to search through a large corpus of documents. The value space was likewise left open. It could, for instance, contain gradients rather than objective values. Therefore the NFL Identification Theorem applies equally to gradient-based methods as much as it does to black-box optimizers. Going further, NFL in this form applies not only to search and optimization, but to control as well. In particular, as path independence shows, NFL is the situation in control where the space of available control actions (the search domain) has no effect on the sensor observations (points in the value space). An NFL fitness measure then represents an *uncontrollable environment*. NFL can also be thought of in game theoretic terms as an *unwinnable game*; this perspective is explored in the next chapter. As a final point, the concept of path independence is not limited to the case where the objectives are static. With appropriate extensions to the definition of NFL, these results also carry over to stochastic and dynamic objective functions as well.

The above results have characterized class P-NFL, but they have not established whether any fitness measures belonging to this class exist. This is the question to which we now turn.

## 12.3 Existence and Construction of NFL

NFL fitness measures exist for every search domain and value space, as stated in Theorem 12.10 below. What is more interesting, however, is the nature of such fitness measures. Two main categories are developed in this section: almost surely constant NFL fitness measures, and fitness measures with identical coordinates constructed via the Kolmogorov extension theorem. Some objections to the nature of these kinds of fitness measures was raised in Section 12.1.3, and several broad classes of spatial regularity are known to block NFL, such as Lipschitz continuity or square integrability. In what follows, sufficient conditions for path independence are given, and broad vector spaces of NFL fitness measures are compared, concluding that NFL does in fact imply such randomness to a degree that one would almost never expect to observe in practical problems.

### 12.3.1 Construction of NFL Fitness Measures

According to the NFL Identification Theorem, every path independent fitness measure is of class P-NFL. But it is not obvious how to construct path independent fitness measures. In this subsection, it is demonstrated that path independence follows whenever all coordinate projections are identically distributed and mutually independent. This fact was initially proven by English [46, 47] for NFL in finite spaces, but it is extended below for arbitrary Hausdorff spaces. Critically, it implies that NFL priors exist in general.

Consider the collection of coordinate projections $\{F_x \mid x \in X\}$ such that $F_x : Y^X \to Y$ is given by $F_x(u) = u(x)$. Each $F_x$ is both Borel and Baire measurable. Given a

Borel (or Baire) fitness measure $\mathbb{F}$, each $F_x$ induces a measure $\mathbb{F}_x$ such that for any Borel subset $E$ of $Y$, $\mathbb{F}_x(E) = \mathbb{F}(\{u \mid u(x) \in E\})$. Furthermore, for any subset $D$ of $X$, a projection $F_D : Y^X \to Y^D$ is defined by $F_D(u) = (u(x))_{x \in D}$. Then $F_D$ is always Borel measurable and is Baire measurable if $D$ has at most countable cardinality. These projections also induce Borel measures $\mathbb{F}_D$ such that for an Borel subset $E$ of $Y^D$,

$$\mathbb{F}_D(E) = \mathbb{F}(\{u \mid \exists v \in E \, s.t. \forall x \in D, \, v(x) = u(x)\}). \tag{12.20}$$

When $D$ is finite, $\mathbb{F}_D$ is called a *finite-dimensional distribution* of $\mathbb{F}$. Next, two useful properties of $\mathbb{F}_x$ and $\mathbb{F}_D$ are introduced.

**Definition 12.17 (Identically Distributed Coordinates).** A Borel (or Baire) fitness measure $\mathbb{F}$ has identically distributed coordinates if for all $x \in X$ the map $x \to \mathbb{F}_x$ is constant.

**Definition 12.18 (Mutually Independent Coordinates).** A Borel (or Baire) fitness measure $\mathbb{F}$ has mutually independent coordinates if for any finite $D \subseteq X$ and any collection of Borel sets $\{E_x \mid x \in D\}$ of $Y$, the measure $\mathbb{F}_D$ factorizes as

$$\mathbb{F}_D\left(\prod_{x \in D} E_x\right) = \prod_{x \in D} \mathbb{F}_x(E_x). \tag{12.21}$$

Finite subsets are used in the definition of mutual independence to avoid a definition of uncountable products and because these are sufficient to prove Theorem 12.8 for Baire fitness measures and even most Borel fitness measures.

Fitness measures of class P-NFL can be constructed by guaranteeing tightness along with mutually independent and identically distributed coordinates. There is one limitation at present in that Borel fitness measures of class P-NFL can only be constructed in this way when $Y^X$ is a Mařík space.

**Theorem 12.8 (Construction of Path Independence).** *If a Baire fitness measure $\mathbb{F}$ has identically distributed and mutually independent coordinates and all of the finite-dimensional distributions of $\mathbb{F}$ are tight, then $\mathbb{F}$ is path independent over $\mathcal{T}$ and therefore of class P-NFL on $\mathcal{T}$. The same holds for a Borel fitness measure if $Y^X$ is a Mařík space.*

*Proof.* Suppose $\mathbb{F}$ satisfies the conditions. Let $D$ be any finite subset of $X$. Mutually independent and identically distributed coordinates together imply that for any collection of Borel sets $\{E_x \mid x \in D\}$ of $Y$,

$$\mathbb{F}_D\left(\prod_{x \in D} E_x\right) = \prod_{x \in D} \mathbb{F}_x(E_x) = \prod_{x \in D} \mathbb{F}_{x_0}(E_x), \tag{12.22}$$

where $x_0$ is a fixed element of $X$ independent of $D$. Importantly, Equation 12.22 uniquely determines $\mathbb{F}_D$ based on Carathéodory's extension theorem. Then $\{\mathbb{F}_D \mid D \subseteq X, D \, \text{finite}\}$ is a consistent family of finite-dimensional distributions. Each $\mathbb{F}_D$ is a coordinate restriction of a tight measure and is thus tight. By the Kolmogorov

Extension Theorem, there is a unique Baire measure $\mathbb{F}_X$ on $Y^X$ that extends the $\mathbb{F}_D$, and it must hold that $\mathbb{F}_X = \mathbb{F}$ since $\mathbb{F}_X$ is unique and its finite-dimensional distributions agree with those of $\mathbb{F}$.

Now let $\pi : X \to X$ be any permutation of $X$. For any subset $D$ of $X$, define $\pi(D) = \{\pi(x) \mid x \in D\}$. Define an isomorphism from $\psi_\pi : Y^X \to Y^{\pi(X)}$ so that $\psi_\pi(v) = v \circ \pi$, and note that the restriction $\psi_\pi^D$ of $\psi_\pi$ to $Y^D$ has range $Y^{\pi(D)}$. For arbitrary subsets $E$ of $Y^D$, define $D_\pi(E) = \{\psi_\pi^D(v) \mid v \in E\}$. Revisiting Equation 12.22, notice that for the collection $\{E_x \mid x \in D\}$ above,

$$D_\pi\left(\prod_{x\in D} E_x\right) = \prod_{x\in\pi(D)} E_{\pi^{-1}(x)} = \prod_{x\in D} E_x \quad \text{so} \quad \mathbb{F}_{\pi(D)} \circ D_\pi\left(\prod_{x\in D} E_x\right) = \prod_{x\in D} \mathbb{F}_{x_0}(E_x).$$

Due to uniqueness, it holds in general that $\mathbb{F}_D = \mathbb{F}_{\pi(D)} \circ D_\pi$.

It remains to identify the distribution $\mathbb{W}_{\mathbb{F}}^z$ with $\mathbb{F}_I$ for some $I \subseteq X$. The set $I$ need not be finite; it will in fact have $|I| = |\mathfrak{T}_X|$. All such $\mathbb{F}_I$ are uniquely defined as Kolmogorov extensions of some subset of $\{\mathbb{F}_D \mid D \subseteq X, D\,\text{finite}\}$. The extension theorem is necessary because $I$ may be uncountable and uncountable projections are not Baire measurable.

Now fix any $I \subseteq X$ such that $|I| = |\mathfrak{T}_X|$ (which may be strictly smaller than $|X|$). Choose $z \in X^{\mathfrak{T}_X}$ to be any non-repeating sequence such that $z_t \in I$ for all $t \in \mathfrak{T}_X$. It should be clear that $z$ exhausts $I$, i.e., $I = \{z_t \mid t \in \mathfrak{T}_X\}$. For any Baire event $E$ of $Y^{\mathfrak{T}_X}$, a Baire event $I_z(E)$ of $Y^I$ can be defined such that

$$I_z(E) = \left\{\tilde{v} \in Y^I \mid \exists v \in E\, s.t. \forall x \in I\, \exists t \in \mathfrak{T}_X, \text{with}\, z_t = x \,\text{and}\, \tilde{v}(x) = v_t\right\},$$

which pairs each element $v \in Y^{\mathfrak{T}_X}$ with an element $\tilde{v} \in Y^I$ that agrees with it. Consequently, $\mathbb{W}_{\mathbb{F}}^z = \mathbb{F}_I \circ I_z$ since

$$\begin{aligned}\mathbb{W}_{\mathbb{F}}^z(E) &= \mathbb{F}(\{u \mid \exists v \in E\, s.t. \forall t \in \mathfrak{T}_X,\ v_t = u(z_t)\}) \\ &= \mathbb{F}(\{u \mid \exists \tilde{v} \in I_z(E)\, s.t. \forall x \in I,\ \tilde{v}(x) = u(x)\}) = \mathbb{F}_I(I_z(E)).\end{aligned} \tag{12.23}$$

Finally, for any permutation $\pi$ of $X$, let $\pi(z) = (\pi(z_t))_{t\in\mathfrak{T}_X}$. Observe that

$$I_{\pi(z)} = D_\pi \circ I_z.$$

By Equation 12.23 and the fact that $\mathbb{F}_I = \mathbb{F}_{\pi(I)} \circ D_\pi$,

$$\mathbb{W}_{\mathbb{F}}^{\pi(z)} = \mathbb{F}_{\pi(I)} \circ I_{\pi(z)} = \mathbb{F}_{\pi(I)} \circ D_\pi \circ I_z = \mathbb{F}_I \circ I_z = \mathbb{W}_{\mathbb{F}}^z.$$

Now for any non-repeating $\tilde{z} \in X^{\mathfrak{T}_X}$ there is some permutation $\pi$ on $X$ such that $\tilde{z} = \pi(z)$, and thus $\mathbb{W}_{\mathbb{F}}^z = \mathbb{W}_{\mathbb{F}}^{\tilde{z}}$. That is, $\mathbb{F}$ is path independent over $\mathfrak{T}$ and therefore of class P-NFL on $\mathfrak{T}$. To extend this result to tight Borel fitness measures, if $Y^X$ is Mařík then $\mathbb{F}_X$ extends uniquely to $\mathbb{F}$. Then $\mathbb{F}_I$ may be assumed to be a tight Borel measure whenever $\mathbb{F}$ is, and the result follows.

Since mutually independent and identically distributed coordinates together imply path independence, one wonders whether these properties are also necessary. In fact, identically distributed coordinates are a necessary consequence of path independence, proven in Theorem 12.9, but mutual independence is not, proven in Theorem 12.13 below.

**Theorem 12.9.** *Every path independent Borel (or Baire) fitness measure has identically distributed coordinates.*

*Proof.* Suppose that $\mathbb{F}$ is path independent without identically distributed coordinates. Then there exist $x, y \in X$ and a Borel subset $E$ of $Y$ such that $\mathbb{F}_x(E) \neq \mathbb{F}_y(E)$. Let $z^x, z^y \in X^{\mathcal{T}_X}$ be non-repeating such that $z_1^x = x$ and $z_1^y = y$. But now consider the set $G = \{v \in Y^{\mathcal{T}_X} \mid v_1 \in E\}$, which is chosen so that $\Phi^{-1}_{W^{z^x}}(G) = \{u \mid u(x) \in E\}$, and

$$\mathbb{W}^{z^x}_{\mathbb{F}}(G) = \mathbb{F}_x(E) \neq \mathbb{F}_y(E) = \mathbb{W}^{z^y}_{\mathbb{F}}(G), \tag{12.24}$$

which achieves the desired contradiction.

### 12.3.2 Existence of P-NFL Fitness Measures

Theorem 12.8 provides a simple way to construct Baire fitness measures of class P-NFL for any Hausdorff search space $X$ and fitness space $Y$. If in addition $Y^X$ is Mařík, then a Borel fitness measure can also be constructed.

The theorem works by using Kolmogorov extension to transform a tight Borel measure on $Y$ into a Baire fitness measure over $Y^X$ that has identically distributed and mutually independent coordinates. This procedure generates the two examples of NFL fitness measures given above. An almost surely constant NFL fitness measure is obtained by placing a degenerate probability measure on $Y$, which is necessarily tight and Borel measurable.The constant fitness measure obviously has identically distributed and mutually independent coordinates, so it is path independent.

Non-degenerate measures represent more interesting case that can be intuitively described. Suppose that $Y = \mathbb{R}$ and each objective value $h_t^Y$ is chosen from a standard normal distribution independent of the search history $h_X$ and independent also of the values $h_s^Y$ for $s \neq t$. That is, the search method chooses a point, and then the search problem randomly assigns an objective value to that point that has no relationship to its previous choices or to the search point proposed. Thus the probability of the observing certain fitness values is then independent of the search points proposed, satisfying the intuitive criterion for probabilistic NFL. This fitness measure has mutually independent and identically distributed coordinates. It is thus path independent by Theorem 12.8 and has the NFL property by Theorem 12.7. Furthermore, the fitness measure in question does exist as a consequence of the Kolmogorov extension theorem and can be computationally approximated. A formal statement and proof of the general case is given next.

**Theorem 12.10 (Existence of P-NFL).** *Let $\mathcal{T}$ be an arbitrary index set. Then there exists a Baire fitness measure of class P-NFL on $\mathcal{T}$ for any search domain $X$. If $Y^X$ is Mařík (Theorem 5.2), there exists a Borel fitness measure of class P-NFL on $\mathcal{T}$.*

*Proof.* Assume $\mathbb{Q}$ is a tight Borel measure on $Y$. Such a measure always exists, since for all $y \in Y$, the degenerate measure $\mathbb{Q}_y(A) = \mathbb{1}_A(y)$ is a tight Borel measure. For each $x \in X$, define $\mathbb{F}_x = \mathbb{Q}$. For any finite subset $D$ of $X$, define $\mathbb{F}_D$ to be the unique measure such that for any collection $\{E_x \mid x \in D\}$ of Borel subsets of $Y$,

$$\mathbb{F}_D\left(\prod_{x\in D} E_x\right) = \prod_{x\in D} \mathbb{F}_x(E_x) = \prod_{x\in D} \mathbb{Q}(E_x). \tag{12.25}$$

Each $\mathbb{F}_D$ is tight since it is a product of tight measures. Thus the collection given by $\{\mathbb{F}_D \mid D \subseteq X,\ D\,\text{finite}\}$ is a consistent family of finite-dimensional distributions for which each member is tight. By the Kolmogorov Extension Theorem, there is a unique Baire fitness measure $\mathbb{F}$ that extends this collection.

$\mathbb{F}$ has mutually independent and identically distributed coordinates. Therefore $\mathbb{F}$ is path independent over $\mathcal{T}$ and of class P-NFL on $\mathcal{T}$ by Theorem 12.8.

If $Y^X$ is Mařík, then $\mathbb{F}$ may be extended to a unique regular Borel fitness measure $\hat{\mathbb{F}}$. The measure $\hat{\mathbb{F}}$ still has mutually independent and identically distributed coordinates since it agrees with $\mathbb{F}$ on all finite projections. Therefore $\hat{\mathbb{F}}$ is path independent over $\mathcal{I}$ and of class P-NFL on $\mathcal{I}$, again by Theorem 12.8.

According to Theorem 12.10, P-NFL fitness measures exist for a broad class of search domains and fitness spaces. Some of these are listed as examples below. Although the index set $\mathcal{T}$ is arbitrary in Theorem 12.10, the class P-NFL is only defined with respect to Borel settings (where $\mathcal{T}$ must be first-countable, e.g. $\mathbb{R}$) and Baire settings (where $\mathcal{T}$ must be countable). For larger settings the unique stopping times $T_t$ cannot be proven measurable for larger index sets. If the definition of P-NFL were expanded to other settings, Theorem 12.10 should still be valid, since it does not rely on the unique stopping times directly.

## *12.3.3 NFL Vector Spaces*

From Theorem 12.10, P-NFL fitness measures exist. But there is more to conclude from it than mere existence. A Borel (or Baire) fitness measure is defined as a Borel (or Baire) probability measure on the space of objective functions. Thus the space of fitness measures with finite total variation is the familiar Banach space of unsigned measures on $Y^X$, either $\mathcal{M}\left[Y^X, \mathcal{B}[Y^X]\right]$ for Borel fitness measures or $\mathcal{M}\left[Y^X, \mathcal{B}a[Y^X]\right]$ for Baire fitness measures. As it turns out, NFL naturally carves out a Banach subspace. In this subsection, the relationships of the properties defined in this chapter are studied in relationship to these Banach subspaces. The definitions of path independence, identically distributed coordinates, and mutually independent coordinates apply to finite signed measures and in some cases carve out further Banach subspaces.

**Theorem 12.11.** *For any index set $\mathcal{T}$, the path independent Borel (or Baire) fitness measures form a closed vector subspace of the Borel (or Baire) fitness measures with identically distributed coordinates, which is in turn a closed vector subspace of all generalized Borel (or Baire) fitness measures, $\mathcal{M}[Y^X, \mathcal{B}_{YX}]$ (or $\mathcal{M}[Y^X, \mathcal{B}a_{YX}]$).*

*Proof.* **Identically distributed coordinates**. Suppose $\mu$ and $\nu$ are Borel (or Baire) fitness functions with identically distributed coordinates. For $\alpha \in \mathbb{R}$, $\alpha\mu$ has identically distributed coordinates. For $x, y \in X$, it holds that $(\mu + \nu)_x = \mu_x + \nu_x = \mu_y + \nu_y = (\mu + \nu)_y$. If $\mu_n \to \mu$ in the total variation norm and $(\mu_n)_x = (\mu_n)_y$ for all $n$, then $\mu_x = \mu_y$. So Borel (or Baire) fitness measures with identically distributed coordinates form a closed subspace of the fitness measures.

**Path independence**. Suppose $\mu$ and $\nu$ are path independent Borel (or Baire) fitness measures. It is obvious that for $\alpha \in \mathbb{R}$, $\alpha\mu$ is path independent. Also, $\mu + \nu$ is path independent since for any non-repeating $z, \tilde{z} \in X^{\mathcal{T}_X}$,

$$\mathbb{W}^z_{\mu+\nu} = \mathbb{W}^z_\mu + \mathbb{W}^z_\nu = \mathbb{W}^{\tilde{z}}_\mu + \mathbb{W}^{\tilde{z}}_\nu = \mathbb{W}^{\tilde{z}}_{\mu+\nu}. \tag{12.26}$$

Now suppose $\mu_n \to \mu$ in the total variation norm. In this case for all non-repeating $z \in X^{\mathcal{T}_X}$, $\mathbb{W}^z_{\mu_n} \to \mathbb{W}^z_\mu$, and since the sequence on the left is identical for each $z$, the limit is likewise identical. That is, the path independent generalized fitness measures form a closed vector subspace of the set of all generalized fitness measures. Since every path independent fitness measure has identically distributed coordinates (the proof of Theorem 12.9 does not require $\mathbb{F}$ to be a probability measure), the subspace of path independent generalized fitness measures is a closed subspace of such measures.

Denote the closed vector subspace of $\mathcal{M}\left[Y^X, \mathcal{B}_{YX}\right]$ consisting of Borel fitness measures that have identically distributed coordinates by $\mathcal{ID}^{\mathcal{B}}_{X,Y}$, and let $\mathcal{ID}^{\mathcal{B}a}_{X,Y}$ be defined analogously for generalized Baire fitness measures. Let $\mathcal{PI}^{\mathcal{T},\mathcal{B}}_{X,Y}$ be the closed vector subspace of $\mathcal{ID}^{\mathcal{B}}_{X,Y}$ consisting of generalized Borel fitness measures that are path independent over $\mathcal{T}$, and similarly define $\mathcal{PI}^{\mathcal{B}a,\mathcal{T}}_{X,Y}$. It is of interest to know whether the subspaces are proper. Although it is obvious that $\mathcal{M}\left[Y^X, \mathcal{B}_{YX}\right] \neq \mathcal{ID}^{\mathcal{B}}_{X,Y}$ whenever $X$ and $Y$ each have at least two elements, it needs to be shown that $\mathcal{ID}^{\mathcal{B}}_{X,Y} \neq \mathcal{PI}^{\mathcal{B},\mathcal{T}}_{X,Y}$. The inequality will be proven for many but not all cases below, using the following definition.

**Definition 12.19 (Equipartitionable).** A tuple $(X,Y)$ is *equipartitionable* if $X$ and $Y$ can be partitioned into $X_1, X_2 \subseteq X$ and $Y_1, Y_2 \subseteq Y$ with the following properties:

1. $X_1, X_2, Y_1, Y_2$ are all nonempty Borel sets in their respective topologies.
2. There exist bijections $\phi$ from $X_1$ to $X_2$ and $\psi$ from $Y_1$ to $Y_2$.
3. There exists a tight measure $\mathbb{Q}$ on $Y$.
4. For $G$ Borel in $Y_1$, the measure $\mathbb{Q}$ has $\mathbb{Q}(\psi(G)) = \mathbb{Q}(G)$.

5. For $G$ Borel in $Y_2$, the measure $\mathbb{Q}$ has $\mathbb{Q}(\psi^{-1}(G)) = \mathbb{Q}(G)$.

If $(X,Y)$ is equipartitionable, the measure $\mathbb{Q}$ has $\mathbb{Q}(Y_1) = \mathbb{Q}(Y_2)$ because $\psi(Y_1) = Y_2$. As usual, $\psi(G) = \{y_2 \in Y_2 \mid \exists y_1 \in Y_1 \, s.t. \, \psi(y_1) = y_2\}$. In particular, $([0,1],\mathbb{R})$, $(\mathbb{N},\mathbb{R})$, and $(\{0,1\}^n,\mathbb{R})$ are all equipartitionable. Search spaces with an odd finite size, however, are not equipartionable. This wrinkle can probably be removed with a revised proof. When $(X,Y)$ are equipartitionable, then path independent fitness measures are properly contained.

**Theorem 12.12.** *If the search domain contains at least three elements and is equipartitionable together with the value space, and if the index set contains at least two elements, then $\mathcal{PI}^{\mathcal{B}a,\mathcal{T}}_{X,Y}$ is a proper subspace of $\mathcal{ID}^{\mathcal{B}a}_{X,Y}$. If in addition $Y^X$ is Mařík (Theorem 5.2), then $\mathcal{PI}^{\mathcal{B},\mathcal{T}}_{X,Y}$ is also a proper subspace of $\mathcal{ID}^{\mathcal{B}}_{X,Y}$. That is, there exists at least one path dependent fitness measure with identically distributed coordinates.*

*Proof.* Given the equipartition from Definition 12.19, construct path independent Baire fitness measures $\mathbb{F}_1$ on $Y_1^{X_1}$ and $\mathbb{F}_2$ on $Y_2^{X_2}$ using the restrictions

$$\mathbb{Q}|_{Y_1}(G) = \frac{\mathbb{Q}(G \cap Y_1)}{\mathbb{Q}(Y_1)} \quad \text{and} \quad \mathbb{Q}|_{Y_2}(G) = \frac{\mathbb{Q}(G \cap Y_2)}{\mathbb{Q}(Y_2)}$$

for Theorem 12.10. The fitness measures $\mathbb{F}_1$ and $\mathbb{F}_2$ will now be combined in a way that destroys path independence while retaining identically distributed coordinates by placing probability one on functions that either 1) take values in $Y_1$ on $X_1$ and in $Y_2$ on $X_2$ or 2) take values in $Y_2$ on $X_1$ and in $Y_1$ on $Y_2$. If values in $Y_1$ are observed for search points in $X_1$, then points in $X_2$ will yield values in $Y_2$ with probability one. By strategically crossing $X_1$ and $X_2$, a path can obtain different value histories.

Let $j,k$ range over $\{1,2\}$. To permit generalizations, let $\psi_{11} = \psi_{22}$ and $\phi_{11} = \phi_{22}$ be the identity functions. Let $\psi_{12} = \psi$, $\psi_{21} = \psi^{-1}$, $\phi_{12} = \phi$, and $\phi_{21} = \phi^{-1}$ using $\phi$ and $\psi$ from Definition 12.19. For any Baire event $E$ of $Y^X$,

$$E_{jk} = \left\{ u \in Y_j^{X_j} \,\middle|\, \exists v \in E \, s.t. \forall x \in X_j, \; \psi_{jk} \circ u \circ \phi_{kj}(x) = v(x) \right\}$$

defines four images of the set $E$ under different conditions. The composition $\psi_{jk} \circ u \circ \phi_{kj}$ moves search points from $X_j$ to $X_k$, takes a fitness value in $Y_k$, and transfers the fitness to $Y_j$. If $j = k$, these operations do nothing. Define $\mathbb{F}$ by

$$\mathbb{F}(E) = \Big( \; \mathbb{F}_1(E_{11}) \, \mathbb{F}_2(E_{22}) \Big) \, \mathbb{Q}(Y_1) + \Big( \; \mathbb{F}_1(E_{12}) \, \mathbb{F}_2(E_{21}) \Big) \, \mathbb{Q}(Y_2).$$

The first term of $\mathbb{F}$ says that with probability $\mathbb{Q}(Y_1)$ fitness functions take values in $Y_1$ on $X_1$ and in $Y_2$ on $X_2$. The second term says that with probability $\mathbb{Q}(Y_2)$ fitness functions take values in $Y_2$ on $X_1$ and in $Y_1$ on $X_2$.

It needs to be shown that $\mathbb{F}$ has identically distributed coordinates but is path dependent. For any Borel event $G$ in $Y$ and any $x \in X$,

$$\left\{u \in Y^X \,\middle|\, u(x) \in G\right\}_{jk} = \begin{cases} \left\{u \in Y_j^{X_j} \,\middle|\, \psi_{jk} \circ u \circ \phi_{kj}(x) \in G \cap Y_j\right\} & \text{if} x \in X_k \\ Y_j^{X_j} & \text{otherwise.} \end{cases}$$

Thus if $x \in X_1$, the coordinate projection has

$$\begin{aligned} \mathbb{F}_x(G) &= \mathbb{F}_1\left(\left\{u \in Y_1^{X_1} \,\middle|\, u(x) \in G \cap Y_1\right\}\right) \mathbb{Q}(Y_1) \\ &\quad + \ \mathbb{F}_2\left(\left\{u \in Y_2^{X_2} \,\middle|\, \psi^{-1} \circ u \circ \phi(x) \in G \cap Y_2\right\}\right) \mathbb{Q}(Y_2) \\ &= \mathbb{Q}(G \cap Y_1) + \mathbb{Q}(\psi(G \cap Y_2)) = \mathbb{Q}(G), \end{aligned}$$

using the fact that $\psi$ preserves $\mathbb{Q}$-measure and that each $\mathbb{F}_j$ has coordinates identically distributed according to $\mathbb{Q}|_{Y_j}$. Similarly, if $x \in X_2$, it follows that $\mathbb{F}_x(G) = \mathbb{Q}(\psi^{-1}(G \cap Y_1)) + \mathbb{Q}(G \cap Y_2) = \mathbb{Q}(G)$. So $\mathbb{F}$ has identically distributed coordinates.

Now choose $x_1^1, x_1^2 \in X_1$ and $x_2 \in X_2$ and let $z^1, z^2 \in X^{\mathcal{T}_X}$ be non-repeating so that $z_1^1 = x_1^1$, $z_2^1 = x_2$, $z_1^2 = x_1^1$ and $z_2^2 = x_1^2$. In this way, $z^1$ first checks the fitness domain for $X_1$, followed by $X_2$. The sequence $z^2$ starts in $X_1$ and then checks another point in $X_1$. It is possible to choose such points because $X$ has at least three elements. Now

$$\mathbb{F}\left(\left\{u \,\middle|\, W_1^{z^1}(u) \in Y_1, W_2^{z^1}(u) \in Y_2\right\}\right) = \mathbb{F}\left(\left\{u \,\middle|\, u(x_1^1) \in Y_1, u(x_2) \in Y_2\right\}\right).$$

To expand this equation, set $x_2^2 = x_2$ and define $Y_{jk}$ by

$$Y_{jk} = \left\{u \,\middle|\, u(x_1^1) \in Y_1, u(x_2) \in Y_2\right\}_{jk} = \left\{u \in Y_j^{X_j} \,\middle|\, \psi_{jk} \circ u \circ \phi_{kj}(x_j^j) \in Y_k\right\}.$$

Observe that on the one hand, $Y_{jk}$ is empty if $j \neq k$ so that

$$\mathbb{F}_1(Y_{12}) = \frac{\mathbb{Q}(\psi(Y_1 \cap Y_2))}{\mathbb{Q}(Y_1)} = 0 \quad \text{and} \quad \mathbb{F}_2(Y_{21}) = \frac{\mathbb{Q}\left(\psi^{-1}(Y_2 \cap Y_1)\right)}{\mathbb{Q}(Y_2)} = 0,$$

and on the other hand,

$$\mathbb{F}_1(Y_{11}) = \frac{\mathbb{Q}(Y_1 \cap Y_1)}{\mathbb{Q}(Y_1)} = 1 \quad \text{and} \quad \mathbb{F}_2(Y_{22}) = \frac{\mathbb{Q}(Y_2 \cap Y_2)}{\mathbb{Q}(Y_2)} = 1.$$

Consequently,

$$\mathbb{F}\left(\left\{u \,\middle|\, W_1^{z^1}(u) \in Y_1, W_2^{z^1}(u) \in Y_2\right\}\right) = \mathbb{Q}(Y_1) = \frac{1}{2}.$$

By contrast, functions that take values in $Y_1$ on $x_1^1 \in X_1$ take values in $Y_1$ on $x_1^2 \in X_1$. Thus, with probability one, $W_2^{z^2}(u) = u(x_1^2) \notin Y_2$, and so

$$\mathbb{F}\left(\left\{u \,\middle|\, W_1^{z^2}(u) \in Y_1, W_2^{z^2}(u) \in Y_2\right\}\right) = 0.$$

That is, $\mathbb{F}$ is path dependent, and the path independent Borel fitness measures are properly contained. By construction, $\mathbb{F}$ is a Baire fitness measure. If $Y^X$ is Mařík, it can be uniquely extended to a Borel fitness measure.

So in important cases, not all fitness measures with identically distributed coordinates are path independent. There also exist Baire fitness measures of class P-NFL that do not have mutually independent coordinates.

**Theorem 12.13.** *If the search domain has at least two elements, the set of Baire fitness measures with mutually independent coordinates is not closed under vector operations. Furthermore, there exists a Baire fitness measure that is path independent (and hence of class P-NFL) but does not have mutually independent coordinates. If $Y^X$ is Mařík, both statements are true of Borel fitness measures as well.*

*Proof.* Let $E$ be any Borel set in $Y$ with $E \neq Y$. Choose $\alpha, \beta \in (0,1)$ so that

$$2\alpha\beta \neq \alpha^2 + \beta^2. \tag{12.27}$$

Let $\mathbb{Q}^1$ and $\mathbb{Q}^2$ be any tight measures with $\mathbb{Q}^1(E) = \alpha$, and $\mathbb{Q}^2(E) = \beta$.

Pick $x, y \in X$, and construct $\mathbb{F}_1$ and $\mathbb{F}_2$ by Theorem 12.8 to be path independent Baire fitness measures with mutually independent coordinates distributed identically according to $\mathbb{Q}^1$ and $\mathbb{Q}^2$, respectively. The weighted sum $\mathbb{F} = \frac{1}{2}\mathbb{F}^1 + \frac{1}{2}\mathbb{F}^2$ is a Baire fitness measure. Set $D = \{x, y\}$. Then

$$\mathbb{F}_D(E \times E) = \frac{1}{2}\mathbb{F}^1_x(E)\mathbb{F}^1_y(E) + \frac{1}{2}\mathbb{F}^2_x(E)\mathbb{F}^2_y(E) = \frac{1}{2}\left(\alpha^2 + \beta^2\right) \tag{12.28}$$

using $\mathbb{F}_D$ and $\mathbb{F}_x$ from Definition 12.18. To have mutual independence,

$$\mathbb{F}_D(E \times E) = \left[\frac{1}{2}(\alpha + \beta)\right]^2 = \frac{1}{4}\left(\alpha^2 + 2\alpha\beta + \beta^2\right), \tag{12.29}$$

which together with Equation 12.28 violates Equation 12.27. That is, $\mathbb{F}$ does not have mutually independent coordinates, and so mutual independence is not closed under vector operations. However, path independence is closed under vector operations by Theorem 12.11, so $\mathbb{F}$ is path independent. Thus $\mathbb{F}$ is path independent without mutually independent coordinates. By Theorem 12.7, it is also of class P-NFL. If $Y^X$ is Mařík, then $\mathbb{F}^1$ and $\mathbb{F}^2$ can be extended to Borel measures without losing path independence, and so in a Mařík space all of the above statements apply to Borel fitness measures as well.

One final NFL topic not yet addressed is the ultimate relationship between set-theoretic and probabilistic NFL. It was mentioned above that Theorem 12.2 shows that under set-theoretic NFL deterministic optimizers produce the same performance vectors, whereas under probabilistic NFL, the performance vectors share the same distribution as stated in Theorems 12.4 and 12.6. Given that the path independent fitness measures form a Banach space, one might wonder whether subsets of objectives that are closed under permutation somehow generate this Banach space, for example, by mapping into a set of vectors that span the space. For finite c.u.p. sets, it is clear how to obtain a fitness measure, but for transfinite c.u.p. sets, there is no obvious relationship. This issue is left open for future inquiries.

## 12.4 Conclusion

NFL theory is now two decades old. Much was initially made of the idea that "good" algorithms for some problem domain are necessarily "bad" for other problem domains, but over time this claim has become less shocking in part due to familiarity but also due to the reality that many of the problems incurring "bad" performance are uninteresting from a practical perspective. No search method suffices for all problems, yet many are good enough for general purposes because the problems of interest possess some general regularities that can be exploited.

This chapter has reviewed the literature around NFL, including set-theoretic and probabilistic varieties, introducing the concept of path independence as a equivalent property to NFL. The fact that identically distributed and mutually independent fitness measures necessarily imply path independence made it possible to construct NFL in a wide variety of spaces using the Kolmogorov extension theorem. Nonetheless, the independence properties of NFL are strict enough that they can be easily disrupted by imposing simple properties on the space of objectives, such as Lipschitz continuity or square-integrability. Since physical reality imposes many regularities, it is reasonable to expect that general-purpose learning is in fact possible, though it will always be less efficient than a solution that is well-adapted to a particular problem.

This excursion into NFL expanded on the concept of NFL as *equal average performance* across problems. Perhaps the more important feature of the NFL idea, though, is the notion of problem-solution alignment. That is, rather than merely understanding fitness measures that induce equal average performance, it might be more beneficial to understand which algorithms are well-aligned with each problem. Then one might hope to understand the cost one *pays for good performance* by specializing the search method. These ideas are taken up in the next chapter.

# Chapter 13
# The Information Maximization Principle

Path independence implies that samples from an NFL fitness measure are chaotic and unpredictable. In this chapter we discuss how the predictability of non-NFL fitnesses can be exploited to optimize over search methods. First, we introduce fitness measures with some degree of predictability, emphasizing that such fitness measures can cover a wide range of settings for search and optimization. Next, *standard* search trajectors that extend standard search generators are introduced as a replacement for static search trajectors so that subsequent results can be simplified and applied to a wider range of search and optimization settings. For the final step in this chapter, the V-performance is defined to extend performance criteria from previous chapters to this new setting. This class of performance measurements ultimately induces a relationship between the fitness measures and search trajectors that will be explored further in Chapter 17. In this chapter, Bellman-type equations are applied to demonstrate the optimal form for exploiting the predictability within the fitness measure. This leads to the *information maximization principle*, which states that the best way to choose points to optimize a problem against a given value preference is to choose the points that yield the greatest amount of information. The chapter concludes with a discussion of how to practically implement this principle.

## 13.1 Predictability in Fitness Measures

Because path independence is a strong condition, there exist many ways in which it can be violated. A path-dependent fitness measure can have only a few predictable paths floating in a sea of chaos, or it can be so structured that knowledge of the value at one point in the search domain determines the values at nearly all other points. This section briefly discusses the former case, known as *Almost NFL*, and then considers two more structured cases. The first case, *the diffusion prior*, is based on the Brownian motion and is tightly related to the field of Gaussian processes. Like Lipschitz continuity, it represents a weak form of structure that nonetheless results in path dependence across almost all paths. The second case, the *universal*



A. J. Lockett, *General-Purpose Optimization Through Information Maximization*,
Natural Computing Series, https://doi.org/10.1007/978-3-662-62007-6_13

*prior*, represents a more abstract case of a fitness measure that generates objective function in proportion to their simplicity. The universal prior is related to Occam's razor, and the section will conclude with some remarks regarding the possibility of a *reality prior* that represents the predictability that occurs within nature.

### 13.1.1 Sparse Predictability and Almost NFL

Path dependence means that the map $z \mapsto \mathbb{W}_{\mathbb{F}}^{z}$ is not constant. This map yields the probability measure over value histories in $Y^{\mathfrak{T}}$ that results from applying the fitness measure $\mathbb{F}$ to a non-repeating search history $z$. If the map is constant except for a single search history $z_0$, then $\mathbb{F}$ is path dependent, but not meaningfully so. For every search history other than $z_0$, the observed value history is distributed independently from the search history, including for search histories $z_0'$ that agree with $z_0$ for the first $n$ timesteps, where $n$ is arbitrary. Although $\mathbb{F}$ is not path independent and therefore not NFL, it is practically indistinguishable from NFL. The fitness measure $\mathbb{F}$ might be called *sparsely predictable*. There are clearly more sparsely predictable fitness measures than there are NFL fitness priors.

A related phenomenon has been named *Almost NFL* by Droste et al. [42]. In their conception, one might have a collection $U$ of "hard" objectives in the black-box sense that it provably requires the evaluation of exponentially many search points in order to find the optimum for each $u \in U$. Their example is the so-called *needle in a haystack* class of functions on the hypercube $X = \{0,1\}^n$, for which there is exactly one $x \in X$ such that $u(x) = 1$ and $u(x') = 0$ for all other $x' \neq x$. There are $2^n$ such functions, and any search heuristic requires exponentially many evaluations on average to find the unknown "needle" $x$. Yet for each $u \in U$ there are many functions $u'$ similar to $u$ that can nonetheless be optimized using a polynomial number of evaluations, so that by expanding $U$ to $U'$, one passes from NFL to a polynomially solvable class of objectives. Thus in some senses the boundary of NFL is sharp. This observation is similar in spirit to the result of Igel and Toussaint [79] that the number of c.u.p. subsets (and hence the number of NFL problems in the set-theoretic sense) goes to zero as the size of the search domain grows without bound.

In the end, while NFL and sparse predictability are theoretically important, they are not entirely relevant to solving practical problems. Sparse predictability is not the same as broad structure we observe in real problems. Instead, problems drawn from physics and engineering tend to be subject to strong constraints such as locality or temporality, *i.e.*, spatial limits on the rate of growth in objective values.

Therefore, if one wishes to devise search and optimization methods that are applicable to a broad range of practical problems, there are a typical set of constraints that come into play almost necessarily. Locality has already been mentioned. Another category of constraints emerges from a sort of constructibility in which there exist underlying latent processes that yield similarities among superficially different measurements, such as the basic concept that all matter is made up of quarks, or the fact that magnetism and electricity are tightly coupled through the electromagnetic force. Rather than attempt to formalize these concepts, we simply refer to the collection of all relevant constraints as the *reality prior* and assert that whatever it is, this fitness measure clearly exhibits a substantial degree of regularity, whence NFL does not apply to it in the traditional sense.

### 13.1.2 The Diffusion Prior

The previous section proposed that one should expect the reality prior to possess more regularity than an NFL fitness measure does. It will be worthwhile to give an example of a fitness measure that possesses such regularity while still being strongly non-specific. In this section, the diffusion prior will be discussed as an example of a fitness measure satisfying one of the desired qualities in a prior, that of locality, or limited spatial growth.

A diffusion prior is defined as a fitness measure over objectives on Euclidean space $\mathbb{R}^n$ based on the Brownian motion. The Brownian Motion assumes random expansion in space at a rate equal to the square root of the time elapsed. It is a well-studied mathematical object that plays a role in mathematical theories ranging from physics to finance [32, 86]. The Brownian motion can be defined as a prior over all continuous functions, or even as a prior over discontinuous ones. However, it enforces a constraint that if a function has a certain value at one point, then it is likely (but not required) to have similar values at nearby points.

The standard Brownian Motion is a random process in one variable, often denoted by $W = (W_t)_{t\in[0,\infty)}$, such that the increments of the Brownian Motion are normally distributed with a variance the size of the increment, i.e.

$$W_t - W_s \sim \mathcal{N}(W_s, t-s), \; \forall s < t.$$

Typically, the process is started at $W_0 = 0$ for convenience, but the process can be initialized at any point. The phrase "Brownian Motion started at $x$" indicates that $W_0 = x$.

The Brownian Motion can be constructed as a prior over continuous functions (that is, over $C[0,\infty)$) against a $\sigma$-algebra of cylinder sets (see e.g. Karatzas and Shreve [86]). Thus the distribution of the Brownian Motion is a type of fitness measure. A function drawn from this Brownian prior is a continuous trajectory of the Brownian Motion.

The Brownian prior has constant mean ($\mathbb{E}W_t = W_0$), but it is not identically distributed at each point and is path dependent. Such a prior is non-NFL. It is easy to demonstrate this fact because $W_t$ is a martingale, which means

$$\mathbb{E}\left[W_t \mid W_s\right] = W_s \neq W_0 = \mathbb{E}W_t \;\; \text{for } 0 < s < t.$$

As a non-NFL prior, it follows that some optimizers perform better than others on this prior. This fact is also easily demonstrated. Consider the search space given by $t \in [0,1]$. Let $\mathcal{G}_1$ be a deterministic search method that proposes evenly spaced points in order, say,

$$0, \frac{1}{100}, \frac{2}{100}, \frac{3}{100}, \frac{4}{100}, \frac{5}{100}, \ldots,$$

and let $\mathcal{G}_2$ be another deterministic search method that iteratively splits $[0,1]$ with evenly spaced points, i.e

$$0, 1, \frac{1}{2}, \frac{1}{4}, \frac{3}{4}, \frac{1}{8}, \frac{3}{8}, \frac{5}{8}, \frac{7}{8}, \frac{1}{16}, \ldots$$

$\mathcal{G}_2$ has better performance than $\mathcal{G}_1$ on a Brownian prior, because at each step it eliminates a larger proportion of functions from consideration than $\mathcal{G}_2$ by placing more constraints on the available functions.

Define an $n$-dimensional diffusion prior as a functional of one or more Brownian Motions on a bounded subset of $\mathbb{R}^n$. It can be described as a random objective $U$ indexed by $x \in [0,1]^n$ given by

$$U_x = g(W_{1,x_1}, \ldots W_{n,x_n}),$$

where $W_1, \ldots, W_n$ are $n$ independent Brownian Motions, and $g$ is a Borel-measurable function from $\mathbb{R}^n$ to $\mathbb{R}^n$.

Let $\mathbb{P}_U$ be the distribution of $U$. Consider the search space $[0,1]^n$. Then a sample from $\mathbb{P}_U$ is a function over the search space. This prior can be extended to all of $\mathbb{R}^n$ by transforming the space. The diffusion prior $\mathbb{P}_U$ is a functional of the Brownian Motion, and as long as the function $g$ is non-degenerate, $\mathbb{P}_U$ does not have the NFL.

Readers familiar with the use of Gaussian processes in predictive function modeling should notice the similarity. The function $g$ defining the diffusion prior corresponds to the choice of kernel in a Gaussian process. The driving Brownian Motions could also be shifted backwards in time by an arbitrary amount (e.g. $\tilde{W}_{n,t} = W_{n,t+\sigma_0^2}$) to obtain a non-degenerate initial distribution. With this in mind, an optimizer could perform well on a diffusion prior by choosing evaluation points in such a way as to minimize the conditional variance of the objective function given the evaluation points under the kernel $g$. Because of this relationship, a diffusion prior might also be called a Gaussian prior.

The diffusion prior provides an example of a non-NFL prior that can place positive probability on all subsets of continuous functions. As this example proves, the existence of NFL does not imply that successful learning is only possible on specific problems.

### *13.1.3 The Universal Prior*

The previous section proved that general-purpose non-NFL fitness measures exist. But what sort of fitness measure is the reality prior? Some conjectures have been previously offered on this topic in the context of inductive inference [149, 36, 31, 150, 77, 142]. Ideas concerning a universal prior have typically centered on the modernized version of Occam's razor, which says that when one is faced with competing hypotheses that explain some data, the simplest hypothesis is most likely to be correct.

This research is rooted in the idea that the universe is generated by a computable program, an idea that goes back to Zuse, and even further back to Leibniz [142]. Given that the universe has an observed state $x$, then according to Occam's razor, the most likely program computing the universe is the shortest program that computes $x$. In 1964, Solomonoff [149, 150] proposed a universal measure over bit sequences conforming to this principle, given by

$$\mathbb{P}'_M(x) = \sum_{\substack{\text{program prefixes } p \\ \text{that compute } x}} 2^{-|p|}. \tag{13.1}$$

Similar measures with a basis in information theory were subsequently proposed by Chaitin and Cover that model the entropy of the observation [36, 31].

Solomonoff's measure is enumerable but not computable because of the halting problem [150, 142]. Computable variants have been proposed based on Minimum Description Length and Kolmogorov complexity [36, 94, 142]. More recently, Schmidhüber [142] proposed the Speed Prior in 2002 after work by Hutter [77] exhibited an algorithm to enumerate all programs that produce an output prefix after a fixed of number of steps. The Speed Prior is similar to Solomonoff's measure in spirit, but accounts for the computation time as well, making it computable in the limit. Schmidhuber provided an algorithm for computing the Speed Prior in finite time within a given tolerance.

Viewed as probabilities over objectives (as program subroutines), neither Schmidhüber's nor Solomonoff's measures could possibly have the NFL property, nor could any similar measure that prioritizes observations on the basis of compressibility. Lattimore and Hutter proved this fact in 2013 [98]. NFL requires that the future be incomputable on the basis of the past, and compressible programs necessarily encode computable regularities. It seems likely that formal results could be derived to demonstrate this claim.

Solomonoff, Schmidhüber, and others have proposed that some variant of the universal prior is in fact the *reality prior* that governs all problems derived from physical reality as suggested at the beginning of this section. Naturally, there is no empirical way to refute this claim in the absolute. Instead, without committing to the idea of the universal prior, it is useful enough to accept that are a few classes of problem constraints that tend to show up in most domains, such as spatiality, temporality, or constructibility. Even across domains that share some kind of spatial constraint, the form of the constraint may be different. Nonetheless, the central idea that most problems are predictable in some aspect rules out NFL. If the reality prior (or priors) conforms to the NFL property, then any universal measure of reality cannot be based on Occam's razor, since NFL assumes that complex programs are substantially more common than simple ones. Therefore, if Occam's razor is accepted, general-purpose learning must be possible.

## 13.2 Standard Search Trajectors and V-performance

In the last several chapters, the theory of search processes has been built on static search trajectors. However, for reasons relating to the details of proving the Information Maximization Principle in the next section, it will now be preferable to introduce a new type of search trajector based on the standard search generators introduced in Chapter 6. These trajectors are also more general in that they can represent search methods in settings beyond static optimization, including stochastic and dynamic objectives. With these trajectors, a revised concept of performance is also needed, which will be called the V-performance, since it is based on a *value function* analogous to the kernel of the previously introduced performance criteria.

### *13.2.1 Search Trajectors*

A *search trajector* on a search domain $X$ with value space $Y$ and temporal index set $\mathcal{T}$ is a function $\mathcal{A} : Y^{\mathcal{T}} \to \mathcal{M}\left[X^{\mathcal{T}}, \Sigma_{X^{\mathcal{T}}}\right]$, where $\Sigma_{X^{\mathcal{T}}}$ is a $\sigma$-algebra on $X^{\mathcal{T}}$. If $\Sigma_{X^{\mathcal{T}}} = \mathcal{B}_{X^{\mathcal{T}}}$, then $\mathcal{A}$ is called a *Borel search trajector.* If $\Sigma_{X^{\mathcal{T}}} = \mathcal{B}a_{X^{\mathcal{T}}}$, then $\mathcal{A}$ is called a *Baire search trajector*. A search trajector $\mathcal{A}$ is proper if $\mathcal{A}[w]$ is a probability measure for all $w \in Y^{\mathcal{T}}$. The Banach space of measure-valued functionals $\mathcal{MF}_{Y^{\mathcal{T}},X^{\mathcal{T}}}$ consists entirely of search trajectors with finite norm.

Search trajectors are related to search generators through an analogue of Equation 9.4, namely,

$$\mathcal{G}[h](B) = \mathbb{E}_{\mathcal{A}[h'_Y]}\left[\mathbb{1}_B \circ Z_{t+1} \mid \mathcal{Z}_t\right](h'_X), \tag{13.2}$$

where $B \subseteq X$ is any measurable set, $t = |h|$ is the length of history prefix $h$, and $h' \in (X \times Y)^{\mathcal{T}}$ is any complete history such that $h \sqsubseteq h'$. Then, as with $\mathbf{G}$ and $\mathbf{E}$ before, we can define a set $\mathbf{E}'$ consisting of pairs $(\mathcal{G}, \mathcal{A})$ that satisfy Equation 13.2 for all $h$, $w$, $B$, and $h'_X \sqsupseteq h_X$. This set is a function from a subset of search generators into (but not onto) the set of search trajectors, and so we can define an extension operator $\mathbf{E}'(\cdot)$ that maps extensible search generators into search trajectors using Equation 13.2.

The concept of *extensible* search generators requires some explanation. As in Section 9.1, there are four criteria required for extensibility: (1) measurability, (2) tightness; (3) a consistent family of finite-dimensional distributions; and (4) a common upper bound on the total variation norm of this family. The requisite definitions are now stated for standard search generators. It is here assumed that $X$ is Borel measurable and $\mathcal{T} = \mathbb{N}$ is discrete.

**Definition 13.1 (Integrable Measurability).** A search generator $\mathcal{G}$ is *integrably measurable* if for all $t \in \mathcal{T}$ and all $h \in (X \times Y)^{[t]}$, the map

$$h \mapsto \int W(x)\, \mathcal{G}[h](dx)$$

is measurable for all random variables $W : X \to \mathbb{R}$.

Integrable measurability is not as strict a condition as it seems. For example, all continuous search generators are necessarily integrably measurable, since the map in the definition is continuous for continuous search generators and continuity implies measurability. Although no proof will be offered, given the difficulty of constructing unmeasurable functions, it seems likely that all search generators defined in this book are integrably measurable.

The family of finite-dimensional distributions is defined recursively given a fixed value history $y \in Y^{\mathcal{T}}$ so that $\mathcal{A}_y^{\emptyset} = \mathcal{G}[\emptyset]$ is a measure on $X^{[1]}$ and for $t \in \mathcal{T}$ and Borel sets $A \subseteq X^{[t]}$ and $B \subseteq X$,

$$\mathcal{A}_y^{[t]}(A \times B) = \int_A \mathcal{G}[z](B)\, \mathcal{A}_y^{[t-1]}(dz),$$

which depends on integrable measurability with $W = \mathbb{1}_B$. Due to the Carathéodory Extension Theorem, this equation uniquely defines $\mathcal{A}_y^{[t]}$ for all $t \in \mathcal{T}$ and $y \in Y^{\mathcal{T}}$. Finally, for any finite set $\mathcal{K} \subseteq \mathcal{T}$, let $t_{\max} = \max \mathcal{K}$ and define $\mathcal{A}_y^{\mathcal{K}}$ so that for any Borel $(A_i)_{i \in \mathcal{K}}$ and $(B_i)_{i \in t_{\max}}$ with $B_i = A_i$ if $i \in \mathcal{K}$ and $B_i = X$ otherwise,

$$\mathcal{A}_y^{\mathcal{K}}\left(\prod_i A_i\right) = \mathcal{A}_y^{[t_{\max}]}\left(\prod_i B_i\right),$$

which uniquely defines $\mathcal{A}_y^{\mathcal{K}}$ as a measure on $X^{\mathcal{K}}$. So $\{\mathcal{A}_y^{\mathcal{K}} \mid \mathcal{K} \text{ is a finite subset of } \mathcal{T}\}$ is a consistent family of finite-dimensional distribution.

**Definition 13.2 (Tightness).** A search generator is tight on a value history $y$ if the measure $\mathcal{A}_y^{\mathcal{K}}$ derived from it is tight for every finite $\mathcal{K} \subseteq \mathcal{T}$. If a search generator is tight on all value histories, it is simply said to be tight.

**Definition 13.3 (Eventual Boundedness).** A search generator is eventually bounded on a value history $y$ if the family of finite-dimensional distributions

$$\{\mathcal{A}_y^{\mathcal{K}} \mid \mathcal{K} \text{ is a finite subset of } \mathcal{T}\}$$

is bounded, *i.e.*, if there exists $N < \infty$ with $\left\|\mathcal{A}_y^{\mathcal{K}}\right\| < N$ for all $\mathcal{K}$. If a search generator is eventually bounded on all value histories, then it is simply said to be eventually bounded.

It is important to note that every proper search generator is eventually bounded.

**Definition 13.4 (Extensibility).** A search generator $\mathcal{G}$ is said to be extensible if there exists a search trajector $\mathcal{A}$ with $(\mathcal{G}, \mathcal{A}) \in \mathbf{E}'$.

**Proposition 13.1.** *Every tight, integrably measurable, eventually bounded search generator is extensible.*

*Proof.* Kolmogorov extension of the family $\{\mathcal{A}_y^{\mathcal{K}} \mid \mathcal{K} \text{ is a finite subset of } \mathcal{T}\}$ yields a Baire measure $\mathcal{A}_y^{\mathcal{T}}$ over $X^{\mathcal{T}}$. This measure can be extended uniquely to a regular Borel measure if $X^{\mathcal{T}}$ is Mařík. Consequently, the map $y \mapsto \mathcal{A}_y^{\mathcal{T}}$ is a Baire search trajector which may in some cases by extended to a Borel search trajector, and in either case Equation 13.2 is satisfied.

Not every search trajector results from an extensible search generator. As pointed out in Proposition 6.13, the search generators enforce a black-box property in that they cannot "see" future objective values. Thus a new definition of the black-box property is required for search trajectors.

**Definition 13.5 (Black-box Property).** Let $\phi_{[t]} : X^{\mathcal{T}} \to X^{[t]}$ be the canonical projection to the search points up to time $t$. For all value histories $y$, let $\mathcal{A}[y] \circ \phi_{[t]}$ be the push-forward measure on $X^{[t]}$, which can be used to observe the search history up to time $t$. A search trajector $\mathcal{A}$ has the black-box property if for all $t \in \mathcal{T}$ and for all $y, w \in Y^{\mathcal{T}}$, $\mathcal{A}[y] \circ \phi_{[t]} = \mathcal{A}[w] \circ \phi_{[t]}$ whenever $y_s = w_s$ for all $s < t$.

**Proposition 13.2.** *Suppose* $(\mathcal{G}, \mathcal{A}) \in \mathbf{E}'$. *Then* $\mathcal{A}$ *has the black-box property.*

*Proof.* Suppose not. Then for some $y, w \in Y^{\mathcal{T}}$ with $y_s = w_s$ for all $s < t$, $\mathcal{A}[y] \circ \phi_{[t]} \neq \mathcal{A}[w] \circ \phi_{[t]}$. But then it follows for some measurable set $B$ that

$$\mathbb{E}_{\mathcal{A}[y]} \left[ \mathbb{1}_B \circ Z_{t+1} \mid \mathcal{Z}_t \right](z) \neq \mathbb{E}_{\mathcal{A}[w]} \left[ \mathbb{1}_B \circ Z_{t+1} \mid \mathcal{Z}_t \right](z)$$

and so with $h$ such that $h_X \sqsubseteq z$ and $h_Y \sqsubseteq y$, $\mathcal{G}[h](B) \neq \mathcal{G}[h](B)$ by Equation 13.2, which is a contradiction.

This proposition implies that the function $\mathbf{E}'(\cdot)$ defined by Equation 13.2 is a surjective function from some superset of the extensible search generators onto the black-box search trajectors. As with the static search trajectors, there may be many search generators that correspond to each search trajector, since the conditional expectation is only unique $\mathcal{A}[h_Y]$-almost everywhere.

The distinction between search trajectors and static search trajectors was discussed briefly in Section 3.3.1. The key point is that a search trajector only observes $Y^{\mathcal{T}}$, which is actually observed, whereas static search trajectors depend on $Y^X$, which may include many search points that will never be observed.

To formalize the relationship between static search trajectors and standard search trajectors, the sets $\mathbf{E}'$, $\mathbf{G}$, and $\mathbf{E}$ can be used in conjunction to define a set $\mathbf{T}$ of pairs $(\mathcal{A}, \mathcal{A}')$ where $\mathcal{A}$ is a search generator, $\mathcal{A}'$ is a static search generator, and there exists $(\mathcal{G}, \mathcal{G}') \in \mathbf{G}$ such that $\mathbf{E}'(\mathcal{G}) = \mathcal{A}$ and $\mathbf{E}(\mathcal{G}') = \mathcal{A}'$. In this case, there is again a function $\mathbf{T}(\cdot)$ that maps all of $\mathcal{MF}_{Y^{\mathcal{T}}, X^{\mathcal{T}}}$ onto the set $\mathbf{E}\left(\mathcal{BB}_{X,Y}^{\mathcal{T}}\right)$ of static search trajectors with the black-box property.

**Proposition 13.3.** *A static search trajector has the black-box property if and only if it is the image under* $\mathbf{T}(\cdot)$ *of some standard search trajector with the black-box property.*

*Proof.* Suppose a static search generator $\mathcal{A}'$ is the image under $\mathbf{T}(\cdot)$ of a search generator $\mathcal{A}$. That is, $\mathcal{A}' = \mathbf{T}(\mathcal{A})$. Then there exist a static search generator $\mathcal{G}'$ and a standard search generator $\mathcal{G}$ with $\mathcal{G}' = \mathbf{G}(\mathcal{G})$, $\mathbf{E}'(\mathcal{G}) = \mathcal{A}$, and $\mathbf{E}(\mathcal{G}') = \mathcal{A}'$. By Proposition 6.13, $\mathcal{G}'$ has the black-box property, whence $\mathcal{A}'$ and $\mathcal{A}$ have the black-box property as well.

If a proper static search trajector $\mathcal{A}'$ has the black-box property, then by definition $\mathcal{A}' = \mathbf{E}(\mathcal{G}')$ for some static black-box search generator $\mathcal{G}'$. By Proposition 6.13, $\mathcal{G}'$ is the image of some standard search generator under $\mathbf{G}(\cdot)$. Call this search generator $\mathcal{G}$ so that $\mathbf{G}(\mathcal{G}) = \mathcal{G}'$.

Examining the construction in Proposition 6.13, $\mathcal{G}$ is extensible. For a given history prefix $z \in (X \times Y)^{\mathfrak{T}}$, $\mathcal{G}[z] = \mathbf{0}$ if $z$ is not objectifiable, which means that all non-objectifiable history traces extend to the zero trajector. All objectifiable trajectories extend for $\mathcal{G}$ just as they do for $\mathcal{G}'$. Therefore, there exists $\mathcal{A} = \mathbf{E}'(\mathcal{G})$, where either $\mathcal{A}[w] = 0$ or $\mathcal{A}[w]$ is a probability measure. $\mathcal{A}$ has the black-box property because it is the extension of a search generator. It follows that $\mathcal{A}' = \mathbf{T}(\mathcal{A})$.

The overall situation is summed up in the following diagram, in which single arrows indicate the presence of a total function, and doubled arrows indicate a partial function that is surjective on its domain.

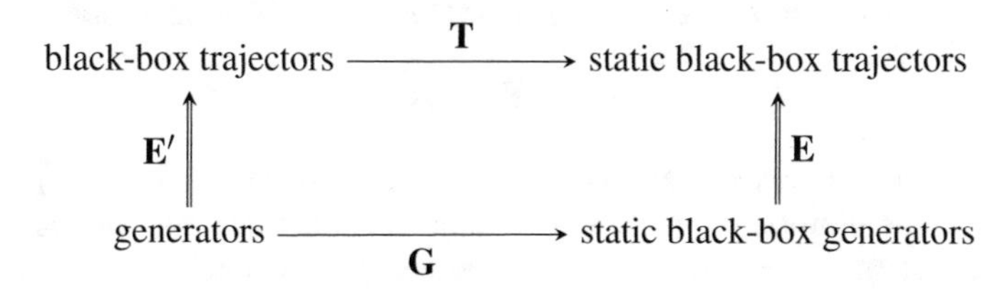


We now proceed to develop a notion of performance on a fitness measure suitable for these trajectors. The first step is to characterize the how history traces are generated from trajectors and fitness measures.

### 13.2.2 Measures over the History Trace

The V-performance defined below averages performance over history traces in proportion to their likelihood. Therefore, a measure is needed to permit observation of history traces. Equation 12.8 defined a measure $\mathcal{A} \times \mathbb{F}$ on $X^{\mathfrak{T}} \times Y^X$ that represents the joint probability over search histories and objectives based on static search trajectors and static fitness measures. This measure generates the interaction between a search method and a randomly selected static objective function. A revised definition for $\mathcal{A} \times \mathbb{F}$ is now constructed for the case where $\mathcal{A}$ is a standard search trajector; this measure is constructed on the space of history traces $(X \times Y)^{\mathfrak{T}}$.

Rather than using Equation 12.8, which requires the definition of *setwise measurability* for trajectors, the measure $\mathcal{A} \times \mathbb{F}$ will be built up by interleaving the progressive operation of the search generator with a conditionalized version of the fitness measure, to be called a *jagged value generator*. Doing so will make it possible to fall back on the requirement of integrable measurability, which will be critical to proving Theorem 13.1.

The fitness measure $\mathbb{F}$ is defined on $Y^X$, and the jagged value generator will be a function $\mathcal{G}^* : \mathcal{H}^{*,\mathcal{T}}_{X,Y} \to \mathcal{M}_Y$ that takes *jagged* histories to unsigned measures on the value space $Y$. A jagged history $h^*$ is a pair $(z,w)$ with $z \in X^{[t+1]}$ and $w \in Y^{[t]}$. This represents what the fitness measure observes at time $t$; it is given a sequence of $t$ search points with values previously assigned, along with one extra search point $z_{t+1}$ for which a value is to be selected. The collection of jagged histories is then the set

$$\mathcal{H}^{*,\mathcal{T}}_{X,Y} = \bigcup_{t \in T} X^{[t+1]} \times Y^{[t]},$$

which can be topologized just as $\mathcal{H}^{\mathcal{T}}_{X,Y}$ was. With $h^* = (z,w)$, the notation $h^*_X$ will be used to refer to $z$ and $h^*_Y$ will refer to $w$.

The appropriate conditioning to generate a jagged history is determined by canonical projections. Given $x \in X$, define the canonical projection $U_x : Y^X \to Y$ by $U_x(u) = u(x)$. Then $U_x$ is necessarily Borel- and Baire-measurable. For $z \in X^{[t]}$, let $\mathcal{U}_z$ be the random variable generated by $(U_{z_s})_{s \le t}$. Now define the *jagged value generator* $\mathcal{G}^*$ so that for all $z \in X^{[t]}$, $x \in X$, $w \in Y^{[t]}$, and $B \subseteq Y$ measurable,

$$\mathcal{G}^*[(z \| x, w)](B) = \mathbb{E}_{\mathbb{F}}\left[\mathbb{1}_B \circ U_x \mid \mathcal{U}_z\right](u), \tag{13.3}$$

where $\|$ is sequence concatenation and $u$ is any objective such that $u(z_s) = w_s$ for all $s \le t$. This definition is not complete, since conditional expectations are only defined almost surely. However, at least one complete definition of $\mathcal{G}^*$ satisfying Equation 13.3 does exist.

Using a triple $(\mathcal{A}, \mathbb{F}, \mathcal{T})$ consisting of a search trajector, a fitness measure, and a temporal index set, a measure $\mathcal{A} \times \mathbb{F}$ can be constructed using Kolmogorov extension. The usual requirements apply: measurability, tightness, consistency of a finite-dimensional family, and eventual boundedness. In order to satisfy these requirements, the search generator $\mathcal{A}$ is assumed to have $\mathcal{A} = \mathbf{E}'(\mathcal{G})$ for some extensible search generator $\mathcal{G}$, and the further conditions are required on the fitness measure $\mathbb{F}$ and its jagged value generator. Throughout this discussion, $\mathcal{T}$ is assumed to be at most countable.

**Definition 13.6 (Setwise Measurable).** A jagged value generator will be called *setwise measurable* if the map

$$x \mapsto \mathcal{G}^*[(z \| x, w)](B)$$

is measurable for all $t \in \mathcal{T}$, $z \in X^{[t]}$, $w \in Y^{[t]}$, and $B \subseteq Y$ measurable. A fitness measure $\mathbb{F}$ will be called setwise measurable if there exists a setwise measurable jagged generator satisfying Equation 13.3 for $\mathbb{F}$.

Given a setwise measurable jagged value generator $\mathcal{G}^*$ and an extensible search generator $\mathcal{G}$, define the first step

$$\mathcal{A} \times \mathbb{F}^{[1]}(E \times F) = \int_E \mathcal{G}^* [(x, \emptyset)] (F) \, \mathcal{G}[\emptyset](dx), \tag{13.4}$$

where $E \subseteq X$ and $F \subseteq Y$ are both measurable in their respective spaces. Note that this equation depends on the integrable measurability of $\mathcal{G}$. By Carathéodory, this defines a measure $\mathcal{A} \times \mathbb{F}^{[1]}$ over $X \times Y$. As a recursive step, define for $t > 1$, $D \in (X \times Y)^{[t]}$, $E \subseteq X$, and $F \subseteq Y$,

$$\mathcal{A} \times \mathbb{F}^{[t]} (D \times (E \times F)) = \int_D \int_E \mathcal{G}^* [(z \| x, w)] (F) \, \mathcal{G}[z](dx) \, \mathcal{A} \times \mathbb{F}^{[t-1]}(d(z, w)), \tag{13.5}$$

which, via Carathéodory, provides complete definitions for $\mathcal{A} \times \mathbb{F}^{[t]}$ as a measure on $(X \times Y)^{[t]}$ for all $t \in \mathcal{T}$.

As before, using $\mathcal{K} \subseteq \mathcal{T}$ finite and $t_{\max} = \max \mathcal{K}$, one can define

$$\mathcal{A} \times \mathbb{F}^{\mathcal{K}} \left( \prod_i A_i \right) = \mathcal{A} \times \mathbb{F}^{[t_{\max}]} \left( \prod_i B_i \right)$$

where each $A_i \subseteq X \times Y$ is measurable and $B_i = A_i$ if $i \in \mathcal{K}$, else $B_i = X$. Using Carathéodory one final time, a consistent finite-dimensional family of distributions $\{\mathcal{A} \times \mathbb{F}^{\mathcal{K}} \mid \mathcal{K} \subseteq \mathcal{T} \text{ finite}\}$ has been defined.

**Definition 13.7 (Tightness).** A triple $(\mathcal{A}, \mathbb{F}, \mathcal{T})$ is tight if each $\mathcal{A} \times \mathbb{F}^{\mathcal{K}}$ defined above is a tight measure. A jagged value generator $\mathcal{G}^*$ is tight if $\mathcal{G}^*[(z, w)]$ is a tight measure for all $z$ and $w$.

**Proposition 13.4.** *A triple $(\mathcal{A}, \mathbb{F}, \mathcal{T})$ is tight if $\mathcal{A} = \mathbf{E}'(\mathcal{G})$ for some tight search generator $\mathcal{G}$ and if there exists a tight jagged value generator $\mathcal{G}^*$ satisfying Equation 13.3 for $\mathbb{F}$.*

**Definition 13.8 (Eventual Boundedness).** A triple $(\mathcal{A}, \mathbb{F}, \mathcal{T})$ is eventually bounded if there exists $N < \infty$ such that $\|\mathcal{A} \times \mathbb{F}^{\mathcal{K}}\| < N$ for every finite $\mathcal{K} \subseteq \mathcal{T}$.

**Proposition 13.5.** *A triple $(\mathcal{A}, \mathbb{F}, \mathcal{T})$ is eventually bounded if $\mathcal{A} = \mathbf{E}'(\mathcal{G})$ for some eventually bounded search generator $\mathcal{G}$.*

*Proof.* Because the jagged value generator $\mathcal{G}^*$ is chosen to extend to $\mathbb{F}$, $\mathbb{F}$ cannot cause $\mathcal{A} \times \mathbb{F}^{\mathcal{K}}$ to grow without bound, or else $\mathbb{F}$ would not be a finite measure. Since $\mathcal{G}$ is eventually bounded, the total variation of $\mathcal{A} \times \mathbb{F}^{\mathcal{K}}$ is eventually bounded as well.

**Definition 13.9 (Standard Baire Setting).** A triple $(\mathcal{A}, \mathbb{F}, \mathcal{T})$ consisting of a Baire search trajector $\mathcal{A}$, a Baire fitness measure $\mathbb{F}$, and a temporal index set $\mathcal{T}$ is called a *standard Baire setting* if $\mathcal{A} = \mathbf{E}'(\mathcal{G})$ for an extensible search generator $\mathcal{G}$, $\mathbb{F}$ has a tight, setwise measurable jagged value generator, and $\mathcal{T}$ is at most countable.

**Proposition 13.6.** *Every standard Baire setting corresponds to a unique Baire measure $\mathcal{A} \times \mathbb{F}$ on the space of history traces $(X \times Y)^{\mathcal{T}}$.*

*Proof.* The family of finite dimensional distributions $\{\mathcal{A} \times \mathbb{F}^{\mathcal{K}} \mid \mathcal{K} \subseteq \mathcal{T} \text{ finite}\}$ is consistent, tight, and eventually bounded and hence extends uniquely to the desired Baire measure.

Although $\mathcal{A} \times \mathbb{F}$ has been defined based on extensible search generators, this definition can be expanded to a potentially broader set of search trajectors using linearity. That is, if $\mathcal{A} = \alpha\mathcal{A}' + \beta\mathcal{A}''$ for $\alpha, \beta \in \mathbb{R}$ with $\mathcal{A}'$ and $\mathcal{A}''$ search trajectors that are generated by extensible search generators, then by linear extension one can define

$$\mathcal{A} = \alpha(\mathcal{A}' \times \mathbb{F}) + \beta(\mathcal{A}'' \times \mathbb{F}),$$

so that whenever $\mathcal{A} \times \mathbb{F}$ is defined, it is defined for at least an entire linear subspace of all search trajectors. The same fact holds for $\mathbb{F} = \alpha\mathbb{F}' + \beta\mathbb{F}''$ irrespective of setwise measurability for $\mathbb{F}$.

The measure $\mathcal{A} \times \mathbb{F}$ over history traces can now be used in order to define a concept of performance suitable for search trajectors.

### *13.2.3 The V-Performance*

Chapter 10 introduced a performance criterion $\psi(\mathcal{G}, u)$ that maps a static search generator $\mathcal{G}$ and an objective $u$ to a real number by integrating with respect to a kernel $V : Y^X \to (X^{\mathcal{T}} \to \mathbb{R})$. The focus on *generators* instead of trajectors meant that the performance criterion was non-linear, and the limitation to a single objective ignores average performance. In the present setting, we will now focus on performance as a property of the interaction of a search trajector $\mathcal{A}$ and a fitness measure $\mathbb{F}$.

**Definition 13.10 (V-Performance).** Let $(\mathcal{A}, \mathbb{F}, \mathcal{T})$ be any triple such that $\mathcal{A} \times \mathbb{F}$ exists, and let $V : (X \times Y)^{\mathcal{T}} \to \mathbb{R}$ be measurable. Then the *V-performance* of $\mathcal{A}$ on $\mathbb{F}$ is given by

$$\langle \mathcal{A}, \mathbb{F} \rangle_V = \mathbb{E}_{\mathcal{A} \times \mathbb{F}}[V] = \int_{(X \times Y)^{\mathcal{T}}} V(h)\, \mathcal{A} \times \mathbb{F}(dh), \tag{13.6}$$

and the function $V$ is called a *value function*.

The value function applies to history traces, in contrast to the kernel of a performance criterion $\psi$, which is indexed to objectives. The term *value function* is derived from control theory, where it describes the objective value of a particular state. Here, the relevant state is the history trace $h$.

It is important to address what can and cannot be represented using this value function. Given the form $V(h)$, the value can only be computed based on the search history $h_X$ and the value history $h_Y$. Because this value function does not depend on the objective $u$, it cannot observe properties of the objective, such as the optimum $u_*$, which figures prominently in several of the performance criteria introduced in Chapter 10 based on commonly used benchmarking techniques.

The optimum $u_*$ is not necessarily measurable even in settings where $u$ can be observed. If $\mathbb{F}$ is a Baire fitness measure and both the temporal index set $\mathcal{T}$ and the search domain $X$ are uncountable, then $u_*$ is not generally measurable, nor are other function properties such as continuity or boundedness. There remain cases in which $u_*$ can nevertheless be known. For instance, the value of $u_*$ might be built in to $\mathbb{F}$, *e.g.*, by fixing $u_* = 0$. If $u_*$ is a known quantity, then it can also be built into $V$.

The primary use of performance criteria such as the average runtime or average final error is for measuring performance on benchmarks. In this case, one has a finite collection of known functions comprising the benchmark. The optima on these benchmarks are known, typically by construction or because they have closed-form analytic solutions. Although useful, benchmarks can be exploited by their very nature. If there are $N$ benchmarks that are substantially distinct from each other, then the exact benchmark being optimized can determined by evaluating at most $N-1$ search points. In fact, if the benchmarks are almost nowhere equal to each other, then the exact benchmark can be determined by just one search point evaluation. The second search point can then be selected as the known optimum of the benchmark. The best average runtime over these benchmarks is thus less than 2, and this solution would in fact be discovered by the information maximization principle introduced in the next section. Benchmarks only work if the algorithm is designed with no knowledge of the benchmark functions.

The point here is that the difficulty of observing $u_*$ is not purely academic, nor is it the result of an arbitrary choice. Real-world optimization problems commonly do not have known optima, or else the problem would not be posed as an optimization. Instead, one has metrics that one can measure and aggregate into a concept of utility.

In fact, the idea of a *value-dependent* performance criterion that was needed to express No Free Lunch theorems corresponds closely to a value function in the sense of the V-performance. We can obtain the same result by defining a value function $V(x,y)$ to be value-dependent if $V(x,y) = V(z,y)$ for all $x,z \in X^{\mathfrak{T}}$. If $\mathcal{A}$ and $\mathcal{A}'$ are any two standard search trajectors such that $\mathbf{T}(\mathcal{A})$ and $\mathbf{T}(\mathcal{A}')$ are setwise measurable and eventually non-repeating and $\mathbb{F}$ is a path independent fitness measure, then by Theorems 12.6 and 12.7, we have the NFL result

$$\langle \mathcal{A}, \mathbb{F} \rangle_V = \langle \mathcal{A}', \mathbb{F} \rangle_V$$

subject to the assumption that $\mathcal{A} \times \mathbb{F} = \mathbf{T}(\mathcal{A}) \times \mathbb{F}$. One subtle reality that value dependence highlights is that if the utility of a search path depends on the nature of the search point, then NFL is invalidated. One example of this case occurs when search points incur differing computational costs.

The notation $\langle \mathcal{A}, \mathbb{F} \rangle_V$ suggests that the V-performance is a bilinear form, but in fact it is not. Although the V-performance depends linearly on both $V$ and $\mathcal{A} \times \mathbb{F}$, the maps $\mathcal{A} \mapsto \langle \mathcal{A}, \mathbb{F} \rangle_V$ and $\mathbb{F} \mapsto \langle \mathcal{A}, \mathbb{F} \rangle_V$ are not generally linear. This result points to the nonlinearity of the operator $\mathbf{T}(\cdot)$. The map

$$(\mathbf{T}[\mathcal{A}], \mathbb{F}) \mapsto \int_{X^{\mathfrak{T}} \times Y^X} V(z, u(z))\, \mathcal{A}[u](dx) \mathbb{F}(du)$$

is obviously bilinear as a function of $\mathbf{T}[\mathcal{A}]$ and $\mathbb{F}$. Unfortunately, the static search generators are not sufficient to express the full range of problems that will be examined in Chapter 17, and the additional power that comes from this range is worth the loss of linearity.

A key assumption below is that the value function is bounded. Boundedness is primarily used below to reverse the order of integration and sums using Fubini's Theorem from real analysis; a similar result, known as Tonelli's Theorem, allows one to reverse the order of integration if the integrand is nonnegative. Incorporating other details, it will be seen that unbounded value functions $V$ can be used in some cases provided that the value function is everywhere positive. If $V(z,u) > 0$ for all pairs $(z,u)$, then $V$ is said to be *strictly positive* for this purpose.

The V-performance represents the overall performance of a search method on a search problem. Thus far, although the search trajector abstraction can represent broader forms of search and optimization, such as stochastic or dynamic optimization, the fitness measure represents a more limited form of search problems. These limits can be removed by introducing a more general representation of search problems, which will be explored further in Chapter 17, when the V-performance will be updated accordingly. At present, within the static setting, it will be shown that the optimal search trajector for the V-performance on a fixed fitness measure can be derived in many cases.

## 13.3 The Information Maximization Principle

Given that there exist non-NFL function priors that represent broad problem classes, the obvious question is: How can path dependence be exploited to make search and optimization more efficient? To answer this question, a search process can be reinterpreted as a control problem in which the value history is controlled by the choice of search points. NFL represents the case in which the value history is uncontrollable, but in all other cases, at least some control is possible. The remainder of this chapter explores how techniques from control theory – Bellman equations [20] in particular – can be applied in order to specify a search method that is optimal with respect to a given problem class. The fitness measure together with a performance criterion provides a *model* of the problem class and the search objective. If this model is changed, the optimal search method may change as well, and so such search methods will be called *model-optimal*. The form of this optimal search method can be described as choosing the search history at each step to maximize the information available about the objective based on the value history, hence the name *Information Maximization Principle* will be applied to this result.

### 13.3.1 Conditional V-Performance

The standard approach in control theory deconstructs a control policy into a series of steps in which the optimal control can be chosen dependent on the previous steps and an estimate of the value of the next step. For the V-performance, these steps are determined as conditional expectations given a history prefix.

Throughout what follows, the value function $V$ will need to be either bounded or strictly positive. A fitness measure $\mathbb{F}$ with a tight, setwise-measurable jagged value generator is used as a source of objectives for evaluation; some theorems will require it to be nonnegative. The class of search methods under consideration consists of all proper search trajectors from search domain $X$ to value space $Y$ over the temporal index set $\mathcal{T}$ such that $(\mathcal{A},\mathbb{F},\mathcal{T})$ is a standard Baire setting. Recall that the space $Y$ represents that information that the search method is allowed to see after proposing a point in the search domain $X$. The only limitation placed on $X$ and $Y$ is that both must be Hausdorff spaces that are measured with the Borel $\sigma$-algebra. Thus the information provided by $Y$ could be a function evaluation, a gradient, or something else.

In order to apply Bellman equations, we need to measure the progress of a search method on a random objective up to some time. As a random variable on $X^{\mathcal{T}} \times Y^{\mathcal{T}}$, $V$ contains information about the complete history. A conditional expectation over this space is used to restrict the available information down to a finite timestep.

Recall that $Z = (Z_t)_{t\in\mathcal{T}}$ is the optimization process as defined in Equation 9.2. Its natural filtration $(\mathcal{Z}_t)_{t\in\mathcal{T}}$ was defined in Equation 9.3. Previously, $Z_t$ was defined as a random variable on $X^{\mathcal{T}}$, but now the underlying measure space is $(X\times Y)^{\mathcal{T}}$. Thus we retain the same notation, but define $Z_t : (X\times Y)^{\mathcal{T}} \to X$ so that $Z_t(h) = h_t^X$, which then implies that $\mathcal{Z}_t$ is a sub-$\sigma$-algebra of $\mathcal{B}a_{(X\times Y)^{\mathcal{T}}}$ rather than just $\mathcal{B}a_{X^{\mathcal{T}}}$ alone.

In addition to $Z_t$, based on the same analogy to the history process of Equation 12.10, define $W_t : (X\times Y)^{\mathcal{T}} \to Y$ so that $W_t(h) = h_t^Y$, which generates a stochastic process $W = (W_t)_{t\in\mathcal{T}}$ and its natural filtration $(\mathcal{W}_t)_{t\in\mathcal{T}}$. Then $\mathcal{Z}_t$ contains information about the search paths up to time $t$, and $\mathcal{W}_t$ contains information about the objective obtained up to time $t$. Define for all $t,s\in T$

$$\mathcal{Z}_t \otimes \mathcal{W}_s = \sigma\left((Z_\tau)_{\tau\leq t}, (W_\kappa)_{\kappa\leq s}\right) = \sigma\left(\{A\cap B \mid A\in\mathcal{Z}_t, B\in\mathcal{W}_s\}\right)$$

as the $\sigma$-algebra generated by the search history up to time $t$ and the value history up to time $s$. Given a standard Baire setting $(\mathcal{A},\mathbb{F},\mathcal{T})$, the conditional expectation

$$\mathbb{E}_{\mathcal{A}\times\mathbb{F}}\left[V \mid \mathcal{Z}_t\otimes\mathcal{W}_s\right] \tag{13.7}$$

is well-defined for $t,s\in\mathcal{T}$, and it is a random variable on $X^{\mathcal{T}}\times Y^{X}$. Also define $\mathcal{Z}_0$ and $\mathcal{W}_0$ as the trivial $\sigma$-algebra $\left\{\emptyset,(X\times Y)^{\mathcal{T}}\right\}$ so that

$$\mathbb{E}_{\mathcal{A}\times\mathbb{F}}\left[V \mid \mathcal{Z}_t\otimes\mathcal{W}_0\right] = \mathbb{E}_{\mathcal{A}\times\mathbb{F}}\left[V \mid \mathcal{Z}_t\right] \quad\text{and}\quad \mathbb{E}_{\mathcal{A}\times\mathbb{F}}\left[V \mid \mathcal{Z}_0\otimes\mathcal{W}_0\right] = \langle\mathcal{A},\mathbb{F}\rangle_V.$$

Equation 13.7 is a random variable on $(X\times Y)^{\mathcal{T}}$, which is to say that it is a function from $(z,w)$ to the reals. However, due to the coordinate restrictions arising from $\mathcal{Z}_t\otimes\mathcal{W}_s$, we have

$$\mathbb{E}_{\mathcal{A}\times\mathbb{F}}\left[V \mid \mathcal{Z}_t\otimes\mathcal{W}_s\right](z,w) \;=\; \mathbb{E}_{\mathcal{A}\times\mathbb{F}}\left[V \mid \mathcal{Z}_t\otimes\mathcal{W}_s\right](z',w') \tag{13.8}$$

whenever $z_\tau = z'_\tau$ for all $\tau \leq t$ and $w_\kappa = w'_\kappa$ for all $\kappa \leq s$. Using non-standard notation, we may write for $(x,y) \in X^{[t]} \times Y^{[s]}$

$$\mathbb{E}_{\mathcal{A}\times\mathbb{F}}\left[V \mid \mathcal{Z}_t \otimes \mathcal{W}_s\right](x,y)$$

to represent the quantity in Equation 13.8. The more traditional notation

$$\mathbb{E}_{\mathcal{A}\times\mathbb{F}}\left[V \mid (Z_\tau)_{\tau\leq t} = x, (W_\kappa)_{\kappa\leq s} = y\right],$$

is cumbersome and unclear and is therefore avoided.

### 13.3.2 Optimal Completion

Using the conditional expectation above, the V-performance can in many cases be progressively decomposed into a telescoping sum. Each term in this sum can be optimized independently under the assumption that later terms have likewise been optimized. These optimized terms will be referred to as the *optimal completion* of a history prefix.

When optimizing over search methods, the search methods and search problems must form a standard Baire setting. A fitness measure $\mathbb{F}$ will be called *admissible* if there exist $\mathcal{A}$ and $\mathcal{T}$ such that $(\mathcal{A},\mathbb{F},\mathcal{T})$ is a standard Baire setting. A fitness measure is admissible if it has an extensible jagged value generator. Similarly, a search trajector will be called admissible if it is proper[1] and if there exist $\mathbb{F}$ and $\mathcal{T}$ such that $(\mathcal{A},\mathbb{F},\mathcal{T})$. A search trajector is admissible if it is the extension of a extensible proper search generator, that is, a generator that is tight, integrably measurable, and eventually bounded. Every admissible search trajector has the black box property. By invoking a standard Baire setting, it is assumed that $\mathcal{T}$ is at most countable.

**Definition 13.11.** The V-performance is progressively decomposable over countable $\mathcal{T}$ if there exists a sequence of value functions $(V_t)_{t\in\mathcal{T}}$ such that

$$\langle\mathcal{A},\mathbb{F}\rangle_V = \sum_{t\in\mathcal{T}} \langle\mathcal{A},\mathbb{F}\rangle_{V_t} \tag{13.9}$$

and each $V_t$ is a $\mathcal{Z}_t \otimes \mathcal{W}_{t-1}$-measurable random variable defined independently of $\mathcal{A}$.

Progressive decomposability is not a particularly exclusive property. All continuous value functions can be progressively decomposed. The following series of definitions and propositions identify sufficient criteria for value functions to be progressively decomposable. Although necessary conditions will not be identified, intuitively, for a value function to fail to be progressively decomposable, the value of

[1] Recall that a search trajector or any measure-valued functional is proper if it yields a probability measure on all inputs.

$V(z,w)$ must depend on the entire infinite sequences $z$ and $w$ and cannot be approximated by any finite subset of their coordinates. Although it will not be proven, it seems unlikely that such a situation is consistent with the measurability of the value function based on the product topology.

**Definition 13.12 (Sequential Transition).** Given any two sequences $\alpha, \alpha' \in A^{\mathcal{T}}$, the *sequential transition* from $\alpha'$ to $\alpha$ is the sequence of sequences $(\alpha^t)_{t\in\mathcal{T}}$ with $\alpha^t \in A^{\mathcal{T}}$ defined so that

$$\alpha_s^t = \begin{cases} \alpha_s \text{ if } s \le t \\ \alpha_s' \text{ if } s > t \end{cases}$$

This sequence transitions from $\alpha'$ to $\alpha$ one index at a time. The notation $\alpha^0 = \alpha'$ is used to denote the starting state of this sequence.

**Definition 13.13 (Progressively Convergent).** A value function $V$ is progressively convergent at $z \in X^{\mathcal{T}}$ and $w \in Y^{\mathcal{T}}$ if there exist $z' \in X^{\mathcal{T}}$ and $w' \in Y^{T}$ such that the sequential transitions $(z^t)_{t\in\mathcal{T}}$ and $(w^t)_{t\in\mathcal{T}}$ from $z'$ to $z$ and $w'$ to $w$ yield

$$V(z,w) = \lim_{t\to\sup\mathcal{T}} V(z^t, w^{t-1}).$$

If there exists a pair $(z', w')$ that makes $V$ progressively convergent at all $(z,w)$, then $V$ is simply called progressively convergent.

**Proposition 13.7.** *Suppose $\mathcal{T}$ is finite. Then any value function $V$ is progressively convergent if it is independent of the final value, i.e., for all $z,w,w'$, $V(z,w) = V(z,w')$ whenever $w_t = w_t'$ for all $t < \sup\mathcal{T}$.*

**Proposition 13.8.** *Suppose $\mathcal{T}$ is countably infinite. Then a value function $V$ is progressively convergent at $(z,w)$ if there exists a neighborhood $N$ of $V(z,w)$ whose pre-image under $V$ contains a neighborhood of $(z,w)$. If every pair $(z,w)$ has such a neighborhood, then $V$ is progressively convergent.*

*Proof.* Suppose is not progressively convergent at $(z,w)$. Then by the topological definition of limits there exist $(z', w') \in X^{\mathcal{T}} \times Y^{\mathcal{T}}$, some neighborhood $N$ of $V(z,w)$, and some $T \in \mathcal{T}$ such that for all $t \ge T$, $V(z^t, w^{t-1}) \notin N$ where $(z^t)_{t\in\mathcal{T}}$ and $(w^t)_{t\in\mathcal{T}}$ are the sequential transitions from $z'$ to $z$ and $w'$ to $w$.

Let $N' \subseteq V^{-1}(N)$ be a neighborhood of $(z,w)$. Let $(N_x^i)_{i\in\mathcal{I}}$ be a neighborhood base for $x \in X$ and similarly let $(N_y^j)_{j\in\mathcal{I}'}$ be a neighborhood base for $y \in Y$. Then by the structure of the product topology,

$$\left\{ \left( \prod_{\tau=1}^{s} N_{z_\tau}^{i_\tau} \times X^{\{\kappa|\kappa>s\}} \right) \times \left( \prod_{\tau=1}^{s'} N_{w_\tau}^{j_\tau} \times Y^{\{\kappa|\kappa>s'\}} \right) \,\middle|\, s, s' \in \mathcal{T}, i \in \mathcal{I}^{[s]}, j \in \mathcal{I}'^{[s']} \right\}$$

is a neighborhood base for $(z,w)$. Consequently, $N'$ contains some subset, call it $N''$, of this form with $(z,w) \in N''$. But then there exists $t > \max\{s, s', T\}$ such that $(z^t, w^{t-1}) \in N''$, whence $V(z^t, w^{t-1}) \in N$, contradicting the original claim and proving that $V$ is progressively convergent at $(z,w)$.

The preceding argument did not depend on the nature of $(z', w')$, and hence if there exists such a neighborhood $N = N(z, w)$ satisfying the conditions for every $(z, w)$, then it follows that $V$ is progressively convergent.

**Proposition 13.9.** *Suppose $\mathcal{T}$ is countably infinite. Then every continuous value function $V$ is progressively convergent.*

*Proof.* For any $(z, w)$, if $N$ is a neighborhood of $V(z, w)$ then its preimage is a neighborhood of $(z, w)$, whence $V$ is progressively convergent by Proposition 13.8.

**Proposition 13.10.** *If a value function $V$ is progressively convergent, then the V-performance is progressively decomposable.*

*Proof.* Fix $z' \in X^{\mathcal{T}}$ and $w' \in Y^{\mathcal{T}}$ as reference points. The proof is by construction based on sequential transitions. For a given $z \in X^{\mathcal{T}}$ and $w \in Y^{\mathcal{T}}$, let $(z^t)_{t \in \mathcal{T}}$ and $(w^t)_{t \in \mathcal{T}}$ be the sequential transitions from $z'$ to $z$ and $w'$ to $w$, respectively. Let

$$V_1(z, w) = V(z^1, w^0) \quad \text{and} \quad V_{t+1}(z, w) = V(z^{t+1}, w^t) - V(z^t, w^{t-1}). \tag{13.10}$$

By the progressive continuity of $V$, we have $V(z^t, w^{t-1}) \to V(z, w)$. Also,

$$\begin{aligned}\sum_{t \in \mathcal{T}} V_t(z, w) &= V(z^1, w^0) + \sum_{t > 1} \left\{ V(z^t, w^{t-1}) - V(z^{t-1}, w^{t-2}) \right\} \\ &= \lim_{t \to \sup \mathcal{T}} V(z^t, w^{t-1}) = V(z, w),\end{aligned}$$

which implies Equation 13.9. The final step is to note that each $V_t(z, w)$ is necessarily $\mathcal{Z}_t \otimes \mathcal{W}_{t-1}$-measurable since it only depends on $z$ up to time $t$ and $w$ up to time $t-1$. Therefore, the V-performance is progressively decomposable.

The optimal completion can now be defined based on a progressive decomposition of the value function.

**Definition 13.14 (Optimal Completion).** Suppose $V$ is progressively decomposable. The *optimal completion* of a history prefix $(x, y) \in X^{[s]} \times Y^{[s]}$ is the best possible performance on a fitness measure $\mathbb{F}$ at any time step $t \in \mathcal{T}$, defined by

$$\hat{V}_{\mathbb{F}}(x, y) = \inf_{\mathcal{A}} \sum_{t > s} \mathbb{E}_{\mathcal{A} \times \mathbb{F}} \left[ V_t \mid \mathcal{Z}_s \otimes \mathcal{W}_s \right](x, y), \tag{13.11}$$

where the infimum is taken over proper black-box search trajectors $\mathcal{A}$ such that $\mathcal{A} = \mathbf{E}(\mathcal{G})$ for some extensible search generator $\mathcal{G}$.[2]

This formulation assumes that low values of $V$ are better; otherwise the supremum may be used. If $s = 0$ then $X^0 \times Y^0$ is a one element set; call this element $(\emptyset, \emptyset)$. Then $\hat{V}_{\mathbb{F}}(\emptyset, \emptyset) = \inf_{\mathcal{A}} \langle \mathcal{A}, \mathbb{F} \rangle_V$. Thus $\hat{V}_{\mathbb{F}}(\emptyset, \emptyset)$ is the optimal $V$-performance on $\mathbb{F}$ over all proper black-box search trajectors, and $\hat{V}_{\mathbb{F}}(x, y)$ is the best performance possible

[2] Recall that a trajector is proper if it yields a probability measure for all inputs.

once $x$ and $y$ have been observed. A series of theorems will now show that the optimal completion may be computed step by step over time. Consequently, the best choices for an optimization method may be determined one step at a time.

**Theorem 13.1.** *Suppose $V$ is progressively decomposable and either bounded or strictly positive. For any admissible search trajector $\mathcal{A}$, for all $N > 0$, and all $(x,y) \in X^{[s]} \times Y^{[s]}$ with $s \in \mathcal{T}$,*

$$\hat{V}_{\mathbb{F}}(x,y) \leq \sum_{t=s+1}^{s+N} \mathbb{E}_{\mathcal{A}\times\mathbb{F}}\left[V_t \,|\, \mathcal{Z}_s \otimes \mathcal{W}_s\right](x,y) \tag{13.12}$$
$$+ \quad \mathbb{E}_{\mathcal{A}\times\mathbb{F}}\left[\hat{V}_{\mathbb{F}}\left((Z_\tau)_{\tau\leq s+N}, (W_\kappa)_{\kappa\leq s+N}\right) \middle| \, \mathcal{Z}_s \otimes \mathcal{W}_s\right](x,y).$$

*Proof.* Fix $(x,y)$ and $\delta > 0$. For each $(r,w) \in X^{[N]} \times Y^{[N]}$, choose $\mathcal{A}^{\delta}_{r,w}$ so that

$$\hat{V}_{\mathbb{F}}(x\|r, y\|w) + \delta > \sum_{t>s+N} \mathbb{E}_{\mathcal{A}\times\mathbb{F}}\left[V_t \,|\, \mathcal{Z}_s \otimes \mathcal{W}_s\right](x\|r, y\|w), \tag{13.13}$$

where $\|$ represents sequence concatenation. By assumption each $\mathcal{A}^{\delta}_{r,w} = \mathbf{E}(\mathcal{G}^{\delta}_{r,w})$ for some proper, extensible search generator $\mathcal{G}^{\delta}_{r,w}$. Define $\tilde{\mathcal{G}}$ so that

$$\tilde{\mathcal{G}}[h] = \begin{cases} \mathcal{G}[h] & \text{if} \, |h| < s+N \\ \mathcal{G}^{\delta}_{r,w}[h] & \text{if} \, x\|r \sqsubseteq h_X \text{ and } y\|w \sqsubseteq h_Y, \end{cases} \tag{13.14}$$

where $h \sqsubseteq h'$ for sequences means that $h$ is a prefix of $h'$. Note that $\tilde{G}$ is everywhere defined. It is also tight and integrably measurable since each of $\mathcal{G}$ and $\mathcal{G}^{\delta}_{r,w}$ are tight and integrably measurable. Eventual boundedness follows from the fact that each of these objects are proper, and so $\tilde{\mathcal{G}}$ is extensible. Therefore there exists admissible $\tilde{A} = \mathbf{E}(\tilde{\mathcal{G}})$ that extends $\tilde{\mathcal{G}}$. Now then,

$$\hat{V}_{\mathbb{F}}(x,y) \leq \sum_{t=s+1}^{\infty} \mathbb{E}_{\tilde{A}\times\mathbb{F}}\left[V_t \mid \mathcal{Z}_s \otimes \mathcal{W}_s\right](x,y) \tag{13.15}$$
$$= \sum_{t=s+1}^{s+N} \mathbb{E}_{\tilde{A}\times\mathbb{F}}\left[V_t \mid \mathcal{Z}_s \otimes \mathcal{W}_s\right](x,y) + \sum_{t=s+N+1}^{\infty} \mathbb{E}_{\tilde{A}\times\mathbb{F}}\left[V_t \mid \mathcal{Z}_s \otimes \mathcal{W}_s\right](x,y).$$

Using the properties of conditional expectation, for $t > s+N$,

$$\mathbb{E}_{\tilde{A}\times\mathbb{F}}\left[V_t \mid \mathcal{Z}_s \otimes \mathcal{W}_s\right] = \mathbb{E}_{\tilde{A}\times\mathbb{F}}\left[\mathbb{E}_{\tilde{A}\times\mathbb{F}}\left[V_t \mid \mathcal{Z}_{s+N} \otimes \mathcal{W}_{s+N}\right] \mid \mathcal{Z}_s \otimes \mathcal{W}_s\right]. \tag{13.16}$$

Summing on both sides, pushing the sum inside the expectation (as we may since $V$ is positive or bounded), and applying Equation 13.13,

$$\sum_{t=s+N+1}^{\infty} \mathbb{E}_{\tilde{A}\times\mathbb{F}}\left[V_t \mid \mathcal{Z}_s \otimes \mathcal{W}_s\right](x,y) \tag{13.17}$$
$$< \delta + \mathbb{E}_{\tilde{A}\times\mathbb{F}}\left[\hat{V}_{\mathbb{F}}\left((Z_\tau)_{\tau\leq s+N}, (W_\kappa)_{\kappa\leq s+N}\right) \middle| \, \mathcal{Z}_s \otimes \mathcal{W}_s\right](x,y).$$

Plugging this back into Equation 13.15 and letting $\delta$ go to zero proves the theorem for $\tilde{\mathcal{A}}$. The proof is completed by observing that $\mathcal{A}$ can be replaced by $\tilde{\mathcal{A}}$ in Equation 13.12 without change, since $V_t$ by definition does not depend on $\mathcal{A}$ or $\mathcal{A}'$ beyond time step $t \le s+N$, and $\mathcal{A}$ and $\tilde{\mathcal{A}}$ are identical up to that point by construction.

**Theorem 13.2.** *Suppose $V$ is progressively decomposable and either bounded or strictly positive. For any admissible fitness measure $\mathbb{F}$, for all $N > 0$ and all $(x,y) \in (X \times Y)^{[s]}$,*

$$\hat{V}_{\mathbb{F}}(x,y) = \inf_{\mathcal{A}} \left\{ \sum_{t=s+1}^{s+N} \mathbb{E}_{\mathcal{A}\times\mathbb{F}}\left[V_t \mid \mathcal{Z}_s \otimes \mathcal{W}_s\right](x,y) \right. \tag{13.18}$$
$$\left. + \ \mathbb{E}_{\mathcal{A}\times\mathbb{F}}\left[\hat{V}_{\mathbb{F}}\left((Z_\tau)_{\tau\le s+N}, (W_\kappa)_{\kappa\le s+N}\right) \middle| \mathcal{Z}_s \otimes \mathcal{W}_s\right](x,y) \right\},$$

*with the infimum taken over admissible search trajectors.*

*Proof.* The fact that the right-hand side is an upper bound is shown in Theorem 13.1. To see that it is also a lower bound, fix $\delta > 0$ and choose $\mathcal{A}^\delta$ so that

$$\hat{V}_{\mathbb{F}}(x,y) + \delta \quad > \sum_{t=s+1}^{\infty} \mathbb{E}_{\mathcal{A}^\delta\times\mathbb{F}}\left[V_t \mid \mathcal{Z}_s \otimes \mathcal{W}_s\right](x,y). \tag{13.19}$$

Split this sum at $t = s+N+1$ as in Equation 13.15. As in Equation 13.16, we can expand the conditional expectation to contain two conditionals. We can also push the sum inside the expectation if $V$ is either bounded (Fubini's Theorem) or strictly positive (Tonelli's Theorem), so that

$$\begin{aligned}
&\sum_{t=s+N+1}^{\infty} \mathbb{E}_{\mathcal{A}^\delta\times\mathbb{F}}\left[V_t \mid \mathcal{Z}_s \otimes \mathcal{W}_s\right](x,y) \\
&\quad = \mathbb{E}_{\mathcal{A}^\delta\times\mathbb{F}}\left[\sum_{t=s+N+1}^{\infty} \mathbb{E}_{\mathcal{A}^\delta\times\mathbb{F}}\left[V_t \mid \mathcal{Z}_{s+N}\otimes\mathcal{W}_{s+N}\right] \middle| \mathcal{Z}_s \otimes \mathcal{W}_s\right](x,y) \\
&\quad \ge \mathbb{E}_{\mathcal{A}^\delta\times\mathbb{F}}\left[\hat{V}_{\mathbb{F}}((Z_\tau)_{\tau=1}^{s+N}, (W_\kappa)_{\kappa=1}^{s+N}) \middle| \mathcal{Z}_s \otimes \mathcal{W}_s\right](x,y)
\end{aligned}$$

by the definition of $\hat{V}_{\mathbb{F}}$. Therefore,

$$\begin{aligned}
\hat{V}_{\mathbb{F}}(x,y) + \delta > &\sum_{t=s+1}^{s+N} \mathbb{E}_{\mathcal{A}^\delta\times\mathbb{F}}\left[V_t \mid \mathcal{Z}_s \otimes \mathcal{W}_s\right](x,y) \\
&+ \ \mathbb{E}_{\mathcal{A}^\delta\times\mathbb{F}}\left[\hat{V}_{\mathbb{F}}\left((Z_\tau)_{\tau=1}^{s+N}, (W_\kappa)_{\kappa=1}^{s+N}\right) \middle| \mathcal{Z}_s \otimes \mathcal{W}_s\right](x,y).
\end{aligned}$$

Letting $\delta$ become arbitrarily small proves the theorem.

**Corollary 13.1.** *For any admissible fitness measure $\mathbb{F}$ with an extensible jagged value generator $\mathcal{G}^*$, for all $(x,y) \in (X \times Y)^{[s]}$, and for $V$ progressively decomposable and either bounded or strictly positive,*

$$\hat{V}_{\mathbb{F}}(x,y) = \inf_{\alpha \in \mathcal{P}_X^{\text{tight}}} \left\{ \int_X V_{s+1}(x \| x_{s+1}, y)\, \alpha(dx_{s+1}) \right.$$
$$\left. + \int_X \int_Y \hat{V}_{\mathbb{F}}(x \| x_{s+1}, y \| y_{s+1})\; \mathcal{G}^*[x \| x_{s+1}, y](dy_{s+1})\, \alpha(dx) \right\}, \tag{13.20}$$

*where $\mathcal{P}_X^{\text{tight}}$ is the set of all tight Borel probability measures on $X$.*

*Proof.* Letting $N = 1$ in Theorem 13.2 gives $\hat{V}_{\mathbb{F}}(x,y) =$

$$\inf_{\mathcal{A}} \left\{ \mathbb{E}_{\mathcal{A}\times\mathbb{F}} \left[ V_{s+1}\left( (Z_\tau)_{\tau=1}^{s+1}, (W_\kappa)_{\kappa=1}^{s} \right) + \hat{V}_{\mathbb{F}}\left( (Z_\tau)_{\tau=1}^{s+1}, (W_\kappa)_{\kappa=1}^{s+1} \right) \Big| \mathcal{Z}_s \otimes \mathcal{W}_s \right] (x,y) \right\} \tag{13.21}$$

where $V_{s+1}$ is treated as a function over $X^{[s+1]} \times Y^{[s]}$ based on its dependency structure. Equation 13.21 depends on exactly one step of a deterministic optimizer $\mathcal{A}$. For each admissible $\mathcal{A}$, there is a probability measure $\alpha_{\mathcal{A}} \in \mathcal{P}_X^{\text{tight}}$ such that $Z_{s+1}$ is distributed according to $\alpha_{\mathcal{A}}$ given $x$ and $y$, and this measure determines $V_{s+1}$ as well when $x$ and $y$ are given. The right term under the conditional expectation likewise depends only on $\alpha_{\mathcal{A}}$ and $\mathbb{F}$. Thus the conditional expectations can be dropped given $\alpha_{\mathcal{A}}$. Further, for every measure in $\alpha \in \mathcal{P}_X^{\text{tight}}$, there is some admissible $\mathcal{A}$ with $\mathcal{A}(x,y) = \alpha$, since $\alpha$ does not by itself prevent integrable measurability. So the infimum can be taken over $\mathcal{P}_X^{\text{tight}}$ rather than over search trajectors. Applying these facts yields Equation 13.20.

These theorems have demonstrated that the value of optimal completion at each time step does not depend on the entire definition of an optimizer, but only on its definition at one step. Because the optimal completion is the best performance possible for each history, optimal decision-making can be achieved one step at a time, discussed next.

### 13.3.3 Model-Optimal Optimization

Corollary 13.1 provides a way of minimizing the optimal completion of the $V$-performance without reference to a particular optimizer. The terms under the infimum in Equation 13.20 can be used to derive $\varepsilon$-optimal search methods. A proper black-box search trajector $\mathcal{A}$ is *$\varepsilon$-optimal* for $\varepsilon > 0$ on a fitness measure $\mathbb{F}$ and a bounded value function $V$ whenever

$$\langle \mathcal{A}, \mathbb{F} \rangle_V - \inf_{\mathcal{A}'} \langle \mathcal{A}', \mathbb{F} \rangle_V < \varepsilon, \tag{13.22}$$

where the infimum is taken over all admissible search trajectors. This difference can be computed as the sum of stepwise errors. Define a function for the term under the infimum in Equation 13.20,

$$\hat{V}_{\mathbb{F}}(x_{s+1},x,y) = V_{s+1}(x\|x_{s+1},y) + \int_Y \hat{V}_{\mathbb{F}}\left(x\|x_{s+1},y\|y_{s+1}\right)\, \mathcal{G}^*[x\|x_{s+1},y](dy_{s+1}) \tag{13.23}$$

where $\mathcal{G}^*$ is the jagged value generator of $\mathbb{F}$. The purpose of this equation is to express the optimal completion under the assumption that $x_{s+1}$ is chosen as the next search point. Then Corollary 13.1 intuitively says that the optimal completion performance is found by choosing the optimal distribution from which to sample the next search point. Notice that the last term of Equation 13.23 is by definition equivalent to

$$\mathbb{E}_{\mathcal{A}\times\mathbb{F}}\left[\hat{V}_{\mathbb{F}}\left((Z_\tau)_{\tau=1}^{s+1},(W_\kappa)_{\kappa=1}^{s+1}\right)\,\middle|\, \mathcal{Z}_{s+1}\otimes\mathcal{W}_s\right](x\|x_{s+1},y).$$

Represent the change in error after choosing search point $x_{s+1}$ by

$$\mathrm{err}_{\mathbb{F},V}(x_{s+1},x,y) = \hat{V}_{\mathbb{F}}(x_{s+1},x,y) - \hat{V}_{\mathbb{F}}(x,y), \tag{13.24}$$

which is called the *one-step error gain*. Then we have the following theorem.

**Theorem 13.3.** *For $V$ progressively decomposable and either bounded or strictly positive, an admissible search trajector $\mathcal{A}$ has $V$-performance error bounded below by the sum over time of the expected one-step error gain,*

$$\langle\mathcal{A},\mathbb{F}\rangle_V - \inf_{\mathcal{A}'}\langle\mathcal{A}',\mathbb{F}\rangle_V = \sum_{t\in\mathcal{T}} \mathbb{E}_{\mathcal{A}\times\mathbb{F}}\left[\mathrm{err}_{\mathbb{F},V}(Z_t,(Z_\tau)_{\tau=1}^{t-1},(W_\kappa)_{\kappa=1}^{t-1})\right]. \tag{13.25}$$

*Proof.* It will be shown by induction that for $V$ bounded or strictly positive,

$$\hat{V}_{\mathbb{F}}(\emptyset,\emptyset) = \mathbb{E}_{\mathcal{A}\times\mathbb{F}}\left[\sum_{t\in\mathcal{T}}\left\{V_t((Z_\tau)_{\tau=1}^{t},(W_\kappa)_{\kappa=1}^{t-1}) - \mathrm{err}_{\mathbb{F},V}((Z_\tau)_{\tau=1}^{t-1},(W_\kappa)_{\kappa=1}^{t-1})\right\}\right], \tag{13.26}$$

which is equivalent to Equation 13.25 using Equation 13.9. The induction inserts a term $\hat{V}_{\mathbb{F}}\left((Z_\tau)_{\tau=1}^{t},(W_\kappa)_{\kappa=1}^{t}\right)$ under the expectation and truncates the sum at time $t$. For the base case $t=1$, by expanding the definition it holds that

$$\mathbb{E}_{\mathcal{A}\times\mathbb{F}}\left[V_{\mathbb{F}}(Z_1,\emptyset,\emptyset)\right] = \mathbb{E}_{\mathcal{A}\times\mathbb{F}}\left[V_1(Z_1,\emptyset) + \hat{V}_{\mathbb{F}}(Z_1,W_1)\right].$$

Moving the left side over to the right, adding $\hat{V}_{\mathbb{F}}(\emptyset,\emptyset)$ on both sides, and substituting the definition of err yields

$$\hat{V}_{\mathbb{F}}(\emptyset,\emptyset) = \mathbb{E}_{\mathcal{A}\times\mathbb{F}}\left[V_1(Z_1,\emptyset) + \hat{V}_{\mathbb{F}}(Z_1,W_1) - \mathrm{err}_{\mathbb{F},V}(Z_1,\emptyset,\emptyset)\right],$$

proving the base case. Suppose for the inductive case that

$$\hat{V}_{\mathbb{F}}(\emptyset,\emptyset) = \mathbb{E}_{\mathcal{A}\times\mathbb{F}}\left[\sum_{s=1}^{t}\left\{V_s((Z_\tau)_{\tau=1}^{s-1},(W_\kappa)_{W=1}^{s-1}) - \mathrm{err}_{\mathbb{F},V}(Z_s,(Z_\tau)_{\tau=1}^{s-1},(W_\kappa)_{W=1}^{s-1})\right\}\right.$$
$$\left. + \; \hat{V}_{\mathbb{F}}((Z_\tau)_{\tau=1}^{t},(W_\kappa)_{\kappa=1}^{t})\right]. \tag{13.27}$$

Taking the expectation in $\mathcal{A} \times \mathbb{F}$ on both sides of Equation 13.24, expanding $\hat{V}_{\mathbb{F}}(x_{s+1}, x, y)$ by the definition in Equation 13.23, and rearranging terms obtains $\mathbb{E}_{\mathcal{A}\times\mathbb{F}}\left[\hat{V}_{\mathbb{F}}((Z_\tau)_{\tau=1}^t, (W_\kappa)_{\kappa=1}^t)\right] =$

$$\mathbb{E}_{\mathcal{A}\times\mathbb{F}}\left[V_{t+1}((Z_\tau)_{\tau=1}^{t+1}, (W_\kappa)_{\kappa=1}^{t}) + \hat{V}_{\mathbb{F}}((Z_\tau)_{\tau=1}^{t+1}, (W_\kappa)_{\kappa=1}^{t+1}) \right. \tag{13.28}$$
$$\left. - \operatorname{err}_{\mathbb{F},V}(Z_{t+1}, (Z_\tau)_{\tau=1}^{t}, (W_\kappa)_{\kappa=1}^{t})\right].$$

Plugging Equation 13.28 into Equation 13.27 shows that Equation 13.28 holds with $t$ replaced by $t+1$, and the induction is complete since $\hat{V}_{\mathbb{F}}(\emptyset, \emptyset) =$

$$\underset{\mathcal{A}\times\mathbb{F}}{\mathbb{E}}\left[\sum_{t\in\mathcal{T}}\left\{V_t((Z_\tau)_{\tau=1}^{t}, (W_\kappa)_{\kappa=1}^{t-1}) - \operatorname{err}_{\mathbb{F},V}(Z_t, (Z_\tau)_{\tau=1}^{t-1}, (W_\kappa)_{\kappa=1}^{t-1})\right\} + \hat{V}_{\mathbb{F}}(Z, W)\right].$$

Equivalence with Equation 13.26 is established by observing that the sum in Equation 13.11 is vacuously zero when $s \geq \sup \mathcal{T}$ so that $\hat{V}_{\mathbb{F}}(Z, W) = 0$, which completes the proof.

Based on the result of Theorem 13.3, it is clear that a search method is only $\varepsilon$-optimal if the infinite sum of the one step error gain converges to something less than $\varepsilon$. For all $\eta > 0$, define a set $\mathcal{V}_\eta[x, y]$ of $\eta$-optimal choices for $x_{s+1}$ by

$$\mathcal{V}_\eta[x, y] = \left\{z \in X \mid \operatorname{err}_{\mathbb{F},V}(z, x, y) < \eta\right\}, \tag{13.29}$$

and let $\overline{\mathcal{V}}_\eta[x, y]$ be defined in the same way with $\leq$ in place of $<$. Underneath an expectation with respect to the measure $\mathcal{A} \times \mathbb{F}$, we may write $\mathcal{V}_\eta^t$ and $\overline{\mathcal{V}_\eta^t}$ to mean $\mathcal{V}_\eta\left[(Z_\tau)_{\tau=1}^{t-1}, (W_\kappa)_{\kappa=1}^{t-1}\right]$ and $\overline{\mathcal{V}}_\eta\left[(Z_\tau)_{\tau=1}^{t-1}, (W_\kappa)_{\kappa=1}^{t-1}\right]$, respectively. The next corollary summarizes the nature of an $\varepsilon$-optimal search method in terms of the one-step error gain.

**Corollary 13.2.** *An admissible search trajector $\mathcal{A}$ is $\varepsilon$-optimal on an admissible fitness measure $\mathbb{F}$ and $V$ progressively decomposable and either bounded or strictly positive if and only if the sequence $(\eta_t)$ defined by*

$$\eta_t = \mathbb{E}_{\mathcal{A}\times\mathbb{F}}\left[\operatorname{err}_{\mathbb{F},V}(Z_t, (Z_\tau)_{\tau=1}^{t-1}, (W_\kappa)_{\kappa=1}^{t-1})\right] \tag{13.30}$$

*has $\sum_{t\in\mathcal{T}} \eta_t < \varepsilon$, and if $\mathcal{A}$ is deterministic then $Z_t \in \overline{\mathcal{V}_{\eta_t}^t} \setminus \mathcal{V}_{\eta_t}^t$ almost surely.*

*Proof.* The corollary is a direct consequence of the definitions and Theorem 13.3.

The meaning of these results is that an $\varepsilon$-optimal search trajector distributes its errors at each step in a way that sums below $\varepsilon$. For deterministic search trajectors, something stronger has been said: An $\varepsilon$-optimal deterministic search trajector $\mathcal{A}$ must choose all points within the set $\overline{\mathcal{V}_{\eta_t}^t}[x, y]$ for some sequence $(\eta_t)$ with $\sum_t \eta_t < \varepsilon$, except possibly on some set of histories that have $\mathcal{A} \times \mathbb{F}$-measure zero.

Suppose now that the error $\varepsilon$ is reduced to zero. If a search trajector is $\varepsilon$-optimal for all $\varepsilon > 0$, then it is said to be *model-optimal* on an objective $\mathbb{F}$ and a value function $V$. It is not clear that model-optimal optimizers always exist, but if they do then their form can be expressed as a formula. This formula will be termed the Information Maximization Principle. To explain the notation used below, for any $t \in \mathfrak{T}$ and any $a \in X^{[t]}$, define $U_a(u) = (u(a_s))_{s \leq t}$, which is a random variable on $Y^X$. Let $\mathfrak{U}_a$ is the $\sigma$-algebra generated by the random variable $U_a \in Y^{[t]}$.

**Information Maximization Principle:** A model-optimal optimizer makes choices so that for a search history $x \in X^{[t]}$ with objective values $y \in Y^{[t]}$,

$$x_{t+1}^{\text{opt}} \in \operatorname*{arg\,inf}_{x_{t+1} \in X} \mathbb{E}_{\mathbb{F}} \left[ V_{t+1}(x \| x_{t+1}, U_{x \| x_{t+1}}) \middle| \mathfrak{U}_x \right](y). \tag{13.31}$$

The choice of $x_{t+1}^{\text{opt}}$ is not necessarily unique. A more concise expression can be stated using the one-step error gain,

$$x_{t+1}^{\text{opt}} \in \operatorname*{arg\,min}_{x_{t+1} \in X} \operatorname{err}(x_{t+1}, x, y). \tag{13.32}$$

Choosing points according to Equations 13.31 or 13.32 if possible guarantees that the sequence of search choices is model-optimal.

The name *Information Maximization Principle* is justified because the model-optimal choice inherently requires knowledge of the fitness measure $\mathbb{F}$, and it uses this information through the value filtration $\mathfrak{U}_x$ in order to select the points that are most likely to provide the best overall outcome given the values observed thus far. In this way, use of the conditional expectation permits the correct balance of exploration and exploitation during search by maximizing its use of the available information.

## 13.4 Discussion of the Information Maximization

The Information-Maximization Principle applies Bellman equations to optimization in order to determine the optimal search method given that the underlying distribution of search problems is known. In this section similar approaches to optimal search choices are reviewed. The practicality of computing model-optimal search points is discussed, along with heuristics for choosing search points inspired by the information maximization principle.

### *13.4.1 Related Methods*

Bellman equations were introduced several decades ago for control, and consequently similar approaches have been followed previously.

Regarding the general problem formulation, Droste, Jansen, and Wegener [44] introduced the *black-box complexity* of a class of objective functions $\mathfrak{F}$ on a class of optimizers $\mathfrak{A}$ as the least value obtained by some optimizer for the worst expected runtime on any problem. That is,

$$\mathcal{BBC}(\mathfrak{A}, \mathfrak{F}) = \inf_{\mathcal{A} \in \mathfrak{A}} \sup_{u \in \mathfrak{F}} \mathbb{E}_{\mathcal{A}[u]} \left[ V(Z, u(Z)) \right]. \tag{13.33}$$

Several general results have been obtained in this framework [101, 40]. Although these authors have not explored connections to Bellman equations, the present results may nonetheless be useful for determining bounds on some problems. the notable difference here is that the V-performance studies the average performance of a search method over a random objective, whereas the black-box complexity is a lower bound over all problems.

Two results in control theory are worth mentioning. Hutter [78] published a theory of *universal artificial intelligence* in which a reinforcement learner named AIXI chooses actions to maximize expected future rewards averaged according to a *universal prior*. For this purpose, it is assumed that the likelihood of a particular environment is inversely proportional to the length of the shortest program computing it [149]. AIXI seeks to avoid model bias by weighting future outcomes accordingly. The formulae that result are close to Equation 13.31. The AIXI model only applies to discrete settings. Issues of computability are rigorously accounted for, and the model is proven computable in the limit [78]. If the results in this paper are restricted to computability, the two results may be equivalent.

The *knowledge gradient* was studied by Frazier et al. and Powell and Ryzhov [56, 129]. A learner is presented with multiple information sources (e.g., the search domain of an optimization problem). The knowledge gradient is the marginal value of the information gained by measuring any particular source; it is similar to the one-step error gain from Equation 13.24. They show that a policy that maximizes the knowledge gradient is asymptotically optimal under certain conditions [56]. The knowledge gradient is used with a Gaussian process as an objective measure. Their objective functions are stochastic in contrast to the static objectives studied here. The experimental success of the knowledge gradient methods on problems with sparse information suggests that the theory of information maximization is practically applicable.

These two related methods illustrate the value of connecting search performance to control theory through the principle of information maximization.

### 13.4.2 Challenges and Practical Application

One interesting aspect of model-optimal optimizers is that Equation 13.31 contains an inner optimization loop. Due to issues of infinite regress, this optimizer is likely to be uncomputable in many cases. Methods that perform an inner optimization have been proposed previously; the inner optimization loop can be computationally expensive [141]. Given the computational difficulties, it is important to consider approximations and heuristics based on information maximization.

If a model of the fitness measure $\mathbb{F}$ is available, a sampling approach can be used to approximate Equation 13.31 for some value functions. If the value function is the sum of weighted errors, *i.e.*, $V(z,w) = \sum_{t=1}^{\infty} \alpha_t \inf_{s \leq t} w_s$ for convergent $(\alpha_t)$, future errors become practically irrelevant beyond a fixed time horizon. A forest of future paths could be sampled using $\mathbb{F}$ to select the point with the best expected future performance. Rolet et al. have previously used a similar Monte Carlo method and observed faster convergence times [134].

However, the model $\mathbb{F}$ may not be known in advance. It must then be chosen to have properties that appear to coincide with observed search and optimization problems. For example, the AIXI model in particular was designed to be agnostic with respect to model uncertainty [78]. The Gaussian model used with the knowledge gradient averages out uncertainty based on the Central Limit Theorem [129].

Even if the model $\mathbb{F}$, it may not be feasible to compute it or to optimize against it. Common probability distributions cannot be computed exactly, even with a truly random source. It would be interesting to know the conditions under which the model-optimal optimizers could be approximated computably. In general, however, the fact that one can only compute with pseudorandom numbers today does not preclude the truly random computing in the future. Similarly, there may exist special cases that can be solved analytically.

The model-optimal optimizers developed above apply to static objectives. Similar results hold for stochastic and dynamic objective functions. This topic will be addressed briefly in Chapter 17 after the introduction of a broader set of search problems. In that case, it will be found that the same results apply almost verbatim.

### 13.4.3 Martingale Optimization

The Information Maximization Principle suggests a new approach to optimization that will be termed *martingale optimization*. A martingale optimizer generates an optimization process that is a martingale with respect to the filtration generated by the evaluation history. In this approach, the results of objective evaluation as an information source that the optimizer can use to choose which points to evaluate. The following paragraphs outline one possible martingale optimizer that can be used if a fitness measure $\mathbb{F}$ is known.

For the sake of discussion, optimization can be treated as a kind of two-player zero-sum game in the sense of von Neumann [117]. In this game, the first player is an optimization method, which chooses a search point. The second player is an optimization problem, which assigns a value to each search point chosen by the first player. The value function for this game observes the sequence of choices from both players, $z \in X^{\mathcal{T}}$ for the optimizer and $w \in Y^{\mathcal{T}}$ for the optimization problem, and assigns a value $V(z,w)$ to the end result.

Suppose now that result of the game is the minimal value obtained within the first $m$ unique search points chosen by the optimizer, corresponding to the performance criterion $\zeta_{T_m}$ in Chapter 11. Thus the optimizing player can spend $m-1$ moves to develop a good model of the objective function around the optimum, and then one final move to guess the minimal point based on the model.

Reviewing Equation 13.31 and applying the Information Maximization Principle, the optimal final move given prior search choices $x$ with values $y$ is given by

$$x_m^{\text{opt}} \in \underset{x_m \in X}{\arg\min}\ \mathbb{E}_{\mathbb{F}}\left[U(x_m) \middle| \mathfrak{U}_x\right](y). \tag{13.34}$$

which is the most rational estimate of the true minimum given the information in $\mathfrak{U}_x$. The initial moves serve to prepare $\mathfrak{U}_x$ so that it holds as much useful information as possible. The random variable $U$ here is the identity function on $Y^X$, so that $U(x_m)$ is the value of the objective at the search point $x_m$.

Computing Equation 13.34 requires an estimate of the objective value $U(x_m)$ at each possible point $x_m$ in the search domain. Suppose that the optimization method maintains an estimate $U_t(x')$ approximating the random objective at each time step $t$; the natural estimator is of course $U_t(x') = \mathbb{E}_{\mathbb{F}}\left[U(x') \mid \mathfrak{U}_z\right](w)$ for the prior history $(z,w) in X^{[t]} \times Y^{[t]}$. In order to maximize the relevant information in $\mathfrak{U}_x$ during the first $m-1$ moves, the first player could attempt to minimize the variance of the estimator $U_t(x)$, since variance represents uncertainty and the estimate is most likely to be mistaken in areas where its variance is high. However, merely minimizing the variance is not enough. In some places, the variance may be high, but the nearby values of the objective function are so large that the true minimum of the function is highly unlikely to reside in that region.

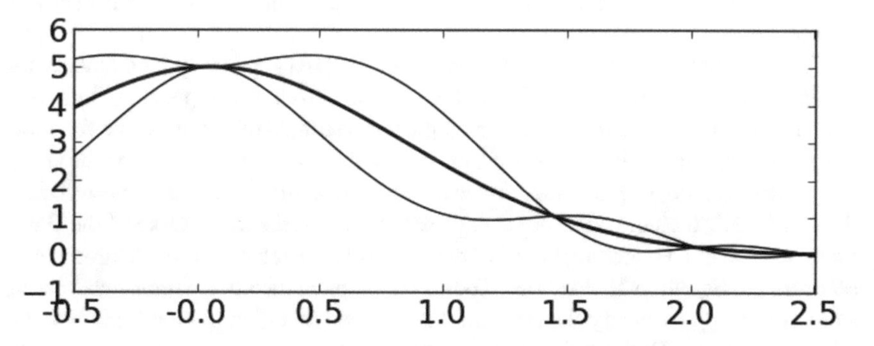


**Fig. 13.1** An example showing an estimated objective function such that the variance of the estimate increases with distance from the control points. The highest variance occurs at $x = 0.75$, but a new optimum is much more likely to be found near 2.25. When optimizing an objective function, merely reducing variance is not enough; the objective values must be taken into account as well. The optimal point to choose should trade off between reducing variance and improving the objective value.

This situation is visualized in the context of a Gaussian process in Figure 13.1. In this case, it is more profitable to minimize the variance in regions where the objective value is low. Thus there is a tradeoff between removing uncertainty in general and removing uncertainty near the expected minimum. Proportional sampling provides one way to address this tradeoff. Let $L_t$ be the lower variance-adjusted estimate of $U_t$ given history $(z,w) \in X^{[t]} \times Y^{[t]}$,

$$L_t(x') = U_t(x') - \alpha\sqrt{\mathrm{Var}\left[U(x') \mid \mathfrak{U}_z\right](w)}$$

for some $\alpha$, with $\alpha = 1$ being the first standard deviation. Then, for $t < m$, the optimizing player can choose $Z_t$ with probability

$$\mathbb{P}\left(Z_t \in dx'\right) \propto \exp\left(\frac{-L_t(x')}{T}\right), \tag{13.35}$$

where $T$ is a factor that controls the sensitivity of the strategy to different values of $L_t$. The probability in Equation 13.35 chooses points proportionally according to the variance-adjusted estimate $L_t$. It balances the choice of points near known good values against the need to increase evaluations in regions with high variance. This probability is similar to the Boltzmann distribution used by simulated annealing.

In summary, then, the proposed strategy for the optimizing player is to choose $m-1$ points in succession according to Equation 13.35 followed by a single point chosen according to Equation 13.34. This strategy is an information-maximizing strategy. It attempts to develop a useful set of information near the apparent optima and then makes its best guess at the end. Such a strategy takes advantage of the control that the optimizing player has over which points can be evaluated. While the tradeoff between exploration and exploitation encoded in Equation 13.35 may not be optimal, it seems plausible that this strategy or a similar one could perform well in practice.

Curiosity Search, introduced by Schaul et al. [141], partially implements the strategy described above. In Curiosity Search, the current set of points and its evaluations are used to construct a Gaussian process estimate of the objective function. The next evaluation point is selected by an internal optimization routine on the Gaussian process. If the objective function is drawn from a diffusion prior as described in Section 13.1.2, then the conditional expectation of $U$ is the mean value of the Gaussian process with a matching kernel. Thus Curiosity Search selects evaluation points according to Equation 13.34 above. To fully implement the information-maximizing strategy for $\zeta_{T_m}$, Curiosity Search would need to be modified to intentionally minimize the variance. Without doing so, however, this method still obtains good results on optimization.

In order to implement the variance-minimizing strategy above, it must be possible to compute or approximate both $\mathbb{E}\left[U(x_m) \mid \mathfrak{U}_x\right]$ and $\mathrm{Var}\left[U(x') \mid \mathfrak{U}_z\right]$. Thus this strategy is still not completely specified, and a computable fitness measure $\mathbb{F}$ is needed in order to instantiate it. In the next chapter, a simpler strategy named *evolutionary annealing* is proposed that ignores the conditional variance and assumes that the function prior is a random mixture sieve. This simpler strategy can be fully specified and efficiently implemented, as demonstrated in the next chapter.

## 13.5 Conclusion

The NFL Identification Theorem showed that the NFL property implies path independence. Path independence in turn implies that learning and prediction are impossible. Given that learning and prediction are observed in reality, one must conclude that common classes of search problems are not subject to NFL. That is, general-purpose optimizers exist, and it makes sense to search for them. This concept was discussed from a pragmatic and philosophical point of view in Section 13.1, where it was conjectured that NFL produces hypothesis that necessarily violate Occam's razor. Thus, if smaller problem descriptions are more likely, then effective general-purpose black-box optimizers exist.

When the function prior is non-NFL, then performance varies over search methods and problems. In the case where the problem class is constrained by some fixed fitness measure, one wishes to know which search methods will perform best on the given problems. The Information Maximization Principle addresses this question in detail using principles of control theory and reveals that the optimal search choices are selected according to Equation 13.31. A theoretical strategy inspired by this idea was presented in Section 13.4.3 based on the Information Maximization Principle. In the next chapter, a concrete class of optimizers is proposed that implements aspects of this information-maximizing approach.

# Chapter 14
# The Evolutionary Annealing Method

In Chapter 13, an optimization method was shown to achieve its best performance on a given problem by making full use of the information about the objective function obtained from function evaluations, and martingale optimizers were proposed as a consequence. In this chapter, *evolutionary annealing* is proposed as a practical martingale optimizer, *i.e.*, a general-purpose optimization technique that efficiently uses past evaluations in order to select new evaluation points. Like simulated annealing, evolutionary annealing is a meta-technique that can be applied to many optimization tasks. This chapter introduces the basic algorithm and shows theoretically that instances of the algorithm converge to the global optimum under certain conditions. The next two chapters evaluate this approach experimentally in two specific spaces, finite-dimensional Euclidean space and neural networks.

## 14.1 Foundation

In the previous chapter, a martingale optimizer was defined as any optimizer whose optimization process is a martingale with respect to the evaluation history. Evolutionary annealing, introduced in this chapter, is a martingale optimizer. Much like the strategy described in Section 13.4.3, it chooses points proportionally to their expected objective value. It also adopts several simplifying assumptions that make it possible to implement the algorithm efficiently. The conditional variance is not used to select new points, and the particular form of the conditional expectation is determined by an implementation-specific family of mutation operators. Given the discussion of globally optimal optimizers in Chapter 13, it may seem disappointing that the practical algorithm proposed here does not implement many of the concepts suggested by the theory. However, it does apply the core ideas of martingale optimization, and the resulting optimizer is competitive with other state-of-the-art optimizers despite its limitations. The success of evolutionary annealing suggests that future work on developing efficient ways to implement more aspects of the Information Maximization Principle is likely to be rewarded.



A. J. Lockett, *General-Purpose Optimization Through Information Maximization*,
Natural Computing Series, https://doi.org/10.1007/978-3-662-62007-6_14

### 14.1.1 Martingales vs. Markov Chains

Martingale optimization stands in contrast to optimization methods based on the convergence of Markov chains. Simulated annealing, for instance, converges globally in some instances because its sequence of accepted points generates an irreducible, aperiodic Markov chain that satisfies the principle of detailed balance [90]. Most evolutionary algorithms are also Markov; the population for each generation is constructed stochastically from only the population in the prior generation. As a result, these algorithms can discover and then forget high-quality regions within the search domain. They can therefore miss crucial information from the past, resulting in suboptimal performance.

This problem can be alleviated by selecting new evaluation points based on the entire pool of previously observed solutions. A genetic algorithm with non-Markovian selection can in principle become trapped in local optima by prematurely focusing on a narrow region of the search space. Evolutionary annealing combines genetic algorithms and simulated annealing using martingales in a manner that prevents this premature focus, resulting in an evolutionary algorithm that takes advantage of the full information gathered from the entire history of function evaluations. Evolutionary annealing solidly outperforms both genetic algorithms and simulated annealing, and compares favorably with the bank of stochastic optimization methods tested in Chapter 11.

### 14.1.2 Characteristics of Evolutionary Annealing

Evolutionary annealing is a global optimization algorithm for Borel-measurable spaces that can be alternately viewed as a genetic algorithm with non-Markovian selection or as a method for performing simulated annealing without the Metropolis sampler. Evolutionary annealing introduces two annealed selection operators, exploiting a connection between the average effect of proportional selection and the annealed Boltzmann distributions used in simulated annealing. Although many genetic algorithms have previously employed the Boltzmann distribution for selection (e.g. [61, 82, 110]), evolutionary annealing is distinct from these approaches in that it can select any member of any prior population and does so using information generated by a sequence of refining partitions of the search domain. Evolutionary annealing is distantly related to Estimation of Distribution Algorithms (EDAs), since it builds a global model of the annealing distributions for the fitness function (see Section 2.7.2, [122, 111]). However, whereas EDAs build models based solely on the best members of the immediately prior generation, evolutionary annealing maintains a martingale model based on the entire history of observation. By leveraging the information acquired from function evaluations, evolutionary annealing builds an increasingly refined estimate of the fitness function that allows it to locate the global optimum. To illustrate this process, the progress of an example run of evolutionary annealing in a two-dimensional space is shown in Figure 14.1.

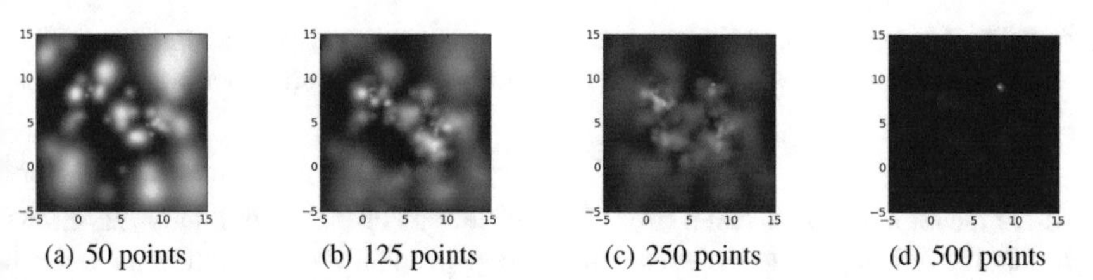

(a) 50 points (b) 125 points (c) 250 points (d) 500 points

**Fig. 14.1** Example run of evolutionary annealing on Shekel's Foxholes in two dimensions (shown in Figure 11.1(e)). Images are heat maps displaying the estimated probability density of evolutionary annealing, that is, the probability that each point will occur in the next generation of evolutionary annealing. White areas are more probable, and dark areas are less probable. Successive frames show how the probability density changes once 50, 125, 250, and 500 points have been evaluated. The resulting distribution increasingly models the fitness function; comparison with Figure 11.1(e) confirms that after 500 evaluations, evolutionary annealing has focused on the true global optimum.

Theoretically, evolutionary annealing converges asymptotically to the true global optima of the fitness function. The proof is given in Section 14.3.6. Experimentally, evolutionary annealing converges at a controlled rate as demonstrated on the twelve global optimization benchmarks from Chapter 11. Because of its efficient use of information gained from evaluations, evolutionary annealing performs well in a comparison with the other optimization methods evaluated in Chapter 11, specifically simulated annealing (SA), differential evolution (DE), evolution strategies with correlated matrix adaption (CMA-ES), particle swarm optimization (PSO), the real-coded Bayesian optimization algorithm (rBOA), a real-coded genetic algorithm (rGA), the Nelder-Mead algorithm (NM), a basic generating set search (GSS), and conjugate gradient descent (CG).

### *14.1.3 Expected Proportional Selection*

Evolutionary annealing builds on concepts from simulated annealing and evolutionary algorithms (discussed in Section 2.5.1, Section 2.6, and Chapter 7). There is an interesting theoretical connection between genetic algorithms and simulated annealing that motivates the global selection mechanism of evolutionary annealing. This connection is exposed by trivial manipulations of a previous result of Mühlenbein and Mahnig [110], as will be discussed in this subsection.

Many genetic algorithms employ *proportional selection*, where individuals in the prior population are selected proportionally to their observed fitness (see Section 7.2.4). Much like simulated annealing, proportional selection sharpens the fitness function implicitly with each generation, so that on averaging over population trajectories the selection operator asymptotically places probability one on the optima of the fitness function. Following Mühlenbein and Mahnig [110], proportional selection at the $n^{th}$ population is given by $\mathcal{S}^n_u(x) \propto u(x) N^{n-1}_x$, where $u$ is the fitness function, $\mathcal{S}_u^n(x)$ is the probability of selecting $x$ at time $t$, and $N_x^n$ is a random variable indicating the number of copies of the solution $x$ in the population at time $n$. Taking the expectation,

$$\mathbb{E}\left[\mathcal{S}_u^n(x)\right] \propto u(x)\mathbb{E}\left[N_x^{n-1}\right], \tag{14.1}$$

where the expectation is taken over all possible runs of the evolutionary algorithm, so that the left-hand side represents the average probability of selecting a fixed $x$ at time step $n$, with the average taken over all possible evolutionary runs. This average selection probability is here termed *expected proportional selection*. It differs from proportional selection in that expected proportional selection may assign positive probability to any point in the search domain. To understand this rule intuitively, imagine running all possible evolutionary trajectories simultaneously; then expected proportional selection describes which proportion of runs choose $x$ to fill a slot in the next population at time $n$.

Suppose that expected proportional selection is treated as a standalone evolutionary algorithm; that is, each successive population is simply sampled with $\mathbb{P}(x) = \mathbb{E}[\mathcal{S}_u^n(x)]$ for all $n$. If the initial population ($n = 0$) is selected uniformly at random, then $\mathbb{E}\left[N_x^0\right]$ is independent of $x$ since each $x$ is equally likely, so

$$\mathbb{E}\left[\mathcal{S}_u^1(x)\right] \propto u(x).$$

When a population of size $K$ is selected by expected proportional selection, then $N_x^n/K$ is just the proportion of the population taking the value $x$. Consequently, $\mathbb{E}[\mathcal{S}_u^n(x)] = \mathbb{E}[N_x^n]/K$, which is simply what it means to sample the population from $\mathbb{E}[\mathcal{S}_u^n(x)]$. Using Equation 14.1, this fact implies that $\mathbb{E}\left[\mathcal{S}_u^2(x)\right] \propto u(x)\mathbb{E}[N_x^1] \propto u(x)^2$. This logic can be repeated recursively to yield $\mathbb{E}[S_u^n(x)] \propto u(x)^n$. Thus expected proportional selection sharpens the fitness function. Introducing $g(x) \equiv -\log(u(x))$,

$$\begin{aligned}\mathbb{E}\left[S_u^n(x)\right] &\propto \exp(-g(x))^n \\ &= \exp\left(-\frac{1}{n^{-1}}g(x)\right).\end{aligned} \tag{14.2}$$

Comparing Equation 4.5 to Equation 14.2, expected proportional selection is found to have an annealing distribution on $-\log u$ with cooling schedule $T_n = n^{-1}$. Since the logarithm is monotonic, the maxima of $u$ are the minima of $g$.

Expected proportional selection is not a feasible selection rule, because it requires total knowledge of the fitness function a priori. If such knowledge were possible, there would be no need for iterative optimization. The optima would already be known. Expected proportional selection could be estimated by averaging over the trajectories of several different runs of a genetic algorithm, but the number of trajectories required for a good estimate would be intractably large. Genetic algorithms with proportional selection can be viewed as an approximation of expected proportional selection.

Evolutionary annealing exploits the theoretical relationship between simulated annealing and genetic algorithms to create a hybridized algorithm that merges qualities of both algorithms, as will be described next.

## 14.2 The Evolutionary Annealing Approach

This section defines the evolutionary annealing algorithm. The formal context and notation are introduced first, followed by the algorithmic details.

### 14.2.1 Formal Context and Assumptions

As elsewhere in this text, the search domain $X$ is assumed to be a topological space with a given Hausdorff topology, and then $(X, \mathcal{B}_X)$ is the measurable space generated by the Borel $\sigma$-algebra on this topology on $X$. The value space will be fixed as $Y = \mathbb{R}$, meaning that the functions to be optimized are real-valued functions. In order to provide grounding for the algorithms, evolutionary annealing is defined with respect to a base measure $\lambda$ on $(X, \mathcal{B}_X)$ that is finite and positive on all open sets. The objective function $u : X \to \mathbb{R}$ is to be minimized, and we assume that $u$ has all necessary integrability properties required by the formulae that follow. Primarily, $\exp(-u/T)$ must be $\lambda$-integrable for $T > 0$.

Since evolutionary annealing is a population-based algorithm, the content and notation of Chapter 7 is used. Thus $H = (P_n)_{n\in\mathcal{T}/K}$ will represent the population process, the sequence of stochastic populations paired with their fitness values, for a fixed population size $K$, where the index set $\mathcal{T}/K$ indexes populations of size $K$ by taking each successive $K$ items from the temporal index set $\mathcal{T}$, which is assumed here to be at most countable. Each population $P_n$ contains $K$ pairs, with each pair denoted as $P_i^k = (x,y)$ for the $k^{th}$ member of the $n^{th}$ population where $x$ is a search point and $y$ is a fitness value. Thus for any positive integer $m$ and any history trace $h \in \mathcal{H}_{X,Y}^{[mK]}$, there is a corresponding population process $H(h) \in \mathcal{H}_{X^K,Y^K}^{[m]}$ and an inverse projection $\text{traj}_K(H(h)) = h$.

For evolutionary annealing, the selection rule will be defined with respect to all unique search points in the population process $H$ up to a given time. Let $A_n$ represent the set of all search points proposed up to and including the $n^{th}$ population, $A_n = \bigcup_{j\leq n,k} \left\{x \mid P_j^k = (x,y)\right\}$. With these definitions, the basic algorithm can be defined.

### 14.2.2 Basic Algorithm

Evolutionary annealing consists of selection and variation phases. The population $P_{n+1}$ is sampled one individual at a time in these two stages. In the selection phase, an element $a \in A_n$ is selected with probability

$$p_n(a) = \xi_n \exp\left(-\frac{u(a)}{T_n}\right)\lambda\left(E_n^a\right), \tag{14.3}$$

where $T_n$ is a cooling schedule, $\xi_n$ is a normalizing factor computed with respect to the set $A_n$, and $\lambda\left(E_n^a\right)$ is the measure of a region surrounding the point $a$, discussed below. This selection mechanism will be termed *annealed proportional selection* based on the relationship between expected proportional selection and annealing described in the prior section. Expressed as a search generator using the formalisms introduced in Chapters 6 and 7, annealed proportional selection may be written as

$$\mathcal{APS}\langle T\rangle[h](\{a\}) = \mathbb{1}_{A_n}(a)\, p_n(a), \tag{14.4}$$

where $n = \lfloor |h|/K \rfloor$ is the current population, $T = (T_n)_{n=1}^{\infty}$ is the cooling schedule and $\mathbb{1}_{A_n}$ is used to ensure that the set $A_n$ has probability one, as required by the formal definition of a selection rule. The primary distinction of $\mathcal{APS}$ is that it can select any member of any prior population; that is, it ranges over $A_n$ and not just $P_n$.

For the variation phase, evolutionary annealing requires a family of probability distributions $\{\nu_n^x\}_{x\in X}$ used to mutate selected points, so that given a selected point $x$, $\nu_n^x$ is used to vary $x$ at time $n$. The choice of mutation distributions is essentially arbitrary, although the convergence theorems that follow will restrict this choice. In Euclidean space, Gaussians can be used, centered at $x$ and with the variation as a hyperparameter $\sigma_n$. In binary spaces, individual bits can be flipped with a probability dependent on $n$. The particular mutation distributions should be chosen based on the needs of the problem at hand; a mutation distribution whose shape is well matched with the objective function will converge much faster than one that is not. The choice of mutation distribution determines the function prior with respect to which evolutionary annealing is best aligned in the sense of Chapters 12 and 17. Some results for a specific instantiation of evolutionary annealing with real vectors will be discussed in Section 15.2. The family of mutation distributions defines a mutation operator in the terminology of Chapter 7 through the equation

$$\mathcal{V}[h](A) = \nu_n^x(A)$$

where $n = \lfloor |h|/K \rfloor$ and $x = h_{-1}^X$ is the individual to be mutated.

Summing up, evolutionary annealing selects an individual $a \in A_n$ with probability $p_n(a)$, and mutates it according to $\nu_n^a$ in order to generate a new member of the population. That is, each individual in the population at time $n+1$ is sampled according to

$$P_n^k \sim \sum_{a\in A_n} p_n(a)\nu_n^a(dx). \tag{14.5}$$

Thus evolutionary annealing samples its populations from a sequence of mixture distributions with one mixing point located at each individual from prior populations. In this way, the selection is non-Markovian; the selected individual could come from any previous generation. The mixture probabilities $p_n(a)$ are chosen according to the annealing formula in Equation 14.3.

Equation 14.5 may be recognized as a convolution, and so evolutionary annealing with annealed proportional selection may be written as

$$\mathcal{EA}\langle T,\mathcal{V}\rangle = \mathcal{APS}\langle T\rangle \star \mathcal{V}, \tag{14.6}$$

reflecting a dependence on the cooling schedule and the choice of mutation distributions.

Intuitively, as the number of mixing points increases and the variance of the mutation distribution decreases, the mixture distribution in Equation 14.5 converges to the annealing distribution $\mathcal{B}_n^u$ in Equation 4.5. It is commonly known that mixtures of Gaussians can model any sufficiently smooth distribution arbitrarily well if enough mixing points are used. It is also true that mixture distributions in general can model any probability measure arbitrarily well subject to certain conditions. A specific proof of convergence for evolutionary annealing is offered in Section 14.3.6; Theorem 14.1 states that evolutionary annealing converges in probability to the optima of $u$ under certain conditions. Therefore $P_n$ is successively sampled from better and better approximations to $\mathcal{B}_n^u$, and as $n \to \infty$, the population sequence $P_n$ will increasingly focus on the optima of $u$. The rate of convergence will be taken up in Section 14.3.7.

---

**Algorithm 2** Evolutionary Annealing Algorithm

---

$N$, the number of generations
$K$, sample points (population size) per generation
$\left(P_1^k\right)_{k=1}^K$, the initial random population
$A_0 \leftarrow \emptyset$, all points from all generations
**for** $n \leftarrow 1$ to $N$ **do**
  $A_n \leftarrow \bigcup_k P_n^k \cup A_{n-1}$
  $p_n \leftarrow \mathit{prepare}\,(A_n)$
  **for** $k \leftarrow 1$ to $K$ **do**
    $y \leftarrow \mathit{sample}\,(p_n)$
    $P_{n+1}^k \leftarrow$ a sample from $\nu_n^y$
  **end for**
**end for**

---

A high-level algorithm for evolutionary annealing over $N$ generations is shown in Algorithm 2. The algorithm depends on two subroutines, *prepare* and *sample*. The subroutine *prepare* builds data structures to support efficient sampling of the quantity $p_n$ from Equation 14.3. The subroutine *sample* samples from $p_n$ using the prepared data structures. Through the use of highly precise approximations as described in Section 15.1.2, both *prepare* and *sample* can be implemented to run in time logarithmic in the population size and the number of generations. The specific implementations of *prepare* and *sample* used in the experiments utilize the methods of Section 15.1.2. The *prepare* routine adds nodes to the trees described in that section and propagates the components of Equations 15.1 and 15.7 up the tree. The *sample* routine employs Equations 15.1 and 15.7 to traverse the tree down from the root in order to select a previously evaluated point. Assuming that sampling $\nu_n^a$ and computing $\lambda\left(E_n^a\right)$ do not add to the complexity, the overall algorithm has performance $O\left(NK \log NK\right)$.

In order to make evolutionary annealing concrete, the cooling schedule must be determined. In light of [65], a default choice for the cooling schedule is given by

$T_n^{-1} = \eta \log n$. Here $\eta$ is a learning rate that scales the fitness function and thereby controls the aggressiveness of selection. A high learning rate focuses selection on the few best individuals and may restrict exploration of the space. A low learning rate allows promiscuous selection, slowing down refinement of previously discovered solutions but increasing the probability of escaping a local minimum. Again following [65], a possible value for $\eta$ is $1/d$ where $d$ is the largest depth of a local minima relative to its surroundings in the fitness landscape. In more complex spaces, different cooling schedules could be considered. There may also be a benefit to linking the variance of the mutation distribution to the cooling schedule, so that as the probability of selecting the current best individual decreases, the variance also decreases to enable refined exploration of the immediate region around the current best. The effect of parameter settings is explored further in Section 14.3.7.

The region weight $\lambda\left(E_n^a\right)$ is present in Equation 14.3 to avoid a particular scenario of premature convergence. Once a good solution is discovered, evolutionary annealing will devote increasing resources to exploring the neighborhood of that point. If these points are also good, then the probability of selecting more points in the same region will increase in a feedback loop. Within a few generations, almost all points selected will come from the immediate environment of these good points. If there is a local minimum in the vicinity, evolutionary annealing would likely become entrapped in that region. The region weight is intended to serve as a measure of how many individuals have been previously sampled in the region surrounding the point $a$. The sets $E_n^a$ partition $X$ around points in $A_n$, the total population so far. Such a partition can be computed in logarithmic time in many spaces. These partitions also play an important role in the convergence proof in Section 14.3.6.

### 14.2.3 Partitioning the Space

To demonstrate convergence of evolutionary annealing, each of the mixing points $a \in A_n$ will be considered representative of a particular region of the search space $X$. Each successive set $A_n$ will be associated with a partition $\{E_n^a\}_{a \in A_n}$ of disjoint sets such that $X = \bigcup_{a \in A_n} E_n^a$ and $a \in E_n^a$ for all $n$. The topology on $X$ is assumed to generate enough Borel sets to support such partitions based on any finite collection of points in $X$. The partitioning set $E_n^a$ is the same as the one that appears in Equation 14.3.

Provided that there exists a computable algorithm to split any set containing two distinct points into two disjoint sets each of which contains exactly one of the points, then the partitions can be stored in a binary tree, and if the splitting algorithm does not depend on the population size of the number of generations, the computational complexity of maintaining a partitioning tree is logarithmic on average.

Algorithm 3 partitions any Borel measure space over a Hausdorff topology given a function for dividing a partition region between two separate points in the region. A partition is represented as a binary tree, with the root representing the entire space $X$ and each branch partitioning $X$ into two sets. The algorithm is initialized with a

**Algorithm 3** Algorithm to Generate a Partition Based on Grid Points

$\{x_m\}_{m=1}^M \subseteq X$, the mixing points
$\mathcal{T} \leftarrow \{X\}$, the partition tree
$k(i) \leftarrow \emptyset$ for all $i = 1,\ldots,M$, node assignment function
**for** $m \leftarrow 1$ to $M$ **do**
  $N \leftarrow$ the leaf node in $\mathcal{T}$ such that $x_m \in N$
  **if** $\exists j \neq m$ s.t. $k(j) = N$ **then**
    $N_0, N_1 \leftarrow separate(x_j, x_m, N)$
    $\mathcal{T} \leftarrow \mathcal{T} \cup \{N_0, N_1\}$
    $k(j) \leftarrow N_0$, $k(m) \leftarrow N_1$
  **else**
    $k(m) \leftarrow N$
  **end if**
**end for**

sequence of points $\{x_m\}_{m=0}^M \subseteq X$ to be partitioned (the mixing points), a tree $\mathcal{T}$ with $X$ as the root node, and an assignment function $k$ such that $k(m)$ is the leaf node of the tree assigned to the point $x_m$, or $\emptyset$ if no assignment has been made. The algorithm then loops through the mixing points, splitting the space where necessary to ensure that each leaf node contains exactly one mixing point. The algorithm relies on *separate*, a domain-specific subroutine to split an existing set. At the end of each iteration of the algorithm's main loop, each leaf node is assigned to exactly one mixing point. When a new mixing point is added, *separate* partitions the leaf node to which it belongs into two new leaf nodes, each containing only one mixing point. The process of adding a single new mixing point to the tree requires only a tree traversal, so that at each generation, updating the partition requires $O(K \log NK)$ time, where $NK$ is the number of points at the $N^{th}$ generation.

In the vector space $\mathbb{R}^d$, one possible definition of the function *separate* can be described easily. Suppose that $X$ is bounded above by a set of coefficients $\{u_i\}_{i=1}^d$ and below by $\{\ell_i\}_{i=1}^d$ so that $X$ has a rectangular shape with $x \in X$ if and only if $\ell_i \leq x_i \leq u_i$ for all $i = 1,\ldots d$. Each node in the partition tree will restrict the upper or lower coefficient for exactly one coordinate, say $j$. In this case, each set $E_n^a$ in the partition is defined as an axis-aligned hyperrectangle on finitely many coordinates, with each step in the traversal of the partitioning tree adding a new coordinate value for some side of the hyperrectangle. So $E_n^a$ can be represented as two vectors, $u^a$ for its upper bounds, and $\ell^a$ for its lower bounds. Given the point $a \in X$ and a second point $x \in E_n^a$, $E_n^a$ can be separated as follows. Let $k = \text{argmax}_{i \leq d} |a_i - x_i|$; $k$ is the index at which the rectangle $E_n^a$ will be split. Suppose $a_k > x_k$ for the sake of simplicity; the opposite situation is handled analogously. Initialize $u^x \leftarrow u^a$ and $\ell^x \leftarrow \ell^a$. Then set $\ell_k^a \leftarrow \frac{1}{2}(a_k + x_k)$ and $u_k^x \leftarrow \frac{1}{2}(a_k + x_k)$. The regions $E_{n+1}^a$ and $E_{n+1}^x$ defined by these boundary vectors are then disjoint if the upper boundary is strict. The result of this partitioning method in $\mathbb{R}^2$ is shown in Figure 14.2.

In an infinite-dimensional vector space, *e.g.*, $\mathbb{R}^\infty$, a similar partitioning method can be used by selecting a maximum dimension $D$ to stand in the place of $d$ in the algorithm of the last paragraph. In this case, it is possible for two distinct vectors to have arbitrarily many identical coefficients, and no computable algorithm can locate

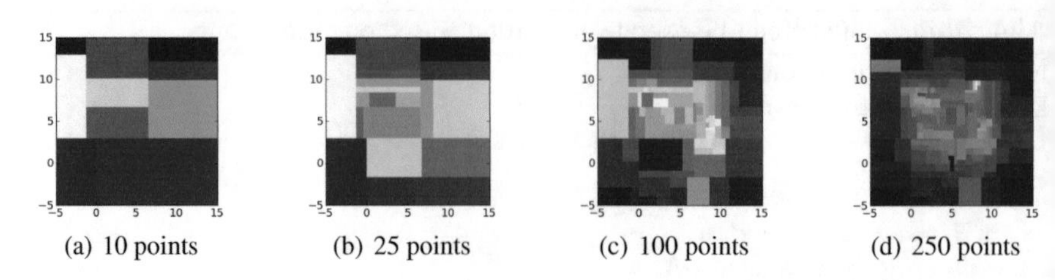

(a) 10 points (b) 25 points (c) 100 points (d) 250 points

**Fig. 14.2** Progression of partition regions $\{E_n^a\}$ on Shekel's Foxholes during the run of evolutionary annealing in Figure 14.1 in two dimensions. Images are heat maps displaying the selection probability of each region; light regions have a higher selection probability. Successive frames show how the partition regions gradually model the shape of the fitness function after 10, 25, 100, and 250 points have been evaluated.

the coefficients in which they differ. This situation is of theoretical more than practical concern, however, and can be ignored in most cases. Separation algorithms of this type can be implemented efficiently in many spaces of interest. Section 15.1.2 discusses how these partition mechanisms can also be used to implement the subroutines *prepare* and *sample* from Algorithm 2.

## 14.2.4 Annealed Tournament Selection

Annealed proportional selection as given in Equation 14.3 is a proportional selection rule; individuals are selected according to their proportion of the overall fitness. Proportional selection has a well-known drawback that also applies to annealed proportional selection. For example, suppose that the fitness function $u$ has a minimal value of 0, and consider the selection probabilities for the points $x, y$ with $u(x) = 0.01$ and $u(y) = 0.001$ at temperature $T_n = 5$. Assume $\lambda\left(E_n^x\right) = \lambda\left(E_n^y\right) = 1$. Then $p_n(y)/p_n(x) = 1.0018$. That is, $x$ is almost equally as likely to be selected as $y$, even though $y$ is a whole order of magnitude closer to the optimum. Thus the more precise solution is no more likely to be selected than rougher solutions close to the optimum, which makes refinement of solutions near a local or global optimum sluggish. These intuitions are confirmed by the experimental results in Chapter 15; annealed proportional selection converges within 0.1 of the optimal fitness without difficulty, but then fails to attain accuracy within 0.001 in most cases.

To address this weakness of proportional selection in genetic algorithms, tournament and ranking selection were introduced (cf. [157]). These methods select among individuals according to their fitness rank in the population rather than according to their raw fitness. For tournament selection, the best individual is selected with some probability $q$, termed the *selection pressure*. If the best individual is not selected, then the second best individual is chosen with probability $q$. Thus the probability of selecting the $n^{th}$-ranked individual of the population is proportional to $q(1-q)^{n-1}$.

A similar concept can be used to define *annealed tournament selection*, a non-Markovian version of tournament selection. Annealed tournament selection replaces Equation 14.3 by

$$p_n(a) = \xi_n \, q^{1/T_n} \left(1 - q^{1/T_n}\right)^{r(a)} \, \lambda\left(E_n^a\right), \tag{14.7}$$

where $q$ is the selection pressure, and $r(a)$ is the fitness rank of $a$ in $A_n$ starting with 0. Annealed tournament selection uses a cooling schedule $T_n$ so that the rank becomes increasingly significant with each generation, with the ultimate result that the top-ranked individual is selected at zero temperature. The main difference from standard tournament selection is that each individual must be ranked against all other individuals from all prior generations. As a consequence, the selection pressure must be much lower. For this paper, the value of $q$ was fixed at 0.025. Rather than varying $q$, the learning rate $\eta$ in the cooling schedule can be varied to achieve the same effect.

In terms of the formalism advanced in this book, annealed tournament selection is given by

$$\mathcal{ATS}\langle T\rangle\,[h](\{a\}) = \mathbb{1}_{A_n}(a)\, p_n(a),$$

where $n = \lfloor |h|/K \rfloor$ and $p_n$ comes from Equation 14.7. Evolutionary annealing with annealed tournament selection may be defined as

$$\mathcal{EA}_{\text{tour}}\langle T, \mathcal{V}\rangle = \mathcal{ATS}\langle T\rangle \star \mathcal{V}.$$

With the basic algorithm defined, it can now be shown that evolutionary annealing converges to the global optimum under some conditions for annealed proportional selection. These results will be confirmed experimentally in Chapter 11.

## 14.3 Convergence Properties

Subject to a reasonable set of conditions, evolutionary annealing with proportional selection converges in probability to the set of optimal points for the fitness function. These conditions include: (1) the partitions must eventually refine to arbitrary granularity; (2) the mutation variance must asymptotically decrease faster than the partition size; (3) the cooling schedule and the partitions must induce quasi-differentiability properties on the annealing distributions around the mixing points; and (4) the fitness function must be essentially bounded and not too irregular in the immediate neighborhood of the optima. With these conditions satisfied, evolutionary annealing converges to the optima. The convergence rate is highly sensitive to the partitioning method, the cooling schedule, and the variance decay; the interaction of these parameters remains a subject of inquiry.

In this section, the preceding concepts are made rigorous, and a proof of convergence for evolutionary annealing is provided, after which the rate of convergence is discussed. Several symbols are used, and their meanings are summarized in Table 14.1 to aid the reader.

### 14.3.1 The Stochastic Population Process

In Section 14.2.2, evolutionary annealing was described as generating a stochastic population process. In order to prove convergence, the distribution of this process needs to be analyzed. The members of the first population are independent and identically distributed according to $\mathbb{G}_1 = \lambda/\lambda(X)$. Each element $P^k_{n+1}$ in the $(n+1)^{th}$ population is independently and identically distributed according to

$$\mathbb{G}_n(dP^k_{n+1}) = \sum_{a \in A_n} p_n(a)\, \nu^a_n(dP^k_{n+1}), \tag{14.8}$$

and so the population $P_{n+1}$ as a whole is distributed as

$$\prod_{k=1}^{K} \mathbb{G}_n(dP^k_{n+1}), \tag{14.9}$$

where this product distribution is on the measurable space $\left(X^K, \mathcal{B}_{X^K}\right)$. The distribution $\mathbb{G}_n$ is conditional on the value of the set $A_n = \bigcup_{i=1}^n P_i$, which is generated by sampling $\mathbb{G}_1$ up to $\mathbb{G}_{n-1}$ in succession. This dependency will be written as $\mathbb{G}_n(dP^k_{n+1} \mid P_1, \ldots, P_n)$. The distribution of the entire process can be built from a family of distributions $\{\mathcal{G}^n_u\}_{n \in \mathbb{N}}$ of the form

$$\mathcal{G}^n_u(dP_1, \ldots, dP_n) = \prod_{m=1}^{n} \prod_{k=1}^{K} \mathbb{G}_m\left(P^k_m \mid P_1, \ldots, P_{m-1}\right), \tag{14.10}$$

obtained by separating the joint distribution of $P_1, \ldots, P_n$ into a product of conditional probabilities.

The set $\{\mathcal{G}^n_u\}_{n \in \mathbb{N}}$ can be used as a base to construct a consistent family of finite-dimensional distributions, and the Kolmogorov Extension Theorem ([94]) guarantees that there exists a unique distribution over infinite sequences on $X^K$ that extends this family. Let $\mathcal{G}_u$ be this distribution, where the subscript $u$ has been used to indicate that the identity of the distribution depends on the fitness function. Then the stochastic population process $(P_n)_{n \in \mathbb{N}}$ is distributed according to $\mathcal{G}_u$ for evolutionary annealing.

The convergence proof depends on certain properties of the population process that hold almost surely under the distributions $\mathcal{G}_u$. A property $\mathcal{P}$ is said to hold almost surely in $\mathcal{G}_u$ (or $\mathcal{G}_u$-a.s.) if $\mathcal{G}_u(\mathcal{P}) = 1$, that is, if $\mathcal{P}$ has probability one under $\mathcal{G}_u$. These properties will be of interest because of their effect on the finite sets $A_n$. Lemma 14.2 below provides a tool for extending properties that hold $\lambda$-a.e. to properties that hold almost surely in $\mathcal{G}_u$. This issue is taken up again in Section 14.3.6; definitions pertaining to the fitness function are introduced next.

**Table 14.1** Table of symbols relating to the convergence proof

| Symbol | Meaning |
|---|---|
| $X$ | The search domain, a Hausdorff (separated) topological space |
| $\mathcal{B}_X$ | The Borel $\sigma$-algebra on $X$ |
| $\lambda$ | A measure on $(X, \mathcal{B}_X)$ that is positive on open sets |
| $u$ | A $\lambda$-integrable fitness function with finite minimum |
| $u_*$ | The minimum of the fitness function (the essential infimum) |
| $X_\varepsilon$ | The set of $\varepsilon$-optimal points in $X$, $\{x : \lvert u(x) - u_*\rvert < \varepsilon\}$ |
| $A_n$ | The set of observed individuals at generation $n$ |
| $A$ | The limiting set of observed individuals as $n \to \infty$; $A = \bigcup_n A_n$ |
| $T_n$ | A cooling schedule, $T_n \downarrow 0$ |
| $E_n^a$ | A set containing $a \in A_n$, with $\{E_n^a\}_{a \in A_n}$ partitioning $X$ |
| $\lambda(E_n^a)$ | The volume of the partition set $E_n^a$ under the measure $\lambda$ |
| $\xi_n$ | A normalizing factor for $p_n(a)$, computed over $A_n$ |
| $p_n(a)$ | The selection probability for $a \in A_n$, $p_n(a) = \xi_n^{-1} \exp(\frac{-u(x)}{T_n})\lambda(E_n^a)$ |
| $\nu_n^a$ | The mutation distribution around a point $a \in A_n$ at generation $n$ |
| $\mathbb{G}_n$ | The distribution of evolutionary annealing at generation $n$ |
| $g_n$ | The annealing density for the fitness function at temperature $T_n$ |
| $\eta_n$ | The normalizer for $g_n$ |
| $h_n$ | The denormalized version of $g_n$, *i.e.* $h_n = \eta_n g_n$ |
| $\mathcal{A}_n$ | The annealing distribution for the fitness function at temperature $T_n$ |
| $\mathcal{A}$ | The limit of the annealing distributions under the total variation norm |
| $g_n^\lambda$ | The neighborhood average of $g_n$ on $E_n^a$ |
| $\lambda_n^a$ | A measure on $E_n^a$ given by $\lambda_n^a(B) = \lambda(B \cap E_n^a)/\lambda(E_n^a) = \lambda(B \mid E_n^a)$ |
| $\mathcal{G}_u$ | Probability over population histories for evolutionary annealing on $u$ |
| $P_m^k$ | The $k^{th}$ member of the $m^{th}$ population |

### *14.3.2 Suitable Fitness and the Annealing Densities*

As mentioned above, the convergence proof requires conditions on the partition sequence, the mutation variance, the annealing distributions, and the fitness function. In the present context, the minimum $u_*$ of a $\lambda$-measurable fitness function $u$ is intended to mean the essential infimum of the function with respect to the measure $\lambda$; that is, $u_*$ is the lower bound of the function $u$ over all measurable sets with positive $\lambda$-measure. Note that this definition is slightly distinct from prior interpretations of $u_*$, in which this notation is assumed to be the minimum in an absolute sense.

In order for convergence to be attained, the fitness function must not be too irregular. In particular, it must be measurable and integrable, its essential infimum must exist, and the boundary of the $\varepsilon$-optimal sets must have $\lambda$-measure zero. The last point needs some clarification. For any $\varepsilon > 0$, define $X_\varepsilon \equiv \{x : u(x) < u_* + \varepsilon\}$. $X_\varepsilon$ includes all points in $X$ that come within $\varepsilon$ of the optimum. It follows from the measurability of $u$ that $X_\varepsilon$ is Borel-measurable for all $\varepsilon > 0$. If $X_\varepsilon$ is so jagged that its boundary possesses positive $\lambda$-mass, the shape of $X_\varepsilon$ can never be well approximated by a countable sequence of estimates; in this case, evolutionary annealing cannot be expected to succeed. A fitness function will be called *suitable* when the fitness function meets the criteria described in this paragraph.

**Definition 14.1 (Suitable Fitness Function).** A fitness function $u$ is termed *suitable* whenever the following three conditions hold:

1. The function $u$ is $\lambda$-measurable and $\lambda$-integrable.
2. The essential infimum exists, i.e. $u_* > -\infty$.
3. There exists a constant $\gamma > 0$ s.t. for all $\varepsilon \in (0, \gamma)$, $\lambda(\partial X_\varepsilon) = 0$.

In addition, the *essential supremum* of the fitness function needs to be finite for the results below. The essential supremum on a set is the upper bound of the function on subsets with positive measure, written $\operatorname{ess\,sup}_{z \in B} u(B)$. If both the essential supremum and the essential infimum are finite on the whole domain, then a function is *essentially bounded*, i.e. $\operatorname{ess\,sup}_X |u| < \infty$. The essential supremum is used below to limit the asymptotic variation of the annealing distributions.

Intuitively, evolutionary annealing converges to the essential infimum because it approximately samples from the annealing distributions with respect to the measure $\lambda$. Specifically, define

$$h_n(x) = \exp\left(\frac{-u(x)}{T_n}\right),$$

and let $\eta_n = \int_X h_n d\lambda$ be a normalizing factor for $h_n$. Then define

$$g_n(x) = \frac{\exp(-u(x)/T_n)}{\int_X \exp(-u(x)/T_n)\, \lambda(dx)} = \frac{h_n}{\eta_n}$$

so that $g_n$ is the density of an annealing distribution generalized to the space $(X, \mathcal{B}_X, \lambda)$, i.e. $\int_X g_n d\lambda = 1$. It is important to show that the denominator $\eta_n$ is well-defined for suitable $u$. Note that $u$ and $u - u_*$ both produce the same $g_n$ due to normalization, so one may assume without loss of generality that $u \geq 0$ $\lambda$-a.e. when working with $g_n$ below. In that case, $|h_n(x)| \leq 1$, and so $\eta_n = \int_X |h_n(x)| \lambda(dx) \leq \lambda(X) < \infty$. Define the annealing distributions by

$$\mathcal{A}_n(B) = \int_B g_n(x) \lambda(dx)$$

so that $\mathcal{A}_n(X) = 1$ and consider the limiting distribution $\mathcal{A} \equiv \lim_n \mathcal{A}_n$, meaning $\mathcal{A}_n(B) \to \mathcal{A}(B)$ for all $B \in \mathcal{B}_X$. The functions $g_n$ are positive, and therefore $\mathcal{A}_n$ is a probability measure. The definition of $\mathcal{A}$ implies $\mathcal{A}$ is positive and $\mathcal{A}(X) = 1$, so $\mathcal{A}$ is a probability measure as well. In particular, $\mathcal{A}$ assigns measure zero to all non-optimal points of $u$.

As an observation, $g_n$ is invariant whenever $u$ is shifted by a constant factor, a fact which is useful for the proofs below.

**Proposition 14.1.** *The annealing density depends on the objective only up to a constant factor; denoting by $g_n^u$ the annealing density for an objective $u$, it holds that $g_n^u = g_n^z$ whenever $z(x) = u(x) + C$ for some constant $C$.*

*Proof.* Notice that $\exp(-z(x)/T_n) = e^{-C} \exp(-u(x)/T_n)$. When $g_n^z$ is normalized, the constant factor $e^{-C}$ is cancelled out, since it appears in both the numerator and the denominator. That is, $g_n^u = g_n^z$.

Because of this observation, the objective function may be shifted in order to make $\exp(-u(x)/T_n)$ positive or negative as necessary.

### 14.3.3 Uniform Compatibility

In order to guarantee that the mixture distributions used by evolutionary annealing are capable of approximating $g_n$, it is necessary that the densities $g_n$ do not vary too quickly, i.e., that the fitness function does not oscillate wildly between infinitesimally close points. Formally, this property depends on the fitness function, the cooling schedule, and the partition sequence.

**Definition 14.2 (Compatibility).** The triple $(u, \{T_n\}_{n\in\mathbb{N}}, \{\{E_n^a\}_{a\in A_n}\}_{n\in\mathbb{N}})$ consisting respectively of a suitable fitness function, a cooling schedule with $T_n > 0$ and $T_n \to 0$, and a partition sequence of Borel-measurable sets is said to be *compatible* if for each $a \in A = \bigcup_n A_n$,

$$\operatorname*{ess\,sup}_{z\in E_n^a} |g_n(a) - g_n(z)| \to 0 \tag{14.11}$$

as $n \to \infty$, and this triple is *uniformly compatible* if the convergence is uniform over $A$ (i.e., the rate of convergence is independent of $a$ for all $a \in A$).

**Proposition 14.2.** *The triple $(u, \{T_n\}_{n\in\mathbb{N}}, \{\{E_n^a\}_{a\in A_n}\}_{n\in\mathbb{N}})$ is compatible if $u$ is essentially bounded,* $\operatorname{ess\,inf} u \le u(a) \le \operatorname{ess\,sup} u$ *for all $a \in A = \bigcup_n A_n$, and*

$$\frac{e^{R/T_n}}{T_n} \operatorname*{ess\,sup}_{z\in E_n^a} |u(a) - u(z)| \to 0, \tag{14.12}$$

*where $\eta_n = \int_X \exp(-u(x)/T_n)\,\lambda(dx)$ and $R = \operatorname{ess\,sup} u - \operatorname{ess\,inf} u$ is the essential range of $u$. Uniform compatibility also holds if the convergence is uniform with respect to $A$.*

*Proof.* Fix $\varepsilon > 0$. Define $k_n(x) = \exp(-x/T_n)$, and notice that $k_n''(x) > 0$ so that $k(x)$ is convex. By convexity, $k_n(x) - k_n(y) \ge k_n'(x)(x-y)$ and hence, using this inequality twice,

$$\frac{u(z)-u(a)}{T_n} e^{-u(a)/T_n} \le e^{-u(a)/T_n} - e^{-u(z)/T_n} \le \frac{u(z)-u(a)}{T_n} e^{-u(z)/T_n}.$$

Collapsing these inequalities, we find that

$$\left| e^{-u(a)/T_n} - e^{-u(z)/T_n} \right| \le \left| \frac{u(z)-u(a)}{T_n} \right| \exp\left( -\frac{u(a)\wedge u(z)}{T_n} \right).$$

Consequently, taking the essential supremum and dividing by the normalizing factor $\eta_n$,

$$\operatorname*{ess\,sup}_{z \in E_n^a} |g_n(a) - g_n(z)| \leq \frac{1}{T_n} \operatorname*{ess\,sup}_{z \in E_n^a} \left\{ |u(a) - u(z)| \frac{\exp\left(-\frac{u(a) \wedge u(z)}{T_n}\right)}{\eta_n} \right\}.$$

Observe that $\eta_n^{-1} \leq \lambda(X)^{-1} \exp(\operatorname{ess\,sup} u / T_n)$. Next, separating the essential supremum and applying the bound $u \geq \operatorname{ess\,inf} u$, we have

$$\operatorname*{ess\,sup}_{z \in E_n^a} \left\{ \frac{\exp\left(-\frac{u(a) \wedge u(z)}{T_n}\right)}{\eta_n} \right\} \leq \frac{e^{-\frac{\operatorname{ess\,inf} u}{T_n}}}{\eta_n} \leq \lambda(X)^{-1} e^{R/T_n}.$$

The proof is completed by noting that the factor $\lambda(X)^{-1}$ is constant and may be discarded. Thus, under the given conditions, compatibility or uniform compatibility depends only on the convergence in Equation 14.12.

This proposition implies that uniform compatibility can be obtained if the temperature decreases sufficiently slower than the partitions so that the factor $e^{R/T_n}/T_n$ is annihilated asymptotically. If $T_n^{-1} = \frac{1}{R} \log n$, the default for evolutionary annealing, then this factor reduces to $\frac{1}{R} n \log n$.

More detailed sufficient conditions for uniform compatibility can be stated if the search space is uniformizable. The intuitive significance of this phrase is that there must be a notion of neighborhood size that is held in common among all points in the search domain. A set of neighborhoods of the same size is called an *entourage*, formally described as a subset of $X \times X$; in a metric space, the set of balls with a fixed radius around each point forms an entourage. A uniform neighborhood for $x \in X$ can be generated from an entourage $U$; this is denoted by $U_x = \{y : (x,y) \in U\}$.

The conditions for uniform compatibility also depend on the concept of nicely shrinking sets. Nicely shrinking sets are a vanishing sequence of sets, each of which possesses some interior volume. Uniform compatibility requires partitions that generate a mesh of nicely shrinking sets that shrink uniformly, as defined below.

**Definition 14.3 (Nicely Shrinking Sets).** A sequence of Borel-measurable sets $\{E_n^a\}_{n \in \mathbb{N}}$ *shrinks nicely* around a point $a$ if (1) for all $n$, $a$ lies in the interior of $E_n^a$ (i.e. there exists an open set $O \subseteq E_n^a$ such that $a \in O$), and (2) for any open set $O$ containing $a$, there is an $N < \infty$ such that $E_n^a \subseteq O$ for all $n > N$.

**Definition 14.4 (Nice Partition Sequence).** A partition sequence $\{\{E_n^a\}_{n \in \mathbb{N}}\}_{a \in A_m}$ is *nice* if for all $m$ and all $a \in A_m$, $\{E_n^a\}_{n \geq m}$ shrinks nicely to $a$. The partition sequence is *uniformly nice* if additionally for any entourage $U$, there is an $N$ such that for all $m$, all $a \in A_m$, and all $n$ with $n > N$ and $n \geq m$, $E_n^a \subseteq U_a$.

**Definition 14.5 (Uniformly Nice).** An evolutionary annealing algorithm is *almost surely uniformly nice* on an objective $u$ if it produces a uniformly nice partition sequence with $\mathcal{G}_u$ probability one.

A uniformly nice partition sequence will be uniformly compatible with a suitable fitness function whose growth is bounded on small sets of uniform size, as long as the cooling schedule schedule decreases slower than the $\lambda$-measure of the partitions, based on the results from Proposition 14.2. These two conditions are made concrete in the next two definitions. The first definition is standard terminology in complexity analysis, and the second generalizes the idea of Lipschitz continuity to the current setting.

**Definition 14.6 (Asymtotic Annihilation).** Given two function $p(n)$ and $q(n)$, $q$ *asymptotically annihilates* $p$ if $\lim_{n\to\infty}\frac{p(n)}{q(n)} = 0$, written $p(n) = o(q(n))$.

**Definition 14.7 (Essentially Lipschitz Continuous).** An objective $u$ defined on a uniformizable search domain $X$ is *essentially Lipschitz continuous* on a set $A$ if there is a constant $c > 0$ such that for any entourage $U \subseteq X \times X$ in the uniformity for $X$, for all $a \in A$, and for any open set $O$ containing $a$ in its interior, $O \subset U_a$ implies

$$\operatorname*{ess\,sup}_{z\in O} |u(a) - u(z)| < c\lambda(U_a \setminus \{a\}). \tag{14.13}$$

If the set $A$ has full $\lambda$ measure, then $u$ is *essentially Lipschitz continuous almost everywhere*.

The condition in Equation 14.13 is a generalized version of Lipschitz continuity. If the search domain is the real line, $X = \mathbb{R}$, and $u$ is essentially Lipschitz continuous almost everywhere on $\mathbb{R}$, then suppose $O$ is a subset of the interval $I = (x-\varepsilon, y+\varepsilon)$ that contains $x$ and $y$. The set $I$ is a projection of the uniform structure onto $X$, and the Lebesgue measure results in $\lambda(I\setminus\{x\}) = y - x + 2\varepsilon$, and so the condition implies that for almost every $x, y$, it holds that $|u(y) - u(x)| < c|y-x| + 2c\varepsilon$. Taking $\varepsilon$ to zero recovers the standard condition for Lipschitz continuity almost everywhere. The set $U_a$ is used to replace the set $O$ on the right hand side of the inequality because $U_a$ is of uniform size, and can thus stand as a proxy for the diameter of the set $O$, which may be irregularly shaped.

The preceding definitions can be combined to give a set of sufficient conditions for uniform compatibility. These conditions are easier to check than the definition, since they separate out the properties required on the objective function, the partition sequence, and the cooling schedule.

**Proposition 14.3.** *A triple $(u, \{T_n\}, \{\{E_n^a\}\})$ is uniformly compatible if*

1. *The search domain $X$ is uniformizable.*
2. *The partition sequence $\{\{E_n^a\}\}$ is uniformly nice.*
3. *The objective function is essentially bounded on $X$, and $u(a)$ is bounded by the essential bound for all $a \in A$.*
4. *The objective $u$ is essentially Lipschitz continuous on $A = \bigcup_n A_n$.*
5. *If $R = \operatorname{ess\,sup} u - \operatorname{ess\,inf} u$, then* $\sup_{a\in A_n} \inf_{\{U : E_n^a \subseteq U_a\}} \lambda(U_a \setminus \{a\}) = o\left(T_n e^{-R/T_n}\right)$.

*Proof.* By Proposition 14.2, we only need to show that Equation 14.12 holds uniformly. Since the partition sequence is uniformly nice, then for any $U$ in the uniformity, for all $n$ large enough and all $a \in A_n$, $E_n^a \subseteq U_a$. Combining Equation 14.13 with Equation 14.12 yields

$$\frac{e^{R/T_n}}{T_n} \operatorname*{ess\,sup}_{z \in E_n^a} |u(a) - u(z)| \quad < \quad c\lambda(U_a \setminus \{a\}) \frac{e^{R/T_n}}{T_n} \tag{14.14}$$

for any $U$ in the uniformity such that $U_a$ contains $E_n^a$. This inequality completes the proof, since the right hand side converges to zero at a rate independent of $a$ by assumption.

Proposition 14.3 provides a set of sufficient conditions that make it easier to check uniform compatibility for an evolutionary annealing algorithm. These conditions are sufficient, not necessary; there may be weaker conditions that also guarantee uniform compatibility. Necessary conditions are currently unknown and remain the subject of inquiry. Based on the proposition above, uniform compatibility can be obtained for essentially bounded objective functions that are also essentially Lipschitz continuous almost everywhere. In this case, with probability one, the objective will also be bounded and essentially Lipschitz continuous on the set $A = \bigcup_n A_n$ produced by evolutionary annealing, so long as $\lambda$ is absolutely continuous with respect to the mutation distributions. The details are discussed below.

If the partition sequence is uniformly nice, the cooling schedule may then be set dynamically during execution to satisfy the conditions using

$$\frac{e^{R/T_n}}{T_n} = \left( \sup_{a \in A_n} \inf_{\{U : E_n^a \subseteq U_a\}} \lambda(U_a \setminus \{a\}) \right)^{\beta - 1} \tag{14.15}$$

for arbitrary $\beta > 0$. As can be seen, the cooling schedule must converge at a rate that is exponentially slower than the rate at which the individual partition regions vanish. It will be seen that the partition regions already shrink exponentially slowly, so that the rate of convergence on the cooling schedule must be double exponentially slow in order to guarantee that the global optimum is reached in the limit. However, even if the global optimum is hard to find, local optima may still be found with faster cooling schedule.

A more difficult question is to determine when the evolutionary annealing algorithm produces a uniformly nice partition sequence. The fitness function and the cooling schedule are specified prior to running evolutionary annealing, but the partition sequence is determined by the population history obtained from evolutionary annealing as it runs. The actual partition sequence obtained from evolutionary annealing is stochastic, and therefore some addition terminology is needed to take this fact into account.

**Definition 14.8 (Almost Surely Uniformly Compatible).** An evolutionary annealing algorithm with a cooling schedule $\{T_n\}$ is *almost surely uniformly compatible* with a suitable fitness function $u$ if the triple $(u, \{T_n\}, \{\{E_n^a\}\})$ is uniformly compatible $\mathcal{G}_u$-a.s.

**Definition 14.9 (Neighborhood Average).** Given a decreasing sequence of sets $\{E_n^a\}_{n \in \mathbb{N}}$ each containing the point $a$ and a sequence of $\lambda$-integrable functions

$\{g_n\}_{n\in\mathbb{N}}$ on a measure space $(X, \mathcal{B}_X, \lambda)$, the *neighborhood average* of $g_n$ on $E_n^a$ is given by

$$g_n^\lambda(a) \equiv \lambda\left(E_n^a\right)^{-1} \int_{E_n^a} g_n d\lambda. \tag{14.16}$$

**Definition 14.10 (Approximation by Neighborhood Average).** On a measure space $(X, \mathcal{B}_X, \lambda)$, a sequence of $\lambda$-integrable functions $\{g_n\}_{n\in\mathbb{N}}$ is *approximated by its neighborhood average* at a point $a$ if for any decreasing sequence of sets $\{E_n^a\}_{n\in\mathbb{N}}$ containing the point $a$,

$$\left|g_n(a) - g_n^\lambda(a)\right| \to 0. \tag{14.17}$$

The approximation is uniform over a set $A \subseteq X$ if the rate of convergence is independent of $a$ for all $a \in A$.

**Proposition 14.4.** *Suppose a triple $(u, \{T_n\}_{n\in\mathbb{N}}, \{\{E_n^a\}_{n\in\mathbb{N}}\}_{a\in A_n})$ is compatible. Then the annealing densities $\{g_n\}_{n\in\mathbb{N}}$ are approximated by their neighborhood averages at each $a \in A = \bigcup_n A_n$. If the triple is uniformly compatible, then the approximation is uniform.*

*Proof.* This proposition follows from the Dominated Convergence Theorem, since

$$\left|g_n(a) - g_n^\lambda(a)\right| \le \frac{1}{\lambda(E_n^a)} \int_{E_n^a} |g_n(a) - g_n(z)|\, \lambda(dz) \le \operatorname*{ess\,sup}_{z\in E_n^a} |g_n(a) - g_n(z)| \to 0. \tag{14.18}$$

The approximation is uniform whenever the convergence on the right is uniform.

**Proposition 14.5.** *Suppose a triple $(u, \{T_n\}_{n\in\mathbb{N}}, \{\{E_n^a\}_{n\in\mathbb{N}}\}_{a\in A_n})$ is uniformly compatible. The ratio $g_n^\lambda / g_n$ converges to one uniformly if*

1. *The search domain $X$ is uniformizable.*
2. *The partition sequence $\{\{E_n^a\}\}$ is uniformly nice.*
3. *The objective function is essentially Lipschitz continuous on $A = \bigcup_n A_n$.*
4. $\sup_{a\in A_n} \inf_{\{U : E_n^a \subseteq U_a\}} \lambda(U_a \setminus \{a\}) = o(T_n)$.

*Proof.* Fix $\varepsilon > 0$ independent of $a$. Note that

$$\begin{aligned}\left|1 - \frac{g_n^\lambda(a)}{g_n(a)}\right| &= \left|1 - \frac{\exp(u(a)/T_n)}{\lambda(E_n^a)} \int_{E_n^a} \exp\left(\frac{-u(z)}{T_n}\right) \lambda(dz)\right| \\ &= \left|1 - \frac{1}{\lambda(E_n^a)} \int_{E_n^a} \exp\left(\frac{u(a) - u(z)}{T_n}\right) \lambda(dz)\right|. \end{aligned} \tag{14.19}$$

Thus it suffices to show that

$$\frac{1}{\lambda(E_n^a)} \int_{E_n^a} \exp\left(\frac{u(a) - u(z)}{T_n}\right) \lambda(dz) \to 1, \tag{14.20}$$

independent of $a$. Because the partition sequence is uniformly nice, for any $U$ in the uniformity and $n$ large, $E_n^a \subseteq U_a$. So we have

$$\frac{1}{\lambda(E_n^a)}\int_{E_n^a}\exp\left(\frac{u(a)-u(z)}{T_n}\right)\lambda(dz) \leq \exp\left(\frac{\operatorname{ess\,sup}_{z\in E_n^a}|u(a)-u(z)|}{T_n}\right)$$
$$\leq \exp\left(\frac{c\,\lambda(U_a\setminus\{a\})}{T_n}\right). \tag{14.21}$$

The opposite inequality holds with a negative sign inside the exponent. The desired result follows from the fact that the quantity inside the exponent converges to zero independent of $a$ in either case.

If the neighborhood average $g_n^\lambda$ of a sequence $g_n$ approximates the values of the sequence at a point, then the neighborhood average can be used as a proxy for the function at that point. If the approximation is uniform, then it can be applied under an increasing sum, such as appears in the definition of annealed proportional selection in Equation 14.3. Notice that $p_n(a)$ from that equation is normalized by the sum $\xi_n$. In the convergence proof, this normalizing sum will need to be replaced by the integral normalizer $\eta_n$. The following lemma uses the neighborhood average to prove that these normalizers converge and are thus asymptotically interchangeable.

**Lemma 14.1.** *Given a uniformly compatible triple* $(u,\{T_n\},\{\{E_n^a\}\})$, *if*

1. *The search domain* $X$ *is uniformizable;*
2. *The partition sequence* $\{\{E_n^a\}\}$ *is uniformly nice;*
3. *The objective function is essentially Lipschitz continuous on* $A=\bigcup_n A_n$*; and*
4. $\sup_{a\in A_n}\inf_{\{U:E_n^a\subseteq U_a\}}\lambda(U_a\setminus\{a\})=o(T_n)$,

*then it holds that* $|1-\xi_n/\eta_n|\to 0$.

*Proof.* Fix $\varepsilon>0$. Suppose without loss of generality that $\lambda(X)=1$ and $u\geq 0$ $\lambda$-a.e. Observe that $\xi_n=\sum_{a\in A_n}h_n(a)\,\lambda(E_n^a)$ from Equation 14.3, and that $\eta_n$ and $\xi_n$ are bounded above by 1 since $u\geq 0$ $\lambda$-a.e. and $\lambda(X)=1$. By Proposition 14.4, the ratio $g_n^\lambda/g_n=h_n^\lambda/h_n$ converges to one uniformly on $A$. It follows that $h_n/h_n^\lambda\to 1$, also uniformly. Notice that

$$\eta_n=\sum_{a\in A_n}\int_{E_n^a}h_n(z)\lambda(dz)=\sum_{a\in A_n}h_n^\lambda(a)\lambda(E_n^a). \tag{14.22}$$

For $n$ large independent of $a$, $h_n(a)<(1+\varepsilon)\,h_n^\lambda(a)$ and $h_n(a)>(1-\varepsilon)\,h_n^\lambda(a)$. Hence,

$$\xi_n=\sum_{a\in A_n}h_n(a)\lambda(E_n^a)<(1+\varepsilon)\sum_{a\in A_n}h_n^\lambda(a)\,\lambda(E_n^a)=(1+\varepsilon)\,\eta_n, \tag{14.23}$$
$$\xi_n=\sum_{a\in A_n}h_n(a)\lambda(E_n^a)>(1-\varepsilon)\sum_{a\in A_n}h_n^\lambda(a)\,\lambda(E_n^a)=(1-\varepsilon)\,\eta_n, \tag{14.24}$$

where a suitable $n$ exists because $h_n(a)/h_n^\lambda(a)$ converges uniformly on $A$. These equations imply that for $n$ large, $\xi_n/\eta_n<1+\varepsilon$ and $\xi_n/\eta_n>1-\varepsilon$. That is, $|1-\xi_n/\eta_n|<\varepsilon$, and the conclusion follows.

The lemmas in this section require the size of the partition regions to shrink faster than the temperature, expressed by the condition

$$\sup_{a\in A_n} \inf_{\{U:E_n^a\subseteq U_a\}} \lambda(U_a\setminus\{a\}) = o\left(T_n e^{-R/T_n}\right).$$

The following definition expresses this as a property of the evolutionary annealing algorithm as a whole.

**Definition 14.11 (Almost Surely Shrinks Faster).** An evolutionary algorithm *almost surely shrinks faster* than a sequence $(k_n)_{n\in\mathbb{N}}$ on an objective $u$ if

$$\sup_{a\in A_n} \inf_{\{U:E_n^a\subseteq U_a\}} \lambda(U_a\setminus\{a\}) = o(k_n)$$

with $\mathcal{G}_u$-probability one.

In the convergence theorem below, it is required that the evolutionary annealing algorithm almost surely shrinks faster than the cooling schedule so that the conditions of Lemma 14.1 are met, which include uniform compatibility. In order to guarantee uniform compatibility if it is not already given, the evolutionary annealing algorithm would need to shrink faster than $T_n e^{-R/T_n}$ as well to satisfy Proposition 14.3.

To sum up, uniform compatibility is implied by essential boundedness of the fitness function together with essential Lipschitz continuity and a cooling schedule that decreases at a rate sufficiently slower than the diameter of the partition regions. Under these same conditions, uniform compatibility in turn yields that the fitness function is uniformly approximated by its neighborhood average. As a consequence, the normalizing sum $\xi_n$ can be exchanged for the integral normalizer $\eta_n$. These facts are crucial to the convergence proof in Section 14.3.6. There may be weaker conditions that also yield uniform compatibility and uniform approximation by the neighborhood average; necessary conditions for these properties remain a subject of inquiry.

### 14.3.4 Fine Partitions

Uniform compatibility guarantees that the fitness function does not oscillate chaotically within the partitions, but it does not imply that the partitions provide adequate coverage of the optimal regions of the search domain. This quality is ensured by requiring the partitions to be arbitrarily fine. A fine partition sequence can model any closed region of the search space within arbitrarily small error.

**Definition 14.12 (Fine Partitions).** The partition sequence $\{\{E_n^a\}_{a\in A_n}\}_{n\in\mathbb{N}}$ is *fine* if for any closed set $F\subseteq X$ and any $\varepsilon>0$, there is an $N$ such that for all $n>N$, there is a subset $C_n\subseteq A_n$ for which $\{E_n^a\}_{a\in C_n}$ covers $F$ (i.e., $F\subseteq\bigcup_{a\in C_n} E_n^a$) and $\lambda(F)+\varepsilon>\sum_{a\in C_n}\lambda(E_n^a)$.

In particular, for a given $\varepsilon$, fine partitions will be used to bound the probability mass of the partition regions containing the essential boundary of $X_\varepsilon$, which is a closed set of measure zero. An evolutionary annealing algorithm must be almost surely fine in order to converge.

**Definition 14.13 (Almost Surely Fine).** An evolutionary annealing algorithm is *almost surely fine* on an objective $u$ if it produces a fine partition sequence with $\mathcal{G}_u$-probability one.

The relationship between partition sequences that are fine and those that are uniformly compatible deserves some discussion. In order to obtain a fine partition sequence, an algorithm must exhaustively search the domain in the limit, but the resulting partitions can still be shaped in such a way that the variation of the annealing density within the partition regions is not asymptotically bounded. On the other hand, a uniformly compatible partition sequence guarantees arbitrarily small fitness variation in the partition regions, but does not necessarily cover each region of the search domain. The convergence of evolutionary annealing requires both properties.

At this time, sufficient conditions to obtain almost surely fine partitions are unknown. Preliminary work suggests that such conditions exist in Euclidean space, at least, but determining the exact requirements is left as future work.

### *14.3.5 Conditions on the Mutation Distributions*

The final two conditions for convergence pertain to the mutation distributions for evolutionary annealing. First, the mutation distributions must be well matched with the base measure $\lambda$ in the sense that sets of $\lambda$-measure zero must also have $\nu_n^a$-measure zero. This property is known as *absolute continuity* of $\nu_n^a$ with respect to $\lambda$. Additionally, mutation must increasingly focus within the partition sequence.

**Definition 14.14 (Increasing Focus).** The mutation distributions $\{\{\nu_n^a\}\}$ are *increasingly focused* within the partition sequence $\{\{E_n^a\}_{a\in A_n}\}_{n\in\mathbb{N}}$ if for all $a \in A = \bigcup_n A_n$, $\nu_n^a(E_n^a) \to 1$ as $n \to \infty$. If the convergence is uniform over $A$, then the mutation distributions are *uniformly focused*; i.e., for all $\varepsilon > 0$, there exists an $N$ such that for all $m$ and all $a \in A_m$, $n > N$ and $n \geq m$ implies $\nu_n^a(E_n^a) > 1 - \varepsilon$.

This requirement of uniform focus is most easily satisfied by construction. As a trivial example, the choice of $\nu_n^a(B) = \lambda(B \cup E_n^a)/\lambda(E_n^a)$ is uniformly focused. In other cases, the uniform convergence $\nu_n^a(E_n^a) \to 1$ can be built into the definition of $\nu_n^a$ by tying the variance of the mutation distribution to the $\lambda$-measure of $E_n^a$. If the partition regions shrink at a uniform rate, then uniform focus follows by construction. The condition of uniform focus need not hold absolutely; it is sufficient for it to hold almost surely in $\mathcal{G}_u$.

Notice that the requirement that the mutation distributions be uniformly focused and that the partition sequence be fine are in competition with each other. Increasing focus requires the mutation variance to shrink quickly enough, but if it shrinks too quickly, the resulting partition sequence may not be fine.

This discussion of the mutation distributions completes the theoretical background in order to prove convergence, which is taken up next.

### *14.3.6 Convergence Proof*

The five conditions outlined in the previous subsections are sufficient to force asymptotic convergence of evolutionary annealing to the true global optimum. Reviewing, these conditions are: (1) an essentially bounded suitable fitness function that is also essentially Lipschitz continuous $\lambda$-a.e., (2) almost sure uniform compatibility, (3) an almost surely uniformly nice and fine partition sequence, (4) uniformly focused and absolutely continuous mutation distributions, and (5) a sufficiently slow cooling schedule. These five properties induce convergence of the observed fitness value to the essential infimum in probability.

Before stating the convergence theorem, some important lemmas are introduced. The first one is used to extend properties that hold $\lambda$-a.e. to properties that hold $\mathcal{G}_u$-a.s.

**Lemma 14.2.** *Given any property $\mathcal{P} \subseteq X$ that holds $\lambda$-almost everywhere, then for all $n$, an evolutionary annealing algorithm has $A_n \subseteq \mathcal{P}$ almost surely in $\mathcal{G}_u$ if its mutation distributions are absolutely continuous with respect to $\lambda$.*

*Proof.* Fix $n$. The initial population is sampled from $\lambda/\lambda(X)$. $\mathcal{P}$ holds $\lambda$-a.e., so with probability one, $A_1 \subseteq \mathcal{P}$.

Suppose evolutionary annealing generates $m$ populations with $A_m \subseteq P$. Then the next population, $A_{m+1} \setminus A_m$, is sampled from $\mathbb{G}_m = \sum_{a \in A_m} p_m(a) \nu_m^a$. Observe that $\mathcal{P}$ holds almost surely for $\nu_m^a$, since $\nu_m^a$ is absolutely continuous with respect to $\lambda$. Therefore $\mathbb{G}_m(\mathcal{P}) = 1$. Thus whenever $A_m \subseteq \mathcal{P}$, $A_{m+1} \subseteq \mathcal{P}$ with probability one.

Chaining these events together for $m \leq n$, the event $A_n \subseteq \mathcal{P}$ has probability one under $\mathcal{G}_u^n$. Since $\mathcal{G}_u$ is consistent with $\mathcal{G}_u^n$, it follows that $A_n \subseteq \mathcal{P}$ $\mathcal{G}_u$-a.s.

The next lemma replaces the mutation distributions with the conditional probability of $\lambda$ given the current partition, which is easier to handle theoretically. For convenience, define $\lambda_n^a(B) \equiv \lambda(B \cap E_n^a) / \lambda(E_n^a)$ for $B \in \mathcal{B}[X]$. If $\lambda(X) = 1$, then $\lambda_n^a(B)$ is the conditional probability of $B$ given $E_n^a$, i.e. $\lambda_n^a(B) = \lambda(B \mid E_n^a)$.

**Lemma 14.3.** *Suppose $u$ is a suitable fitness function and let $\varepsilon \in (0, \gamma)$. Let $\mathbb{G}_n$ be defined as above and assume that (1) the mutation distributions are uniformly focused, and (2) the partition sequence is fine. Assume also that $A_n \cap \partial X_\varepsilon = \emptyset$ for all $n$. Define $\tilde{\mathbb{G}}_n(X_\varepsilon) = \sum_{a \in A_n} p_n(a) \lambda_n^a(X_\varepsilon)$. Then $|\mathbb{G}_n(X_\varepsilon) - \tilde{\mathbb{G}}_n(X_\varepsilon)| \to 0$.*

*Proof.* Fix $\delta > 0$. Assume without loss of generality that $\lambda(X) = 1$; otherwise $\tilde{\lambda} = \lambda/\lambda(X)$ may be used. Since $\nu_n^a$ is uniformly focused within $E_n^a$, for $n$ sufficiently large, $\nu_n^a(X_\varepsilon \setminus E_n^a) < \delta/3$ independent of $a$.

Next, the portion of the $E_n^a$ that intersects the essential boundary of $X_\varepsilon$ will be bounded in size. Note that $\partial X_\varepsilon$ may include disconnected components of zero measure with empty interiors. Thus we consider the boundary of $\mathring{X}_\varepsilon$, which discards any disconnected components of $\partial X_\varepsilon$. Let $B$ be this boundary, consisting of the limit points of the interior of $X_\varepsilon$. Since $B \subseteq \partial X_\varepsilon$, $\lambda(B) = 0$. The set $B$ is closed. Since the partition is fine, for each $n$ sufficiently large, there is a set $C_n$ such that $\{E_n^a\}_{a\in C_n}$ covers $B$ with

$$\sum_{a\in C_n} \lambda(E_n^a) < \lambda(B) + \frac{\delta}{6} = \frac{\delta}{6}. \tag{14.25}$$

For $a \in C_n$, it holds that $\xi_n \geq \exp(-u(a)/T_n)\lambda(X) = \exp(-u(a)/T_n)$ and so

$$\begin{aligned}\sum_{a\in C_n} p_n(a)\,|\nu_n^a(X_\varepsilon \cap E_n^a) - \lambda_n^a(X_\varepsilon)| &\leq 2\sum_{a\in C_n} p_n(a)\\ &\leq 2\sum_{a\in C_n} \xi_n^{-1}\exp\left(-\frac{u(a)}{T_n}\right)\lambda(E_n^a)\\ &\leq 2\sum_{a\in C_n}\lambda(E_n^a) < \frac{\delta}{3}.\end{aligned} \tag{14.26}$$

For $a \in A_n \setminus C_n$, either $\mathring{X}_\varepsilon \cap E_n^a = E_n^a$ or $\mathring{X}_\varepsilon \cap E_n^a = \emptyset$. Since $\varepsilon < \gamma$, the measure of $\partial X_\varepsilon$ can be ignored, and so for $a \in A_n \setminus C_n$, either $\lambda_n^a(X_\varepsilon) = 0$ or $\lambda_n^a(X_\varepsilon) = 1$. Similarly, for $a \notin C_n$, $n$ can be chosen independent of $a$ so that $\nu_n^a(X_\varepsilon \cap E_n^a)$ comes within $\delta/3$ of either 0 or 1, since $\nu_n^a$ is uniformly focused on $a$ and either $\nu_n^a(X_\varepsilon \cap E_n^a) = \nu_n^a(E_n^a)$ or $\nu_n^a(X_\varepsilon \cap E_n^a) = 0$, depending on whether $a \in \mathring{X}_\varepsilon$. Therefore, $|\nu_n^a(X_\varepsilon \cap E_n^a) - \lambda_n^a(X_\varepsilon)| < \frac{\delta}{3}$ independent of $a$, and

$$\begin{aligned}\left|\mathbb{G}_n(X_\varepsilon) - \tilde{\mathbb{G}}_n(X_\varepsilon)\right| &\leq \sum_{a\in A_n} p_n(a)\,|\nu_n^a(X_\varepsilon) - \lambda_n^a(X_\varepsilon)|\\ &\leq \sum_{a\in A_n} p_n(a)\,\nu_n^a(X_\varepsilon \setminus E_n^a)\\ &\quad + \sum_{a\in A_n\setminus C_n} p_n(a)\,|\nu_n^a(X_\varepsilon \cap E_n^a) - \lambda_n^a(X_\varepsilon)|\\ &\quad + \sum_{a\in C_n} p_n(a)\,|\nu_n^a(X_\varepsilon \cap E_n^a) - \lambda_n^a(X_\varepsilon)|\\ &< \frac{\delta}{3} + \frac{\delta}{3} + \frac{\delta}{3} = \delta,\end{aligned} \tag{14.27}$$

and the conclusion follows.

The final lemma extends the uniform compatibility of $\{\{E_n^a\}\}$ to the partition of $X_\varepsilon$ given by $\{\{X_\varepsilon \cap E_n^a\}_{a\in A_n}\}_{n\in\mathbb{N}}$.

**Lemma 14.4.** *Suppose that the triple $(u, \{T_n\}, \{\{E_n^a\}\})$ is uniformly compatible on $X$. Let $\varepsilon \in (0,\gamma)$ and suppose $a$ is in the interior of $X_\varepsilon$. Then it holds that the triple $(u, \{T_n\}, \{\{X_\varepsilon \cap E_n^a\}_{a\in A_n \cap X_\varepsilon}\}_{n\in\mathbb{N}})$ is uniformly compatible.*

*Proof.* The result follows immediately from

$$\lim \operatorname*{ess\,sup}_{z \in X_\varepsilon \cap E_n^a} |g_n(a) - g_n(z)| \leq \lim \operatorname*{ess\,sup}_{z \in E_n^a} |g_n(a) - g_n(z)| = 0. \tag{14.28}$$

**Theorem 14.1.** *The fitness of a sample taken from an evolutionary annealing algorithm with annealed proportional selection converges in probability to the essential infimum of any essentially bounded suitable fitness function that is essentially Lipschitz continuous almost everywhere provided that the algorithm almost surely shrinks faster than its cooling schedule, is almost surely uniformly nice and fine, and has mutation distributions that are almost surely uniformly focused within the partitions and absolutely continuous with respect to the base measure.*

*Proof.* Fix $\varepsilon, \delta > 0$ with $\varepsilon < \gamma$. Without loss of generality, assume $\lambda(X) = 1$; if not, $\tilde{\lambda} \equiv \lambda / \lambda(X)$ will satisfy this equality. Also without loss of generality, suppose $u \geq 0$ $\lambda$-a.e.; otherwise $\tilde{u} = u - u_*$ may be used. Let $P$ be a population history sampled from $\mathcal{G}_f$. The conclusion will follow if $\mathbb{P}\left(|u(P_n^k) - u_*| < \varepsilon\right) > 1 - \delta$ for any $k$. Observe that

$$\mathbb{P}\left(|u(P_n^k) - u_*| < \varepsilon\right) = \int_{C_{n-1}} \mathbb{G}_n(X_\varepsilon \mid P_1, \ldots, P_{n-1})\, \mathcal{G}_u^{n-1}(dP_1, \ldots, dP_{n-1}), \tag{14.29}$$

where $C_{n-1}$ is the $(n-1)$-prefix of any set $C$ such that $\mathcal{G}_u(C) = 1$. That is,

$$C_{n-1} = \{z \in X^{K(n-1)} \mid \exists c \in C \text{ s.t. for } 1 \leq m < n, c_m = z_m\}. \tag{14.30}$$

Note that $C$ is a measurable set of infinite sequences on $X^K$. It will be shown for a particular $C$ that for all $c \in C$ and all $n > N_0$,

$$\mathbb{G}_n(X_\varepsilon) = \mathbb{G}_n(X_\varepsilon \mid P_1 = c_1, \ldots, P_n = c_n) < 1 - \delta, \tag{14.31}$$

and the proof will follow. Note that the sequence of partitions produced by a particular sequence $c$ is assumed to be deterministic, and that the set $A_n$ is a deterministic function of $c$, i.e. $A_n = A_n(c)$. Because $\mathcal{A}(X_\varepsilon) = 1$ for all $\varepsilon$, it is sufficient to prove that $|\mathbb{G}_n(X_\varepsilon) - \mathcal{A}(X_\varepsilon)| < \delta$ for large $n$.

In order to carry the proof, the desired set $C$ must satisfy the following properties for each $c \in C$:

1. The sequence of partitions generated by $c$ is uniformly nice and fine;
2. The mutation distributions are uniformly focused;
3. The cooling schedule has $\sup_{a \in A_n} \inf_{\{U : E_n^a \subseteq U_a\}} \lambda(U_a \setminus \{a\}) = o(T_n)$ for all $n$;
4. For all $n$ and all $a \in A_n(c)$, $|u(a)| < \operatorname{ess\,sup}_X |u|$;
5. For all $n$, $A_n(c) \cap \partial X_\varepsilon = \emptyset$;
6. For all $n$ and all $a \in A_n(c)$, $0 \leq u(a) < \infty$;
7. The fitness function $u$ is essentially Lipschitz continuous on $A = \bigcup_n A_n$.

Define sets $C_i$ so that $C_i$ is the largest set satisfying the $i^{th}$ property above. The sets $C_1$, $C_2$ and $C_3$ have $\mathcal{G}_u$-probability one by assumption. The sets $C_4$, $C_5$, $C_6$, and $C_7$

each have $\mathcal{G}_u$-probability one by Lemma 14.2. For $C_4$, use $\mathcal{P} = X \setminus \partial X_\varepsilon$. For $C_5$, use $\mathcal{P} = \{x | u(x) \leq \operatorname{ess\,sup}_X |u|\}$. For $C_6$, use $\mathcal{P} = \{x | 0 \leq u(x) < \infty\}$. For $C_7$, $\mathcal{P} = \{x | u \text{ is essentially Lipschitz continuous at } \{x\}\}$. Let $C = \bigcap_i C_i$. Then $\mathcal{G}_u(C) = 1$. As a consequence, we may assume below that properties listed above hold for all $n$.

By Lemma 14.3, for $n$ sufficiently large, $|\mathbb{G}_n(X_\varepsilon) - \tilde{\mathbb{G}}_n(X_\varepsilon)| < \frac{\delta}{2}$. It will be shown that $|\tilde{\mathbb{G}}_n(X_\varepsilon) - \mathcal{A}(X_\varepsilon)| \to 0$. The argument is based on the fact that $\tilde{\mathbb{G}}_n$ is an approximate martingale and uses a series of conditional expectations. Here and below, the notation $\mathbb{1}_B = \mathbb{1}_B(x)$ is defined as

$$\mathbb{1}_B(x) = \begin{cases} 1 \text{ if} x \in B \\ 0 \text{ otherwise} \end{cases}. \tag{14.32}$$

The fact that $\int_X \mathbb{1}_B \, d\lambda = \lambda(B)$ is used frequently. The expression $Y\mathbb{1}_B$ is shorthand for $Y(x)\mathbb{1}_B(x)$, as is common in the study of stochastic processes.

Define a random process $Y_n(x) = \sum_{a \in A_n} \mathbb{1}_{E_n^a}(x) g_n(a)$. Observe that

$$\begin{aligned} \left|\tilde{\mathbb{G}}(X_\varepsilon) - \mathbb{E}(Y_n \mathbb{1}_{X_\varepsilon})\right| &= \left| \sum_{a \in A_n} \left[ p_n(a)\lambda_n^a(X_\varepsilon) - \int_{E_n^a} \mathbb{1}_{X_\varepsilon}(z) g_n(a) \lambda(dz) \right] \right| \\ &\leq \sum_{a \in A_n} |p_n(a)\lambda_n^a(X_\varepsilon) - g_n(a)\lambda(X_\varepsilon \cap E_n^a)| \\ &\leq \sum_{a \in A_n} |p_n(a) - g_n(a)\lambda(E_n^a)| \, \lambda_n^a(X_\varepsilon) \\ &\leq \sum_{a \in A_n} |p_n(a) - g_n(a)\lambda(E_n^a)| \\ &\leq \left|1 - \frac{\xi_n}{\eta_n}\right| \sum_{a \in A_n} p_n(a) = \left|1 - \frac{\xi_n}{\eta_n}\right| < \frac{\delta}{8}, \end{aligned} \tag{14.33}$$

by an application of Lemma 14.1 together with Proposition 14.3. Here we have relied on the fact that $\frac{g_n(a)\lambda(E_n^a))}{p_n(a)} = \frac{\xi_n}{\eta_n}$ since only the normalizers differ.

Let $\{\mathcal{E}_n^{A_n}\}$ be the filtration generated by the sequence of partitions $\{E_n^a\}$. Now consider the process generated by conditioning $Y_n \mathbb{1}_{X_\varepsilon}$ on $\{\mathcal{E}_n^{A_n}\}$,

$$Y_n^\varepsilon = \mathbb{E}\left[Y_n \mathbb{1}_{X_\varepsilon} \mid \mathcal{E}_n^{A_n}\right] = \sum_{a \in A_n} \mathbb{1}_{E_n^a}(x) g_n(a) \lambda_n^a(X_\varepsilon). \tag{14.34}$$

Note that $\mathbb{E}[Y_n^\varepsilon] = \mathbb{E}[Y_n \mathbb{1}_{X_\varepsilon}]$ by the properties of conditional expectations. Let $Z_n(x) = g_n(x)$, and define $Z_n^\varepsilon = \mathbb{E}\left[Z_n \mathbb{1}_{X_\varepsilon} \mid \mathcal{E}_n^{A_n}\right]$ similarly to $Y_n^\varepsilon$, so that

$$Z_n^\varepsilon = \sum_{a \in A_n} \mathbb{1}_{E_n^a}(x) \int_{X_\varepsilon} g_n(x)\lambda_n^a(dx) = \sum_{a \in A_n} \frac{\mathbb{1}_{E_n^a}(x)}{\lambda(E_n^a)} \int_{X_\varepsilon \cap E_n^a} g_n(x)\lambda(dx), \tag{14.35}$$

observing that

$$\mathbb{E}[Z_n^\varepsilon] = \mathbb{E}[Z_n 1_{X_\varepsilon}] = \mathcal{A}_n(X_\varepsilon). \tag{14.36}$$

It holds that $Y_n^\varepsilon(x)$ and $Z_n^\varepsilon(x)$ converge together in mean. To prove this statement, there are three types of points to consider: (1) points inside a covering set for the boundary of $\mathring{X}_\varepsilon$, (2) points located away from $X_\varepsilon$, and (3) points solidly inside of $X_\varepsilon$. For the first case, as in the proof of Lemma 14.3, let $B$ be the boundary of $\mathring{X}_\varepsilon$, a closed subset of $\partial X_\varepsilon$ with disconnected components of measure zero removed. Because the partition sequence is fine, for $n$ sufficiently large, there is a set $F_n$ such that $B \subseteq \bigcup_{a\in F_n} E_n^a$ and $\sum_{a\in F_n} \lambda(E_n^a) < \lambda(B) + \frac{\delta}{16} = \frac{\delta}{16}$.

For the second case, let $z \in E_n^{\tilde{a}}$ with $\tilde{a} \notin X_\varepsilon$ and $\tilde{a} \notin F_n$. Then $\mathring{X}_\varepsilon \cap E_n^{\tilde{a}} = \emptyset$, so $Y_n^\varepsilon(z) = Z_n^\varepsilon(z) = 0$ since $\lambda(\partial X_\varepsilon \cap E_n^{\tilde{a}}) = 0$. For the third case, in which $\tilde{a} \in X_\varepsilon$ and $\tilde{a} \notin F_n$, then $\tilde{a} \in \mathring{X}_\varepsilon$ by assumption, and by Lemma 14.4 together with Proposition 14.4, $g_n$ is uniformly approximated by its neighborhood average on $\{\{X_\varepsilon \cap E_n^a\}_{a\in A_n \cap X_\varepsilon}\}_{n\in\mathbb{N}}$. So in the third case, for $n$ large and independent of $\tilde{a}$,

$$
\begin{aligned}
|Y_n^\varepsilon(z) - Z_n^\varepsilon(z)| &\leq \left| g_n(\tilde{a})\,\lambda_n^{\tilde{a}}(X_\varepsilon) - \frac{1}{\lambda(E_n^{\tilde{a}})} \int_{X_\varepsilon \cap E_n^{\tilde{a}}} g_n(x)\,\lambda(dx) \right| \\
&= \lambda_n^{\tilde{a}}(X_\varepsilon) \left| g_n(\tilde{a}) - \frac{1}{\lambda(X_\varepsilon \cap E_n^{\tilde{a}})} \int_{X_\varepsilon \cap E_n^{\tilde{a}}} g_n(x)\,\lambda(dx) \right| \\
&< \frac{\delta}{16},
\end{aligned}
\tag{14.37}
$$

where the final line follows because $\lambda_n^{\tilde{a}} \leq 1$. Uniform convergence is critical, since otherwise the size of $n$ would depend on $\tilde{a}$. Putting these three cases together,

$$
\begin{aligned}
\mathbb{E}\,|Y_n^\varepsilon - Z_n^\varepsilon| &\leq \sum_{a\in C_n} \left| g_n(a)\lambda(X_\varepsilon \cap E_n^a) - \int_{X_\varepsilon \cap E_n^a} g_n(z)\lambda(dz) \right| \\
&\quad + \sum_{a\in A_n\setminus C_n} \int_{E_n^a} |Y_n^\varepsilon(z) - Z_n^\varepsilon(z)|\,\lambda(dz) \\
&< \sum_{a\in C_n} \left| g_n(a) - \operatorname*{ess\,sup}_{z\in E_n^a} g_n(z) \right| \lambda(E_n^a) + \frac{\delta}{16} \sum_{a\in A_n\setminus C_n} \lambda(E_n^a) \\
&< \frac{\delta}{16} + \frac{\delta}{16} = \frac{\delta}{8},
\end{aligned}
\tag{14.38}
$$

where the last line follows from the fact that $|g_n(a) - \operatorname*{ess\,sup}_{z\in E_n^a} g_n(z)| \leq 1$. Thus for $n$ sufficiently large,

$$
|\mathbb{E}_n(Y_n \mathbb{1}_{X_\varepsilon}) - \mathcal{A}_n(X_\varepsilon)| = |\mathbb{E}(Y_n^\varepsilon) - \mathbb{E}(Z_n^\varepsilon)| \leq \mathbb{E}\,|Y_n^\varepsilon - Z_n^\varepsilon| < \frac{\delta}{8}. \tag{14.39}
$$

That is,

$$
\begin{aligned}
\left|\tilde{\mathbb{G}}_n(X_\varepsilon) - \mathcal{A}_n(X_\varepsilon)\right| &\leq \left|\tilde{\mathbb{G}}_n(X_\varepsilon) - \mathbb{E}(Y_n \mathbb{1}_{X_\varepsilon})\right| + \left|\mathbb{E}(Y_n \mathbb{1}_{X_\varepsilon}) - \mathcal{A}_n(X_\varepsilon)\right| \\
&< \frac{\delta}{8} + \frac{\delta}{8} = \frac{\delta}{4}
\end{aligned}
\tag{14.40}
$$

Putting these facts together, for $n$ sufficiently large,

$$\begin{aligned} |\mathbb{G}_n(X_\varepsilon) - \mathcal{A}(X_\varepsilon)| &\leq \left|\mathbb{G}_n(X_\varepsilon) - \tilde{\mathbb{G}}(X_\varepsilon)\right| + \left|\tilde{\mathbb{G}}_n(X_\varepsilon) - \mathcal{A}_n(X_\varepsilon)\right| + |\mathcal{A}_n(X_\varepsilon) - \mathcal{A}(X_\varepsilon)| \\ &< \frac{\delta}{2} + \frac{\delta}{4} + \frac{\delta}{4} = \delta, \end{aligned} \tag{14.41}$$

completing the proof.

In sum, Theorem 14.1 shows that evolutionary annealing with annealed proportional selection is guaranteed to converge asymptotically arbitrarily close to the minima of the fitness function provided that the fitness function is not too irregular and the cooling schedule and variance decay are not too aggressive. Less can be said regarding the rate of convergence, as will be discussed next.

### *14.3.7 Convergence Rates*

An examination of the proof of Theorem 14.1 shows that there are three basic sources of approximation error: (1) the variance of the mutation distribution, (2) the accuracy of the neighborhood average, and (3) the speed of convergence for the annealing distributions, due to the cooling schedule. Of these, the variance and the cooling schedule are under the direct control of the practitioner. Implicitly, these two factors also control the accuracy of the neighborhood average. In order to set the cooling schedule and variance decay to maximize the rate of convergence, the effects of these three error sources must be carefully considered.

The first source of error is due to the difference $|\nu_n^a(X_\varepsilon) - \lambda_n^a(X_\varepsilon)|$. Convergence occurs because both of these measures $\nu_n^a$ and $\lambda_n^a$ asymptotically become point masses, the former because variance decays and the latter because the mixing points eventually fill the search space. To minimize error, these two measures should be kept as close as possible for mixing points in the vicinity of the optima. As the algorithm begins to focus on a small group of optima, the partitions in that region will become smaller, and the variance of the mutation distribution should decrease at a similar rate. Notably, however, decreasing the variance also reduces the probability that the global optimum will be discovered if it has not already and if it is located sufficiently far from the current regions of focus. Also, when a new and better local optimum is discovered after the search has already focused on other local optima, the exploration of the new optimum will proceed slowly if the variance has already decayed substantially. Therefore it may make sense to scale the variance to reflect the size of the partition region for the mixing point being mutated. In this way, larger variances will typically be employed in unexplored regions, whereas a narrower variance will be used in well explored regions, so that $\nu_n^a$ and $\lambda_n^a$ are well-matched in general. The fact that the mixing points eventually fill the space guarantees that a dynamic and locally scaled variance decay schedule of this type will eventually become increasingly centered as required.

The second source of error pertains to the accuracy of approximating the average value of the annealing distribution by its value at the mixing points, due to the term $\left|g_n^\lambda(x) - g_n(a)\right|$. This error depends strongly on the fitness function. If the fitness function is relatively homogeneous and does not fluctuate at different rates in different regions of the search space, then this source of error strongly reflects the mass of the partition region, $\lambda(E_n^a)$. In a large region, the approximation $g_n(a)\lambda(E_n^a)$ is likely to differ more substantially from $\int_{E_n^a} g_n \, d\lambda$ than it would in a smaller region. Thus this source of error can perhaps be reduced by spreading the mixing points more evenly through the space in order to keep the partition regions uniformly small. This goal can be accomplished by using a high variance at the outset to guarantee full exploration of the space. At later stages, the use of a high variance is at odds with the need to match the mutation distribution $\nu_n^a$ with the measure $\lambda_n^a$, and thus the first two sources of error must be balanced and cannot be mutually eliminated. The accuracy of the average approximation is also affected by the cooling schedule. At high temperature, the approximation must be more accurate because $g_n$ will vary less over the region $E_n^a$ if $T_n$ is large. Lowering the temperature increases this source of error by causing the function $g_n$ to fluctuate more.

The third source of error concerns the speed of convergence of the annealing distributions due to the difference $|g - g_n|$. The faster the cooling schedule takes the temperature to zero, the faster this error will be minimized. But a fast cooling schedule will increase the error due to the neighborhood approximation. The speed of the cooling schedule must balance the need to minimize both the neighborhood approximation error and the annealing convergence error.

Overall, experiments show that a logarithmic cooling schedule, e.g. $T_n^{-1} = \eta \log n$, works well in practice. Experimental results also suggest that the variance should start off quite large and decay exponentially fast. Also, scaling the variance locally based on the size of the partition region for the mixing point being mutated should be effective. The next chapter discusses experimental results along these lines.

## 14.4 Conclusion

Evolutionary annealing was introduced as an optimization strategy that seeks the global optimum by building a probabilistic model based on the objective evaluations of all points produced by the optimization process. Under certain conditions, this model converges to an accurate representation of the objective function near the global optimum, as seen in Theorem 14.1. The proof relies on the fact that the model underlying evolutionary annealing is an approximate Lévy martingale. Such a martingale may be thought of as an estimate of some quantity that improves as more information becomes available. Thus evolutionary annealing is a *martingale method*, a new class of optimization method based on using increasing information to improve optimization. In the case of evolutionary annealing, objective evaluations provide the source of information, and evolutionary annealing leverages this information in order to improve its optimization.

Evolutionary annealing converges in theory on certain fitness functions, but its experimental performance also needs to be analyzed. To do so, evolutionary annealing must be instantiated within a particular search domain. Chapter 15 presents experiments with an instantiation in finite-dimensional Euclidean space, $\mathbb{R}^d$, along with specific implementation details that make it possible to sample evolutionary annealing in logarithmic time with respect to the number of evaluations. In the course of these experiments, it is seen that evolutionary annealing generally performs well. Chapter 16 applies evolutionary annealing to the problem of training neural networks, showing that the basic optimization concept can be expanded to search effectively in complex spaces. These neural networks outperform networks trained by other methods on tasks that require a complex network topology. Successes in both real vectors and neural networks will establish evolutionary annealing as an effective optimization method for practical tasks.

# Chapter 15
# Evolutionary Annealing in Euclidean Space

Evolutionary annealing was developed in the last chapter as a general-purpose optimization technique. This chapter presents an application of evolutionary annealing to the space of finite real vectors. Experiments are performed to compare real-space evolutionary annealing (REA) on the set of benchmarks and algorithms from Chapter 11. REA performs well in general and is complementary to the earlier optimizers tested, outperforming other methods on multimodal objectives with irregular structure. This feature suggests that REA is well-aligned with a mixture sieve prior.

## 15.1 Evolutionary Annealing in Euclidean Space

Evolutionary annealing can be used to search for bit strings, real vectors, neural networks, Bayesian network structures, game strategies, programs, state machines, and any other structure that can be embedded within a suitable measure space. As a baseline evaluation, experiments were performed in finite-dimensional Euclidean space on a set of twelve standard benchmarks from Chapter 11. As before, the domain is a problem-specific hypercube $Q \subseteq \mathbb{R}^d$, with the base measure $\lambda$ being the Lebesgue measure restricted to the Borel $\sigma$-algebra over $Q$, i.e. $\lambda(B) = \int_B dx / \int_Q dx$. The instantiation of evolutionary annealing in Euclidean space is termed Real-Space Evolutionary Annealing (REA), and it has been tested with both annealed proportional selection (REA-P) and annealed tournament selection (REA-T).

### 15.1.1 Instantiation Details

The version of REA in this chapter uses Gaussian mutation distributions with $\nu_n^a = \mathcal{N}\left(a, \sigma_n(a)^2\right)$. The standard deviation $\sigma_n(a)$ is scaled to the area of the partition region with $\sigma_n(a) = \frac{1}{2} w \lambda\left(E_n^a\right)^{1/d}$, where $d$ is the dimension of the problem and $w$ is


A. J. Lockett, *General-Purpose Optimization Through Information Maximization*,
Natural Computing Series, https://doi.org/10.1007/978-3-662-62007-6_15

the width of the space (*i.e.* $\frac{1}{2}w$ is the side length of $Q$). This choice of variance seeks to align the shape of $\nu_n^a$ and $\lambda\left(E_n^a\right)$ as discussed in Section 14.3.7. Specifically, if $E_n^a$ were a hypercube centered at $a$, then the first standard deviation of $\nu_n^a$ would be contained within $E_n^a$.

This implementation of REA does not meet the requirements of Theorem 14.1, but it is on the cusp of doing so. Most importantly, there is no way to determine at this time whether REA generates uniformly nice and fine partitions. It is likely that REA converges too quickly to maintain nice and fine partitions, meaning that REA may fail to find the optima in certain cases, but finds the optima more quickly in others.

The vector-separating algorithm from Section 14.2.3 was used, which does guarantee that each search point $a$ is contained in the interior of the region $E_n^a$ for all $n$. The mutation distributions are absolutely continuous with respect to the Lebesgue measure. On average, $\nu(E_n^a)$ is a nonzero constant less than 1. If the partitions were to reduce in size regularly, and if $\sigma_n(a)$ were multiplied by a decaying factor, say, $e^{-\beta n}$ for $\beta$ close to zero, then $\nu_n^a$ would be increasingly focused ($\nu_n^a(E_n^a) \to 1$). In the experiments that follow, no decay factor was applied for $d = 5$ and $d = 10$. It was not necessary to do so, since the samples in the experiments converged towards a fixed distribution without a decay factor. In 25 dimensions, however, a decay factor of $n^{-\frac{1}{2}}$ (i.e. $\sigma_n(a) = \frac{1}{2}wn^{-\frac{1}{2}}\lambda\left(E_n^a\right)^{1/d}$) was applied in order to achieve faster convergence.

REA can be computationally expensive because of the overhead involved in handling an expanding set of evaluations points. An efficient implementation can be obtained by implementing the sampling routines in a way that requires computation that is only logarithmic in the number of evaluation point. The details are discussed next.

### *15.1.2 Sampling Algorithms for Annealed Selection*

The computational efficiency of evolutionary annealing is primarily determined by the cost of preparing and sampling annealed proportional selection. A naïve approach to computing Equation 14.3 would make the cost of preparing and sampling $p_n$ be linear, since the normalizing factor $\xi_n$ must be computed one element at a time and because sampling from a probability vector typically requires iterating through the vector. In fact, annealed proportional selection can be approximately computed in logarithmic time in the average case by leveraging the partition tree, with most operations occurring in subroutines that guarantee worst-case logarithmic time, as described in Section 14.2.4. The approximation can be made accurate at close to machine-level precision, so that it is sufficiently precise for all practical purposes.

In order to reduce the sampling complexity for evolutionary annealing from linear to logarithmic time, a tree-sampling method is needed for sampling $p_n$. The partition tree provides a tree such that the leaves are associated exactly with the components of $p_n$. The goal, then, is to create a sequence of decisions made along a path through the partition tree such that the decision process assigns probability mass to each complete path in equality with the probability of the leaf at the end of the path under $p_n$.

Let $\nu$ be an internal node of the partition tree. Let $N \subseteq A_n$ be the set of previously observed individuals residing within leaves of the partition tree that are descended from $\nu$. Let $\mu$ be one of the two child nodes of $\nu$, and let $M \subseteq N$ contain the leaf descendants of $\mu$. To extend a probabilistic path that has reached $\nu$, a choice must be made at node $\nu$ whether to add node $\mu$ or its sibling to the path. Suppose the choice is made according to

$$\mathbb{P}(\mu \mid \nu) = \frac{\sum_{x \in M} \alpha(x)^{\log n} \lambda\left(E_n^x\right)}{\sum_{y \in N} \alpha(y)^{\log n} \lambda\left(E_n^y\right)}, \tag{15.1}$$

where $\alpha(x) \equiv \exp(-\eta u(x))$, mirroring Equation 14.3 with cooling schedule $T_n^{-1} = \eta \log n$. Now let $\pi_x$ be a path from the root to the leaf containing the point $x$, and observe that a sequence of decisions made according to Equation 15.1 yields

$$\mathbb{P}(\pi_x) = \prod_{\nu} \mathbb{P}(\text{child}(\nu, \pi_x) \mid \nu) = \xi_n \alpha(x)^{\log n} \lambda\left(E_n^x\right) = p_n(x), \tag{15.2}$$

with $\text{child}(\nu, \pi_x)$ being the child node of $\nu$ on the path $\pi_x$. The next to last equality in Equation 15.2 holds because each successive denominator cancels the numerator of the previous one, leaving only the denominator from the root node, which is equal to $\xi_n^{-1}$, and the numerator from the leaf node, which is $\alpha(x)^{\log n} \lambda\left(E_n^x\right) = \exp(-u(x)/T_n) \lambda\left(E_n^x\right)$. Therefore, sampling a path through the tree starting from the root samples from $p_n$ provided that the decision at each node is made according to Equation 15.1.

The difficulty of this method is that the sum in the numerator of Equation 15.1 must be computed for each node. If the temperature were fixed, then the value of the sum could be stored on each node. The sum only changes when new leaves are inserted, and then only the nodes that are direct ancestors of the inserted node need to adjust their sums, resulting in logarithmic updates to the tree. As long as the temperature does not change, then, the tree-sampling method is logarithmic both to prepare the data structures and to sample them.

It remains to account for changes in temperature without recomputing the numerator of Equation 15.1 at each time step. Introducing $h(T) = \sum_{x \in N} \alpha(x)^T \lambda\left(E_n^x\right)$ to capture the fact that the sum varies with the generation, the problem is that the exponent cannot be pulled out of the sum, meaning that the sum must be recomputed with every change in temperature. However, $h(T)$ is infinitely differentiable in $T$, with $m^{th}$ derivative

$$h^{(m)}(T) = \sum_{x \in N} \alpha(x)^T (\log \alpha(x))^m \lambda\left(E_n^x\right). \tag{15.3}$$

Thus a Taylor approximation is possible, since

$$h(T) = \sum_{m=1}^{\infty} \left( \sum_{x \in N} \frac{(\log \alpha(x))^m}{m!} \alpha(x)^{T_0} \lambda\left(E_n^x\right) \right) (T - T_0)^m . \tag{15.4}$$

The Taylor approximation can be computed by storing a vector of coefficients $t = (t_1 \dots t_m)$ with $t_j \equiv \sum_{x \in N} (\log \alpha(x))^j \alpha(x)^{T_0} \lambda\left(E_n^x\right)$ for all $j \in 1 \dots m$, with a fixed value $T_0$. These vector sums can then be propagated up the tree in logarithmic time, and the sampling method can approximate $h(\log n)$ as needed at each node.

To complete the description of the sampling method, $T_0$ and $m$ must be specified. As a general feature of $h(T)$, the approximation is substantially correct for $T > T_0$ over a larger interval than for $T < T_0$. With $m = 10$, the approximation is highly accurate for $T \in [T_0, T_0 + 1/2]$ but degrades outside that interval. Thus the Taylor coefficients must be recomputed for the entire tree on every interval of $T$ of size $1/2$. For practical purposes, the value of $T_0$ is set to 1 for the first few generations, and then is reset when

$$T = \log n = 3/2, 2, 5/2, \dots$$

This resetting feature is actually not as burdensome as it may sound, and it only needs to be performed logarithmically often, so that the entire procedure of maintaining and sampling the tree still has logarithmic complexity overall. Some example statistics for computation time are shown in Table 15.1. The next section discusses a similar method for sampling annealed tournament selection, and introduces data structures that make it possible to sample annealed selection in average case logarithmic time.

### *15.1.3 Sampling Annealed Tournament Selection*

As with annealed proportional selection, it is not computationally efficient to sample Equation 14.7 directly. In addition, annealed tournament selection introduces the need to sort all previously proposed solutions by fitness. In order to accommodate these issues, a balanced binary tree can be used, called the *score tree*. Like the partition tree, the score tree contains one leaf node per proposed solution; the internal nodes represent the set of nodes in their span. The score tree reorganizes the partition tree so that points with higher fitness are always to the left and points with lower fitness are always to the right. Using standard tree algorithms, the score tree can be balanced in logarithmic time after each insertion.

Annealed tournament selection can be sampled by walking the score tree, making a decision at each node whether to follow the lower- or the higher-ranked branch. The probability at each node will depend on the area represented by the node and the height of the subtree underneath the node. The area of a leaf node can be copied from the partition tree. Both the area and the height can then be propagated up the score tree in logarithmic time after each insertion. In this way, the score tree is also a partition tree. However, the internal nodes of the score tree correspond approximately to the level sets of the fitness function, and thus the regions that they represent can be arbitrarily complex to describe. Therefore, although the score tree defines a partition over the search space, the score tree cannot replace the partition tree, because there is no efficient way to determine whether a point resides in the region represented by an internal node of the score tree. However, the score tree is kept balanced, providing worst-case logarithmic performance.

When sampling annealed tournament selection using the score tree, the decision must be made at each internal node $\nu$ whether to follow the higher- or lower-ranked branch. Let $h+1$ be the height of the subtree under node $\nu$, and assume the tree is perfectly balanced. Then $\nu$ has $2^{h+1}$ leaf nodes in its span. Let $\mu$ be the higher-ranked child node of $\nu$. Suppose further that the nodes spanned by $\nu$ range in rank from $R$ to $R+2^{h+1}-1$, so that the nodes spanned by $\mu$ range in rank from $R$ to $R+2^h-1$. Ignoring the region weight temporarily, a direct application of standard tournament selection yields

$$\mathbb{Q}_T(\mu \mid \nu) = \frac{\sum_{m=0}^{2^h-1} q^{1/T}\left(1-q^{1/T}\right)^{R+m}}{\sum_{j=0}^{2^{h+1}-1} q^{1/T}\left(1-q^{1/T}\right)^{R+j}}. \tag{15.5}$$

Let $\kappa$ be the lower ranked sibling of $\mu$, spanning ranks $R+2^h$ to $R+2^{h+1}-1$. Then the ratio for selecting $\mu$ over $\kappa$ is given by

$$\frac{\mathbb{Q}_T(\mu \mid \nu)}{\mathbb{Q}_T(\kappa \mid \nu)} = \frac{\sum_{m=0}^{2^h-1} q^{1/T}\left(1-q^{1/T}\right)^{R+m}}{\sum_{m=0}^{2^h-1} q^{1/T}\left(1-q^{1/T}\right)^{R+2^h+m}} = \frac{1}{\left(1-q^{1/T}\right)^{2^h}} \equiv \tilde{q}(h,T). \tag{15.6}$$

The function $\tilde{q}(h,T)$ gives the selection preference of the higher branch over the lower branch. Finally, incorporating the region weights, let

$$\mathbb{P}_T(\mu \mid \nu) = \frac{\tilde{q}(h,T)\,\lambda(\mu)}{\tilde{q}(h,T)\,\lambda(\mu) + (1-\tilde{q}(h,T))\,\lambda(\kappa)}, \tag{15.7}$$

where $\lambda(\mu)$ and $\lambda(\kappa)$ are the cumulative weights of the partition regions of the points in the span of $\mu$ and $\kappa$, respectively. Equation 15.7 is normalized and implies $\mathbb{P}_T(\kappa \mid \nu) = 1 - \mathbb{P}_T(\mu \mid \nu)$.

To show that this process does in fact implement annealed tournament selection, notice that

$$\mathbb{P}_T(\mu \mid \nu) \propto \tilde{q}(h,T)\frac{\lambda(\mu)}{\lambda(\nu)}, \quad \mathbb{P}_T(\kappa \mid \nu) \propto \frac{\lambda(\mu)}{\lambda(\nu)}, \tag{15.8}$$

introducing the $\lambda(\nu)$ factor as a proportional constant. Thus for a general path $\pi_x$, recalling that $\tilde{q}(h,T) \propto \mathbb{Q}_T(\mu \mid \nu)$ by definition,

$$\mathbb{P}_T(\text{child}(\nu,\pi_x) \mid \nu) \propto \mathbb{Q}_T(\text{child}(\nu,\pi_x) \mid \nu)\frac{\lambda(\mu)}{\lambda(\nu)}, \tag{15.9}$$

and therefore

$$
\begin{aligned}
\mathbb{P}_{T_n}(\pi_x) &= \prod_{v\in\pi_x} \mathbb{P}_{T_n}(\text{child}(v,\pi_x) \mid v) \\
&\propto \prod_{v\in\pi_x} \mathbb{Q}_{T_n}(\text{child}(v,\pi_x) \mid v) \frac{\lambda(\text{child}(v,\pi_x))}{\lambda(v)} \\
&= \mathbb{Q}_{T_n}(\pi_x) \frac{\lambda(E_n^x)}{\lambda(X)} \\
&\propto p_n(x).
\end{aligned} \tag{15.10}
$$

The last equality holds because the area ratios successively cancel each other, and the last proportionality follows from the fact that $\mathbb{Q}_{T_n}$ was defined to implement tournament selection with selection pressure $q^{1/T_n}$. The ultimate conclusion is that a tree-sampling algorithm with node selection probabilities as given in Equation 15.7 can be used to sample from annealed tournament selection in worst-case logarithmic time.

As a final note on efficiency, notice that sampling in the score tree has worst-case logarithmic time, whereas sampling on the partition tree has average case logarithmic time. Therefore it makes sense to sample annealed proportional selection from the score tree rather than the partition tree. The only additional requirement is that the Taylor coefficients for annealed proportional selection should be propagated up the score tree rather than the partition tree. In this way, regardless of whether tournament or proportional selection is used, the sampling operations of evolutionary annealing require logarithmic time in the worst case.

### *15.1.4 Implementation*

Because evolutionary annealing relies on several data structures, it can be complex to implement. In order to further clarify implementation details and to permit the reproducibility of the experimental results that follow, an open-source implementation was released under the name *pyec* (http://pypi.python.org/pypi/PyEC). This package implements both annealed proportional and tournament selection along with many other popular evolutionary computation methods, including the exact code used to run the experiments described in Section 15.2. This package is intended to encourage further experimentation and evaluation of the evolutionary annealing method beyond the results reported in this book.

Performance statistics for evolutionary annealing were gathered using this implementation in order to demonstrate the actual computational costs of running the algorithm in Table 15.1. These statistics were compiled by averaging results from four runs each of the algorithm using tournament selection on the benchmarks *shekel* and *rastrigin*. Tournament and proportional selection both traverse the score tree when sampling, so the numbers are representative for both selection rules. The columns of Table 15.1 show the average time required for sampling the score tree, for inserting a point into the partition tree, for inserting a point into the ranked score tree, and for the total processing overhead per individual. Each entry shows the average time in milliseconds to process a single individual given a certain number of stored points in the database. The averages are cumulative, so for example the fact that sampling requires 12.9 ms with $100,000$ points in the database means that the average sample time over all $100,000$ individuals was 12.9 ms. As an exception, the total processing time per individual shows the cost per individual averaged over 100 samples. Logarithmic growth in complexity is clear from the table.

Since the implementation details have been fully discussed, the experiments for REA can now be presented.

**Table 15.1** Performance statistics for Evolutionary Annealing on a 2GHz Intel Core 2 Duo processor using the open-source implementation available at http://pypi.python.org/pypi/PyEC. For each number of observed points, the table gives the time in milliseconds for sampling one point, for inserting one point into the partition tree, for inserting one point into the ranked score tree, and for the total processing overhead per function evaluation. Complexity grows logarithmically in the number of points.

| points | sample | partition | rank | total |
|---|---|---|---|---|
| 1,000 | 8.6 | 18.2 | 20.6 | 59.2 |
| 5,000 | 10.5 | 22.1 | 24.7 | 64.5 |
| 10,000 | 11.2 | 24.1 | 26.4 | 68.1 |
| 25,000 | 11.8 | 27.6 | 28.2 | 76.8 |
| 50,000 | 12.4 | 34.0 | 30.4 | 99.2 |
| 100,000 | 12.9 | 47.3 | 32.8 | 113.6 |

## 15.2 Experiments with REA

REA was tested on the twelve benchmarks defined in Table 11.1 using the same methodology as in Chapter 11. The parameters for REA-P and REA-T are the learning rate $\eta$ and the population size $K$. Several values for $\eta$ were tested, shown in Table 15.2 for each benchmark. Preliminary experiments showed that the learning rate influences the performance of REA more than the population size, and thus experiments varying the population size were left for future work. REA-P was not tested in 25 dimensions to conserve computational resources; preliminary experiments showed that REA-T substantially outperformed REA-P in 25 dimensions, just as it does in five and ten dimensions.

### 15.2.1 Experimental Results

As in Chapter 11, all algorithms were run on all benchmarks 200 times for each tested parameter setting. These 200 runs are sufficient to guarantee statistical significance on the estimated success rates for each algorithm at the 95% level within ±0.5% [160]. When a single number is shown as the result of an experiment, that number represents the best value achieved on any parameter setting for that algorithm, unless otherwise stated.

The complete experimental results are included in tabular form in Appendix A. Figures 15.1, 15.2, 15.3, 15.4, and 15.5 show the results for REA-P and REA-T in five dimensions on various performance criteria from Section 10.1, with DE and CMA-ES included for comparison. Results for REA-T in 10 and 25 dimensions on $\zeta_T$ are shown in Figure 15.6 and 15.7.

In short, REA-T, DE, and CMA-ES are the most effective optimizers on this set of benchmarks, with restarted optimizers excluded. REA-T is more effective on problems that are asymmetric, non-separable, and multimodal such as *shekel*, *langerman*, and *whitley*. In Section 15.3, this fact will be discussed in terms of alignment with a particular function prior. DE outperforms REA-T on some but not all radially symmetric problems such as *rastrigin*, *salomon*, and *griewank*. CMA-ES performs particularly well on *rastrigin* and *griewank*. Comparing the two versions of REA, REA-P performs well, but fails to refine solutions near global and local optima. Its performance also degrades in higher dimensions. In contrast, REA-T attains precisely refined solutions, most often at the global optimum, and is therefore the stronger method on these benchmarks.

More specifically, in five dimensions, the results show that REA-P and REA-T are effective at locating the global optima of complex fitness functions. REA-P is successful on most problems at the 0.1 success level, with notable exceptions for *rastrigin* and *schwefel*. For *schwefel*, REA-P actually located the region of the true global optimum on most trials, but was unable to refine these solutions further. For comparison, the failures of CMA-ES and PSO on this benchmark were over an order of magnitude worse and were not in the correct region of the search space. On *rastrigin*, it was not possible to configure REA-P to succeed predictably. The algorithm may succeed at a lower learning rate (e.g. $\eta = 0.001$) with more function evaluations, but an even lower learning rate would further slow down the refinement of the solution.

By contrast, REA-T is very effective at refining points around the optima. In most cases where REA-T came within 0.1 of the optima, it also managed to attain machine-level precision. The exceptions to this statement primarily involved local optima with fitness values close to those of the true optimum (i.e. *salomon*, *langerman* and *griewank*). In the case of *rastrigin*, tournament selection even helped REA-T escape local optima in several cases, so that it attained the true global optimum more often than REA-P.

In higher dimensions, all of the algorithms had trouble attaining the global optimum. However, a review of the errors in Table A.41 shows that REA-T was competitive with the others. In preliminary trials, REA-P failed on *whitley* and *rosenbrock* as a consequence of numeric issues. In both of these problems, the region of the search space containing reasonable fitness values (e.g. $u(x) < 100$) is small relative to the overall area, and in higher dimensions this region becomes exponentially smaller. Annealed proportional selection overflows on large fitness values (Equation 14.3) and must therefore be capped, so the probability that REA-P selects any particular point is effectively constant. This problem can be overcome by using a very small learning rate, but then REA-P would not be able to converge once the feasible region is attained. Because annealed tournament selection is only sensitive to the fitness rank of points, REA-T does not suffer from numeric issues and continues to perform relatively well on *whitley* and *rosenbrock* even in higher dimensions. It is possible that with lower learning rates, REA-T could perform even better in 25 dimensions.

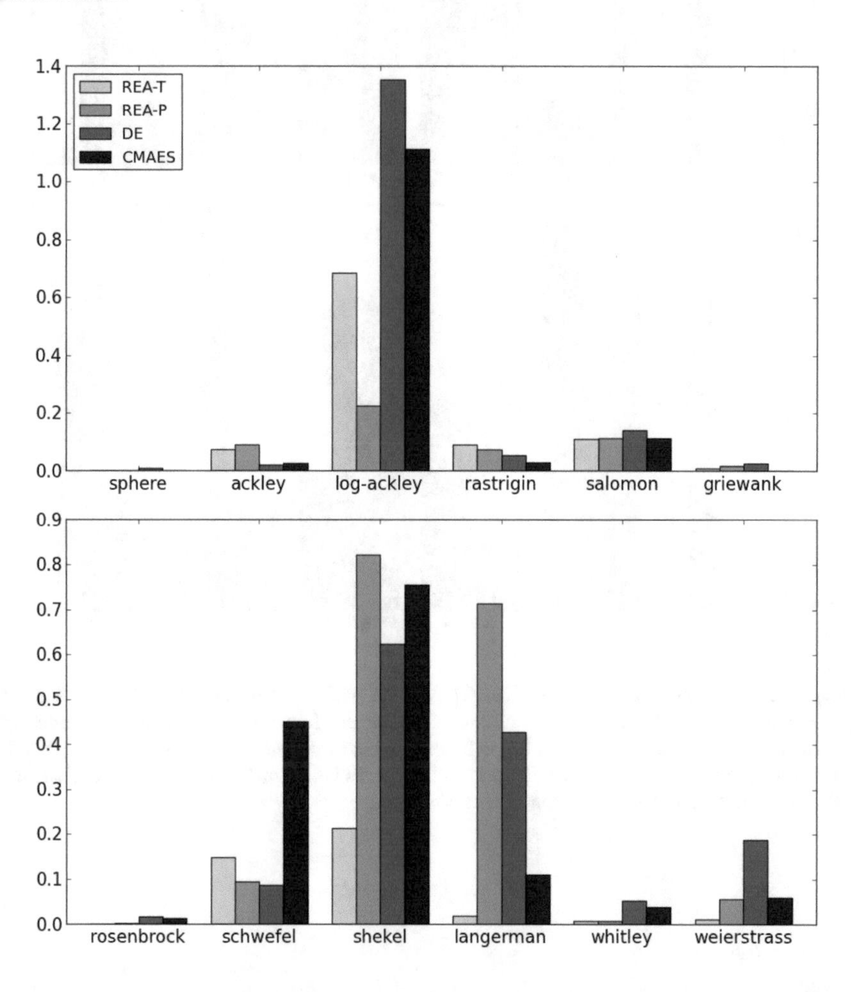


**Fig. 15.1** Performance of REA, DE, and CMA-ES in five dimensions on the average unweighted error $\phi_1$ (scaled), as reported in Figure 11.2 for all optimizers. Lower values are better. All four optimizers are generally comparable on this criterion, with REA-T performing best on *salomon*, *rosenbrock*, *shekel*, *langerman*, *whitley*, and *weierstrass*. REA-T generally has lower error than REA-P, although REA-P also performs well in five dimensions.

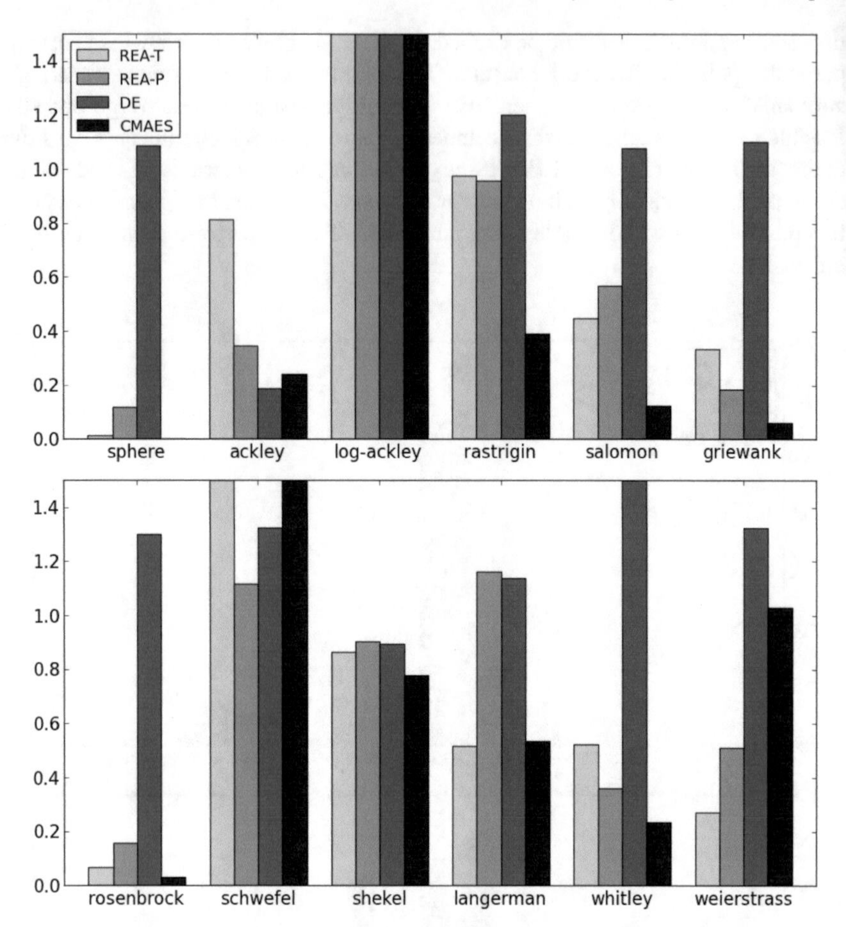


**Fig. 15.2** Performance of REA, DE, and CMA-ES in five dimensions on the average weighted error $\phi_2$ (scaled), as reported in Figure 11.3 for all optimizers. Lower values are better. The criterion $\phi_2$ emphasizes early errors, and thus favors faster converging optimizers such as CMA-ES. REA-T converges at the same rate as CMA-ES in several cases, and often makes less early errors than DE. REA-T is the best on *langerman* and *weierstrass* for $\phi_2$.

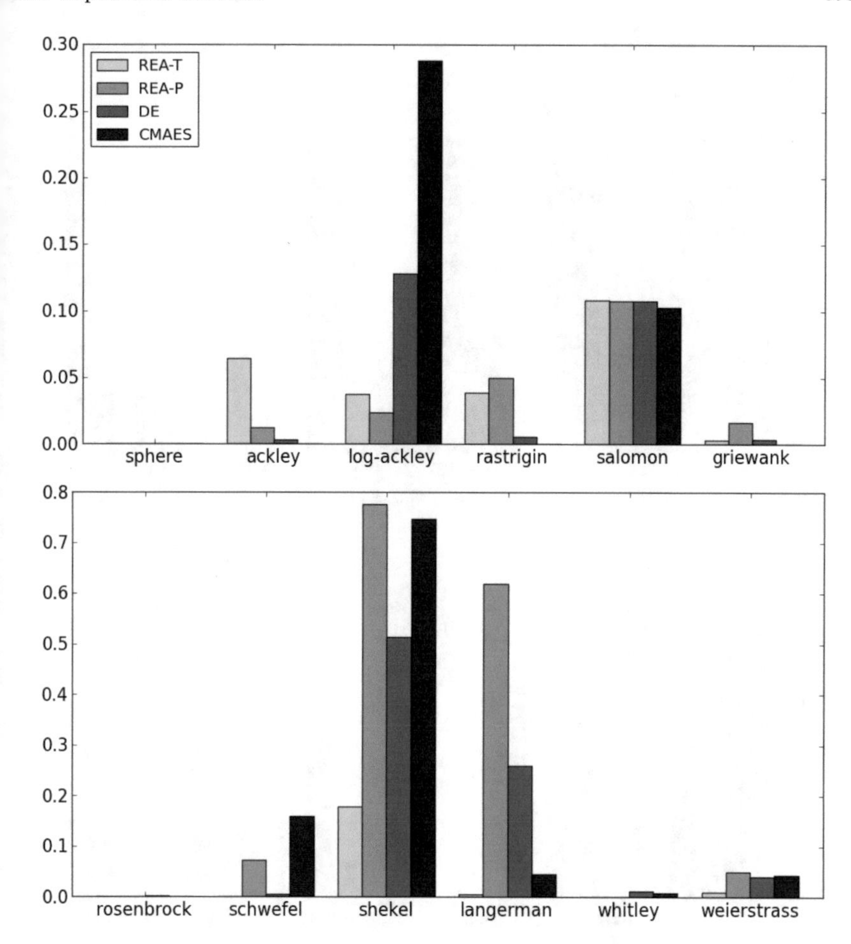


**Fig. 15.3** Performance of REA, DE, and CMA-ES in five dimensions on the average final error $\zeta_{T_{250,000}}$ (scaled), as reported in Figure 11.4 for all optimizers. Lower values are better. REA-T performs best on *schwefel*, *shekel*, *langerman*, *whitley*, and *weierstrass*, with lower average error at the end of evaluation.

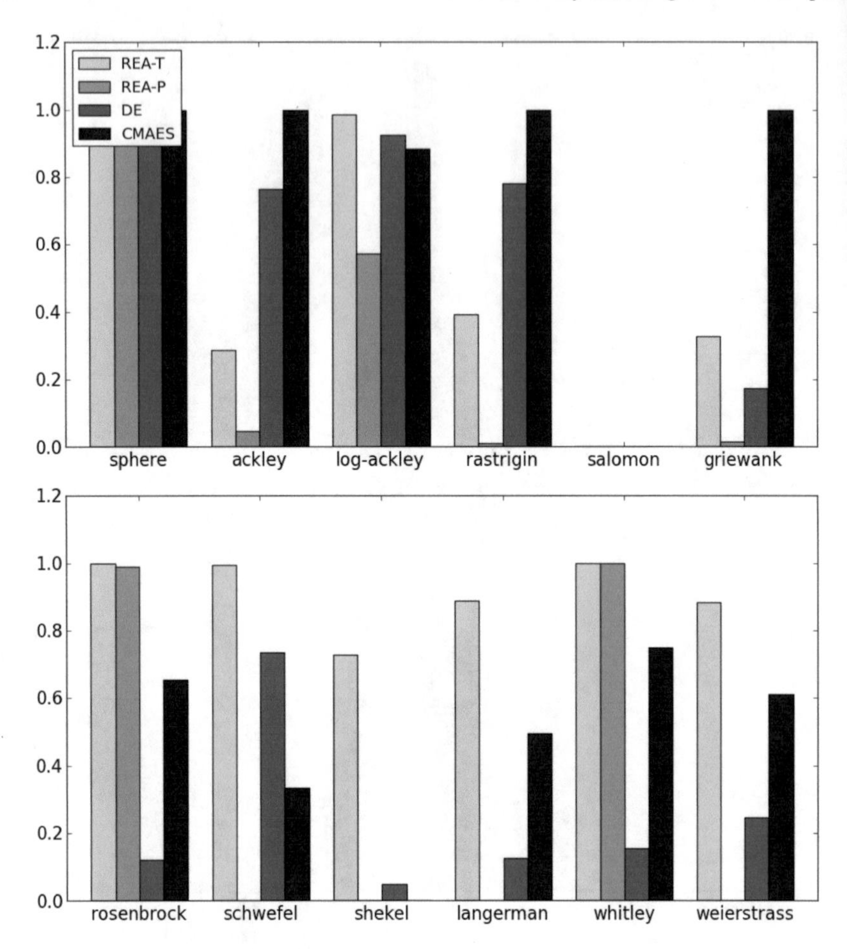


**Fig. 15.4** Performance of REA, DE, and CMA-ES in five dimensions on the success probability $\sigma_\varepsilon^N$ with $\varepsilon = 0.01$ and $N = 250,000$, as reported in Figure 11.7 for all optimizers. Higher values are better. REA-T generally performs best on the irregular problems, such as *shekel*, *langerman*, and *whitley*.

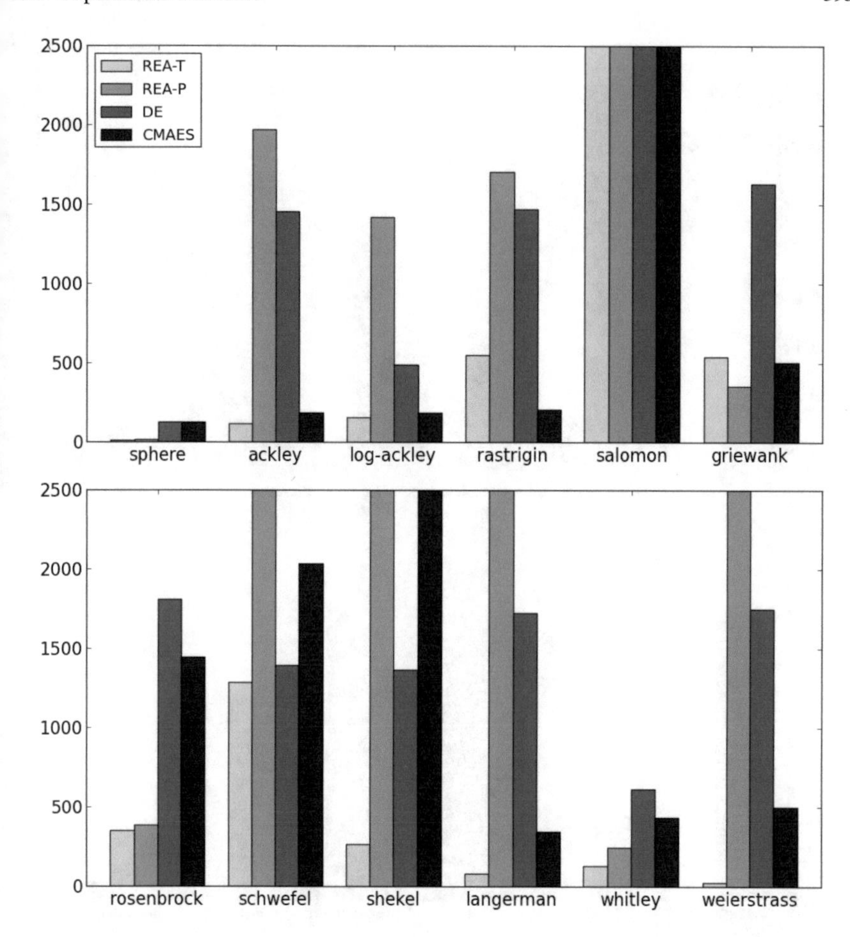


**Fig. 15.5** Performance of REA, DE, and CMA-ES in five dimensions on the average hitting time $\frac{1}{100}\hat{\psi}_{\varepsilon}^{N}$ with $\varepsilon = 0.01$ and $N = 250,000$, as reported in Figure 11.6 for all optimizers. Lower values are better; 2500 is maximum value. REA-T converges at about the same rate as CMA-ES when successful, suggesting that its performance could be boosted by restarting. In general, REA-T is more accurate than CMA-ES.

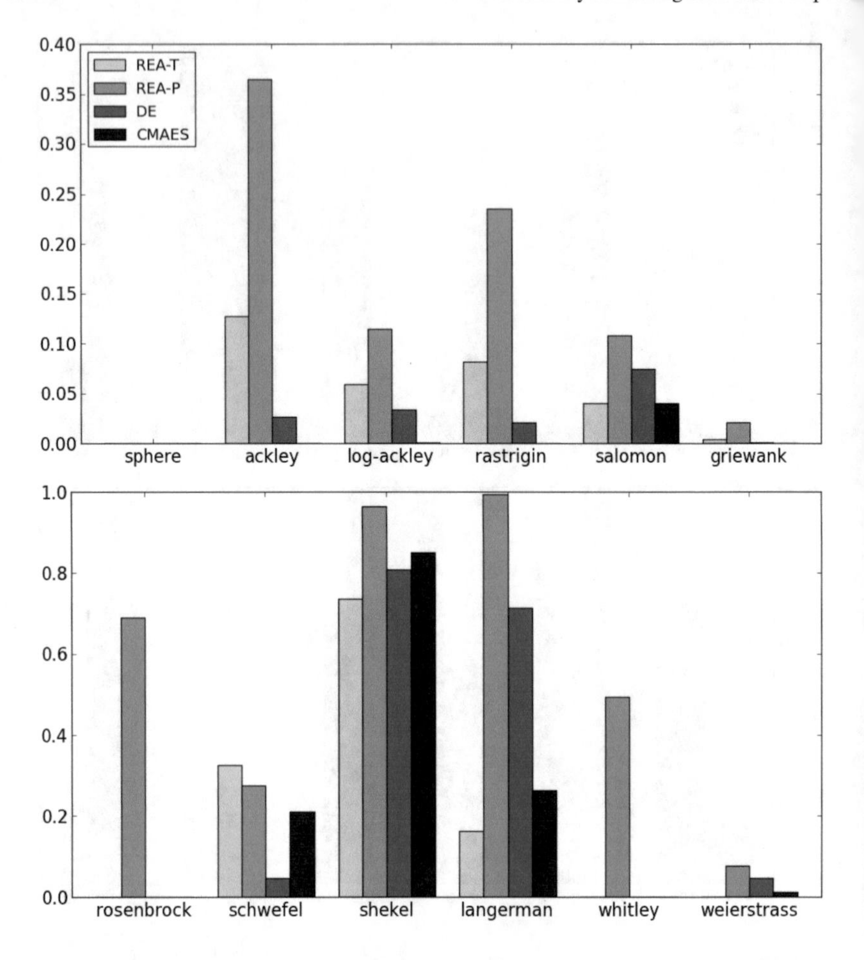


**Fig. 15.6** Performance of REA, DE, and CMA-ES in 10 dimensions on the average final error $\zeta_{T_{250,000}}$ (scaled). The performance of REA-P degrades as the dimension increases. REA-T performs best on several problems, including *salomon*, *shekel*, *langerman*, and *weierstrass*.

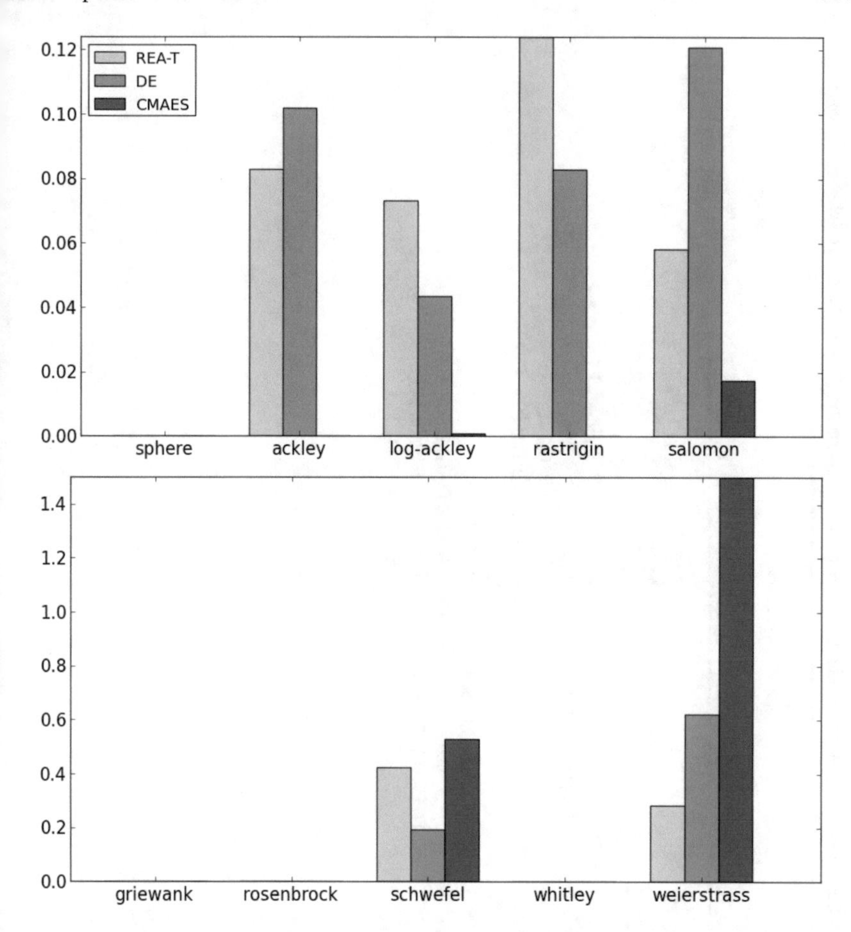


**Fig. 15.7** Performance of REA, DE, and CMA-ES in 25 dimensions on the average final error $\zeta_{T_{250,000}}$ (scaled). REA-P performed substantially worse and was omitted. All algorithms perform relatively well on *sphere*, *griewank*, *rosenbrock*, and *whitley*. REA-T performs best among these three methods on *weierstrass*.

**Table 15.2** Learning rates $\eta$ for REA-P and REA-T tested in the experiments. Lower values yield higher success probability at the cost of slower convergence.

| | REA-P | | REA-T | | |
|---|---|---|---|---|---|
| Benchmark | $d = 5$ | $d = 10$ | $d = 5$ | $d = 10$ | $d = 25$ |
| sphere | 10 | 1, 10 | 10 | 1, 10 | 0.1, 1,10 |
| ackley | 0.25 | 0.25, 1 | 0.25 | 0.25, 1 | 0.05, 0.25, 1 |
| log-ackley | 0.25 | 0.25, 1 | 0.05, 0.25 | 0.25, 1 | 0.05, .25, 1 |
| whitley | 0.1 | 0.25, 1 | 0.05, 0.25 | 0.25, 1 | 0.05, 0.25, 1 |
| shekel | 0.1, 0.25 | 0.1, 1 | 0.1, 0.5, 1.0, 5.0 | 0.1, 1 | – |
| rosenbrock | 1 | 1, 5 | 5 | 1, 5 | 0.1, 1,5 |
| rastrigin | 0.01, 0.1 | 0.035, 1 | 0.01, 0.035, 0.050, 0.075 | 0.035, 1 | 0.01, 0.035, 1 |
| salomon | 2 | 1, 2 | 2 | 1, 2 | 0.1, 1, 2 |
| langerman | 0.1, 0.5 | 0.25, 1 | 0.1, 0.5, 1.0, 5.0 | 0.25, 1 | – |
| schwefel | 0.015 | 0.001, 0.01 | 0.001 | 0.001, 0.01 | 0.0001, 0.001, 0.01 |
| griewank | 1, 10 | 0.1, 1 | 0.025, 0.1, 0.25, 0.5 | 0.1, 1 | 0.01, 0.1, 1 |
| weierstrass | 5 | 1, 5 | 5 | 1, 5 | 0,1, 1, 5 |

In higher dimensions, all of the algorithms had trouble attaining the global optimum. However, a review of the errors in Table A.41 shows that REA-T was competitive with the others. In preliminary trials, REA-P failed on *whitley* and *rosenbrock* as a consequence of numeric issues. In both of these problems, the region of the search space containing reasonable fitness values (e.g. $u(x) < 100$) is small relative to the overall area, and in higher dimensions this region becomes exponentially smaller. Annealed proportional selection overflows on large fitness values (Equation 14.3) and must therefore be capped, so the probability that REA-P selects any particular point is effectively constant. This problem can be overcome by using a very small learning rate, but then REA-P would not be able to converge once the feasible region is attained. Because annealed tournament selection is only sensitive to the fitness rank of points, REA-T does not suffer from numeric issues and continues to perform relatively well on *whitley* and *rosenbrock* even in higher dimensions. It is possible that with lower learning rates, REA-T could perform even better in 25 dimensions.

Figure 15.8 shows the progression of the success probability and Figure 15.9 the magnitude of the error as a function of the number of evaluations for REA-T with different learning rates on selected benchmarks. As the learning rate is decreased, REA-T converges slower and succeeds more often. Thus there is a trade-off between the number of evaluations and solution quality. A higher learning rate can be used to reduce the number of evaluations, but at the cost of reducing the probability of success. Notice that the shape of the graph remains remarkably constant in Figure 15.8 while the learning rate changes, suggesting that the success probability changes smoothly and predictably as a function of the learning rate and the number of evaluations.

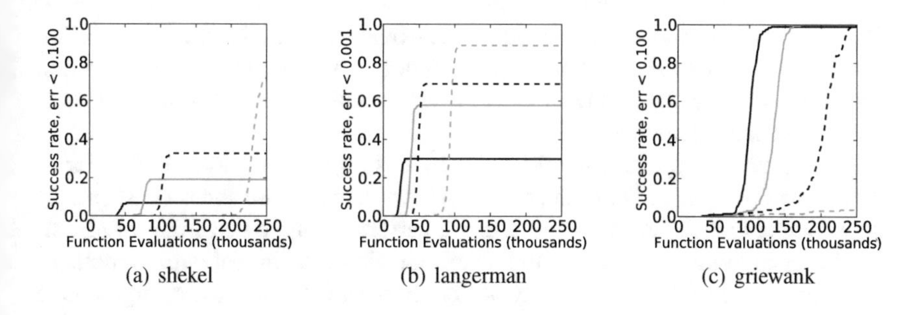


**Fig. 15.8** Success probabilities for REA-T on selected benchmarks in five dimensions for four different learning rates. Decreasing the learning rate improves the success probability overall but requires more fitness evaluations.

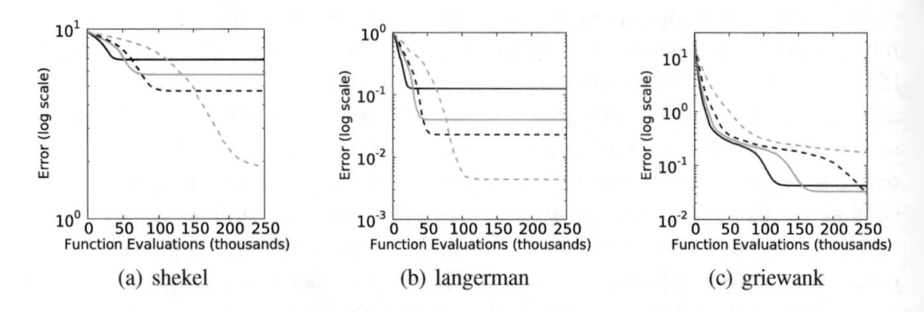


(a) shekel (b) langerman (c) griewank

**Fig. 15.9** Average error rates for REA-T on selected benchmarks in five dimensions for four different learning rates. The black solid line is the average error for the largest learning rate in Table 15.2; the grey solid line is the second largest; the black dotted line is the third largest, and the grey dotted line is the smallest learning rate. Decreasing the learning rate thus reduces error overall at the cost of increased error in early generations.

## *15.2.2 Analysis of REA Results*

The experimental results in Section 15.2 favor evolutionary annealing, especially with annealed tournament selection. There are some generalizations that may be drawn from the results. First, REA-T is generally better than REA-P for optimization and is thus the preferred implementation for Euclidean space. Second, REA is most successful relative to other algorithms on problems that do not possess an easily identifiable structure, such as *langerman* and especially *shekel*. The reason is that REA does not assume a particular problem structure in its definition. This observation is discussed further in Section 15.3. In structured domains, such as *sphere*, REA may use more function evaluations than would otherwise be necessary to eliminate the possibility that the current best solution is a local optimum. However, in unstructured environments, these extra function evaluations help REA avoid becoming trapped in local optima.

Among the non-restarted algorithms, REA-T is most comparable to DE in terms of optimization quality. DE is an elegant and simple algorithm and is consequently more computationally efficient than REA-T, performing up to two orders of magnitude faster in terms of per-generation overhead. However, in real-world problems, the computation of fitness values typically far outweighs the cost of algorithmic overhead. The overhead of REA is generally unrelated to the fitness function being optimized, so in domains where the fitness takes a long time to compute, the use of REA will not add substantially to the overall computation time.

Also, the results on the benchmarks suggest that DE and REA-T are complementary, with REA-T being preferable on highly unstructured problems, and DE performing better on problems with some degree of symmetry around the optimum. In practice, there are many real-world problems both with and without symmetry. If the degree of structure is not known, and fitness can be calculated quickly, a reasonable approach is to test DE first and use REA-T if DE fails.

All of the restarted algorithms (NM-R, GSS-R, and CMA-ES-R) generally performed as well or better than REA-T on most benchmarks, with the notable exception of *langerman*. Restarting after convergence is a form of boot-strapping that can augment the probability of success. For example, if an algorithm has a 5% chance of success, but converges after $1,000$ evaluations, then by running the algorithm 100 times, that 5% success rate can be boosted to 99.4%. To benefit from numerous restarts, an algorithm must obtain a positive success rate quickly. For REA, if the learning rate $\eta$ is set at a high level (e.g. $> 1$), then REA-T will converge quickly. If this convergence can be measured, then REA-T can be restarted to boost its success rate as well. Such an extension is an interesting direction for future work.

In contrast to the other successful optimizers, evolutionary annealing is well-defined in any suitable measure space. Thus evolutionary annealing can be used to search for neural networks, game strategies, Bayesian network structure and many other problem domains where it is unclear how DE, CMA-ES, NM, GSS, or PSO might be applied. In fact, preliminary experiments have been performed in all these problem domains with promising results.

The benchmark set also shows that REA performs well on problems to which it should not be particularly well-suited, at least while using Gaussian variation. For instance, separable problems such as *schwefel* and *weierstrass* can be more efficiently solved by searching in only one dimension. The optimizer rGA succeeds on *schwefel* by using recombination to cross-pollinate correct components, and DE succeeds by sharing component-level information among the different members of its population through its unique crossover mechanism. In contrast, REA must learn each component separately. While this aspect of REA could be improved for *schwefel* by implementing a mutation distribution that employs crossover, it is nonetheless promising that REA is able to learn the correct value for all components independently without using excessively more function evaluations than the other algorithms.

Given that REA-T is designed to search a space exhaustively for the global optimum, it might be expected to perform worse than more greedy algorithms in higher dimensional spaces. The results show that the opposite is true: REA-T still performs among the best algorithms tested even in 25 dimensions. One reason is the addition of the decay factor $n^{-\frac{1}{2}}$; without this decay factor, REA-T failed to find good solutions in 25 dimensions. To see why, consider that in $d$ dimensions, $2^d$ evaluations must be performed in order to cut the average side length of a partition region $E_n^a$ in half. Thus the variance $\sigma_n(a)$ reduces exponentially slowly in higher dimension. The decay factor forces evolutionary annealing to focus only on the most promising solutions. In this way, evolutionary annealing can obtain good solutions in reasonable time for higher dimensions at the cost of global optimality.

In Chapter 16, evolutionary annealing will be applied to neural networks, and that application will benefit from the results of the experiments in this chapter. The purpose of defining evolutionary annealing at the chosen level of abstraction is to provide a means for developing new algorithms to search in complex spaces without having to reinvent the underlying evolutionary apparatus from whole cloth. Evolutionary annealing provides convergence guarantees as well as heuristics for setting learning parameters for a wide variety of search domains.

More work remains to be done to establish the rate of convergence for evolutionary annealing beyond the heuristics provided in Section 14.3.6. For example, maximum likelihood estimates of mixture distributions with increasing mixing points are known to approximate continuous distributions at a relatively fast rate of $C\left(\frac{\log n}{n}\right)^{0.25}$ [58]. The distributions employed in evolutionary annealing are not the same, but similar performance may be hoped for on continuous fitness functions. Also, theoretical work needs to be done to find sufficient conditions on the cooling schedule and mutation distributions to make an evolutionary annealing algorithm exhaustive.

Ultimately, the success of evolutionary annealing must be determined by experimentation in real-world applications. It is difficult to predict in advance whether evolutionary annealing will be successful in such applications, but the results on benchmarks make it clear that evolutionary annealing is worthy of consideration as a method for global optimization in general-purpose domains. The next section analyzes the types of problems on which evolutionary annealing should perform best.

## 15.3 Problem Alignment

REA was presented as a general optimization technique for arbitrary problems in Euclidean space. However, in line with the discussion in Chapters 12 and 17, it may be expected that there are function priors on which REA performs better than other optimizers. In fact, the experimental results just presented provide evidence for this sort of alignment. REA outperforms other optimizers on irregular, multimodal objectives such as *langerman*, *shekel*, and *whitley*. So what is the natural function prior corresponding to REA, *i.e.*, what sorts of problems play to REA's strengths?

REA is a model-building optimizer. At each time step, REA builds a probability distribution defined by

$$\mathbb{P}(dx) = \xi_n \sum_{a \in A_n} \frac{\lambda(E_n^a)}{\sigma_n(a)\,(2\pi)^{d/2}} \exp\left(-\frac{u(a)}{T_n} - \frac{|x-a|^2}{2\sigma_n(a)^2}\right)\lambda(dx). \tag{15.11}$$

This distribution is a mixture of Gaussians, where the number of mixing points increases with each time step and the variance is a decreasing function of the number of mixing points. In light of Section 13.4.3, REA might be expected to perform best when the conditional expectation under the prior has a shape similar to the $\lambda$-density in Equation 15.11.

In the initial generations, the distribution in Equation 15.11 has relatively few large modes distributed broadly through the search space. As the number of function evaluations increases, the modes become smaller, but the points explored become closer to each other, since REA emphasizes exploration within the modes it has already discovered. Thus REA expects to discover secondary modes distributed across the modes already known. Extrapolating out to infinite time, the well-aligned function prior should prefer functions with a fractal structure that results from the composition of many Gaussian modes overlaid in tight clusters and distributed sparsely throughout the search space. The location of large modes might appear as though drawn from a Dirichlet prior at various levels of refinement. With high probability, the existing modes would be maintained, and with low probability a new mode would be sampled.

As an example of this kind of prior, consider the following iterative sampling scheme for functions over $\mathbb{R}^1$ on the interval $[-10,10]$. First, choose 10 points $x_1^0,\ldots,x_{10}^0$ distributed uniformly over the interval. Assign a value $F(x_1^0)$ uniformly at random on $[0,1]$. Then for $n$ greater than zero, sample $x_1^n,\ldots,x_{10}^n$ from a mixture distribution with density

$$G_n(x) = \xi_n \sum_{a\in\bigcup_{i=1}^n\bigcup_j\{x_j^i\}} \frac{1}{1.035^{-n}\sqrt{2\pi}} \exp\left(F(a) - \frac{|x-a|^2}{2\times 1.035^{-2n}}\right), \qquad (15.12)$$

where $\xi_n$ is a normalizing factor. Notice the similarity to Equation 15.11, except for the area-sensitive variance. Consider each $G_n$ as a random objective, and notice that the conditional expectation of $G_n$ based on the first $10m$ points obeys the equation

$$\mathbb{E}\left[G_n \mid F(x_j^i), i\le m\right] = C \sum_{a\in\bigcup_{i=1}^m\bigcup_j\{x_j^i\}} \frac{1}{1.035^{-n}\sqrt{2\pi}} \exp\left(F(a) - \frac{|x-a|^2}{2\times 1.035^{-2n}}\right),$$

where $C$ is a constant reflecting the normalizing factor and the expectation of the remaining $10(n-m)$ terms from the sum in $G_n$. Annealed selection (proportional or tournament) is unchanged by the addition of a constant multiplier, since such terms are normalized out. Thus the annealed selection rules are approximately martingales on the sequence of priors given by $G_n$, supposing that the information contained in $F(x_j^i)$ for $i\le m$ is the same or similar to the information in $G_1,\ldots,G_m$. Loosely, then, it seems that evolutionary annealing implements the Information Maximization Principle for the prior $G_\infty = \lim_n G_n$, with the final step of that strategy being unnecessary, since the expected minimum is one of the $x_j^i$. Figure 15.10 shows an objective sampled from the $G_n$ at $G_{25}$, $G_{50}$, $G_{75}$ as an example of the sorts of objectives that might be generated by this procedure. These images show the type of function on which REA should perform well. Note that if the fixed variance decay in $G_n$ were replaced with REA's variance-sensitive decay, the generated objectives would probably be slightly smoother, especially in regions where the $x_j^i$ are more sparse. Examining the progression of $G_n$ reveals the fractal nature of such a function prior.

Examining the heat maps in Figure 11.1, it may be seen that the procedure from the last paragraph most accurately describes the benchmarks *langerman*, *shekel*, and to some extent *whitley*. The benchmark *langerman* in particular fits the description. It has four large modes, two of which overlap, with rugged and detailed volcano-

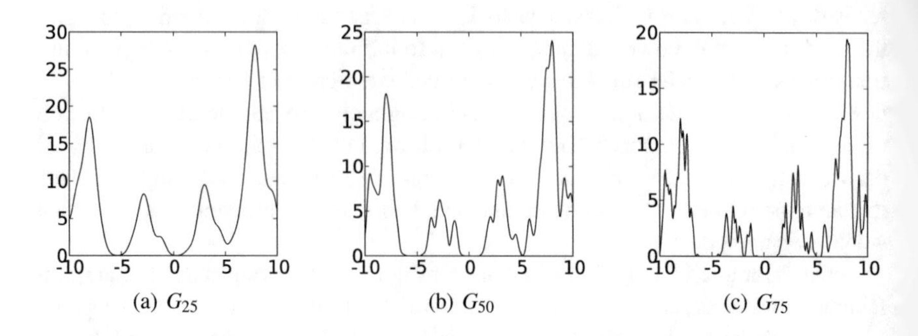


**Fig. 15.10** Examples of priors drawn from $G_{25}$, $G_{50}$, and $G_{75}$ using Equation 15.12. The random objective $G_\infty$ may be well aligned with REA, supposing a fixed, decaying variance was used in place of REA's area-sensitive variance decay. An area-sensitive decay would be more smooth, particularly in sparse regions. The fractal structure of $G_n$ is clearly evident in these images. This function roughly resembles *langerman*, on which REA-T performs best among all optimizers tested, suggesting that REA is well aligned with this random objective.

like structures at the top of each mode. Thus the intuition about what kinds of priors REA might prefer is confirmed by the experiments, and REA should be considered as a good alternative for optimizing multimodal functions with generally irregular structure.

## 15.4 Conclusion

In this chapter, REA was applied to optimization in finite-dimensional Euclidean space. Details were offered for an efficient implementation of annealed selection. Experiments on the benchmarks from Chapter 11 showed that REA performs well in comparison to other optimization methods, particularly with annealed tournament selection. The performance profile of REA was found to be complementary to the performance of the other optimizers that were tested. This complementarity results from the alignment of REA to function priors that generate irregular objectives with good fitness distributed sparsely among different modes at several fractal levels.

Euclidean space is a common target for optimization, but evolutionary annealing can be applied to other spaces as well. The next chapter studies evolutionary annealing as a tool for optimizing neural networks.

# Chapter 16
# Neuroannealing

In the previous chapter, evolutionary annealing was applied to bounded subsets of Euclidean space to demonstrate that an information maximizing approach to optimization is both feasible and effective. An important feature of evolutionary annealing is that it can be applied to any measurable space. In this chapter, evolutionary annealing is employed to search a space of recurrent artificial neural networks; this approach to learning neural networks will be termed *neuroannealing*. Neuroannealing is compared to a successful neuroevolution method, NEAT, and is shown to perform better on certain kinds of problems, in particular, those that require large neural networks with deep structure. As will be discussed in this chapter, neuroannealing is able to discover more complex solutions than NEAT because it retains all previously tested solutions, allowing it to pass through regions with lower objective values in order to reach the solutions.

## 16.1 Evolving Recurrent Neural Networks

Recurrent neural networks (RNNs) are a flexible class of parameterized nonlinear dynamic functions. In a supervised setting, the dynamics of an RNN can be learned using techniques such as Backpropagation Through Time [139]. However, RNNs are often applied to control tasks, where a supervised learning signal is not generally available. In such cases, the two most prominent approaches for training neural controllers are reinforcement learning [156, 164, 140, 168] and evolutionary computation, termed *neuroevolution* [52, 173, 109, 63, 153, 79, 64].

Recently, reinforcement learning methods based on deep learning have become prominent. In these methods, a network with thousands of neurons and millions of parameters is trained on for a control task using the backpropagation algorithm [108, 102, 147]. The success of these methods is largely orthogonal to the experimental control tasks in this chapter, which are characterized by dozens rather than hundreds or thousands of input dimensions.



A. J. Lockett, *General-Purpose Optimization Through Information Maximization*,
Natural Computing Series, https://doi.org/10.1007/978-3-662-62007-6_16

Neuroevolution, especially methods such as NEAT [153], have been more effective on these lower dimensional tasks, frequently requiring more than an order of magnitude fewer evaluations to solve exemplar tasks such as pole balancing, discussed below. The reason for this efficiency is that backpropagation moves slowly but methodically through millions of parameters, whereas neuroevolution rapidly explores a somewhat smaller space by using neural networks with hundreds or thousands of parameters as opposed to millions.

This chapter introduces neuroannealing as an application of evolutionary annealing to search a space of RNNs. Evolutionary annealing was compared to several competing optimizers in Chapter 15. In this chapter, neuroannealing will be compared experimentally to NEAT in order to demonstrate the benefits of the annealed selection for learning neural networks.

### *16.1.1 RNN Basics*

An RNN consists of a set of artificial *neurons*, or *nodes*, connected by artificial *synapses*, or *links*, with a signal-modulating *weight*. A subset of the nodes, termed the *input nodes*, are used as sensors to observe external state. A disjoint subset, the *output nodes*, are treated as the network's output signal. The remaining nodes are referred to as *hidden nodes*. In addition, most RNNs use a *bias* on each node to predispose the neuron to be more or less easily activated. Computation in an RNN proceeds by propagating an input signal through the synapses until equilibrium, and then measuring the *activation* or *excitation* of the output nodes. An RNN is characterized by the fact that the network graph, formed by taking the neurons as nodes and the synapses as edges, may contain loops. A neural network without loops is termed a *feedforward neural network*.

An RNN is determined by its connectivity and its weights. The *network topology* refers to the particular pattern of connectivity within a network. It is not the same as a topology of a space, although the two are distantly related. Two disconnected neurons may be described as being connected with a zero weights, and thus one mathematical representation of an RNN uses a pair of weight matrices, one for connections to the input (the input weights), and one for connections among the hidden and output nodes (the hidden weights). For an RNN with $N$ inputs, $H$ hidden nodes and $M$ outputs, the input weight matrix $I$ is an $(H+M)\times N$ matrix, the hidden weight matrix $W$ is an $(H+M)\times(H+M)$ matrix, the bias $b$ is an $M$-dimensional vector. Collectively, $I$, $W$ and $b$ constitute the parameters of an RNN.

The state of the RNN is a vector in $\mathbb{R}^{H+M}$ that assigns a real number to each output and hidden node. If $x_n$ is the state of a network, then given an input $u_{n+1}$, the next state $x_{n+1}$ is computed as

$$x_{n+1} = \sigma\left(Iu_{n+1} + Wx_n + b\right), \tag{16.1}$$

where $\sigma$ is a nonlinear *activation function*, often called a *squashing function* because it is usually intended to compress the neuron state within a small finite range. Typical activation functions are the hyperbolic tangent, $\sigma(x) = \tanh(x)$, and the logistic function, $\sigma(x) = (1+\exp(-x))^{-1}$. The hyperbolic tangent compresses activation values into $[-1,1]$, and the logistic compresses them to $[0,1]$. In this chapter, neuroannealing uses the hyperbolic tangent, and NEAT uses the logistic function. There is no significant difference between the two in terms of computing power.

A feedforward neural network with enough nodes and sufficiently precise weights can approximate any integrable real function [38]. Discrete-time RNNs are strictly more powerful than feedforward networks. In terms of computational theory, every binary language is decidable by some RNN with real weights, meaning that RNNs are capable of performing tasks that a Turing Machine cannot [146]. This result remains true even if the RNN is only run for a finite number of steps [30]. With rational weights, RNNs are at least as powerful as Turing Machines [146].

As dynamical systems, most RNNs are Lyapunov-stable and converge to equilibrium exponentially fast [18], meaning that their neural activations tend towards a static equilibrium in very few steps when the inputs are fixed. Thus an RNN with random weights and no inputs cannot generally compute an arbitrary time sequence. However, a specially constructed RNN can generate limit cycles [138]. Such limit cycles are induced by a chain of neurons arranged in a singly-connected loop; in such a chain, the activation is passed along each neuron, generating a time-varying source that does not depend on the inputs. This fact is utilized in developing the neuroannealing approach.

### *16.1.2 Neuroevolution*

The term *neuroevolution* describes the process of applying evolutionary algorithms to search a space of neural networks to find a network that optimizes some fitness criterion. In this approach, the parameters of a neural network are encoded inside of one or more artificial genes, which are then selected and mutated to form new networks. Early work focused on networks with fixed size and topology [118, 52, 167, 173, 34]. With this assumption, it is straightforward to encode a neural network as a vector in $\mathbb{R}^C$ where $C$ is the number of connections in the network.

Subsequent work resulted in methods for evolving networks one neuron at a time, averaging over the performance of different networks to estimate the value of particular parameters. Such methods include SANE [109], ESP [63], and CoSyNE [64]. All of these methods use a fixed number of hidden nodes.

In a different vein, NeuroEvolution of Augmenting Topologies (NEAT) was introduced as a neuroevolution algorithm that seeks to produce only those hidden nodes that improve the overall fitness of a recurrent neural network [152, 153]. NEAT has been widely applied to several experimental settings with success [104, 105]. Later in this chapter, neuroannealing will be compared experimentally with NEAT, and so some discussion of the algorithmic details is necessary.

NEAT is initialized with a population of networks that contain no hidden nodes and no recurrent links. These networks consist only of input nodes directly connected to the output nodes. In successive generations, NEAT uses proportional selection to choose a pair of network parents. It then applies crossover (either intermediate or multipoint crossover). After crossover, a network may undergo one or more modifications, either adding a node, adding a link, or mutating an existing weight. Weight mutation applies a relatively large Gaussian ($\sigma \approx 2$) to the current weight. New links are added with a small random weight. When a new node is added, it replaces an existing link between any two connected nodes. In this case, two new connections are also added. One connection is added from the source of the link to the new node with a weight of 1.0. A second connection is added from the new node to the target of the original link, copying the original weight. The general purpose of these details is to preserve network function as much as possible. If a structural mutation substantially impairs the performance of a network, then the new mutation will be immediately ejected from the population. Adding nodes and links using the method above increases the chance that the new network will survive.

NEAT has several additional features that improve its performance. *Speciation* segregates the population of networks into subgroups based on the similarity of topology and weights, and crossover is restricted so that both parents are usually but not always drawn from the same species. The measure of similarity can be tightened or relaxed. The use of species in NEAT preserves suboptimal solution candidates that are different from existing solutions during reproduction, promoting more thorough exploration of network topologies. Species are allowed to go extinct after a fixed number of generations with no improvement. Additionally, NEAT marks each new structural feature (i.e. a new node or connection) with a unique identifier, so that when crossover is applied to networks with different structures, the shared structures can be properly aligned. NEAT also uses *elitism*, retaining the best member of the last population, except when the species containing the best member of the population goes extinct. There are many other details required to describe NEAT fully, and they make a difference in how well NEAT performs. It is thus difficult to give complete mathematical account of NEAT's behavior. The source code for NEAT is publicly available, and this code was used to test NEAT in the experiments below.

### 16.1.3 Evaluating the Performance of NEAT

Neuroevolution in general has been shown to perform well in control tasks, such as controlling a finless rocket in flight [62] or generating a gait for a multi-legged robot [159]. Experiments by Stanley showed that NEAT performs well in a number of domains, including pole-balancing, board games, obstacle avoidance in driving simulations, and control of virtual robots [154]. In general, NEAT quickly locates small and efficient recurrent networks that solve a task.

NEAT does not always perform well, however. The failure modes of NEAT were studied by Kohl [91], who found that NEAT's performance tends to degrade with the complexity of the problem, as determined by the total variation of the problem. A neural network defines a map between input states and output states. Such a map will be termed a *state-action map*. Neuroevolution searches through the space of neural networks as a tractable proxy for the space of state-action maps. Let $\Pi$ be any partition of the search space consisting of hyperrectangles, and suppose that action space is metric. The *variation* of a state-action map on a hyperrectangle $H$ is the largest distance between the action values at any two corners of $H$. The *total variation* of a state-action map over the partition $\Pi$ is the sum of the variation of the map on each hyperrectangle in $\Pi$. The total variation of a state-action map is the supremum over all partitions consisting of hyperrectangles. The total variation of a problem is the infimum of the total variation of state-action maps that solve it.

Kohl showed that the performance of NEAT degrades as the total variation of the problem increases, a property that he termed *fracture* [91]. Kohl exhibited several problems with fractured state spaces, such as recognizing concentric spirals, implementing a multiplexer for address-based lookup, and robotic keepaway soccer. He also proposed a solution using radial basis function nodes that improved NEAT's performance on these domains. As Kohl observed, when NEAT does succeed in fractured domains, the successful networks tend to be larger, allowing them to encode higher complexity that reflects the fractured problem domain. Kohl's solution works because the mix of radial basis functions with sigmoidal nodes allows compact networks to exhibit more complex behavior. Kohl also experimented with cascaded networks in which the existing weights of the network are frozen and new sigmoidal nodes are added, which was also successful. In contrast, neuroannealing allows all weights in a network to change throughout training.

### 16.1.4 Experimental Hypothesis

In contrast to Kohl's approach, neuroannealing does not add radial basis functions and does not freeze weights, but is still able to find more complex solutions to fractured problems. It is able to do so because it retains information from all prior evaluations in order to generate new candidate solutions.

In order to move from a simple network with reasonably good performance on the objective to a complex network with better performance, an optimization method must either make all structural changes to the network in one step, or else it must make a series of incremental changes, each of which may degrade the objective value of the network. As a population-Markov optimizer, NEAT discards previously evaluated networks that fail to improve fitness. The speciation mechanism used by NEAT preserves novel structure for a period of time, but any network that does not improve performance is eventually eliminated. The probability that the required intermediate steps are preserved in the population under NEAT therefore decreases exponentially with the number of steps required.

In neuroannealing, the intermediate solutions remain in the pool of previously observed networks, and thus it is possible to discover more complex networks that achieve higher fitness. As a result, neuroannealing should outperform NEAT on fractured problems. However, the fact that more exploration is performed around sub-optimal points imposes a cost in terms of the number of evaluations. When NEAT succeeds, it should succeed faster than neuroannealing.

This hypothesis will be tested experimentally on concentric spirals, multiplexers, double pole-balancing, and automated currency trading. First, the application of evolutionary annealing to the space of neural networks is described.

## 16.2 Evolutionary Annealing for Neural Networks

In order to apply evolutionary annealing to the space of neural networks, three components must be defined: (1) a base measure over neural networks, (2) an algorithm for partitioning sets of neural networks, and (3) a sequence of mutation distributions likely to improve the objective value of a network. This section proposes a particular approach to defining these components that is collectively termed *neuroannealing*. First, the concept of a layer of nodes is introduced as a building block for RNNs, and then each of the three components are described in turn.

### *16.2.1 Layered RNNs*

Neuroannealing searches the space of RNNs for the optimal networks to solve an objective. In order to generate different network topologies, neuroannealing stochastically adds and removes new links and nodes to existing networks. In addition, neuroannealing organizes nodes into layers and provides mutation operators to add and remove entire layers of neurons. A layer is a group of nodes such that within a layer, all nodes are of the same type, either inputs, outputs, or hidden nodes. In a layered RNN, links interconnect neural layers, so that two nodes are connected if and only if their respective layers are connected. Links between two layers are associated with a weight matrix containing the connection strengths between the nodes in each layer.

The concept of layers is standard when training neural networks using supervised techniques. Every layered RNN corresponds exactly to a basic RNNs described above. Layers merely add a conceptual separation that is useful for computational efficiency, since it reduces the number of weights that must be stored and multiplied. In neuroannealing, layers also play a role in allowing the structure of the network to expand in useful ways. Neuroannealing probabilistically inserts layers that are designed to store the prior state of another layer, providing a natural way for RNNs to develop an otherwise improbable memory.

To represent a NEAT RNN as a layered network, each node can be assigned to its own layer. The effect on neuroannealing's optimization ability can be tested by enforcing this property on all proposed networks.

### 16.2.2 Base Measure for RNNs

Section 5.4 demonstrated how to construct a topology over a space of neural networks along with an example measure. This section follows that pattern for the class of RNNs described in the paragraphs above, which differs only in that all neurons are assumed to use a hyperbolic tangent activation.

The base measure over RNNs used by neuroannealing is a sum of simpler measures built on top of each other. The space of layered RNNs can be partitioned according to the following four features: (1) the number of layers $\ell$, (2) the number of nodes in each layer $s$, (3) the connectivity pattern among the links $c$, and (4) the weight values $w$. A layered RNN representation can be identified exactly by the tuple $(\ell, s, c, w)$. The base measure will be constructed by addressing each of these items in reverse. The construction of the measure is an important aspect of evolutionary annealing. Since the value of the base measure appears in the selection probability for the next population of networks (Equation 14.7), networks that are preferred by the base measure will be explored more thoroughly. In general, the driving force behind the decisions below is to emphasize smaller, less complex networks without penalizing extra structure too severely.

The first three criteria above comprise the network topology. If $\ell$, $s$, and $c$ are all fixed, then an RNN may be described completely by listing its weights and biases. There are a fixed number of weights and biases, and so an RNN with a given topology may be treated as a vector in $\mathbb{R}^C$ where $C = C(c)$ is the number of weights and biases. At this point, one could place a bound on the magnitude of the weight and use the Lebesgue measure, as was done for REA. Instead, neuroannealing utilizes a Gaussian measure to allow unbounded weights with a preference for small weights. For a given $\ell$, $s$, and $c$, then, the measure over RNNs matching this profile is given by

$$\lambda_{\ell,s,c}(A) = \int_A \exp\left(-\frac{x^2}{2\gamma^2}\right) dx$$

for $A \in \mathcal{B}[\mathbb{R}^C]$. The factor $\gamma$ is termed the *space scale*; it reflects the average absolute correlation between connected nodes. A good default for the space scale is $\gamma = 1$.

Next, networks with the same number of layers and layer sizes but different connectivity are handled. The connectivity pattern $c$ can be represented as a binary string of size $L = N^2$ where $N$ is the total number of nodes in the network, $N = \sum_i s_i$. $L$ is the number of possible links. Let $n(c) = \sum_i c_i$ be the number of actual links in $c$. Given $\ell$ and $s$, there are exactly $2^L$ distinct connectivity patterns. Let $P$ be the set of such patterns. A set $A$ of RNN representations with different connectivity patterns may be partitioned into a finite family of sets $\{A_c\}_{c \in P}$, separating out RNNs by connectivity. A measure over such sets is given by

$$\lambda_{\ell,s}(A) = \sum_{c \in P} \frac{1}{n(c)} \binom{L}{n(c)} \lambda_{\ell,s,c}(A_c).$$

Here the factor $1/n(c)$ is applied to prefer networks with lower connectivity, and hence fewer parameters. The factor $\binom{L}{n(c)}$ is added to emphasize networks that have about half of the possible number of links. The combined effect of the two parameters prefers smaller networks that possess a reasonable number of links.

If only the number of layers is fixed, the number of sizes $s$ is a vector of positive integers greater than one with dimension $\ell$. Networks with smaller layer sizes are preferable, but layers of size one should not be emphasized too strongly, or else neuroannealing will not consider larger layer sizes. This balance was accomplished by weighting each size profile inversely to the total number of nodes in the network. There are countably many possible layer sizes, and these can be enumerated. Let $S$ be the set of size profiles, and define

$$\lambda_\ell(A) = \sum_{s\in S} \frac{1}{\sum_i s_i} \lambda_{\ell,s}(A_s),$$

where $A_s$, like $A_c$ in the last paragraph, decomposes $A$ according to size profiles. It is notable that $\lambda_\ell$ is not finite, unlike $\lambda_{\ell,s}$ and $\lambda_{\ell,s,w}$. First, there are many size profiles with equivalent sums, and second $\sum 1/k = \infty$ even if there were not. The theory of evolutionary annealing only applies to finite measures. A finite measure over size profiles can be obtained by capping the total size of the network with some large value. In practice, the experiments in this chapter never produced a network larger than 256 nodes, and so this value was used as a maximum network size.

The final step in developing the base measure over RNN representations is to place a measure over the number of layers. This number is an integer greater than one. As with sizes, a set of RNNs may be decomposed according to the number of layers, so that for a given set of RNNs $A$, the set $A_\ell$ is the subset of $A$ with $\ell$ layers. Then a measure over arbitrary layered RNNs is given by

$$\lambda(A) = \sum_{\ell=2}^{\infty} \frac{1}{\ell} \lambda_\ell(A_\ell). \tag{16.2}$$

Putting it all together,

$$\lambda(A) = \sum_{\ell=2}^{\infty} \frac{1}{\ell} \sum_{s\in S} \frac{1}{\sum_i s_i} \sum_{c\in P} \frac{1}{n(c)} \binom{L}{n(c)} \int_{A_{\ell,s,c}} \exp\left(-\frac{x^2}{2\gamma^2}\right) dx.$$

Once again, this measure is not finite, but a finite measure can be obtained by bounding the size of the network at some large value. In the experiments that follow, the number of layers was bounded above by 256; more than 20 layers were rarely observed.

The base measure induces a $\sigma$-algebra over the space of RNNs whose structure primarily reflects the underlying Euclidean space of the weights. The space of network topologies is discrete and countable, and so a topology for the space of RNNs can be defined as a countable product topology based on the Euclidean topology over the weights.

### 16.2.3 Redundant Representations

The space of RNNs is treated as a proxy for searching a functional space. Due to Cybenko's density result [38], the space of RNNs spans at least $L^1[\mathbb{R}^d]$. This relationship is not a formal isomorphism, however. There may be many RNNs that compute the same function, even if the number of hidden nodes is fixed. Equivalent RNNs can be generated by swapping the connection strengths between equivalently connected RNNs. Thus even at a basic level, RNN representations are not unique in the sense of computing equivalent functions. The addition of layers introduces further potential for different representations of the same function.

This non-uniqueness does not prevent searching for neural networks, but it is an issue to consider in neuroannealing for two reasons. First, the space of RNN representations contains numerous redundancies. Functions with many representations will be assigned greater mass than functions with fewer representations, meaning that neuroannealing will be more likely to select such functions. Second, the objective value of different representations of equivalent functions is the same, and an information-maximizing optimizer should utilize this information in order to narrow the search space as quickly as possible. In spite of these two observations, it is not clear that removing such redundancies is necessary or beneficial. A non-redundant measure might improve performance when searching among complex networks, but such a measure would potentially be more complex to compute.

Given the base measure $\lambda$, the next question is how to leverage this measure in order to partition the space of RNNs.

### 16.2.4 Partitioning Networks

Evolutionary annealing works by partitioning the search space at increasingly fine resolution one point at a time. There are many ways in which such partitioning could be done. Neuroannealing extends the basic partitioning algorithm in Algorithm 3 (Section 14.2.3) to account for differences in network topology. For this purpose, the partition tree is conceptually stratified into four sections, one for each of the four levels used to define the base measure in Section 16.2.2.

The stratification can be best understood by starting with the node-separation algorithm. Given two networks $x_1$ and $x_2$ and a set $A$, neuroannealing must create disjoint sets $A_1$ and $A_2$ such that $x_1 \in A_1$ and $x_2 \in A_2$. The networks can be decomposed so that $x_i = (\ell_i, s_i, c_i, w_i)$ for $i = 1,2$. If $\ell_1 \neq \ell_2$, then compute the midpoint $\tilde{\ell} = \lceil \frac{\ell_1+\ell_2}{2} \rceil$, and let $A_1$ be the set of networks in $A$ with less that $\tilde{\ell}$ layers, and let $A_2 = A \setminus A_1$. This process is a straightforward application of the vector separation method in Section 14.2.3. If $\ell_1 = \ell_2$ but $s_1 \neq s_2$, then the vector separation method can be applied to the size vectors $s_1$ and $s_2$. The same approach can also be applied if $c_1 \neq c_2$, and finally if $w_1 \neq w_2$. This approach to separation assumes a hierarchy of separation levels, so that $\ell$ is separated first, then $s$, then $c$, and finally $w$.

Provided that any traversal through the partition tree from the root respects the ordering of this hierarchy, the tree will correspond to a valid partition. If the ordering is violated, for example, by separating on $w$ at a higher node in the tree, by $\ell$ at a lower level, and then by $w$ at the leaf, then the regions contained in distinct branches of the tree may overlap, with deleterious results. Thus a traversal through the tree must be stratified. Any separation on $\ell$ must occur first, then separation on $s$, and so on.

Algorithm 3 can be modified to support this stratification by allowing separation at nodes other than the leaves. The network partitioning algorithm for neuroannealing locates the first separating boundary for the new network. If this node is a leaf, then the algorithm proceeds as before using the separation algorithm from this section. But if this boundary occurs at an internal node, then a new internal node must be created, and the point being inserted must be separated from every node under the span of the boundary node. In order to make this approach possible, each node in the partition tree must be marked with the representation $(\ell, s, c, w)$ that was used to create the node and the index of the tuple that was most recently used to separate the node. Note that the portion of this representation that creates the boundary is shared among all points under the space of the boundary node. For example, if the boundary occurs at $s$, so that $s' \neq s$ where $s'$ is the size profile of the network being inserted, then it holds that every node underneath the boundary shares the size profile $s$. By separating $s'$ from $s$ using the vector separation algorithm, the inserted network is partitioned away from every node under the internal boundary node.

The hierarchical partitioning algorithm is given in Algorithm 4. To add a new point to an existing tree, the tree is traversed from the root so long as the representations agree up to the marked separation index. The first node containing the inserted point that disagrees on some part of the representation up to the marked separation index is the boundary node, and is chosen for partitioning. This node is separated as described in the previous paragraph. The new internal node is marked with the separation index at which the inserted point first disagreed. The branch containing the existing nodes is unchanged. The inserted point is assigned to the other branch and is marked with the representation of the inserted point and separation index 4.

The basic partitioning algorithm introduced in Section 14.2.3 and used for Euclidean space in Chapter 15 maintained a partition tree that represents the entire area of the search space. In contrast, the hierarchical partitioning method only represents the area of the network topologies discovered at each point during execution. When neuroannealing is initialized, the area of the first topology inserted into the tree is used to compute the area of the whole tree for sampling purposes. Thus if the first point is $x_1 = (\ell_1, s_1, c_1, w_1)$, the partition tree is assigned the initial area $\lambda_{\ell_1,s_1,c_1}(X_{\ell_1,s_1,c_1})$. Whenever a point with a distinct topology is encountered, say, $x_2 = (\ell_2, s_2, c_2, w_2)$, then the new node for this topology is assigned the area $\lambda_{\ell_2,s_2,c_2}(X_{\ell_2,s_2,c_2})$. Thus the total area of the partition tree is increased whenever a new topology is inserted. This increase is ignored for the purpose of sampling, as though the area of the new topology had always been present, uniformly distributed among the existing leaf nodes. Since sampling from the tree is normalized, this effect is invisible.

**Algorithm 4** Algorithm to Generate a Partition of RNNs

$\{x_m\}_{m=1}^M \subseteq X$, the observed networks as $(\ell, s, c, w)$ tuples
$\mathcal{T} \leftarrow \{X\}$, the partition tree
$k(i) \leftarrow \emptyset$ for all $i = 1, \dots, M$, node assignment function
$\mu(\{X\}) \leftarrow (0,0,0,0)$, the node marking function
$idx(\{X\}) = 4$, the node separation index function
**for** $m \leftarrow 1$ to $M$ **do**
  $N \leftarrow$ highest node in $\mathcal{T}$ s.t. $x_m \in N$ and $\exists i \leq idx(N)\, s.t.\, \mu(N)_i \neq x_{m,i}$
  **if** $\exists j \neq m$ s.t. $k(j) = N$ **then**
    $N_0, N_1 \leftarrow separate\,(x_j, x_m, N)$
    $\mathcal{T} \leftarrow \mathcal{T} \cup \{N_0, N_1\}$
    $k(j) \leftarrow N_0$, $k(m) \leftarrow N_1$
    $\mu(N_0) \leftarrow \mu(N)$, $\mu(N_1) \leftarrow x_m$
    $idx(N_0) \leftarrow idx(N)$, $idx(N_1) \leftarrow 4$
    $idx(N) \leftarrow$ the minimum $i$ s.t. $x_{m,i} \neq \mu(N_0)_i$
  **else**
    $k(m) \leftarrow N$
    $\mu(N) = x_m$
    $idx(N) \leftarrow 4$
  **end if**
**end for**

Because the area of new topologies is only added to the partition when a new topology appears, the new area only needs to be propagated up the partition tree. Insertion into the score tree is done as for any other point. Thus the approach of adding new area as topologies are discovered avoids an otherwise troublesome problem of reallocating area from existing nodes in the partition and score tree.

As a result, when a new topology appears, it immediately acquires substantial area, forcing some exploration of the new topology. This effect parallels the use of speciation in NEAT, but is a natural mathematical property of the hierarchical partitioning method.

The hierarchical partitioning algorithm can be easily generalized to other search spaces where there is a hierarchy of criteria useful for partitioning points.

### *16.2.5 Network Mutations*

Once neuroannealing has selected a network to mutate, a sequence of mutations is applied to modify the network. Eight types of mutation are employed, in the following order: (1) uniform crossover, (2) addition of a hidden layer, (3) removal of a hidden layer, (4) addition of a node to a hidden layer, (5) removal of a node from a hidden layer, (6) addition of a link between any two unconnected layers, (7) removal of an existing link, and (8) mutation of the weights with an area-sensitive Gaussian.

After selecting a network, neuroannealing applies crossover with probability 0.5. Crossover combines two networks to form a third network that shares properties of the two parents. When crossover is used in neuroannealing, a second network is selected independently of the first using annealed tournament selection. The structure of the networks is aligned according to the indices of their layers, then the weights from any shared links are recombined using either uniform crossover as in Equation 7.20 with probability 0.6 or intermediate crossover as in Equation 7.29 with probability 0.4. The combined network retains the topology of the first parent, but integrates weights and biases from the second parent where they share structure.

In the context of neural networks, crossover is useful because networks are naturally modular. A subset of weights or structure from a network can increase the objective value of the network independent of the other network parameters. Ideally, crossover would be performed by identifying different modules within the network and creating a new network by recombining the modules from successful networks. In the present case, it is not clear how to identify such modules, and so neuroannealing randomly chooses weights from one or the other parent. Experiments suggest that the use of crossover on about half of the population improves neuroannealing.

Whether or not crossover is applied, further mutations are attempted in the order presented below. Only one such mutation is allowed. Once a layer, node, or link has been added or removed, no further structural changes are permitted.

First, neuroannealing adds a *chained layer* to a network with probability 0.01. A chained layer is a layer of hidden nodes that copies an existing layer of the network and adds two links. The first link runs from the copied layer to the chain layer with the identity matrix as the link weight matrix. The second link connects to a random layer in the network other than the chain layer, including possibly the copied layer. If the copied layer was already connected to the target layer, then the weights are also copied from the existing to the new link. Otherwise, the new weights are sampled from a Gaussian with configurable variance $\hat{\sigma}^2$, defaulting to $\hat{\sigma}^2 = 0.01$. A chain layer preserves the prior state of the copied layer into the next step. This mutation was intended to allow the creation of limit cycles within the network, in accordance with the results in [138]. Successive chain layers can quickly add a short-term memory to the RNN that would otherwise be difficult to attain randomly.

Next, if no chain layer was added, neuroannealing deletes a random hidden layer and all of its associated links with probability 0.01. Removing layers allows unneeded structure to be culled once good solutions are located.

If no modifications are made to the network layers, a node is added to a random hidden layer with probability 0.01. The weights and bias for the new node are sampled from a Gaussian using the same variance $\hat{\sigma}^2$ as described above for new layers. The new node's connections are determined by the existing links over the layered structure. If no node is added, a node is removed from a random hidden layer with probability 0.01, and all of its connections are deleted.

The next two structural mutations alter the network connections if no layers or nodes have been mutated. A new link is added between two random layers with probability 0.025. Nothing is done if the randomly selected layers are already connected. Any new weights are sampled from the same Gaussian as is used for adding nodes and layers, with variance $\hat{\sigma}^2$. If no link is added, a random link is removed with probability 0.025. Once again, the link is removed by selecting two random layers. If the layers are not connected, nothing is done. Link removal is performed this way so that it is less likely that links will be removed from a sparsely connected network.

If no structural mutations have been performed, then the existing weights of the network are randomly modified with probability 0.5 using a Gaussian that reflects the structure of the current partition of the space. The partition tree is traversed to obtain the current upper and lower boundaries on the weights of the potentially recombined network. The upper and lower boundaries are used to determine distinct variances for each weight or bias. Let $u$ and $\ell$ be the upper and lower partition boundaries for the network's weights. Because the weight space is unbounded, these vectors may be infinite on either side. When the upper and lower boundaries are finite, the desired standard deviation for each parameter is half the distance between the upper and lower boundaries. To account for unbounded weights, $u$ and $\ell$ are modified by using the cumulative distribution of the Gaussian,

$$\Phi_\gamma(z) = \frac{1}{\sqrt{2\pi}\gamma} \int_{-\infty}^{z} \exp\left(-\frac{x^2}{\gamma^2}\right) dz, \tag{16.3}$$

reflecting the warping of the weight space that is also applied by the base measure of Section 16.2.2. The standard deviation for mutating each weight or bias is then given by

$$\sigma_{n,i} = \frac{\Phi_\gamma(u_i) - \Phi_\gamma(\ell_i)}{2\log n}, \tag{16.4}$$

where $n$ is the number of the generation and $i$ is the index of the component within the weight and bias vector as used for partitioning in Section 16.2.4. Each weight or bias is mutated independently. Scaling the variance in this way preserves well-explored parameters, for which the distance between the upper and lower boundaries is small, while forcing exploration of parameters that have not been partitioned much. The extra logarithmic factor is used to compel faster convergence in higher dimensional spaces, as was done in the 25-dimensional experiments on REA in Chapter 15.

### *16.2.6 Neuroannealing Instantiation*

With the previous subsections in mind, the complete neuroannealing algorithm can be stated. Neuroannealing is evolutionary annealing in the space of layered RNNs with annealed tournament selection using using the base measure from Section 16.2.2 and the hierarchical partitioning algorithm of Section 16.2.4. Selected networks are mutated using the chain of mutations described in Section 16.2.5. The hidden and output layers of the RNNs uses hyperbolic tangent activations.

The initial population of networks is sampled as follows. All initial networks have the same topology, which consists of a single input layer and a single output layer, with the input layer fully connected to the output layer. Within this topology, the initial weights and biases are chosen uniformly at random inside $[-\hat{\sigma}, \hat{\sigma}]$ where $\hat{\sigma}^2$ is the variance to be used when adding layers, nodes, and links. At initialization, the weights are intended to be small so that the activation can quickly change with new mutations, promoting fast exploration of the space.

As presented, neuroannealing has four parameters that must be configured: (1) the population size $K$, (2) the learning rate $\eta$, (3) the space scale $\gamma$, and (4) the standard deviation of the components, $\hat{\sigma}$. Based on experiments, a reasonable set of defaults is $K = 50$, $\eta = 0.1$, $\gamma = 1.0$, and $\hat{\sigma} = 0.1$. The defaults work well for all of the experiments below except for double pole-balancing with reduced inputs, where the values $K = 50$, $\eta = 0.025$, $\gamma = 2.5$ and $\hat{\sigma} = 0.25$ were used instead.

With the algorithm fully described, a set of experiments will be presented to compare neuroannealing with NEAT.

## 16.3 Neuroannealing Experiments

Experiments were performed in four domains, two in which NEAT performs well, and two in which it does not. The domains and experiments are described below.

### *16.3.1 Experimental Setup*

For the experiments in this section, except as noted otherwise, both neuroannealing and NEAT were run for up to 1,000 generations with a population size of 50, totaling 50,000 evaluations. The parameters for NEAT were set according to the defaults distributed with the publicly available C++ package, except for non-Markov double-pole balancing, where they were set to match the best published performance [154]. In contrast to previous experiments, each of the tasks below is stated as a maximization problem. To maximize with neuroannealing, the ranking used for tournament selection simply sorts from highest to lowest score rather than the opposite.

Results are reported for each experiment using the performance criteria of Chapter 11: success probability ($\sigma_\varepsilon^N$), hitting time on success ($\hat{\psi}_\varepsilon^N$), final error ($\zeta_{T_n}$), average error ($\phi_1$), and weighted average error ($\phi_2$). The three error-based performance criteria are scaled between 0 and 1 where possible. The error threshold $\varepsilon$ was chosen separately for each task and is given in the table. Each task is now described in turn along with its experimental results. The results for all experiments are also compiled and presented in Appendix A as Table A.51.

### 16.3.2 Double Pole-Balancing

The double pole-balancing task is a control problem in which two poles are attached to a moving cart with hinges. The first pole is 1m in length with mass 0.1kg, and the second is 0.1m with mass 0.01kg. The 10kg cart moves along a track 4.8 meters in length, and must balance the two poles simultaneously by keeping them with 36 degrees of vertical. A motor is attached to the cart that outputs a force up to 10N in either direction along the track at each point in time. Interactions are assumed to be frictionless. This physical system is simulated using a fourth order Runge-Kutta method with state updates every 0.2 seconds. The system starts with the cart in the middle of the track, the smaller pole upright, and the larger pole at 4 degrees from vertical. A successful controller must remain on the track and keep both poles within the tolerance for $100,000$ steps, or about half an hour of real time. The physical equations for the system and further details of the simulation can be found in the literature [62, 154, 64].

The neural network is tasked with controlling the direction of the force and is queried after each state update. Six state variables are available: the position and velocity of the cart, $x$ and $\dot{x}$, the angle and angular velocity of the first pole, $\theta_1$ and $\dot{\theta}_1$, and the angle and angular velocity of the second pole, $\theta_2$ and $\dot{\theta}_2$. There are two versions of this task. In the first version, all six variables are provided to the network, and the network output is scaled to $[-10, 10]$ and applied as the force. This Markov version of the problem can be solved without any hidden nodes. A second, more difficult version of the task provides only the position and angles to the network, requiring the network to infer the velocities over time. This non-Markov version can be solved with as few as two hidden nodes [154].

**Table 16.1** Published results for selected methods on both versions of the Double Pole-Balancing task, as given by Gomez et al. [64]. Reported quantity is the average number of evaluations before success, with failed trials excluded (i.e., $\hat{\psi}_0^N$, $N = 100,000$ for new results). Results for neuroannealing are new (as indicated by the asterisks), as well as the results for NEAT (determined experimentally using the parameters published by Stanley [154]).

| Method | Markov | non-Markov |
|---|---|---|
| SANE | 12,600 | 262,700 |
| Q-MLP | 10,582 | – |
| Neuroannealing | *7,767 | *7,499 |
| ESP | 3,800 | 7,374 |
| NEAT | *1,819 | *4,676 |
| CMA-ES | 895 | 3,521 |
| CoSyNE | 954 | 1,249 |

The objective value of a network for double pole-balancing with or without velocities is the number of steps for which the cart remains on the track with the poles upright. The version with the velocities is termed *Markov*, and the version just the position and angles is termed *non-Markov*.

For NEAT, the parameters for non-Markov double-pole balancing were chosen to match those used by Stanley [154]. Most importantly, the population size for NEAT was increased to 150. For neuroannealing, the settings $K = 50$, $\eta = 0.025$, $\gamma = 2.5$, and $\hat{\sigma} = .25$ were used in place of the defaults to promote larger weights and more thorough exploration of the space.

The Markov version of the task has been solved by a number of methods, including reinforcement learning (Q-MLP), and both versions have been solved by neuroevolution methods (SANE, ESP, NEAT, CMA-ES, CoSyNE) [64]. Due to the relatively low dimensionality of the problem, deep reinforcement learning methods, such as policy gradients or deep Q-networks, still require tens of thousands of evaluations. The non-Markov task is more challenging and has so far only been solved through neuroevolution. The number of network evaluations required to solve the problem is shown in Table 16.1. As these results together with Table 16.2 show, neuroannealing is able to solve the pole-balancing task, but requires twice as many evaluations as NEAT does. Neuroannealing takes longer because it searches more thoroughly around previously observed solutions. While such a search is not particularly useful on this problem, it turns out to be valuable on the next two tasks.

### 16.3.3 Multiplexers

A multiplexer is a circuit that selects one of several input lines using a binary address. Multiplexers are used to implement computer memory circuits and are easily implemented in hardware. The function of a multiplexer is difficult for a network to learn because it requires the use of a large percentage of the binary input space. A single perceptron can only distinguish a fraction of the binary numbers, and thus multiple neurons must be used in concert to solve the multiplexer problem. As a result, methods like NEAT have difficulty discovering the required complexity [91].

The experiments below test the ability of neuroannealing to learn multiplexers with four different inputs. Mux12 has one address line and four binary inputs. Mux24 uses two address lines and four binary inputs. Mux35 has three address lines and five binary inputs, while Mux36 has three address lines and six inputs. The versions with three address lines use less than the possible eight data inputs in order to simplify the task for neural networks. The task in each case is to learn a network that reads the binary address lines and outputs the binary input at the specified address line. The data inputs are numbered in the standard binary order. Figure 16.1 shows a visualization of the multiplexer problem, taken from Kohl [91].

**Table 16.2** Results of neural network experiments on both versions of the double pole-balancing task using the performance criteria of Chapter 11: success probability ($\sigma_\varepsilon^N$), hitting time on success ($\hat{\psi}_\varepsilon^N$), final error ($\zeta_T$), average error ($\phi_1$), and weighted average error ($\phi_2$). The first standard deviation is provided where possible; the value for $\sigma_\varepsilon^N$ is accurate within $\pm 0.005$ with $p < 0.05$. The error threshold $\varepsilon$ was chosen separately for each task and is given in the table. Neuroannealing succeeds on both tasks, but requires more objective evaluations than NEAT.

Neuroannealing

| Task | $\varepsilon <$ | $\sigma_\varepsilon^N$ | $\hat{\psi}_\varepsilon^N$ | $\zeta_T$ | $\phi_1$ | $\phi_2$ |
|---|---|---|---|---|---|---|
| Double Pole (Markov) | 1.0 | 0.845 | 7,767 $\pm$ 4,871 | 0.154 $\pm$ 0.360 | 0.203 $\pm$ 0.342 | 0.991 $\pm$ 0.071 |
| Double Pole (non-Markov) | 1.0 | 0.960 | 7,499 $\pm$ 3,157 | 0.039 $\pm$ 0.195 | 0.163 $\pm$ 0.181 | 0.998 $\pm$ 0.006 |

NEAT

| Task | $\varepsilon <$ | $\sigma_\varepsilon^N$ | $\hat{\psi}_\varepsilon^N$ | $\zeta_T$ | $\phi_1$ | $\phi_2$ |
|---|---|---|---|---|---|---|
| Double Pole (Markov) | 1.0 | 1.000 | 1,819 $\pm$ 2,276 | 0.000 $\pm$ 0.000 | 0.018 $\pm$ 0.044 | 0.566 $\pm$ 0.472 |
| Double Pole (non-Markov) | 1.0 | 1.000 | 4,676 $\pm$ 2,107 | 0.000 $\pm$ 0.000 | 0.012 $\pm$ 0.012 | 0.742 $\pm$ 0.415 |

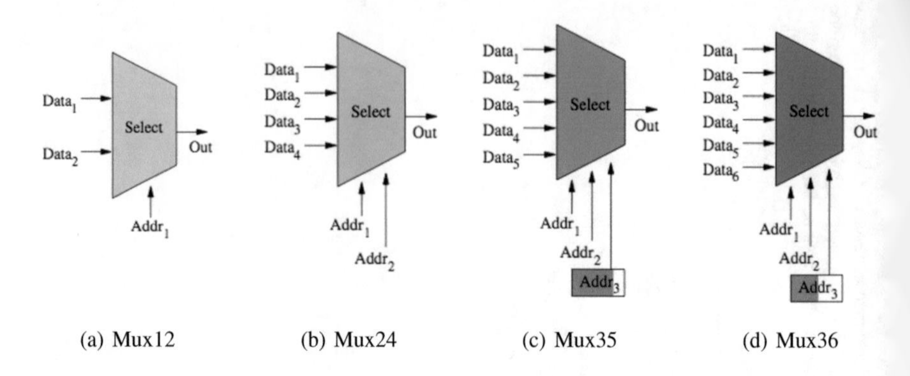


**Fig. 16.1** The multiplexer learning problem. The correct output is determined by the value of the data input at the address specified. With three address bits (Mux35 and Mux36), not all addresses were used to simplify the problem for a neural network.

The objective function (i.e. the fitness function) sums the error at each feasible address and data input. The network outputs are scaled to $[0,1]$ for this purpose. If $\text{net}(a,d)$ is the scaled output of the network for an address $a$ and a data input $d$ and $d_a$ is the addressed data, the objective is given by

$$f(\text{net}) = \sum_{a,d} |d_a - \text{net}(a,d)|. \tag{16.5}$$

Importantly, the objective function is structured to maximize the error and hence the learning signal; in practice, it is sufficient to measure the results by checking whether the net output exceeds a threshold.

The results in Table 16.3 show that neuroannealing performs better than NEAT on the multiplexer problems. On 13% of all runs, neuroannealing completely solves Mux12, whereas NEAT was unable to find a solution after 200 runs. The best solution discovered by neuroannealing for Mux24 was also completely correct, although the average solution achieved a fitness of 0.75 against an average of 0.72 for NEAT. On the versions of the problem with three address lines, Mux35 and Mux36, neuroannealing similarly performed well, with an average fitness of 0.72 and 0.70, compared to an average fitness of 0.68 and 0.65 for NEAT. The best fitness in 200 trials for neuroannealing on Mux35 was 0.97, and on Mux36 it was 0.92. The best networks on this task were larger than those produced by NEAT. Typical solutions for neuroannealing used 4-6 layers with about 20 nodes. Thus neuroannealing is able to solve the multiplexer problems better than NEAT because it is able to discover more complex networks with high objective values that NEAT is unable to reach. The next task, learning concentric spirals, reinforces this point.

### 16.3.4 Concentric Spirals

In the Concentric Spirals problem [127], the state space is divided into two interlocking spirals, one "black" and the other "white", and the task is to identify whether each point in the space falls inside of the black or white spiral [127, 91]. The black spiral is determined by 97 points, given in polar coordinates by

$$r_i = \frac{6.5}{104}\left(104 - i\right), \quad \theta_i = \frac{\pi}{16} i \tag{16.6}$$

for $i = 0, \ldots, 96$. The white spiral is defined by inverting the sign of $r_i$. The 194 points are shown in Figure 16.2(a). The spate space is then divided between the two spirals by classifying each point to match the closest spiral. The resulting state space is shown in Figure 16.2(b). An evenly spaced $100 \times 100$ grid was overlaid on the state space over the region $\left[-6.5, 6.5\right]^2$, and the resulting $10,000$ points were classified in this manner.

The neural network has two inputs and one output. The Cartesian coordinates of the state space are passed to the network as input, and the single output should read 1.0 for black, and 0.0 for white. For this experiment, the objective function summed the errors at each output for every point on the $100 \times 100$ grid, scaled between 0.0 and 1.0. Thus the sigmoidal outputs of NEAT were used directly, and the hyperbolic tangent outputs of neuroannealing were shifted and scaled as required. If $\mathrm{net}(x,y)$ is the scaled output of the network for the given Cartesian coordinates and $c(i,j)$ is the correct classification for position $(i,j)$ on the grid, then the objective function (i.e. the fitness function) is

$$f(\mathrm{net}) = \sum_{i=1}^{100} \sum_{j=1}^{100} \left| c(i,j) - \mathrm{net}(x_i, x_j) \right|. \tag{16.7}$$

It is possible to score a fitness of 0.67 on this problem by learning a correctly angled hyperplane on the state space. To achieve higher scores, the network must learn the spiral structure. Concentric spirals tests the ability of a network to distinguish nearby points in the state space that should be classified differently. In Kohl's terms, the state space is fractured. Such a task requires networks with many nodes to represent the space, which were shown by Kohl to be difficult for NEAT to discover [91].

Experiments were performed for both neuroannealing and NEAT for $1,000$ generations with a population size of 50 and 200 trials. As expected, NEAT performed poorly, rarely exceeding the basic hyperplane solution with fitness 0.67. By contrast, neuroannealing outperformed the hyperplane approximation on about half of the runs, correctly classifying 69% of the points on the average. Complete results are in Table 16.4.

Figure 16.3 shows the learned classifications from several runs of neuroannealing. Over time, neuroannealing eventually discovers solutions that correspond to a spiral shape on the state space. Such solutions generally correspond to larger net-

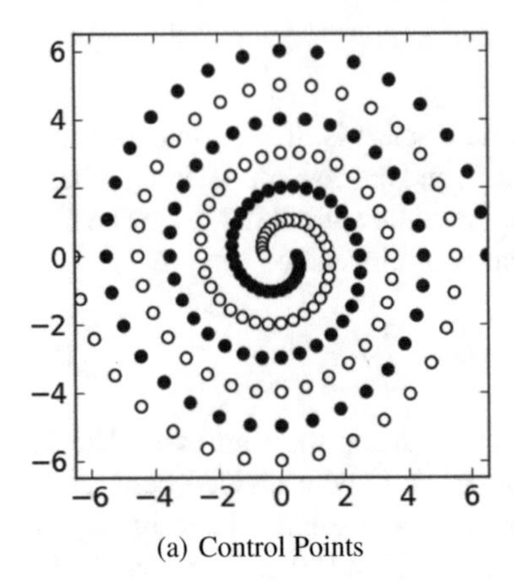


(a) Control Points

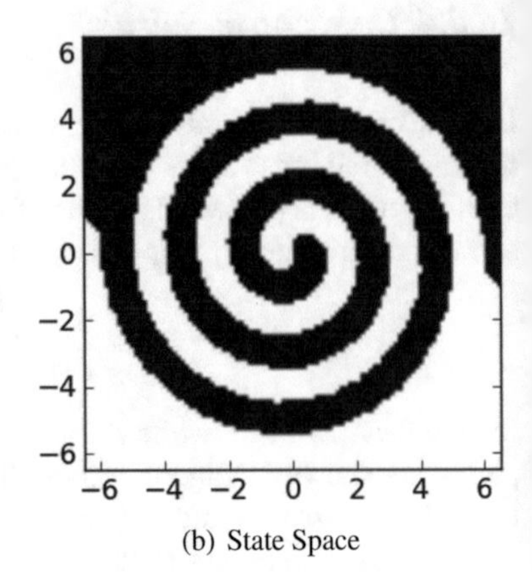


(b) State Space

**Fig. 16.2** Illustration of the Concentric Spirals Problem, in which points must be correctly classified as belonging to interlaced black and white spirals. The left panel shows the 197 control points used to define the problem, and the right panel shows the state space divided according to whether a black or white point is closer. The percentage of correct classifications on the points in the $100 \times 100$ grid in the right panel was used for training neural networks.

works. Only one of the solutions shown in the figure comes from a network with less than 20 nodes. Networks in the figure generally consisted of $4-7$ layers: The largest network, with 77 nodes, had a chained layer of size 37 that allowed correct classification of 30 extra points more than the network without the chained layer. As the networks become larger, they are better able to model the concentric spirals, but the learning progress slows down because larger networks have higher dimension. Neuroannealing was still improving at the end of 50,000 evaluations (1,000 generations), and it is possible that much better networks would have been discovered with more evaluations. In general, it may be conjectured that neuroannealing is more capable of discovering complex solutions in part because annealed selection allows it to follow suboptimal intermediate steps to arrive at more complex optima.

**Table 16.3** Results of neural network experiments on the multiplexer problem using the performance criteria of Chapter 11. The error threshold $\varepsilon$ for each task is given in the table. Neuroannealing outperforms NEAT on the multiplexer problems in terms of final error; these results are statistically significant ($p < 0.01$).

Neuroannealing

| Task | $\varepsilon <$ | $\sigma_\varepsilon^N$ | $\hat{\psi}_\varepsilon^N$ | $\zeta_T$ | $\phi_1$ | $\phi_2$ |
|---|---|---|---|---|---|---|
| Multiplexer, $1 \times 2$ | 0.010 | 0.130 | 15,376 ± 15,930 | 0.108 ± 0.042 | 0.113 ± 0.033 | 0.176 ± 0.050 |
| Multiplexer, $2 \times 4$ | 0.200 | 0.047 | 19,833 ± 10,351 | 0.247 ± 0.037 | 0.252 ± 0.025 | 0.329 ± 0.023 |
| Multiplexer, $3 \times 5$ | 0.250 | 0.028 | 20,566 ± 15,509 | 0.285 ± 0.013 | 0.287 ± 0.013 | 0.363 ± 0.017 |
| Multiplexer, $3 \times 6$ | 0.300 | 0.036 | 17,675 ± 12,449 | 0.305 ± 0.013 | 0.308 ± 0.011 | 0.385 ± 0.012 |

NEAT

| Task | $\varepsilon <$ | $\sigma_\varepsilon^N$ | $\hat{\psi}_\varepsilon^N$ | $\zeta_T$ | $\phi_1$ | $\phi_2$ |
|---|---|---|---|---|---|---|
| Multiplexer, $1 \times 2$ | 0.010 | 0.000 | 50,000 ± 0,000 | 0.166 ± 0.027 | 0.180 ± 0.012 | 0.187 ± 0.000 |
| Multiplexer, $2 \times 4$ | 0.200 | 0.000 | 50,000 ± 0,000 | 0.279 ± 0.001 | 0.282 ± 0.001 | 0.300 ± 0.008 |
| Multiplexer, $3 \times 5$ | 0.250 | 0.000 | 50,000 ± 0,000 | 0.322 ± 0.001 | 0.325 ± 0.001 | 0.340 ± 0.007 |
| Multiplexer, $3 \times 6$ | 0.300 | 0.000 | 50,000 ± 0,000 | 0.348 ± 0.003 | 0.351 ± 0.002 | 0.367 ± 0.005 |

**Table 16.4** Results of neural network experiments on the concentric spirals problem using the performance criteria of Chapter 11. The error threshold $\varepsilon$ is given in the table. Neuroannealing outperforms NEAT substantially in terms of final error; this result is statistically significant ($p < 0.01$).

Neuroannealing

| Task | $\varepsilon <$ | $\sigma_\varepsilon^N$ | $\hat{\psi}_\varepsilon^N$ | $\zeta_T$ | $\phi_1$ | $\phi_2$ |
|---|---|---|---|---|---|---|
| Concentric Spirals | 0.300 | 0.261 | 21,687 ± 7,834 | 0.310 ± 0.021 | 0.317 ± 0.014 | 0.333 ± 0.001 |

NEAT

| Task | $\varepsilon <$ | $\sigma_\varepsilon^N$ | $\hat{\psi}_\varepsilon^N$ | $\zeta_T$ | $\phi_1$ | $\phi_2$ |
|---|---|---|---|---|---|---|
| Concentric Spirals | 0.300 | 0.000 | 50,000 ± 0,000 | 0.331 ± 0.000 | 0.331 ± 0.000 | 0.332 ± 0.000 |

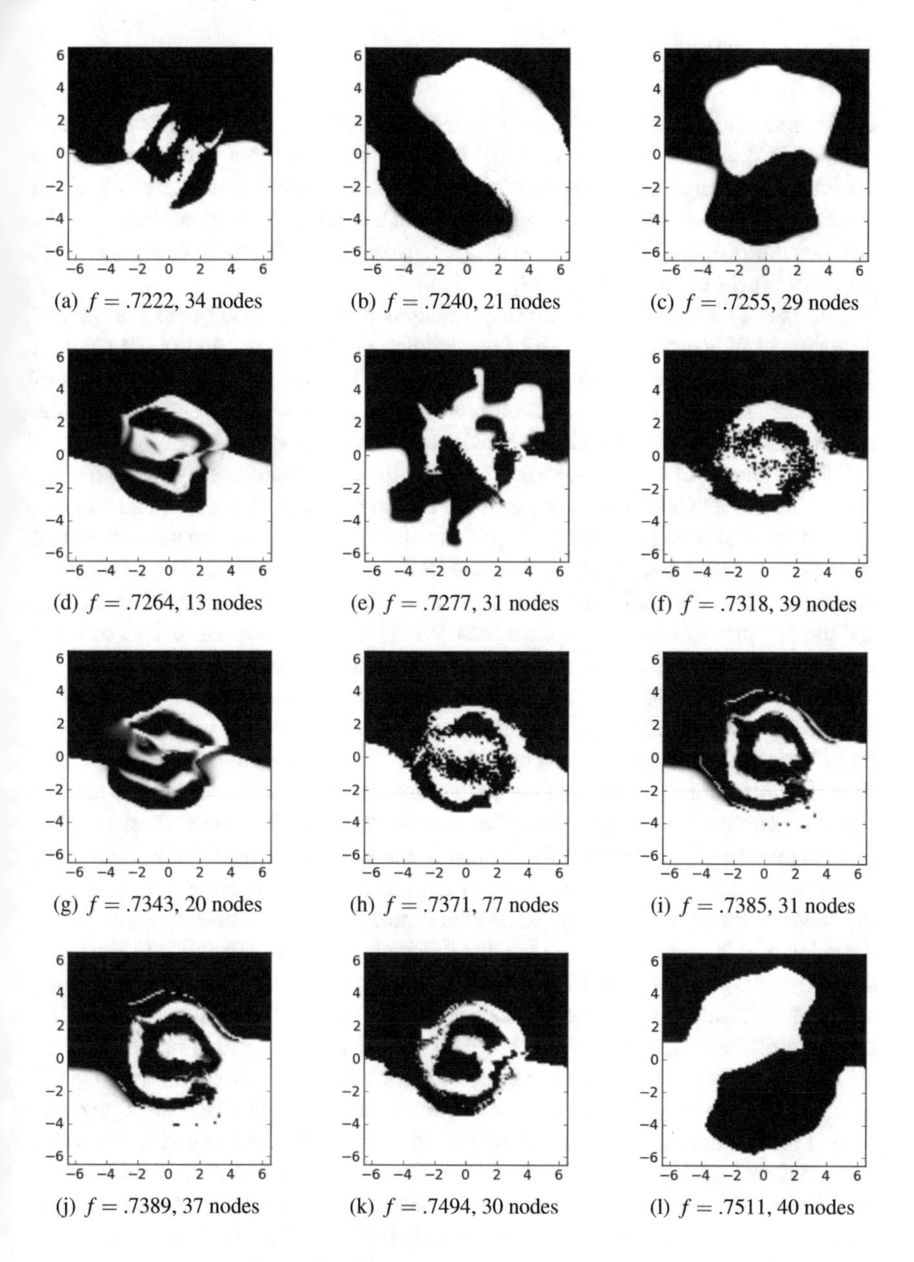

(a) $f = .7222$, 34 nodes (b) $f = .7240$, 21 nodes (c) $f = .7255$, 29 nodes

(d) $f = .7264$, 13 nodes (e) $f = .7277$, 31 nodes (f) $f = .7318$, 39 nodes

(g) $f = .7343$, 20 nodes (h) $f = .7371$, 77 nodes (i) $f = .7385$, 31 nodes

(j) $f = .7389$, 37 nodes (k) $f = .7494$, 30 nodes (l) $f = .7511$, 40 nodes

**Fig. 16.3** State space classification for the concentric spirals problem as learned by neuroannealing. Objective values and network sizes are shown for each solution. More accurate solutions require larger networks. Neuroannealing is able to discover these solutions, whereas NEAT does not.

### 16.3.5 Currency Trading

Both neuroannealing and NEAT were also tested on the task of automated currency trading. In this task, a neural network is presented with input data derived from the hourly exchange rate between two currencies. The network must decide which currency to hold each hour and with how much leverage. The complexity of the task is unknown, but it is of interest as a real-world problem where optimization of neural networks could prove useful.

The task of an automated currency trader is to progressively read a sequence of technical indicators for an arbitrary currency exchange rate and output trading decisions. Performance on this task was tested using a dataset with six months of hourly exchange rate data for nine different currency pairs. Complete details of the task and the associated datasets are provided in Appendix B.

The sequence of technical indicators for this experiment consists of ten real-valued inputs derived from the exponential moving average at five, 20, and 50 periods, the relative strength index at 14 periods, the fast and slow stochastics at 14 and three periods respectively, the width of the Bollinger Bands, the position of the closing price within the Bollinger Bands, the absolute difference between the opening and closing price, and the difference between the high and low price. Each of these indicators were scaled to remove the price details, as described in Appendix B.

In addition to these ten technical indicators, three trading inputs were used to describe the automated trader's current position. The first trading input specifies the trader's current long position as a percentage of the possible long value if the trader were operating at full leverage. This input is zero if the trader's position is currently short or neutral. The second trading input gives the trader's current short position as a percentage of the possible short value in the same way. The third trading input provides the length of time the trader's current position has been held under an exponent. If the number of time steps the current position has been open is $\ell$, then the value of this input is $\exp(-\ell)$, so that this input exponentially tends toward zero the longer the position is held. These three trading inputs allow the network to be aware of the status of its trades, which is necessary since not every network decision can be implemented, and the simulator may impose a margin call after a bad trade.

The networks for this task have three outputs. The output values are normalized so that the three outputs total to 1. The first output is for buy decisions, and the second for sell decisions. The third output represents risk aversion and is only used to normalize the other two. If the normalized value of first output exceeds the second by 0.05, a buy decision is entered, or a long position is held. If the normalized value of the second output exceeds the first by 0.05, a sell decision is entered, or a short position is held. Otherwise, the current position is liquidated. The amount of leverage is decided as a percentage of the possible leverage (up to 50:1 in currency trading) based on the absolute difference of the normalized buy and sell signals.

The objective value of a currency trader is determined by the account value after trading each of the nine currency pairs in succession for six months. The network starts with an account value of 1 and is queried once per hour on historical data.

**Table 16.5** Results of neural network experiments on the currency trading task using the performance criteria of Chapter 11. For currency trading, the global optimal value are unknown, and the values for $\sigma_\varepsilon^N$ and $\hat{\psi}_\varepsilon^N$ were found by averaging over trials on which the final account value was greater than 250. Neuroannealing and NEAT both perform well, but neuroannealing performs slightly better, achieving higher success probability, higher final account values, and the largest overall account value.

Neuroannealing

| Task | $\varepsilon <$ | $\sigma_\varepsilon^N$ | $\hat{\psi}_\varepsilon^N$ | $\zeta_T$ | $\phi_1$ | $\phi_2$ |
|---|---|---|---|---|---|---|
| Currency Trading | $f > 250$ | 0.749 | 20,054 ± 11,189 | 31016.331 ± 55094.212 | 10904.020 ± 20980.836 | 0.930 ± 0.284 |

NEAT

| Task | $\varepsilon <$ | $\sigma_\varepsilon^N$ | $\hat{\psi}_\varepsilon^N$ | $\zeta_T$ | $\phi_1$ | $\phi_2$ |
|---|---|---|---|---|---|---|
| Currency Trading | $f > 250$ | 0.028 | 29,425 ± 8,095 | 43.365 ± 103.898 | 15.910 ± 48.626 | 0.033 ± 0.105 |

Objective values less than 1 represent an overall loss, and values above 1 correspond to a gain. Six months of trading on nine pairs equates to four and a half years of trading time, so a 10% annual gain would result in a fitness of $1.1^{4.5} = 1.54$. A fitness of 10 implies an annual gain of 66%.

Table 16.5 presents the results for this task. Since the true optimum is unknown, the raw fitness values were reported rather than the error. A fitness of 250 or more was considered a success. In the columns for $\zeta_T$, $\phi_1$, and $\phi_2$, the final fitness, average fitness, and weighted average fitness were reported instead of the standard values for these performance criteria.

On this task, neuroannealing achieved a substantially higher average fitness than NEAT over 200 trials. Neuroannealing attained a fitness of $31,016.331$ on average, versus 43.365 for NEAT. Due to the definition of the fitness, the account value grows exponentially when a neural network trader succeeds, accounting for the large numbers. Neuroannealing also posted higher account values more frequently than NEAT, as reflected in the value for $\sigma_{\varepsilon}^{N}$ in Table A.51, which shows that neuroannealing attained an account value of 250 or more on 74.9% of all runs, versus 2.8% for NEAT. This difference is statistically significant. The highest fitness value discovered by NEAT was 803 for NEAT, compared with the average fitness for neuroannealing at $31,016$. Both neuroannealing and NEAT performed well on this task overall. It is unclear whether these results will generalize to real-world trading contexts, since it is possible that both algorithms are overfitting the data. Nonetheless, higher objective values suggest greater success in this task. Of the two methods, neuroannealing performs considerably better.

## 16.4 Neuroannealing Discussion

The experiments show that neuroannealing is an effective method for training neural networks in three different domains: multiplexers, concentric spirals, and currency trading. Neuroannealing works well on these problems because it searches more thoroughly through complex networks and is not constrained by population size. Annealed selection makes it possible for neuroannealing to attempt more ways of increasing network complexity without forgetting previously successful solutions. This property allows neuroannealing to step through regions of suboptimal fitness in order to find successful complex networks. When simple solutions exist, neuroannealing usually finds them, because it searches simple networks first. When complexity is required, however, neuroannealing considers progressively more complex solutions.

In double pole-balancing, neuroannealing does not find solutions as quickly as NEAT, ESP, or CoSyNE, but it does solve the problem. This success is achieved despite the fact that neuroannealing is designed to focus on thorough optimization rather than speed. Neuroannealing is a robust optimizer even in domains where NEAT performs well. Interestingly, neuroannealing performed better in the more difficult, non-Markov version of the task, in which the velocities had to be inferred from successive positions of the cart and poles. This fact supports the claim that neuroannealing improves learning on more difficult tasks, while introducing an overhead on simpler tasks.

On the multiplexer problems and on concentric spirals, neuroannealing performs substantially better than NEAT because it is more capable of discovering complex networks. The size of these networks can exceed those of NEAT networks by a full order of magnitude, as demonstrated by the networks with up to 77 nodes in Figure 16.3. As noted by Kohl [91], these problems require complexity in order to be solved, and neuroannealing is able to deliver.

Neuroannealing thus demonstrates the power of leveraging information in order to drive optimization. As an instance of evolutionary annealing, neuroannealing proves that evolutionary annealing can work well in at least some high-dimensional domains and further reinforces the value of the annealed selection methods.

Future research into neuroannealing could focus on determining the effect of the various mutation operators and tuning their parameters. In addition, the good use of chain layers suggest that there may be other large-scale agglomerative combination methods for constructing large neural networks from known modular components. One approach in this direction would be to refine the partition method so that partitions reflect network behavior. Modular networks could then be constructed by merging networks from different behavioral regions. Similar work with NEAT has already yielded valuable results in this direction [100].

For evolutionary annealing in general, the principle that partition regions should reflect meaningful distinctions in the search domain is one that deserves further considerations. The current partitions based on axis-parallel hyperrectangles are a rudimentary tool that could be substantially refined.

Also, it can be argued that methods like DE and CMA-ES are effective because they compress the prior search history into a very brief and compact form. By contrast, evolutionary annealing performs no compression and uses the entire evaluation history. It is likely that there is some useful middle ground. Perhaps the results of evolutionary annealing can be achieved while retaining a summarized version of the evaluation history.

Furthermore, the mutation operators for an evolutionary annealing instance will be more successful if they are better aligned with the function prior from which problems are drawn. In neuroannealing as in NEAT, the progressive addition of network structure is intended to mirror the principle of Minimum Description Length, aligning the search methodology with the general assumptions about the nature of the universal prior, as discussed in Chapter 13. The success of both NEAT and neuroannealing in a wide range of domains bolsters this assumption. Still, it seems that neither neuroannealing nor NEAT fully implements the principles of modularity and locality to a satisfactory degree at this time. More work needs to be performed to quantify and implement these principles.

## 16.5 Conclusion

Neuroannealing was shown to be an effective optimizer in diverse domains, including pole-balancing, multiplexers, concentric spirals and currency trading. In fractured domains, neuroannealing solidly outperforms NEAT due to its ability to discover larger networks with higher objective values. These results demonstrate that neuroannealing is an effective method for optimizing neural networks.

Evolutionary annealing is based on an information-maximizing heuristic. It has been tested in Euclidean space and neural network with positive results. It is expected that similar information-maximizing optimizers will continue to prove their usefulness in the future.

# Chapter 17
# The Optimization Game

The majority of this book has thus far focused on characterizing search methods rather than search problems. Search problems, when presented, have primarily been constrained to static optimization problems, in particular, through fitness measures that allow stochastic selection of a static problem. In Chapter 12, the question of alignment between search problems and solutions was prominently raised in the context of No Free Lunch; NFL itself was described as a type of search problem most closely aligned with blind random search. Chapter 13 presented how methods of control theory could be used to find the optimal search method for a given fitness measure. The existence of such a method suggests that the best-aligned solution to a problem can be thought of as a function of the problem. Change the problem, and the solution should change too.

This concept of alignment can be described as a duality between search methods and problems. To explore the full extent of this duality, the types of search problems considered must be opened far wider than previously considered, so that the search setting becomes a zero-sum, two-player game between an *optimization player* who controls the search method, and a *fitness player*, termed an *adversarial fitness*, who assigns fitness or objective value to search points selected by the optimization player. By admitting such general problems, it becomes possible to represent not only static optimization problems, but also stochastic, dynamic, and even adaptive problems. This final category is critical to appreciating evolutionary methods, since adaptation is the key problem characteristic that motivates the choice of evolutionary solutions.

If the fitness player is allowed to choose any strategy, it can obviously frustrate the efforts of any optimizer by playing a No Free Lunch strategy, so the actual performance of an optimizer is not of interest here. Nonetheless, the flexibility of the setting makes it possible to draw broad conclusions about the structure of the space of search methods. For example, every measure of performance that is based on averaging scored search histories corresponds to a game with some adversarial fitness. In addition, the NFL results of Chapter 12 and the Bellman equations of Chapter 13 can be shown to hold for much larger classes of problems using these adversarial fitnesses.



A. J. Lockett, *General-Purpose Optimization Through Information Maximization*,
Natural Computing Series, https://doi.org/10.1007/978-3-662-62007-6_17

This chapter introduces the formal underpinnings of the Optimization Game and connects this game to the previous formalisms based on search generators and trajectors. The result is a dualistic theory of paired stochastic processes in which search and optimization are merely two reflexes of an abstract mathematical system. This existence of this theory is the essence of this book, and it suggests a mathematization of stochastic programs to be discussed further in Chapter 18.

## 17.1 Adversarial Optimization

To begin, suppose that the search and fitness space are both finite, and imagine optimization to be a two-player zero-sum game between an optimizer and a fitness function. Throughout this chapter, we refer to *optimization* rather than *search* since it is in optimization where the game-like iteration is most apparent, but the abstraction introduced below applies even more generally than just search and optimization, and the choice of words here is illustrative rather than definitive. Likewise, an objective function is replaced by the fitness player to indicate that this player is broader in scope than a mere function, but the underlying significance lies in the mathematical abstractions rather than in the word choice.

### *17.1.1 Optimization as a Zero-Sum Game*

In game theory [117], a game consists of one or more players; here we are interested in games with two players, an optimizer and a fitness player. Each player chooses a strategy, say, $\sigma_X$ for the optimizer and $\sigma_Y$ for the fitness, and each player receives a value based on both choices, e.g. $V_X(\sigma_X, \sigma_Y)$ for the first player and $V_Y(\sigma_X, \sigma_Y)$ for the second. For a zero-sum game, $V_Y$ is defined so that $V_Y(\sigma_X, \sigma_Y) = -V_X(\sigma_X, \sigma_Y)$, *i.e.*, the two values sum to zero.

For turn-based games, one distinguishes between the *normal form* and the *extended form* of a game. In the extended form, each player makes a series of choices that can depend on the choices made by the other player. This form is often represented as a *game tree*, in which alternating levels of the tree indicate the alternating turns of each player. From the perspective of the extended form, each player makes a sequence of choices that are dependent on the choices of the other player. This extended form corresponds to the notion of a *search generator* or *jagged value generator* as introduced in previous chapters. By contrast, in the normal form, each player chooses a complete strategy in a single decision at the outset of the game; these are the strategies $\sigma_X$ and $\sigma_Y$ mentioned above. The normal form and extended form are generally equivalent for turn-based games. Under this equivalence, the strategy $\sigma_X$ can be thought of as a function or a program that determines the first player's decisions at each step, conditioned on the opponent's responses. The normal form corresponds to the *search trajectors* that were introduced in Chapter 13.

Now consider the process of optimizing a fitness function as a game between an optimization method and a fitness function. In each turn of the optimization game, the optimization player first chooses a search point, and then the fitness player chooses a fitness value to assign to that search point. The turns continue on indefinitely, although an external observer might terminate the game if certain stopping criteria are met. For a fixed static fitness function, the responses of the fitness player are determined by the choice of search points. However, if the fitness player is allowed to take arbitrary actions, a more complex situation arises in which the fitness values can even be chosen to frustrate the optimization process. This type of fitness player will be called *adversarial.* An adversarial fitness subsumes a wide variety of optimization settings including static, stochastic, dynamic, and adaptive fitness functions.

### 17.1.2 Optimization and Fitness Players

For simplicity, suppose that the search space $X$ and the fitness domain $Y$ (the value space) are both finite. In the optimization game, the *optimizer*'s moves consist of sequences of search points. The *adversarial fitness* makes moves that consist of sequences of fitness values. Each player is allowed to observe the strategy played by the other player in a limited fashion corresponding to a black-box assumption, as described below. Each player's moves are also randomized according to some probability distribution, so that the overall context is stochastic.

For the normal form of the game, this situation can be formalized by identifying each player as a function that maps the *observed* move of the other player to a probability distribution over the player's own moves. Note that a move here includes the player's entire history – the sequence of all search points chosen by the optimizer and the sequence of all fitness values chosen by the adversarial fitness. For now, suppose that the game is played for $T$ turns, where $T$ is some fixed finite number, so that $T$ search points are proposed and $T$ fitness values selected in response. This setting is identical to that of Chapter 3, except that the fitness function now has the same degree of freedom as the optimizer.

The observed move for the optimizer is a sequence of search points $z = (z_n)_{n=1}^T$ such that $z_n \in X$ is a single search point for each natural number $n$. The observed move for the adversarial fitness is a sequence of fitness values $y = (y_n)_{n=1}^T$, where $y_n$ is a single fitness value assigned to the search point $z_n$. An optimizer $\mathcal{A}$ is function that takes a sequence $y$ of fitness values as above and assigns to it a probability distribution that can be sampled to obtain the sequence $z$ of search points above. Conversely, an adversarial fitness $\mathcal{F}$ is a function that takes a sequence $z$ of search points and assigns to it a probability distribution that can be sampled to obtain the sequence $y$ of fitness values. Thus we may write that the search sequence has $z \sim \mathcal{A}[y]$, and the fitness sequence has $y \sim \mathcal{F}[z]$. Plainly, these objects – both the optimizer and the fitness – are search trajectors. As in Chapter 10, a black-box assumption is required, defined in the following subsections.

This formal setup is relatively simple. An optimizer observes a value history and uses it to sample a search history. An adversarial fitness observes a search history and uses it to sample a value history. From this pair of trajectories, generators can be obtained as was done in Chapter 13. But these objects are not their own source; instead, they are drawn from the context of a game or games that constitute their *raison d'être*. The next few subsections define this game, and show how trajectors and generators are derived from it.

### 17.1.3 Measures over Game Histories

Consider the extended form of the optimization game, which corresponds to the concept of iterative optimization methods. An optimizer proposes a search point that is scored by the fitness, and the previous points and their scores are used to determine the next search point. The optimizer can therefore be represented as a search generator, which therefore corresponds to the extended form of the game. The adversarial fitness, on the other hand, receives a proposed search point and scores it in the context of the previous points and scores. Thus an adversarial fitness is represented by a jagged value generator as discussed in Chapter 13.

The optimization game is built by interleaving the operation of the optimizer and the generator. In fact, the text in Section 13.2.2 built up a distribution $\mathcal{A} \times \mathbb{F}$ over history traces in exactly this manner. This construction process required a number of extraneous definitions: *integrable measurability*, *setwise measurability*, *tightness*, and *eventual boundedness*. These complexities can be avoided by specifying the optimization game itself as the primary object.

**Definition 17.1 (Optimization Game).** Given a Hausdorff search domain $X$, a Hausdorff value space $Y$, and a strictly ordered temporal index set $\mathcal{T}$, an *optimization game* $\mathcal{A} \times \mathcal{F}$ is a finite signed measure on the space of history traces $(X \times Y)^{\mathcal{T}}$ measured according to a $\sigma$-algebra $\Sigma$. If $\Sigma$ is the Borel $\sigma$-algebra, then $\mathcal{A} \times \mathcal{F}$ is called a *Borel optimization game*. If $\Sigma$ is the Baire $\sigma$-algebra, then $\mathcal{A} \times \mathcal{F}$ is called a *Baire optimization game*. If $\mathcal{A} \times \mathcal{F}$ is a probability measure, then it is called a *proper optimization game*.

In this definition, the optimization game has been denoted as $\mathcal{A} \times \mathcal{F}$, where $\mathcal{A}$ will eventually represent an optimization player and $\mathcal{F}$ an adversarial fitness. However, the optimization game has been defined independently from these two objects, from which they will now be derived. In this way, restrictions on the two objects $\mathcal{A}$ and $\mathcal{F}$ are derived from their respective roles in the optimization game. Even further, the temporal index set need not be well-ordered in this definition, but only strictly ordered, so that one might assign $\mathcal{T}$ to be the real line, for instance, in order to consider infinite and continuous search paths. Thus the setting of the optimization game as a signed measure over history traces abstracts away many details while capturing the essence of a search process.

Players are extracted from the game in two stages. In the first stage below, conditional expectations are used to extract proper search trajectors. In the second stage (Section 17.3.2), these proper players are linearly projected to construct players for all possible games. Define $\phi_X : (X \times Y)^{\mathcal{T}} \to X^{\mathcal{T}}$ and $\phi_Y : (X \times Y)^{\mathcal{T}} \to Y^{\mathcal{T}}$ to be the canonical projections onto $X^{\mathcal{T}}$ and $Y^{\mathcal{T}}$, so that $\phi_X(h) = h_X$ and $\phi_Y(h) = h_Y$. Suppose these projections are both measurable. Let $\Phi_X = \sigma(\phi_X)$ be the $\sigma$-algebra generated by $\phi_X$, and let $\Phi_Y = \sigma(\phi_Y)$ be the $\sigma$-algebra generated by $\phi_Y$. Then we have the following definition.

**Definition 17.2 (Proper Game Players).** Suppose $\mathcal{A} \times \mathcal{F}$ is a proper optimization game such that $\phi_X$ and $\phi_Y$ are measurable functions. If for two search trajectors $\mathcal{A} : Y^{\mathcal{T}} \to \mathcal{M}[X^{\mathcal{T}}, \Phi_X]$ and $\mathcal{F} : X^{\mathcal{T}} \to \mathcal{M}[Y^{\mathcal{T}}, \Phi_Y]$,

$$\mathcal{A}[y](A) = \mathbb{E}_{\mathcal{A}\times\mathcal{F}}\left[\mathbb{1}_{\phi_X^{-1}(A)} \,\middle|\, \Phi_Y\right](y), \quad \text{and} \tag{17.1}$$

$$\mathcal{F}[z](B) = \mathbb{E}_{\mathcal{A}\times\mathcal{F}}\left[\mathbb{1}_{\phi_Y^{-1}(B)} \,\middle|\, \Phi_X\right](z) \tag{17.2}$$

for all $y \in Y^{\mathcal{T}}$, all measurable $A \subseteq X^{\mathcal{T}}$, all $z \in X^{\mathcal{T}}$, and all $B \subseteq Y^{\mathcal{T}}$, then the pair $(\mathcal{A}, \mathcal{F})$ is called a pair of proper players of the game $\mathcal{A} \times \mathcal{F}$. $\mathcal{A}$ is an *optimization player* for the game, and $\mathcal{F}$ is an *adversarial fitness*.

There are several things to notice in this definition. First of all, the measurability properties of the players is determined by the measurability of the game through the projections $\phi_X$ and $\phi_Y$. Thus if $\mathcal{A} \times \mathcal{F}$ is a Baire game, but the temporal index set $\mathcal{T}$ is uncountable, then $\mathcal{A} \times \mathcal{F}$ has no players, because $\phi_X$ and $\phi_Y$ are unmeasurable. This feature forces Baire games to have countable time $\mathcal{T}$ and Baire search trajectors as players. Likewise, Borel games may have uncountable $\mathcal{T}$, *e.g.*, continuous time, but this forces the players to be Borel search trajectors.

Secondly, the conditional expectation is not unique, so a given game can have many optimization players and adversarial fitnesses, each of which differ from each other on sets of history traces that have $\mathcal{A} \times \mathcal{F}$-measure zero. To understand this, imagine a two-player game of poker in which one player always folds immediately after the cards are dealt. The differences between the opponents of this player can only be seen on turns where the universal folder is the second to bet. If two different opponents both behave the same way on the first turn, but differently later, there is no way to distinguish them in a game against a universal folder. That effect is precisely what is meant by saying that an optimization game may have many players; the nature of the interaction within a particular game may hide many distinguishing features of the players.

Finally, it should be noted that every proper optimization game has a pair of proper players and these players can be chosen to be proper, justifying the name. The proof of these facts is straightforward.

**Proposition 17.1.** *Every proper optimization game has a pair of proper players, and these players can be chosen to be proper search trajectors.*

*Proof.* By the existence of the conditional expectation, there exist functions

$$\mathcal{A}''[y](A) = \mathbb{E}_{\mathcal{A}\times\mathcal{F}}\left[\mathbb{1}_{\phi_X^{-1}(A)} \,\middle|\, \Phi_Y\right](y) \quad \text{and} \quad \mathcal{F}''[z](B) = \mathbb{E}_{\mathcal{A}\times\mathcal{F}}\left[\mathbb{1}_{\phi_Y^{-1}(B)} \,\middle|\, \Phi_X\right](z).$$

It must be shown that $\mathcal{A}''[y]$ and $\mathcal{F}''[z]$ so defined are finite signed measures, which means that they are $\sigma$-additive. Let $(A_n)_{n\in\mathbb{N}}$ be any countable collection of disjoint measurable subsets of $X^{\mathcal{T}}$. Then almost everywhere in $\mathcal{A}\times\mathcal{F}|_{\Phi_Y}$

$$\mathcal{A}''[y]\left(\bigcup_n A_n\right) \stackrel{a.e.}{=} \mathbb{1}_{\phi_X^{-1}(\bigcup_n A_n)}(z,y) \stackrel{a.e.}{=} \sum_n \mathbb{1}_{\phi_X^{-1}(A_n)}(z,y) \stackrel{a.e.}{=} \sum_n \mathcal{A}''[y](A_n), \tag{17.3}$$

which implies that $\mathcal{A}''[y]$ is a signed measure for almost every $y$. This result can be repeated for for $\mathcal{F}''[z]$. Let $\mu_X$ and $\mu_Y$ be arbitrary probability measures on $X^{\mathcal{T}}$ and $Y^{\mathcal{T}}$ respectively, and define

$$\mathcal{A}'[y] = \begin{cases} \mathcal{A}''[y] \text{ if } \mathcal{A}''[y] \text{ is a signed measure} \\ \mu_X \quad \text{otherwise} \end{cases}$$

$$\mathcal{F}'[z] = \begin{cases} \mathcal{F}''[z] \text{ if } \mathcal{F}''[y] \text{ is a signed measure} \\ \mu_Y \quad \text{otherwise.} \end{cases}$$

Now both $\mathcal{A}'$ and $\mathcal{F}'$ are measure-valued functionals that satisfy the conditional expectations above. Because $\mathcal{A}\times\mathcal{F}$ is finite, $|\mathcal{A}'[y](A)| \leq \|\mathcal{A}\times\mathcal{F}\| < \infty$ and similarly for $|\mathcal{F}'[z](B)|$. Therefore $\mathcal{A}'$ and $\mathcal{F}'$ are search trajectors with finite norm less than $2\|\mathcal{A}\times\mathcal{F}\|$.

$\mathcal{A}\times\mathcal{F}$ is proper, so $\|\mathcal{A}\times\mathcal{F}\| = 1$. Because the conditional expectations above are then integrals of a non-negative function over a non-negative measure, we have $\mathcal{A}'[y](A) \geq 0$ and $\mathcal{F}'[z](B) \geq 0$ for almost every $y$ and $z$. Because $\|\mathcal{A}\times\mathcal{F}\| = 1$, $\mathcal{A}'[y](X^{\mathcal{T}}) = 1$ and $\mathcal{F}'[z](Y^{\mathcal{T}}) = 1$ for almost every $y$ and $z$, which implies that both $\mathcal{A}'[y]$ and $\mathcal{F}'[z]$ are probability measures for almost every $y$ and $z$. Let $\mathbb{P}_X$ be any probability measure on $X^{\mathcal{T}}$, and define a proper search trajector $\mathcal{A}$ so that $\mathcal{A}[y] = \mathcal{A}'[y]$ if $\mathcal{A}'[y]$ is a probability measure and $\mathcal{A}[y] = \mathbb{P}_X$ otherwise. Define a proper search trajector $\mathcal{F}$ similarly, so that for some probability measure $\mathbb{P}_Y$ on $Y^{\mathcal{T}}$, $\mathcal{F}[z] = \mathcal{F}'[z]$ wherever $\mathcal{F}'[z]$ is a probability measure and $\mathcal{F}[z] = \mathbb{P}_Y$ otherwise. Then $\mathcal{A}$ and $\mathcal{F}$ are proper search trajectors satisfying Equations 17.1 and 17.2, which completes the proof.

Although the definition of the proper game players suffices to extract search trajectors from a proper optimization game, these players may still violate the black-box properties introduced in Chapters 9 and 13, meaning that the game players may be able to "see the future" and, perhaps more importantly, that the game has no starting point and no order of play. This loophole is removed in the next section.

## 17.2 Enforcing the Black-box Properties

The black-box properties guarantee that an optimization game can be played one step at a time, which ensures that the normal-form game can be converted to an extended-form game. Since every game has both forms, it will not surprise that every proper game can be derived as an interaction of two players with black-box properties, proven as Proposition 17.2 below. The black-box properties pertain primarily to the concept of search and value generators, and so this section unpacks the relationship between these objects, games, and trajectors, both deconstructing and reconstructing the game. In this process, it should be remembered that the black-box properties here have a more general meaning than just the traditional notion of oracle access to a fitness function; the traditional concept is recovered when the value space $Y$ represents the evaluation of a fitness function. In the adversarial setting, many broader forms of optimization can be studied, and some of these, including stochastic or dynamic fitness functions, are enumerated in this section.

### *17.2.1 From Games to Generators*

The trajectors $\mathcal{A}$ and $\mathcal{F}$ described by Equations 17.1 and 17.2 and constructed in Proposition 17.1 represent the game strategies of two proper players in normal form. The question now is how to obtain search generators that represent the same strategies in extended form. The challenge is that a naïve derivation of these generators is ambiguous, unless the black-box properties are imposed, which prevents players from making decisions at time $t$ based on the opponent's move subsequent to time $t$.

Search generators can also be obtained from the optimization game as conditional expectations over measurable projections just as in Equation 13.2. Define $Z_t$ and $W_t$ as stochastic processes as in Section 13.3.1. Then $Z_t$ is the search point at time $t$ and $W_t$ is the fitness value at time $t$. The process $(Z_t)_{t\in\mathcal{T}}$ is regarded as a process on $X^{\mathcal{T}}$ and $(W_t)_{t\in\mathcal{T}}$ as a process on $Y^{\mathcal{T}}$. The variables $Z_t$ and $W_t$ are canonical projections and are typically measurable. Define filtrations

$$\mathcal{Z}_{<t} = \sigma\left((Z_\tau)_{\tau<t}\right) \quad \text{and} \quad \mathcal{W}_{<t} = \sigma\left((W_\tau)_{\tau<t}\right).$$

A search generator and a jagged value generator may be obtained by setting

$$\mathcal{G}[h](A) = \mathbb{E}_{\mathcal{A}[h'_Y]}\left[\mathbb{1}_A \circ Z_t \mid \mathcal{Z}_{<t}\right](h'_X) \qquad \text{and} \tag{17.4}$$

$$\mathcal{G}^*[h_X \| x, h_Y](B) = \mathbb{E}_{\mathcal{F}[h''_X]}\left[\mathbb{1}_B \circ W_t \mid \mathcal{W}_{<t}\right](h''_Y), \tag{17.5}$$

where $x \in X$, $A \in \mathcal{B}_X$, $B \in \mathcal{B}_Y$, $h \in \mathcal{H}^{\mathcal{T}}_{X,Y}$ is a history prefix defined for all times less than $t$, *i.e.*, $|h| = |\{\tau \in \mathcal{T} \mid \tau < t\}|$, and $h', h'' \in (X \times Y)^{\mathcal{T}}$ are history traces such that $h'_\tau = h''_\tau = h_\tau$ for $\tau < t$ and additionally $h''^X_t = x$.

Notice, though, that Equation 17.4 and 17.5 are potentially ambiguous, and not just on sets of measure zero. If there are multiple histories $h'$ that agree with $h$ but for which $\mathcal{A}[h'_Y]$ differs, then $\mathcal{G}[h]$ will be unequal to itself. Clearly, this ambiguity is not acceptable. As before, the solution to this ambiguity is to impose the black-box properties. The black-box property for search trajectors was given in Definition 13.5. A similar definition, the jagged black-box property, is now given to account for the fact that the adversarial fitness can see the decisions of the optimization player one step farther, up to the current time. The subtle difference in the two definitions is the structure of their time dependence; for the black-box property, $y_s = w_s$ for $s < t$, whereas for the jagged black-box property, $z_s = w_s$ for $s \leq t$.

**Definition 17.3 (Jagged Black-box Property).** Let $\phi^Y_{[t]} : Y^{\mathcal{T}} \to Y^{[t]}$ be the canonical projection to the search points up to time $t$. For all search histories $z$, let $\mathcal{F}[z] \circ \phi^Y_{[t]}$ be the push-forward measure on $Y^{[t]}$, which can be used to observe the value history up to time $t$. A search trajector $\mathcal{F}$ has the *jagged black-box property* if for all $t \in \mathcal{T}$ and for all $z, w \in X^{\mathcal{T}}$, $\mathcal{F}[z] \circ \phi^Y_{[t]} = \mathcal{F}[w] \circ \phi^Y_{[t]}$ whenever $z_s = w_s$ for all $s \leq t$.

**Definition 17.4 (Black-Box Game).** A proper optimization game $\mathcal{A} \times \mathcal{F}$ has the *black-box property* if there exists a pair of players $(\mathcal{A}, \mathcal{F})$ for the game such that $\mathcal{A}$ has the black-box property and $\mathcal{F}$ has the jagged black-box property.

Now it is expected from game theory that the extended form and the normal form of a game are equivalent, so it would be problematic if there existed optimization games for which no extended form could be given. Fortunately, by deriving the players from the game rather than taking the players as absolute, it can be shown that *every* proper optimization game admits the black-box property, as expected. In order to prove this fact, it is first shown that a lack of ambiguity in Equations 17.4 and 17.5 is equivalent to the black-box properties.

**Lemma 17.1.** *Equation 17.4 is unambiguous if and only if $\mathcal{A}$ has the black-box property, and Equation 17.5 is unambiguous if and only if $\mathcal{F}$ has the jagged black-box property.*

*Proof.* Suppose that $\mathcal{A}$ has the black-box property and let $y, w \in Y^{\mathcal{T}}$ and $t \in \mathcal{T}$ be such that $y_s = w_s$ for $s < t$. First, since the integrands and conditioning are independent of $Z_s$ for $s > t$, irrespective of the black-box property

$$\mathbb{E}_{\mathcal{A}[y]}\left[\mathbb{1}_A \circ Z_t \mid \mathcal{Z}_{<t}\right](z) \quad = \quad \mathbb{E}_{\mathcal{A}[y]\circ\phi^X_{[t]}}\left[\mathbb{1}_A \circ Z_t \mid \mathcal{Z}_{<t}\right](z_{[t]})$$

for all $z \in X^{\mathcal{T}}$ where $z_{[t]} = (z_s)_{s \leq t}$. Then since $\mathcal{A}[y] \circ \phi^X_{[t]} = \mathcal{A}[w] \circ \phi^X_{[t]}$, for all $z, z' \in X^{\mathcal{T}}$ with $z_s = z'_s$ for $s < t$ it follows that

$$\mathbb{E}_{\mathcal{A}[y]}\left[\mathbb{1}_A \circ Z_t \mid \mathcal{Z}_{<t}\right](z) \quad = \quad \mathbb{E}_{\mathcal{A}[w]}\left[\mathbb{1}_A \circ Z_t \mid \mathcal{Z}_{<t}\right](z'),$$

where the interchangeability of $z$ and $z'$ results from the fact that $z_{[t]} = z'_{[t]}$. But this equation implies that Equation 17.4 is unambiguous, as desired.

In the opposite direction, suppose Equation 17.4 is unambiguous. It will be proven by induction that $\mathcal{A}$ has the black-box property. Again let $y,w \in Y^{\mathcal{T}}$. If $t=1$, then $\mathcal{A}[y]\circ\phi_{[t]}^X = \mathcal{A}[w]\circ\phi_{[t]}^X = \mathcal{G}[\emptyset]$. If $t>1$, $y_s = w_s$ for $s<t$ and $\mathcal{A}[y]\circ\phi_{[1,t)}^X = \mathcal{A}[w]\circ\phi_{[1,t)}^X$, then for any measurable sets $B\subseteq X^{[1,t)}$ and $A\subseteq X$,

$$\begin{aligned}
\mathcal{A}[y]\circ\phi_{[t]}^X(B\times A) &= \mathbb{E}_{\mathcal{A}[y]\circ\phi_{[1,t)}^X}\left[\mathbb{1}_B\circ\phi_{[1,t)}\,\mathbb{E}_{\mathcal{A}[y]\circ\phi_{[t]}^X}\left[\mathbb{1}_A\circ Z_t \mid \mathcal{Z}_{<t}\right]\right]\\
&= \mathbb{E}_{\mathcal{A}[w]\circ\phi_{[1,t)}^X}\left[\mathbb{1}_B\circ\phi_{[1,t)}\,\mathbb{E}_{\mathcal{A}[w]\circ\phi_{[t]}^X}\left[\mathbb{1}_A\circ Z_t \mid \mathcal{Z}_{<t}\right]\right]\\
&= \mathcal{A}[w]\circ\phi_{[t]}^X(B\times A),
\end{aligned}$$

where $w$ can be substituted for $y$ in one case due to the induction hypothesis and in the other due to Equation 17.4. That is, $\mathcal{A}[y]\circ\phi_{[t]}^X = \mathcal{A}[w]\circ\phi_{[t]}^X$, so that $\mathcal{A}$ has the black-box property.

The proof of equivalence for the jagged black-box property and the unambiguity of Equation 17.5 is similar and left as an exercise for the reader.

**Proposition 17.2.** *Every proper game has the black-box property.*

*Proof.* The proof strategy is to show that a game has players for which Equations 17.4 and 17.5 are unambiguous and then invoke Lemma 17.1.

First, let $\mathcal{A}\times\mathcal{F}$ be a proper optimization game. Let $\tilde{Z}_t$ be the projection from $(X\times Y)^{\mathcal{T}}$ to $X$ such that $Z_t = \tilde{Z}_t\circ\phi_X$ and similarly for $\tilde{W}_t$ with $W_t = \tilde{W}_t\circ\phi_Y$. Both are processes on $(X\times Y)^{\mathcal{T}}$. Define two filtrations,

$$\tilde{\mathcal{Z}}_t\otimes\tilde{\mathcal{W}}_{<t} = \sigma\left((\tilde{Z}_\tau)_{\tau\le t},(\tilde{W}_\tau)_{\tau<t}\right) \quad\text{and}\quad \tilde{\mathcal{Z}}_{<t}\otimes\tilde{\mathcal{W}}_{<t} = \sigma\left((\tilde{Z}_\tau)_{\tau\le t},(\tilde{W}_\tau)_{\tau<t}\right).$$

Next, define measure-valued functionals $\tilde{\mathcal{G}}_t$ and $\tilde{\mathcal{G}}_t^*$ so that

$$\begin{aligned}
\tilde{\mathcal{G}}_t[h'](A) &= \mathbb{E}_{\mathcal{A}\times\mathcal{F}}\left[\mathbb{1}_A\circ\tilde{Z}_t \mid \tilde{\mathcal{Z}}_{<t}\otimes\tilde{\mathcal{W}}_{<t}\right](h') \qquad\text{and}\\
\tilde{\mathcal{G}}_t^*[h'](B) &= \mathbb{E}_{\mathcal{A}\times\mathcal{F}}\left[\mathbb{1}_B\circ\tilde{W}_t \mid \tilde{\mathcal{Z}}_t\otimes\tilde{\mathcal{W}}_{<t}\right](h'),
\end{aligned}$$

for any $x\in X$, any measurable $A\subseteq X$ and $B\subseteq Y$, and any complete history trace $h'\in(X\times Y)^{\mathcal{T}}$. $\tilde{\mathcal{G}}_t[h']$ and $\tilde{\mathcal{G}}_t^*[h']$ can in fact be chosen to be probability measures for the same reasons as in the proof of Proposition 17.1, and this choice will now be assumed so that $\tilde{\mathcal{G}}$ and $\tilde{\mathcal{G}}^*$ are proper.

Let $h''$ be a history trace that agrees with $h'$ up to time $t$ so that $h'_s = h''_s$ for $s<t$. If $\tilde{\mathcal{G}}_t[h'](A) = \tilde{\mathcal{G}}_t[h''](A)$ for all measurable $A$ and $\mathcal{A}\times\mathcal{F}$-almost every such pair $(h',h'')$, then $\tilde{\mathcal{G}}$ can be used to define a search generator. But for any set $C\in\tilde{\mathcal{Z}}_{<t}\otimes\tilde{\mathcal{W}}_{<t}$, $h'\in C$ implies $h''\in C$, since $C$ is generated by pre-image with respect to the canonical projections $Z_s$ and $W_s$ for $s<t$. Hence $\tilde{\mathcal{G}}_t$ and by a similar argument $\tilde{\mathcal{G}}_t^*$ can be chosen so that

$$\mathcal{G}\left[(h_s)_{s<t}\right](A) = \tilde{\mathcal{G}}_t[h](A) \quad\text{and}\quad \mathcal{G}^*\left[\left(h_s^X\right)_{s\le t},\left(h_s^Y\right)_{s<t}\right](B) = \tilde{\mathcal{G}}_t^*[h](A)$$

unambiguously define a proper search generator and a proper jagged value generator, respectively. Unlike Equations 17.4 and 17.5, this definition is unambiguous due to the restrictions on both $\tilde{Z}_t$ and $\tilde{W}_t$, whereas in those equations only one or the other was restricted.

The next step is to show that these generators are equivalent to those in Equations 17.4 and 17.5 for some choice of optimization player $\mathcal{A}$ and adversarial fitness function $\mathcal{F}$. For almost every $h'$ with respect to the restricted measure $\mathcal{A}\times\mathcal{F}|_{\tilde{\mathcal{Z}}_{<t}\otimes\tilde{\mathcal{W}}_{<t}}$, the definition of conditional expectation implies

$$\mathcal{G}\left[(h'_s)_{s<t}\right](A) \overset{a.e.}{=} \mathbb{1}_A\circ\tilde{Z}_t(h') = \mathbb{1}_A\circ Z_t(h'_X).$$

At the same time, for any value history $h'_Y$ and almost every $h'_X$ over $\mathcal{A}[h'_Y]$,

$$\mathbb{1}_A\circ Z_t(h'_X) \overset{a.e.}{=} \mathbb{E}_{\mathcal{A}[h'_Y]}\left[\mathbb{1}_A\circ Z_t \mid \mathcal{Z}_{<t}\right](h'_X).$$

Now $\mathcal{A}[h'_Y]$ is itself a conditional expectation with respect to the $\sigma$-algebra $\Phi_Y$, so the phrase "almost every $h'_X$ over $\mathcal{A}[h'_Y]$" is equivalent to the phrase "almost every $h'$ over $\tilde{\mathcal{Z}}_{<t}\otimes\Phi_Y$", since $\tilde{\mathcal{Z}}_{<t}=\sigma\left((\tilde{Z}_\tau)_{\tau<t}\right)$ implies $\mathcal{Z}_{<t}=\left\{\phi_X(B)\,\middle|\,B\in\tilde{\mathcal{Z}}_{<t}\right\}$.[1] Going further, $\tilde{\mathcal{Z}}_{<t}\otimes\tilde{\mathcal{W}}_{<t}$ is a sub-$\sigma$-algebra of $\tilde{\mathcal{Z}}_{<t}\otimes\Phi_Y$, so that equality for almost every $h'$ over $\tilde{\mathcal{Z}}_{<t}\otimes\Phi_Y$ implies equality for almost every $h'$ over $\tilde{\mathcal{Z}}_{<t}\otimes\tilde{\mathcal{W}}_{<t}$. Putting it together,

$$\mathcal{G}\left[(h'_\tau)_{\tau<t}\right](A) \overset{a.e.}{=} \mathbb{E}_{\mathcal{A}[h'_Y]}\left[\mathbb{1}_A\circ Z_t \mid \mathcal{Z}_{<t}\right](h'_X).$$

That is, $\mathcal{G}$ satisfies Equation 17.4 unambiguously. Thus by Lemma 17.1 $\mathcal{A}$ necessarily has the black-box property.

A similar argument can be used to prove that Equation 17.5 is unambiguous for some $\mathcal{F}$ satisfying Equation 17.2 so that $(\mathcal{A},\mathcal{F})$ is a pair of proper players for the proper game $\mathcal{A}\times\mathcal{F}$, which therefore has the black-box property.

As a consequence of Proposition 17.2, every pair of proper players for a given proper game satisfies the black-box property almost everywhere and can be replaced with a version that satisfies it everywhere. Observe now that by beginning with a definition of the optimization game, we have been able to derive definitions of (equivalence classes of) optimization players and adversarial fitness functions that automatically satisfy the black-box property. This fact justifies the previous choice to define standard search generators in Chapter 6 and jagged value generators in Chapter 13 in a form that automatically respected black-box assumptions.

As a reminder, in this context "black-box" does not necessarily imply oracle access to an objective function, but only the ability to observe elements of the search domain $X$ and the value space $Y$ in their proper temporal order. The standard definition of black-box optimization as oracle access to an objective function is recovered when $Y$ is constrained by the choice of the adversarial fitness $\mathcal{F}$ to be the output of

[1] There is a small detail that this equivalence is only true for *almost every* $h'_Y$, but in this case $\mathcal{A}$ can be chosen so that it does indeed apply for all $h'_Y$, since it is itself a conditional expectation that is only defined almost everywhere in $\mathcal{A}\times\mathcal{F}|_{\Phi_Y}$.

some static, stochastic, or dynamic objective. In the next subsection, some different types of adversarial fitness are reviewed, which will help to clarify this point.

### *17.2.2 Adversarial Fitness Players*

Thus far, an adversarial fitness may not seem particularly useful for optimization. After all, if the fitness can make any decision whatsoever about the values sent to the optimizer, then successful optimization is impossible. The fitness can always confound the optimizer; the No Free Lunch theorems prove as much. However, the idea of an adversarial fitness is useful because it subsumes a wide array of search and optimization scenarios and connects search to game theory. Adversarial fitnesses can be used to characterize any of the following special cases, each of which represents a common practical optimization setting.

**Static Fitness.** Static optimization is the case where there is a single objective function $u : X \to Y$. The goal of the optimizer is to find the minimum or maximum of $u$. Let $z$ be an arbitrary search history, and let $y$ be the value history with $y_t = u(z_t)$ for all $t \in \mathcal{T}$. In this case, an adversarial fitness $\mathcal{F}^u$ can be defined to place probability one on $y$ given $z$, i.e., $\mathcal{F}^u[z](B) = \mathbb{1}_B(y)$, which, due to the Hausdorff property, implies that $y$ is the only value history that confers nonzero probability.

**Randomized Static Fitness.** Static optimization can be performed with a randomly selected objective function, as in Chapters 12 and 13. Suppose that $\mathbb{F}$ is a fitness measure over $Y^X$. For each search history $z$, consider the projection $\phi_z : Y^X \to Y^{\mathcal{T}}$ given by $\phi_z(u) = (u(z_t))_{t\in\mathcal{T}}$. The function $\phi_z$ induces a push-forward measure $\mathbb{F} \circ \phi_z$ on the value sequence space $Y^{\mathcal{T}}$. Then the map $z \mapsto \mathbb{F} \circ \phi_z$ defines an adversarial fitness function.

**Stochastic Fitness.** A randomized static fitness function randomly selects the fitness function, but once the fitness function is selected, the same response must be given to successive search queries at the same search point. A stochastic fitness removes this restriction but requires that the fitness at a point have the same distribution at all times.

A stochastic fitness can be defined by including multiple sources of randomness. Suppose $\Omega$ is some Hausdorff space and consider the space of extended fitness functions $Y^{X\times\Omega}$. If $u \in Y^{X\times\Omega}$, then it can happen that $u(x, \omega) \neq u(x, \omega')$ for $\omega, \omega' \in \Omega$. By randomly selecting $\omega$ independently of the search point, $u$ can be thought of as a stochastic fitness function; the optimizer chooses $x$, but $\omega$ is chosen at random when the fitness value is queried.

So suppose that $\mathbb{S}$ is a probability measure over $\Omega \times Y^{X\times\Omega}$, which is a measurable space. Now for each search history $z \in X^{\mathcal{T}}$, consider the expanded projection $\hat{\phi}_z : \Omega \times Y^{X\times\Omega} \to Y^{\mathcal{T}}$ given by $\hat{\phi}_z(\omega, u) = \big(u(z_t, \omega)\big)_{t\in\mathcal{T}}$. This projection is measurable so $\mathbb{S} \circ \hat{\phi}_z$ is defined, and $z \mapsto \mathbb{S} \circ \hat{\phi}_z$ is an adversarial fitness.

**Dynamic Fitness.** A dynamic fitness function is allowed to change over time. There are many possible definitions of dynamism. We give a brief example that generalizes a stochastic fitness to $Y^{\mathcal{T}\times X\times\Omega}$, where the fitness function has been expanded again to take an index parameter. Let $\mathbb{D}$ be a measure on $\Omega \times Y^{\mathcal{T}\times X\times\Omega}$. For each search history $z \in X^{\mathcal{T}}$, the expanded projection $\tilde{\phi}_z(\omega,u) = (u(t,z_t,\omega))_{t\in\mathcal{T}}$ again induces a measure $\mathbb{D}\circ\tilde{\phi}_z$. The map $z \mapsto \mathbb{D}\circ\tilde{\phi}_z$ is again an adversarial fitness.

A randomized static fitness is a subcase of a stochastic fitness, which is a subcase of a dynamic fitness. An potentially larger category is that of adaptive fitness functions that respond dynamically to the selected search points.

**Adaptive Fitness.** To generalize dynamic fitnesses further, suppose $\mathbb{A}$ is a measure on $\Omega \times Y^{\mathcal{H}_X^{\mathcal{T}}\times X\times\Omega}$. The projection becomes $\bar{\phi}_z(\omega,u) = (u((z_\tau)_{\tau<t},z_t,\omega))_{t\in\mathcal{T}}$, and $z \mapsto \mathbb{A}\circ\bar{\phi}_z$ is an adversarial fitness that takes into account the entire search history before deciding upon a stochastically varying value to assign to the current search point.

Now that it has been shown how an adversarial fitness function corresponds to a number of important optimization scenarios, we explore how the optimization game arises as an interaction of players.

### *17.2.3 Games from Players*

In Section 17.1.3, the optimization player and adversarial fitness were derived from an optimization game. Since the game only describes a particular interaction of two players, this derivation does not address how the game arises from independently selected strategies. Given a search trajector $\mathcal{A}$, we wish to compare two games $\mathcal{A}\times\mathcal{F}$ and $\mathcal{A}\times\mathcal{F}'$ for distinct search trajectors $\mathcal{F}$ and $\mathcal{F}'$. It is clear that each game may reveal different facets of $\mathcal{A}$ to the extent that the equivalence classes of optimization players for each game may differ. This section discusses how search trajectors $\mathcal{A}$ and $\mathcal{F}$ chosen independently generate a unique optimization game in the case where the temporal index set is countable.

Invoking Proposition 17.2, the most interesting choices are the case when $\mathcal{A}$ has the black-box property and $\mathcal{F}$ has the jagged black-box property, since such trajectors are sufficient to generate all proper optimization games.

Given these choices, Equations 17.4 and 17.5 define a search generator $\mathcal{G}$ that generates $\mathcal{A}$ and a jagged value-generator $\mathcal{G}^*$ that generates $\mathcal{F}$. Using this pair of generators, the optimization game is produced by recapitulating the development of Section 13.2.2 with minor alterations. First of all, it must be demonstrated that $\mathcal{G}$ is integrably measurable (Definition 13.1) and that $\mathcal{G}^*$ is setwise measurable (Definition 13.6). In order to use the Kolmogorov extension theorem, both $\mathcal{G}$ and $\mathcal{G}^*$ must further be tight. Eventual boundedness is guaranteed by the fact that all these generators and trajectors are proper by assumption.

Tightness of generators has already been defined. Recall that if the search domain $X$ is a complete, separable metric space (such as a finite space, Euclidean space, the space of bounded continuous functions, or the space of square integrable functions), then every measure and hence every generator is tight. Outside of these spaces, tightness must be demonstrated.

When the temporal index set $\mathcal{T}$ is countable, the construction of $\mathcal{A} \times \mathcal{F}$ from generators is straightforward. Let 1 represent the least element in $\mathcal{T}$ and define $\mathcal{A} \times \mathcal{F}^{[1]}$ and $\mathcal{A} \times \mathcal{F}^{[t]}$ for $t > 1$ using analogues of Equation 13.4 and Equation 13.5, adapted to adversarial fitnesses. Specifically, for measurable sets $D \subseteq (X \times Y)^{[t-1]}$, $E \subseteq X$ and $F \subseteq Y$,

$$\mathcal{A} \times \mathcal{F}^{[1]}(E \times F) = \int_E \mathcal{G}^*[\emptyset \| x, \emptyset](F)\, \mathcal{G}[\emptyset](dx) \quad \text{and}$$

$$\mathcal{A} \times \mathcal{F}^{[t]}(D \times (E \times F)) = \int_D \int_E \mathcal{G}^*[h_X \| x, h_Y](F)\, \mathcal{G}[h](dx)\, \mathcal{A} \times \mathcal{F}^{[t-1]}(dh)$$

define $\mathcal{A} \times \mathcal{F}^{[1]}$ and $\mathcal{A} \times \mathcal{F}^{[t]}$ through Carathéodory extension.

From these measures, $\mathcal{A} \times \mathcal{F}^{[t]}$ is defined for all $t \in \mathcal{T}$. For all finite $\mathcal{K} \subseteq \mathcal{T}$, $\mathcal{K}$ has a greatest element, say, $t_{\max}$. Define a projection $\phi^{\mathcal{K}}_{X \times Y}(z) = (z_t)_{t \in \mathcal{K}}$ on the product space $(X \times Y)^{[t_{\max}]}$, which is measurable. Let $\mathcal{A} \times \mathcal{F}^{\mathcal{K}} = \mathcal{A} \times \mathcal{F} \circ \phi^{\mathcal{K}}_{X \times Y}$, and then $\{\mathcal{A} \times \mathcal{F}^{\mathcal{K}} | \mathcal{K} \subset \mathcal{T} \text{ finite}\}$ is a consistent family of finite distributions. A game $\mathcal{A} \times \mathcal{F}$ is obtained from Kolmogorov extension of these measures. Because $\mathcal{T}$ is countable, this game is both Baire and Borel, since the two are identical in this case.

One might wish to consider how to generate a game when $\mathcal{T}$ is not countable, but to do so one needs generators for intervals as well as a mechanism to guarantee consistency among intervals, and so no mechanism for constructing such games from generators is presented here. The following theorem, proven by the previous paragraphs, sums up countable case.

**Theorem 17.1.** *Suppose $\mathcal{A}$ is proper black-box search trajector, $\mathcal{F}$ is a proper search trajector with the jagged black-box property, and $\mathcal{T}$ is at most countable. If the search generator defined by Equation 17.4 is integrably measurable, the jagged value generator defined by Equation 17.5 is setwise measurable, and both are tight, then there is a unique, proper Baire and Borel optimization game $\mathcal{A} \times \mathcal{F}$ on $(X \times Y)^{\mathcal{T}}$, constructed above.*

Thus, subject to the restrictions above, a tuple $(\mathcal{A}, \mathcal{F}, \mathcal{T})$ of an optimization player, an adversarial fitness, and a temporal index set yields a unique optimization game $\mathcal{A} \times \mathcal{F}$. There may still be some other tuple $(\mathcal{A}', \mathcal{F}', \mathcal{T})$ such that $\mathcal{A} \times \mathcal{F} = \mathcal{A}' \times \mathcal{F}'$, but this game is unique in the sense that there is no other game that may be generated from $\mathcal{A}$ and $\mathcal{F}$ in the same way.

Furthermore, if $(\mathcal{A}, \mathcal{F})$ is a pair of proper players for $\mathcal{A} \times \mathcal{F}$ by Definition 17.2 and $(\mathcal{A}, \mathcal{F}, \mathcal{T})$ satisfies them premises of Theorem 17.1, then $\mathcal{A} \times \mathcal{F}$ is indeed the game constructed by that theorem, as stated next.

**Theorem 17.2.** *Suppose $\mathcal{A}$ is proper black-box search trajector, $\mathcal{F}$ is a proper search trajector with the jagged black-box property, and $\mathcal{T}$ is at most countable. If these objects satisfy the conditions of Theorem 17.1, then $(\mathcal{A}, \mathcal{F})$ is a pair of proper players for the game $\mathcal{A} \times \mathcal{F}$ constructed in that theorem.*

*Proof.* The goal is to prove that for all histories $h$, all $A \in \Phi_X$ and all $B \in \Phi_Y$,

$$\mathcal{A}[h_Y](A) = \mathop{\mathbb{E}}_{\mathcal{A}\times\mathcal{F}}\left[\mathbb{1}_{\phi_X^{-1}(A)}\middle|\Phi_Y\right](h) \quad \text{and} \quad \mathcal{F}[h_X](B) = \mathop{\mathbb{E}}_{\mathcal{A}\times\mathcal{F}}\left[\mathbb{1}_{\phi_Y^{-1}(B)}\middle|\Phi_X\right](h). \tag{17.6}$$

The proof is accomplished by demonstrating that the finite projections are the same and appealing to the uniqueness of Kolmogorov extensions.

Now $\Phi_Y$ is the $\sigma$-algebra generated by $\phi_Y(h)$, which is invariant with respect to $h_X$. Thus the function

$$\mathcal{A}' : h_Y \mapsto \mathop{\mathbb{E}}_{\mathcal{A}\times\mathcal{F}}\left[\mathbb{1}_{\phi_X^{-1}(A)}\middle|\Phi_Y\right](\iota(h_Y))$$

is well defined for any injection $\iota : Y^{\mathcal{T}} \to (X \times Y)^{\mathcal{T}}$, and it is a search trajector with finite norm as discussed in the proof of Proposition 17.1. The goal is to show that $\mathcal{A}$ and $\mathcal{A}'$ have the same finite projections, which will be true if they produce the same generators.

The generator for $\mathcal{A}$ is given by Equation 13.2. The generator for $\mathcal{A}'$ is given by Equation 17.4. The equivalence of these two formulations has already been shown in the proof of Proposition 17.2, and thus $\mathcal{A} \overset{a.e.}{=} \mathcal{A}'$. The same argument can be applied to $\mathcal{F}$, whence the theorem is true.

In this section, the optimization game and its players have been defined, both in normal and extended form. It has been shown how to build players from a game and a game from players, and the concept of a *black-box* game has been elucidated. Adversarial fitnesses have been introduced and explained with examples. The next section addresses the broader question of how performance can be evaluated in optimization games, and how problems and solutions are related in this context.

## 17.3 Evaluating Performance of Optimization Games

Chapter 13 addressed the question of performance of an optimizer with respect to a probability distribution over objective functions. The main tool of that chapter, the V-performance, can be extended trivially to measure the performance of optimization players against an adversarial fitness. This section summarizes the results of this generalization and places a capstone on the subject matter of this book.

### *17.3.1 V-Performance Revisited*

In game theory, a game is always tied to a concept of value that determines the winner. This value is a real number and is a function of the game history. We have already introduced the V-performance in the context of a fitness measure in Definition 13.10. It is a simple generalization to define the V-performance for adversarial fitnesses as well in order to provide the needed concept of the value of a game.

**Definition 17.5 (Generalized V-Performance).** Let $(\mathcal{A},\mathcal{F})$ be a pair of proper players for the proper optimization game $\mathcal{A}\times\mathcal{F}$, and let $V:(X\times Y)^{\mathcal{T}}\to\mathbb{R}$ be bounded and measurable. Then the V-performance of $\mathcal{A}$ on $\mathcal{F}$ is

$$\langle\mathcal{A},\mathcal{F}\rangle_V=\mathbb{E}_{\mathcal{A}\times\mathcal{F}}[V]=\int_{(X\times Y)^{\mathcal{T}}}V(h)\,\mathcal{A}\times\mathcal{F}(dh), \tag{17.7}$$

and the function $V$ is called a *value function*.

Everything that was said about the V-performance for fitness measures in Chapter 13 applies verbatim to the V-performance for adversarial fitnesses. This statement includes Theorem 13.3 and Corollary 13.2, so that the model-optimal choices of $\mathcal{A}$ given $\mathcal{F}$ are already well defined. These statements are sufficiently important that the generalized theorems are reformulated here. The proofs and definitions involve little more than replacing $\mathbb{F}$ with $\mathcal{F}$.

**Theorem 17.3.** *Suppose the temporal index set $\mathcal{T}$ is countable. For $V$ progressively decomposable and either bounded or strictly positive, an admissible search trajector $\mathcal{A}$ has $V$-performance error bounded below by the sum over time of the expected one-step error gain,*

$$\langle\mathcal{A},\mathcal{F}\rangle_V-\inf_{\mathcal{A}'}\langle\mathcal{A}',\mathcal{F}\rangle_V=\sum_{t\in\mathcal{T}}\mathbb{E}_{\mathcal{A}\times\mathcal{F}}\left[\mathrm{err}_{\mathcal{F},V}(Z_t,(Z_\tau)_{\tau=1}^{t-1},(W_\kappa)_{\kappa=1}^{t-1})\right]. \tag{17.8}$$

**Corollary 17.1.** *Suppose the temporal index set $\mathcal{T}$ is countable. An admissible search trajector $\mathcal{A}$ is $\varepsilon$-optimal on an admissible adversarial fitness $\mathcal{F}$ and $V$ progressively decomposable and either bounded or strictly positive if and only if the sequence $(\eta_t)$ defined by*

$$\eta_t=\mathbb{E}_{\mathcal{A}\times\mathcal{F}}\left[\mathrm{err}_{\mathcal{F},V}(Z_t,(Z_\tau)_{\tau=1}^{t-1},(W_\kappa)_{\kappa=1}^{t-1})\right] \tag{17.9}$$

*has $\sum_{t\in\mathcal{T}}\eta_t<\varepsilon$, and if $\mathcal{A}$ is deterministic then $Z_t\in\overline{\mathcal{V}^t_{\eta_t}}\setminus\mathcal{V}^t_{\eta_t}$ almost surely.*

This theorem and its corollary reiterate that for a fixed adversarial fitness $\mathcal{F}$ and evaluation scenario $V$, a near-optimal search point can be chosen at each point in time based strictly on the previous search points and the resulting values.

As defined, the V-performance depends on both $\mathcal{F}$ and $V$. At the cost of admitting non-proper adversarial fitness functions, $V$ can be absorbed into $\mathcal{F}$. The absorption process to remove $V$ from the V-performance is defined as a function on proper adversarial fitnesses, $T^*_V:\mathcal{PF}_{X^{\mathcal{T}},Y^{\mathcal{T}}}\to\mathcal{MF}_{X^{\mathcal{T}},Y^{\mathcal{T}}}$, defined so that for all $z\in X^{\mathcal{T}}$ and measurable $B\subseteq Y^{\mathcal{T}}$

$$T^*_V\mathcal{F}[z](B)=\int_B V(z,y)\,\mathcal{F}[z](dy)\quad\text{and}\quad\mathcal{A}\times T^*_V\mathcal{F}(H)=\int_H V(h)\,\mathcal{A}\times\mathcal{F}(dh).$$

The game $\mathcal{A}\times T^*_V\mathcal{F}$ is well defined but not necessarily proper.

This game can also be constructed from $\mathcal{A}$ and $T^*_V\mathcal{F}$ by Kolmogorov extension. Suppose $V\geq 0$. Define $V^*_0(\emptyset,\emptyset)=1$ and for $z\in X^{[t]}$, $y\in Y^{[t]}$

$$V_t^*(z,y) \;=\; \begin{cases} 0 \quad \text{if } V_{t-1}^*(z_{[t-1]},y_{[t-1]}) = 0 \\ \dfrac{\mathbb{E}_{\mathcal{F}[h_X]}\left[V \mid \mathcal{W}_t\right](h_Y)}{V_{t-1}^*(z_{[t-1]},y_{[t-1]})} \;=\; \dfrac{\mathbb{E}_{\mathcal{A}\times\mathcal{F}}\left[V \mid \tilde{\mathcal{Z}}_t \otimes \tilde{\mathcal{W}}_t\right](h)}{V_{t-1}^*(z_{[t-1]},y_{[t-1]})} & \text{otherwise,} \end{cases}$$

where $z_{[t]} = (z_s)_{s\leq t}$, $y_{[t]} = (y_s)_{s\leq t}$, and $h$ satisfies $h_s^X = z_s$ and $h_s^Y = y_s$ for all $s \leq t$. Define a jagged value generator $T_V^*\mathcal{G}^*$ by

$$T_V^*\mathcal{G}^*[z,y](B) = \int_B V_t^*(z,y\|w)\; \mathcal{G}^*[z,y](dw),$$

which, due to the boundedness and measurability of $V$, is tight, setwise-measurable, and eventually bounded whenever $\mathcal{G}^*$ is tight, setwise-measurable, and proper. Also, the finite projection of $T_V^*\mathcal{F}[z]$ up to time $t$ is $T_V^*\mathcal{F}_z^{[t]}(B) =$

$$\begin{aligned} \int_B \prod_{s\leq t} T_V^*\mathcal{G}^*[z_{[t]},y_{[t-1]}](dy_t) &= \int_B \mathbb{E}_{\mathcal{F}[z]}\left[V \mid \mathcal{W}_t\right](y) \prod_{s\leq t} \mathcal{G}^*[z_{[t]},y_{[t-1]}](dy_t) \\ &= \int_{\tilde{B}} \mathbb{E}[V|\mathcal{W}_t]\; \mathcal{F}[z](dy) \;=\; \int_{\tilde{B}} V(z,y)\; \mathcal{F}[z](dy), \end{aligned}$$

where $B \subseteq Y^{[t]}$ and $\tilde{B} \in \mathcal{W}_t$ is the set whose image up to time $t$ is $B$. Thus in every case where $\mathcal{G}^*$ extends to $\mathcal{F}$ by Kolmogorov extension, $T_V^*\mathcal{G}^*$ extends to $T_V^*\mathcal{F}$ as well.

Furthermore, the finite projection of the game $\mathcal{A} \times T_V^*\mathcal{F}$ up to time $t$ is

$$\begin{aligned} \mathcal{A} \times T_V^*\mathcal{F}^{[t]}(H) &= \int_H \prod_{s=1}^{t} T_V^*\mathcal{G}^*[z_{[s]},y_{[s-1]}](dy_s)\; \mathcal{G}[z_{[s-1]},y_{[s-1]}](dz_s) \\ &= \int_H \mathop{\mathbb{E}}_{\mathcal{A}\times\mathcal{F}}\left[V \mid \tilde{\mathcal{Z}}_t \otimes \tilde{\mathcal{W}}_t\right] \prod_{s=1}^{t} \mathcal{G}^*[z_{[s]},y_{[s-1]}](dy_s)\; \mathcal{G}[z_{[s-1]},y_{[s-1]}](dz_s), \\ &= \int_{\tilde{H}} V(h)\; \mathcal{A} \times \mathcal{F}(dh), \end{aligned}$$

where $\tilde{H}$ is the element of $\tilde{\mathcal{Z}}_t \otimes \tilde{\mathcal{W}}_t$ whose image under projection is $H$. This equation implies that whenever $\mathcal{A} \times \mathcal{F}$ can be constructed from $\mathcal{A}$ and $\mathcal{F}$ by Theorem 17.1, an analogous construction can produce $\mathcal{A} \times T_V^*\mathcal{F}$ from $\mathcal{A}$, $\mathcal{F}$, and $V$ by Kolmogorov extension.

For nonnegative $V$, if can simply be observed that there are measurable nonnegative functions $V_+$ and $V_-$ such that $V = V_+ - V_-$ and $T_V^* = T_{V_+}^* - T_{V_-}^*$. Thus the conclusions of the last two paragraphs are true for all measurable and bounded value functions $V$, not just nonnegative ones.

To take this argument to its full extent, suppose that $\mathcal{F}'$ is any search trajector with finite norm having the jagged-black-box property. Let $(P_z, N_z)$ be the Hahn decomposition of $\mathcal{F}'[z]$. Define

$$V_{\mathcal{F}}(z,y) = \frac{1}{2}\left\{\mathcal{F}'[z](P_z)\mathbb{1}_{P_z}(y) + \mathcal{F}'[z](N_z)\mathbb{1}_{N_z}(y)\right\}, \tag{17.10}$$

which is measurable and bounded, since $\mathcal{F}'$ has finite norm. Let

$$V'_{\mathcal{F}}(h) = \begin{cases} 0 & \text{if } V_{\mathcal{F}}(h) = 0 \\ 1/V_{\mathcal{F}}(h) & \text{otherwise} \end{cases} \quad \text{so that} \quad \mathcal{F}[z](B) = \int_B V'_{\mathcal{F}}(z,y)\mathcal{F}'[z](dy)$$

is a proper search trajector such that $T^*_{V_{\mathcal{F}}}\mathcal{F} = \mathcal{F}'$. The following proposition has now been proven.

**Proposition 17.3.** *The map $(V,\mathcal{F}) \mapsto T^*_V\mathcal{F}$ is a surjective function from the bounded, measurable functions and the jagged-black-box proper search trajectors onto the vector space of jagged-black-box search trajectors with finite norm.*

The purpose of introducing $T^*_V$ was to reduce from the V-performance to the 1-performance, which is done in the next proposition.

**Proposition 17.4.** *Given any value function $V$ and any jagged-black-box proper search trajector $\mathcal{F}$, there exists a jagged-black-box search trajector $\mathcal{F}'$ such that*

$$\langle \mathcal{A},\mathcal{F}\rangle_V = \mathcal{A} \times \mathcal{F}'\left((X\times Y)^{\mathcal{T}}\right) \equiv \langle \mathcal{A},\mathcal{F}'\rangle .$$

*Further, for each $\mathcal{A}$ the map $\mathcal{F}' \mapsto \langle \mathcal{A},\mathcal{F}'\rangle$ is a function from the vector space of jagged-black-box search trajectors with finite norm into the real numbers.*

*Proof.* The result is obvious from $\mathcal{F}' = T^*_V\mathcal{F}$ and Proposition 17.3.

The previous derivations have shown that all evaluation scenarios for optimization players that rely on averaging some bounded, measurable function of the game history can be described in terms of an optimization game against an adversarial fitness that is not necessarily proper. In proving this fact, the set of optimization games that could be derived from search trajectors was expanded to include an entire vector space of adversarial fitness functions. This same procedure can be continued for the optimization players as well.

### 17.3.2 Performance of Non-Proper Players

As shown in the last subsection, an optimization game can constructed from a proper optimization player and an entire vector space of adversarial fitnesses. It is now shown that this same process can be applied to optimization players as well, so that a player-game map can be defined as an operator among three vector spaces: (1) finite, black-box search trajectors, (2) finite, jagged-black-box search trajectors, and (3) optimization games.

We will construct an operator $T_V : \mathcal{PF}_{Y^{\mathcal{T}},X^{\mathcal{T}}} \to \mathcal{MF}_{Y^{\mathcal{T}},X^{\mathcal{T}}}$ such that $\langle \mathcal{A},\mathcal{F}\rangle_V = \langle T_V\mathcal{A},\mathcal{F}\rangle_1$. Define for $y \in Y^{\mathcal{T}}$, measurable $A \subseteq X^{\mathcal{T}}$, $H \subseteq (X\times Y)^{\mathcal{T}}$

$$T_V\mathcal{A}[y](A) = \int_A V(z,y)\,\mathcal{A}[y](dz) \quad \text{and} \quad T_V\mathcal{A}\times\mathcal{F}(H) = \int_H V(h)\,\mathcal{A}\times\mathcal{F}(dh).$$

As with $T_V^*$, if a search generator $\mathcal{G}$ extends to $\mathcal{A}$ by Kolmogorov extension, a search generator $T_V\mathcal{G}$ can be defined to extend to $T_V\mathcal{A}$, this time by integrating

$$V_t(z,y) = \begin{cases} 0 \quad \text{if } V_{t-1}(z_{[t-1]}, y_{[t-2]}) = 0 \\ \dfrac{\mathbb{E}_{\mathcal{A}[h_Y]}[V \mid \mathcal{Z}_t](h_X)}{V_{t-1}(z_{[t-1]}, y_{[t-2]})} = \dfrac{\mathbb{E}_{\mathcal{A}\times\mathcal{F}}\left[V \mid \tilde{\mathcal{Z}}_t \otimes \tilde{\mathcal{W}}_{t-1}\right](h)}{V_{t-1}^*(z_{[t-1]}, y_{[t-2]})} & \text{otherwise,} \end{cases}$$

with $V_0(\emptyset,\emptyset) = 1$ and again assuming nonnegative $V$. Notice, though, that the conditioning on $\mathcal{A}\times\mathcal{F}$ has $\tilde{\mathcal{W}}_{t-1}$ rather than $\tilde{\mathcal{W}}_t$. This is a consequence of the fact that the optimization player plays first, without seeing the adversarial fitness' choice. A similar issue was addressed by the concept of *progressive convergence* (Definition 13.13) in Chapter 13. A similar definition will be introduced here. The condition in this definition is obviously true for $V_t^*$ but can fail to hold for $V_t$.

**Definition 17.6.** A bounded, measurable value function $V$ is jaggedly convergent if $\lim_{t\to\infty} V_t(z_{[t]}, y_{[t-1]}) = V(z,y)$.

Without recapitulating the entire development, the search generator

$$T_V\mathcal{G}[h](A) = \int_A V_{t+1}(h^X_{[t]} \| x, h^Y_{[t]})\, \mathcal{G}[h](dx)$$

for $h \in \mathcal{H}^{\mathcal{T}}_{X,Y}$ and measurable $A \subseteq X$ extends $T_V\mathcal{A}$ and can be used to construct $T_V\mathcal{A}\times\mathcal{F}$ by Kolmogorov extension when $V$ is jaggedly convergent and nonnegative. For bounded and measurable $V$, the construction can be performed twice with $V_+$ and $V_-$ and then added.

Suppose $\mathcal{A}'$ is any black-box search trajector that extends from a tight, integrably measurable, eventually bounded search generator. Then as before one can define from Hahn decompositions $(P_y, N_y)$ for $\mathcal{A}[y]$

$$V_{\mathcal{A}}(z,y) = \mathcal{A}'[y](P_y)\mathbb{1}_{P_y}(z) + \mathcal{A}'[y](N_y)\mathbb{1}_{N_y}(z), \tag{17.11}$$

which is measurable and bounded, since $\mathcal{A}'$ has finite norm. This $V_{\mathcal{A}}$ is also jaggedly convergent due to the black-box property on $\mathcal{A}$. Let

$$V'_{\mathcal{A}}(h) = \begin{cases} 0 & \text{if } V_{\mathcal{A}}(h) = 0 \\ 1/V_{\mathcal{A}}(h) & \text{otherwise} \end{cases} \quad \text{so that} \quad \mathcal{A}[y](A) = \int_A V'_{\mathcal{A}}(z,y)\mathcal{A}'[y](dz),$$

and $\mathcal{A}$ is a proper search trajector such that $T_{V_{\mathcal{A}}}\mathcal{A} = \mathcal{A}'$. The result of this discussion is the following proposition.

**Proposition 17.5.** *The map $(V,\mathcal{A}) \mapsto T_V\mathcal{A}$ is a surjective function from the bounded, measurable, jaggedly convergent functions and the set of black-box proper search trajectors onto the vector space of black-box search trajectors with finite norm.*

The operators $T_V$ and $T_V^*$ together provide a way to define players for optimization games that are not necessarily proper, and as the discussion above shows, in many cases games can be constructed directly from players.

**Definition 17.7 (Normalized Form).** A proper search trajector $\mathcal{A}'$ is a *normalized form* of a search trajector $\mathcal{A}$ if there exists a value function $V$ such that $\mathcal{A} = T_V \mathcal{A}'$, and the value function $V$ is called a *normalizer* of $\mathcal{A}$.

**Definition 17.8 (Generalized Game Players).** A pair $(\mathcal{A}, \mathcal{F})$ of search trajectors is a pair of players for an optimization game $\mathcal{A} \times \mathcal{F}$ if there exist normalized forms $\mathcal{A}'$ and $\mathcal{F}'$ of $\mathcal{A}$ and $\mathcal{F}$, respectively, such that $(\mathcal{A}', \mathcal{F}')$ is a pair of proper players for a game $\mathcal{A}' \times \mathcal{F}'$ such that

$$\mathcal{A} \times \mathcal{F}(H) = \int_H V_{\mathcal{A}}(h) V_{\mathcal{F}}(h)\, \mathcal{A}' \times \mathcal{F}'(dh) \tag{17.12}$$

where $V_{\mathcal{A}}$ and $V_{\mathcal{F}}$ are the normalizers for $\mathcal{A}$ and $\mathcal{F}$ yielding $\mathcal{A}'$ and $\mathcal{F}'$.

The definition of game players does not require the value functions $V_{\mathcal{A}}$ and $V_{\mathcal{F}}$ to conform to Equations 17.11 and 17.10, which allows for a more general definition of players. It is worthwhile, however, to name the results of these two functions as *canonical* if for no other reason than to point out that every search trajector has at least one normalized form. Since the two equations are identical apart from reversing the search and value space, the following definition applies equally to both $\mathcal{A}'$ and $\mathcal{F}'$.

**Definition 17.9 (Canonical Normalized Form).** The *canonical normalized form* $\mathcal{A}_{\text{can}}$ of a search trajector $\mathcal{A}$ from $X$ to $Y$ is the normalized form of $\mathcal{A}$ that is obtained by the normalizer

$$V(z,y) = \mathcal{A}[y](P_y)\mathbb{1}_{P_y}(z) + \mathcal{A}[y](N_y)\mathbb{1}_{N_y}(z)$$

as in Equation 17.11 with $(P_y, N_y)$ the Hahn decomposition of $\mathcal{A}[y]$ for each $y$.

Using normalized forms, a game can be constructed from a pair $(\mathcal{A}, \mathcal{F})$ be normalizing, extending, and then denormalizing again. In this case, by extending, we are referring directly to Kolmogorov extension of a tight, measurable, proper search generator, and hence the phrase "extends in the Kolmogorov sense" is used to distinguish such extension from mere membership in the set $\mathbf{E}'$ from Section 13.2.2.

**Theorem 17.4.** *Suppose that $\mathcal{A}$ is a black-box search trajector from $X$ to $Y$ with a normalized form that extends in the Kolmogorov sense from a tight, integrably measurable proper search generator and that $\mathcal{F}$ is a jagged-black-box search trajector from $Y$ to $X$ with a normalized form that extends in the Kolmogorov sense from a tight, setwise-measurable proper jagged value generator. Then the pair $(\mathcal{A}, \mathcal{F})$ is a pair of players for a unique optimization game $\mathcal{A} \times \mathcal{F}$.*

*Proof.* Let $\mathcal{A}'$ and $\mathcal{F}'$ be the Kolmogorov-extensible normalized forms of $\mathcal{A}$ and $\mathcal{F}$, respectively. By Theorems 17.1 and 17.2 there is a game $\mathcal{A}' \times \mathcal{F}'$ such that $(\mathcal{A}', \mathcal{F}')$ are proper players for the game. Setting $V_{\mathcal{A}}$ and $V_{\mathcal{F}}$, respectively, to be the normalizers of $\mathcal{A}$ and $\mathcal{F}$ with respect to $\mathcal{A}'$ and $\mathcal{F}'$, Equation 17.12 can be taken as a definition of a game $\mathcal{A} \times \mathcal{F}$ for which Definition 17.8 is satisfied. Uniqueness follows from the uniqueness of $\mathcal{A}' \times \mathcal{F}'$.

Theorem 17.4 maps many pairs of players to an optimization game, but it is incomplete in the sense that there remain pairs of players that can not be mapped to a game. Define a set

$$\mathbf{P} = \{((\mathcal{A},\mathcal{F}),\mathcal{A}\times\mathcal{F}) \mid (\mathcal{A},\mathcal{F}) \text{ is a pair of players for } \mathcal{A}\times\mathcal{F}\},$$

which captures the relationships between players and games in Definition 17.8. Ideally, $\mathbf{P}$ would be the graph of a total function. Although it may in fact be so, this text will not prove it. However, it is easy to see that if it is a function, it is surjective. That is, every game has players.

**Theorem 17.5.** *There exists a pair of players for every optimization game.*

*Proof.* Let $\mu$ be an optimization game and note that by Hahn-Jordan decomposition and normalization, $\mu = \alpha\mathbb{P} - \beta\mathbb{Q}$ for probability measures $\mathbb{P},\mathbb{Q}$ and positive real numbers $\alpha,\beta$. Let $(P,N)$ be the Hahn decomposition and define

$$V'_\mu(h) = \frac{1}{2\alpha}\mathbb{1}_P(h) - \frac{1}{2\beta}\mathbb{1}_N(h) \text{ so that } \bar{\mu}(H) = \int_H V'_\mu(h)\,\mu(h) = \frac{1}{2}(\mathbb{P}(H)+\mathbb{Q}(H))$$

is a probability measure. Then $\bar{\mu}$ is a proper game and has a pair of proper players $(\mathcal{A},\mathcal{F})$ by Proposition 17.1, and these players can be chosen to have black-box properties by Proposition 17.2. Setting $V_\mu = 2\alpha\mathbb{1}_P + 2\beta\mathbb{1}_N$, it holds that $T^*_{V_\mu}\mathcal{F}$ is a search trajector from $Y$ to $X$, and since $V_\mathcal{A} = 1$ and $V_{T^*_{V_\mu}\mathcal{F}} = V_\mu$, it follows that $(\mathcal{A}, T_{V_\mu}\mathcal{F})$ is a pair of players for the game

$$H \mapsto \int_H V_\mu(h)\,\bar{\mu}(dh) = \int_H V_\mu(h)V'_\mu(h)\,\mu(h) = \mu(H\cap P) + \mu(H\cap N) = \mu(H),$$

which completes the proof.

The proof of Theorem 17.5 also clearly demonstrates that every optimization game has many different pairs of players, since any essentially distinct pair $(V_\mathcal{A}, V_\mathcal{F})$ such that $V_\mu = V_\mathcal{A} V_\mathcal{F}$ yields a distinct pair of players $(T_{V_\mathcal{A}}\mathcal{A}, T^*_{V_\mathcal{F}}\mathcal{F})$. This fact opens the possibility that if one is careful about zero sets, one can in fact fix $V_\mathcal{A}$ and then choose $V_\mathcal{F}$ to play a given game.

**Definition 17.10 (Zero Set).** The *zero set* of a search trajector $\mathcal{A}$ from $X$ to $Y$ is the subset of $\Sigma$ given by

$$\mathscr{Z}[\mathcal{A}] = \{H \in \Sigma \mid |\mathcal{A}[y]|(\phi_X(H)) = 0 \text{ for every } y \in \phi_Y(H)\}.$$

**Definition 17.11 (Compatibility).** A search trajector $\mathcal{A}$ is said to be *compatible* with a game $\mu$ if $|\mu|(H) = 0$ for all $H \in \mathscr{Z}[\mathcal{A}]$.

**Proposition 17.6.** *A search trajector $\mathcal{A}$ is compatible with an optimization game $\mu$ if and only if $\mu(H) = 0$ for all $H \in \mathscr{Z}_\mu[\mathcal{A}]$ given by*

$$\mathscr{Z}_\mu[\mathcal{A}] = \{H \in \Sigma \mid |\mathcal{A}[y]|(\phi_X(H)) = 0 \textit{ for } \mu\textit{-almost every } y \in \phi_Y(H)\}.$$

*Proof.* The "if" direction is a consequence of the fact that $\mathscr{Z}[\mathcal{A}] \subseteq \mathscr{Z}_\mu[\mathcal{A}]$, so assume that $\mathcal{A}$ is compatible with $\mu$; *i.e.*, $|\mu|(H) = 0$ for all $H \in \mathscr{Z}[\mathcal{A}]$. Let $H' \in \mathscr{Z}_\mu[\mathcal{A}]$ be arbitrary, and it must be shown that $|\mu|(H') = 0$. Then there is a set $H'' \subseteq H'$ with $|\mu|(H'') = 0$ on which $|\mathcal{A}[y]|(\phi_X(H')) \neq 0$ for $y \in \phi_Y(H'')$, but for every $y \in \phi_Y(H' \setminus H'')$, $|\mathcal{A}[y]|(\phi_X(H')) = |\mathcal{A}[y]|(\phi_X(H' \setminus H'')) = 0$. It follows that $H' \setminus H'' \in \mathscr{Z}[\mathcal{A}]$, so that $|\mu|(H' \setminus H'') = 0$. But then it holds that $|\mu|(H') = |\mu|(H'') + |\mu|(H' \setminus H'') = 0$ as desired.

The preceding definitions and proposition were stated in terms of a trajector $\mathcal{A}$, but it should be noted that in fact these definitions and proposition apply equally to the adversarial fitness, so that a search trajector $\mathcal{F}$ from $Y$ to $X$ is compatible with a game on $(X \times Y)^{\mathcal{T}}$ if that game assigns zero probability to every set in $\mathscr{Z}[\mathcal{F}]$, the zero set of $\mathcal{F}$, as in the next propositions and theorem.

**Proposition 17.7.** *Let a search trajector* $\mathcal{A}$ *with canonical normalizer* $V_{\mathcal{A}}$. *Then* $\mathscr{Z}[\mathcal{A}_{\text{can}}] = \mathscr{Z}[\mathcal{A}]$.

*Proof.* Note that $V_{\mathcal{A}}$ is always nonzero since $P_y \cup N_y = X^{\mathcal{T}}$. Thus $|\mathcal{A}[y]|(\phi_X(H)) = 0$ implies that $|\mathcal{A}_{\text{can}}[y]|(\phi_X(H)) = 0$ and *vice versa*. Consequently $\mathscr{Z}[\mathcal{A}_{\text{can}}] = \mathscr{Z}[\mathcal{A}]$.

**Proposition 17.8.** *If* $(\mathcal{A}, \mathcal{F})$ *is a pair of players for a game* $\mu$, *then* $\mathcal{A}$ *and* $\mathcal{F}$ *are both compatible with* $\mu$.

*Proof.* Let $V_{\mathcal{A}}$ be the canonical normalizer of $\mathcal{A}$, and similarly let $V_{\mathcal{F}}$ be the canonical normalizer of $\mathcal{F}$. Suppose that $H \in \mathscr{Z}[\mathcal{A}_{\text{can}}] = \mathscr{Z}[\mathcal{A}]$. Then let $\mu' = \mathcal{A}_{\text{can}} \times \mathcal{F}_{\text{can}}$ be the proper game played by $\mathcal{A}_{\text{can}}$ and $\mathcal{F}_{\text{can}}$. Then

$$\begin{aligned}\mu'(H) = \mathbb{E}_{\mu'}\left[\mathbb{E}_{\mu'}\left[\mathbb{1}_H \mid \Phi_Y\right]\right] &\leq \mathbb{E}_{\mu'}\left[\mathbb{1}_{\phi_Y^{-1}(\phi_Y(H))} \mathbb{E}_{\mu'}\left[\mathbb{1}_{\phi_X^{-1}(\phi_x(H))} \middle| \Phi_Y\right]\right] \\ &= \mathbb{E}_{\mu'}\left[\mathbb{1}_{\phi_Y^{-1}(\phi_Y(H))} \mathcal{A}_{\text{can}}[\phi_Y](\phi_X(H))\right] = 0.\end{aligned}$$

Since $\mu'$ is a probability measure and $\mathbb{1}_H$ is positive, it follows that $\mu'(H) = 0$ and so $\mu(H) = \int_H V_{\mathcal{A}} V_{\mathcal{F}} d\mu' = 0$. Nearly identical reasoning also shows that if $H \in \mathscr{Z}[\mathcal{F}]$ then $\mu(H) = 0$. Thus both $\mathcal{A}$ and $\mathcal{F}$ are compatible with $\mu$.

**Theorem 17.6.** *Suppose* $\mu$ *is an optimization game and* $\mathcal{A}'$ *is a search trajector compatible with it. Then there exists a search trajector* $\mathcal{F}'$ *such that* $(\mathcal{A}', \mathcal{F}')$ *is a pair of players for* $\mu$. *Conversely, if* $\mathcal{F}'$ *is a search trajector compatible with* $\mu$, *then there exists a search trajector* $\mathcal{A}'$ *such that* $(\mathcal{A}', \mathcal{F}')$ *is a pair of players for* $\mu$.

*Proof.* The theorem is proven for $\mathcal{A}'$ given; when $\mathcal{F}'$ is given the reasoning is identical. Let $V_\mu$ and $(\mathcal{A}, \mathcal{F})$ be as in the proof of Theorem 17.5. Define

$$V_{\mathcal{A}}(h) = \frac{d\mathcal{A}'[h_Y]}{d\mathcal{A}[h_Y]}(h_X) \quad \text{and} \quad V_{\mathcal{F}}(h) = \begin{cases} 0 & \text{if } V_{\mathcal{A}}(h) = 0 \\ V_\mu(h)/V_{\mathcal{A}}(h) & \text{otherwise,} \end{cases}$$

and then set $\mathcal{F}' = T_{V_{\mathcal{F}}} \mathcal{F}$. By Definition 17.8, the pair $(\mathcal{A}', \mathcal{F}')$ is a pair of players for $\mu$ if $V_{\mathcal{A}}(h) V_{\mathcal{F}}(h) = V_\mu(h)$ for $\mu$-almost every $h$.

Using the definitions above, the only way that $V_\mu \neq V_\mathcal{A} V_\mathcal{F}$ is if there is a set $H$ with $|\mu|(H) \neq 0$ on which $V_\mathcal{A}(h) = 0$ but $V_\mu(h) \neq 0$. But then $|\mathcal{A}'[h_Y]|(\phi_X(H)) = 0$ for $\mu$-almost every $h_Y$ in $\phi_Y(H)$. Thus $H \in \mathscr{Z}_\mu[\mathcal{A}]$ and by the compatibility of $\mathcal{A}$ and $\mu$ together with Proposition 17.6, $|\mu|(H) = 0$. By contradiction, $V_\mu = V_\mathcal{A} V_\mathcal{F}$ almost everywhere in $\mu$, completing the proof.

Theorem 17.6 intuitively means that a search trajector can play any optimization game except for those that require it to make decisions it refuses to make. In order to play the games it is willing to play, it merely needs a suitable opponent. Returning to the poker example, the player who always folds will never be able to play a winning game, but a player who has a nonzero probability of taking each possible action can observe every possible poker game given enough opportunities.

As was shown in the last subsection, the V-performance $\langle \mathcal{A}, \mathcal{F} \rangle_V$ generates every possible evaluation scenario that is obtained by averaging some bounded, measurable function of the history trace, and it is equal to the overall measure of a game between $\mathcal{A}$ and a generalized adversarial fitness $T_V^* \mathcal{F}$. In this subsection, it has been added that a particular optimization player corresponding to some black-box search trajector, can play every game that it chooses to play, with only the games assigning positive value to some member of its zero set being excluded.

### *17.3.3 Nonlinearity of the Player-Game Map*

Definition 17.2 defined the players of a proper optimization game as conditional expectations with respect to subspace projections. Together with this definition, Theorem 17.1 generates a two-way relationship between pairs of proper search trajectors $(\mathcal{A}, \mathcal{F})$ and proper games $\mathcal{A} \times \mathcal{F}$. The previous subsection extended these concepts to arbitrary search trajectors.

The relationship is defined more broadly when passing from games to players; in this case the temporal index set can be transfinite, the $\sigma$-algebra can be Borel, and the entire vector space of search trajectors can potentially participate. Nonetheless, it should be noticed that both the spaces of search trajectors as well as the space of games are vector spaces. Even in the constructive process for generating games from players, the underlying properties are linear: tightness, measurability, and black-box properties.

Given the profusion of linear properties and vector spaces, one might expect to find linear relationships between games and players. Unfortunately, such an extension is impossible while preserving the characterization of players as conditional expectations, because conditional expectations are not linear with respect to the integrating measure.

To demonstrate this fact with an example, consider the unit interval $[0,1]$. Let $p(x) = 2x$ and $q(x) = 2 - 2x$. Let $\mathbb{P}$ be the measure with density $p$ and $\mathbb{Q}$ the measure with density $q$, as we may since both $p$ and $q$ integrate to one. Let $\mathbb{S} = (1-\alpha)\mathbb{P} + \alpha\mathbb{Q}$ for $\alpha \in [0,1]$, which implies that $\mathbb{S}$ has density $s(x) = 2\alpha + 2x(1-2\alpha)$. Let $g(x) = 1$ for $x \leq 1/4$ and $1/2$ otherwise. Conditioning on $x < 1/2$,

$$\begin{aligned} \mathbb{P}(x \mid x < \tfrac{1}{2}) &= 8x & & & \mathbb{E}_{\mathbb{P}}\left[g \mid x < \tfrac{1}{2}\right] &= \tfrac{5}{8} \\ \mathbb{Q}(x \mid x < \tfrac{1}{2}) &= \tfrac{8-8x}{3} & &\text{whence} & \mathbb{E}_{\mathbb{Q}}\left[g \mid x < \tfrac{1}{2}\right] &= \tfrac{19}{24} \\ \mathbb{S}(x \mid x < \tfrac{1}{2}) &\doteq \tfrac{8\alpha+8x(1-2\alpha)}{1+2\alpha} & & & \mathbb{E}_{\mathbb{S}}\left[g \mid x < \tfrac{1}{2}\right] &= \tfrac{1}{1+2\alpha}\left(\tfrac{7}{4}\alpha + \tfrac{5}{8}\right). \end{aligned}$$

So $\mathbb{E}_{\mathbb{S}}\left[g \mid x < \frac{1}{2}\right]$ is a non-linear function of $\mathbb{E}_{\mathbb{P}}\left[g \mid x < \frac{1}{2}\right]$ and $\mathbb{E}_{\mathbb{Q}}\left[g \mid x < \frac{1}{2}\right]$ when varying $\alpha$. In this case, the relationship is concave, but another example could have been chosen to make it convex or neither. The general relationship for any $\sigma$-algebra $\mathcal{K}$ in any measurable space is

$$\mathbb{E}_{\mathbb{S}}[g \mid \mathcal{K}] = (1-\alpha)\,\mathbb{E}_{\mathbb{P}}[g \mid \mathcal{K}]\,\frac{d\mathbb{P}}{d\mathbb{S}} + \alpha\,\mathbb{E}_{\mathbb{Q}}[g \mid \mathcal{K}]\,\frac{d\mathbb{Q}}{d\mathbb{S}}.$$

This fact does not support a linear extension of the relationship between games and players, nor even a convex or concave one. The partial map $(\mathcal{A},\mathcal{F}) \mapsto \mathcal{A} \times \mathcal{F}$ outlined by Definition 17.8 and encapsulated in the set **P** is inherently nonlinear. This nonlinearity results from the fact that conditioning essentially renormalizes a probability distribution, and the renormalization process is nonlinear by its nature.

The nonlinearity of this player-game map has implications. Firstly, any linear extension of the map to all tuples $(\mathcal{A},\mathcal{F})$ in some pair of vector spaces would induce a contradiction; whatever games are played by non-proper players therefore cannot be generated by linear operations on trajectors. Thus in general it is not true that $(\mathcal{A}+\mathcal{A}',\mathcal{F})$ is a pair of proper players for the summed game $\mathcal{A}\times\mathcal{F}+\mathcal{A}'\times\mathcal{F}$.

Secondly, and more importantly, nonlinearity leaves open the possibility that in sufficiently large spaces with sufficiently large temporal index sets (at least countable), stochastic optimization methods may outperform deterministic optimization methods on some problems. If the player-game map were linear in both arguments or even just convex, the Banach-Alaoglu Theorem could potentially be invoked to show that the unit ball of the search trajectors is weak-star compact. Then, since the V-performance is linear with respect to the optimization game, the Krein-Milman theorem would imply that every linear or weak-star convex functional takes its extreme value on the extreme points of a compact set, which for the proper search trajectors comprises the deterministic or atomistic search trajectors, as discussed in Chapter 6.

Wrapping up this section, the V-performance has been generalized to adversarial fitnesses, and it has been shown that, in fact, the value function $V = 1$ is sufficient to generate all possible versions of the V-performance, at the cost of introducing games against non-black-box adversarial fitnesses that can intuitively see the future. Using value functions, players have been defined for optimization games in which both the players and the games are not necessarily proper and may not satisfy black-box properties. It has also been shown that many of these players and games can be constructed from search generators and jagged value generators using Kolmogorov extension. Finally, Theorem 17.6 shows that a search trajector can play any game as long as its sets of measure zero align with those of the game, simultaneously providing a mechanism for constructing the adversarial fitness that generates that game.

This section has thus elaborated the relationship between players and games substantially. The final section of this chapter will now address the crucial question of alignment and duality between optimization players and adversarial fitnesses, between solutions and problems.

## 17.4 Problem-Solution Alignment in Optimization Games

The purpose of defining the optimization game and its players is to provide a general theoretical framework within which interesting questions about optimization or search performance can be posed and answered. This section poses some of these questions and provides some initial answers.

### *17.4.1 Defining Alignment*

Supposing that the V-performance represents a reasonable way to characterize the performance of a search or optimization process, relying as it does on assigning values to each history trace, according to the results of the previous section it suffices to represent a search or optimization problem as an arbitrary adversarial fitness, a search trajector from the value space $Y$ to the search domain $X$. This adversarial fitness carries the concept of value, the $V$ of V-performance, within itself, so that performance is just the overall measure of a game played against this adversarial fitness. Thus an adversarial fitness, jagged-black-box or otherwise, represents a *search* or *optimization problem*, for which a solution is sought among the black-box proper search trajectors.

The obvious metric of alignment between an optimization player $\mathcal{A}$ and an adversarial fitness $\mathcal{A}$ is the quantity $\langle \mathcal{A}, \mathcal{F} \rangle - \inf_{\mathcal{A}'} \langle \mathcal{A}', \mathcal{F} \rangle$. If $V$ is a normalizer of $\mathcal{F}$, then $V$ is plainly bounded, since $\mathcal{F}$ has finite norm by assumption. If $V$ is also progressively decomposable, then Theorem 17.3 and Corollary 17.1 demonstrate the properties that an optimization player must have to obtain alignment to $\mathcal{F}$ within an $\varepsilon > 0$ tolerance.

If one has a correct model of $\mathcal{F}$ and if $\mathcal{F}$ is well-behaved, $\varepsilon$-alignment might be achieved directly from Corollary 17.1 by Monte Carlo methods, that is, by sampling $\mathcal{F}$ some number of steps into the future and choosing the next search point as the search point whose outcomes are best. By "well-behaved", one means that $\mathcal{F}$ itself does not look arbitrarily far ahead, leading the Monte Carlo method into a trap in which the best samples are necessarily misleading. Given that infinite time is involved, there is no way to guarantee good behavior except by stipulating it. Stated otherwise, the general question of alignment with a problem, that is, with an adversarial fitness, is only interesting if the space of problems and the space of solutions is constrained somehow.

For this reason, when defining alignment, we consider a problem set $\mathcal{P} \subseteq \mathcal{MF}_{X^{\mathcal{T}},Y^{\mathcal{T}}}$ of search trajectors from $Y$ to $X$ and a solution set $\mathcal{S} \subseteq \mathcal{MF}_{Y^{\mathcal{T}},X^{\mathcal{T}}}$ of search trajectors from $X$ to $Y$. The problem set $\mathcal{P}$ contains all allowed adversarial fitnesses, and the solution set $\mathcal{S}$ contains the allowed optimization players. In any case, for all $\mathcal{A}$ and $\mathcal{F}$ both with finite norm, $|\langle \mathcal{A}, \mathcal{F} \rangle| \leq \|\mathcal{A}\| \, \|\mathcal{F}\| < \infty$, justifying the following definitions.

One reasonable constraint is that the adversarial fitness should have a normalized form that is a jagged-black-box search trajector for which the normalizer is progressively decomposable. This scenario is sufficiently broad to cover all of the adversarial fitnesses discussed in Section 17.2.2, since each of these classes only allow dependence on prior search points in order to assign a value.

Further possible constraints are to specify any one of the five classes of adversarial fitnesses defined in Section 17.2.2: static, randomized, stochastic, dynamic, and adaptive, and to require a particular value function, such as bounded versions of the running time or the final or averaged value. Each of these constraints incorporates the broader constraint from the previous paragraph and represents a point of attack for analyzing alignment.

**Definition 17.12 (Admissibility of Problem-Solution Pairs).** A problem-solution pair $(\mathcal{P}, \mathcal{S})$ consisting of a problem set and a solution set is admissible if for all $\mathcal{A} \in \mathcal{S}$ and all $\mathcal{F} \in \mathcal{P}$, $\mathcal{A}$ is proper and the pair $(\mathcal{A}, \mathcal{F})$ is a pair of players for an optimization game $\mathcal{A} \times \mathcal{F}$.

**Definition 17.13 (Alignment).** Given an admissible problem-solution pair $(\mathcal{P}, \mathcal{S})$, the alignment of a search trajector $\mathcal{A} \in \mathcal{S}$ with a search trajector $\mathcal{F} \in \mathcal{P}$ is $\mathfrak{A}(\mathcal{A}, \mathcal{F}) = 1$ if $\sup_{\mathcal{A}' \in \mathcal{S}} \langle \mathcal{A}', \mathcal{F} \rangle = \inf_{\mathcal{A}' \in \mathcal{S}} \langle \mathcal{A}', \mathcal{F} \rangle$ and otherwise

$$\mathfrak{A}(\mathcal{A}, \mathcal{F}) = \frac{\langle \mathcal{A}, \mathcal{F} \rangle - \inf_{\mathcal{A}' \in \mathcal{S}} \langle \mathcal{A}', \mathcal{F} \rangle}{\sup_{\mathcal{A}' \in \mathcal{S}} \langle \mathcal{A}', \mathcal{F} \rangle - \inf_{\mathcal{A}' \in \mathcal{S}} \langle \mathcal{A}', \mathcal{F} \rangle}.$$

The function $\mathfrak{A}$ computing the alignment is called the *alignment gauge* for $\mathcal{P}$ and $\mathcal{S}$, and it necessarily takes values in $[0, 1]$.

The restriction to proper optimization players prevents the alignment from depending on scaling effects from $\mathcal{A}$; it is reasonable to do so, since only proper optimization players can be sampled in a probabilistic sense. The adversarial fitness, on the other hand, may be non-proper, primarily in order to represent an absorbed value function.

In the degenerate situation where $\sup_{\mathcal{A}' \in \mathcal{S}} \langle \mathcal{A}', \mathcal{F} \rangle = \inf_{\mathcal{A}' \in \mathcal{S}} \langle \mathcal{A}', \mathcal{F} \rangle$ for some fixed $\mathcal{F}$, the alignment has $\mathfrak{A}(\mathcal{A}, \mathcal{F}) = 1$ for all $\mathcal{A} \in \mathcal{S}$. This situation evokes the concept of Weak NFL, a term that has previously been used to describe the situation where the average performance on a problem is constant across optimization method with respect to some particular performance criterion. This observation is captured in the next definition.

**Definition 17.14 (Adversarial Weak No Free Lunch).** An adversarial fitness function $\mathcal{F}$ has the *weak NFL property* relative to a solution set $\mathcal{S}$ if $\sup_{\mathcal{A}' \in \mathcal{S}} \langle \mathcal{A}', \mathcal{F} \rangle =$

$\inf_{\mathcal{A}' \in \mathcal{S}} \langle \mathcal{A}', \mathcal{F} \rangle$. A problem set $\mathcal{P}$ is said to have the weak NFL property relative to $\mathcal{S}$ if there exists $\mathcal{F} \in \mathcal{P}$ with $\mathcal{F} \neq 0$ such that $\mathcal{F}$ has the weak NFL property relative to $\mathcal{S}$.

Recall that the strong version of No Free Lunch guarantees not just constant performance but a constant probability distribution over value histories. This topic is taken up next.

### *17.4.2 Adversarial No Free Lunch*

Chapter 12 addressed the case where the adversarial fitness is a randomized static fitness satisfying No Free Lunch requirements, for which *all* black-box optimization players are perfectly aligned with the adversarial fitness. In that chapter, NFL as a property of optimization histories was reduced to path independence, which is a property of optimization problems without referring to optimization methods at all. In this subsection, these results will be extended to the optimization game, where path independence remains equivalent to NFL given sufficiently wide solution sets. It is then shown that for stochastic fitness function, NFL prior still exist.

Considering the proof of equivalence between NFL and path independence, one should notice that there are two key elements. First is the avoidance of repeating search histories, which is specific to the setting of a randomized static fitness function, and second is the use of deterministic optimization methods that ignore the optimization objective. By generalizing these two elements and adapting the setting from static search trajectors to standard search trajectors, these proofs can be expanded to give a general proof of equivalence between NFL and path independence in the optimization game.

**Definition 17.15 (Adversarial P-NFL).** A proper adversarial fitness function $\mathcal{F}$ has the P-NFL property relative to a solution set $\mathcal{S}$ if the problem-solution pair $(\{\mathcal{F}\}, \mathcal{S})$ is admissible and the function $\mathcal{A} \mapsto (\mathcal{A} \times \mathcal{F}) \circ \phi_Y$ is constant on $\mathcal{S}$. A problem set $\mathcal{P}$ is said to have the P-NFL property relative to $\mathcal{S}$ if there exists $\mathcal{F} \in \mathcal{P}$ such that $\mathcal{F}$ has the P-NFL property relative to $\mathcal{S}$.

The NFL property of Chapter 12 is recovered from this definition by considering the solution set $\mathcal{S}$ of almost surely non-repeating, black-box proper optimization players together with the problem set of randomized fitness functions, and Theorem 12.10 demonstrated that this problem set has the NFL property for this solution set, regardless of the search space or value domain. The P-NFL property implies Weak P-NFL, subject to value dependence.

**Definition 17.16 (Value Dependence).** A value function $V : (X \times Y)^{\mathcal{T}} \to \mathbb{R}$ is value-dependent if $V(h) = V(h')$ whenever $h_Y = h'_Y$. In this case, $\hat{V} : Y^{\mathcal{T}} \to \mathbb{R}$ indicates the function such that $\hat{V}(h_Y) = V(h)$ for all $h \in (X \times Y)^{\mathcal{T}}$.

**Theorem 17.7 (Strong NFL Implies Weak NFL).** *Suppose a search trajector $\mathcal{F}$ from $Y$ to $X$ has the P-NFL property relative to a solution set $\mathcal{S}$. If $\mathcal{F}'$ is any search trajector such that $\mathcal{F}$ is a normalized form of $\mathcal{F}'$ with a value-dependent normalizer $V_{\mathcal{F}}$, then $\mathcal{F}'$ has the weak P-NFL property relative to $\mathcal{S}$.*

*Proof.* Since $(\{\mathcal{F}\},\mathcal{S})$ is an admissible problem-solution pair, for all $\mathcal{A} \in \mathcal{S}$, $(\mathcal{A},\mathcal{F})$ is a pair of proper players for a proper game $\mathcal{A}\times\mathcal{F}$. Therefore $(\mathcal{A},\mathcal{F}')$ is a pair of players for the game $\mathcal{A}\times\mathcal{F}'$ defined by $\mathcal{A}\times\mathcal{F}'(H) = \int_H V_{\mathcal{F}}\, d\mathcal{A}\times\mathcal{F}$. So $(\{\mathcal{F}'\},\mathcal{S})$ is an admissible problem-solution pair.

Because $V_{\mathcal{F}}$ is value-dependent, it is $\Phi_Y$-measurable. Consequently, $\mathop{\mathbb{E}}_{\mathcal{A}\times\mathcal{F}}[V_{\mathcal{F}}] = \mathop{\mathbb{E}}_{(\mathcal{A}\times\mathcal{F})\circ\phi_Y}\left[\hat{V}_{\mathcal{F}}\right]$. For any $\mathcal{A},\mathcal{A}' \in \mathcal{S}$, the P-NFL property implies

$$\left\langle \mathcal{A},\mathcal{F}'\right\rangle = \mathop{\mathbb{E}}_{(\mathcal{A}\times\mathcal{F})\circ\phi_Y}\left[\hat{V}_{\mathcal{F}}\right] = \mathop{\mathbb{E}}_{(\mathcal{A}'\times\mathcal{F})\circ\phi_Y}\left[\hat{V}_{\mathcal{F}}\right] = \left\langle \mathcal{A}',\mathcal{F}'\right\rangle,$$

so $\sup_{\mathcal{A}'\in\mathcal{S}}\langle\mathcal{A}',\mathcal{F}\rangle = \inf_{\mathcal{A}'\in\mathcal{S}}\langle\mathcal{A}',\mathcal{F}\rangle$ and $\mathcal{F}'$ has the weak P-NFL property.

Thus the P-NFL property is a stronger property than weak P-NFL, since a P-NFL adversarial fitness forces all games to produce the same distribution over value histories across a bank of optimization methods. Weak P-NFL, by contrast, merely requires that the average evaluation of a history trace be insensitive to the choice of optimization method. The caveat is that if evaluation depends on the search history, P-NFL does not imply weak P-NFL.

The same strategy as in Chapter 12 is used to show that P-NFL is equivalent to path independence and P-NFL adversarial fitnesses exist.

**Definition 17.17 (Path Independence).** A proper adversarial fitness function $\mathcal{F}$ is path independent on a set $A \subseteq X^{\mathcal{T}}$ of search histories if $\mathcal{F}|_A$ is a constant function, *i.e.*, there exists a probability measure $\mathbb{P}$ on $Y^{\mathcal{T}}$ such that for all $z \in A$, $\mathcal{F}[z] = \mathbb{P}$. If $\mathcal{F}$ is not path independent, it is path dependent.

**Proposition 17.9.** *If $\mathcal{A}$ is a proper, deterministic (Definition 6.9) and value-agnostic (Definition 6.21) search trajector from $X$ to $Y$ and $\mathcal{F}$ is a proper search trajector from $Y$ to $X$, there exist $z \in X^{\mathcal{T}}$ and a game $\mathcal{A}\times\mathcal{F}$ with $(\mathcal{A}\times\mathcal{F})\circ\phi_Y = \mathcal{F}[z]$ for which $(\mathcal{A},\mathcal{F})$ is a pair of proper players.*

*Proof.* By determinism, for each $y \in Y^{\mathcal{T}}$ there exists $z_y \in X^{\mathcal{T}}$ such that $\mathcal{A}[y](A) = \mathbb{1}_A(z_y)$. Since $\mathcal{A}$ is value-agnostic, the map $y \mapsto z_y$ is constant. Let $z$ be this constant value. Suppose $(\mathcal{A},\mathcal{F})$ plays a game $\mathcal{A}\times\mathcal{F}$. For $B \in Y^{\mathcal{T}}$,

$$\begin{aligned}\mathcal{A}\times\mathcal{F}(\phi_Y^{-1}(B)) &= \mathbb{E}_{\mathcal{A}\times\mathcal{F}}[\mathcal{F}[\phi_X](B)] = \mathbb{E}_{\mathcal{A}\times\mathcal{F}}\left[\mathbb{E}_{\mathcal{A}\times\mathcal{F}}[\mathcal{F}[\phi_X](B) \mid \Phi_X]\right] \\ &= \mathbb{E}_{\mathcal{A}\times\mathcal{F}}\left[\mathbb{E}_{\mathcal{A}[\phi_Y]}[\mathcal{F}[\phi_X](B)]\right] = \mathcal{F}[z](B).\end{aligned}$$

It remains to show that $\mathcal{A}\times\mathcal{F}$ exists. For $A \in X^{\mathcal{T}}$, let $\mathcal{A}\times\mathcal{F}(A\times B) = \mathbb{1}_z(A)\mathcal{F}[z](B)$, which defines $\mathcal{A}\times\mathcal{F}$ as a probability measure by Carathéodory extension. Then $\mathop{\mathbb{E}}_{\mathcal{A}\times\mathcal{F}}\left[\mathbb{1}_{\phi_X^{-1}(A)} \middle| \Phi_Y\right](h) = \mathbb{1}_A(z) = \mathcal{A}[h_Y](A)$ and $\mathop{\mathbb{E}}_{\mathcal{A}\times\mathcal{F}}\left[\mathbb{1}_{\phi_Y^{-1}(B)} \middle| \Phi_X\right](h) = \mathcal{F}[z](B)$.

**Definition 17.18 (Deterministically Complete Solution Set).** A solution set $\mathcal{S}$ is *deterministically complete* with respect to a set $A \subseteq X^{\mathfrak{T}}$ of search histories if for every $\mathcal{A} \in \mathcal{S}$, $\phi_X^{-1}(X^{\mathfrak{T}} \setminus A) \in \mathcal{Z}[\mathcal{A}]$ and for all $z \in A$, there exists $\mathcal{A}_z \in \mathcal{S}$ such that for all $y \in Y^{\mathfrak{T}}$, $\mathcal{A}_z[y](A) = \mathbb{1}_A(z)$. I.e., for all $z \in A$, $\mathcal{S}$ contains a deterministic, value-agnostic search trajector whose deterministic core is $z$.

**Lemma 17.2.** *Suppose $\mathcal{S}$ is a deterministically complete solution set with respect to a set A. If an adversarial fitness function has the P-NFL property relative to $\mathcal{S}$, then it is path independent on A.*

*Proof.* If $\mathcal{F}$ has the P-NFL property relative to $\mathcal{S}$ but is path dependent on $A$, then there exist search histories $z_1, z_2 \in A$ such that $\mathcal{F}[z_1] \neq \mathcal{F}[z_2]$. But since $\mathcal{S}$ is deterministically complete on $A$, there exist $\mathcal{A}_{z_1}, \mathcal{A}_{z_2} \in \mathcal{S}$ such that $z_1$ and $z_2$ respectively are the deterministic cores of each. Therefore by the P-NFL property and Proposition 17.9 applied to $\mathcal{A}_{z_1}$ and $\mathcal{A}_{z_2}$,

$$\mathcal{F}[z_1] = (\mathcal{A}_{z_1} \times \mathcal{F}) \circ \phi_Y = (\mathcal{A}_{z_2} \times \mathcal{F}) \circ \phi_Y = \mathcal{F}[z_2],$$

which is a contradiction. So $\mathcal{F}$ is path independent with respect to $A$.

**Proposition 17.10.** *Suppose $(\mathcal{A}, \mathcal{F})$ is a proper pair of players for a proper optimization game $\mathcal{A} \times \mathcal{F}$. If $\mathcal{F}$ is path independent on a set A, then for any $z_0 \in A$ and $B \subseteq Y^{\mathfrak{T}}$*

$$\mathcal{A} \times \mathcal{F}(A \times B) = \int_B \mathcal{A}[y](A)\, \mathcal{F}[z_0](dy).$$

*Proof.* First note that for any $h_0$ with $h_0^X = z_0 \in A$,

$$\int_B \mathcal{A}[y](A)\, \mathcal{F}[z_0](dy) = \mathop{\mathbb{E}}_{\mathcal{A}\times\mathcal{F}} \left[ \mathbb{1}_{\phi_Y^{-1}(B)} \mathop{\mathbb{E}}_{\mathcal{A}\times\mathcal{F}} \left[ \mathbb{1}_{\phi_X^{-1}(A)} \middle|\, \Phi_Y \right] \middle|\, \Phi_X \right] (h_0)$$

Next, since the right hand side is constant for all $h_0 \in A \times B$,

$$\int_B \mathcal{A}[y](A)\, \mathcal{F}[z_0](B) = \mathop{\mathbb{E}}_{\mathcal{A}\times\mathcal{F}} \left[ \mathbb{1}_{\phi_Y^{-1}(B)} \mathop{\mathbb{E}}_{\mathcal{A}\times\mathcal{F}} \left[ \mathbb{1}_{\phi_X^{-1}(A)} \middle|\, \Phi_Y \right] \right] = \mathop{\mathbb{E}}_{\mathcal{A}\times\mathcal{F}} \left[ \mathbb{1}_{\phi_Y^{-1}(B)} \mathbb{1}_{\phi_X^{-1}(A)} \right],$$

where the final quantity is equal to $\mathcal{A} \times \mathcal{F}(A \times B)$ by definition.

**Lemma 17.3.** *Suppose $\mathcal{S}$ is a deterministically complete solution set with respect to a set A. If an adversarial fitness function $\mathcal{F}$ is path independent on A and $(\{\mathcal{F}\}, \mathcal{S})$ is an admissible problem-solution pair, then $\mathcal{F}$ has the P-NFL property relative to $\mathcal{S}$.*

*Proof.* Fix $B \subseteq Y^{\mathfrak{T}}$ and $\mathcal{A} \in \mathcal{S}$. Because $\mathcal{S}$ is deterministically complete on $A$, it follows that $\phi_X^{-1}(X^{\mathfrak{T}} \setminus A) \in \mathcal{Z}[\mathcal{A}]$. By Proposition 17.8, $\mathcal{A} \times \mathcal{F}$ is compatible with $\mathcal{A}$. Consequently, $\mathcal{A} \times \mathcal{F}(\phi_Y^{-1}(B)) = \mathcal{A} \times \mathcal{F}(A \times B)$ since $(X^{\mathfrak{T}} \setminus A) \times B$ has measure zero. Then $\mathcal{A}[y](A) = \mathcal{A}[y](X^{\mathfrak{T}}) = 1$ since $\phi_X^{-1}(X^{\mathfrak{T}} \setminus A) \in \mathcal{Z}[\mathcal{A}]$. Proposition 17.10 then implies

$$\mathcal{A} \times \mathcal{F}(\phi_Y^{-1}(B)) = \int_B \mathcal{A}[y](A)\, \mathcal{F}[z_0](dy) = \mathcal{F}[z_0](B)$$

for any $z_0 \in A$. That is, $\mathcal{F}$ has the P-NFL property relative to $\mathcal{S}$.

**Theorem 17.8 (NFL Identification Theorem).** *Suppose $\mathcal{S}$ is a deterministically complete solution set with respect to a set A. A proper adversarial fitness $\mathcal{F}$ has the P-NFL property if and only if $\mathcal{F}$ is path independent on A and $(\{\mathcal{F}\}, \mathcal{S})$ is an admissible problem-solution pair.*

*Proof.* If $\mathcal{F}$ has the P-NFL property, then this property implies that $(\{\mathcal{F}\}, \mathcal{S})$ is admissible, and Lemma 17.2 implies that $\mathcal{F}$ is path independent on $A$. Conversely, if $\mathcal{F}$ is path independent on $A$ and $(\{\mathcal{F}\}, \mathcal{S})$ is admissible, Lemma 17.3 completes the proof.

Theorem 17.8 generalizes Theorem 12.7 to the optimization game and provides a tool to demonstrate the existence of NFL in a wider variety of optimization and search settings. The existence of NFL in the adversarial fitness case is immediate from the definition of path independence. One needs only a probability distribution $\mathbb{P}$ over $Y^{\mathcal{T}}$, and then by setting $\mathcal{F}[z] = \mathbb{P}$ for all $z \in X^{\mathcal{T}}$ a path independent adversarial fitness is immediately obtained. Such probability distributions always exist, and thus NFL always exists as well.

**Theorem 17.9 (NFL Exists).** *Suppose $\mathcal{S}$ is a deterministically complete solution set with respect to a set A. There exists a proper adversarial fitness function $\mathcal{F}$ with the P-NFL property relative to the solution set*

$$\mathcal{S}' = \{\mathcal{A} \in \mathcal{S} \mid (\mathcal{A}, \mathcal{F}) \textit{ is a pair of proper players for some proper game } \mathcal{A} \times \mathcal{F}\}.$$

*Proof.* As stated in the text, let $\mathbb{P}$ be an arbitrary probability measure on $Y^{\mathcal{T}}$ and define $\mathcal{F}[z] = \mathbb{P}$ for all $z \in X^{\mathcal{T}}$, and then $\mathcal{F}$ is path independent with respect to every subset of $X^{\mathcal{T}}$. Furthermore, $(\{\mathcal{F}\}, \mathcal{S}')$ is an admissible problem-solution pair. Notice that by Proposition 17.9, if $\mathcal{A}_z \in \mathcal{S}$ is a proper, deterministic, and value-agnostic, then $(\mathcal{A}_z, \mathcal{F})$ is a pair of proper players for a proper game. Therefore, every proper, deterministic, and value-agnostic search trajector in $\mathcal{S}$ is also in $\mathcal{S}'$, whence $\mathcal{S}'$ is deterministically complete. By Theorem 17.8, $\mathcal{F}$ has the P-NFL property relative to $\mathcal{S}'$.

In Chapter 12, the concept of identically distributed and mutually independent coordinates was needed in order to prove the existence of NFL. These properties were necessary because that chapter considered distributions over $Y^X$ instead of $Y^{\mathcal{T}}$; identical distributions and mutually independent coordinates guarantee that mapping from $X^{\mathcal{T}} \times Y^X \to Y^{\mathcal{T}}$ does not introduce dependence on the search history.

This existence theorem guarantees the existence of an adversarial fitness with the P-NFL property but does not guarantee that the adversarial fitness has any particular properties. For example, if one wishes to consider only adversarial fitnesses that are derived from random static fitnesses as in Section 17.2.2 and as studied in Chatper 12, then path independence of the adversarial fitness can only result when repeating histories are excluded or the adversarial fitness is constant. Otherwise, repetitions in the search history induce path dependence, because the same search point must be assigned the same value.

Whereas NFL for randomized static fitnesses requires one to exclude repeating search histories to obtain path independence, stochastic fitnesses as described in Section 17.2.2 can have the NFL property even over repeating search histories. Below is a proof based again on identically distributed and mutually independent coordinates.

**Definition 17.19 (Identically Distributed and Mutually Independent).** Let $\mathbb{S}$ be a probability measure on $\Omega \times Y^{X\times\Omega}$. Define $S_x : \Omega \times Y^{X\times\Omega} \to Y$ so that $S_x(\omega, u) = u(x,\omega)$ and $S_D : \Omega \times Y^{X\times\Omega} \to Y^D$ for $D \subseteq X$ finite so that $S_D(\omega, u) = (u(x,\omega))_{x\in D}$. Let $\mathbb{S}_x = \mathbb{S} \circ S_x$ and $\mathbb{S}_D = \mathbb{S} \circ S_D$. Then $\mathbb{S}$ is said to have identically distributed coordinates if $\mathbb{S}_x = \mathbb{S}_{x'}$ for all $x, x' \in X$, and $\mathbb{S}$ has mutually independent coordinates if $\mathbb{S}_D\left(\prod_{x\in D} E_x\right) = \prod_{x\in D} \mathbb{S}_x(E_x)$.

**Lemma 17.4.** *Let $\mathcal{F}$ be an adversarial fitness derived from a stochastic fitness measure $\mathbb{S}$ on $\Omega \times Y^{X\times\Omega}$ as described in Section 17.2.2. If $\mathbb{S}$ has identically distributed and mutually independent coordinates and $\mathbb{S}_x$ is tight for each $x \in X$, then $\mathcal{F}$ is path independent on $X^{\mathcal{T}}$.*

*Proof.* If $\mathbb{S}$ has identically distributed and mutually independent coordinates and $\mathbb{S}_x$ is tight, by Kolmogorov extension there is a probability measure $\mathbb{S}_X$ on $Y^X$ whose finite dimensional distributions are given by $\mathbb{S}_D$ for finite $D$. It is necessary to pass from a measure on $Y^X$ to a measure on $Y^{\mathcal{T}}$. The proof of Theorem 12.8 uses permutations for this purpose since the search histories there are non-repeating. With repeating histories allowed, permutations are not appropriate, and a more direct route may be followed. Instead, note

$$\mathcal{F}[z](B) = \mathbb{S}_X\left(\left\{u \in Y^X \mid \exists y \in B \text{ s.t. } \forall t \in \mathcal{T}, y_t = u(z_t)\right\}\right),$$

which is analogous to Equation 12.23. Let $\mathcal{K} \subset \mathcal{T}$ be finite and $\phi_{\mathcal{K}} : Y^{\mathcal{T}} \to Y^{\mathcal{K}}$ the canonical projection from $Y^{\mathcal{T}}$ to $Y^{\mathcal{K}}$. Let $\mathcal{B} = (B_k)_{k\in\mathcal{K}}$ be a sequence of measurable subsets of $Y$, *i.e.*, $B_k \subseteq Y$. Then for $D(z) = (z_k)_{k\in\mathcal{K}}$

$$\mathcal{F}[z]\left(\phi_{\mathcal{K}}^{-1}\left(\prod_{k\in\mathcal{K}} B_k\right)\right) = \mathbb{S}_{D(z)}\left(\prod_{k\in\mathcal{K}} B_k\right) = \prod_{k\in\mathcal{K}} \mathbb{S}_{x_0}(B_k),$$

where $x_0 \in X$ is arbitrary. Thus for all $z, z' \in X^{\mathcal{T}}$, all finite $\mathcal{K} \subseteq \mathcal{T}$, and all $\mathcal{B}$,

$$\mathcal{F}[z]\left(\phi_{\mathcal{K}}^{-1}\left(\prod_{k\in\mathcal{K}} B_k\right)\right) = \mathcal{F}[z']\left(\phi_{\mathcal{K}}^{-1}\left(\prod_{k\in\mathcal{K}} B_k\right)\right)$$

and so $\mathcal{F}[z] = \mathcal{F}[z']$. Therefore $\mathcal{F}$ is path independent on $X^{\mathcal{T}}$.

**Theorem 17.10.** *Let $\mathcal{S}$ be the solution set containing all proper black-box search trajectors from $X$ to $Y$. There exists a proper adversarial fitness $\mathcal{F}$ derived from a stochastic fitness measure that has the P-NFL property with respect to*

$$\mathcal{S}' = \{\mathcal{A} \in \mathcal{S} \mid (\mathcal{A}, \mathcal{F}) \textit{ is a proper pair of players for a game } \mathcal{A} \times \mathcal{F}\}.$$

*Proof.* $\mathcal{S}$ is deterministically complete, since for each $z \in X^{\mathcal{T}}$ there exists a proper, deterministic and value-agnostic search trajector $\mathcal{A}_z$ whose deterministic core is $z$, and by Proposition 17.9, for any proper adversarial fitness $\mathcal{F}$, $(\mathcal{A}_z, \mathcal{F})$ is a pair of proper players for a game $\mathcal{A}_z \times \mathcal{F}$. Every deterministic, value-agnostic search trajector has the black-box property, and so $\mathcal{A}_z \in \mathcal{S}'$.

Let $\mathbb{P}$ be any probability measure over $\Omega$, and let $\mathbb{Q}$ be any tight probability measure over $Y$, *e.g.* $\mathbb{Q}(B) = \mathbb{1}_B(y_0)$ for $B \subseteq Y$ and some fixed $y_0 \in Y$. Let $\tilde{D} \subseteq X \times \Omega$ be finite and $\phi_{\tilde{D}} : Y^{X\times\Omega} \to Y^{\tilde{D}}$ the canonical projection. Choose $(B_{x,\omega})_{(x,\omega)\in\tilde{D}}$ with $B_{x,\omega} \subseteq Y$. Define $\mathbb{S}^{\mathbb{Q}}$ for $E \subseteq \Omega$ and $(B_{x,\omega})_{(x,\omega)\in\tilde{D}}$ by

$$\mathbb{S}^{\mathbb{Q}}\left(E \times \phi_{\tilde{D}}^{-1}\left(\prod_{(x,\omega)\in\tilde{D}} B_{x,\omega}\right)\right) \quad = \quad \mathbb{P}(E) \prod_{(x,\omega)\in\tilde{D}} \mathbb{Q}(B_{x,\omega}),$$

which is a complete definition of $\mathbb{S}^{\mathbb{Q}}$ by Kolmogorov extension. Observe that $\mathbb{S}^{\mathbb{Q}}_x(B) = \mathbb{Q}(B)$ and that for finite $D \subseteq X$, $\mathbb{S}^{\mathbb{Q}}_D\left(\prod_{x\in D} B_x\right) = \prod_{x\in D} \mathbb{S}^{\mathbb{Q}}_x(B_x)$. That is, $\mathbb{S}^{\mathbb{Q}}$ has identically distributed and mutually independent coordinates.

Lemma 17.4 proves that $\mathcal{F}$ derived from $\mathbb{S}^{\mathbb{Q}}$ is path independent on $X^{\mathcal{T}}$, and Theorem 17.8 establishes that $\mathcal{F}$ has the P-NFL property on $\mathcal{S}$.

It has thus been shown that the NFL results of Chapter 12 also hold for the optimization game in arbitrary spaces. It has also been shown specifically that the problem set of stochastic fitnesses has the NFL property with respect to the solution set of all proper black-box optimization methods. Theorem 17.7 further connects this result back to the concept of alignment based on the V-performance as in Definition 17.13.

### 17.4.3 Practically Optimizing Alignment

The key concept from the NFL results in the last subsection is that whenever an adversarial fitness is non-constant, meaning whenever $\mathcal{F}[z] \neq \mathcal{F}[z']$ for some $z, z' \in X^{\mathcal{T}}$, there is an opportunity for some optimization method to exploit these differences in order to achieve an advantage on some performance criterion. In this final subsection, we consider how these differences may be exploited.

In fact, quite plainly, deterministic and value-agnostic optimization methods are sufficient to exploit this advantage; one need only choose $z_\varepsilon \in X^{\mathcal{T}}$ such that

$$\left| \mathbb{E}_{\mathcal{F}[z_\varepsilon]}\left[\hat{V}\right] - \inf_z \mathbb{E}_{\mathcal{F}[z]}\left[\hat{V}\right] \right| < \varepsilon$$

to minimize a value-dependent value function $V$ within tolerance $\varepsilon$. Assuming value dependence, $\mathbb{E}_{\mathcal{A}_z\times\mathcal{F}}[V] = \mathbb{E}_{\mathcal{F}[z]}\left[\hat{V}\right]$ by Proposition 17.9. Therefore one can trivially conclude that for a fixed adversarial fitness $\mathcal{F}$, there is a deterministic, value-agnostic search trajector $\mathcal{A}_{z_\varepsilon}$ that attains near-optimal value against the adversarial fitness.

Of course, this method of choosing $\mathcal{A}_{z\varepsilon}$ is cheating. The result is only valid for one adversarial fitness, and finding it requires either exhaustive search or complete knowledge of the adversarial fitness. For practical reasons, one wishes to discover optimization methods that perform robustly across a wide variety of realistic problem scenarios under the assumption that the exact scenarios cannot be known in advance.

The result of Theorem 17.3 already provides a high-level approach to efficient search and optimization, and given the breadth of the scenarios incorporated into the optimization game, one should not expect broader results without constraining the problem. The challenges of implementing the results of this theorem include the need to model the adversarial fitness and value function accurately and the requirement to perform a nested optimization over all possible rollouts of the future trajectory. Except in highly constrained environments, these challenges can only be addressed approximately or intuitively at present.

Chapters 14, 15, and 16 followed a similarly inspired approach based on martingale optimization as introduced in Section 13.4.3. The fitness function was modeled as piecewise constant over region partitions determined by the current history trace, and the overall optimization was a variant of simulated annealing. Rather than relying on nested optimization, however, the search point selection step was performed by sampling the annealed model. Also, the value function was not incorporated into the search point selection.

Chapter 13 laid out several realistic scenarios whose constraints suggest ways of aligning optimization methods with problems. For example, the diffusion prior (Section 13.1.2) imposed a constraint of weak locality, in which the observed values are more likely to be close for search points that are close. The universal prior (Section 13.1.3) addressed the constraint of compressibility or representation by a short program. Many if not most problems encountered in practice exhibit both locality and compressibility. The two are not distinct; locality is a form of compressibility, since it enables prediction based on interpolation of nearby points.

In general, effective search and optimization strategies leverage available constraints in order to restrict the search to points that are likely to yield desirable values. In the highly constrained case of optimizing convex, differentiable functions, the global optimum can be quickly located by estimating gradients and following them. If these same methods are applied to functions that are non-convex but still differentiable, only local optima are found by gradient methods. In other practical scenarios, such as when evaluations are stochastic and the underlying objective function is non-convex and non-differentiable in some places, gradient methods can fail entirely. The success of gradient methods comes entirely from the strong constraints placed on the objective function.

Any effective general-purpose optimization method must make assumptions at least as specific as those of the universal prior, namely, that the problem being solved has at least some predictable structure, and that simpler structures are more likely than complex structures.

From this observation, one can propose a rubric for problem solving that seeks first to answer broad questions about the problem and then gradually funnels down to one of a few scenarios upon which a well performing method can be applied. The first of such questions might include:

1. Is the objective function deterministic or stochastic?
2. Is the objective function static or dynamic?
3. Is the objective function adaptive?

From these questions one can then delve further. For example, if the objective is deterministic, then one might ask next whether it is continuous, and then whether it is differentiable. All of the scenarios above occur in practical problems, and at a high level, the type of solution appropriate to each case is widely divergent. For differentiable functions, gradient methods are appropriate. For adaptive objective function – that is, for full adversarial fitnesses – some form of adaptation must be employed by the optimization method; it is here that evolutionary methods become appropriate.

Many researchers focus on either particular problem classes or particular solution types, often forgetting that other problem classes and solution types exist. For example, someone accustomed to problems expressed as differentiable formulae may have difficulty believing that evolutionary methods can be used to optimize effectively in any scenario. However, if the problem being solved changes in unexpected and unpredictable ways, the solution must be able to adapt as well. This case occurs when the problem incorporates the decisions or actions of another agent. Adaptive problems are thus commonly encountered when dealing with living or thinking things, such as in social sciences, management, government, economics, games, and biology.

Conversely, when highly adaptive methods are applied to highly structured problems, there is no reason to believe that they will work efficiently. It is wasteful not to employ known structure within a search or optimization process. As has been known at least since the discovery of No Free Lunch, not every optimization method solves every problem.

And yet real problems are not random problems; they come from somewhere. They are derived from fundamental physical laws with sometimes more and sometimes less layers of indirection. Consequently, there is hope that general-purpose methods can be found for solving a wide range of problems. Given that the one of the primary emergent features of the real world is in fact adaptation within structure, one might even conjecture that biological evolution is just such a general-purpose method, a method that constructs machines to solve new problems as they develop, ultimately building thinking machines who conceptualize and subdivide problems to discover new and innovative solutions. In this process, new adaptations result, new problems are formed, and old problems recede or change.

Ultimately, a formalism as abstract as the optimization game is as much as tool for conceiving new problems and solutions as it is a unification of existing problems. Its primary usage is to subdivide problems and solutions, identifying potential matches and mismatches. Embracing convex optimization, model-building machine learning, evolutionary methods, reinforcement learning, and game theory, the optimization game provides a roof for a great warehouse of problems and solution. It remains to provide better theoretical tools for building rooms within this warehouse and expressing more refined pathways from problems to solutions.

## 17.5 Conclusion

This chapter introduced the concept of the optimization game as a general framework for understanding search and optimization problems and how they interact with solutions. By allowing the problems themselves to be treated as agents, the whole scenario of search and optimization was expanded into a zero-sum game. The unfolding of this game was encapsulated as a probability distribution over history traces containing all decisions made in the course of the game. In a wide range of cases, it was shown that by modeling the game agents as search trajectors, the probability distribution encapsulating the game could be derived as a function of these search trajectors, which themselves can be derived from the sequential decisions over the course of the game represented by search generators. Thus the optimization game represents a theory of both solutions and problems, as opposed to previous chapters, in which the problem was mostly assumed to be static and taken as a given.

The kinds of problems represented by adversarial fitness include static, dynamic, stochastic, and adaptive objective functions. At this level of generality, one might not expect specific results, and yet it is still possible to observe that the optimal solution to a given problem is described by Bellman-type equations as in Theorem 17.3. Given the generality of the problems described, however, little practical advantage is provided by this result without somehow constraining the range of problems and solutions.

Furthermore, No Free Lunch theorems naturally extend to even this general setting through the concept of path independence, and the existence of NFL can even be proven for stochastic optimization if not also for broader problem classes. Yet the very nature of path independence implies that general-purpose optimization is possible unless the problem is independent of all search actions. As stated in Chapter 12, the world of NFL is an impractical world of either extreme flatness or random fuzz. Since neither of these properties seem to hold broadly in the real world, there must exist one or more general-purpose optimization methods that solve a wide range of problems effectively, even if they are less efficient on particular problems with known structure.

In sum, this chapter has been about taming the wide variety of problems and solutions encountered by across the spectrum of optimization. Though there remains much that is still wild and untamed, many basic principles of how to approach problems and what kinds of problems exist flow naturally from observing the rough structure of the optimization game. Future investigations may reveal further structure still; these investigations are now left to the reader.

# Chapter 18
# Conclusion

The common thread running through this book is the concept that search and optimization problems and solutions themselves can be analyzed and manipulated as mathematical objects. In this context, the main contributions of this book are (1) the formalization of iterative stochastic optimization methods leading to theoretical results such as the No Free Lunch Identification Theorems and the Information-Maximization Principle, including the introduction of the optimization game as a template for formalizing search and optimization problems; (2) harnessing the Information Maximization Principle in order to generate effective new optimization methods such as evolutionary annealing that operate in a wide range of topological spaces from bit strings to Euclidean space to neural networks; and (3) an analysis of optimization performance in many settings that suggests general-purpose optimizers exist for real-world problems. One overarching theme is the mathematization of computer programs within an abstract setting, for which the formalization of stochastic optimization is merely one instance and to which the traditional mathematical toolkit, including analytic and algebraic tools, may be applied. The other primary theme is the potential for general-purpose optimization in real-world settings based on information maximization. This chapter discusses the three points above and identifies potential avenues to solidify and expand them into new areas.

## 18.1 Bridging Optimization and Functional Analysis

While optimization has been studied for centuries, the analytic relationships among the distinct optimization methodologies proliferating over the last century have previously received little attention. In this book, arbitrary optimizers for a variety of problem settings have been studied from a functional analytic point of view, with some surprising results. This work represents one attempt towards bridging the gulf between the practice of global optimization and the fields of advanced mathematics. Many other approaches might be as successful or more so, but the present approach is validated by the breadth and depth of the results presented in the previous chap-



A. J. Lockett, *General-Purpose Optimization Through Information Maximization*,
Natural Computing Series, https://doi.org/10.1007/978-3-662-62007-6_18

ters. In this section, the accomplishments of the formal approach of this book are reviewed, and some alternative formalizations are briefly considered. Finally, the concept of mathematizing computer programs is discussed conceptually.

### 18.1.1 Significance of the Formal Approach

Chapter 3 presented a formalization of iterative stochastic optimization methods in finite spaces based on the search points they propose given a history of search points and their assigned values, a point of view that has its origins in the initial characterizations of No Free Lunch. Chapter 6, extended these search generators to Hausdorff topological spaces, revealing that the space of iterative stochastic optimization methods represented by search generators is a closed, convex subset of a complete normed vector space. Interestingly, this result still holds true when optimization methods are formalized based on the infinite sequence of evaluation points they propose for a given objective, as shown in Chapter 9, invoking a connection with normalized versus extended forms in game theory, developed further in Chapter 17.

The profusion of optimization methods and the clear distinctions in the way they have been presented would lead an observer to initially conclude that methods such as gradient descent and Monte Carlo optimization are unrelated. The vector space result instead implies that between any two optimization methods there is a line in generator space that smoothly transforms one method into the other. This fact was explored experimentally in Chapter 11, where it was shown that in many cases the performance of the optimizers along that line changes continuously as well.

The search generator concept covers a wide range of search and optimization methods. Chapter 7 demonstrated that the most common population-based optimization methods can be expressed naturally as static search generators. The formalization even makes it possible to compare methods directly in a mathematical setting, as was done in Theorem 7.1, where the $(1+1)$–ES was shown to be the norm limit of Simulated Annealing in Euclidean space.

Chapters 8 and 10 harnessed static search generators as a vehicle for mathematical analysis by proving the conditions under which optimization methods respond continuously to changes in the search history or the fitness function. As shown in those chapters, genetic algorithms, evolution strategies, swarm optimizers, differential evolution, and stochastic gradient descent are continuous in most circumstances, especially on trajectories of unambivalent fitness value. This continuity even carries over to performance criteria, so that the performance of most popular methods changes continuously along with the objective or the optimization method.

In Chapters 13 and 17, several extensions of these concepts were introduced focused on search trajectors as the extended behavior of search generators to represent how search and optimization methods generate entire histories all at once. The relationship between static and standard search generators and trajectors was explored, and these objects were used to define the optimization game, which extends many key results to stochastic, dynamic, and adaptive search and optimization problems.

Throughout this work, the assumption that optimization methods could be modeled as measure-valued functionals or, viewed otherwise, as conditional probability distributions, has proved to be a mathematically fertile perspective and will likely continue to do so. The large number of theoretical and practical results presented in this book demonstrates the value and significance of this approach.

### *18.1.2 An Alternative Characterization: Density Maps*

From the beginning of Chapter 6, a particular norm was chosen and remained fixed throughout the text. This choice cast the space of measure-valued functionals as a Banach space. Alternative characterizations of the space of optimizers are possible and may be useful for obtaining further results. To demonstrate the truth of this claim, an example involving density maps is presented that leads naturally to a Hilbert space, also known as an inner product space. This subsection develops the concept of trajector density maps to show how this Hilbert space is connected to the space of black-box search trajectors, and then presents the inner product as a result of fixing a base measure on $X^{\mathfrak{T}}$ along with an auxiliary measure on $Y^{\mathfrak{T}}$, which has consequences extending into the optimization game as well.

For measure-valued functionals, arbitrary finite signed measures were allowed. But suppose that one wishes to study only optimization methods that are absolutely continuous with respect to a fixed measure $\mu$ on the search domain. In Euclidean space, one might choose the Lebesgue measure – the standard measure of volume in $\mathbb{R}^n$ – to play this role. Each optimization method can then be described as a Radon-Nikodym derivative capturing the deviation of the method from this base measure. The usefulness of such a base measure should be apparent from Chapters 14 to 16, where just such a base measure was used to generate a family of optimization methods.

This concept of deviation from a base measure is called a *density map* in recognition of its fundamental similarity to probability densities. The derivation relies on *Lebesgue decomposition*. Recall that using Radon-Nikodym derivatives, any two finite signed measures $\nu$ and $\mu$ can be related using

$$\nu(A) = \int_A \left\{ \frac{d\nu}{d\mu} \right\} d\mu$$

for some function $\frac{d\nu}{d\mu}$ defined $\mu$-almost everywhere, under the assumption that $|\mu(A)| = 0$ implies $|\nu(A)| = 0$, a property known as *absolute continuity*, written $\mu << \nu$ and previously used in Chapter 14. If absolute continuity is violated, a maximal measurable set $S$ can be found for which $|\nu|(S) > 0$ but $|\mu|(S) = 0$. The signed measure $\nu$ can then be subdivided into two new measures,

$$\nu_{\parallel}^{\mu}(A) = \int_A \mathbb{1}_{S^c}\, d\nu \text{ so that } \nu_{\parallel}^{\mu} << \mu, \text{ and } \quad \nu_{\perp}^{\mu}(A) = \int_A \mathbb{1}_S\, d\nu \text{ so that } \nu_{\perp}^{\mu} \perp \mu,$$

where $S^c$ is the complement of $S$. The notation $\delta \perp \mu$ indicates *singularity* and means that $|\delta|(A) > 0$ if and only if $|\mu|(A) = 0$. Then $\nu = \nu_{\parallel}^{\mu} + \nu_{\perp}^{\mu}$, and

$$\nu(A) = \int_A \left\{ \frac{d\nu_{\parallel}^{\mu}}{d\mu}(x) \right\} \mu(dx) + \nu_{\perp}^{\mu}(A),$$

which permits an alignment of $\nu$ with $\mu$ where both have mass, simultaneously yielding an extra term for $\nu$ where $\mu$ is zero. The measure $\nu_{\parallel}^{\mu}$ is the non-singular component of $\nu$ and the measure $\nu_{\perp}^{\mu}$ is the singular component of $\nu$, both with respect to $\mu$.

Suppose $\mu$ is a finite signed measure on the space of search histories $X^{\mathfrak{T}}$. For any search trajector $\mathcal{A}$, define a restricted search trajector $\mathcal{A}_{\parallel}^{\mu}$ that restricts to the non-singular component of the Lebesgue decomposition with respect to $\mu$; that is,

$$\mathcal{A}_{\parallel}^{\mu}[y] = (\mathcal{A}[y])_{\parallel}^{\mu}.$$

With this definition $\mathcal{A}_{\parallel}^{\mu}[y] << \mu$ for all $y \in Y^{\mathfrak{T}}$. Consider the map

$$\partial\mathcal{A} : \mu \mapsto \left( h \mapsto \frac{d\mathcal{A}_{\parallel}^{\mu}[h_Y]}{d\mu}(h_X) \right).$$

The map $\partial\mathcal{A}$ maps optimization games to functions over complete history traces defined by a family of Radon-Nikodym derivatives, one for each value history. This map will be called the *trajector density map* of $\mathcal{A}$, since the function $\partial\mathcal{A}(\mu)$ can be viewed as the density function of the trajector $\mathcal{A}$ with respect to the measure $\mu$. For notation, let $\partial_\mu \mathcal{A} = \partial\mathcal{A}(\mu)$, which embodies the intuition that this quantity is a kind of derivative with respect to $\mu$.

In light of the optimization game of Chapter 17, the same concept can be applied to adversarial fitnesses with the measure $\xi$ being defined on the space of value histories $Y^{\mathfrak{T}}$ rather than on search histories. Thus the notation $\partial_\xi \mathcal{F}$ might be used in this case to represent a trajector density map for an adversarial fitness. Now whenever $\mathcal{A}[y] << \mu$ and $\mathcal{F}[z] << \xi$,

$$\mathcal{A}[y](A) = \int_A \partial_\mu \mathcal{A}(z,y)\, \mu(dz) \quad \text{and} \quad \mathcal{F}[z](B) = \int_B \partial_\xi \mathcal{F}(z,y)\, \xi(dy),$$

which suggests that in fact one can start with a trajector density map and then derive a search trajector in either case. In order to do so, the nature of acceptable trajector density maps must be clarified.

**Definition 18.1 (Measure-Consistent).** Any $\rho : \mathcal{M}[X^{\mathfrak{T}}] \to \left((X \times Y)^{\mathfrak{T}} \to \mathbb{R}\right)$ is *measure-consistent* in $X$ if for any pair of measures $\mu$ and $\nu$ with $\nu << \mu$,

$$\rho(\mu)(h) = \rho(\nu)(h)\,\frac{d\nu}{d\mu}(h_X)$$

for each $h_Y$ and for all $h_X$ in a set of full $\nu$-measure. If the same statement holds with $X$ and $Y$ interchanged, then $\rho$ is measure-consistent in $Y$.

**Definition 18.2 (Globally Integrable).** Any $\rho : \mathcal{M}[X^{\mathcal{T}}] \to \left((X \times Y)^{\mathcal{T}} \to \mathbb{R}\right)$ is *globally integrable* in $X$ if $\rho(\mu)$ is measurable for all $\mu$ and there exists $0 \leq M < \infty$ such that for any measure $\mu \in \mathcal{M}[X^{\mathcal{T}}]$ and any $h_Y \in Y^{\mathcal{T}}$, $\int |\rho(\mu)(h)|\,|\mu|(dh_X) \leq M$. If the same statement holds with $X$ and $Y$ interchanged, then $\rho$ is globally integrable in $Y$.

**Definition 18.3 (Trajector Density Map).** A function $\rho$ with signature $\mathcal{M}[X^{\mathcal{T}}] \to \left((X \times Y)^{\mathcal{T}} \to \mathbb{R}\right)$ is a trajector density map in $X$ if $\rho$ is measure-consistent and globally integrable in $X$. Similarly, $\rho$ is a *trajector density map* in $Y$ if it is measure-consistent and globally integrable in $Y$.

Measure consistency in $X$ guarantees for $\nu, \mu$ with $\nu << \mu$ and $\mu << \nu$,

$$\int_A \rho(\mu)(h)\,\mu(dh_X) = \int_A \rho(\nu)(h)\,\nu(dh_X)$$

for all measurable $A \subseteq X^{\mathcal{T}}$, so that $\rho$ generates a measure in its own right, independent of the integrating measure $\mu$ or $\nu$, except for their null sets. Global integrability guarantees that there is an upper bound (the $M$ of the definition) on the value these integrals across all measures.

**Proposition 18.1.** *The maps $\partial\mathcal{A}$ and $\partial\mathcal{F}$ defined above are trajector density maps in $X$ and $Y$ respectively.*

*Proof.* Suppose $\nu << \mu$, and let $h_Y$ be arbitrary. Let $S_\mu$ be the singular set of $\mu$ with respect to which $\nu$, that is, the largest measurable set for which $|\mu|(S_\mu) \neq 0$ but $|\nu|(S_\mu) = 0$. For any measurable $A$,

$$\mathcal{A}_{\|}^{\nu}[h_Y](A \setminus S_\mu) = \int_{A \setminus S_\mu} \partial_\nu \mathcal{A}(h)\,\nu(dh_X) = \int_{A \setminus S_\mu} \partial_\mu \mathcal{A}(h)\,\mu(dh_X),$$

where the second equality holds because $\mathcal{A}_{\|}^{\nu}[h_Y] = \mathcal{A}_{\|}^{\mu}[h_Y]$ for any subset of $X^{\mathcal{T}} \setminus S_\mu$. This equation establishes measure consistency in $X$ since $X^{\mathcal{T}} \setminus S_\mu$ has full $\nu$-measure. Measure consistency in $Y$ can be established analogously for $\partial\mathcal{F}$. Global integrability of $\partial\mathcal{A}$ is established by

$$\int_A \left|\partial_\mu \mathcal{A}(h)\right|\,|\mu|(dh_X) = \left|\mathcal{A}_{\|}^{\mu}[h_Y](A)\right| \leq |\mathcal{A}[h_Y](A)| \leq \|\mathcal{A}\| < \infty,$$

and a similar equation holds for $\partial\mathcal{F}$.

**Lemma 18.1.** *For every trajector density map $\rho$ in $X$, there is a unique search trajector $\mathcal{A}$ from $X$ to $Y$ with finite norm such that $\partial\mathcal{A} = \rho$.*

*Proof.* Consider the map over measurable sets $A$ and measures $\mu$ given by

$$\mathcal{R}_{h_Y} : (A, \mu) \mapsto \int_A \rho(\mu)(h)\, \mu(dh_X)$$

for some $h_Y \in Y^{\mathcal{T}}$. Being defined by an integral, the map $\mathcal{R}_{h_Y}$ is obviously $\sigma$-additive for fixed $\mu$, and so the map $A \mapsto \mathcal{R}_{h_Y}(A, \mu)$ is a measure. This measure will be called $\mathcal{A}^{\mu}_{\mathcal{R}}[h_Y]$, since it is absolutely continuous with respect to $\mu$. Global integrability of $\rho$ implies that this measure is finite when $\mu$ is.

The dependence on the measure $\mu$ must now be removed. By global integrability, there exists $0 \leq M < \infty$ such that $\|\mathcal{A}^{\mu}_{\mathcal{R}}[h_Y]\| \leq M$. Therefore it is possible to define for each $A$ and $h_Y$

$$\mathcal{A}[h_Y](A) = \sup_{\mu} \left\{ \left(\mathcal{A}^{\mu}_{\mathcal{R}}[h_Y]\right)_+(A) \right\} - \sup_{\mu} \left\{ \left(\mathcal{A}^{\mu}_{\mathcal{R}}[h_Y]\right)_-(A) \right\},$$

where the $+$ and $-$ subscripts indicate the positive and negative components of the Hahn-Jordan decomposition for the measure $\mathcal{A}^{\mu}_{\mathcal{R}}[h_Y]$, respectively.

The goal now is to show that $\mathcal{A}$ is a search trajector with finite norm. Finiteness is immediate from global integrability. For $\sigma$-additivity, first note that for any disjoint collection $(A_n)_{n\in\mathbb{N}}$,

$$\sup_{\mu} \left(\mathcal{A}^{\mu}_{\mathcal{R}}[h_Y]\right)_+ \left(\bigcup_n A_n\right) = \sup_{\mu} \sum_n \left(\mathcal{A}^{\mu}_{\mathcal{R}}[h_Y]\right)_+(A_n) \leq \sum_n \sup_{\mu} \left(\mathcal{A}^{\mu}_{\mathcal{R}}[h_Y]\right)_+(A_n). \tag{18.1}$$

Secondly, for $\varepsilon > 0$, suppose $(\mu_n)_{n\in\mathbb{N}}$ is some collection of measures such that

$$\sum_n \left(\mathcal{A}^{\mu_n}_{\mathcal{R}}[h_Y]\right)_+(A_n) > \sum_n \sup_{\mu} \left(\mathcal{A}^{\mu}_{\mathcal{R}}[h_Y]\right)_+(A_n) - \varepsilon.$$

The sum on the left is a limit, and so for every $\varepsilon' > 0$ there is some $N$ with

$$\sum_{n=1}^{N} \left(\mathcal{A}^{\mu_n}_{\mathcal{R}}[h_Y]\right)_+(A_n) > \sum_n \left(\mathcal{A}^{\mu_n}_{\mathcal{R}}[h_Y]\right)_+(A_n) - \varepsilon' > \sum_n \sup_{\mu} \left(\mathcal{A}^{\mu}_{\mathcal{R}}[h_Y]\right)_+(A_n) - \varepsilon - \varepsilon'.$$

Now define a measure $\mu'$ such that $\mu'(H) = \sum_{n=1}^{N} \mu_n(H \cap A_n)$. This measure is finite, since it is a finite sum of finite measures, and

$$\sup_{\mu} \sum_n \left(\mathcal{A}^{\mu}_{\mathcal{R}}[h_Y]\right)_+(A_n) \geq \sum_n \left(\mathcal{A}^{\mu'}_{\mathcal{R}}[h_Y]\right)_+(A_n) = \sum_{n=1}^{N} \left(\mathcal{A}^{\mu_n}_{\mathcal{R}}[h_Y]\right)_+(A_n).$$

Letting first $\varepsilon$ and then $\varepsilon'$ go to zero, it follows that

$$\sup_{\mu} \sum_n \left(\mathcal{A}^{\mu}_{\mathcal{R}}[h_Y]\right)_+(A_n) \geq \sum_n \sup_{\mu} \left(\mathcal{A}^{\mu}_{\mathcal{R}}[h_Y]\right)_+(A_n),$$

which implies $\mathcal{A}[h_Y](\bigcup_n A_n) = \sum_n \mathcal{A}[h_Y](A_n)$ when taken together with Equation 18.1 and the fact that the same process can be applied to $\left(\mathcal{A}^{\mu}_{\mathcal{R}}[h_Y]\right)_-$. So $\mathcal{A}[h_Y]$ is $\sigma$-additive with finite norm bounded above by $2M < \infty$, and $\mathcal{A}$ is a search trajector with finite norm.

The claim that $\partial\mathcal{A} = \rho$ is proven based on measure consistency in $X$. By construction $\rho(\mu)(h) = \frac{d\mathcal{A}^{\mu}_{\mathcal{R}}[h_Y]}{d\mu}(h_X)$, and therefore $\partial\mathcal{A} = \rho$ if $\mathcal{A}^{\mu}_{\parallel} = \mathcal{A}^{\mu}_{\mathcal{R}}$. Let $S_\mu$ be the singular set of $(\mathcal{A}[h_Y])_+$ with respect to $\mu$ so that

$$\left(\mathcal{A}^{\mu}_{\parallel}[h_Y]\right)_+(A) = (\mathcal{A}[h_Y])_+(A \setminus S_\mu)$$

Let $S^c_\mu$ be the complement of $S_\mu$. If for any set $A \subseteq S^c_\mu$ the measure $\left(\mathcal{A}^{\mu}_{\mathcal{R}}\right)_+$ has $\left(\mathcal{A}^{\mu}_{\mathcal{R}}\right)_+(A) = \sup_{\mu'}\left(\mathcal{A}^{\mu'}_{\mathcal{R}}\right)_+(A)$, the desired result will follow.

For all measures $\mu'$, $\left(\mathcal{A}^{\mu'}_{\mathcal{R}}[h_Y]\Big|_{S^c_\mu}\right)_+ << \left(\mathcal{A}^{\mu}_{\mathcal{R}}[h_Y]\right)_+$. Otherwise, there would be some nonempty set $S' \subseteq S^c_\mu$ with $\left(\mathcal{A}^{\mu'}_{\mathcal{R}}[h_Y]\right)_+(S') > 0$ but $\left(\mathcal{A}^{\mu}_{\mathcal{R}}[h_Y]\right)_+(S') = 0$ so that $S' \subseteq S_\mu$, which contradicts the fact that $S' \cap S_\mu = \emptyset$. By Lebesgue decomposition, for $A \subseteq S^c_\mu$,

$$\left(\mathcal{A}^{\mu}_{\mathcal{R}}[h_Y]\right)_+(A) = \left(\mathcal{A}^{\mu'}_{\mathcal{R}}[h_Y]\right)_+(A) + \left(\mathcal{A}^{\mu}_{\mathcal{R}}[h_Y]\right)_{+,\perp}(A),$$

which implies that $\left(\mathcal{A}^{\mu'}_{\mathcal{R}}[h_Y]\right)_+(A) \leq \left(\mathcal{A}^{\mu}_{\mathcal{R}}[h_Y]\right)_+(A)$ for all $\mu$. That is,

$$\left(\mathcal{A}^{\mu}_{\mathcal{R}}[h_Y]\right)_+(A) = \sup_{\mu'}\left\{\left(\mathcal{A}^{\mu'}_{\mathcal{R}}[h_Y]\right)_+(A)\right\} = \left(\mathcal{A}^{\mu}_{\parallel}[h_Y]\right)_+(A) = (\mathcal{A}[h_Y])_+(A)$$

for all $A \subseteq S^c_\mu$. Since the same reasoning applies to the negative components with subscript $-$, the conclusion follows that $\mathcal{A}^{\mu}_{\parallel} = \mathcal{A}^{\mu}_{\mathcal{R}}$ and $\partial\mathcal{A} = \rho$.

Suppose there were another search trajector $\mathcal{A}'$ such that $\partial\mathcal{A}' = \rho$. Then it would follow that $\mathcal{A}'^{\mu}_{\parallel}[h_Y](A) = \int_A \rho(\mu)(h)\mu(dh_X) = \mathcal{A}^{\mu}_{\parallel}[h_Y](A)$. Since this is true for all $\mu$, it follows that $\mathcal{A}' = \mathcal{A}$, and so $\mathcal{A}$ is unique.

**Lemma 18.2.** *For every trajector density map $\rho$ in $Y$, there is a unique search trajector $\mathcal{F}$ from $Y$ to $X$ with finite norm such that $\partial\mathcal{F} = \rho$.*

*Proof.* The proof is similar to that of Lemma 18.1, *mutatis mutandis.*

**Theorem 18.1.** *The space of trajector density maps in $X$ and $Y$ are both isomorphic to the space of search trajectors with finite norm.*

*Proof.* This theorem combines the statements of Proposition 18.1 and Lemmas 18.1 and 18.2.

**Theorem 18.2.** *The space of trajector density maps in X or Y is a vector space under pointwise addition and scalar multiplication, and the vector spaces of trajector density maps and search trajectors are isomorphic with equivalent norm topologies.*

*Proof.* If $\rho$ and $\rho'$ are two trajector density maps, then $(\rho+\rho')(\mu)(h) = \rho(\mu)(h) + \rho'(\mu)(h)$. Measure consistency in $X$ or $Y$ follows from the definition, and if $M$ and $M'$ are the global integration bounds for $\rho$ and $\rho'$ respectively, then $\int |(\rho + \rho')(\mu)(h)||\mu_X[h_Y]|(dh_Y) \leq M + M'$, proving global integrability. It follows that $\rho + \rho'$ is a trajector density map. For $\lambda \in \mathbb{R}$, $(\lambda\rho)(\mu)(h) = \lambda(\rho(\mu)(h))$, and again measure consistency and global integrability are immediate, so that $\lambda\rho$ is a trajector density map. Commutative, distributive, and associative properties follow trivially from the same properties of scalar multiplication and addition, so that the space of trajector density maps is a vector space.

If $\mathcal{A}$ and $\mathcal{A}'$ are search trajectors with finite norm and $\lambda, \lambda' \in \mathbb{R}$, then $\partial(\lambda\mathcal{A} + \lambda'\mathcal{A}') = \lambda\partial\mathcal{A} + \lambda'\partial\mathcal{A}'$ by the linearity of the Radon-Nikodym derivative, which follows from the linearity of the integral. Thus the vector spaces agree as claimed.

Based on Theorem 18.1, the topology of the search trajectors can be imposed on the space of trajector density maps, which produces a norm topology over density maps with norm

$$\|\rho\| = \sup_{\mu, h_Y} \int |\rho(\mu)(h)||\mu|(dh_X)$$

so that $\|\partial\mathcal{A}\| = \|\mathcal{A}\|$.

Theorem 18.2 shows that in fact the trajector density maps and the search trajectors produce identical Banach spaces under pointwise addition and scalar multiplication. Trajector density maps are thus just another way of seeing search trajectors.

Coming to the point of this subsection, suppose that $\mu$ is a fixed measure on $X^{\mathcal{T}}$ and $\xi$ is a fixed measure on $Y^{\mathcal{T}}$. Let $\mu \times \xi$ be the product measure on $X^{\mathcal{T}} \times Y^{\mathcal{T}}$ given by $\mu \times \xi(A \times B) = \mu(A)\nu(B)$. Now consider the map $\mathcal{A} \mapsto \partial_\mu\mathcal{A}$ defined on all search trajectors, and use this map to define an equivalence relation $\mathcal{A} \sim \mathcal{A}'$ if $\partial_\mu\mathcal{A}(z,y) \stackrel{a.e}{=} \partial_\mu\mathcal{A}'(z,y)$ for $\mu \times \xi$-almost every $(z,y)$, which implies that $\mathcal{A}^\mu_\parallel[y] = \mathcal{A}'^\mu_\parallel[y]$ for $\nu$-almost every $y$. Let $[\mathcal{A}]$ represent the equivalence class of $\mathcal{A}$ and define the space of these equivalence classes as

$$\mathcal{D}[\mu,\xi] = \left\{ [\mathcal{A}] \,\middle|\, \mathcal{A} \in \mathcal{MF}_{Y^{\mathcal{T}},X^{\mathcal{T}}} \right\}.$$

Each element of $\mathcal{D}[\mu,\xi]$ has $\mathcal{A}[y]$ absolutely continuous with respect to $\mu$ for $\nu$-almost every $y \in Y^{\mathcal{T}}$, and $\partial_\mu\mathcal{A}$ is determined by the pointwise Radon-Nikodym derivatives with respect to $\mu$. This last fact makes it possible to define an inner product on $\mathcal{D}[\mu,\xi]$.

**Definition 18.4 (Density Product).** Given base measure $\mu$ on $X^{\mathcal{T}}$ and an auxiliary measure $\xi$ on $Y^{\mathcal{T}}$, the *density product* of two search trajectors $\mathcal{A}$ and $\mathcal{B}$ with respect to $\mu$ and $\xi$ is the double integral

$$(\mathcal{A},\mathcal{B})_{\mu,\xi} = \int_{Y^{\mathcal{T}}} \int_{X^{\mathcal{T}}} \partial_{\mu}\mathcal{A}(z,y)\, \partial_{\mu}\mathcal{B}(z,y)\, \mu(dz)\, \xi(dy).$$

**Theorem 18.3.** *For each positive base measure $\mu$ on $X^{\mathcal{T}}$ and each positive auxiliary measure $\xi$ on $Y^{\mathcal{T}}$, the density product with respect to $\mu$ and $\xi$ is an inner product on $D[\mu,\xi]$, which is thus a Hilbert space.*

*Proof.* An inner product is a function taking two vectors to a scalar such that the function is bilinear and symmetric and the inner product of a vector with itself is nonnegative and equal to zero only for the zero vector. The integral definition of the density product combined with Theorem 18.2 guarantees bilinearity and symmetry. For any search trajector $\mathcal{A}$,

$$(\mathcal{A},\mathcal{A})_{\mu,\xi} = \int_{Y^{\mathcal{T}}} \int_{X^{\mathcal{T}}} \left(\partial_{\mu}\mathcal{A}(z,y)\right)^2\, \mu(dz)\, \xi(dy) \geq 0.$$

Note that this equation relies on the premise that $\mu$ and $\xi$ are both positive. Furthermore, if $(\mathcal{A},\mathcal{A})_{\mu,\xi} = 0$, then $\partial_{\mu}\mathcal{A}(z,y) = 0$ for $\mu \times \xi$ almost every $(z,y)$. It follows that $\mathcal{A} \in [\mathbf{0}]$, which completes the proof.

Hilbert spaces have many salutary properties. Firstly, a Hilbert space is reflexive, meaning that each such space is its own topological dual. The topological dual is the space of continuous linear functions on a topological space. Essentially, for each $\mathcal{A}$, the inner product map $\mathcal{B} \mapsto (\mathcal{A},\mathcal{B})_{\mu,\xi}$ is a continuous linear function over search trajectors, and this formulation encompasses *every* continuous linear function; this is the Riesz Representation Theorem.

Secondly, the closed unit ball of every Hilbert space is weakly compact, which further implies that the set of proper members of $\mathcal{D}[\mu,\xi]$ – that is, the set of $[\mathcal{A}] \in \mathcal{D}[\mu,\xi]$ such that $\mathcal{A}^{\mu}_{\parallel}[y]$ is a probability measure for $\xi$-almost every $y$ – is a weakly compact set because it is a closed subset of a weakly compact set. Since this set is also convex, the Krein-Milman Theorem implies that it is the closed convex hull of its extreme points. Theorem 6.4 showed that for proper measure-valued functionals, the extreme points are the proper deterministic functionals. It is likely that this point of view can be extended to $\mathcal{D}[\mu,\xi]$ as well, so that the extreme points of the proper elements of $\mathcal{D}[\mu,\xi]$ would be the image of the deterministic proper search trajectors taking values on points of positive $\mu$-density. The relevance of this last point comes from the Bauer Minimum Principle, which depends on the Krein-Milman Theorem and states that the every convex continuous functional takes its extreme values over a convex compact set at the extreme points of the set.

Revisiting the optimization game of Chapter 17, suppose that $\mu$ and $\xi$ are both probability measures and let $\mathcal{A}$ and $\mathcal{F}$ be search trajectors from $X$ to $Y$ and $Y$ to $X$, respectively. Then $(\mathcal{A}^{\mu}_{\parallel}, \mathcal{F}^{\xi}_{\parallel})$ is a pair of players for the game

$$(\mathcal{A} \times \mathcal{F})_{\mu,\xi}(H) = \int_{H} \partial_{\mu}\mathcal{A}(h)\, \partial_{\xi}\mathcal{F}(h)\, \mu(dh_X)\, \xi(dh_Y),$$

where the value function is just $V = \partial_\mu \mathcal{A} \times \partial_\xi \mathcal{F}$. This game is invariant over the Cartesian product $[\mathcal{A}] \times [\mathcal{F}]$. That is, there is a function $([\mathcal{A}],[\mathcal{F}]) \mapsto (\mathcal{A} \times \mathcal{F})_{\mu,\xi}$ that assigns games to players and is defined on all of $\mathcal{D}_{X,Y}[\mu,\xi] \times \mathcal{D}_{Y,X}[\xi,\mu]$, where the $X$ and $Y$ subscripts are added to indicate that the left-hand side is based on equivalence classes of search trajectors from $X$ to $Y$ whereas the right-hand side is based on search trajectors from $Y$ to $X$. This function is surjective onto the vector subspace of optimization games that is absolutely continuous with respect to $\mu \times \xi$. Thus by passing to $\mathcal{D}[\mu,\xi]$, a complete player-game map is obtained, whereas no such map has thus far been obtained for the space of all search trajectors.

Going further, a performance criterion can be defined analogous to the V-performance,

$$\langle \mathcal{A}, \mathcal{F} \rangle_{\mu,\xi} = (\mathcal{A} \times \mathcal{F})_{\mu,\xi} \left( (X \times Y)^{\mathcal{T}} \right).$$

This performance criterion is as powerful as the V-performance on the subset of optimization games that is absolutely continuous with respect to $\mu \times \xi$, because any value function can be absorbed into $\partial_\mu \mathcal{F}$ directly.

Unlike the V-performance on search trajectors, this new criterion is a bilinear form due to Theorem 18.2. Since it is bilinear, we can define for each $\mathcal{F}$ a map $\mathcal{A} \mapsto \langle \mathcal{A}, \mathcal{F} \rangle_{\mu,\xi}$, and this map is a continuous linear functional, where continuity follows from operator theory given that the optimization game is finite and therefore bounded. Continuous linear functionals are also convex, and thus the Bauer Minimum Principle implies that this functional takes its extreme values on the extreme points of each closed (weakly) compact convex set. Since these extreme points are likely the image of the deterministic proper search trajectors, it follows that deterministic optimization should perform best relative to fixed $\mu$ and $\nu$, if this analysis holds under further scrutiny.

Thus the inner product space $\mathcal{D}[\mu,\xi]$ is an alternative way of analyzing the space of optimizers, one that places a geometry over optimizers and clearly permits stronger theoretical results. This Hilbert space is only one additional formalism; another formalization based on Wasserstein (Earth Mover) distance has other desirable properties, such as continuous deterministic trajectors. There remains wide opportunity to extend these formalizations to better understand search and optimization problems.

### 18.1.3 Reasoning About Computational Agents

A computational agent in a stimulus-response setting such as that of Section 1.1 is a type of program that takes a stimulus as input and generates a response. Traditionally, theoretical analysis of programs has been based on formalisms such as Turing Machines, boolean circuits, and other finite automata. These formalisms are intended to describe discrete, sequential computation. However, in realistic problems, the relationship between stimulus and response is stochastic. Even when this relationship is deterministic, stochastic programs may still provide efficient and easily implemented approaches to solving such problems. Traditional computational theory has been extended to address such methods using a random bit-flipping instruction, and computational learning theory has flourished for analyzing probability in finite systems using the concept of probably approximate correctness (PAC). Yet there is a broader context in which a stimulus-response agent plays a kind of probabilistic game against an external environment, and discrete theories of programs are not particularly adept at modeling the stochastic interaction between a program and a computation context.

The search trajector formalism is directly suited to the analysis of this game, as shown in Chapter 17 with the development of the optimization game; it might be that the space $\mathcal{D}[\mu,\xi]$ from the last subsection is even better suited to this analysis. The underlying concept is that a program, whether stochastic or deterministic, can be viewed as a unitary mathematical object in its own right independent of the sequence of steps or rules it follows. These objects have spatial features, such as the vector space structures of search generators and trajectors. These spatial features can be used to reason about broad groups of programs, for example, to choose from among several programs a convex combination of programs that is best suited to a particular problem (see Section 18.2.1 below for this convex control task). In addition, these spatial objects can be combined to reason about the interaction between the program and its computational environment, which can also be viewed as a kind of program; this is the message of the optimization game.

Specifically, let $\mathcal{A}$ be a search trajector from $X$ to $Y$ representing an agent operating in an environment $\mathcal{E}$, which is likewise modeled as a search trajector for $Y$ to $X$. In this case, the domain $X$ consists of actuator commands or program outputs while the space $Y$ represents sensor inputs. Then if there exists an optimization game $\mathcal{A}\times\mathcal{E}$ such that $(\mathcal{A},\mathcal{E})$ is a pair of players for the game. This agent is a policy in the sense of reinforcement learning. If there exists a search generator $\mathcal{G}$ that extends to $\mathcal{A}$, then $\mathcal{G}$ represents how the agent synthesizes its sensor history in order to issue a set of controls for a given time. If there further exists a jagged value generator $\mathcal{G}^*$ that extends to $\mathcal{E}$, then $\mathcal{G}^*$ represents the sensor response at the time subsequent to the action.

This scenario brings several aspects of the theory of search trajectors into focus. Firstly, generators and trajectors are useful in domains beyond just search and optimization; for example, they describe how agents interact with environments in stimulus-response fashion. Secondly, it emphasizes why Bellman equations show up as the method for optimizing over optimization methods in Chapter 13. Thirdly, for an arbitrary robot placed into a real-world environment, the environment clearly takes on the role of the *reality prior* described in Section 13.1.3.

This last point bears expounding. No Free Lunch theorems show that if one wishes to solve all optimization problems, one may as well perform blind random search, since all methods work equally well. At the other extreme, if one wishes to solve one particular problem, then clearly the choice of method should be tuned to that one problem, based on the principle of problem-solution alignment. But in reality, one wishes to solve many problems, all of which arise from the real world as an environment. The process by which humans solve these problems includes imposing constraints, isolating and ignoring irrelevant factors, proposing convenient representations, and otherwise manipulating the problem to make it tractable with known methods.

When researchers speak of *problems*, they usually refer to the output of this problem-solving process: a constrained, isolated, and tractable problem with a convenient representation. They often ignore the far more complicated process of generating that crystalized form of the problem to which a known optimization technique is applied. Yet it is the human problem-solving process itself that represents a solution to the various problems imposed by the reality prior, and advances in artificial intelligence should provide better general-purpose solutions for various instantiations of the reality prior. Viewed this way, speaking of problem-solution alignment in the NFL sense will not suffice. The problem itself must be refined to generate a form for which one can reason about such alignment, and this problem refinement is itself a sort of interaction with an environment derived from the real world. In other words, the production of solvable problems is itself an artifact of a broader problem solving process, and it is in fact the manifestations of this broader problem that artificial intelligence researchers truly wish to solve.

In sum, the approach of this book provides new mathematical tools for treating programmatic approaches to solving problems as spatial objects. These tools add to the repertoire of analysis tools for understanding computation, search, optimization, reinforcement learning, and more. By thinking of these problems in terms of search generators and trajectors as well as in terms of the optimization game, new aspects of how to solve real-world problems become apparent. These techniques can be used by researchers in order to propose new ways of solving general-purpose problems, and they provide a framework within which to analyze the results.

## 18.2 Improved Optimization Methods

The analytic elegance of studying search trajectors and adversarial fitnesses are clear, but the original purpose of this inquiry was to improve optimization methods. Three methods were proposed along these line for developing new optimization methods that may be appropriate in certain scenarios. In this section we review these three approaches: (1) Convex Control of Optimizer Porfolios; (2) Benchmark-driven Algorithm Selection; and (3) Martingale Optimization.

### *18.2.1 Convex Control of Optimizer Portfolios*

Chapter 6 proved that convex combinations of computable optimizers are also computable optimizers. Chapter 10 demonstrated that performance varies nonlinearly as the convex combination changes, and some of the results in Chapter 11 even suggested that convex combinations may outperform any of the optimizers being combined.

In light of these facts, can one choose a good optimizing strategy by convexly combining existing optimizers? One way to do so is to test several strategies and allocate resources to the strategies that perform best. This approach is termed *convex control* of optimizers; it is explored theoretically in this section.

Suppose that there is a finite set of search generators $\mathfrak{G} = \{\mathcal{G}_1, \ldots, \mathcal{G}_m\}$, each of which is known to perform well on a range of optimization problems. Such a set will be termed an *optimizer portfolio*, and it might include general-purpose methods such as simulated annealing, hill climbing with random restarts, or differential evolution. The convex control problem can be stated as follows: Given a set of optimizers, a fitness measure $\mathbb{F}$, and a performance criterion $\psi$, choose a time-varying probability vector $\alpha(t) = (\alpha_1(t), \ldots, \alpha_m(t))$ with $\sum_i \alpha_i(t) = 1$ for all $t \in \mathcal{T}$ such that the search generator $\mathcal{G}_\alpha$ defined by $\mathcal{G}_\alpha[h] = \sum_i \alpha_i(t)\mathcal{G}_i[h]$ for $h \in (X \times Y)^{[t]}$ minimizes $\tilde{\alpha} \mapsto \mathbb{E}_{u \sim \mathbb{F}}[\psi(\mathcal{G}_{\tilde{\alpha}}, u)]$.

At each time step, $\mathcal{G}_\alpha$ is a convex combination over the optimizer set $\mathfrak{G}$. As a function of time, $\mathcal{G}_\alpha$ moves along a path in generator space contained in the convex span of $\mathfrak{G}$. Because the probability vector $\alpha(t)$ changes with time, $\mathcal{G}_\alpha$ itself cannot be expressed directly as a convex combination over $\mathfrak{G}$.

The convex control problem can be addressed from two perspectives. In the first, the goal is to find a single, stable convex combination that is adapted to a given objective function. In the second, a dynamic control procedure is sought that makes the most efficient use of the optimizer portfolio. Both of these directions are interesting lines of research for future work.

For now, suppose that the probability vector $\alpha$ does not vary with time, i.e. $\alpha = (\alpha_1, \ldots, \alpha_m)$ independent of $t$. Then $\mathcal{G}_\alpha$ is a single convex combination over $\mathfrak{G}$ contained within the convex span of $\mathcal{G}$. A convex combination of optimizers can be regarded as a choice over optimizers. At each time step, the probability vector $\alpha$ is sampled to choose one of the $\mathcal{G}_i$, and then $\mathcal{G}_i$ is sampled to choose the next evaluation point. Call the sequence of such choices a *control history* of $\mathcal{G}_\alpha$ , e.g. $\mathcal{G}_3\mathcal{G}_1\mathcal{G}_4\mathcal{G}_1\mathcal{G}_2\mathcal{G}_2\mathcal{G}_2\mathcal{G}_3\ldots$. The set of all control histories of $\mathcal{G}_\alpha$ may be regarded as the set of optimization strategies available to $\mathcal{G}_\alpha$ within the Optimization Game of Chapter 13.

Since the fitness measure $\mathbb{F}$ is fixed, the adversarial fitness player is a chance node in game theory jargon, and the set of histories of $\mathcal{G}_\alpha$ can be thought of as tracing out an $m$-ary game tree in which the chance nodes are omitted. A fundamental question is whether convex combinations in the span of $\mathfrak{G}$ can outperform the best individual element in $\mathfrak{G}$. If $\mathfrak{G}$ contains just two members, e.g. $\mathcal{G}_1$ and $\mathcal{G}_2$, then the game tree is a binary tree. The question then reduces to whether one of the two homogeneous control histories, $\mathcal{G}_1\mathcal{G}_1\mathcal{G}_1\mathcal{G}_1\ldots$ or $\mathcal{G}_2\mathcal{G}_2\mathcal{G}_2\mathcal{G}_2\ldots$, outperforms all other histories.

Each choice in this binary tree can be represented as a zero if the left branch is followed, and a one if the right branch is followed. A single history contains infinitely many such choices. Thus the set of histories corresponds to a binary representation of the real numbers between zero and one and is therefore uncountable.

On this basis, it would be surprising if the two homogeneous control histories were the only two interesting ones from the perspective of performance.

As discussed in Section 10.2.1, the performance criterion $\psi$ is non-linear over search generators. If it were linear, then only the homogeneous histories could be optimal. Because it is non-linear, it is possible that one of the uncountably many internal histories could perform best.

Research on applying algorithm portfolios to optimization has been performed by Silverthorn and Miikkulainen [148] with promising results. The discussion above provides further theoretical basis for this research and places it within the context of general-purpose optimization.

### *18.2.2 Benchmark-Driven Algorithm Selection*

Chapters 8 and 10 demonstrated that many if not most optimization methods are continuous almost everywhere as the objective function changes, and that this continuity carries through to performance as well. Consequently, most optimization methods perform similarly on similar objectives. It is therefore a reasonable strategy to maintain a range of optimization methods and their performance on a set of benchmarks and to select an optimization method to use on a new objective function based on the best-performing optimization method for the most similar benchmark.

This possibility was mentioned in passing in Chapter 10, and Chapter 11 provided just such a benchmarked set of common optimization methods. The complete results of these experiments are contained in the appendices of this book, and could be used to implement a basic form of this approach to algorithm selection.

In addition to selecting among substantially different kinds of optimization methods, one could used the benchmark-driven method to select hyperparameters for a given class of optimizer, using interpolation to guess the likely performance of a parameterized optimization method based on its performance on similar benchmark objectives.

This approach does have a significant weakness, however, in that the space of objective functions is vast, while the space of optimization methods is even larger. Nonetheless, many interesting objective functions can be well approximated by lower-dimensional parameterized forms. Furthermore, when one selects an optimization method, it is usually drawn from less than a dozen popular approaches to optimization, so that the high dimensionality of the space of optimizers is not such an issue in many cases.

### *18.2.3 Martingale Optimization*

Chapter 13 proposed the Information Maximization Principle, that the optimal optimizer should fully utilize all available information about the objective function.

This principle was used to derive evolutionary annealing as a martingale optimization method in Chapter 14. There remains substantial future work to implement the theory of information maximization. Several improvements to the evolutionary annealing method are possible as well.

The theorems of Chapter 13 demonstrated that $\varepsilon$-optimal choices could be made based on Bellman equations given knowledge of a fitness measure generating the objective function. But this result was based on having a model of the fitness measure and on being able to choose search points based on an infinite lookahead against this model. In some situations, it may be possible to solve for the optimal search points, either analytically when the integrals are tractable or using Monte Carlo methods when the lookahead values are well-behaved for a particular fitness measure. Nonetheless, the information maximization principle reveals just how difficult optimization can be in the most general case. One must have full knowledge of the problem being optimized, and then one must perform a nested optimization in order to obtain the best search point at each step.

Evolutionary annealing was introduced as a tractable optimization procedure based on the information maximization principle. Since it is tractable, it is necessarily approximate and involves substantial simplifications. However, there are many ways in which the evolutionary annealing method of Chapter 14 could be improved.

In evolutionary annealing, a partitioning method was used to determine how annealed selection apportions probability mass among previously observed evaluation points. For Euclidean space, REA employed axis-parallel hyperrectangles as an efficiently computable partitioning approach (Chapter 15). Neuroannealing generalized this partition method to use hierarchical partitioning on a larger space, but at its base, neural networks were also partitioned using axis-parallel hyperrectangles to separate the weights (Chapter 16).

What if the partitions of the search space could be arranged to match the natural structure of the objective function? Such *semantic partitions* might be better able to locate the optima of an objective function by allocating probability more efficiently among the different regions. In a way, hierarchical partitioning as used by neuroannealing is a rudimentary step in this direction. Additionally, it might be possible to recognize fractal structure within semantic regions, and to propagate this structure across different partition regions in order to build a more accurate model of the objective function. For example, the RNN space in neuroannealing contains redundant network representations. If the similarities between two network topologies could be identified, then the objective evaluations from one network topology could be used to estimate the fitness structure of networks in the other network topology without additional objective evaluations. This type of approach could substantially improve the accuracy of evolutionary annealing.

It was mentioned in Chapter 14 that evolutionary annealing has been tested in several domains, including bit strings, structure-learning for Bayesian networks, and game-playing strategies. In each of these domains, competitive results were obtained on benchmark problems. Nonetheless, there is substantial effort involved in applying evolutionary annealing to a new domain. A partitioning method must be developed, along with a base measure and a set of effective mutation distributions.

Further experiments in other domains will promote the development of a generalized methodology for instantiating these objects.

Evolutionary annealing retains the complete results for every objective evaluation it performs. The requirements to store this data are manageable, but they also introduce substantial overhead. In addition, computing the next point with evolutionary annealing requires logarithmic rather than constant time in terms of the number of previously evaluated points. It would be desirable to reduce or eliminate this overhead where possible.

As a martingale method, evolutionary annealing is primarily concerned with preserving the full information provided by prior evaluations. However, it is possible that the complete information or a nearly complete approximation can be achieved by compressing the previously evaluated points into a smaller representation. A compressed representation would also have the benefit of generalizing the information learned from the previous evaluations so that objective evaluations are chosen more efficiently in regions where they are more likely to improve performance. Compressed information representations provide another direction for improving evolutionary annealing.

Evolutionary annealing was formulated in the context of static objective functions, and it assumes that fitness evaluations are permanent. But if the objective function is dynamic or stochastic, some alterations are required. In the case of stochastic domains, selection of partition regions could be performed in a way that selects a larger region higher in the tree containing several points, effectively averaging over the points in order to avoid committing to a point with spuriously optimal fitness. Alternatively, one could avoid partitioning a region until several evaluations within the region had been performed, so that the average evaluation on the region could be used rather than attaching a single search point to each partition region.

In adaptive environments, the compression of information might become especially important if the dynamics of the environment are predictable. In this case, it might work well to extract a set of invariant principles governing the dynamics and to use these principles in conjunction with the observed objective values to determine which points to explore next. If the environment is unpredictable, then some form of strategic forgetting may be helpful. Evolutionary annealing with strategic forgetting would become similar to evolutionary algorithms in which individuals have a "lifetime" that might span several generations.

Aside from improvements to evolutionary annealing, the martingale optimization approach of Section 13.4.3 assumed that the variance of fitness evaluations could be used in addition to the mean evaluation. This variance was ignored in the development of evolutionary annealing, but correctly accounting for variance could produce a more powerful method.

Overall, the martingale optimization approach presents a fresh perspective on optimization that yields several interesting paths for future work. Methods such as semantic partitioning and information compression could improve on evolutionary annealing. Mutation operators for evolutionary annealing that capture the regularities of general-purpose function priors may also produce substantial advances in optimization technology that are more capable of searching high-dimensional spaces to find elegant solutions for important problems. Advanced martingale optimization methods may make better approximations of the information-maximization principle possible as well.

## 18.3 Final Thoughts: General-Purpose Optimization

Optimization tasks are ubiquitous throughout the engineering disciplines. This observation is especially true of artificial intelligence and machine learning, where nearly every problem is expressed in terms of searching for a solution that is optimal according to some criterion. Thus the study of optimization is central to the quest for a strong artificial intelligence.

This book has studied the relationships among optimization methods at a general level by examining the probability distribution over the sequence of evaluation points produced by the optimization process. This study produced several results that perhaps seemed unintuitive at the outset. Optimizers are vectors. There is a well-defined objective measure of distance between any two optimizers. Between any two optimizers there is an entire spectrum of optimizers, and in most cases behavior and performance changes smoothly along this line. The discovery of these facts was made possible by analyzing the optimization process as a mathematical object.

The results presented only scratch the surface of what is possible to achieve using such an analysis. Future work on convex control of optimization portfolios could provide a way to allocate resources automatically to the best optimizer for a particular problem. The study of performance-based linear projections could even make it possible to analytically construct optimizers that are well-aligned with specific problem classes. Accurate approximations of optimizer methods may enable the implementation of near-optimal optimizers based on the information-maximization principle. Each of these topics can be explored by applying mathematical disciplines such as functional analysis and topology to the space of optimization methods as described in this book.

One of the most important immediate contributions of this book is the No Free Lunch Identification Theorems, which demonstrated that in static and stochastic optimization settings as well as many others the No Free Lunch concept is equivalent to path independence. That is, No Free Lunch makes a prior assumption that objective evaluations at one search point provide no information whatsoever about the value of the objective at any other point. Since it is patently absurd to claim that objective evaluations in real-world problems are completely uncorrelated, this theorem refutes the claim that no effective general-purpose optimization algorithms exist for problems emanating from the real world. A successful general-purpose optimizer should therefore structure its search so as to prioritize solutions that are more likely to be correct. In line with the principle of Occam's razor, simpler solutions should be preferred over more complex ones. For real-world problems, physical principles such a locality, smoothness, periodicity, and fractal structure should be used to guide the optimization process efficiently.

The construction of the diffusion prior demonstrated mathematically that there exist very general problem classes on which certain optimization strategies outperform others. The recognition that the No Free Lunch theorems do not preclude general-purpose optimization led to the articulation of the Information Maximization Principle, based on a proof from Bellman equations that the optimal optimization method for any particular problem class is the one that makes the full use of the information obtained from function evaluations. Explicitly information-maximizing optimizers form a new class of martingale-based optimization methods that deserve further theoretical and experimental study.

As an initial step in this direction, this book has proposed evolutionary annealing, which samples the same distribution as simulated annealing but replaces the Markov-based Metropolis algorithm with a martingale representation that successively partitions the search domain. A proof of asymptotic global convergence stated the conditions under which evolutionary annealing can be expected to find the true global optimum. More importantly, experiments with real vectors and neural networks demonstrated the effectiveness of evolutionary annealing as a practical optimization method. Neuroannealing in particular was structured in such a way as to implement the information-maximization principle in accordance with Occam's razor, preferring simple network solutions over more complex ones until the simple networks have been ruled out.

Future research into information-maximizing optimizers will find more compact ways of partitioning the search space to represent the knowledge obtained through the optimization process. They will leverage the physical principles that govern the natural world in order to develop more effective means for choosing the next evaluation points. And they should ultimately outperform existing methods by a wide margin in more complex domains.

Much of the analysis in this book pertained to the case of randomized static optimization, in which there is a fixed objective function that is selected stochastically at the outset of optimization. This setting is fairly limited in that it does not account for stochastic or dynamic evaluations. The introduction of the optimization game made it possible to study a much wider array of optimization settings using the formal approach of this book. In particular, key results such as the No Free Lunch Identification Theorem and the Information Maximization Principle were shown to hold in these expanded spaces based on adversarial fitnesses. In this chapter, a further refinement to this approach has been suggested that produces a Hilbert space of optimization methods in which a bilinear map can be derived relating players in this Hilbert space to optimization games. The study of the optimization game as presented in this book is preliminary, but it demonstrates the claim that the formal approach has wide application and that there remain many theoretical and practical results that can be derived from it.

The greatest known optimizer at present is the human brain. By methodically applying scientific principles in conjunction with a creative instinct, human researchers have utilized knowledge gained from experience to construct increasingly refined and accurate models of the natural world. The exact nature of this creative instinct is poorly understood at present, but it may be surmised that human creativity comprises a set of hidden mental operations that project past observations into highly probable future states that accord with a core set of fundamental physical principles. If this hypothesis is true, then the most important endeavor in the search for a general artificial intelligence is to identify these fundamental principles. Once enumerated, these principles can be used to construct a general-purpose information-maximizing optimization method capable of human-level discoveries.

Thus the study of all optimization methods taken together has reinforced a fundamental insight regarding the nature of learning and artificial intelligence. General-purpose learners can be effective to the degree that their assumptions and biases reflect the physical laws of their environment. The study of artificial intelligence must in fact be a study of the abstract pillars of reality. By incorporating these principles, it should be possible to develop general-purpose learners of increasing capability and true intelligence.

# Appendix A
# Performance Experiment Results

This appendix contains the results for the performance experiments in Chapters 11 and 15 in tabular form (Tables A.1 to A.51). Tables are presented for each of the performance criteria $\sigma_\varepsilon^N$ (success probability with threshold $\varepsilon$), $\hat{\psi}_\varepsilon^N$ (number of evaluations until success), $\zeta_{T_m}$ (average error after $m$ evaluations), $\phi_1$ (average error over all evaluations), and $\phi_2$ (weighted average error over all evaluations) in order on the following pages (see Chapter 11 for details). Tables are grouped by the dimension of the experiments, with $d = 5$, 10, and 25. Additionally, the results for the neural network experiments in Chapter 16 are aggregated in a single table at the end of this appendix.

The values of $\zeta_{T_m}$, $\phi_1$, and $\phi_2$ are scaled as described in Chapter 11. The scaling factors for each benchmark are listed in Table A.1. Additionally, variances are provided for $\zeta_{T_m}$, $\phi_1$, and $\phi_2$ in separate tables so that statistical significance can be checked. All values are based on 200 trials. The variances for $\hat{\psi}_\varepsilon^N$ were not given, since these averages were only computed for successful trials, and the number of successful trials varies in every case. The estimated values of $\sigma_\varepsilon^N$ are accurate up to $\pm 0.005$ with $p < 0.05$. Any values greater than $100,000$ appeared only in rare cases and were written simply as "$\infty$".

Some of the algorithms were run with different parameters. The specific parameters are shown in the second column of the table. For CMA-ES and CMA-ES-R, the parameter is the population size. For DE, the parameters are given as CR / F, so that ".2/.9" means a crossover rate of .2 and a learning rate of .9. For PSO, the parameters are listed as $\omega/\phi_g$ in the same way. The algorithms REA-P and REA-T were run with different learning rates as shown in Table 15.2, reproduced in this appendix as Table A.2 for convenience. The different results for each learning rates are shown order from top to bottom for each benchmark, matching the order of the values in Table 15.2 from left to right. That is, the smallest learning rates are given at the top, and the largest learning rates at the bottom. Extra entries in the table were marked with "–".



A. J. Lockett, *General-Purpose Optimization Through Information Maximization*,
Natural Computing Series, https://doi.org/10.1007/978-3-662-62007-6

**Table A.1** Scaling factors used for the scaled variants of $\zeta_{T_m}$, $\phi_1$, and $\phi_2$ by dimension $d$.

| | $d = 5$ | $d = 10$ | $d = 25$ |
|---|---|---|---|
| sphere | 1.247 | 12.552 | 84.481 |
| ackley | 2.442 | 4.059 | 5.509 |
| log-ackley | 0.393 | 24.250 | 184.941 |
| whitley | 43.998 | 20,726.140 | 5,018,118.903 |
| shekel | 10.472 | 10.247 | – |
| rosenbrock | 100.254 | 2,926.197 | 64,226.113 |
| rastrigin | 16.567 | 70.078 | 272.727 |
| salomon | 0.927 | 2.453 | 5.740 |
| langerman | 0.746 | 0.965 | – |
| schwefel | 99.070 | 187.058 | 270.419 |
| griewank | 5.125 | 43.479 | 290.635 |
| weierstrass | 3.248 | 10.275 | 34.156 |

**Table A.2** Learning rates $\eta$ for REA-P and REA-T tested in the experiments. Lower values yield higher success probability at the cost of slower convergence.

| | REA-P | | REA-T | | |
|---|---|---|---|---|---|
| Benchmark | $d = 5$ | $d = 10$ | $d = 5$ | $d = 10$ | $d = 25$ |
| sphere | 10 | 1, 10 | 10 | 1, 10 | 0.1, 1,10 |
| ackley | 0.25 | 0.25, 1 | 0.25 | 0.25, 1 | 0.05, 0.25, 1 |
| log-ackley | 0.25 | 0.25, 1 | 0.05, 0.25 | 0.25, 1 | 0.05, .25, 1 |
| whitley | 0.1 | 0.25, 1 | 0.05, 0.25 | 0.25, 1 | 0.05, 0.25, 1 |
| shekel | 0.1, 0.25 | 0.1, 1 | 0.1, 0.5, 1.0, 5.0 | 0.1, 1 | – |
| rosenbrock | 1 | 1, 5 | 5 | 1, 5 | 0.1, 1,5 |
| rastrigin | 0.01, 0.1 | 0.035, 1 | 0.01, 0.035, 0.050, 0.075 | 0.035, 1 | 0.01, 0.035, 1 |
| salomon | 2 | 1, 2 | 2 | 1, 2 | 0.1, 1, 2 |
| langerman | 0.1, 0.5 | 0.25, 1 | 0.1, 0.5, 1.0, 5.0 | 0.25, 1 | – |
| schwefel | 0.015 | 0.001, 0.01 | 0.001 | 0.001, 0.01 | 0.0001, 0.001, 0.01 |
| griewank | 1, 10 | 0.1, 1 | 0.025, 0.1, 0.25, 0.5 | 0.1, 1 | 0.01, 0.1, 1 |
| weierstrass | 5 | 1, 5 | 5 | 1, 5 | 0,1, 1, 5 |

**Table A.3** Results on performance criterion $\sigma_{\varepsilon}^{N}$ with $\varepsilon = 0.100$ and $N = 250,000$ in 5 dimensions.

| $\sigma_{\varepsilon}^{N}$ | | sphr. | ack. | lg-ack. | whit. | shek. | rosen. | rastr. | sal. | lange. | schw. | grie. | weier. |
|---|---|---|---|---|---|---|---|---|---|---|---|---|---|
| CG | | 1.000 | 0.000 | 0.000 | 0.000 | 0.005 | 0.825 | 0.000 | 0.000 | 0.000 | 0.005 | 0.000 | 0.000 |
| CG-R | | 1.000 | 0.000 | 0.000 | 0.005 | 0.995 | 1.000 | 0.010 | 1.000 | 0.990 | 0.180 | 0.000 | 0.000 |
| CMA-ES | 100 | 0.999 | 0.669 | 0.613 | 0.003 | 0.000 | 0.027 | 0.413 | 0.915 | 0.400 | 0.020 | 0.999 | 0.580 |
| | 750 | 1.000 | 1.000 | 0.890 | 0.625 | 0.000 | 0.300 | 1.000 | 1.000 | 0.860 | 0.110 | 1.000 | 0.755 |
| | 1250 | 1.000 | 1.000 | 0.875 | 0.750 | 0.000 | 0.905 | 1.000 | 1.000 | 0.900 | 0.200 | 1.000 | 0.695 |
| | 2500 | 1.000 | 1.000 | 0.820 | 0.245 | 0.000 | 0.795 | 1.000 | 1.000 | 0.900 | 0.345 | 1.000 | 0.255 |
| CMA-ES-R | 100 | 1.000 | 1.000 | 1.000 | 0.025 | 0.000 | 0.125 | 1.000 | 1.000 | 1.000 | 0.390 | 1.000 | 1.000 |
| | 750 | 1.000 | 0.905 | 0.235 | 0.600 | 0.000 | 0.000 | 0.980 | 1.000 | 0.510 | 0.080 | 0.080 | 0.030 |
| | 1250 | 1.000 | 0.000 | 0.010 | 0.605 | 0.000 | 0.000 | 0.670 | 1.000 | 0.015 | 0.020 | 0.000 | 0.000 |
| | 2500 | 0.730 | 0.000 | 0.000 | 0.064 | 0.000 | 0.000 | 0.175 | 0.064 | 0.020 | 0.025 | 0.000 | 0.000 |
| DE | .2/.9 | 1.000 | 1.000 | 0.160 | 0.005 | 0.010 | 0.005 | 0.195 | 0.095 | 0.155 | 0.100 | 0.015 | 0.000 |
| | .2/.2 | 1.000 | 1.000 | 0.925 | 0.175 | 0.110 | 0.135 | 0.920 | 1.000 | 0.305 | 0.785 | 1.000 | 0.485 |
| | .9/.2 | 1.000 | 1.000 | 0.925 | 0.100 | 0.045 | 0.420 | 0.005 | 1.000 | 0.220 | 0.000 | 0.890 | 0.000 |
| | .9/.9 | 0.170 | 1.000 | 0.000 | 0.000 | 0.000 | 0.000 | 0.000 | 0.000 | 0.000 | 0.000 | 0.000 | 0.000 |
| GSS | | 1.000 | 0.000 | 0.000 | 0.069 | 0.085 | 0.775 | 0.005 | 0.064 | 0.020 | 0.420 | 0.000 | 0.095 |
| GSS-R | | 1.000 | 0.480 | 0.025 | 0.650 | 0.755 | 0.740 | 0.095 | 0.771 | 0.054 | 1.000 | 0.085 | 0.680 |
| NM | | 1.000 | 0.020 | 0.000 | 0.054 | 0.045 | 0.750 | 0.000 | 0.000 | 0.000 | 0.015 | 0.405 | 0.375 |
| NM-R | | 1.000 | 0.185 | 0.940 | 1.000 | 1.000 | 1.000 | 0.175 | 0.135 | 0.040 | 0.855 | 1.000 | 1.000 |
| PSO | –0.5/2 | 0.820 | 0.000 | 0.000 | 0.059 | 0.000 | 0.000 | 0.010 | 0.075 | 0.000 | 0.105 | 0.000 | 1.000 |
| | 1/2 | 0.000 | 0.000 | 0.000 | 0.000 | 0.000 | 0.000 | 0.000 | 0.000 | 0.000 | 0.000 | 0.000 | 1.000 |
| REA-P | R1 | 1.000 | 0.984 | 1.000 | 1.000 | 0.000 | 0.989 | 0.000 | 1.000 | 0.005 | 0.261 | 0.693 | 0.904 |
| | R2 | – | – | – | – | 0.000 | – | 0.291 | – | 0.000 | – | 0.412 | – |
| REA-T | R1 | 1.000 | 0.285 | 0.985 | 1.000 | 0.728 | 1.000 | 0.391 | 0.995 | 1.000 | 0.995 | 1.000 | 0.960 |
| | R2 | – | – | 0.840 | 0.825 | 0.326 | – | 0.226 | – | 0.959 | – | 1.000 | – |
| | R3 | – | – | – | – | 0.195 | – | 0.195 | – | 0.920 | – | 1.000 | – |
| | R4 | – | – | – | – | 0.069 | – | 0.040 | – | 0.680 | – | 0.990 | – |
| SA | | 1.000 | 0.000 | 0.000 | 0.000 | 0.000 | 1.000 | 0.115 | 0.190 | 0.125 | 0.000 | 0.000 | 0.000 |
| rBOA | | 1.000 | 1.000 | 0.965 | 0.000 | 0.000 | 0.000 | 1.000 | 1.000 | 0.100 | 0.000 | 1.000 | 0.000 |
| rGA | | 1.000 | 0.000 | 0.020 | 0.685 | 0.020 | 0.069 | 0.010 | 1.000 | 0.495 | 0.520 | 0.075 | 0.000 |

**Table A.4** Results on performance criterion $\sigma_\varepsilon^N$ with $\varepsilon = 0.010$ and $N = 250,000$ in 5 dimensions.

| $\sigma_\varepsilon^N$ | | sphr. | ack. | lg-ack. | whit. | shek. | rosen. | rastr. | sal. | lange. | schw. | grie. | weier. |
|---|---|---|---|---|---|---|---|---|---|---|---|---|---|
| CG | | 1.000 | 0.000 | 0.000 | 0.000 | 0.005 | 0.825 | 0.000 | 0.000 | 0.000 | 0.005 | 0.000 | 0.000 |
| CG-R | | 1.000 | 0.000 | 0.000 | 0.005 | 0.995 | 1.000 | 0.010 | 1.000 | 0.830 | 0.180 | 0.000 | 0.000 |
| CMA-ES | 100 | 0.999 | 0.665 | 0.613 | 0.003 | 0.000 | 0.004 | 0.413 | 0.000 | 0.209 | 0.020 | 0.678 | 0.265 |
| | 750 | 1.000 | 1.000 | 0.885 | 0.625 | 0.000 | 0.140 | 1.000 | 0.000 | 0.430 | 0.110 | 0.995 | 0.610 |
| | 1250 | 1.000 | 1.000 | 0.875 | 0.750 | 0.000 | 0.655 | 1.000 | 0.000 | 0.495 | 0.200 | 1.000 | 0.485 |
| | 2500 | 1.000 | 1.000 | 0.820 | 0.245 | 0.000 | 0.345 | 1.000 | 0.000 | 0.300 | 0.335 | 1.000 | 0.015 |
| CMA-ES-R | 100 | 1.000 | 1.000 | 1.000 | 0.025 | 0.000 | 0.020 | 1.000 | 0.000 | 0.995 | 0.380 | 1.000 | 0.995 |
| | 750 | 0.845 | 0.250 | 0.090 | 0.390 | 0.000 | 0.000 | 0.435 | 0.000 | 0.205 | 0.000 | 0.015 | 0.000 |
| | 1250 | 0.845 | 0.000 | 0.010 | 0.320 | 0.000 | 0.000 | 0.190 | 0.000 | 0.000 | 0.000 | 0.000 | 0.000 |
| | 2500 | 0.160 | 0.000 | 0.000 | 0.030 | 0.000 | 0.000 | 0.080 | 0.000 | 0.000 | 0.005 | 0.000 | 0.000 |
| DE | .2/.9 | 0.995 | 0.660 | 0.005 | 0.000 | 0.000 | 0.000 | 0.000 | 0.000 | 0.049 | 0.000 | 0.000 | 0.000 |
| | .2/.2 | 1.000 | 0.765 | 0.925 | 0.155 | 0.049 | 0.030 | 0.780 | 0.000 | 0.125 | 0.735 | 0.175 | 0.245 |
| | .9/.2 | 1.000 | 0.580 | 0.875 | 0.075 | 0.040 | 0.120 | 0.005 | 0.000 | 0.015 | 0.000 | 0.000 | 0.000 |
| | .9/.9 | 0.000 | 0.700 | 0.000 | 0.000 | 0.000 | 0.000 | 0.000 | 0.000 | 0.000 | 0.000 | 0.000 | 0.000 |
| GSS | | 1.000 | 0.000 | 0.000 | 0.069 | 0.085 | 0.700 | 0.005 | 0.000 | 0.005 | 0.420 | 0.000 | 0.095 |
| GSS-R | | 1.000 | 0.480 | 0.025 | 0.650 | 0.755 | 0.675 | 0.095 | 0.000 | 0.045 | 1.000 | 0.000 | 0.675 |
| NM | | 1.000 | 0.020 | 0.000 | 0.054 | 0.045 | 0.750 | 0.000 | 0.000 | 0.000 | 0.015 | 0.005 | 0.355 |
| NM-R | | 1.000 | 0.185 | 0.940 | 1.000 | 1.000 | 1.000 | 0.175 | 0.000 | 0.025 | 0.855 | 0.975 | 1.000 |
| PSO | –0.5/2 | 0.160 | 0.000 | 0.000 | 0.049 | 0.000 | 0.000 | 0.000 | 0.015 | 0.000 | 0.105 | 0.000 | 1.000 |
| | 1/2 | 0.000 | 0.000 | 0.000 | 0.000 | 0.000 | 0.000 | 0.000 | 0.000 | 0.000 | 0.000 | 0.000 | 1.000 |
| REA-P | R1 | 1.000 | 0.045 | 0.572 | 1.000 | 0.000 | 0.989 | 0.000 | 0.000 | 0.000 | 0.000 | 0.015 | 0.000 |
| | R2 | – | – | – | – | 0.000 | – | 0.010 | – | 0.000 | – | 0.000 | – |
| REA-T | R1 | 1.000 | 0.285 | 0.985 | 1.000 | 0.728 | 1.000 | 0.391 | 0.000 | 0.889 | 0.995 | 0.326 | 0.885 |
| | R2 | – | – | 0.840 | 0.825 | 0.326 | – | 0.226 | – | 0.688 | – | 0.095 | – |
| | R3 | – | – | – | – | 0.195 | – | 0.195 | – | 0.580 | – | 0.090 | – |
| | R4 | – | – | – | – | 0.069 | – | 0.040 | – | 0.300 | – | 0.040 | – |
| SA | | 1.000 | 0.000 | 0.000 | 0.000 | 0.000 | 0.120 | 0.000 | 0.000 | 0.000 | 0.000 | 0.000 | 0.000 |
| rBOA | | 1.000 | 1.000 | 0.010 | 0.000 | 0.000 | 0.000 | 1.000 | 0.855 | 0.000 | 0.000 | 1.000 | 0.000 |
| rGA | | 1.000 | 0.000 | 0.020 | 0.255 | 0.000 | 0.000 | 0.000 | 0.000 | 0.245 | 0.010 | 0.000 | 0.000 |

**Table A.5** Results on performance criterion $\sigma_\varepsilon^N$ with $\varepsilon = 0.001$ and $N = 250,000$ in 5 dimensions.

| $\sigma_\varepsilon^N$ | | sphr. | ack. | lg-ack. | whit. | shek. | rosen. | rastr. | sal. | lange. | schw. | grie. | weier. |
|---|---|---|---|---|---|---|---|---|---|---|---|---|---|
| CG | | 1.000 | 0.000 | 0.000 | 0.000 | 0.005 | 0.825 | 0.000 | 0.000 | 0.000 | 0.005 | 0.000 | 0.000 |
| CG-R | | 1.000 | 0.000 | 0.000 | 0.005 | 0.995 | 1.000 | 0.010 | 1.000 | 0.820 | 0.170 | 0.000 | 0.000 |
| CMA-ES | 100 | 0.999 | 0.665 | 0.609 | 0.003 | 0.000 | 0.000 | 0.413 | 0.000 | 0.209 | 0.018 | 0.339 | 0.100 |
| | 750 | 1.000 | 1.000 | 0.885 | 0.625 | 0.000 | 0.069 | 1.000 | 0.000 | 0.430 | 0.105 | 0.970 | 0.460 |
| | 1250 | 1.000 | 1.000 | 0.875 | 0.750 | 0.000 | 0.490 | 1.000 | 0.000 | 0.495 | 0.200 | 0.985 | 0.365 |
| | 2500 | 1.000 | 1.000 | 0.820 | 0.245 | 0.000 | 0.145 | 1.000 | 0.000 | 0.295 | 0.290 | 1.000 | 0.000 |
| CMA-ES-R | 100 | 1.000 | 1.000 | 1.000 | 0.025 | 0.000 | 0.010 | 1.000 | 0.000 | 0.995 | 0.370 | 1.000 | 0.000 |
| | 750 | 0.190 | 0.000 | 0.015 | 0.145 | 0.000 | 0.000 | 0.030 | 0.000 | 0.085 | 0.000 | 0.005 | 0.000 |
| | 1250 | 0.415 | 0.000 | 0.000 | 0.064 | 0.000 | 0.000 | 0.025 | 0.000 | 0.000 | 0.000 | 0.000 | 0.000 |
| | 2500 | 0.025 | 0.000 | 0.000 | 0.010 | 0.000 | 0.000 | 0.059 | 0.000 | 0.000 | 0.000 | 0.000 | 0.000 |
| DE | .2/.9 | 0.425 | 0.000 | 0.000 | 0.000 | 0.000 | 0.000 | 0.000 | 0.000 | 0.010 | 0.000 | 0.000 | 0.000 |
| | .2/.2 | 1.000 | 0.005 | 0.925 | 0.150 | 0.040 | 0.000 | 0.645 | 0.000 | 0.010 | 0.705 | 0.005 | 0.185 |
| | .9/.2 | 1.000 | 0.000 | 0.660 | 0.059 | 0.035 | 0.015 | 0.005 | 0.000 | 0.000 | 0.000 | 0.000 | 0.000 |
| | .9/.9 | 0.000 | 0.000 | 0.000 | 0.000 | 0.000 | 0.000 | 0.000 | 0.000 | 0.000 | 0.000 | 0.000 | 0.000 |
| GSS | | 1.000 | 0.000 | 0.000 | 0.069 | 0.085 | 0.660 | 0.005 | 0.000 | 0.005 | 0.420 | 0.000 | 0.095 |
| GSS-R | | 1.000 | 0.480 | 0.025 | 0.650 | 0.755 | 0.635 | 0.095 | 0.000 | 0.045 | 1.000 | 0.000 | 0.675 |
| NM | | 1.000 | 0.020 | 0.000 | 0.054 | 0.045 | 0.750 | 0.000 | 0.000 | 0.000 | 0.015 | 0.000 | 0.350 |
| NM-R | | 1.000 | 0.185 | 0.940 | 1.000 | 1.000 | 1.000 | 0.175 | 0.000 | 0.025 | 0.855 | 0.430 | 1.000 |
| PSO | –0.5/2 | 0.020 | 0.000 | 0.000 | 0.035 | 0.000 | 0.000 | 0.000 | 0.005 | 0.000 | 0.105 | 0.000 | 1.000 |
| | 1/2 | 0.000 | 0.000 | 0.000 | 0.000 | 0.000 | 0.000 | 0.000 | 0.000 | 0.000 | 0.000 | 0.000 | 1.000 |
| REA-P | R1 | 1.000 | 0.000 | 0.000 | 1.000 | 0.000 | 0.135 | 0.000 | 0.000 | 0.000 | 0.000 | 0.000 | 0.000 |
| | R2 | – | – | – | – | 0.000 | – | 0.000 | – | 0.000 | – | 0.000 | – |
| REA-T | R1 | 1.000 | 0.285 | 0.985 | 1.000 | 0.728 | 0.995 | 0.391 | 0.000 | 0.889 | 0.995 | 0.045 | 0.845 |
| | R2 | – | – | 0.840 | 0.825 | 0.326 | – | 0.226 | – | 0.688 | – | 0.020 | – |
| | R3 | – | – | – | – | 0.195 | – | 0.195 | – | 0.580 | – | 0.005 | – |
| | R4 | – | – | – | – | 0.069 | – | 0.040 | – | 0.300 | – | 0.010 | – |
| SA | | 0.100 | 0.000 | 0.000 | 0.000 | 0.000 | 0.000 | 0.000 | 0.000 | 0.000 | 0.000 | 0.000 | 0.000 |
| rBOA | | 1.000 | 1.000 | 0.000 | 0.000 | 0.000 | 0.000 | 1.000 | 0.850 | 0.000 | 0.000 | 1.000 | 0.000 |
| rGA | | 1.000 | 0.000 | 0.010 | 0.015 | 0.000 | 0.000 | 0.000 | 0.000 | 0.240 | 0.000 | 0.000 | 0.000 |

**Table A.6** Results on performance criterion $\frac{1}{100}\hat{\psi}_{\varepsilon}^{N}$ with $\varepsilon = 0.100$ and $N = 250,000$ in 5 dimensions.

| $\frac{1}{100}\hat{\psi}_{\varepsilon}^{N}$ | | sphr. | ack. | lg-ack. | whit. | shek. | rosen. | rastr. | sal. | lange. | schw. | grie. | weier. |
|---|---|---|---|---|---|---|---|---|---|---|---|---|---|
| CG | | 0.0 | – | – | – | 3.0 | 16.6 | – | – | – | 1.0 | – | – |
| CG-R | | 0.0 | – | – | 873.0 | 545.6 | 21.0 | 1995.5 | 78.2 | 532.5 | 1138.0 | – | – |
| CMA-ES | 100 | 4.9 | 21.8 | 21.5 | 19.0 | – | 224.3 | 18.0 | 12.8 | 14.8 | 107.0 | 18.2 | 64.8 |
| | 750 | 87.9 | 149.6 | 151.0 | 197.0 | – | 1127.4 | 169.5 | 102.6 | 135.6 | 356.3 | 190.9 | 390.0 |
| | 1250 | 204.9 | 339.5 | 347.6 | 413.5 | – | 1135.6 | 373.8 | 224.9 | 287.2 | 703.4 | 417.8 | 835.2 |
| | 2500 | 614.8 | 1071.6 | 1078.6 | 1235.2 | – | 2207.2 | 1146.7 | 741.1 | 868.4 | 1752.8 | 1324.1 | 2185.7 |
| CMA-ES-R | 100 | 4.9 | 28.8 | 37.5 | 1540.6 | – | 1322.6 | 52.7 | 20.1 | 100.1 | 1165.1 | 18.2 | 71.7 |
| | 750 | 200.6 | 915.7 | 1006.5 | 1174.8 | – | – | 690.6 | 104.5 | 1178.9 | 1476.2 | 1440.2 | 1524.8 |
| | 1250 | 409.1 | – | 318.5 | 1395.8 | – | – | 1174.3 | 261.6 | 1916.3 | 1749.7 | – | – |
| | 2500 | 1268.3 | – | – | 1828.8 | – | – | 1515.0 | 1021.1 | 1743.7 | 1920.0 | – | – |
| DE | .2/.9 | 225.6 | 355.7 | 1786.7 | 1138.0 | 2306.5 | 1656.0 | 1896.3 | 1687.1 | 1415.5 | 1874.9 | 1840.0 | – |
| | .2/.2 | 65.4 | 311.7 | 331.4 | 486.2 | 1265.6 | 834.9 | 1051.2 | 571.3 | 1315.0 | 1204.8 | 383.3 | 1592.2 |
| | .9/.2 | 68.3 | 338.6 | 1067.5 | 1080.0 | 1647.2 | 1611.9 | 1612.0 | 874.9 | 1269.4 | – | 1285.6 | – |
| | .9/.9 | 1249.6 | 351.7 | – | – | – | – | – | – | – | – | – | – |
| GSS | | 5.0 | – | – | 11.5 | 11.1 | 339.3 | 11.0 | 10.4 | 6.7 | 10.8 | – | 8.5 |
| GSS-R | | 4.9 | 1165.1 | 1078.7 | 932.2 | 1064.3 | 348.5 | 1144.0 | 951.7 | 203.6 | 235.7 | 1227.6 | 609.0 |
| NM | | 0.4 | 1.2 | – | 1.2 | 1.0 | 3.9 | – | – | – | 1.6 | 1.4 | 0.0 |
| NM-R | | 0.4 | 951.1 | 697.9 | 190.2 | 137.4 | 8.0 | 1028.9 | 1171.2 | 1314.7 | 798.8 | 10.7 | 5.9 |
| PSO | –0.5/2 | 705.2 | – | – | 1144.0 | – | – | 1242.5 | 854.1 | – | 137.5 | – | 0.0 |
| | 1/2 | – | – | – | – | – | – | – | – | – | – | – | 0.0 |
| REA-P | R1 | 8.6 | 1284.3 | 146.0 | 150.5 | – | 140.1 | – | 255.3 | 1865.0 | 1182.3 | 68.7 | 118.7 |
| | R2 | – | – | – | – | – | – | 1113.3 | – | – | – | 1495.2 | – |
| REA-T | R1 | 7.7 | 104.2 | 138.7 | 124.3 | 229.1 | 100.1 | 515.7 | 26.3 | 63.3 | 1133.2 | 395.3 | 19.9 |
| | R2 | – | – | 67.9 | 63.4 | 101.3 | – | 257.1 | – | 34.6 | – | 199.6 | – |
| | R3 | – | – | – | – | 76.0 | – | 148.3 | – | 27.5 | – | 132.8 | – |
| | R4 | – | – | – | – | 42.8 | – | 56.6 | – | 16.8 | – | 99.2 | – |
| SA | | 6.7 | – | – | – | – | 237.3 | 1369.0 | 1456.2 | 1182.8 | – | – | – |
| rBOA | | 7.7 | 37.5 | 753.6 | – | – | – | 36.9 | 23.6 | 1349.6 | – | 15.2 | – |
| rGA | | 26.1 | – | 37.2 | 840.8 | 135.5 | 411.2 | 251.5 | 320.7 | 1113.1 | 620.0 | 1531.6 | – |

**Table A.7** Results on performance criterion $\frac{1}{100}\hat{\psi}_{\varepsilon}^{N}$ with $\varepsilon = 0.010$ and $N = 250,000$ in 5 dimensions.

| $\frac{1}{100}\hat{\psi}_{\varepsilon}^{N}$ | | sphr. | ack. | lg-ack. | whit. | shek. | rosen. | rastr. | sal. | lange. | schw. | grie. | weier. |
|---|---|---|---|---|---|---|---|---|---|---|---|---|---|
| CG | | 0.0 | – | – | – | 3.0 | 16.6 | – | – | – | 1.0 | – | – |
| CG-R | | 0.0 | – | – | 873.0 | 548.0 | 21.0 | 1995.5 | 206.4 | 914.6 | 1138.0 | – | – |
| CMA-ES | 100 | 7.3 | 25.3 | 24.5 | 20.0 | – | 159.5 | 20.7 | – | 17.6 | 110.2 | 24.0 | 95.2 |
| | 750 | 126.9 | 187.2 | 185.2 | 211.9 | – | 1288.6 | 207.2 | – | 162.9 | 419.0 | 230.5 | 502.4 |
| | 1250 | 296.3 | 435.6 | 419.6 | 437.4 | – | 1447.1 | 462.0 | – | 345.2 | 785.3 | 508.5 | 1055.0 |
| | 2500 | 951.3 | 1349.8 | 1296.7 | 1287.7 | – | 2364.1 | 1436.7 | – | 1060.4 | 2036.9 | 1530.3 | 2283.3 |
| CMA-ES-R | 100 | 7.3 | 32.6 | 40.3 | 1542.8 | – | 974.0 | 55.4 | – | 298.3 | 1168.4 | 53.2 | 421.2 |
| | 750 | 914.5 | 1112.4 | 1386.0 | 1199.4 | – | – | 1109.0 | – | 1227.7 | – | 1527.3 | – |
| | 1250 | 1111.6 | – | 349.5 | 1447.9 | – | – | 1177.0 | – | – | – | – | – |
| | 2500 | 1714.8 | – | – | 1783.3 | – | – | 1575.0 | – | – | 1825.0 | – | – |
| DE | .2/.9 | 939.4 | 1487.0 | 2436.0 | – | – | – | – | – | 1698.1 | – | – | – |
| | .2/.2 | 126.0 | 1459.5 | 492.1 | 617.7 | 1367.1 | 862.1 | 1470.5 | – | 1728.8 | 1399.5 | 1630.2 | 1750.7 |
| | .9/.2 | 167.7 | 1516.2 | 1455.8 | 1409.8 | 1719.0 | 1811.7 | 1740.0 | – | 1396.3 | – | – | – |
| | .9/.9 | – | 1550.5 | – | – | – | – | – | – | – | – | – | – |
| GSS | | 7.6 | – | – | 13.3 | 14.4 | 330.6 | 14.0 | – | 7.0 | 13.4 | – | 13.7 |
| GSS-R | | 7.6 | 1168.1 | 1080.5 | 934.4 | 1066.9 | 331.8 | 1146.6 | – | 233.1 | 238.2 | – | 623.5 |
| NM | | 0.7 | 1.2 | – | 1.3 | 1.2 | 4.5 | – | – | – | 2.0 | 2.0 | 0.0 |
| NM-R | | 0.7 | 951.5 | 698.3 | 190.3 | 137.7 | 8.6 | 1029.1 | – | 1278.6 | 799.0 | 565.3 | 6.5 |
| PSO | –0.5/2 | 1087.0 | – | – | 1149.1 | – | – | – | 1511.3 | – | 257.4 | – | 0.0 |
| | 1/2 | – | – | – | – | – | – | – | – | – | – | – | 0.0 |
| REA-P | R1 | 17.2 | 1971.0 | 1423.9 | 246.6 | – | 385.4 | – | – | – | – | 357.6 | – |
| | R2 | – | – | – | – | – | – | 1707.0 | – | – | – | – | – |
| REA-T | R1 | 11.8 | 114.7 | 156.4 | 133.4 | 264.8 | 351.9 | 553.3 | – | 80.6 | 1287.3 | 541.3 | 28.9 |
| | R2 | – | – | 77.4 | 68.3 | 117.0 | – | 274.9 | – | 41.7 | – | 251.6 | – |
| | R3 | – | – | – | – | 88.0 | – | 160.8 | – | 32.2 | – | 160.4 | – |
| | R4 | – | – | – | – | 50.7 | – | 62.3 | – | 19.1 | – | 118.1 | – |
| SA | | 165.4 | – | – | – | – | 1439.0 | – | – | – | – | – | – |
| rBOA | | 10.8 | 44.0 | 812.0 | – | – | – | 39.3 | 659.5 | – | – | 51.7 | – |
| rGA | | 34.9 | – | 78.7 | 1278.5 | – | – | – | – | 1127.6 | 1877.0 | – | – |

**Table A.8** Results on performance criterion $\frac{1}{100}\hat{\psi}_{\varepsilon}^{N}$ with $\varepsilon = 0.001$ and $N = 250,000$ in 5 dimensions.

| $\frac{1}{100}\hat{\psi}_{\varepsilon}^{N}$ | | sphr. | ack. | lg-ack. | whit. | shek. | rosen. | rastr. | sal. | lange. | schw. | grie. | weier. |
|---|---|---|---|---|---|---|---|---|---|---|---|---|---|
| CG | | 0.0 | – | – | – | 3.0 | 16.6 | – | – | – | 1.0 | – | – |
| CG-R | | 0.0 | – | – | 873.0 | 548.0 | 21.0 | 1995.5 | 206.4 | 969.5 | 1170.5 | – | – |
| CMA-ES | 100 | 9.7 | 29.2 | 26.1 | 21.0 | – | – | 23.0 | – | 20.2 | 42.7 | 35.9 | 177.4 |
| | 750 | 164.3 | 243.3 | 223.4 | 230.3 | – | 1619.7 | 244.8 | – | 193.1 | 411.5 | 266.7 | 617.1 |
| | 1250 | 386.6 | 561.8 | 506.1 | 471.0 | – | 1573.0 | 553.6 | – | 400.5 | 881.9 | 592.0 | 1275.1 |
| | 2500 | 1244.5 | 1785.6 | 1555.9 | 1392.3 | – | 2400.8 | 1755.0 | – | 1288.1 | 2294.8 | 1805.0 | – |
| CMA-ES-R | 100 | 9.7 | 36.3 | 42.9 | 1545.6 | – | 1012.5 | 57.9 | – | 300.9 | 1203.4 | 131.1 | – |
| | 750 | 1163.3 | – | 1694.6 | 1205.7 | – | – | 1588.5 | – | 1392.2 | – | 1095.0 | – |
| | 1250 | 1393.7 | – | – | 1574.7 | – | – | 1275.0 | – | – | – | – | – |
| | 2500 | 1960.0 | – | – | 1787.5 | – | – | 1791.6 | – | – | – | – | – |
| DE | .2/.9 | 1831.0 | – | – | – | – | – | – | – | 1932.5 | – | – | – |
| | .2/.2 | 193.1 | 1661.0 | 639.9 | 685.5 | 1309.2 | – | 1658.9 | – | 1767.5 | 1513.9 | 2499.0 | 1884.6 |
| | .9/.2 | 273.1 | – | 1543.2 | 1507.7 | 1696.1 | 1979.6 | 1862.0 | – | – | – | – | – |
| | .9/.9 | – | – | – | – | – | – | – | – | – | – | – | – |
| GSS | | 10.3 | – | – | 15.7 | 16.8 | 373.3 | 16.0 | – | 10.0 | 16.0 | – | 18.5 |
| GSS-R | | 10.2 | 1172.1 | 1084.0 | 936.4 | 1069.6 | 349.5 | 1149.8 | – | 234.8 | 240.9 | – | 628.5 |
| NM | | 1.0 | 1.5 | – | 1.5 | 1.6 | 4.9 | – | – | – | 2.3 | – | 0.0 |
| NM-R | | 1.0 | 951.9 | 698.5 | 190.6 | 137.9 | 9.0 | 1029.5 | – | 1278.8 | 799.3 | 1118.1 | 6.6 |
| PSO | –0.5/2 | 1905.5 | – | – | 1237.4 | – | – | – | 1993.0 | – | 676.1 | – | 0.0 |
| | 1/2 | – | – | – | – | – | – | – | – | – | – | – | 0.0 |
| REA-P | R1 | 166.2 | – | – | 848.6 | – | 1691.5 | – | – | – | – | – | – |
| | R2 | – | – | – | – | – | – | – | – | – | – | – | – |
| REA-T | R1 | 15.8 | 126.6 | 174.1 | 141.7 | 298.9 | 838.1 | 585.3 | – | 92.6 | 1447.6 | 527.1 | 37.7 |
| | R2 | – | – | 86.4 | 72.5 | 131.6 | – | 290.4 | – | 48.1 | – | 253.2 | – |
| | R3 | – | – | – | – | 98.8 | – | 170.4 | – | 37.7 | – | 149.0 | – |
| | R4 | – | – | – | – | 56.2 | – | 68.2 | – | 23.4 | – | 119.5 | – |
| SA | | 1311.0 | – | – | – | – | – | – | – | – | – | – | – |
| rBOA | | 13.2 | 47.9 | – | – | – | – | 42.2 | 661.1 | – | – | 59.0 | – |
| rGA | | 94.8 | – | 387.0 | 1592.0 | – | – | – | – | 1110.6 | – | – | – |

**Table A.9** Results on performance criterion $\zeta_{T_m}$ (scaled) with $m = 25,000$ in 5 dimensions.

| $\zeta_{T_m}$ (scaled) | | sphr. | ack. | lg-ack. | whit. | shek. | rosen. | rastr. | sal. | lange. | schw. | grie. | weier. |
|---|---|---|---|---|---|---|---|---|---|---|---|---|---|
| CG | | 0.000 | 2.323 | 168.967 | 14.178 | 0.789 | 1.069 | 2.555 | 3.831 | 1.290 | 2.298 | 13.171 | 3.038 |
| CG-R | | 0.000 | 0.918 | 38.969 | 0.307 | 0.416 | 0.000 | 0.401 | 0.035 | 0.630 | 0.676 | 1.536 | 1.659 |
| CMA-ES | 100 | 0.000 | 0.026 | 1.110 | 0.089 | 0.762 | 0.012 | 0.046 | 0.112 | 0.261 | 0.978 | 0.001 | 0.060 |
| | 750 | 0.000 | 0.000 | 0.289 | 0.016 | 0.748 | 0.015 | 0.000 | 0.106 | 0.070 | 0.653 | 0.000 | 0.555 |
| | 1250 | 0.036 | 0.390 | 10.236 | 0.312 | 0.839 | 0.062 | 0.246 | 0.108 | 0.422 | 1.187 | 0.166 | 1.305 |
| | 2500 | 0.254 | 0.675 | 26.277 | 0.469 | 0.858 | 0.229 | 0.565 | 0.498 | 0.802 | 1.051 | 0.349 | 1.113 |
| CMA-ES-R | 100 | 0.000 | 0.000 | 0.013 | 0.038 | 0.736 | 0.012 | 0.001 | 0.107 | 0.041 | 0.564 | 0.000 | 0.006 |
| | 750 | 0.057 | 0.387 | 26.086 | 0.333 | 0.865 | 0.061 | 0.035 | 0.107 | 0.711 | 0.652 | 0.282 | 0.424 |
| | 1250 | 0.123 | 0.693 | 20.884 | 0.367 | 0.865 | 0.128 | 0.247 | 0.166 | 0.787 | 1.068 | 0.389 | 1.090 |
| | 2500 | 0.264 | 0.685 | 27.786 | 0.485 | 0.864 | 0.227 | 0.549 | 0.519 | 0.915 | 1.051 | 0.383 | 1.130 |
| DE | .2/.9 | 0.075 | 0.057 | 12.595 | 0.354 | 0.794 | 0.096 | 0.250 | 0.454 | 0.678 | 0.234 | 0.242 | 0.494 |
| | .2/.2 | 0.000 | 0.052 | 1.432 | 0.103 | 0.693 | 0.018 | 0.151 | 0.194 | 0.645 | 0.277 | 0.034 | 0.461 |
| | .9/.2 | 0.001 | 0.055 | 7.202 | 0.215 | 0.702 | 0.024 | 0.395 | 0.212 | 0.600 | 0.807 | 0.073 | 0.951 |
| | .9/.9 | 0.383 | 0.055 | 34.254 | 0.525 | 0.856 | 0.338 | 0.690 | 0.666 | 0.795 | 0.918 | 0.528 | 1.000 |
| GSS | | 0.000 | 2.356 | 18.674 | 0.103 | 0.718 | 0.005 | 0.312 | 0.478 | 0.872 | 0.309 | 0.098 | 0.755 |
| GSS-R | | 0.000 | 0.128 | 12.708 | 0.062 | 0.666 | 0.005 | 0.222 | 0.297 | 0.806 | 0.123 | 0.074 | 0.499 |
| NM | | 0.000 | 0.212 | 77.499 | 0.239 | 0.745 | 0.040 | 1.643 | 2.728 | 1.281 | 1.482 | 1.398 | 0.291 |
| NM-R | | 0.000 | 0.118 | 3.471 | 0.010 | 0.125 | 0.000 | 0.162 | 0.460 | 1.005 | 0.279 | 0.003 | 0.000 |
| PSO | −0.5/2 | 0.232 | 0.338 | 37.586 | 0.476 | 0.892 | 0.334 | 1.195 | 0.620 | 1.145 | 0.427 | 0.396 | 0.000 |
| | 1/2 | 4.590 | 0.340 | 91.646 | 132.687 | 0.937 | 9.609 | 2.163 | 2.018 | 1.289 | 0.897 | 4.086 | 0.000 |
| REA-P | R1 | 0.000 | 0.219 | 0.226 | 0.000 | 0.863 | 0.002 | 0.625 | 0.109 | 0.858 | 0.108 | 0.016 | 0.064 |
| | R2 | – | – | – | – | 0.866 | – | 0.087 | – | 0.867 | – | 0.049 | – |
| REA-T | R1 | 0.000 | 0.064 | 0.037 | 0.000 | 0.180 | 0.000 | 0.360 | 0.108 | 0.005 | 0.601 | 0.034 | 0.010 |
| | R2 | – | – | 0.402 | 0.003 | 0.450 | – | 0.075 | – | 0.030 | – | 0.006 | – |
| | R3 | – | – | – | – | 0.547 | – | 0.067 | – | 0.053 | – | 0.006 | – |
| | R4 | – | – | – | – | 0.660 | – | 0.119 | – | 0.169 | – | 0.008 | – |
| SA | | 0.005 | 0.815 | 26.249 | 0.326 | 0.862 | 0.003 | 0.141 | 0.311 | 0.618 | 0.681 | 0.108 | 0.646 |
| rBOA | | 0.000 | 0.000 | 0.435 | 0.194 | 0.811 | 0.024 | 0.000 | 0.065 | 0.609 | 2.046 | 0.000 | 1.127 |
| rGA | | 0.000 | 0.435 | 16.841 | 0.066 | 0.756 | 0.015 | 0.216 | 0.108 | 0.941 | 0.136 | 0.046 | 0.533 |

**Table A.10** Variance for performance criterion $\zeta_{T_m}$ (scaled) with $m = 25,000$ in 5 dimensions.

| $Var(\zeta_{T_m})$ (scaled) | | sphr. | ack. | lg-ack. | whit. | shek. | rosen. | rastr. | sal. | lange. | schw. | grie. | weier. |
|---|---|---|---|---|---|---|---|---|---|---|---|---|---|
| CG | | 0.000 | 0.412 | 58.034 | 141.565 | 0.079 | 10.672 | 1.106 | 1.283 | 0.045 | 0.853 | 7.080 | 0.562 |
| CG-R | | 0.000 | 0.208 | 14.311 | 0.093 | 0.319 | 0.000 | 0.169 | 0.061 | 0.522 | 0.266 | 0.909 | 0.231 |
| CMA-ES | 100 | 0.000 | 0.038 | 1.483 | 0.085 | 0.037 | 0.006 | 0.046 | 0.020 | 0.227 | 0.443 | 0.001 | 0.086 |
| | 750 | 0.000 | 0.000 | 0.817 | 0.022 | 0.026 | 0.003 | 0.000 | 0.006 | 0.120 | 0.342 | 0.000 | 0.304 |
| | 1250 | 0.017 | 0.071 | 3.803 | 0.040 | 0.039 | 0.020 | 0.078 | 0.011 | 0.160 | 0.217 | 0.031 | 0.172 |
| | 2500 | 0.114 | 0.089 | 5.905 | 0.060 | 0.046 | 0.101 | 0.132 | 0.107 | 0.155 | 0.179 | 0.084 | 0.153 |
| CMA-ES-R | 100 | 0.000 | 0.000 | 0.195 | 0.042 | 0.008 | 0.005 | 0.010 | – | 0.082 | 0.345 | 0.000 | 0.008 |
| | 750 | 0.051 | 0.218 | 8.272 | 0.128 | 0.023 | 0.049 | 0.036 | 0.001 | 0.174 | 0.331 | 0.106 | 0.309 |
| | 1250 | 0.104 | 0.090 | 7.908 | 0.074 | 0.021 | 0.057 | 0.086 | 0.133 | 0.164 | 0.218 | 0.090 | 0.172 |
| | 2500 | 0.115 | 0.091 | 5.727 | 0.061 | 0.025 | 0.098 | 0.141 | 0.108 | 0.161 | 0.214 | 0.087 | 0.159 |
| DE | .2/.9 | 0.051 | 0.033 | 5.104 | 0.071 | 0.059 | 0.054 | 0.096 | 0.127 | 0.143 | 0.136 | 0.052 | 0.132 |
| | .2/.2 | 0.000 | 0.031 | 1.209 | 0.052 | 0.143 | 0.010 | 0.068 | 0.062 | 0.156 | 0.183 | 0.011 | 0.113 |
| | .9/.2 | 0.001 | 0.033 | 2.787 | 0.057 | 0.123 | 0.009 | 0.117 | 0.063 | 0.106 | 0.238 | 0.018 | 0.164 |
| | .9/.9 | 0.174 | 0.032 | 6.976 | 0.083 | 0.030 | 0.169 | 0.150 | 0.159 | 0.152 | 0.200 | 0.149 | 0.142 |
| GSS | | 0.000 | 0.415 | 10.806 | 0.071 | 0.224 | 0.008 | 0.186 | 0.335 | 0.296 | 0.378 | 0.039 | 0.514 |
| GSS-R | | 0.000 | 0.046 | 6.740 | 0.052 | 0.251 | 0.007 | 0.109 | 0.163 | 0.326 | 0.198 | 0.028 | 0.458 |
| NM | | 0.000 | 0.103 | 49.791 | 0.181 | 0.168 | 0.361 | 0.958 | 1.213 | 0.071 | 0.619 | 4.643 | 0.397 |
| NM-R | | 0.000 | 0.046 | 3.318 | 0.019 | 0.259 | 0.000 | 0.081 | 0.168 | 0.252 | 0.198 | 0.001 | 0.000 |
| PSO | –0.5/2 | 0.253 | 0.066 | 11.900 | 0.159 | 0.022 | 0.476 | 0.400 | 0.290 | 0.150 | 0.225 | 0.225 | 0.000 |
| | 1/2 | 2.090 | 0.070 | 19.158 | 245.385 | 0.014 | 7.659 | 0.454 | 0.473 | 0.022 | 0.070 | 1.876 | 0.000 |
| REA-P | R1 | 0.000 | 0.045 | 0.492 | 0.008 | 0.022 | 0.009 | 0.146 | 0.008 | 0.169 | 0.129 | 0.010 | 0.130 |
| | R2 | – | – | – | – | 0.026 | – | 0.044 | – | 0.167 | – | 0.011 | – |
| REA-T | R1 | 0.000 | 0.043 | 0.305 | 0.000 | 0.291 | 0.000 | 0.092 | 0.007 | 0.018 | 0.183 | 0.008 | 0.057 |
| | R2 | – | – | 0.922 | 0.009 | 0.313 | – | 0.040 | – | 0.085 | – | 0.003 | – |
| | R3 | – | – | – | – | 0.271 | – | 0.044 | – | 0.116 | – | 0.003 | – |
| | R4 | – | – | – | – | 0.184 | – | 0.057 | – | 0.217 | – | 0.004 | – |
| SA | | 0.002 | 0.112 | 6.018 | 0.048 | 0.026 | 0.010 | 0.063 | 0.103 | 0.146 | 0.237 | 0.023 | 0.113 |
| rBOA | | 0.000 | 0.000 | 0.236 | 0.014 | 0.024 | 0.006 | 0.000 | 0.047 | 0.128 | 0.361 | 0.000 | 0.158 |
| rGA | | 0.000 | 0.150 | 10.581 | 0.053 | 0.121 | 0.014 | 0.116 | 0.001 | 0.329 | 0.167 | 0.010 | 0.077 |

**Table A.11** Results on performance criterion $\zeta_{T_m}$ (scaled) with $m = 100,000$ in 5 dimensions.

| $\zeta_{T_m}$ (scaled) | | sphr. | ack. | lg-ack. | whit. | shek. | rosen. | rastr. | sal. | lange. | schw. | grie. | weier. |
|---|---|---|---|---|---|---|---|---|---|---|---|---|---|
| CG | | 0.000 | 2.323 | 168.967 | 14.178 | 0.789 | 1.069 | 2.555 | 3.831 | 1.290 | 2.298 | 13.171 | 3.038 |
| CG-R | | 0.000 | 0.690 | 24.186 | 0.208 | 0.118 | 0.000 | 0.225 | 0.001 | 0.153 | 0.418 | 0.626 | 1.416 |
| CMA-ES | 100 | 0.000 | 0.026 | 1.101 | 0.086 | 0.762 | 0.012 | 0.046 | 0.112 | 0.261 | 0.969 | 0.001 | 0.056 |
| | 750 | 0.000 | 0.000 | 0.288 | 0.013 | 0.748 | 0.005 | 0.000 | 0.106 | 0.069 | 0.392 | 0.000 | 0.044 |
| | 1250 | 0.000 | 0.000 | 0.333 | 0.008 | 0.758 | 0.001 | 0.000 | 0.105 | 0.045 | 0.271 | 0.000 | 0.066 |
| | 2500 | 0.006 | 0.150 | 2.185 | 0.188 | 0.781 | 0.031 | 0.027 | 0.104 | 0.078 | 0.811 | 0.083 | 0.942 |
| CMA-ES-R | 100 | 0.000 | 0.000 | 0.000 | 0.014 | 0.735 | 0.006 | 0.000 | 0.107 | 0.001 | 0.293 | 0.000 | 0.002 |
| | 750 | 0.012 | 0.104 | 17.128 | 0.168 | 0.840 | 0.027 | 0.004 | 0.106 | 0.444 | 0.156 | 0.173 | 0.124 |
| | 1250 | 0.022 | 0.598 | 17.158 | 0.163 | 0.841 | 0.053 | 0.016 | 0.105 | 0.591 | 0.281 | 0.300 | 0.219 |
| | 2500 | 0.123 | 0.581 | 21.118 | 0.353 | 0.837 | 0.065 | 0.091 | 0.383 | 0.623 | 0.807 | 0.300 | 0.846 |
| DE | .2/.9 | 0.008 | 0.011 | 3.346 | 0.189 | 0.654 | 0.030 | 0.074 | 0.287 | 0.537 | 0.037 | 0.131 | 0.239 |
| | .2/.2 | 0.000 | 0.011 | 0.190 | 0.018 | 0.616 | 0.009 | 0.018 | 0.109 | 0.485 | 0.050 | 0.007 | 0.136 |
| | .9/.2 | 0.000 | 0.011 | 0.571 | 0.081 | 0.608 | 0.004 | 0.228 | 0.108 | 0.444 | 0.450 | 0.021 | 0.686 |
| | .9/.9 | 0.213 | 0.012 | 25.813 | 0.442 | 0.823 | 0.178 | 0.510 | 0.494 | 0.642 | 0.676 | 0.369 | 0.842 |
| GSS | | 0.000 | 2.356 | 18.674 | 0.103 | 0.718 | 0.003 | 0.312 | 0.478 | 0.872 | 0.309 | 0.098 | 0.755 |
| GSS-R | | 0.000 | 0.074 | 6.954 | 0.021 | 0.449 | 0.004 | 0.123 | 0.177 | 0.790 | 0.002 | 0.054 | 0.287 |
| NM | | 0.000 | 0.212 | 77.499 | 0.239 | 0.745 | 0.040 | 1.643 | 2.728 | 1.281 | 1.482 | 1.398 | 0.291 |
| NM-R | | 0.000 | 0.088 | 0.812 | 0.000 | 0.000 | 0.000 | 0.084 | 0.302 | 0.744 | 0.101 | 0.001 | 0.000 |
| PSO | –0.5/2 | 0.091 | 0.338 | 29.919 | 0.392 | 0.888 | 0.128 | 0.798 | 0.403 | 1.082 | 0.320 | 0.267 | 0.000 |
| | 1/2 | 4.586 | 0.340 | 91.117 | 132.687 | 0.937 | 9.609 | 2.163 | 2.018 | 1.289 | 0.893 | 4.064 | 0.000 |
| REA-P | R1 | 0.000 | 0.077 | 0.047 | 0.000 | 0.829 | 0.001 | 0.435 | 0.107 | 0.699 | 0.082 | 0.016 | 0.053 |
| | R2 | – | – | – | – | 0.837 | – | 0.056 | – | 0.700 | – | 0.029 | – |
| REA-T | R1 | 0.000 | 0.064 | 0.037 | 0.000 | 0.178 | 0.000 | 0.038 | 0.108 | 0.005 | 0.009 | 0.003 | 0.010 |
| | R2 | – | – | 0.402 | 0.003 | 0.450 | – | 0.055 | – | 0.030 | – | 0.004 | – |
| | R3 | – | – | – | – | 0.547 | – | 0.067 | – | 0.053 | – | 0.006 | – |
| | R4 | – | – | – | – | 0.660 | – | 0.119 | – | 0.169 | – | 0.008 | – |
| SA | | 0.002 | 0.691 | 19.443 | 0.275 | 0.828 | 0.000 | 0.058 | 0.169 | 0.478 | 0.391 | 0.078 | 0.500 |
| rBOA | | 0.000 | 0.000 | 0.213 | 0.175 | 0.790 | 0.019 | 0.000 | 0.031 | 0.454 | 2.046 | 0.000 | 0.938 |
| rGA | | 0.000 | 0.434 | 16.837 | 0.034 | 0.756 | 0.005 | 0.214 | 0.107 | 0.582 | 0.135 | 0.033 | 0.459 |

**Table A.12** Variance for performance criterion $\zeta_{T_m}$ (scaled) with $m = 100,000$ in 5 dimensions.

| $Var(\zeta_{T_m})$ (scaled) | | sphr. | ack. | lg-ack. | whit. | shek. | rosen. | rastr. | sal. | lange. | schw. | grie. | weier. |
|---|---|---|---|---|---|---|---|---|---|---|---|---|---|
| CG | | 0.000 | 0.412 | 58.034 | 141.565 | 0.079 | 10.672 | 1.106 | 1.283 | 0.045 | 0.853 | 7.080 | 0.562 |
| CG-R | | 0.000 | 0.159 | 9.245 | 0.068 | 0.252 | 0.000 | 0.098 | 0.010 | 0.285 | 0.213 | 0.424 | 0.194 |
| CMA-ES | 100 | 0.000 | 0.038 | 1.482 | 0.084 | 0.037 | 0.006 | 0.046 | 0.020 | 0.226 | 0.443 | 0.001 | 0.085 |
| | 750 | 0.000 | 0.000 | 0.817 | 0.018 | 0.026 | 0.004 | 0.000 | 0.006 | 0.120 | 0.251 | 0.000 | 0.106 |
| | 1250 | 0.000 | 0.000 | 0.906 | 0.016 | 0.036 | 0.001 | 0.000 | 0.010 | 0.087 | 0.202 | 0.000 | 0.107 |
| | 2500 | 0.003 | 0.074 | 1.783 | 0.049 | 0.054 | 0.008 | 0.019 | 0.011 | 0.066 | 0.177 | 0.019 | 0.221 |
| CMA-ES-R | 100 | 0.000 | 0.000 | 0.000 | 0.011 | 0.003 | 0.003 | 0.000 | 0.004 | 0.005 | 0.180 | 0.000 | 0.000 |
| | 750 | 0.013 | 0.155 | 8.885 | 0.153 | 0.024 | 0.009 | 0.005 | 0.005 | 0.265 | 0.136 | 0.069 | 0.050 |
| | 1250 | 0.031 | 0.081 | 6.749 | 0.120 | 0.028 | 0.019 | 0.016 | 0.009 | 0.094 | 0.212 | 0.055 | 0.093 |
| | 2500 | 0.071 | 0.084 | 4.677 | 0.077 | 0.029 | 0.025 | 0.073 | 0.107 | 0.098 | 0.190 | 0.050 | 0.310 |
| DE | .2/.9 | 0.007 | 0.008 | 2.240 | 0.069 | 0.189 | 0.016 | 0.044 | 0.085 | 0.155 | 0.035 | 0.046 | 0.068 |
| | .2/.2 | 0.000 | 0.007 | 0.666 | 0.015 | 0.253 | 0.007 | 0.024 | 0.015 | 0.189 | 0.096 | 0.003 | 0.063 |
| | .9/.2 | 0.000 | 0.007 | 0.800 | 0.045 | 0.234 | 0.004 | 0.078 | 0.007 | 0.177 | 0.211 | 0.005 | 0.117 |
| | .9/.9 | 0.090 | 0.009 | 5.714 | 0.059 | 0.039 | 0.083 | 0.109 | 0.116 | 0.084 | 0.181 | 0.083 | 0.119 |
| GSS | | 0.000 | 0.415 | 10.806 | 0.071 | 0.224 | 0.007 | 0.186 | 0.335 | 0.296 | 0.378 | 0.039 | 0.514 |
| GSS-R | | 0.000 | 0.039 | 3.283 | 0.028 | 0.334 | 0.007 | 0.051 | 0.062 | 0.331 | 0.023 | 0.021 | 0.452 |
| NM | | 0.000 | 0.103 | 49.791 | 0.181 | 0.168 | 0.361 | 0.958 | 1.213 | 0.071 | 0.619 | 4.643 | 0.397 |
| NM-R | | 0.000 | 0.034 | 1.264 | 0.002 | 0.000 | 0.000 | 0.044 | 0.095 | 0.225 | 0.123 | 0.000 | 0.000 |
| PSO | –0.5/2 | 0.081 | 0.066 | 9.366 | 0.141 | 0.021 | 0.100 | 0.386 | 0.229 | 0.166 | 0.180 | 0.085 | 0.000 |
| | 1/2 | 2.092 | 0.070 | 19.117 | 245.385 | 0.014 | 7.659 | 0.454 | 0.473 | 0.022 | 0.075 | 1.898 | 0.000 |
| REA-P | R1 | 0.000 | 0.039 | 0.020 | 0.000 | 0.036 | 0.007 | 0.104 | 0.000 | 0.123 | 0.121 | 0.010 | 0.128 |
| | R2 | – | – | – | – | 0.046 | – | 0.037 | – | 0.104 | – | 0.007 | – |
| REA-T | R1 | 0.000 | 0.043 | 0.305 | 0.000 | 0.292 | 0.000 | 0.032 | 0.007 | 0.018 | 0.017 | 0.001 | 0.057 |
| | R2 | – | – | 0.922 | 0.009 | 0.313 | – | 0.036 | – | 0.085 | – | 0.002 | – |
| | R3 | – | – | – | – | 0.271 | – | 0.044 | – | 0.116 | – | 0.003 | – |
| | R4 | – | – | – | – | 0.184 | – | 0.057 | – | 0.217 | – | 0.004 | – |
| SA | | 0.001 | 0.094 | 4.782 | 0.043 | 0.040 | 0.000 | 0.043 | 0.050 | 0.166 | 0.196 | 0.018 | 0.082 |
| rBOA | | 0.000 | 0.000 | 0.090 | 0.021 | 0.022 | 0.007 | 0.000 | 0.046 | 0.167 | 0.361 | 0.000 | 0.133 |
| rGA | | 0.000 | 0.150 | 10.581 | 0.032 | 0.122 | 0.004 | 0.116 | 0.000 | 0.450 | 0.167 | 0.007 | 0.060 |

**Table A.13** Results on performance criterion $\zeta_{T_m}$ (scaled) with $m = 250,000$ in 5 dimensions.

| $\zeta_{T_m}$ (scaled) | | sphr. | ack. | lg-ack. | whit. | shek. | rosen. | rastr. | sal. | lange. | schw. | grie. | weier. |
|---|---|---|---|---|---|---|---|---|---|---|---|---|---|
| CG | | 0.000 | 2.323 | 168.967 | 14.178 | 0.789 | 1.069 | 2.555 | 3.831 | 1.290 | 2.298 | 13.171 | 3.038 |
| CG-R | | 0.000 | 0.572 | 17.835 | 0.154 | 0.003 | 0.000 | 0.158 | 0.000 | 0.016 | 0.278 | 0.379 | 1.268 |
| CMA-ES | 100 | 0.000 | 0.026 | 1.101 | 0.085 | 0.762 | 0.012 | 0.046 | 0.112 | 0.261 | 0.969 | 0.001 | 0.055 |
| | 750 | 0.000 | 0.000 | 0.288 | 0.013 | 0.748 | 0.004 | 0.000 | 0.106 | 0.069 | 0.392 | 0.000 | 0.044 |
| | 1250 | 0.000 | 0.000 | 0.333 | 0.008 | 0.758 | 0.000 | 0.000 | 0.105 | 0.045 | 0.271 | 0.000 | 0.044 |
| | 2500 | 0.000 | 0.000 | 0.473 | 0.023 | 0.755 | 0.000 | 0.000 | 0.103 | 0.046 | 0.158 | 0.000 | 0.084 |
| CMA-ES-R | 100 | 0.000 | 0.000 | 0.000 | 0.008 | 0.732 | 0.003 | 0.000 | 0.107 | 0.000 | 0.154 | 0.000 | 0.001 |
| | 750 | 0.004 | 0.015 | 10.552 | 0.050 | 0.823 | 0.021 | 0.001 | 0.104 | 0.268 | 0.035 | 0.129 | 0.091 |
| | 1250 | 0.003 | 0.529 | 14.654 | 0.049 | 0.824 | 0.038 | 0.005 | 0.104 | 0.517 | 0.048 | 0.259 | 0.149 |
| | 2500 | 0.056 | 0.513 | 17.045 | 0.236 | 0.819 | 0.034 | 0.033 | 0.326 | 0.526 | 0.228 | 0.254 | 0.228 |
| DE | .2/.9 | 0.001 | 0.003 | 0.831 | 0.096 | 0.514 | 0.020 | 0.017 | 0.215 | 0.441 | 0.005 | 0.065 | 0.144 |
| | .2/.2 | 0.000 | 0.003 | 0.189 | 0.011 | 0.606 | 0.008 | 0.005 | 0.107 | 0.340 | 0.035 | 0.003 | 0.040 |
| | .9/.2 | 0.000 | 0.003 | 0.128 | 0.046 | 0.581 | 0.002 | 0.144 | 0.107 | 0.259 | 0.275 | 0.013 | 0.548 |
| | .9/.9 | 0.143 | 0.003 | 21.453 | 0.405 | 0.795 | 0.119 | 0.424 | 0.416 | 0.584 | 0.557 | 0.303 | 0.750 |
| GSS | | 0.000 | 2.356 | 18.674 | 0.103 | 0.718 | 0.002 | 0.312 | 0.478 | 0.872 | 0.309 | 0.098 | 0.755 |
| GSS-R | | 0.000 | 0.043 | 4.809 | 0.006 | 0.215 | 0.001 | 0.090 | 0.138 | 0.768 | 0.000 | 0.039 | 0.239 |
| NM | | 0.000 | 0.212 | 77.499 | 0.239 | 0.745 | 0.040 | 1.643 | 2.728 | 1.281 | 1.482 | 1.398 | 0.291 |
| NM-R | | 0.000 | 0.067 | 0.148 | 0.000 | 0.000 | 0.000 | 0.057 | 0.237 | 0.585 | 0.033 | 0.000 | 0.000 |
| PSO | –0.5/2 | 0.044 | 0.338 | 24.066 | 0.331 | 0.883 | 0.080 | 0.475 | 0.268 | 1.000 | 0.286 | 0.220 | 0.000 |
| | 1/2 | 4.586 | 0.340 | 91.059 | 132.687 | 0.937 | 9.609 | 2.163 | 2.018 | 1.289 | 0.893 | 4.064 | 0.000 |
| REA-P | R1 | 0.000 | 0.012 | 0.023 | 0.000 | 0.777 | 0.000 | 0.329 | 0.107 | 0.620 | 0.073 | 0.016 | 0.050 |
| | R2 | – | – | – | – | 0.811 | – | 0.049 | – | 0.638 | – | 0.020 | – |
| REA-T | R1 | 0.000 | 0.064 | 0.037 | 0.000 | 0.178 | 0.000 | 0.038 | 0.108 | 0.005 | 0.001 | 0.003 | 0.010 |
| | R2 | – | – | 0.402 | 0.003 | 0.450 | – | 0.055 | – | 0.030 | – | 0.004 | – |
| | R3 | – | – | – | – | 0.547 | – | 0.067 | – | 0.053 | – | 0.006 | – |
| | R4 | – | – | – | – | 0.660 | – | 0.119 | – | 0.169 | – | 0.008 | – |
| SA | | 0.001 | 0.605 | 15.293 | 0.244 | 0.802 | 0.000 | 0.026 | 0.125 | 0.339 | 0.267 | 0.065 | 0.430 |
| rBOA | | 0.000 | 0.000 | 0.140 | 0.164 | 0.778 | 0.017 | 0.000 | 0.012 | 0.314 | 2.046 | 0.000 | 0.840 |
| rGA | | 0.000 | 0.433 | 16.835 | 0.015 | 0.756 | 0.003 | 0.213 | 0.107 | 0.389 | 0.135 | 0.028 | 0.408 |

**Table A.14** Variance for performance criterion $\zeta_{T_m}$ (scaled) with $m = 250,000$ in 5 dimensions.

| $Var(\zeta_{T_m})$ (scaled) | | sphr. | ack. | lg-ack. | whit. | shek. | rosen. | rastr. | sal. | lange. | schw. | grie. | weier. |
|---|---|---|---|---|---|---|---|---|---|---|---|---|---|
| CG | | 0.000 | 0.412 | 58.034 | 141.565 | 0.079 | 10.672 | 1.106 | 1.283 | 0.045 | 0.853 | 7.080 | 0.562 |
| CG-R | | 0.000 | 0.137 | 7.385 | 0.063 | 0.046 | 0.000 | 0.072 | 0.000 | 0.065 | 0.176 | 0.240 | 0.172 |
| CMA-ES | 100 | 0.000 | 0.038 | 1.482 | 0.083 | 0.037 | 0.006 | 0.046 | 0.020 | 0.226 | 0.444 | 0.001 | 0.085 |
| | 750 | 0.000 | 0.000 | 0.817 | 0.018 | 0.026 | 0.003 | 0.000 | 0.006 | 0.120 | 0.251 | 0.000 | 0.106 |
| | 1250 | 0.000 | 0.000 | 0.906 | 0.016 | 0.036 | 0.001 | 0.000 | 0.010 | 0.087 | 0.202 | 0.000 | 0.091 |
| | 2500 | 0.000 | 0.000 | 1.049 | 0.016 | 0.053 | 0.001 | 0.000 | 0.013 | 0.067 | 0.135 | 0.000 | 0.077 |
| CMA-ES-R | 100 | 0.000 | 0.000 | 0.000 | 0.007 | 0.021 | 0.002 | 0.000 | 0.004 | 0.002 | 0.132 | 0.000 | 0.000 |
| | 750 | 0.004 | 0.019 | 8.982 | 0.096 | 0.039 | 0.004 | 0.001 | 0.012 | 0.258 | 0.064 | 0.060 | 0.032 |
| | 1250 | 0.006 | 0.075 | 5.660 | 0.088 | 0.030 | 0.012 | 0.006 | 0.011 | 0.130 | 0.070 | 0.046 | 0.045 |
| | 2500 | 0.045 | 0.081 | 4.417 | 0.104 | 0.032 | 0.012 | 0.031 | 0.087 | 0.128 | 0.168 | 0.043 | 0.106 |
| DE | .2/.9 | 0.001 | 0.002 | 0.706 | 0.049 | 0.286 | 0.011 | 0.015 | 0.071 | 0.206 | 0.004 | 0.032 | 0.044 |
| | .2/.2 | 0.000 | 0.002 | 0.666 | 0.011 | 0.267 | 0.006 | 0.017 | 0.000 | 0.231 | 0.087 | 0.001 | 0.039 |
| | .9/.2 | 0.000 | 0.002 | 0.522 | 0.032 | 0.271 | 0.004 | 0.058 | 0.000 | 0.152 | 0.174 | 0.004 | 0.105 |
| | .9/.9 | 0.062 | 0.002 | 5.050 | 0.051 | 0.050 | 0.048 | 0.101 | 0.097 | 0.091 | 0.151 | 0.060 | 0.122 |
| GSS | | 0.000 | 0.415 | 10.806 | 0.071 | 0.224 | 0.006 | 0.186 | 0.335 | 0.296 | 0.378 | 0.039 | 0.514 |
| GSS-R | | 0.000 | 0.038 | 1.887 | 0.012 | 0.314 | 0.004 | 0.044 | 0.048 | 0.330 | 0.000 | 0.014 | 0.460 |
| NM | | 0.000 | 0.103 | 49.791 | 0.181 | 0.168 | 0.361 | 0.958 | 1.213 | 0.071 | 0.619 | 4.643 | 0.397 |
| NM-R | | 0.000 | 0.034 | 0.588 | 0.000 | 0.000 | 0.000 | 0.034 | 0.072 | 0.169 | 0.081 | 0.000 | 0.000 |
| PSO | –0.5/2 | 0.046 | 0.066 | 7.733 | 0.148 | 0.021 | 0.055 | 0.289 | 0.157 | 0.198 | 0.187 | 0.055 | 0.000 |
| | 1/2 | 2.092 | 0.070 | 19.026 | 245.385 | 0.014 | 7.659 | 0.454 | 0.473 | 0.022 | 0.075 | 1.898 | 0.000 |
| REA-P | R1 | 0.000 | 0.009 | 0.010 | 0.000 | 0.112 | 0.003 | 0.075 | 0.000 | 0.090 | 0.114 | 0.010 | 0.127 |
| | R2 | – | – | – | – | 0.050 | – | 0.035 | – | 0.075 | – | 0.004 | – |
| REA-T | R1 | 0.000 | 0.043 | 0.305 | 0.000 | 0.292 | 0.000 | 0.032 | 0.007 | 0.018 | 0.016 | 0.001 | 0.057 |
| | R2 | – | – | 0.922 | 0.009 | 0.313 | – | 0.036 | – | 0.085 | – | 0.002 | – |
| | R3 | – | – | – | – | 0.271 | – | 0.044 | – | 0.116 | – | 0.003 | – |
| | R4 | – | – | – | – | 0.184 | – | 0.057 | – | 0.217 | – | 0.004 | – |
| SA | | 0.000 | 0.089 | 4.097 | 0.040 | 0.061 | 0.000 | 0.026 | 0.034 | 0.170 | 0.159 | 0.014 | 0.067 |
| rBOA | | 0.000 | 0.000 | 0.062 | 0.024 | 0.024 | 0.007 | 0.000 | 0.032 | 0.149 | 0.361 | 0.000 | 0.104 |
| rGA | | 0.000 | 0.150 | 10.581 | 0.023 | 0.122 | 0.001 | 0.116 | 0.000 | 0.462 | 0.167 | 0.006 | 0.061 |

**Table A.15** Results on performance criterion $\phi_1$ (scaled) in 5 dimensions.

| $\phi_1$ (scaled) | | sphr. | ack. | lg-ack. | whit. | shek. | rosen. | rastr. | sal. | lange. | schw. | grie. | weier. |
|---|---|---|---|---|---|---|---|---|---|---|---|---|---|
| CG | | 0.000 | 2.323 | 168.968 | 19.012 | 0.789 | 1.381 | 2.555 | 3.831 | 1.290 | 2.298 | 13.172 | 3.038 |
| CG-R | | 0.000 | 0.711 | 26.518 | 17.445 | 0.145 | 0.005 | 0.251 | 0.021 | 0.218 | 0.450 | 0.839 | 1.429 |
| CMA-ES | 100 | 0.000 | 0.027 | 1.111 | 0.087 | 0.762 | 0.012 | 0.047 | 0.112 | 0.261 | 0.974 | 0.001 | 0.060 |
| | 750 | 0.016 | 0.030 | 1.468 | 0.039 | 0.755 | 0.022 | 0.029 | 0.123 | 0.109 | 0.482 | 0.019 | 0.175 |
| | 1250 | 0.033 | 0.073 | 3.142 | 0.066 | 0.773 | 0.036 | 0.062 | 0.142 | 0.141 | 0.450 | 0.043 | 0.314 |
| | 2500 | 0.067 | 0.221 | 8.189 | 0.180 | 0.788 | 0.072 | 0.165 | 0.205 | 0.279 | 0.543 | 0.117 | 0.677 |
| CMA-ES-R | 100 | 0.000 | 0.000 | 0.032 | 0.018 | 0.735 | 0.007 | 0.001 | 0.107 | 0.016 | 0.319 | 0.000 | 0.008 |
| | 750 | 0.032 | 0.144 | 17.035 | 0.169 | 0.841 | 0.045 | 0.036 | 0.122 | 0.464 | 0.259 | 0.191 | 0.213 |
| | 1250 | 0.056 | 0.600 | 17.902 | 0.174 | 0.841 | 0.076 | 0.075 | 0.147 | 0.619 | 0.364 | 0.311 | 0.355 |
| | 2500 | 0.138 | 0.582 | 21.621 | 0.348 | 0.838 | 0.101 | 0.197 | 0.399 | 0.662 | 0.619 | 0.307 | 0.696 |
| DE | .2/.9 | 0.036 | 0.021 | 5.220 | 0.207 | 0.644 | 0.058 | 0.106 | 0.309 | 0.552 | 0.087 | 0.147 | 0.278 |
| | .2/.2 | 0.011 | 0.019 | 1.354 | 0.053 | 0.634 | 0.023 | 0.055 | 0.143 | 0.484 | 0.109 | 0.027 | 0.187 |
| | .9/.2 | 0.009 | 0.020 | 2.601 | 0.106 | 0.623 | 0.017 | 0.248 | 0.141 | 0.428 | 0.480 | 0.041 | 0.710 |
| | .9/.9 | 0.243 | 0.021 | 26.708 | 0.465 | 0.823 | 0.207 | 0.527 | 0.513 | 0.663 | 0.699 | 0.395 | 0.855 |
| GSS | | 0.000 | 2.356 | 18.674 | 0.103 | 0.718 | 0.003 | 0.312 | 0.478 | 0.872 | 0.309 | 0.098 | 0.755 |
| GSS-R | | 0.000 | 0.172 | 7.178 | 0.023 | 0.391 | 0.004 | 0.128 | 0.184 | 0.784 | 0.031 | 0.051 | 0.308 |
| NM | | 0.000 | 0.212 | 77.499 | 0.239 | 0.745 | 0.040 | 1.643 | 2.728 | 1.281 | 1.482 | 1.398 | 0.291 |
| NM-R | | 0.000 | 0.088 | 1.538 | 0.004 | 0.034 | 0.000 | 0.099 | 0.327 | 0.751 | 0.132 | 0.002 | 0.000 |
| PSO | –0.5/2 | 0.107 | 0.338 | 29.792 | 0.398 | 0.887 | 0.162 | 0.782 | 0.412 | 1.071 | 0.341 | 0.279 | 0.000 |
| | 1/2 | 4.590 | 0.340 | 91.196 | 132.701 | 0.937 | 9.609 | 2.163 | 2.018 | 1.289 | 0.894 | 4.069 | 0.000 |
| REA-P | R1 | 0.000 | 0.089 | 0.226 | 0.008 | 0.822 | 0.001 | 0.456 | 0.113 | 0.712 | 0.093 | 0.018 | 0.057 |
| | R2 | – | – | – | – | 0.835 | – | 0.073 | – | 0.721 | – | 0.032 | – |
| REA-T | R1 | 0.000 | 0.075 | 0.721 | 0.011 | 0.213 | 0.000 | 0.111 | 0.109 | 0.018 | 0.148 | 0.018 | 0.010 |
| | R2 | – | – | 0.683 | 0.008 | 0.458 | – | 0.093 | – | 0.035 | – | 0.011 | – |
| | R3 | – | – | – | – | 0.552 | – | 0.090 | – | 0.056 | – | 0.010 | – |
| | R4 | – | – | – | – | 0.662 | – | 0.127 | – | 0.170 | – | 0.010 | – |
| SA | | 0.003 | 0.697 | 19.848 | 0.278 | 0.828 | 0.001 | 0.077 | 0.201 | 0.472 | 0.420 | 0.081 | 0.518 |
| rBOA | | 0.000 | 0.002 | 0.318 | 0.177 | 0.792 | 0.019 | 0.003 | 0.034 | 0.456 | 2.046 | 0.000 | 0.958 |
| rGA | | 0.003 | 0.435 | 16.944 | 0.046 | 0.757 | 0.013 | 0.216 | 0.115 | 0.596 | 0.137 | 0.037 | 0.464 |

**Table A.16** Variance for performance criterion $\phi_1$ (scaled) in 5 dimensions.

| $Var(\phi_1)$ (scaled) | | sphr. | ack. | lg-ack. | whit. | shek. | rosen. | rastr. | sal. | lange. | schw. | grie. | weier. |
|---|---|---|---|---|---|---|---|---|---|---|---|---|---|
| CG | | 0.000 | 0.412 | 58.034 | 141.302 | 0.079 | 10.644 | 1.106 | 1.283 | 0.045 | 0.853 | 7.080 | 0.562 |
| CG-R | | 0.000 | 0.123 | 7.287 | 23.847 | 0.142 | 0.004 | 0.074 | 0.016 | 0.184 | 0.160 | 0.319 | 0.148 |
| CMA-ES | 100 | 0.000 | 0.038 | 1.480 | 0.083 | 0.037 | 0.006 | 0.046 | 0.020 | 0.226 | 0.441 | 0.001 | 0.084 |
| | 750 | 0.005 | 0.002 | 0.795 | 0.018 | 0.025 | 0.007 | 0.005 | 0.006 | 0.114 | 0.229 | 0.003 | 0.100 |
| | 1250 | 0.009 | 0.006 | 0.900 | 0.015 | 0.032 | 0.011 | 0.010 | 0.012 | 0.079 | 0.164 | 0.006 | 0.090 |
| | 2500 | 0.021 | 0.019 | 1.279 | 0.018 | 0.048 | 0.019 | 0.025 | 0.022 | 0.054 | 0.104 | 0.015 | 0.112 |
| CMA-ES-R | 100 | 0.000 | 0.001 | 0.041 | 0.010 | 0.006 | 0.002 | 0.001 | 0.003 | 0.015 | 0.142 | 0.000 | 0.003 |
| | 750 | 0.010 | 0.086 | 7.053 | 0.098 | 0.025 | 0.011 | 0.007 | 0.006 | 0.198 | 0.088 | 0.055 | 0.048 |
| | 1250 | 0.024 | 0.063 | 5.714 | 0.074 | 0.022 | 0.015 | 0.014 | 0.017 | 0.083 | 0.109 | 0.042 | 0.075 |
| | 2500 | 0.049 | 0.065 | 3.610 | 0.057 | 0.023 | 0.024 | 0.038 | 0.078 | 0.080 | 0.127 | 0.040 | 0.130 |
| DE | .2/.9 | 0.013 | 0.007 | 1.554 | 0.046 | 0.172 | 0.017 | 0.031 | 0.064 | 0.138 | 0.031 | 0.034 | 0.050 |
| | .2/.2 | 0.004 | 0.007 | 0.656 | 0.017 | 0.228 | 0.008 | 0.021 | 0.011 | 0.162 | 0.091 | 0.004 | 0.044 |
| | .9/.2 | 0.003 | 0.007 | 0.749 | 0.034 | 0.214 | 0.006 | 0.057 | 0.010 | 0.123 | 0.168 | 0.004 | 0.080 |
| | .9/.9 | 0.068 | 0.008 | 4.486 | 0.049 | 0.031 | 0.064 | 0.082 | 0.090 | 0.070 | 0.135 | 0.063 | 0.099 |
| GSS | | 0.000 | 0.415 | 10.806 | 0.071 | 0.224 | 0.006 | 0.186 | 0.335 | 0.296 | 0.378 | 0.039 | 0.514 |
| GSS-R | | 0.000 | 0.034 | 2.410 | 0.019 | 0.250 | 0.007 | 0.043 | 0.051 | 0.321 | 0.034 | 0.014 | 0.434 |
| NM | | 0.000 | 0.103 | 49.791 | 0.181 | 0.168 | 0.361 | 0.958 | 1.213 | 0.071 | 0.619 | 4.643 | 0.397 |
| NM-R | | 0.000 | 0.030 | 1.143 | 0.004 | 0.037 | 0.000 | 0.034 | 0.072 | 0.159 | 0.089 | 0.003 | 0.000 |
| PSO | –0.5/2 | 0.077 | 0.066 | 7.646 | 0.128 | 0.020 | 0.122 | 0.275 | 0.178 | 0.160 | 0.174 | 0.080 | 0.000 |
| | 1/2 | 2.089 | 0.070 | 18.984 | 245.378 | 0.014 | 7.659 | 0.454 | 0.473 | 0.022 | 0.072 | 1.890 | 0.000 |
| REA-P | R1 | 0.000 | 0.019 | 0.128 | 0.002 | 0.044 | 0.005 | 0.072 | 0.002 | 0.095 | 0.114 | 0.010 | 0.128 |
| | R2 | – | – | – | – | 0.038 | – | 0.034 | – | 0.081 | – | 0.004 | – |
| REA-T | R1 | 0.000 | 0.042 | 0.308 | 0.001 | 0.275 | 0.000 | 0.030 | 0.007 | 0.018 | 0.026 | 0.002 | 0.057 |
| | R2 | – | – | 0.918 | 0.010 | 0.306 | – | 0.034 | – | 0.084 | – | 0.002 | – |
| | R3 | – | – | – | – | 0.266 | – | 0.043 | – | 0.116 | – | 0.003 | – |
| | R4 | – | – | – | – | 0.182 | – | 0.057 | – | 0.216 | – | 0.004 | – |
| SA | | 0.001 | 0.074 | 3.719 | 0.035 | 0.032 | 0.002 | 0.030 | 0.041 | 0.127 | 0.154 | 0.013 | 0.061 |
| rBOA | | 0.000 | 0.001 | 0.088 | 0.018 | 0.020 | 0.006 | 0.002 | 0.035 | 0.120 | 0.361 | 0.000 | 0.097 |
| rGA | | 0.003 | 0.149 | 10.561 | 0.030 | 0.121 | 0.009 | 0.116 | 0.003 | 0.388 | 0.167 | 0.006 | 0.047 |

**Table A.17** Results on performance criterion $\phi_2$ (scaled) in 5 dimensions.

| $\phi_2$ (scaled) | | sphr. | ack. | lg-ack. | whit. | shek. | rosen. | rastr. | sal. | lange. | schw. | grie. | weier. |
|---|---|---|---|---|---|---|---|---|---|---|---|---|---|
| CG | | 0.000 | 2.323 | 169.502 | 1226.341 | 0.789 | 104.733 | 2.555 | 3.831 | 1.290 | 2.481 | 13.448 | 3.038 |
| CG-R | | 0.000 | 1.700 | 109.497 | 3534.461 | 0.735 | 1.584 | 1.356 | 1.307 | 1.218 | 2.213 | 10.130 | 2.480 |
| CMA-ES | 100 | 0.000 | 0.242 | 5.513 | 0.234 | 0.779 | 0.032 | 0.392 | 0.124 | 0.534 | 1.806 | 0.060 | 1.030 |
| | 750 | 0.927 | 0.958 | 46.539 | 0.928 | 0.912 | 0.967 | 1.031 | 0.917 | 1.239 | 1.889 | 0.852 | 1.687 |
| | 1250 | 0.908 | 0.984 | 47.301 | 0.921 | 0.914 | 0.895 | 1.025 | 0.917 | 1.254 | 1.879 | 0.879 | 1.678 |
| | 2500 | 0.623 | 0.903 | 42.169 | 0.653 | 0.897 | 0.604 | 0.905 | 0.805 | 1.198 | 1.701 | 0.709 | 1.416 |
| CMA-ES-R | 100 | 0.000 | 0.241 | 5.719 | 0.231 | 0.784 | 0.032 | 0.368 | 0.125 | 0.517 | 1.811 | 0.059 | 1.019 |
| | 750 | 0.840 | 0.939 | 45.881 | 0.777 | 0.904 | 0.810 | 0.985 | 0.873 | 1.163 | 1.774 | 0.858 | 1.518 |
| | 1250 | 0.802 | 0.958 | 45.670 | 0.799 | 0.906 | 0.757 | 0.965 | 0.883 | 1.206 | 1.841 | 0.808 | 1.484 |
| | 2500 | 0.602 | 0.894 | 41.902 | 0.662 | 0.898 | 0.641 | 0.855 | 0.791 | 1.192 | 1.731 | 0.688 | 1.440 |
| DE | .2/.9 | 1.541 | 0.187 | 57.456 | 2.877 | 0.904 | 1.873 | 1.202 | 1.199 | 1.187 | 1.324 | 1.469 | 1.324 |
| | .2/.2 | 1.426 | 0.198 | 54.959 | 2.977 | 0.900 | 1.687 | 1.200 | 1.245 | 1.174 | 1.341 | 1.372 | 1.361 |
| | .9/.2 | 1.086 | 0.196 | 53.513 | 1.643 | 0.896 | 1.299 | 1.201 | 1.078 | 1.137 | 1.500 | 1.101 | 1.535 |
| | .9/.9 | 1.504 | 0.195 | 59.798 | 2.669 | 0.904 | 1.646 | 1.294 | 1.222 | 1.170 | 1.531 | 1.431 | 1.434 |
| GSS | | 0.000 | 2.357 | 18.888 | 0.142 | 0.720 | 0.035 | 0.317 | 0.478 | 0.873 | 0.310 | 0.118 | 0.761 |
| GSS-R | | 0.000 | 2.289 | 17.773 | 0.140 | 0.747 | 0.048 | 0.311 | 0.455 | 0.852 | 0.348 | 0.115 | 0.716 |
| NM | | 0.000 | 0.212 | 77.499 | 0.239 | 0.745 | 0.040 | 1.643 | 2.728 | 1.281 | 1.482 | 1.398 | 0.291 |
| NM-R | | 0.000 | 0.189 | 41.857 | 0.152 | 0.673 | 0.009 | 0.926 | 1.661 | 1.263 | 1.094 | 0.106 | 0.028 |
| PSO | −0.5/2 | 0.381 | 0.338 | 42.878 | 0.651 | 0.894 | 0.650 | 1.404 | 0.957 | 1.222 | 0.965 | 0.514 | 0.000 |
| | 1/2 | 4.648 | 0.340 | 92.256 | 133.918 | 0.937 | 9.682 | 2.168 | 2.019 | 1.289 | 0.933 | 4.128 | 0.000 |
| REA-P | R1 | 0.117 | 0.346 | 20.960 | 0.361 | 0.905 | 0.156 | 1.288 | 0.569 | 1.161 | 1.117 | 0.563 | 0.511 |
| | R2 | – | – | – | – | 0.906 | – | 0.957 | – | 1.172 | – | 0.185 | – |
| REA-T | R1 | 0.014 | 0.815 | 50.302 | 1.174 | 0.905 | 0.066 | 1.371 | 0.447 | 1.031 | 1.519 | 1.277 | 0.271 |
| | R2 | – | – | 37.816 | 0.523 | 0.893 | – | 1.230 | – | 0.838 | – | 0.747 | – |
| | R3 | – | – | – | – | 0.886 | – | 1.139 | – | 0.740 | – | 0.466 | – |
| | R4 | – | – | – | – | 0.864 | – | 0.976 | – | 0.516 | – | 0.333 | – |
| SA | | 0.049 | 1.180 | 47.480 | 0.596 | 0.906 | 0.051 | 1.399 | 1.506 | 1.162 | 1.424 | 0.179 | 1.089 |
| rBOA | | 0.028 | 0.306 | 9.952 | 0.324 | 0.880 | 0.074 | 0.573 | 0.202 | 1.177 | 2.046 | 0.024 | 1.556 |
| rGA | | 0.865 | 0.855 | 40.262 | 2.884 | 0.889 | 1.281 | 0.618 | 0.906 | 1.245 | 0.554 | 0.804 | 0.963 |

**Table A.18** Variance for performance criterion $\phi_2$ (scaled) in 5 dimensions.

| $Var(\phi_2)$ (scaled) | | sphr. | ack. | lg-ack. | whit. | shek. | rosen. | rastr. | sal. | lange. | schw. | grie. | weier. |
|---|---|---|---|---|---|---|---|---|---|---|---|---|---|
| CG | | 0.000 | 0.412 | 58.596 | 874.804 | 0.079 | 37.801 | 1.106 | 1.283 | 0.045 | 1.006 | 7.060 | 0.562 |
| CG-R | | 0.000 | 0.345 | 52.084 | 1924.897 | 0.067 | 0.405 | 0.598 | 1.001 | 0.239 | 1.056 | 5.079 | 0.363 |
| CMA-ES | 100 | 0.000 | 0.050 | 2.643 | 0.057 | 0.035 | 0.005 | 0.136 | 0.027 | 0.153 | 0.260 | 0.016 | 0.260 |
| | 750 | 0.375 | 0.128 | 9.138 | 0.475 | 0.014 | 0.498 | 0.263 | 0.194 | 0.084 | 0.222 | 0.324 | 0.178 |
| | 1250 | 0.347 | 0.113 | 8.048 | 0.447 | 0.008 | 0.468 | 0.217 | 0.182 | 0.063 | 0.244 | 0.295 | 0.182 |
| | 2500 | 0.260 | 0.107 | 8.226 | 0.165 | 0.017 | 0.288 | 0.214 | 0.173 | 0.117 | 0.265 | 0.235 | 0.211 |
| CMA-ES-R | 100 | 0.000 | 0.055 | 2.715 | 0.059 | 0.035 | 0.004 | 0.149 | 0.026 | 0.163 | 0.278 | 0.016 | 0.276 |
| | 750 | 0.375 | 0.124 | 8.994 | 0.278 | 0.015 | 0.399 | 0.214 | 0.169 | 0.136 | 0.249 | 0.280 | 0.214 |
| | 1250 | 0.304 | 0.115 | 7.661 | 0.307 | 0.012 | 0.382 | 0.226 | 0.170 | 0.106 | 0.219 | 0.260 | 0.213 |
| | 2500 | 0.261 | 0.136 | 8.427 | 0.188 | 0.018 | 0.301 | 0.221 | 0.175 | 0.117 | 0.231 | 0.222 | 0.209 |
| DE | .2/.9 | 0.771 | 0.045 | 11.097 | 3.650 | 0.016 | 1.116 | 0.305 | 0.282 | 0.115 | 0.287 | 0.577 | 0.168 |
| | .2/.2 | 0.678 | 0.049 | 11.159 | 3.536 | 0.027 | 1.055 | 0.305 | 0.296 | 0.125 | 0.313 | 0.537 | 0.204 |
| | .9/.2 | 0.536 | 0.048 | 12.044 | 1.920 | 0.023 | 1.019 | 0.272 | 0.243 | 0.147 | 0.293 | 0.500 | 0.173 |
| | .9/.9 | 0.673 | 0.047 | 10.784 | 2.955 | 0.017 | 0.908 | 0.266 | 0.241 | 0.113 | 0.270 | 0.522 | 0.182 |
| GSS | | 0.000 | 0.416 | 10.780 | 0.075 | 0.221 | 0.247 | 0.185 | 0.334 | 0.296 | 0.378 | 0.039 | 0.513 |
| GSS-R | | 0.000 | 0.481 | 9.274 | 0.078 | 0.181 | 0.328 | 0.164 | 0.369 | 0.309 | 0.382 | 0.040 | 0.474 |
| NM | | 0.000 | 0.103 | 49.791 | 0.181 | 0.168 | 0.361 | 0.958 | 1.213 | 0.071 | 0.619 | 4.643 | 0.397 |
| NM-R | | 0.000 | 0.085 | 35.131 | 0.093 | 0.212 | 0.046 | 0.612 | 0.813 | 0.097 | 0.426 | 1.194 | 0.076 |
| PSO | –0.5/2 | 0.399 | 0.066 | 13.468 | 0.760 | 0.022 | 0.737 | 0.368 | 0.300 | 0.113 | 0.152 | 0.320 | 0.000 |
| | 1/2 | 2.082 | 0.070 | 18.365 | 245.263 | 0.014 | 7.609 | 0.453 | 0.474 | 0.022 | 0.092 | 1.847 | 0.000 |
| REA-P | R1 | 0.411 | 0.078 | 5.882 | 0.044 | 0.015 | 0.185 | 0.276 | 0.147 | 0.123 | 0.312 | 0.256 | 0.302 |
| | R2 | – | – | – | – | 0.015 | – | 0.230 | – | 0.128 | – | 0.035 | – |
| REA-T | R1 | 0.009 | 0.105 | 9.325 | 0.834 | 0.014 | 0.026 | 0.263 | 0.103 | 0.167 | 0.241 | 0.454 | 0.089 |
| | R2 | – | – | 6.246 | 0.073 | 0.021 | – | 0.280 | – | 0.133 | – | 0.228 | – |
| | R3 | – | – | – | – | 0.016 | – | 0.237 | – | 0.125 | – | 0.119 | – |
| | R4 | – | – | – | – | 0.021 | – | 0.216 | – | 0.146 | – | 0.069 | – |
| SA | | 0.031 | 0.160 | 10.611 | 0.280 | 0.014 | 0.160 | 0.584 | 0.446 | 0.149 | 0.273 | 0.035 | 0.172 |
| rBOA | | 0.048 | 0.126 | 4.964 | 0.064 | 0.024 | 0.082 | 0.342 | 0.079 | 0.117 | 0.361 | 0.013 | 0.204 |
| rGA | | 0.612 | 0.181 | 12.626 | 4.793 | 0.024 | 1.041 | 0.193 | 0.267 | 0.089 | 0.228 | 0.450 | 0.140 |

**Table A.19** Results on performance criterion $\sigma_\varepsilon^N$ with $\varepsilon = 1.000$ and $N = 250,000$ in 10 dimensions.

| $\sigma_\varepsilon^N$ | | sphr. | ack. | lg-ack. | whit. | shek. | rosen. | rastr. | sal. | lange. | schw. | grie. | weier. |
|---|---|---|---|---|---|---|---|---|---|---|---|---|---|
| CG | | 1.000 | 0.000 | 0.000 | 0.000 | 0.000 | 0.870 | 0.000 | 0.045 | 1.000 | 0.000 | 0.145 | 0.000 |
| CG-R | | 1.000 | 0.000 | 0.000 | 0.000 | 0.910 | 1.000 | 0.000 | 1.000 | 1.000 | 0.000 | 1.000 | 0.000 |
| CMA-ES | 100 | 1.000 | 1.000 | 0.230 | 0.000 | 0.000 | 0.000 | 0.360 | 1.000 | 1.000 | 0.000 | 1.000 | 0.069 |
| | 750 | 1.000 | 1.000 | 0.820 | 0.005 | 0.000 | 0.000 | 1.000 | 1.000 | 1.000 | 0.005 | 1.000 | 0.600 |
| | 1250 | 1.000 | 1.000 | 0.945 | 0.054 | 0.000 | 0.000 | 1.000 | 1.000 | 1.000 | 0.000 | 1.000 | 0.805 |
| | 2500 | 1.000 | 1.000 | 0.980 | 0.095 | 0.000 | 0.000 | 1.000 | 1.000 | 1.000 | 0.000 | 1.000 | 0.960 |
| CMA-ES-R | 100 | 1.000 | 1.000 | 1.000 | 0.000 | 0.000 | 0.000 | 1.000 | 1.000 | 1.000 | 0.000 | 1.000 | 1.000 |
| | 750 | 1.000 | 1.000 | 1.000 | 0.000 | 0.000 | 0.000 | 1.000 | 1.000 | 1.000 | 0.000 | 1.000 | 0.000 |
| | 1250 | 1.000 | 1.000 | 0.740 | 0.000 | 0.000 | 0.000 | 0.980 | 1.000 | 1.000 | 0.000 | 1.000 | 0.000 |
| | 2500 | 0.995 | 1.000 | 0.085 | 0.000 | 0.000 | 0.000 | 0.090 | 1.000 | 1.000 | 0.000 | 0.195 | 0.000 |
| DE | .2/.9 | 0.999 | 1.000 | 0.000 | 0.000 | 0.000 | 0.000 | 0.000 | 0.995 | 1.000 | 0.011 | 0.010 | 0.000 |
| | .2/.2 | 1.000 | 1.000 | 0.510 | 0.059 | 0.010 | 0.064 | 0.365 | 1.000 | 1.000 | 0.260 | 1.000 | 0.840 |
| | .9/.2 | 1.000 | 1.000 | 0.220 | 0.015 | 0.030 | 0.035 | 0.000 | 1.000 | 1.000 | 0.000 | 1.000 | 0.000 |
| | .9/.9 | 0.005 | 1.000 | 0.000 | 0.000 | 0.000 | 0.000 | 0.000 | 0.015 | 1.000 | 0.000 | 0.000 | 0.000 |
| GSS | | 1.000 | 0.000 | 0.000 | 0.049 | 0.030 | 0.730 | 0.000 | 0.775 | 1.000 | 0.385 | 0.995 | 0.075 |
| GSS-R | | 1.000 | 1.000 | 0.000 | 0.205 | 0.125 | 0.725 | 0.000 | 1.000 | 1.000 | 0.985 | 1.000 | 0.505 |
| NM | | 1.000 | 0.985 | 0.000 | 0.000 | 0.015 | 0.750 | 0.000 | 0.005 | 1.000 | 0.000 | 0.875 | 0.005 |
| NM-R | | 1.000 | 1.000 | 0.005 | 0.105 | 0.940 | 1.000 | 0.000 | 0.245 | 1.000 | 0.000 | 1.000 | 0.905 |
| PSO | –0.5/2 | 1.000 | 0.015 | 0.000 | 0.000 | 0.000 | 0.000 | 0.025 | 0.985 | 1.000 | 0.005 | 0.000 | 1.000 |
| | 1/2 | 0.000 | 0.029 | 0.000 | 0.000 | 0.000 | 0.000 | 0.000 | 0.000 | 1.000 | 0.000 | 0.000 | 1.000 |
| REA-P | R1 | 1.000 | 0.000 | 0.085 | 0.000 | 0.000 | 0.000 | 0.000 | 1.000 | 1.000 | 0.000 | 0.000 | 1.000 |
| | R2 | 1.000 | 0.000 | 0.005 | 0.000 | 0.000 | 0.000 | 0.000 | 1.000 | 1.000 | 0.000 | 0.830 | 0.255 |
| REA-T | R1 | 1.000 | 0.885 | 0.320 | 0.059 | 0.095 | 0.175 | 0.015 | 1.000 | 1.000 | 0.000 | 1.000 | 0.855 |
| | R2 | 1.000 | 1.000 | 0.235 | 0.135 | 0.125 | 0.355 | 0.000 | 1.000 | 1.000 | 0.000 | 1.000 | 1.000 |
| SA | | 1.000 | 0.000 | 0.000 | 0.000 | 0.000 | 0.840 | 0.000 | 1.000 | 1.000 | 0.000 | 0.015 | 0.000 |
| rBOA | | 1.000 | 1.000 | 0.135 | 0.000 | 0.000 | 0.000 | 0.999 | 1.000 | 1.000 | 0.000 | 1.000 | 0.000 |
| rGA | | 1.000 | 0.045 | 0.000 | 0.005 | 0.005 | 0.000 | 0.000 | 1.000 | 1.000 | 0.030 | 0.935 | 0.000 |

**Table A.20** Results on performance criterion $\sigma_{\varepsilon}^{N}$ with $\varepsilon = 0.100$ and $N = 250,000$ in 10 dimensions.

| $\sigma_{\varepsilon}^{N}$ | | sphr. | ack. | lg-ack. | whit. | shek. | rosen. | rastr. | sal. | lange. | schw. | grie. | weier. |
|---|---|---|---|---|---|---|---|---|---|---|---|---|---|
| CG | | 1.000 | 0.000 | 0.000 | 0.000 | 0.000 | 0.870 | 0.000 | 0.005 | 0.000 | 0.000 | 0.000 | 0.000 |
| CG-R | | 1.000 | 0.000 | 0.000 | 0.000 | 0.910 | 1.000 | 0.000 | 1.000 | 0.005 | 0.000 | 0.100 | 0.000 |
| CMA-ES | 100 | 1.000 | 0.220 | 0.190 | 0.000 | 0.000 | 0.000 | 0.110 | 0.080 | 0.010 | 0.000 | 1.000 | 0.000 |
| | 750 | 1.000 | 0.995 | 0.780 | 0.005 | 0.000 | 0.000 | 0.980 | 1.000 | 0.110 | 0.005 | 1.000 | 0.220 |
| | 1250 | 1.000 | 1.000 | 0.935 | 0.054 | 0.000 | 0.000 | 1.000 | 1.000 | 0.085 | 0.000 | 1.000 | 0.580 |
| | 2500 | 1.000 | 1.000 | 0.975 | 0.090 | 0.000 | 0.000 | 1.000 | 1.000 | 0.205 | 0.000 | 1.000 | 0.815 |
| CMA-ES-R | 100 | 1.000 | 1.000 | 0.995 | 0.000 | 0.000 | 0.000 | 0.995 | 0.985 | 0.290 | 0.000 | 1.000 | 0.265 |
| | 750 | 1.000 | 0.015 | 0.005 | 0.000 | 0.000 | 0.000 | 0.000 | 1.000 | 0.595 | 0.000 | 0.000 | 0.000 |
| | 1250 | 1.000 | 0.000 | 0.000 | 0.000 | 0.000 | 0.000 | 0.000 | 1.000 | 0.470 | 0.000 | 0.000 | 0.000 |
| | 2500 | 0.645 | 0.000 | 0.000 | 0.000 | 0.000 | 0.000 | 0.000 | 1.000 | 0.000 | 0.000 | 0.000 | 0.000 |
| DE | .2/.9 | 0.801 | 0.580 | 0.000 | 0.000 | 0.000 | 0.000 | 0.000 | 0.000 | 0.000 | 0.000 | 0.000 | 0.000 |
| | .2/.2 | 1.000 | 0.490 | 0.435 | 0.059 | 0.005 | 0.020 | 0.125 | 0.170 | 0.000 | 0.240 | 0.950 | 0.290 |
| | .9/.2 | 1.000 | 0.455 | 0.160 | 0.015 | 0.030 | 0.000 | 0.000 | 0.040 | 0.000 | 0.000 | 0.049 | 0.000 |
| | .9/.9 | 0.000 | 0.490 | 0.000 | 0.000 | 0.000 | 0.000 | 0.000 | 0.000 | 0.000 | 0.000 | 0.000 | 0.000 |
| GSS | | 1.000 | 0.000 | 0.000 | 0.045 | 0.030 | 0.575 | 0.000 | 0.000 | 0.005 | 0.385 | 0.240 | 0.035 |
| GSS-R | | 1.000 | 0.000 | 0.000 | 0.155 | 0.125 | 0.555 | 0.000 | 0.005 | 0.000 | 0.985 | 0.890 | 0.315 |
| NM | | 1.000 | 0.000 | 0.000 | 0.000 | 0.015 | 0.750 | 0.000 | 0.000 | 0.000 | 0.000 | 0.565 | 0.000 |
| NM-R | | 1.000 | 0.000 | 0.005 | 0.105 | 0.940 | 1.000 | 0.000 | 0.000 | 0.000 | 0.000 | 1.000 | 0.215 |
| PSO | –0.5/2 | 0.460 | 0.000 | 0.000 | 0.000 | 0.000 | 0.000 | 0.000 | 0.000 | 0.000 | 0.000 | 0.000 | 1.000 |
| | 1/2 | 0.000 | 0.000 | 0.000 | 0.000 | 0.000 | 0.000 | 0.000 | 0.000 | 0.000 | 0.000 | 0.000 | 1.000 |
| REA-P | R1 | 1.000 | 0.000 | 0.000 | 0.000 | 0.000 | 0.000 | 0.000 | 0.000 | 0.000 | 0.000 | 0.000 | 0.000 |
| | R2 | 1.000 | 0.000 | 0.005 | 0.000 | 0.000 | 0.000 | 0.000 | 0.000 | 0.000 | 0.000 | 0.000 | 0.000 |
| REA-T | R1 | 1.000 | 0.000 | 0.064 | 0.059 | 0.095 | 0.000 | 0.000 | 1.000 | 0.635 | 0.000 | 0.295 | 0.800 |
| | R2 | 1.000 | 0.000 | 0.170 | 0.115 | 0.125 | 0.000 | 0.000 | 1.000 | 0.410 | 0.000 | 0.000 | 1.000 |
| SA | | 1.000 | 0.000 | 0.000 | 0.000 | 0.000 | 0.005 | 0.000 | 0.000 | 0.000 | 0.000 | 0.000 | 0.000 |
| rBOA | | 1.000 | 1.000 | 0.000 | 0.000 | 0.000 | 0.000 | 0.975 | 1.000 | 0.000 | 0.000 | 1.000 | 0.000 |
| rGA | | 1.000 | 0.000 | 0.000 | 0.000 | 0.000 | 0.000 | 0.000 | 0.005 | 0.000 | 0.000 | 0.000 | 0.000 |

**Table A.21** Results on performance criterion $\sigma_\varepsilon^N$ with $\varepsilon = 0.010$ and $N = 250,000$ in 10 dimensions.

| $\sigma_\varepsilon^N$ | | sphr. | ack. | lg-ack. | whit. | shek. | rosen. | rastr. | sal. | lange. | schw. | grie. | weier. |
|---|---|---|---|---|---|---|---|---|---|---|---|---|---|
| CG | | 1.000 | 0.000 | 0.000 | 0.000 | 0.000 | 0.870 | 0.000 | 0.000 | 0.000 | 0.000 | 0.000 | 0.000 |
| CG-R | | 1.000 | 0.000 | 0.000 | 0.000 | 0.910 | 1.000 | 0.000 | 0.960 | 0.000 | 0.000 | 0.000 | 0.000 |
| CMA-ES | 100 | 1.000 | 0.220 | 0.190 | 0.000 | 0.000 | 0.000 | 0.110 | 0.000 | 0.005 | 0.000 | 0.995 | 0.000 |
| | 750 | 1.000 | 0.995 | 0.780 | 0.005 | 0.000 | 0.000 | 0.980 | 0.000 | 0.054 | 0.005 | 1.000 | 0.035 |
| | 1250 | 1.000 | 1.000 | 0.935 | 0.054 | 0.000 | 0.000 | 1.000 | 0.000 | 0.015 | 0.000 | 1.000 | 0.390 |
| | 2500 | 1.000 | 1.000 | 0.975 | 0.085 | 0.000 | 0.000 | 1.000 | 0.000 | 0.000 | 0.000 | 0.995 | 0.495 |
| CMA-ES-R | 100 | 1.000 | 1.000 | 0.995 | 0.000 | 0.000 | 0.000 | 0.995 | 0.000 | 0.225 | 0.000 | 1.000 | 0.000 |
| | 750 | 0.730 | 0.000 | 0.000 | 0.000 | 0.000 | 0.000 | 0.000 | 0.000 | 0.430 | 0.000 | 0.000 | 0.000 |
| | 1250 | 0.120 | 0.000 | 0.000 | 0.000 | 0.000 | 0.000 | 0.000 | 0.000 | 0.140 | 0.000 | 0.000 | 0.000 |
| | 2500 | 0.000 | 0.000 | 0.000 | 0.000 | 0.000 | 0.000 | 0.000 | 0.000 | 0.000 | 0.000 | 0.000 | 0.000 |
| DE | .2/.9 | 0.000 | 0.000 | 0.000 | 0.000 | 0.000 | 0.000 | 0.000 | 0.000 | 0.000 | 0.000 | 0.000 | 0.000 |
| | .2/.2 | 1.000 | 0.000 | 0.435 | 0.045 | 0.005 | 0.005 | 0.105 | 0.000 | 0.000 | 0.220 | 0.005 | 0.069 |
| | .9/.2 | 1.000 | 0.000 | 0.120 | 0.010 | 0.030 | 0.000 | 0.000 | 0.000 | 0.000 | 0.000 | 0.000 | 0.000 |
| | .9/.9 | 0.000 | 0.000 | 0.000 | 0.000 | 0.000 | 0.000 | 0.000 | 0.000 | 0.000 | 0.000 | 0.000 | 0.000 |
| GSS | | 1.000 | 0.000 | 0.000 | 0.045 | 0.030 | 0.550 | 0.000 | 0.000 | 0.000 | 0.385 | 0.010 | 0.035 |
| GSS-R | | 1.000 | 0.000 | 0.000 | 0.155 | 0.125 | 0.540 | 0.000 | 0.000 | 0.000 | 0.985 | 0.135 | 0.315 |
| NM | | 1.000 | 0.000 | 0.000 | 0.000 | 0.015 | 0.750 | 0.000 | 0.000 | 0.000 | 0.000 | 0.054 | 0.000 |
| NM-R | | 1.000 | 0.000 | 0.005 | 0.105 | 0.940 | 1.000 | 0.000 | 0.000 | 0.000 | 0.000 | 0.980 | 0.110 |
| PSO | –0.5/2 | 0.035 | 0.000 | 0.000 | 0.000 | 0.000 | 0.000 | 0.000 | 0.000 | 0.000 | 0.000 | 0.000 | 1.000 |
| | 1/2 | 0.000 | 0.000 | 0.000 | 0.000 | 0.000 | 0.000 | 0.000 | 0.000 | 0.000 | 0.000 | 0.000 | 1.000 |
| REA-P | R1 | 0.000 | 0.000 | 0.000 | 0.000 | 0.000 | 0.000 | 0.000 | 0.000 | 0.000 | 0.000 | 0.000 | 0.000 |
| | R2 | 0.975 | 0.000 | 0.000 | 0.000 | 0.000 | 0.000 | 0.000 | 0.000 | 0.000 | 0.000 | 0.000 | 0.000 |
| REA-T | R1 | 1.000 | 0.000 | 0.010 | 0.059 | 0.095 | 0.000 | 0.000 | 0.000 | 0.345 | 0.000 | 0.000 | 0.740 |
| | R2 | 1.000 | 0.000 | 0.170 | 0.095 | 0.125 | 0.000 | 0.000 | 0.000 | 0.235 | 0.000 | 0.000 | 1.000 |
| SA | | 0.020 | 0.000 | 0.000 | 0.000 | 0.000 | 0.000 | 0.000 | 0.000 | 0.000 | 0.000 | 0.000 | 0.000 |
| rBOA | | 1.000 | 0.995 | 0.000 | 0.000 | 0.000 | 0.000 | 0.955 | 0.010 | 0.000 | 0.000 | 1.000 | 0.000 |
| rGA | | 1.000 | 0.000 | 0.000 | 0.000 | 0.000 | 0.000 | 0.000 | 0.000 | 0.000 | 0.000 | 0.000 | 0.000 |

**Table A.22** Results on performance criterion $\frac{1}{100}\hat{\psi}_{\varepsilon}^{N}$ with $\varepsilon = 1.000$ and $N = 250,000$ in 10 dimensions.

| $\frac{1}{100}\hat{\psi}_{\varepsilon}^{N}$ | | sphr. | ack. | lg-ack. | whit. | shek. | rosen. | rastr. | sal. | lange. | schw. | grie. | weier. |
|---|---|---|---|---|---|---|---|---|---|---|---|---|---|
| CG | | 0.0 | – | – | – | – | 53.6 | – | 1.2 | 0.0 | – | 57.3 | – |
| CG-R | | 0.0 | – | – | – | 750.1 | 64.5 | – | 57.9 | 0.0 | – | 267.4 | – |
| CMA-ES | 100 | 7.5 | 18.2 | 29.0 | – | – | – | 42.0 | 7.3 | 0.0 | – | 14.9 | 67.5 |
| | 750 | 58.8 | 98.8 | 147.8 | 187.0 | – | – | 160.4 | 56.4 | 0.0 | 405.0 | 84.7 | 402.5 |
| | 1250 | 127.9 | 186.1 | 267.2 | 337.2 | – | – | 287.7 | 125.5 | 0.0 | – | 161.3 | 704.9 |
| | 2500 | 433.8 | 478.0 | 650.0 | 986.8 | – | – | 768.3 | 372.1 | 0.0 | – | 614.8 | 1684.7 |
| CMA-ES-R | 100 | 7.3 | 17.9 | 165.7 | – | – | – | 109.2 | 7.0 | 0.0 | – | 14.8 | 113.3 |
| | 750 | 57.8 | 100.6 | 193.3 | – | – | – | 173.5 | 54.9 | 0.0 | – | 83.3 | – |
| | 1250 | 125.9 | 183.1 | 725.2 | – | – | – | 744.5 | 123.8 | 0.0 | – | 162.0 | – |
| | 2500 | 656.1 | 493.3 | 814.7 | – | – | – | 1602.7 | 387.2 | 0.0 | – | 1314.1 | – |
| DE | .2/.9 | 367.9 | 8.7 | – | – | – | – | – | 690.6 | 0.0 | 2239.5 | 1800.0 | – |
| | .2/.2 | 98.6 | 9.1 | 597.4 | 639.3 | 1394.5 | 894.1 | 1838.8 | 152.1 | 0.0 | 1523.1 | 280.3 | 1655.2 |
| | .9/.2 | 93.2 | 9.7 | 1717.6 | 1888.0 | 993.8 | 1644.4 | – | 116.4 | 0.0 | – | 406.7 | – |
| | .9/.9 | 1752.0 | 8.5 | – | – | – | – | – | 1694.0 | 0.0 | – | – | – |
| GSS | | 12.8 | – | – | 46.0 | 33.1 | 275.6 | – | 8.1 | 0.0 | 27.3 | 31.2 | 15.1 |
| GSS-R | | 13.4 | 353.9 | – | 1157.2 | 1294.7 | 231.4 | – | 66.2 | 0.0 | 496.7 | 33.0 | 1036.6 |
| NM | | 3.4 | 1.2 | – | – | 9.3 | 35.4 | – | 2.0 | 0.0 | – | 4.0 | 2.0 |
| NM-R | | 3.5 | 6.8 | 1055.0 | 1030.0 | 756.3 | 45.7 | – | 1156.5 | 0.0 | – | 11.7 | 875.4 |
| PSO | –0.5/2 | 417.1 | 5.3 | – | – | – | – | 1288.0 | 627.1 | 0.0 | 554.0 | – | 0.0 |
| | 1/2 | – | 1.0 | – | – | – | – | – | – | 0.0 | – | – | 0.0 |
| REA-P | R1 | 114.5 | – | 1539.6 | – | – | – | – | 343.1 | 0.0 | – | – | 1070.2 |
| | R2 | 203.7 | – | 792.0 | – | – | – | – | 156.0 | 0.0 | – | 1515.3 | 139.7 |
| REA-T | R1 | 105.6 | 1849.5 | 2352.0 | 1425.8 | 750.8 | 2258.1 | 2303.0 | 133.6 | 0.0 | – | 481.3 | 60.5 |
| | R2 | 166.7 | 1282.1 | 1443.6 | 2099.3 | 917.0 | 2200.4 | – | 155.1 | 0.0 | – | 562.4 | 105.9 |
| SA | | 3.8 | – | – | – | – | 458.7 | – | 254.0 | 0.0 | – | 1755.3 | – |
| rBOA | | 13.8 | 44.1 | 1470.0 | – | – | – | 326.6 | 2.9 | 0.2 | – | 1.7 | – |
| rGA | | 38.7 | 89.2 | – | 2456.0 | 825.0 | – | – | 42.9 | 0.0 | 1112.1 | 769.8 | – |

**Table A.23** Results on performance criterion $\frac{1}{100}\hat{\psi}_{\varepsilon}^{N}$ with $\varepsilon = 0.100$ and $N = 250,000$ in 10 dimensions.

| $\frac{1}{100}\hat{\psi}_{\varepsilon}^{N}$ | | sphr. | ack. | lg-ack. | whit. | shek. | rosen. | rastr. | sal. | lange. | schw. | grie. | weier. |
|---|---|---|---|---|---|---|---|---|---|---|---|---|---|
| CG | | 0.0 | – | – | – | – | 53.6 | – | 1.0 | – | – | – | – |
| CG-R | | 0.0 | – | – | – | 760.0 | 64.5 | – | 250.8 | 246.0 | – | 1592.7 | – |
| CMA-ES | 100 | 11.7 | 25.0 | 32.6 | – | – | – | 30.5 | 22.3 | 26.0 | – | 24.5 | – |
| | 750 | 74.6 | 152.6 | 174.7 | 202.0 | – | – | 179.6 | 134.7 | 186.5 | 450.0 | 137.7 | 529.0 |
| | 1250 | 149.5 | 271.5 | 312.0 | 362.2 | – | – | 313.8 | 243.0 | 335.0 | – | 244.9 | 901.8 |
| | 2500 | 528.0 | 670.7 | 760.2 | 1080.5 | – | – | 948.3 | 606.7 | 766.4 | – | 842.8 | 2043.8 |
| CMA-ES-R | 100 | 11.4 | 144.7 | 202.5 | – | – | – | 401.0 | 666.4 | 1140.3 | – | 24.4 | 1084.4 |
| | 750 | 73.6 | 1329.6 | 1207.0 | – | – | – | – | 133.4 | 1065.9 | – | – | – |
| | 1250 | 149.2 | – | – | – | – | – | – | 242.5 | 1210.5 | – | – | – |
| | 2500 | 1243.6 | – | – | – | – | – | – | 604.2 | – | – | – | – |
| DE | .2/.9 | 1687.7 | 1732.1 | – | – | – | – | – | – | – | – | – | – |
| | .2/.2 | 200.1 | 1713.0 | 799.1 | 793.9 | 1227.0 | 1250.7 | 1943.6 | 1934.5 | – | 1726.6 | 1382.4 | 2059.4 |
| | .9/.2 | 257.2 | 1700.3 | 1885.5 | 1968.0 | 1188.0 | – | – | 1892.8 | – | – | 1898.1 | – |
| | .9/.9 | – | 1810.0 | – | – | – | – | – | – | – | – | – | – |
| GSS | | 22.0 | – | – | 48.4 | 43.0 | 299.6 | – | – | 20.0 | 36.5 | 48.9 | 23.4 |
| GSS-R | | 22.8 | – | – | 1237.4 | 1304.0 | 278.4 | – | 1872.0 | – | 506.0 | 743.8 | 1125.0 |
| NM | | 4.3 | – | – | – | 10.0 | 41.3 | – | – | – | – | 7.4 | – |
| NM-R | | 4.4 | – | 1058.0 | 1031.0 | 757.3 | 52.3 | – | – | – | – | 34.3 | 1269.5 |
| PSO | –0.5/2 | 1194.1 | – | – | – | – | – | – | – | – | – | – | 0.0 |
| | 1/2 | – | – | – | – | – | – | – | – | – | – | – | 0.0 |
| REA-P | R1 | 1217.1 | – | – | – | – | – | – | – | – | – | – | – |
| | R2 | 315.2 | – | 1601.0 | – | – | – | – | – | – | – | – | – |
| REA-T | R1 | 215.4 | – | 2423.9 | 1543.2 | 942.4 | – | – | 660.4 | 2184.0 | – | 2310.2 | 95.9 |
| | R2 | 346.1 | – | 1542.9 | 2169.0 | 1123.8 | – | – | 819.7 | 968.0 | – | – | 164.4 |
| SA | | 115.5 | – | – | – | – | 1347.0 | – | – | – | – | – | – |
| rBOA | | 22.5 | 275.3 | – | – | – | – | 347.5 | 94.2 | – | – | 99.8 | – |
| rGA | | 59.9 | – | – | – | – | – | – | 2157.0 | – | – | – | – |

**Table A.24** Results on performance criterion $\frac{1}{100}\hat{\psi}_{\varepsilon}^{N}$ with $\varepsilon = 0.010$ and $N = 250{,}000$ in 10 dimensions.

| $\frac{1}{100}\hat{\psi}_{\varepsilon}^{N}$ | | sphr. | ack. | lg-ack. | whit. | shek. | rosen. | rastr. | sal. | lange. | schw. | grie. | weier. |
|---|---|---|---|---|---|---|---|---|---|---|---|---|---|
| CG | | 0.0 | – | – | – | – | 53.6 | – | – | – | – | – | – |
| CG-R | | 0.0 | – | – | – | 767.9 | 64.5 | – | 592.0 | – | – | – | – |
| CMA-ES | 100 | 15.7 | 29.7 | 36.8 | – | – | – | 34.2 | – | 33.0 | – | 28.5 | – |
| | 750 | 90.1 | 173.8 | 195.8 | 217.0 | – | – | 197.4 | – | 244.5 | 480.0 | 155.3 | 678.0 |
| | 1250 | 171.2 | 301.3 | 351.8 | 388.5 | – | – | 338.5 | – | 491.3 | – | 275.4 | 1117.6 |
| | 2500 | 685.5 | 852.6 | 939.1 | 1113.2 | – | – | 1104.8 | – | – | – | 1028.6 | 2347.7 |
| CMA-ES-R | 100 | 15.4 | 149.4 | 206.4 | – | – | – | 404.8 | – | 1054.5 | – | 29.0 | – |
| | 750 | 1042.4 | – | – | – | – | – | – | – | 1118.7 | – | – | – |
| | 1250 | 856.4 | – | – | – | – | – | – | – | 1402.4 | – | – | – |
| | 2500 | – | – | – | – | – | – | – | – | – | – | – | – |
| DE | .2/.9 | – | – | – | – | – | – | – | – | – | – | – | – |
| | .2/.2 | 305.7 | – | 975.8 | 802.6 | 1324.0 | 1415.0 | 2153.1 | – | – | 1817.9 | 2262.0 | 2252.8 |
| | .9/.2 | 428.2 | – | 1961.5 | 1822.5 | 1357.8 | – | – | – | – | – | – | – |
| | .9/.9 | – | – | – | – | – | – | – | – | – | – | – | – |
| GSS | | 31.3 | – | – | 57.3 | 52.0 | 433.5 | – | – | – | 45.6 | 61.0 | 33.4 |
| GSS-R | | 31.9 | – | – | 1246.0 | 1313.1 | 429.1 | – | – | – | 515.4 | 1041.6 | 1113.5 |
| NM | | 5.2 | – | – | – | 11.3 | 44.3 | – | – | – | – | 9.6 | – |
| NM-R | | 5.1 | – | 1060.0 | 1032.0 | 758.3 | 55.4 | – | – | – | – | 564.5 | 1352.5 |
| PSO | −0.5/2 | 1307.8 | – | – | – | – | – | – | – | – | – | – | 0.0 |
| | 1/2 | – | – | – | – | – | – | – | – | – | – | – | 0.0 |
| REA-P | R1 | – | – | – | – | – | – | – | – | – | – | – | – |
| | R2 | 1342.4 | – | – | – | – | – | – | – | – | – | – | – |
| REA-T | R1 | 330.7 | – | 2398.0 | 1655.2 | 1119.4 | – | – | – | 2297.5 | – | – | 132.7 |
| | R2 | 513.6 | – | 1696.7 | 2253.8 | 1333.2 | – | – | – | 1076.3 | – | – | 221.8 |
| SA | | 969.2 | – | – | – | – | – | – | – | – | – | – | – |
| rBOA | | 30.8 | 309.0 | – | – | – | – | 368.7 | 905.0 | – | – | 167.5 | – |
| rGA | | 546.3 | – | – | – | – | – | – | – | – | – | – | – |

**Table A.25** Results on performance criterion $\zeta_{T_m}$ (scaled) with $m = 25,000$ in 10 dimensions.

| $\zeta_{T_m}$ (scaled) | | sphr. | ack. | lg-ack. | whit. | shek. | rosen. | rastr. | sal. | lange. | schw. | grie. | weier. |
|---|---|---|---|---|---|---|---|---|---|---|---|---|---|
| CG | | 0.000 | 1.431 | 6.279 | 0.133 | 0.846 | 0.014 | 1.267 | 2.045 | 1.000 | 1.329 | 0.143 | 1.940 |
| CG-R | | 0.000 | 0.936 | 3.325 | 0.003 | 0.598 | 0.045 | 0.485 | 0.069 | 0.995 | 0.707 | 0.022 | 1.335 |
| CMA-ES | 100 | 0.000 | 0.032 | 0.067 | 0.002 | 0.851 | 0.001 | 0.029 | 0.073 | 0.624 | 0.898 | 0.000 | 0.288 |
| | 750 | 0.000 | 0.000 | 0.010 | 0.002 | 0.852 | 0.002 | 0.000 | 0.040 | 0.403 | 1.070 | 0.000 | 0.733 |
| | 1250 | 0.000 | 0.142 | 0.145 | 0.002 | 0.878 | 0.002 | 0.260 | 0.041 | 0.602 | 1.171 | 0.002 | 1.276 |
| | 2500 | 0.351 | 0.689 | 2.029 | 0.032 | 0.967 | 0.199 | 0.681 | 0.545 | 0.999 | 1.063 | 0.283 | 1.066 |
| CMA-ES-R | 100 | 0.000 | 0.006 | 0.013 | 0.001 | 0.851 | 0.001 | 0.009 | 0.055 | 0.425 | 0.652 | 0.000 | 0.048 |
| | 750 | 0.001 | 0.085 | 0.031 | 0.002 | 0.880 | 0.003 | 0.009 | 0.040 | 0.371 | 0.958 | 0.017 | 0.517 |
| | 1250 | 0.002 | 0.142 | 0.142 | 0.002 | 0.964 | 0.003 | 0.246 | 0.040 | 0.597 | 1.114 | 0.019 | 1.120 |
| | 2500 | 0.365 | 0.687 | 2.072 | 0.033 | 0.967 | 0.204 | 0.669 | 0.544 | 0.999 | 1.084 | 0.292 | 1.072 |
| DE | .2/.9 | 0.127 | 0.105 | 1.337 | 0.008 | 0.959 | 0.082 | 0.357 | 0.527 | 0.985 | 0.444 | 0.147 | 0.587 |
| | .2/.2 | 0.003 | 0.108 | 0.394 | 0.003 | 0.924 | 0.010 | 0.257 | 0.278 | 0.987 | 0.454 | 0.025 | 0.580 |
| | .9/.2 | 0.009 | 0.107 | 0.978 | 0.004 | 0.913 | 0.010 | 0.563 | 0.257 | 0.951 | 0.924 | 0.032 | 1.026 |
| | .9/.9 | 0.539 | 0.107 | 2.646 | 0.125 | 0.966 | 0.369 | 0.785 | 0.739 | 0.999 | 0.984 | 0.548 | 1.040 |
| GSS | | 0.000 | 1.427 | 0.983 | 0.001 | 0.814 | 0.000 | 0.213 | 0.325 | 0.780 | 0.099 | 0.006 | 0.401 |
| GSS-R | | 0.000 | 1.404 | 0.948 | 0.001 | 0.832 | 0.000 | 0.202 | 0.332 | 0.790 | 0.082 | 0.006 | 0.417 |
| NM | | 0.000 | 0.133 | 2.867 | 0.051 | 0.815 | 0.004 | 0.844 | 1.686 | 1.000 | 0.943 | 0.190 | 0.670 |
| NM-R | | 0.000 | 0.131 | 1.140 | 0.460 | 0.505 | 0.000 | 0.326 | 0.782 | 0.999 | 0.646 | 0.000 | 0.161 |
| PSO | –0.5/2 | 0.133 | 0.320 | 1.943 | 0.010 | 0.962 | 0.090 | 0.832 | 0.532 | 0.999 | 0.477 | 0.153 | 0.000 |
| | 1/2 | 2.735 | 0.320 | 4.892 | 64.875 | 0.984 | 8.484 | 1.590 | 1.615 | 1.000 | 0.549 | 2.819 | 0.000 |
| REA-P | R1 | 0.029 | 0.786 | 0.918 | 0.496 | 0.966 | 0.691 | 0.644 | 0.445 | 0.999 | 0.987 | 0.091 | 0.165 |
| | R2 | 0.057 | 0.600 | 1.418 | 0.650 | 0.966 | 0.803 | 0.747 | 0.310 | 0.999 | 0.856 | 0.040 | 0.255 |
| REA-T | R1 | 0.005 | 0.544 | 1.552 | 0.004 | 0.942 | 0.009 | 0.608 | 0.221 | 0.949 | 0.909 | 0.051 | 0.027 |
| | R2 | 0.029 | 0.486 | 1.314 | 0.005 | 0.950 | 0.023 | 0.604 | 0.272 | 0.705 | 0.961 | 0.069 | 0.000 |
| SA | | 0.004 | 0.805 | 1.758 | 0.004 | 0.966 | 0.001 | 0.245 | 0.402 | 0.985 | 0.881 | 0.031 | 0.664 |
| rBOA | | 0.000 | 0.049 | 0.164 | 0.001 | 0.959 | 0.002 | 0.076 | 0.040 | 0.974 | 1.407 | 0.000 | 1.212 |
| rGA | | 0.000 | 0.407 | 0.989 | 0.002 | 0.848 | 0.004 | 0.194 | 0.115 | 0.999 | 0.187 | 0.023 | 0.639 |

**Table A.26** Variance for performance criterion $\zeta_{T_m}$ (scaled) with $m = 25,000$ in 10 dimensions.

| $Var(\zeta_{T_m})$ (scaled) | | sphr. | ack. | lg-ack. | whit. | shek. | rosen. | rastr. | sal. | lange. | schw. | grie. | weier. |
|---|---|---|---|---|---|---|---|---|---|---|---|---|---|
| CG | | 0.000 | 0.184 | 1.532 | 1.710 | 0.040 | 0.198 | 0.359 | 0.619 | – | 0.460 | 0.188 | 0.257 |
| CG-R | | 0.000 | 0.099 | 0.605 | 0.001 | 0.312 | 0.457 | 0.112 | 0.086 | 0.066 | 0.149 | 0.013 | 0.095 |
| CMA-ES | 100 | 0.000 | 0.019 | 0.048 | 0.000 | 0.004 | 0.000 | 0.018 | 0.021 | 0.184 | 0.175 | 0.000 | 0.123 |
| | 750 | 0.000 | 0.002 | 0.021 | 0.000 | 0.002 | 0.000 | 0.001 | 0.000 | 0.216 | 0.135 | 0.000 | 0.219 |
| | 1250 | 0.000 | 0.029 | 0.071 | 0.000 | 0.017 | 0.000 | 0.067 | 0.000 | 0.126 | 0.081 | 0.001 | 0.084 |
| | 2500 | 0.085 | 0.047 | 0.217 | 0.025 | 0.001 | 0.069 | 0.089 | 0.074 | 0.003 | 0.089 | 0.061 | 0.069 |
| CMA-ES-R | 100 | 0.000 | 0.012 | 0.020 | 0.000 | 0.000 | 0.000 | 0.009 | 0.017 | 0.159 | 0.166 | 0.000 | 0.032 |
| | 750 | 0.000 | 0.021 | 0.021 | 0.000 | 0.008 | 0.000 | 0.004 | 0.000 | 0.202 | 0.150 | 0.002 | 0.192 |
| | 1250 | 0.001 | 0.030 | 0.065 | 0.000 | 0.003 | 0.000 | 0.072 | 0.000 | 0.183 | 0.076 | 0.002 | 0.080 |
| | 2500 | 0.082 | 0.047 | 0.222 | 0.023 | 0.001 | 0.067 | 0.087 | 0.061 | 0.000 | 0.082 | 0.067 | 0.070 |
| DE | .2/.9 | 0.061 | 0.020 | 0.259 | 0.007 | 0.004 | 0.041 | 0.083 | 0.093 | 0.022 | 0.109 | 0.053 | 0.090 |
| | .2/.2 | 0.001 | 0.019 | 0.119 | 0.000 | 0.026 | 0.006 | 0.068 | 0.063 | 0.017 | 0.117 | 0.002 | 0.088 |
| | .9/.2 | 0.004 | 0.020 | 0.174 | 0.000 | 0.017 | 0.006 | 0.082 | 0.054 | 0.053 | 0.111 | 0.004 | 0.083 |
| | .9/.9 | 0.133 | 0.021 | 0.280 | 0.083 | 0.001 | 0.142 | 0.093 | 0.086 | 0.003 | 0.078 | 0.140 | 0.065 |
| GSS | | 0.000 | 0.186 | 0.281 | 0.000 | 0.147 | 0.000 | 0.084 | 0.165 | 0.174 | 0.137 | 0.004 | 0.240 |
| GSS-R | | 0.000 | 0.191 | 0.308 | 0.000 | 0.090 | 0.000 | 0.078 | 0.176 | 0.172 | 0.127 | 0.004 | 0.251 |
| NM | | 0.000 | 0.049 | 1.575 | 0.452 | 0.113 | 0.063 | 0.330 | 0.437 | 0.000 | 0.220 | 0.558 | 0.376 |
| NM-R | | 0.000 | 0.046 | 0.695 | 3.420 | 0.320 | 0.000 | 0.125 | 0.201 | 0.001 | 0.144 | 0.000 | 0.070 |
| PSO | –0.5/2 | 0.115 | 0.031 | 0.522 | 0.015 | 0.001 | 0.100 | 0.279 | 0.196 | 0.001 | 0.237 | 0.112 | 0.000 |
| | 1/2 | 0.826 | 0.036 | 0.683 | 69.350 | 0.002 | 4.853 | 0.200 | 0.222 | – | 0.014 | 0.818 | 0.000 |
| REA-P | R1 | 0.009 | 0.057 | 0.194 | 0.586 | 0.001 | 0.444 | 0.082 | 0.064 | 0.003 | 0.074 | 0.017 | 0.030 |
| | R2 | 0.066 | 0.050 | 0.503 | 0.777 | 0.001 | 0.493 | 0.104 | 0.053 | 0.001 | 0.103 | 0.011 | 0.191 |
| REA-T | R1 | 0.005 | 0.044 | 0.194 | 0.000 | 0.011 | 0.004 | 0.081 | 0.047 | 0.042 | 0.086 | 0.013 | 0.062 |
| | R2 | 0.013 | 0.043 | 0.203 | 0.000 | 0.007 | 0.008 | 0.087 | 0.042 | 0.132 | 0.082 | 0.017 | 0.000 |
| SA | | 0.001 | 0.081 | 0.268 | 0.000 | 0.001 | 0.001 | 0.080 | 0.110 | 0.027 | 0.115 | 0.002 | 0.081 |
| rBOA | | 0.000 | 0.062 | 0.051 | 0.000 | 0.002 | 0.000 | 0.092 | 0.000 | 0.031 | 0.129 | 0.000 | 0.080 |
| rGA | | 0.000 | 0.093 | 0.398 | 0.000 | 0.060 | 0.006 | 0.072 | 0.017 | 0.010 | 0.104 | 0.002 | 0.043 |

**Table A.27** Results on performance criterion $\zeta_{T_m}$ (scaled) with $m = 100,000$ in 10 dimensions.

| $\zeta_{T_m}$ (scaled) | | sphr. | ack. | lg-ack. | whit. | shek. | rosen. | rastr. | sal. | lange. | schw. | grie. | weier. |
|---|---|---|---|---|---|---|---|---|---|---|---|---|---|
| CG | | 0.000 | 1.431 | 6.279 | 0.133 | 0.846 | 0.014 | 1.267 | 2.045 | 1.000 | 1.329 | 0.143 | 1.940 |
| CG-R | | 0.000 | 0.817 | 2.645 | 0.002 | 0.249 | 0.000 | 0.359 | 0.009 | 0.994 | 0.552 | 0.010 | 1.237 |
| CMA-ES | 100 | 0.000 | 0.032 | 0.067 | 0.002 | 0.851 | 0.001 | 0.028 | 0.073 | 0.624 | 0.897 | 0.000 | 0.287 |
| | 750 | 0.000 | 0.000 | 0.010 | 0.002 | 0.851 | 0.002 | 0.000 | 0.040 | 0.401 | 0.482 | 0.000 | 0.087 |
| | 1250 | 0.000 | 0.000 | 0.002 | 0.001 | 0.854 | 0.002 | 0.000 | 0.040 | 0.367 | 0.397 | 0.000 | 0.045 |
| | 2500 | 0.000 | 0.000 | 0.001 | 0.001 | 0.854 | 0.002 | 0.000 | 0.040 | 0.264 | 0.983 | 0.000 | 1.005 |
| CMA-ES-R | 100 | 0.000 | 0.000 | 0.000 | 0.001 | 0.851 | 0.001 | 0.001 | 0.041 | 0.320 | 0.492 | 0.000 | 0.018 |
| | 750 | 0.000 | 0.071 | 0.017 | 0.001 | 0.872 | 0.002 | 0.005 | 0.040 | 0.211 | 0.362 | 0.015 | 0.264 |
| | 1250 | 0.001 | 0.120 | 0.041 | 0.002 | 0.961 | 0.003 | 0.012 | 0.040 | 0.299 | 0.373 | 0.017 | 0.353 |
| | 2500 | 0.039 | 0.158 | 0.078 | 0.003 | 0.960 | 0.016 | 0.046 | 0.040 | 0.993 | 0.998 | 0.037 | 1.001 |
| DE | .2/.9 | 0.023 | 0.060 | 0.579 | 0.003 | 0.933 | 0.022 | 0.149 | 0.344 | 0.864 | 0.168 | 0.047 | 0.331 |
| | .2/.2 | 0.000 | 0.061 | 0.035 | 0.000 | 0.846 | 0.002 | 0.072 | 0.117 | 0.872 | 0.155 | 0.003 | 0.225 |
| | .9/.2 | 0.000 | 0.061 | 0.284 | 0.002 | 0.821 | 0.002 | 0.406 | 0.112 | 0.824 | 0.766 | 0.009 | 0.876 |
| | .9/.9 | 0.369 | 0.062 | 2.301 | 0.044 | 0.964 | 0.215 | 0.677 | 0.629 | 0.997 | 0.888 | 0.367 | 0.958 |
| GSS | | 0.000 | 1.427 | 0.983 | 0.001 | 0.814 | 0.000 | 0.213 | 0.325 | 0.780 | 0.099 | 0.006 | 0.401 |
| GSS-R | | 0.000 | 0.094 | 0.724 | 0.000 | 0.780 | 0.000 | 0.145 | 0.187 | 0.610 | 0.016 | 0.002 | 0.208 |
| NM | | 0.000 | 0.133 | 2.867 | 0.010 | 0.815 | 0.004 | 0.844 | 1.686 | 1.000 | 0.943 | 0.190 | 0.670 |
| NM-R | | 0.000 | 0.114 | 0.553 | 0.001 | 0.237 | 0.000 | 0.200 | 0.589 | 0.998 | 0.507 | 0.000 | 0.083 |
| PSO | –0.5/2 | 0.034 | 0.320 | 1.320 | 0.004 | 0.960 | 0.016 | 0.411 | 0.315 | 0.999 | 0.364 | 0.056 | 0.000 |
| | 1/2 | 2.668 | 0.320 | 4.885 | 64.875 | 0.984 | 8.484 | 1.590 | 1.615 | 1.000 | 0.548 | 2.741 | 0.000 |
| REA-P | R1 | 0.008 | 0.682 | 0.178 | 0.495 | 0.965 | 0.691 | 0.479 | 0.263 | 0.997 | 0.880 | 0.046 | 0.098 |
| | R2 | 0.001 | 0.439 | 0.310 | 0.650 | 0.965 | 0.803 | 0.438 | 0.152 | 0.997 | 0.543 | 0.024 | 0.189 |
| REA-T | R1 | 0.000 | 0.321 | 0.672 | 0.002 | 0.762 | 0.001 | 0.410 | 0.040 | 0.472 | 0.689 | 0.015 | 0.027 |
| | R2 | 0.000 | 0.273 | 0.450 | 0.003 | 0.744 | 0.001 | 0.412 | 0.041 | 0.345 | 0.815 | 0.016 | 0.000 |
| SA | | 0.002 | 0.683 | 1.429 | 0.003 | 0.964 | 0.000 | 0.157 | 0.183 | 0.894 | 0.745 | 0.028 | 0.546 |
| rBOA | | 0.000 | 0.001 | 0.079 | 0.001 | 0.955 | 0.002 | 0.002 | 0.040 | 0.917 | 1.407 | 0.000 | 1.117 |
| rGA | | 0.000 | 0.403 | 0.988 | 0.000 | 0.842 | 0.002 | 0.186 | 0.093 | 0.991 | 0.185 | 0.021 | 0.602 |

**Table A.28** Variance for performance criterion $\zeta_{T_m}$ (scaled) with $m = 100,000$ in 10 dimensions.

| $Var(\zeta_{T_m})$ (scaled) | | sphr. | ack. | lg-ack. | whit. | shek. | rosen. | rastr. | sal. | lange. | schw. | grie. | weier. |
|---|---|---|---|---|---|---|---|---|---|---|---|---|---|
| CG | | 0.000 | 0.184 | 1.532 | 1.710 | 0.040 | 0.198 | 0.359 | 0.619 | – | 0.460 | 0.188 | 0.257 |
| CG-R | | 0.000 | 0.085 | 0.435 | 0.000 | 0.336 | 0.000 | 0.087 | 0.021 | 0.066 | 0.107 | 0.005 | 0.082 |
| CMA-ES | 100 | 0.000 | 0.019 | 0.048 | 0.000 | 0.004 | 0.000 | 0.018 | 0.021 | 0.184 | 0.177 | 0.000 | 0.123 |
| | 750 | 0.000 | 0.002 | 0.021 | 0.000 | 0.001 | 0.000 | 0.001 | – | 0.217 | 0.151 | 0.000 | 0.087 |
| | 1250 | 0.000 | 0.000 | 0.010 | 0.000 | 0.009 | 0.000 | 0.000 | 0.000 | 0.203 | 0.132 | 0.000 | 0.062 |
| | 2500 | 0.000 | 0.000 | 0.006 | 0.000 | 0.009 | 0.000 | 0.000 | 0.000 | 0.180 | 0.085 | 0.000 | 0.096 |
| CMA-ES-R | 100 | 0.000 | 0.000 | 0.004 | 0.000 | 0.000 | 0.000 | 0.004 | 0.004 | 0.161 | 0.120 | 0.000 | 0.008 |
| | 750 | 0.000 | 0.016 | 0.005 | 0.000 | 0.004 | 0.000 | 0.001 | 0.000 | 0.169 | 0.108 | 0.001 | 0.032 |
| | 1250 | 0.000 | 0.024 | 0.013 | 0.000 | 0.006 | 0.000 | 0.004 | 0.000 | 0.192 | 0.139 | 0.002 | 0.051 |
| | 2500 | 0.060 | 0.036 | 0.025 | 0.000 | 0.002 | 0.019 | 0.025 | 0.000 | 0.010 | 0.074 | 0.025 | 0.106 |
| DE | .2/.9 | 0.012 | 0.019 | 0.158 | 0.000 | 0.024 | 0.011 | 0.037 | 0.070 | 0.101 | 0.062 | 0.012 | 0.062 |
| | .2/.2 | 0.000 | 0.017 | 0.038 | 0.000 | 0.074 | 0.002 | 0.028 | 0.027 | 0.097 | 0.086 | 0.001 | 0.066 |
| | .9/.2 | 0.000 | 0.019 | 0.121 | 0.000 | 0.120 | 0.000 | 0.065 | 0.026 | 0.107 | 0.118 | 0.002 | 0.085 |
| | .9/.9 | 0.099 | 0.017 | 0.241 | 0.029 | 0.001 | 0.084 | 0.088 | 0.081 | 0.003 | 0.075 | 0.098 | 0.059 |
| GSS | | 0.000 | 0.186 | 0.281 | 0.000 | 0.147 | 0.000 | 0.084 | 0.165 | 0.174 | 0.137 | 0.004 | 0.240 |
| GSS-R | | 0.000 | 0.022 | 0.190 | 0.000 | 0.175 | 0.000 | 0.046 | 0.064 | 0.139 | 0.035 | 0.002 | 0.191 |
| NM | | 0.000 | 0.049 | 1.575 | 0.032 | 0.113 | 0.063 | 0.330 | 0.437 | 0.000 | 0.220 | 0.558 | 0.376 |
| NM-R | | 0.000 | 0.034 | 0.290 | 0.000 | 0.323 | 0.000 | 0.069 | 0.129 | 0.010 | 0.103 | 0.000 | 0.052 |
| PSO | –0.5/2 | 0.033 | 0.031 | 0.383 | 0.000 | 0.000 | 0.015 | 0.220 | 0.122 | 0.004 | 0.195 | 0.035 | 0.000 |
| | 1/2 | 0.885 | 0.036 | 0.694 | 69.350 | 0.003 | 4.853 | 0.200 | 0.222 | – | 0.015 | 0.897 | 0.000 |
| REA-P | R1 | 0.002 | 0.047 | 0.083 | 0.586 | 0.002 | 0.445 | 0.075 | 0.039 | 0.007 | 0.067 | 0.005 | 0.010 |
| | R2 | 0.000 | 0.033 | 0.188 | 0.777 | 0.001 | 0.493 | 0.177 | 0.022 | 0.008 | 0.103 | 0.001 | 0.133 |
| REA-T | R1 | 0.000 | 0.035 | 0.100 | 0.000 | 0.247 | 0.000 | 0.061 | 0.000 | 0.019 | 0.084 | 0.001 | 0.062 |
| | R2 | 0.000 | 0.030 | 0.106 | 0.000 | 0.260 | 0.000 | 0.066 | 0.004 | 0.214 | 0.073 | 0.002 | 0.000 |
| SA | | 0.000 | 0.068 | 0.205 | 0.000 | 0.001 | 0.000 | 0.047 | 0.040 | 0.091 | 0.119 | 0.001 | 0.061 |
| rBOA | | 0.000 | 0.009 | 0.031 | 0.000 | 0.004 | 0.000 | 0.015 | 0.003 | 0.052 | 0.129 | 0.000 | 0.078 |
| rGA | | 0.000 | 0.093 | 0.398 | 0.000 | 0.060 | 0.000 | 0.072 | 0.015 | 0.059 | 0.104 | 0.002 | 0.035 |

**Table A.29** Results on performance criterion $\zeta_{T_m}$ (scaled) with $m = 250,000$ in 10 dimensions.

| $\zeta_{T_m}$ (scaled) | | sphr. | ack. | lg-ack. | whit. | shek. | rosen. | rastr. | sal. | lange. | schw. | grie. | weier. |
|---|---|---|---|---|---|---|---|---|---|---|---|---|---|
| CG | | 0.000 | 1.431 | 6.279 | 0.133 | 0.846 | 0.014 | 1.267 | 2.045 | 1.000 | 1.329 | 0.143 | 1.940 |
| CG-R | | 0.000 | 0.752 | 2.294 | 0.002 | 0.060 | 0.000 | 0.303 | 0.001 | 0.989 | 0.469 | 0.006 | 1.164 |
| CMA-ES | 100 | 0.000 | 0.032 | 0.067 | 0.002 | 0.851 | 0.001 | 0.028 | 0.073 | 0.624 | 0.896 | 0.000 | 0.286 |
| | 750 | 0.000 | 0.000 | 0.010 | 0.002 | 0.851 | 0.002 | 0.000 | 0.040 | 0.401 | 0.482 | 0.000 | 0.087 |
| | 1250 | 0.000 | 0.000 | 0.002 | 0.001 | 0.854 | 0.002 | 0.000 | 0.040 | 0.367 | 0.394 | 0.000 | 0.040 |
| | 2500 | 0.000 | 0.000 | 0.000 | 0.001 | 0.851 | 0.001 | 0.000 | 0.040 | 0.264 | 0.210 | 0.000 | 0.012 |
| CMA-ES-R | 100 | 0.000 | 0.000 | 0.000 | 0.001 | 0.851 | 0.000 | 0.000 | 0.040 | 0.212 | 0.411 | 0.000 | 0.013 |
| | 750 | 0.000 | 0.059 | 0.013 | 0.001 | 0.868 | 0.002 | 0.004 | 0.040 | 0.087 | 0.239 | 0.013 | 0.238 |
| | 1250 | 0.001 | 0.106 | 0.035 | 0.001 | 0.957 | 0.003 | 0.009 | 0.040 | 0.147 | 0.242 | 0.016 | 0.299 |
| | 2500 | 0.008 | 0.144 | 0.067 | 0.002 | 0.958 | 0.007 | 0.022 | 0.040 | 0.982 | 0.324 | 0.026 | 0.389 |
| DE | .2/.9 | 0.005 | 0.026 | 0.275 | 0.002 | 0.887 | 0.011 | 0.076 | 0.242 | 0.767 | 0.047 | 0.029 | 0.209 |
| | .2/.2 | 0.000 | 0.029 | 0.034 | 0.000 | 0.825 | 0.002 | 0.021 | 0.075 | 0.740 | 0.079 | 0.001 | 0.047 |
| | .9/.2 | 0.000 | 0.029 | 0.074 | 0.002 | 0.809 | 0.001 | 0.307 | 0.075 | 0.714 | 0.644 | 0.004 | 0.765 |
| | .9/.9 | 0.282 | 0.029 | 2.130 | 0.021 | 0.962 | 0.151 | 0.620 | 0.573 | 0.993 | 0.821 | 0.296 | 0.910 |
| GSS | | 0.000 | 1.427 | 0.983 | 0.001 | 0.814 | 0.000 | 0.213 | 0.325 | 0.780 | 0.099 | 0.006 | 0.401 |
| GSS-R | | 0.000 | 0.077 | 0.594 | 0.000 | 0.701 | 0.000 | 0.113 | 0.136 | 0.515 | 0.000 | 0.001 | 0.120 |
| NM | | 0.000 | 0.133 | 2.867 | 0.010 | 0.815 | 0.004 | 0.844 | 1.686 | 1.000 | 0.943 | 0.190 | 0.670 |
| NM-R | | 0.000 | 0.102 | 0.337 | 0.000 | 0.040 | 0.000 | 0.154 | 0.495 | 0.997 | 0.432 | 0.000 | 0.047 |
| PSO | –0.5/2 | 0.014 | 0.320 | 0.991 | 0.003 | 0.960 | 0.008 | 0.236 | 0.213 | 0.998 | 0.304 | 0.035 | 0.000 |
| | 1/2 | 2.646 | 0.320 | 4.875 | 64.875 | 0.984 | 8.484 | 1.590 | 1.615 | 1.000 | 0.547 | 2.694 | 0.000 |
| REA-P | R1 | 0.004 | 0.616 | 0.114 | 0.495 | 0.964 | 0.691 | 0.367 | 0.190 | 0.994 | 0.822 | 0.036 | 0.075 |
| | R2 | 0.000 | 0.364 | 0.304 | 0.650 | 0.964 | 0.803 | 0.235 | 0.108 | 0.996 | 0.276 | 0.021 | 0.176 |
| REA-T | R1 | 0.000 | 0.216 | 0.094 | 0.001 | 0.762 | 0.000 | 0.081 | 0.040 | 0.163 | 0.326 | 0.004 | 0.027 |
| | R2 | 0.000 | 0.127 | 0.059 | 0.002 | 0.736 | 0.000 | 0.082 | 0.040 | 0.243 | 0.683 | 0.010 | 0.000 |
| SA | | 0.001 | 0.619 | 1.266 | 0.003 | 0.964 | 0.000 | 0.124 | 0.131 | 0.775 | 0.654 | 0.027 | 0.478 |
| rBOA | | 0.000 | 0.000 | 0.059 | 0.001 | 0.952 | 0.002 | 0.000 | 0.040 | 0.864 | 1.407 | 0.000 | 1.064 |
| rGA | | 0.000 | 0.401 | 0.988 | 0.000 | 0.841 | 0.002 | 0.181 | 0.082 | 0.984 | 0.183 | 0.020 | 0.574 |

**Table A.30** Variance for performance criterion $\zeta_{T_m}$ (scaled) with $m = 250,000$ in 10 dimensions.

| $Var(\zeta_{T_m})$ (scaled) | | sphr. | ack. | lg-ack. | whit. | shek. | rosen. | rastr. | sal. | lange. | schw. | grie. | weier. |
|---|---|---|---|---|---|---|---|---|---|---|---|---|---|
| CG | | 0.000 | 0.184 | 1.532 | 1.710 | 0.040 | 0.198 | 0.359 | 0.619 | – | 0.460 | 0.188 | 0.257 |
| CG-R | | 0.000 | 0.082 | 0.388 | 0.000 | 0.193 | 0.000 | 0.069 | 0.007 | 0.084 | 0.089 | 0.003 | 0.081 |
| CMA-ES | 100 | 0.000 | 0.019 | 0.048 | 0.000 | 0.004 | 0.000 | 0.017 | 0.021 | 0.184 | 0.176 | 0.000 | 0.124 |
| | 750 | 0.000 | 0.002 | 0.021 | 0.000 | 0.000 | 0.000 | 0.001 | – | 0.217 | 0.151 | 0.000 | 0.087 |
| | 1250 | 0.000 | 0.000 | 0.010 | 0.000 | 0.009 | 0.000 | 0.000 | – | 0.203 | 0.131 | 0.000 | 0.061 |
| | 2500 | 0.000 | 0.000 | 0.006 | 0.000 | 0.000 | 0.000 | 0.000 | 0.000 | 0.181 | 0.085 | 0.000 | 0.032 |
| CMA-ES-R | 100 | 0.000 | 0.000 | 0.002 | 0.000 | – | 0.000 | 0.001 | 0.000 | 0.155 | 0.097 | 0.000 | 0.005 |
| | 750 | 0.000 | 0.013 | 0.003 | 0.000 | 0.003 | 0.000 | 0.001 | 0.000 | 0.099 | 0.078 | 0.001 | 0.025 |
| | 1250 | 0.000 | 0.020 | 0.009 | 0.000 | 0.013 | 0.000 | 0.002 | – | 0.133 | 0.084 | 0.001 | 0.031 |
| | 2500 | 0.011 | 0.022 | 0.018 | 0.000 | 0.003 | 0.003 | 0.007 | 0.000 | 0.019 | 0.131 | 0.004 | 0.050 |
| DE | .2/.9 | 0.003 | 0.014 | 0.093 | 0.000 | 0.037 | 0.008 | 0.026 | 0.052 | 0.128 | 0.030 | 0.004 | 0.040 |
| | .2/.2 | 0.000 | 0.015 | 0.038 | 0.000 | 0.127 | 0.001 | 0.015 | 0.022 | 0.137 | 0.073 | 0.000 | 0.046 |
| | .9/.2 | 0.000 | 0.015 | 0.056 | 0.000 | 0.147 | 0.000 | 0.067 | 0.017 | 0.139 | 0.120 | 0.001 | 0.091 |
| | .9/.9 | 0.084 | 0.015 | 0.218 | 0.012 | 0.002 | 0.059 | 0.076 | 0.071 | 0.013 | 0.075 | 0.080 | 0.058 |
| GSS | | 0.000 | 0.186 | 0.281 | 0.000 | 0.147 | 0.000 | 0.084 | 0.165 | 0.174 | 0.137 | 0.004 | 0.240 |
| GSS-R | | 0.000 | 0.014 | 0.149 | 0.000 | 0.252 | 0.000 | 0.034 | 0.035 | 0.084 | 0.007 | 0.001 | 0.169 |
| NM | | 0.000 | 0.049 | 1.575 | 0.032 | 0.113 | 0.063 | 0.330 | 0.437 | 0.000 | 0.220 | 0.558 | 0.376 |
| NM-R | | 0.000 | 0.026 | 0.174 | 0.000 | 0.160 | 0.000 | 0.048 | 0.098 | 0.013 | 0.090 | 0.000 | 0.036 |
| PSO | –0.5/2 | 0.013 | 0.031 | 0.270 | 0.000 | 0.000 | 0.005 | 0.142 | 0.084 | 0.006 | 0.157 | 0.012 | 0.000 |
| | 1/2 | 0.909 | 0.036 | 0.701 | 69.350 | 0.003 | 4.853 | 0.200 | 0.222 | – | 0.014 | 0.966 | 0.000 |
| REA-P | R1 | 0.001 | 0.048 | 0.060 | 0.587 | 0.001 | 0.445 | 0.062 | 0.027 | 0.012 | 0.066 | 0.003 | 0.007 |
| | R2 | 0.000 | 0.028 | 0.189 | 0.777 | 0.001 | 0.493 | 0.090 | 0.017 | 0.008 | 0.089 | 0.001 | 0.127 |
| REA-T | R1 | 0.000 | 0.025 | 0.084 | 0.000 | 0.249 | 0.000 | 0.033 | 0.000 | 0.202 | 0.074 | 0.003 | 0.062 |
| | R2 | 0.000 | 0.049 | 0.041 | 0.000 | 0.279 | 0.000 | 0.042 | 0.000 | 0.220 | 0.074 | 0.002 | 0.000 |
| SA | | 0.000 | 0.063 | 0.182 | 0.000 | 0.001 | 0.000 | 0.039 | 0.025 | 0.119 | 0.115 | 0.001 | 0.060 |
| rBOA | | 0.000 | 0.000 | 0.029 | 0.000 | 0.004 | 0.000 | 0.001 | 0.004 | 0.072 | 0.129 | 0.000 | 0.070 |
| rGA | | 0.000 | 0.093 | 0.399 | 0.000 | 0.061 | 0.000 | 0.071 | 0.006 | 0.081 | 0.104 | 0.002 | 0.035 |

**Table A.31** Results on performance criterion $\phi_1$ (scaled) in 10 dimensions.

| $\phi_1$ (scaled) | | sphr. | ack. | lg-ack. | whit. | shek. | rosen. | rastr. | sal. | lange. | schw. | grie. | weier. |
|---|---|---|---|---|---|---|---|---|---|---|---|---|---|
| CG | | 0.000 | 1.431 | 6.280 | 0.348 | 0.847 | 0.062 | 1.267 | 2.045 | 1.000 | 1.331 | 0.160 | 1.940 |
| CG-R | | 0.000 | 0.827 | 2.719 | 0.003 | 0.263 | 0.090 | 0.372 | 0.036 | 0.994 | 0.573 | 0.027 | 1.241 |
| CMA-ES | 100 | 0.000 | 0.033 | 0.069 | 0.002 | 0.852 | 0.001 | 0.031 | 0.073 | 0.625 | 0.902 | 0.000 | 0.298 |
| | 750 | 0.009 | 0.023 | 0.073 | 0.004 | 0.856 | 0.009 | 0.030 | 0.055 | 0.429 | 0.574 | 0.008 | 0.216 |
| | 1250 | 0.023 | 0.049 | 0.140 | 0.006 | 0.864 | 0.019 | 0.060 | 0.073 | 0.425 | 0.584 | 0.020 | 0.294 |
| | 2500 | 0.061 | 0.133 | 0.375 | 0.008 | 0.881 | 0.037 | 0.151 | 0.130 | 0.425 | 0.665 | 0.054 | 0.579 |
| CMA-ES-R | 100 | 0.000 | 0.003 | 0.006 | 0.001 | 0.852 | 0.001 | 0.005 | 0.044 | 0.316 | 0.511 | 0.000 | 0.034 |
| | 750 | 0.009 | 0.089 | 0.078 | 0.003 | 0.875 | 0.009 | 0.033 | 0.054 | 0.232 | 0.426 | 0.022 | 0.344 |
| | 1250 | 0.022 | 0.153 | 0.170 | 0.005 | 0.961 | 0.017 | 0.068 | 0.072 | 0.330 | 0.483 | 0.034 | 0.474 |
| | 2500 | 0.087 | 0.246 | 0.433 | 0.009 | 0.961 | 0.047 | 0.170 | 0.130 | 0.991 | 0.718 | 0.084 | 0.733 |
| DE | .2/.9 | 0.053 | 0.061 | 0.674 | 0.015 | 0.925 | 0.046 | 0.182 | 0.358 | 0.859 | 0.198 | 0.077 | 0.358 |
| | .2/.2 | 0.014 | 0.062 | 0.168 | 0.009 | 0.855 | 0.016 | 0.107 | 0.151 | 0.853 | 0.201 | 0.020 | 0.250 |
| | .9/.2 | 0.011 | 0.062 | 0.391 | 0.005 | 0.836 | 0.010 | 0.414 | 0.142 | 0.818 | 0.769 | 0.022 | 0.878 |
| | .9/.9 | 0.386 | 0.063 | 2.340 | 0.067 | 0.964 | 0.242 | 0.687 | 0.639 | 0.996 | 0.890 | 0.394 | 0.963 |
| GSS | | 0.000 | 1.427 | 0.985 | 0.001 | 0.815 | 0.000 | 0.213 | 0.325 | 0.780 | 0.099 | 0.007 | 0.402 |
| GSS-R | | 0.000 | 0.250 | 0.703 | 0.000 | 0.763 | 0.000 | 0.142 | 0.187 | 0.598 | 0.021 | 0.003 | 0.209 |
| NM | | 0.000 | 0.133 | 2.867 | 0.048 | 0.815 | 0.009 | 0.844 | 1.686 | 1.000 | 0.943 | 0.191 | 0.670 |
| NM-R | | 0.000 | 0.114 | 0.656 | 0.222 | 0.238 | 0.002 | 0.220 | 0.615 | 0.998 | 0.518 | 0.002 | 0.094 |
| PSO | –0.5/2 | 0.055 | 0.320 | 1.366 | 0.009 | 0.960 | 0.035 | 0.441 | 0.332 | 0.999 | 0.376 | 0.075 | 0.000 |
| | 1/2 | 2.681 | 0.320 | 4.886 | 64.875 | 0.984 | 8.484 | 1.590 | 1.615 | 1.000 | 0.548 | 2.737 | 0.000 |
| REA-P | R1 | 0.022 | 0.688 | 0.362 | 0.496 | 0.965 | 0.691 | 0.486 | 0.287 | 0.996 | 0.888 | 0.062 | 0.127 |
| | R2 | 0.033 | 0.456 | 0.547 | 0.650 | 0.965 | 0.803 | 0.417 | 0.182 | 0.997 | 0.540 | 0.049 | 0.227 |
| REA-T | R1 | 0.012 | 0.342 | 0.740 | 0.006 | 0.800 | 0.011 | 0.351 | 0.091 | 0.518 | 0.642 | 0.034 | 0.037 |
| | R2 | 0.020 | 0.290 | 0.492 | 0.007 | 0.790 | 0.015 | 0.376 | 0.101 | 0.386 | 0.812 | 0.040 | 0.022 |
| SA | | 0.003 | 0.691 | 1.465 | 0.004 | 0.965 | 0.000 | 0.177 | 0.226 | 0.879 | 0.753 | 0.029 | 0.554 |
| rBOA | | 0.000 | 0.015 | 0.101 | 0.002 | 0.955 | 0.002 | 0.027 | 0.041 | 0.914 | 1.407 | 0.000 | 1.121 |
| rGA | | 0.004 | 0.406 | 0.999 | 0.004 | 0.848 | 0.006 | 0.189 | 0.104 | 0.991 | 0.187 | 0.024 | 0.605 |

**Table A.32** Variance for performance criterion $\phi_1$ (scaled) in 10 dimensions.

| $Var(\phi_1)$ (scaled) | | sphr. | ack. | lg-ack. | whit. | shek. | rosen. | rastr. | sal. | lange. | schw. | grie. | weier. |
|---|---|---|---|---|---|---|---|---|---|---|---|---|---|
| CG | | 0.000 | 0.184 | 1.533 | 1.700 | 0.040 | 0.195 | 0.359 | 0.619 | – | 0.460 | 0.183 | 0.257 |
| CG-R | | 0.000 | 0.067 | 0.348 | 0.000 | 0.214 | 0.060 | 0.066 | 0.025 | 0.060 | 0.081 | 0.004 | 0.063 |
| CMA-ES | 100 | 0.000 | 0.019 | 0.047 | 0.000 | 0.004 | 0.000 | 0.018 | 0.021 | 0.184 | 0.174 | 0.000 | 0.122 |
| | 750 | 0.001 | 0.002 | 0.021 | 0.001 | 0.001 | 0.002 | 0.002 | 0.001 | 0.206 | 0.136 | 0.001 | 0.081 |
| | 1250 | 0.004 | 0.001 | 0.014 | 0.003 | 0.009 | 0.005 | 0.004 | 0.002 | 0.183 | 0.106 | 0.002 | 0.062 |
| | 2500 | 0.010 | 0.006 | 0.025 | 0.004 | 0.003 | 0.010 | 0.012 | 0.009 | 0.137 | 0.064 | 0.008 | 0.064 |
| CMA-ES-R | 100 | 0.000 | 0.001 | 0.004 | 0.000 | 0.000 | 0.000 | 0.002 | 0.003 | 0.131 | 0.091 | 0.000 | 0.008 |
| | 750 | 0.001 | 0.012 | 0.005 | 0.001 | 0.003 | 0.001 | 0.002 | 0.001 | 0.118 | 0.075 | 0.002 | 0.026 |
| | 1250 | 0.004 | 0.019 | 0.013 | 0.002 | 0.005 | 0.004 | 0.005 | 0.002 | 0.134 | 0.075 | 0.003 | 0.037 |
| | 2500 | 0.027 | 0.021 | 0.029 | 0.003 | 0.001 | 0.011 | 0.016 | 0.007 | 0.009 | 0.079 | 0.014 | 0.059 |
| DE | .2/.9 | 0.013 | 0.014 | 0.111 | 0.008 | 0.021 | 0.013 | 0.029 | 0.047 | 0.080 | 0.041 | 0.012 | 0.046 |
| | .2/.2 | 0.002 | 0.013 | 0.036 | 0.006 | 0.074 | 0.005 | 0.022 | 0.020 | 0.078 | 0.071 | 0.003 | 0.045 |
| | .9/.2 | 0.002 | 0.014 | 0.075 | 0.002 | 0.102 | 0.003 | 0.049 | 0.017 | 0.090 | 0.097 | 0.003 | 0.066 |
| | .9/.9 | 0.073 | 0.013 | 0.186 | 0.026 | 0.001 | 0.065 | 0.065 | 0.058 | 0.008 | 0.057 | 0.075 | 0.044 |
| GSS | | 0.000 | 0.186 | 0.280 | 0.000 | 0.146 | 0.000 | 0.084 | 0.165 | 0.173 | 0.137 | 0.004 | 0.240 |
| GSS-R | | 0.000 | 0.028 | 0.148 | 0.000 | 0.159 | 0.000 | 0.035 | 0.046 | 0.097 | 0.025 | 0.001 | 0.164 |
| NM | | 0.001 | 0.049 | 1.575 | 0.227 | 0.112 | 0.065 | 0.330 | 0.437 | – | 0.220 | 0.559 | 0.376 |
| NM-R | | 0.000 | 0.031 | 0.235 | 1.930 | 0.207 | 0.023 | 0.053 | 0.099 | 0.009 | 0.080 | 0.008 | 0.037 |
| PSO | –0.5/2 | 0.032 | 0.031 | 0.270 | 0.013 | 0.000 | 0.021 | 0.150 | 0.085 | 0.004 | 0.168 | 0.030 | 0.000 |
| | 1/2 | 0.860 | 0.036 | 0.690 | 69.350 | 0.003 | 4.853 | 0.200 | 0.222 | – | 0.014 | 0.891 | 0.000 |
| REA-P | R1 | 0.003 | 0.031 | 0.064 | 0.586 | 0.001 | 0.444 | 0.048 | 0.022 | 0.007 | 0.050 | 0.004 | 0.008 |
| | R2 | 0.008 | 0.020 | 0.168 | 0.777 | 0.001 | 0.493 | 0.087 | 0.013 | 0.006 | 0.058 | 0.006 | 0.131 |
| REA-T | R1 | 0.002 | 0.015 | 0.050 | 0.001 | 0.188 | 0.002 | 0.041 | 0.006 | 0.060 | 0.044 | 0.003 | 0.062 |
| | R2 | 0.003 | 0.020 | 0.051 | 0.002 | 0.195 | 0.002 | 0.043 | 0.006 | 0.151 | 0.045 | 0.003 | 0.002 |
| SA | | 0.000 | 0.051 | 0.164 | 0.000 | 0.001 | 0.000 | 0.039 | 0.034 | 0.072 | 0.097 | 0.001 | 0.048 |
| rBOA | | 0.000 | 0.011 | 0.028 | 0.000 | 0.003 | 0.000 | 0.018 | 0.002 | 0.044 | 0.129 | 0.000 | 0.061 |
| rGA | | 0.001 | 0.093 | 0.397 | 0.003 | 0.057 | 0.002 | 0.071 | 0.009 | 0.051 | 0.103 | 0.002 | 0.029 |

**Table A.33** Results on performance criterion $\phi_2$ (scaled) in 10 dimensions.

| $\phi_2$ (scaled) | | sphr. | ack. | lg-ack. | whit. | shek. | rosen. | rastr. | sal. | lange. | schw. | grie. | weier. |
|---|---|---|---|---|---|---|---|---|---|---|---|---|---|
| CG | | 0.000 | 1.431 | 6.729 | 24.032 | 0.914 | 2.809 | 1.267 | 2.059 | 1.000 | 1.859 | 1.373 | 1.940 |
| CG-R | | 0.000 | 1.276 | 6.006 | 0.071 | 0.901 | 4.561 | 0.986 | 1.330 | 1.000 | 1.397 | 1.030 | 1.712 |
| CMA-ES | 100 | 0.013 | 0.371 | 0.892 | 0.003 | 0.960 | 0.011 | 0.609 | 0.202 | 0.990 | 1.418 | 0.033 | 1.476 |
| | 750 | 0.702 | 0.858 | 2.765 | 0.261 | 0.975 | 0.601 | 0.939 | 0.798 | 1.000 | 1.384 | 0.623 | 1.441 |
| | 1250 | 0.692 | 0.876 | 2.863 | 0.276 | 0.976 | 0.599 | 0.945 | 0.813 | 1.000 | 1.388 | 0.673 | 1.424 |
| | 2500 | 0.541 | 0.826 | 2.634 | 0.120 | 0.972 | 0.379 | 0.838 | 0.721 | 1.000 | 1.302 | 0.532 | 1.250 |
| CMA-ES-R | 100 | 0.010 | 0.366 | 0.863 | 0.003 | 0.957 | 0.009 | 0.590 | 0.193 | 0.956 | 1.380 | 0.031 | 1.310 |
| | 750 | 0.628 | 0.838 | 2.700 | 0.191 | 0.972 | 0.509 | 0.894 | 0.760 | 1.000 | 1.338 | 0.583 | 1.278 |
| | 1250 | 0.621 | 0.847 | 2.734 | 0.166 | 0.972 | 0.475 | 0.875 | 0.776 | 1.000 | 1.345 | 0.587 | 1.278 |
| | 2500 | 0.552 | 0.819 | 2.632 | 0.117 | 0.972 | 0.398 | 0.826 | 0.729 | 1.000 | 1.314 | 0.537 | 1.242 |
| DE | .2/.9 | 1.190 | 0.227 | 3.491 | 1.937 | 0.972 | 1.451 | 1.056 | 1.088 | 0.999 | 1.129 | 1.152 | 1.140 |
| | .2/.2 | 1.135 | 0.228 | 3.490 | 1.988 | 0.971 | 1.498 | 1.050 | 1.083 | 1.000 | 1.120 | 1.163 | 1.154 |
| | .9/.2 | 0.800 | 0.228 | 3.179 | 0.708 | 0.969 | 0.934 | 1.010 | 0.963 | 0.999 | 1.212 | 0.866 | 1.293 |
| | .9/.9 | 1.116 | 0.225 | 3.561 | 1.577 | 0.971 | 1.417 | 1.085 | 1.031 | 0.999 | 1.216 | 1.097 | 1.241 |
| GSS | | 0.135 | 1.447 | 1.479 | 0.007 | 0.952 | 0.074 | 0.309 | 0.357 | 0.836 | 0.369 | 0.160 | 0.488 |
| GSS-R | | 0.145 | 1.432 | 1.430 | 0.007 | 0.953 | 0.079 | 0.300 | 0.361 | 0.849 | 0.364 | 0.161 | 0.507 |
| NM | | 0.086 | 0.133 | 2.873 | 7.862 | 0.832 | 0.850 | 0.846 | 1.686 | 1.000 | 0.973 | 0.242 | 0.670 |
| NM-R | | 0.074 | 0.140 | 3.277 | 5.600 | 0.810 | 0.457 | 0.829 | 1.478 | 0.999 | 0.963 | 0.229 | 0.455 |
| PSO | –0.5/2 | 0.388 | 0.320 | 2.633 | 0.168 | 0.965 | 0.342 | 1.088 | 0.811 | 1.000 | 0.844 | 0.397 | 0.000 |
| | 1/2 | 2.821 | 0.320 | 4.910 | 64.878 | 0.984 | 8.484 | 1.591 | 1.615 | 1.000 | 0.638 | 2.885 | 0.000 |
| REA-P | R1 | 0.849 | 1.014 | 3.300 | 0.611 | 0.972 | 0.780 | 1.056 | 0.974 | 1.000 | 1.220 | 0.767 | 1.157 |
| | R2 | 1.005 | 0.978 | 3.540 | 0.758 | 0.971 | 0.907 | 1.087 | 0.958 | 1.000 | 1.189 | 1.146 | 1.207 |
| REA-T | R1 | 0.744 | 0.967 | 3.296 | 0.638 | 0.969 | 0.717 | 1.034 | 0.941 | 1.000 | 1.223 | 0.852 | 1.033 |
| | R2 | 0.859 | 0.947 | 3.224 | 0.785 | 0.969 | 0.865 | 1.038 | 0.940 | 0.999 | 1.218 | 0.918 | 1.130 |
| SA | | 0.038 | 1.067 | 2.801 | 0.111 | 0.972 | 0.007 | 1.040 | 1.225 | 1.000 | 1.173 | 0.041 | 0.975 |
| rBOA | | 0.186 | 0.376 | 0.926 | 0.055 | 0.968 | 0.159 | 0.831 | 0.196 | 1.000 | 1.407 | 0.011 | 1.309 |
| rGA | | 0.582 | 0.828 | 2.533 | 1.105 | 0.966 | 0.797 | 0.675 | 0.846 | 1.000 | 0.698 | 0.634 | 0.954 |

**Table A.34** Variance for performance criterion $\phi_2$ (scaled) in 10 dimensions.

| $Var(\phi_2)$ (scaled) | | sphr. | ack. | lg-ack. | whit. | shek. | rosen. | rastr. | sal. | lange. | schw. | grie. | weier. |
|---|---|---|---|---|---|---|---|---|---|---|---|---|---|
| CG | | 0.000 | 0.184 | 2.104 | 1.922 | 0.031 | – | 0.359 | 0.638 | – | 0.739 | 0.000 | 0.257 |
| CG-R | | 0.000 | 0.138 | 1.872 | 0.004 | 0.036 | 0.359 | 0.293 | 0.701 | 0.000 | 0.355 | 0.000 | 0.186 |
| CMA-ES | 100 | 0.004 | 0.032 | 0.105 | 0.000 | 0.001 | 0.003 | 0.078 | 0.032 | 0.010 | 0.092 | 0.002 | 0.097 |
| | 750 | 0.169 | 0.065 | 0.320 | 0.198 | 0.002 | 0.251 | 0.116 | 0.101 | 0.000 | 0.089 | 0.150 | 0.092 |
| | 1250 | 0.162 | 0.054 | 0.301 | 0.219 | 0.002 | 0.242 | 0.089 | 0.092 | 0.000 | 0.085 | 0.140 | 0.087 |
| | 2500 | 0.137 | 0.059 | 0.281 | 0.089 | 0.002 | 0.160 | 0.097 | 0.098 | 0.000 | 0.091 | 0.133 | 0.099 |
| CMA-ES-R | 100 | 0.002 | 0.030 | 0.105 | 0.000 | 0.002 | 0.002 | 0.092 | 0.027 | 0.038 | 0.099 | 0.002 | 0.090 |
| | 750 | 0.154 | 0.063 | 0.285 | 0.134 | 0.001 | 0.219 | 0.102 | 0.096 | 0.000 | 0.094 | 0.138 | 0.092 |
| | 1250 | 0.157 | 0.062 | 0.246 | 0.126 | 0.002 | 0.178 | 0.104 | 0.099 | 0.000 | 0.088 | 0.129 | 0.087 |
| | 2500 | 0.124 | 0.061 | 0.269 | 0.083 | 0.002 | 0.142 | 0.116 | 0.091 | 0.000 | 0.078 | 0.129 | 0.091 |
| DE | .2/.9 | 0.311 | 0.033 | 0.419 | 1.571 | 0.002 | 0.657 | 0.143 | 0.135 | 0.000 | 0.121 | 0.294 | 0.099 |
| | .2/.2 | 0.294 | 0.029 | 0.426 | 1.583 | 0.002 | 0.696 | 0.136 | 0.132 | 0.000 | 0.124 | 0.308 | 0.097 |
| | .9/.2 | 0.269 | 0.035 | 0.380 | 0.709 | 0.002 | 0.535 | 0.130 | 0.137 | 0.000 | 0.112 | 0.245 | 0.081 |
| | .9/.9 | 0.262 | 0.032 | 0.355 | 1.254 | 0.002 | 0.595 | 0.123 | 0.119 | 0.001 | 0.100 | 0.260 | 0.089 |
| GSS | | 0.061 | 0.184 | 0.266 | 0.005 | 0.011 | 0.056 | 0.094 | 0.151 | 0.133 | 0.150 | 0.064 | 0.218 |
| GSS-R | | 0.059 | 0.179 | 0.286 | 0.003 | 0.010 | 0.050 | 0.089 | 0.165 | 0.123 | 0.143 | 0.062 | 0.227 |
| NM | | 0.377 | 0.049 | 1.580 | 39.019 | 0.101 | 3.355 | 0.332 | 0.437 | 0.000 | 0.217 | 0.671 | 0.376 |
| NM-R | | 0.327 | 0.053 | 1.482 | 31.496 | 0.126 | 2.533 | 0.332 | 0.370 | 0.000 | 0.210 | 0.627 | 0.201 |
| PSO | –0.5/2 | 0.256 | 0.031 | 0.524 | 0.429 | 0.003 | 0.326 | 0.158 | 0.183 | 0.000 | 0.180 | 0.214 | 0.000 |
| | 1/2 | 0.744 | 0.036 | 0.668 | 69.348 | 0.002 | 4.853 | 0.200 | 0.221 | – | 0.134 | 0.787 | 0.000 |
| REA-P | R1 | 0.254 | 0.074 | 0.319 | 0.620 | 0.002 | 0.420 | 0.115 | 0.129 | 0.000 | 0.096 | 0.195 | 0.094 |
| | R2 | 0.261 | 0.084 | 0.396 | 0.841 | 0.002 | 0.482 | 0.127 | 0.128 | 0.000 | 0.101 | 0.276 | 0.084 |
| REA-T | R1 | 0.180 | 0.078 | 0.324 | 0.529 | 0.002 | 0.304 | 0.111 | 0.100 | 0.000 | 0.099 | 0.188 | 0.088 |
| | R2 | 0.208 | 0.066 | 0.320 | 0.576 | 0.001 | 0.336 | 0.121 | 0.119 | 0.000 | 0.096 | 0.194 | 0.081 |
| SA | | 0.017 | 0.090 | 0.419 | 0.301 | 0.002 | 0.010 | 0.288 | 0.190 | 0.000 | 0.107 | 0.005 | 0.109 |
| rBOA | | 0.175 | 0.068 | 0.193 | 0.129 | 0.002 | 0.184 | 0.207 | 0.049 | 0.000 | 0.129 | 0.003 | 0.101 |
| rGA | | 0.216 | 0.089 | 0.400 | 1.266 | 0.002 | 0.437 | 0.110 | 0.136 | 0.000 | 0.121 | 0.203 | 0.072 |

**Table A.35** Results on performance criterion $\sigma_\varepsilon^N$ with $\varepsilon = 10.000$ and $N = 250,000$ in 25 dimensions.

| $\sigma_\varepsilon^N$ | | sphr. | ack. | lg-ack. | whit. | rosen. | rastr. | sal. | schw. | grie. | weier. |
|---|---|---|---|---|---|---|---|---|---|---|---|
| CG | | 1.000 | 1.000 | 0.000 | 0.000 | 0.970 | 0.000 | 0.965 | 0.000 | 1.000 | 0.000 |
| CG-R | | 1.000 | 1.000 | 0.000 | 0.000 | 1.000 | 0.000 | 1.000 | 0.000 | 1.000 | 0.000 |
| CMA-ES | 100 | 1.000 | 1.000 | 0.995 | 0.000 | 1.000 | 0.835 | 1.000 | 0.000 | 1.000 | 0.005 |
| | 750 | 1.000 | 1.000 | 1.000 | 0.000 | 0.000 | 1.000 | 1.000 | 0.000 | 1.000 | 0.004 |
| | 1250 | 1.000 | 1.000 | 1.000 | 0.000 | 0.000 | 1.000 | 1.000 | 0.000 | 1.000 | 0.010 |
| | 2500 | 1.000 | 1.000 | 1.000 | 0.000 | 0.000 | 1.000 | 1.000 | 0.000 | 1.000 | 0.000 |
| CMA-ES-R | 100 | 1.000 | 1.000 | 1.000 | 0.000 | 0.000 | 1.000 | 1.000 | 0.000 | 1.000 | 0.080 |
| | 750 | 1.000 | 1.000 | 1.000 | 0.000 | 0.000 | 1.000 | 1.000 | 0.000 | 1.000 | 0.000 |
| | 1250 | 1.000 | 1.000 | 0.000 | 0.000 | 0.000 | 0.000 | 1.000 | 0.000 | 1.000 | 0.005 |
| | 2500 | 1.000 | 1.000 | 0.000 | 0.000 | 0.000 | 0.000 | 1.000 | 0.000 | 1.000 | 0.000 |
| DE | .2/.9 | 1.000 | 1.000 | 0.000 | 0.000 | 0.000 | 0.000 | 1.000 | 0.000 | 0.000 | 0.000 |
| | .2/.2 | 1.000 | 1.000 | 0.705 | 0.005 | 0.005 | 0.030 | 1.000 | 0.000 | 0.680 | 0.000 |
| | .9/.2 | 1.000 | 1.000 | 0.075 | 0.000 | 0.010 | 0.000 | 1.000 | 0.000 | 1.000 | 0.000 |
| | .9/.9 | 0.040 | 1.000 | 0.000 | 0.000 | 0.000 | 0.000 | 1.000 | 0.000 | 0.000 | 0.000 |
| GSS | | 1.000 | 1.000 | 0.000 | 0.005 | 1.000 | 0.000 | 1.000 | 0.950 | 1.000 | 0.985 |
| GSS-R | | 1.000 | 1.000 | 0.000 | 0.000 | 1.000 | 0.000 | 1.000 | 0.990 | 1.000 | 0.990 |
| NM | | 0.985 | 1.000 | 0.000 | 0.000 | 0.465 | 0.000 | 1.000 | 0.000 | 0.400 | 0.000 |
| NM-R | | 0.980 | 1.000 | 0.000 | 0.000 | 0.565 | 0.000 | 1.000 | 0.000 | 0.455 | 0.010 |
| PSO | –0.5/2 | 1.000 | 1.000 | 0.000 | 0.000 | 0.000 | 0.059 | 1.000 | 0.000 | 1.000 | 1.000 |
| | 1/2 | 0.000 | 1.000 | 0.000 | 0.000 | 0.000 | 0.000 | 1.000 | 0.000 | 0.000 | 1.000 |
| REA-P | R1 | 0.000 | 1.000 | 0.000 | 0.000 | 0.000 | 0.000 | 1.000 | 0.000 | 0.000 | 0.000 |
| | R2 | 0.000 | 1.000 | 0.000 | 0.000 | 0.000 | 0.000 | 1.000 | 0.000 | 0.000 | 0.000 |
| REA-T | R1 | 1.000 | 1.000 | 0.255 | 0.000 | 0.000 | 0.000 | 1.000 | 0.000 | 1.000 | 0.545 |
| | R2 | 0.000 | 1.000 | 0.000 | 0.000 | 0.000 | 0.000 | 1.000 | 0.000 | 0.000 | 0.000 |
| | R3 | 0.000 | 1.000 | 0.000 | 0.000 | 0.000 | 0.000 | 1.000 | 0.000 | 0.000 | 0.005 |
| SA | | 1.000 | 1.000 | 0.000 | 0.000 | 0.045 | 0.000 | 1.000 | 0.000 | 1.000 | 0.000 |
| rBOA | | 0.999 | 1.000 | 0.000 | 0.000 | 0.000 | 0.000 | 1.000 | 0.000 | 0.999 | 0.000 |
| rGA | | 1.000 | 1.000 | 0.000 | 0.000 | 0.000 | 0.000 | 1.000 | 0.000 | 1.000 | 0.000 |

**Table A.36** Results on performance criterion $\sigma_\varepsilon^N$ with $\varepsilon = 1.000$ and $N = 250,000$ in 25 dimensions.

| $\sigma_\varepsilon^N$ | | sphr. | ack. | lg-ack. | whit. | rosen. | rastr. | sal. | schw. | grie. | weier. |
|---|---|---|---|---|---|---|---|---|---|---|---|
| CG | | 1.000 | 0.000 | 0.000 | 0.000 | 0.880 | 0.000 | 0.010 | 0.000 | 1.000 | 0.000 |
| CG-R | | 1.000 | 0.000 | 0.000 | 0.000 | 1.000 | 0.000 | 1.000 | 0.000 | 1.000 | 0.000 |
| CMA-ES | 100 | 1.000 | 1.000 | 0.005 | 0.000 | 0.020 | 0.000 | 1.000 | 0.000 | 1.000 | 0.000 |
| | 750 | 1.000 | 1.000 | 0.690 | 0.000 | 0.000 | 0.970 | 1.000 | 0.000 | 1.000 | 0.000 |
| | 1250 | 1.000 | 1.000 | 0.905 | 0.000 | 0.000 | 1.000 | 1.000 | 0.000 | 1.000 | 0.000 |
| | 2500 | 1.000 | 1.000 | 1.000 | 0.000 | 0.000 | 1.000 | 1.000 | 0.000 | 1.000 | 0.000 |
| CMA-ES-R | 100 | 1.000 | 1.000 | 0.305 | 0.000 | 0.000 | 0.035 | 1.000 | 0.000 | 1.000 | 0.000 |
| | 750 | 1.000 | 1.000 | 0.755 | 0.000 | 0.000 | 0.980 | 1.000 | 0.000 | 1.000 | 0.000 |
| | 1250 | 0.059 | 0.000 | 0.000 | 0.000 | 0.000 | 0.000 | 1.000 | 0.000 | 1.000 | 0.000 |
| | 2500 | 0.000 | 0.000 | 0.000 | 0.000 | 0.000 | 0.000 | 1.000 | 0.000 | 1.000 | 0.000 |
| DE | .2/.9 | 0.000 | 1.000 | 0.000 | 0.000 | 0.000 | 0.000 | 0.000 | 0.000 | 0.000 | 0.000 |
| | .2/.2 | 1.000 | 1.000 | 0.010 | 0.000 | 0.000 | 0.000 | 0.860 | 0.000 | 0.380 | 0.000 |
| | .9/.2 | 1.000 | 1.000 | 0.000 | 0.000 | 0.000 | 0.000 | 0.975 | 0.000 | 0.965 | 0.000 |
| | .9/.9 | 0.000 | 1.000 | 0.000 | 0.000 | 0.000 | 0.000 | 0.000 | 0.000 | 0.000 | 0.000 |
| GSS | | 1.000 | 0.010 | 0.000 | 0.005 | 0.665 | 0.000 | 0.225 | 0.770 | 1.000 | 0.610 |
| GSS-R | | 1.000 | 0.985 | 0.000 | 0.000 | 0.625 | 0.000 | 0.504 | 0.960 | 1.000 | 0.665 |
| NM | | 0.965 | 1.000 | 0.000 | 0.000 | 0.215 | 0.000 | 0.000 | 0.000 | 0.140 | 0.000 |
| NM-R | | 0.975 | 1.000 | 0.000 | 0.000 | 0.290 | 0.000 | 0.000 | 0.000 | 0.142 | 0.000 |
| PSO | –0.5/2 | 0.870 | 0.000 | 0.000 | 0.000 | 0.000 | 0.000 | 0.755 | 0.000 | 0.000 | 1.000 |
| | 1/2 | 0.000 | 0.000 | 0.000 | 0.000 | 0.000 | 0.000 | 0.000 | 0.000 | 0.000 | 1.000 |
| REA-P | R1 | 0.000 | 0.000 | 0.000 | 0.000 | 0.000 | 0.000 | 0.000 | 0.000 | 0.000 | 0.000 |
| | R2 | 0.000 | 0.000 | 0.000 | 0.000 | 0.000 | 0.000 | 0.000 | 0.000 | 0.000 | 0.000 |
| REA-T | R1 | 1.000 | 1.000 | 0.000 | 0.000 | 0.000 | 0.000 | 1.000 | 0.000 | 1.000 | 0.000 |
| | R2 | 0.000 | 0.000 | 0.000 | 0.000 | 0.000 | 0.000 | 0.000 | 0.000 | 0.000 | 0.000 |
| | R3 | 0.000 | 0.000 | 0.000 | 0.000 | 0.000 | 0.000 | 0.000 | 0.000 | 0.000 | 0.000 |
| SA | | 1.000 | 0.000 | 0.000 | 0.000 | 0.000 | 0.000 | 0.120 | 0.000 | 0.000 | 0.000 |
| rBOA | | 0.999 | 0.140 | 0.000 | 0.000 | 0.000 | 0.000 | 1.000 | 0.000 | 0.995 | 0.000 |
| rGA | | 1.000 | 0.000 | 0.000 | 0.000 | 0.000 | 0.000 | 1.000 | 0.000 | 0.000 | 0.000 |

**Table A.37** Results on performance criterion $\sigma_\varepsilon^N$ with $\varepsilon = 0.100$ and $N = 250,000$ in 25 dimensions.

| $\sigma_\varepsilon^N$ | | sphr. | ack. | lg-ack. | whit. | rosen. | rastr. | sal. | schw. | grie. | weier. |
|---|---|---|---|---|---|---|---|---|---|---|---|
| CG | | 1.000 | 0.000 | 0.000 | 0.000 | 0.880 | 0.000 | 0.005 | 0.000 | 0.990 | 0.000 |
| CG-R | | 1.000 | 0.000 | 0.000 | 0.000 | 1.000 | 0.000 | 0.930 | 0.000 | 1.000 | 0.000 |
| CMA-ES | 100 | 1.000 | 0.045 | 0.000 | 0.000 | 0.000 | 0.000 | 0.000 | 0.000 | 1.000 | 0.000 |
| | 750 | 1.000 | 1.000 | 0.320 | 0.000 | 0.000 | 0.750 | 0.040 | 0.000 | 1.000 | 0.000 |
| | 1250 | 1.000 | 1.000 | 0.545 | 0.000 | 0.000 | 0.985 | 0.205 | 0.000 | 1.000 | 0.000 |
| | 2500 | 1.000 | 1.000 | 0.820 | 0.000 | 0.000 | 1.000 | 0.725 | 0.000 | 1.000 | 0.000 |
| CMA-ES-R | 100 | 1.000 | 0.640 | 0.045 | 0.000 | 0.000 | 0.000 | 0.000 | 0.000 | 1.000 | 0.000 |
| | 750 | 1.000 | 1.000 | 0.030 | 0.000 | 0.000 | 0.335 | 0.020 | 0.000 | 0.000 | 0.000 |
| | 1250 | 0.000 | 0.000 | 0.000 | 0.000 | 0.000 | 0.000 | 0.000 | 0.000 | 1.000 | 0.000 |
| | 2500 | 0.000 | 0.000 | 0.000 | 0.000 | 0.000 | 0.000 | 0.000 | 0.000 | 1.000 | 0.000 |
| DE | .2/.9 | 0.000 | 0.000 | 0.000 | 0.000 | 0.000 | 0.000 | 0.000 | 0.000 | 0.000 | 0.000 |
| | .2/.2 | 0.990 | 0.000 | 0.005 | 0.000 | 0.000 | 0.000 | 0.000 | 0.000 | 0.025 | 0.000 |
| | .9/.2 | 1.000 | 0.000 | 0.000 | 0.000 | 0.000 | 0.000 | 0.000 | 0.000 | 0.175 | 0.000 |
| | .9/.9 | 0.000 | 0.000 | 0.000 | 0.000 | 0.000 | 0.000 | 0.000 | 0.000 | 0.000 | 0.000 |
| GSS | | 1.000 | 0.000 | 0.000 | 0.005 | 0.615 | 0.000 | 0.000 | 0.770 | 1.000 | 0.345 |
| GSS-R | | 1.000 | 0.000 | 0.000 | 0.000 | 0.570 | 0.000 | 0.000 | 0.960 | 1.000 | 0.430 |
| NM | | 0.955 | 0.000 | 0.000 | 0.000 | 0.155 | 0.000 | 0.000 | 0.000 | 0.005 | 0.000 |
| NM-R | | 0.970 | 0.000 | 0.000 | 0.000 | 0.165 | 0.000 | 0.000 | 0.000 | 0.010 | 0.000 |
| PSO | –0.5/2 | 0.105 | 0.000 | 0.000 | 0.000 | 0.000 | 0.000 | 0.000 | 0.000 | 0.000 | 1.000 |
| | 1/2 | 0.000 | 0.000 | 0.000 | 0.000 | 0.000 | 0.000 | 0.000 | 0.000 | 0.000 | 1.000 |
| REA-P | R1 | 0.000 | 0.000 | 0.000 | 0.000 | 0.000 | 0.000 | 0.000 | 0.000 | 0.000 | 0.000 |
| | R2 | 0.000 | 0.000 | 0.000 | 0.000 | 0.000 | 0.000 | 0.000 | 0.000 | 0.000 | 0.000 |
| REA-T | R1 | 1.000 | 0.000 | 0.000 | 0.000 | 0.000 | 0.000 | 0.000 | 0.000 | 1.000 | 0.000 |
| | R2 | 0.000 | 0.000 | 0.000 | 0.000 | 0.000 | 0.000 | 0.000 | 0.000 | 0.000 | 0.000 |
| | R3 | 0.000 | 0.000 | 0.000 | 0.000 | 0.000 | 0.000 | 0.000 | 0.000 | 0.000 | 0.000 |
| SA | | 0.010 | 0.000 | 0.000 | 0.000 | 0.000 | 0.000 | 0.000 | 0.000 | 0.000 | 0.000 |
| rBOA | | 0.999 | 0.000 | 0.000 | 0.000 | 0.000 | 0.000 | 0.049 | 0.000 | 0.746 | 0.000 |
| rGA | | 0.005 | 0.000 | 0.000 | 0.000 | 0.000 | 0.000 | 0.000 | 0.000 | 0.000 | 0.000 |

**Table A.38** Results on performance criterion $\frac{1}{100}\hat{\psi}_{\varepsilon}^{N}$ with $\varepsilon = 10.000$ and $N = 250,000$ in 25 dimensions.

| $\frac{1}{100}\hat{\psi}_{\varepsilon}^{N}$ | | sphr. | ack. | lg-ack. | whit. | rosen. | rastr. | sal. | schw. | grie. | weier. |
|---|---|---|---|---|---|---|---|---|---|---|---|
| CG | | 0.0 | 0.0 | – | – | 216.1 | – | 4.2 | – | 0.0 | – |
| CG-R | | 0.0 | 0.0 | – | – | 226.3 | – | 4.9 | – | 0.0 | – |
| CMA-ES | 100 | 10.2 | 0.0 | 45.0 | – | 399.3 | 55.6 | 0.0 | – | 14.4 | 173.0 |
| | 750 | 82.0 | 0.0 | 244.3 | – | – | 262.7 | 0.0 | – | 101.4 | 735.0 |
| | 1250 | 144.5 | 0.0 | 387.1 | – | – | 411.7 | 0.0 | – | 171.7 | 1400.0 |
| | 2500 | 303.2 | 0.0 | 750.3 | – | – | 787.1 | 0.0 | – | 362.8 | – |
| CMA-ES-R | 100 | 9.5 | 0.0 | 45.1 | – | – | 72.7 | 0.0 | – | 13.7 | 981.6 |
| | 750 | 32.6 | 0.0 | 119.7 | – | – | 129.0 | 0.0 | – | 46.3 | – |
| | 1250 | 134.3 | 0.0 | – | – | – | – | 0.0 | – | 29.3 | 2104.0 |
| | 2500 | 303.3 | 0.0 | – | – | – | – | 0.0 | – | 14.3 | – |
| DE | .2/.9 | 1097.7 | 0.0 | – | – | – | – | 0.0 | – | – | – |
| | .2/.2 | 146.5 | 0.0 | 1066.1 | 1229.0 | 1778.0 | 2222.1 | 0.0 | – | 1842.4 | – |
| | .9/.2 | 128.0 | 0.0 | 2279.5 | – | 2234.0 | – | 0.0 | – | 911.0 | – |
| | .9/.9 | 2017.6 | 0.0 | – | – | – | – | 0.0 | – | – | – |
| GSS | | 66.7 | 0.0 | – | 210.0 | 650.0 | – | 0.0 | 126.8 | 105.1 | 36.2 |
| GSS-R | | 65.8 | 0.0 | – | – | 622.4 | – | 0.0 | 197.3 | 104.1 | 33.1 |
| NM | | 116.6 | 0.0 | – | – | 488.1 | – | 0.0 | – | 20.5 | 25.0 |
| NM-R | | 105.4 | 0.0 | – | – | 617.8 | – | 0.0 | – | 20.1 | 24.8 |
| PSO | –0.5/2 | 179.6 | 0.0 | – | – | – | 1521.8 | 0.0 | – | 534.2 | 1.3 |
| | 1/2 | – | 0.0 | – | – | – | – | 0.0 | – | – | 0.0 |
| REA-P | R1 | – | 0.0 | – | – | – | – | 0.0 | – | – | – |
| | R2 | – | 0.0 | – | – | – | – | 0.0 | – | – | – |
| REA-T | R1 | 27.0 | 0.0 | 799.1 | – | – | – | 0.0 | – | 74.3 | 121.5 |
| | R2 | – | 0.0 | – | – | – | – | 0.0 | – | – | – |
| | R3 | – | 0.0 | – | – | – | – | 0.0 | – | – | 2434.0 |
| SA | | 2.9 | 0.0 | – | – | 1443.7 | – | 0.0 | – | 5.0 | – |
| rBOA | | 91.6 | 0.0 | – | – | – | – | 0.0 | – | 402.0 | – |
| rGA | | 61.2 | 0.0 | – | – | – | – | 0.0 | – | 76.5 | – |

**Table A.39** Results on performance criterion $\frac{1}{100}\,\hat{\psi}_{\varepsilon}^{N}$ with $\varepsilon = 1.000$ and $N = 250,000$ in 25 dimensions.

| $\frac{1}{100}\,\hat{\psi}_{\varepsilon}^{N}$ | | sphr. | ack. | lg-ack. | whit. | rosen. | rastr. | sal. | schw. | grie. | weier. |
|---|---|---|---|---|---|---|---|---|---|---|---|
| CG | | 0.0 | – | – | – | 217.0 | – | 4.5 | – | 0.0 | – |
| CG-R | | 0.0 | – | – | – | 239.7 | – | 194.0 | – | 0.0 | – |
| CMA-ES | 100 | 18.4 | 39.5 | 62.0 | – | 2210.0 | – | 19.8 | – | 33.0 | – |
| | 750 | 125.3 | 213.1 | 344.8 | – | – | 321.4 | 124.5 | – | 203.9 | – |
| | 1250 | 207.9 | 353.7 | 525.1 | – | – | 475.1 | 208.2 | – | 330.7 | – |
| | 2500 | 406.2 | 710.1 | 943.5 | – | – | 889.6 | 421.3 | – | 626.0 | – |
| CMA-ES-R | 100 | 17.6 | 38.7 | 1189.7 | – | – | 1216.0 | 18.9 | – | 32.2 | – |
| | 750 | 53.8 | 99.0 | 1069.3 | – | – | 689.6 | 54.7 | – | 96.3 | – |
| | 1250 | 299.5 | – | – | – | – | – | 201.0 | – | 61.9 | – |
| | 2500 | – | – | – | – | – | – | 422.0 | – | 33.1 | – |
| DE | .2/.9 | – | 181.3 | – | – | – | – | – | – | – | – |
| | .2/.2 | 361.2 | 179.2 | 1942.5 | – | – | – | 1375.5 | – | 2168.9 | – |
| | .9/.2 | 466.1 | 190.3 | – | – | – | – | 1334.5 | – | 1823.0 | – |
| | .9/.9 | – | 178.6 | – | – | – | – | – | – | – | – |
| GSS | | 131.9 | 68.5 | – | 270.0 | 1193.9 | – | 51.7 | 184.9 | 281.5 | 294.0 |
| GSS-R | | 131.5 | 1497.6 | – | – | 1213.7 | – | 888.6 | 429.0 | 281.9 | 261.3 |
| NM | | 139.5 | 4.1 | – | – | 547.0 | – | – | – | 24.3 | 25.0 |
| NM-R | | 139.5 | 4.1 | – | – | 801.6 | – | – | – | 24.3 | 25.0 |
| PSO | –0.5/2 | 1005.0 | – | – | – | – | – | 1013.9 | – | – | 5.6 |
| | 1/2 | – | – | – | – | – | – | – | – | – | 0.0 |
| REA-P | R1 | – | – | – | – | – | – | – | – | – | – |
| | R2 | – | – | – | – | – | – | – | – | – | – |
| REA-T | R1 | 85.2 | 825.1 | – | – | – | – | 90.5 | – | 499.4 | – |
| | R2 | – | – | – | – | – | – | – | – | – | – |
| | R3 | – | – | – | – | – | – | – | – | – | – |
| SA | | 80.0 | – | – | – | – | – | 2019.1 | – | – | – |
| rBOA | | 303.9 | 1345.8 | – | – | – | – | 38.7 | – | 1515.2 | – |
| rGA | | 118.2 | – | – | – | – | – | 284.4 | – | – | – |

**Table A.40** Results on performance criterion $\frac{1}{100}\hat{\psi}_{\varepsilon}^{N}$ with $\varepsilon = 0.100$ and $N = 250,000$ in 25 dimensions.

| $\frac{1}{100}\hat{\psi}_{\varepsilon}^{N}$ | | sphr. | ack. | lg-ack. | whit. | rosen. | rastr. | sal. | schw. | grie. | weier. |
|---|---|---|---|---|---|---|---|---|---|---|---|
| CG | | 0.0 | – | – | – | 217.0 | – | 4.0 | – | 238.9 | – |
| CG-R | | 0.0 | – | – | – | 239.7 | – | 827.4 | – | 240.2 | – |
| CMA-ES | 100 | 25.4 | 52.8 | – | – | – | – | – | – | 43.0 | – |
| | 750 | 163.8 | 268.1 | 319.2 | – | – | 335.5 | 244.3 | – | 262.7 | – |
| | 1250 | 266.3 | 434.8 | 517.8 | – | – | 530.9 | 421.1 | – | 425.0 | – |
| | 2500 | 503.5 | 849.7 | 991.9 | – | – | 988.0 | 887.5 | – | 794.5 | – |
| CMA-ES-R | 100 | 24.5 | 1030.9 | 1667.8 | – | – | – | – | – | 42.2 | – |
| | 750 | 72.8 | 177.1 | 1368.5 | – | – | 1256.6 | 1167.7 | – | – | – |
| | 1250 | – | – | – | – | – | – | – | – | 78.7 | – |
| | 2500 | – | – | – | – | – | – | – | – | 43.1 | – |
| DE | .2/.9 | – | – | – | – | – | – | – | – | – | – |
| | .2/.2 | 643.0 | – | 1796.0 | – | – | – | – | – | 2366.8 | – |
| | .9/.2 | 823.9 | – | – | – | – | – | – | – | 2390.4 | – |
| | .9/.9 | – | – | – | – | – | – | – | – | – | – |
| GSS | | 197.4 | – | – | 323.0 | 1534.1 | – | – | 250.0 | 336.3 | 459.0 |
| GSS-R | | 197.5 | – | – | – | 1544.2 | – | – | 494.2 | 335.4 | 412.2 |
| NM | | 149.7 | – | – | – | 713.9 | – | – | – | 24.9 | 25.0 |
| NM-R | | 168.7 | – | – | – | 1134.9 | – | – | – | 24.9 | 25.0 |
| PSO | –0.5/2 | 1478.7 | – | – | – | – | – | – | – | – | 7.1 |
| | 1/2 | – | – | – | – | – | – | – | – | – | 0.0 |
| REA-P | R1 | – | – | – | – | – | – | – | – | – | – |
| | R2 | – | – | – | – | – | – | – | – | – | – |
| REA-T | R1 | 183.4 | – | – | – | – | – | – | – | 965.3 | – |
| | R2 | – | – | – | – | – | – | – | – | – | – |
| | R3 | – | – | – | – | – | – | – | – | – | – |
| SA | | 1666.5 | – | – | – | – | – | – | – | – | – |
| rBOA | | 579.2 | – | – | – | – | – | 2029.2 | – | 1849.1 | – |
| rGA | | 349.0 | – | – | – | – | – | – | – | – | – |

**Table A.41** Results on performance criterion $\zeta_{T_m}$ (scaled) with $m = 25,000$ in 25 dimensions.

| $\zeta_{T_m}$ (scaled) | | sphr. | ack. | lg-ack. | whit. | rosen. | rastr. | sal. | schw. | grie. | weier. |
|---|---|---|---|---|---|---|---|---|---|---|---|
| CG | | 0.000 | 1.050 | 2.313 | 0.009 | 0.139 | 0.793 | 1.428 | 1.141 | 0.000 | 1.454 |
| CG-R | | 0.000 | 0.872 | 1.818 | 0.000 | 0.000 | 0.531 | 0.175 | 0.892 | 0.000 | 1.204 |
| CMA-ES | 100 | 0.000 | 0.032 | 0.030 | 0.000 | 0.000 | 0.029 | 0.058 | 0.805 | 0.000 | ∞ |
| | 750 | 0.000 | 0.041 | 0.043 | 0.000 | 0.000 | 0.064 | 0.029 | 1.134 | 0.000 | ∞ |
| | 1250 | 0.003 | 0.287 | 0.330 | 0.000 | 0.001 | 0.493 | 0.111 | 1.127 | 0.005 | ∞ |
| | 2500 | 0.455 | 0.713 | 1.141 | 0.060 | 0.246 | 0.760 | 0.599 | 1.063 | 0.365 | ∞ |
| CMA-ES-R | 100 | 0.000 | 0.027 | 0.023 | 0.000 | 0.000 | 0.022 | 0.061 | 0.739 | 0.000 | ∞ |
| | 750 | 0.000 | 0.012 | 0.015 | 0.000 | 0.000 | 0.006 | 0.036 | 0.946 | 0.001 | ∞ |
| | 1250 | 0.019 | 0.304 | 0.327 | 0.000 | 0.002 | 0.488 | 0.103 | 1.084 | 0.000 | ∞ |
| | 2500 | 0.447 | 0.717 | 1.147 | 0.060 | 0.243 | 0.767 | 0.600 | 1.058 | 0.000 | ∞ |
| DE | .2/.9 | 0.278 | 0.167 | 1.002 | 0.021 | 0.123 | 0.545 | 0.657 | 0.657 | 1.683 | 1.263 |
| | .2/.2 | 0.038 | 0.167 | 0.497 | 0.000 | 0.014 | 0.438 | 0.432 | 0.680 | 0.439 | 0.905 |
| | .9/.2 | 0.049 | 0.168 | 0.713 | 0.000 | 0.013 | 0.644 | 0.373 | 0.987 | 0.352 | 1.186 |
| | .9/.9 | 0.441 | 0.168 | 1.280 | 0.073 | 0.240 | 0.818 | 0.716 | 0.982 | ∞ | ∞ |
| GSS | | 0.000 | 0.869 | 0.554 | 0.000 | 0.000 | 0.277 | 0.253 | 0.013 | 0.003 | 0.047 |
| GSS-R | | 0.000 | 0.861 | 0.537 | 0.000 | 0.000 | 0.282 | 0.285 | 0.017 | 0.003 | 0.041 |
| NM | | 0.026 | 0.088 | 1.266 | 0.178 | 0.094 | 0.636 | 1.331 | 0.709 | – | – |
| NM-R | | 0.025 | 0.087 | 1.244 | 0.198 | 0.061 | 0.637 | 1.084 | 0.698 | – | – |
| PSO | –0.5/2 | 0.075 | 0.300 | 0.735 | 0.000 | 0.015 | 0.560 | 0.341 | 0.680 | 0.071 | 0.000 |
| | 1/2 | 1.782 | 0.297 | 2.264 | 13.294 | 3.635 | 1.319 | 1.363 | 0.416 | 1.795 | 0.000 |
| REA-P | R1 | 0.708 | 0.898 | 1.533 | 0.244 | 0.448 | 0.864 | 0.821 | 0.997 | 0.683 | 1.046 |
| | R2 | 0.627 | 0.877 | 1.558 | 0.231 | 0.449 | 0.892 | 0.821 | 0.992 | 0.707 | 1.060 |
| REA-T | R1 | 0.000 | 0.313 | 0.323 | 0.000 | 0.000 | 0.256 | 0.065 | 0.729 | 0.004 | 0.287 |
| | R2 | 0.639 | 0.873 | 1.476 | 0.224 | 0.441 | 0.874 | 0.810 | 0.987 | 0.654 | 1.010 |
| | R3 | 0.618 | 0.865 | 1.462 | 0.224 | 0.419 | 0.857 | 0.800 | 0.987 | 0.638 | 0.983 |
| SA | | 0.005 | 0.873 | 0.979 | 0.000 | 0.000 | 0.447 | 0.638 | 0.958 | 0.011 | 0.759 |
| rBOA | | 0.021 | 0.282 | 0.326 | 0.000 | 0.004 | 0.568 | 0.088 | 1.178 | 0.078 | 1.169 |
| rGA | | 0.002 | 0.486 | 0.578 | 0.000 | 0.001 | 0.309 | 0.173 | 0.308 | 0.010 | 0.776 |

**Table A.42** Variance for performance criterion $\zeta_{T_m}$ (scaled) with $m = 25,000$ in 25 dimensions.

| $Var(\zeta_{T_m})$ (scaled) | | sphr. | ack. | lg-ack. | whit. | rosen. | rastr. | sal. | schw. | grie. | weier. |
|---|---|---|---|---|---|---|---|---|---|---|---|
| CG | | 0.000 | 0.074 | 0.356 | 0.050 | 0.383 | 0.139 | 0.277 | 0.285 | 0.000 | 0.112 |
| CG-R | | 0.000 | 0.059 | 0.219 | 0.000 | 0.000 | 0.067 | 0.215 | 0.203 | 0.000 | 0.056 |
| CMA-ES | 100 | 0.000 | 0.008 | 0.010 | 0.000 | 0.000 | 0.009 | 0.010 | 0.097 | 0.000 | ∞ |
| | 750 | 0.000 | 0.009 | 0.016 | 0.000 | 0.000 | 0.025 | 0.007 | 0.034 | 0.000 | ∞ |
| | 1250 | 0.000 | 0.013 | 0.022 | 0.000 | 0.000 | 0.035 | 0.009 | 0.034 | 0.000 | ∞ |
| | 2500 | 0.056 | 0.030 | 0.060 | 0.026 | 0.058 | 0.045 | 0.039 | 0.035 | 0.049 | ∞ |
| CMA-ES-R | 100 | 0.000 | 0.004 | 0.112 | 0.017 | 0.006 | 0.148 | 0.022 | 150.146 | 0.000 | ∞ |
| | 750 | 0.000 | 0.001 | 0.055 | 0.015 | 0.008 | 0.018 | 0.007 | 245.633 | 0.000 | ∞ |
| | 1250 | 0.034 | 0.513 | 19.894 | 0.072 | 0.520 | 65.416 | 0.061 | 318.209 | 0.000 | ∞ |
| | 2500 | 17.242 | 2.840 | 244.431 | 21978.049 | 3971.182 | 161.246 | 2.084 | 303.517 | 0.000 | ∞ |
| DE | .2/.9 | 0.056 | 0.018 | 0.099 | 0.014 | 0.038 | 0.059 | 0.063 | 0.062 | 0.237 | 0.059 |
| | .2/.2 | 0.010 | 0.017 | 0.063 | 0.000 | 0.005 | 0.055 | 0.060 | 0.068 | 0.085 | 0.060 |
| | .9/.2 | 0.013 | 0.017 | 0.072 | 0.000 | 0.005 | 0.050 | 0.046 | 0.046 | 0.083 | 0.044 |
| | .9/.9 | 0.089 | 0.017 | 0.110 | 0.049 | 0.084 | 0.054 | 0.064 | 0.049 | ∞ | ∞ |
| GSS | | 0.000 | 0.155 | 0.115 | 0.000 | 0.000 | 0.089 | 0.090 | 0.061 | 0.000 | 0.067 |
| GSS-R | | 0.000 | 0.151 | 0.102 | 0.000 | 0.000 | 0.111 | 0.136 | 0.087 | 0.000 | 0.065 |
| NM | | 0.077 | 0.024 | 0.404 | 0.699 | 0.278 | 0.146 | 0.145 | 0.104 | – | – |
| NM-R | | 0.074 | 0.025 | 0.402 | 0.691 | 0.199 | 0.153 | 0.090 | 0.097 | – | – |
| PSO | –0.5/2 | 0.070 | 0.014 | 0.201 | 0.001 | 0.023 | 0.224 | 0.131 | 0.149 | 0.062 | 0.000 |
| | 1/2 | 0.324 | 0.014 | 0.175 | 4.117 | 0.738 | 0.080 | 0.088 | 0.029 | 0.258 | 0.000 |
| REA-P | R1 | 0.074 | 0.032 | 0.076 | 0.117 | 0.107 | 0.047 | 0.043 | 0.037 | 0.078 | 0.035 |
| | R2 | 0.069 | 0.030 | 0.074 | 0.099 | 0.102 | 0.044 | 0.048 | 0.038 | 0.076 | 0.035 |
| REA-T | R1 | 0.000 | 0.016 | 0.034 | 0.000 | 0.000 | 0.040 | 0.010 | 0.037 | 0.000 | 0.062 |
| | R2 | 0.064 | 0.030 | 0.076 | 0.084 | 0.092 | 0.043 | 0.046 | 0.039 | 0.075 | 0.035 |
| | R3 | 0.069 | 0.035 | 0.076 | 0.084 | 0.089 | 0.044 | 0.049 | 0.036 | 0.068 | 0.035 |
| SA | | 0.001 | 0.051 | 0.095 | 0.000 | 0.000 | 0.098 | 0.096 | 0.047 | 0.001 | 0.065 |
| rBOA | | 0.017 | 0.030 | 0.033 | 0.000 | 0.003 | 0.069 | 0.016 | 0.063 | 0.042 | 0.048 |
| rGA | | 0.000 | 0.057 | 0.124 | 0.000 | 0.000 | 0.041 | 0.020 | 0.064 | 0.000 | 0.029 |

**Table A.43** Results on performance criterion $\zeta_{T_m}$ (scaled) with $m = 100,000$ in 25 dimensions.

| $\zeta_{T_m}$ (scaled) | | sphr. | ack. | lg-ack. | whit. | rosen. | rastr. | sal. | schw. | grie. | weier. |
|---|---|---|---|---|---|---|---|---|---|---|---|
| CG | | 0.000 | 1.050 | 2.313 | 0.001 | 0.030 | 0.793 | 1.428 | 1.141 | 0.000 | 1.454 |
| CG-R | | 0.000 | 0.826 | 1.593 | 0.000 | 0.000 | 0.455 | 0.028 | 0.658 | 0.000 | 1.146 |
| CMA-ES | 100 | 0.000 | 0.032 | 0.030 | 0.000 | 0.000 | 0.029 | 0.058 | 0.805 | 0.000 | ∞ |
| | 750 | 0.000 | 0.000 | 0.005 | 0.000 | 0.000 | 0.001 | 0.028 | 0.583 | 0.000 | ∞ |
| | 1250 | 0.000 | 0.000 | 0.003 | 0.000 | 0.000 | 0.000 | 0.020 | 1.024 | 0.000 | ∞ |
| | 2500 | 0.000 | 0.001 | 0.001 | 0.000 | 0.000 | 0.000 | 0.017 | 1.054 | 0.000 | ∞ |
| CMA-ES-R | 100 | 0.000 | 0.018 | 0.013 | 0.000 | 0.000 | 0.014 | 0.048 | 0.595 | 0.000 | ∞ |
| | 750 | 0.000 | 0.003 | 0.005 | 0.000 | 0.000 | 0.001 | 0.033 | 0.547 | 0.000 | ∞ |
| | 1250 | 0.018 | 0.296 | 0.275 | 0.000 | 0.002 | 0.149 | 0.047 | 0.913 | 0.000 | ∞ |
| | 2500 | 0.036 | 0.362 | 0.416 | 0.000 | 0.006 | 0.361 | 0.120 | 1.052 | 0.000 | ∞ |
| DE | .2/.9 | 0.122 | 0.126 | 0.702 | 0.001 | 0.035 | 0.359 | 0.498 | 0.449 | 1.649 | 1.252 |
| | .2/.2 | 0.000 | 0.125 | 0.068 | 0.000 | 0.001 | 0.213 | 0.207 | 0.441 | 0.179 | 0.728 |
| | .9/.2 | 0.000 | 0.126 | 0.383 | 0.000 | 0.000 | 0.555 | 0.199 | 0.913 | 0.033 | 1.135 |
| | .9/.9 | 0.281 | 0.128 | 1.132 | 0.019 | 0.120 | 0.748 | 0.613 | 0.933 | ∞ | ∞ |
| GSS | | 0.000 | 0.868 | 0.553 | 0.000 | 0.000 | 0.277 | 0.253 | 0.012 | – | – |
| GSS-R | | 0.000 | 0.860 | 0.537 | 0.000 | 0.000 | 0.282 | 0.223 | 0.016 | – | – |
| NM | | 0.012 | 0.088 | 1.265 | 0.166 | 0.083 | 0.635 | 1.331 | 0.692 | – | – |
| NM-R | | 0.010 | 0.087 | 1.145 | 0.156 | 0.050 | 0.571 | 0.988 | 0.673 | – | – |
| PSO | −0.5/2 | 0.018 | 0.300 | 0.493 | 0.000 | 0.002 | 0.257 | 0.202 | 0.562 | 0.020 | 0.000 |
| | 1/2 | 1.681 | 0.297 | 2.234 | 13.294 | 3.635 | 1.314 | 1.363 | 0.412 | 1.701 | 0.000 |
| REA-P | R1 | 0.595 | 0.859 | 1.432 | 0.181 | 0.369 | 0.798 | 0.753 | 0.953 | 0.558 | 0.969 |
| | R2 | 0.567 | 0.833 | 1.463 | 0.171 | 0.381 | 0.838 | 0.744 | 0.945 | 0.630 | 1.029 |
| REA-T | R1 | 0.000 | 0.136 | 0.074 | 0.000 | 0.000 | 0.124 | 0.058 | 0.444 | 0.000 | 0.283 |
| | R2 | 0.528 | 0.823 | 1.369 | 0.113 | 0.311 | 0.805 | 0.740 | 0.943 | 0.548 | 0.920 |
| | R3 | 0.508 | 0.816 | 1.367 | 0.107 | 0.292 | 0.798 | 0.736 | 0.939 | 0.527 | 0.818 |
| SA | | 0.002 | 0.778 | 0.822 | 0.000 | 0.000 | 0.330 | 0.321 | 0.901 | 0.009 | 0.628 |
| rBOA | | 0.000 | 0.236 | 0.246 | 0.000 | 0.000 | 0.432 | 0.041 | 1.178 | 0.006 | 1.168 |
| rGA | | 0.001 | 0.476 | 0.576 | 0.000 | 0.000 | 0.282 | 0.140 | 0.300 | 0.008 | 0.754 |

**Table A.44** Variance for performance criterion $\zeta_{T_m}$ (scaled) with $m = 100,000$ in 25 dimensions.

| $Var(\zeta_{T_m})$ (scaled) | | sphr. | ack. | lg-ack. | whit. | rosen. | rastr. | sal. | schw. | grie. | weier. |
|---|---|---|---|---|---|---|---|---|---|---|---|
| CG | | 0.000 | 0.074 | 0.356 | 0.019 | 0.187 | 0.139 | 0.277 | 0.285 | 0.000 | 0.112 |
| CG-R | | 0.000 | 0.056 | 0.157 | 0.000 | 0.000 | 0.050 | 0.033 | 0.125 | 0.000 | 0.045 |
| CMA-ES | 100 | 0.000 | 0.008 | 0.010 | 0.000 | 0.000 | 0.009 | 0.010 | 0.097 | 0.000 | ∞ |
| | 750 | 0.000 | 0.002 | 0.005 | 0.000 | 0.000 | 0.001 | 0.007 | 0.092 | 0.000 | ∞ |
| | 1250 | 0.000 | 0.000 | 0.004 | 0.000 | 0.000 | 0.000 | 0.005 | 0.065 | 0.000 | ∞ |
| | 2500 | 0.000 | 0.000 | 0.002 | 0.000 | 0.000 | 0.000 | 0.000 | 0.035 | 0.000 | ∞ |
| CMA-ES-R | 100 | 0.000 | 0.002 | 0.040 | 0.015 | 0.006 | 0.064 | 0.014 | 96.977 | 0.000 | ∞ |
| | 750 | 0.000 | 0.000 | 0.008 | 0.015 | 0.008 | 0.001 | 0.006 | 82.606 | 0.000 | ∞ |
| | 1250 | 0.030 | 0.485 | 14.207 | 0.068 | 0.458 | 6.239 | 0.013 | 228.087 | 0.000 | ∞ |
| | 2500 | 0.115 | 0.727 | 32.238 | 0.109 | 2.577 | 36.096 | 0.084 | 299.574 | 0.000 | ∞ |
| DE | .2/.9 | 0.030 | 0.014 | 0.075 | 0.001 | 0.012 | 0.043 | 0.055 | 0.050 | 0.239 | 0.063 |
| | .2/.2 | 0.000 | 0.013 | 0.023 | 0.000 | 0.000 | 0.042 | 0.040 | 0.068 | 0.043 | 0.054 |
| | .9/.2 | 0.000 | 0.013 | 0.067 | 0.000 | 0.000 | 0.047 | 0.033 | 0.055 | 0.053 | 0.045 |
| | .9/.9 | 0.071 | 0.014 | 0.111 | 0.017 | 0.050 | 0.048 | 0.069 | 0.050 | ∞ | ∞ |
| GSS | | 0.000 | 0.156 | 0.115 | 0.000 | 0.000 | 0.089 | 0.090 | 0.061 | – | – |
| GSS-R | | 0.000 | 0.153 | 0.102 | 0.000 | 0.000 | 0.111 | 0.061 | 0.087 | – | – |
| NM | | 0.056 | 0.024 | 0.404 | 0.676 | 0.253 | 0.146 | 0.145 | 0.100 | – | – |
| NM-R | | 0.049 | 0.025 | 0.382 | 0.598 | 0.170 | 0.145 | 0.082 | 0.096 | – | – |
| PSO | –0.5/2 | 0.018 | 0.014 | 0.132 | 0.000 | 0.002 | 0.134 | 0.086 | 0.137 | 0.016 | 0.000 |
| | 1/2 | 0.409 | 0.014 | 0.230 | 4.117 | 0.738 | 0.094 | 0.088 | 0.023 | 0.374 | 0.000 |
| REA-P | R1 | 0.064 | 0.030 | 0.061 | 0.099 | 0.100 | 0.042 | 0.042 | 0.033 | 0.072 | 0.040 |
| | R2 | 0.070 | 0.028 | 0.066 | 0.086 | 0.096 | 0.040 | 0.042 | 0.034 | 0.062 | 0.032 |
| REA-T | R1 | 0.000 | 0.033 | 0.024 | 0.000 | 0.000 | 0.032 | 0.010 | 0.044 | 0.000 | 0.061 |
| | R2 | 0.050 | 0.029 | 0.066 | 0.041 | 0.056 | 0.043 | 0.044 | 0.037 | 0.064 | 0.032 |
| | R3 | 0.052 | 0.029 | 0.060 | 0.042 | 0.057 | 0.043 | 0.040 | 0.037 | 0.058 | 0.047 |
| SA | | 0.000 | 0.057 | 0.078 | 0.000 | 0.000 | 0.068 | 0.050 | 0.049 | 0.000 | 0.072 |
| rBOA | | 0.000 | 0.026 | 0.027 | 0.000 | 0.000 | 0.071 | 0.009 | 0.063 | 0.004 | 0.047 |
| rGA | | 0.000 | 0.056 | 0.124 | 0.000 | 0.000 | 0.042 | 0.011 | 0.063 | 0.000 | 0.025 |

**Table A.45** Results on performance criterion $\zeta_{T_m}$ (scaled) with $m = 250,000$ in 25 dimensions.

| $\zeta_{T_m}$ (scaled) | | sphr. | ack. | lg-ack. | whit. | rosen. | rastr. | sal. | schw. | grie. | weier. |
|---|---|---|---|---|---|---|---|---|---|---|---|
| CG | | 0.000 | 1.050 | 2.313 | 0.001 | 0.030 | 0.793 | 1.428 | 1.141 | 0.000 | 1.454 |
| CG-R | | 0.000 | 0.794 | 1.473 | 0.000 | 0.000 | 0.417 | 0.007 | 0.583 | 0.000 | 1.110 |
| CMA-ES | 100 | 0.000 | 0.032 | 0.030 | 0.000 | 0.000 | 0.029 | 0.058 | 0.805 | 0.000 | ∞ |
| | 750 | 0.000 | 0.000 | 0.005 | 0.000 | 0.000 | 0.001 | 0.028 | 0.577 | 0.000 | ∞ |
| | 1250 | 0.000 | 0.000 | 0.003 | 0.000 | 0.000 | 0.000 | 0.020 | 0.530 | 0.000 | ∞ |
| | 2500 | 0.000 | 0.000 | 0.000 | 0.000 | 0.000 | 0.000 | 0.017 | 0.672 | 0.000 | ∞ |
| CMA-ES-R | 100 | 0.000 | 0.014 | 0.009 | 0.000 | 0.000 | 0.011 | 0.044 | 0.535 | 0.000 | ∞ |
| | 750 | 0.000 | 0.002 | 0.002 | 0.000 | 0.000 | 0.000 | 0.031 | 0.472 | 0.000 | ∞ |
| | 1250 | 0.016 | 0.289 | 0.268 | 0.000 | 0.002 | 0.138 | 0.046 | 0.772 | 0.000 | ∞ |
| | 2500 | 0.034 | 0.355 | 0.412 | 0.000 | 0.006 | 0.342 | 0.117 | 0.879 | 0.000 | ∞ |
| DE | .2/.9 | 0.066 | 0.102 | 0.534 | 0.000 | 0.014 | 0.261 | 0.402 | 0.316 | 1.596 | 1.242 |
| | .2/.2 | 0.000 | 0.103 | 0.043 | 0.000 | 0.000 | 0.082 | 0.136 | 0.193 | 0.027 | 0.623 |
| | .9/.2 | 0.000 | 0.104 | 0.131 | 0.000 | 0.000 | 0.487 | 0.120 | 0.847 | 0.001 | 1.099 |
| | .9/.9 | 0.204 | 0.104 | 1.017 | 0.007 | 0.070 | 0.704 | 0.536 | 0.894 | ∞ | ∞ |
| GSS | | 0.000 | 0.868 | 0.553 | 0.000 | 0.000 | 0.277 | 0.253 | 0.012 | – | – |
| GSS-R | | 0.000 | 0.132 | 0.480 | 0.000 | 0.000 | 0.230 | 0.188 | 0.008 | – | – |
| NM | | 0.004 | 0.088 | 1.264 | 0.046 | 0.032 | 0.634 | 1.331 | 0.680 | – | – |
| NM-R | | 0.004 | 0.087 | 0.996 | 0.050 | 0.021 | 0.502 | 0.934 | 0.637 | – | – |
| PSO | –0.5/2 | 0.006 | 0.300 | 0.375 | 0.000 | 0.001 | 0.147 | 0.133 | 0.507 | 0.009 | 0.000 |
| | 1/2 | 1.623 | 0.297 | 2.232 | 13.294 | 3.635 | 1.313 | 1.363 | 0.411 | 1.651 | 0.000 |
| REA-P | R1 | 0.524 | 0.836 | 1.372 | 0.168 | 0.352 | 0.759 | 0.700 | 0.929 | 0.479 | 0.862 |
| | R2 | 0.544 | 0.800 | 1.411 | 0.159 | 0.363 | 0.807 | 0.693 | 0.912 | 0.582 | 1.009 |
| REA-T | R1 | 0.000 | 0.082 | 0.073 | 0.000 | 0.000 | 0.123 | 0.058 | 0.424 | 0.000 | 0.283 |
| | R2 | 0.463 | 0.791 | 1.305 | 0.071 | 0.238 | 0.769 | 0.692 | 0.914 | 0.483 | 0.803 |
| | R3 | 0.441 | 0.786 | 1.295 | 0.065 | 0.229 | 0.760 | 0.692 | 0.905 | 0.464 | 0.557 |
| SA | | 0.002 | 0.709 | 0.746 | 0.000 | 0.000 | 0.286 | 0.204 | 0.857 | 0.008 | 0.550 |
| rBOA | | 0.000 | 0.208 | 0.204 | 0.000 | 0.000 | 0.358 | 0.027 | 1.178 | 0.001 | 1.165 |
| rGA | | 0.001 | 0.466 | 0.576 | 0.000 | 0.000 | 0.250 | 0.133 | 0.296 | 0.008 | 0.738 |

**Table A.46** Variance for performance criterion $\zeta_{T_m}$ (scaled) with $m = 250,000$ in 25 dimensions.

| $Var(\zeta_{T_m})$ (scaled) | | sphr. | ack. | lg-ack. | whit. | rosen. | rastr. | sal. | schw. | grie. | weier. |
|---|---|---|---|---|---|---|---|---|---|---|---|
| CG | | 0.000 | 0.074 | 0.356 | 0.019 | 0.187 | 0.139 | 0.277 | 0.285 | 0.000 | 0.112 |
| CG-R | | 0.000 | 0.054 | 0.138 | 0.000 | 0.000 | 0.045 | 0.011 | 0.069 | 0.000 | 0.046 |
| CMA-ES | 100 | 0.000 | 0.008 | 0.010 | 0.000 | 0.000 | 0.009 | 0.010 | 0.097 | 0.000 | ∞ |
| | 750 | 0.000 | 0.002 | 0.005 | 0.000 | 0.000 | 0.001 | 0.007 | 0.087 | 0.000 | ∞ |
| | 1250 | 0.000 | 0.000 | 0.004 | 0.000 | 0.000 | 0.000 | 0.005 | 0.076 | 0.000 | ∞ |
| | 2500 | 0.000 | 0.000 | 0.002 | 0.000 | 0.000 | 0.000 | 0.000 | 0.134 | 0.000 | ∞ |
| CMA-ES-R | 100 | 0.000 | 0.001 | 0.019 | 0.014 | 0.006 | 0.042 | 0.011 | 78.474 | 0.000 | ∞ |
| | 750 | 0.000 | 0.000 | 0.002 | 0.015 | 0.008 | 0.000 | 0.006 | 61.078 | 0.000 | ∞ |
| | 1250 | 0.024 | 0.462 | 13.484 | 0.065 | 0.400 | 5.286 | 0.012 | 162.192 | 0.000 | ∞ |
| | 2500 | 0.104 | 0.696 | 31.644 | 0.096 | 2.476 | 32.253 | 0.080 | 209.650 | 0.000 | ∞ |
| DE | .2/.9 | 0.016 | 0.011 | 0.054 | 0.000 | 0.004 | 0.035 | 0.047 | 0.045 | 0.249 | 0.067 |
| | .2/.2 | 0.000 | 0.012 | 0.022 | 0.000 | 0.000 | 0.033 | 0.034 | 0.065 | 0.036 | 0.050 |
| | .9/.2 | 0.000 | 0.011 | 0.064 | 0.000 | 0.000 | 0.052 | 0.023 | 0.061 | 0.000 | 0.048 |
| | .9/.9 | 0.056 | 0.011 | 0.101 | 0.007 | 0.032 | 0.058 | 0.064 | 0.053 | ∞ | ∞ |
| GSS | | 0.000 | 0.156 | 0.115 | 0.000 | 0.000 | 0.089 | 0.090 | 0.061 | – | – |
| GSS-R | | 0.000 | 0.019 | 0.069 | 0.000 | 0.000 | 0.050 | 0.041 | 0.076 | – | – |
| NM | | 0.033 | 0.024 | 0.403 | 0.224 | 0.145 | 0.146 | 0.145 | 0.097 | – | – |
| NM-R | | 0.031 | 0.025 | 0.293 | 0.282 | 0.110 | 0.121 | 0.075 | 0.083 | – | – |
| PSO | –0.5/2 | 0.005 | 0.014 | 0.102 | 0.000 | 0.000 | 0.078 | 0.054 | 0.135 | 0.005 | 0.000 |
| | 1/2 | 0.466 | 0.014 | 0.232 | 4.117 | 0.738 | 0.097 | 0.088 | 0.021 | 0.437 | 0.000 |
| REA-P | R1 | 0.061 | 0.027 | 0.063 | 0.096 | 0.104 | 0.042 | 0.041 | 0.032 | 0.052 | 0.070 |
| | R2 | 0.083 | 0.029 | 0.064 | 0.088 | 0.101 | 0.036 | 0.047 | 0.038 | 0.062 | 0.027 |
| REA-T | R1 | 0.000 | 0.014 | 0.024 | 0.000 | 0.000 | 0.032 | 0.010 | 0.044 | 0.000 | 0.061 |
| | R2 | 0.048 | 0.026 | 0.058 | 0.027 | 0.048 | 0.038 | 0.037 | 0.037 | 0.052 | 0.040 |
| | R3 | 0.047 | 0.026 | 0.056 | 0.023 | 0.043 | 0.041 | 0.038 | 0.035 | 0.051 | 0.102 |
| SA | | 0.000 | 0.057 | 0.066 | 0.000 | 0.000 | 0.065 | 0.029 | 0.051 | 0.000 | 0.080 |
| rBOA | | 0.000 | 0.026 | 0.023 | 0.000 | 0.000 | 0.064 | 0.007 | 0.063 | 0.001 | 0.045 |
| rGA | | 0.000 | 0.056 | 0.124 | 0.000 | 0.000 | 0.037 | 0.010 | 0.063 | 0.000 | 0.026 |

**Table A.47** Results on performance criterion $\phi_1$ (scaled) in 25 dimensions.

| $\phi_1$ (scaled) | | sphr. | ack. | lg-ack. | whit. | rosen. | rastr. | sal. | schw. | grie. | weier. |
|---|---|---|---|---|---|---|---|---|---|---|---|
| CG | | 0.000 | 1.050 | 2.316 | 0.012 | 0.127 | 0.793 | 1.429 | 1.148 | 0.000 | 1.454 |
| CG-R | | 0.000 | 0.829 | 1.617 | 0.000 | 0.000 | 0.464 | 0.069 | 0.687 | 0.000 | ∞ |
| CMA-ES | 100 | 0.000 | 0.036 | 0.034 | 0.000 | 0.000 | 0.036 | 0.059 | 0.814 | 0.000 | ∞ |
| | 750 | 0.013 | 0.039 | 0.060 | 0.002 | 0.008 | 0.053 | 0.051 | 0.733 | 0.010 | ∞ |
| | 1250 | 0.026 | 0.067 | 0.098 | 0.005 | 0.015 | 0.086 | 0.063 | 0.819 | 0.021 | ∞ |
| | 2500 | 0.054 | 0.139 | 0.198 | 0.008 | 0.029 | 0.169 | 0.107 | 0.997 | 0.049 | ∞ |
| CMA-ES-R | 100 | 0.000 | 0.023 | 0.018 | 0.000 | 0.000 | 0.022 | 0.050 | 0.614 | 0.000 | ∞ |
| | 750 | 0.003 | 0.019 | 0.027 | 0.000 | 0.002 | 0.024 | 0.041 | 0.594 | 0.004 | ∞ |
| | 1250 | 0.039 | 0.320 | 0.328 | 0.003 | 0.014 | 0.206 | 0.083 | 0.903 | 0.001 | ∞ |
| | 2500 | 0.084 | 0.406 | 0.511 | 0.008 | 0.034 | 0.421 | 0.186 | 1.013 | 0.000 | ∞ |
| DE | .2/.9 | 0.149 | 0.128 | 0.726 | 0.015 | 0.059 | 0.374 | 0.507 | 0.458 | 1.641 | 1.252 |
| | .2/.2 | 0.024 | 0.129 | 0.174 | 0.007 | 0.015 | 0.229 | 0.241 | 0.417 | 0.197 | 0.738 |
| | .9/.2 | 0.019 | 0.129 | 0.389 | 0.001 | 0.008 | 0.559 | 0.218 | 0.910 | 0.106 | 1.135 |
| | .9/.9 | 0.295 | 0.130 | 1.132 | 0.033 | 0.137 | 0.751 | 0.614 | 0.933 | ∞ | ∞ |
| GSS | | 0.009 | 0.874 | 0.565 | 0.000 | 0.002 | 0.281 | 0.256 | 0.025 | 0.041 | 0.066 |
| GSS-R | | 0.009 | 0.547 | 0.526 | 0.000 | 0.003 | 0.264 | 0.228 | 0.026 | 0.040 | 0.060 |
| NM | | 0.017 | 0.088 | 1.265 | 0.132 | 0.073 | 0.635 | 1.331 | 0.692 | 0.253 | 0.973 |
| NM-R | | 0.015 | 0.087 | 1.121 | 0.150 | 0.044 | 0.559 | 0.994 | 0.668 | 0.230 | 0.956 |
| PSO | –0.5/2 | 0.034 | 0.300 | 0.520 | 0.001 | 0.009 | 0.290 | 0.219 | 0.576 | 0.035 | 0.000 |
| | 1/2 | 1.686 | 0.297 | 2.240 | 13.294 | 3.636 | 1.314 | 1.363 | 0.414 | 1.705 | 0.000 |
| REA-P | R1 | 0.600 | 0.861 | 1.436 | 0.197 | 0.385 | 0.804 | 0.753 | 0.957 | 0.559 | 0.949 |
| | R2 | 0.577 | 0.833 | 1.472 | 0.187 | 0.395 | 0.842 | 0.750 | 0.946 | 0.636 | 1.029 |
| REA-T | R1 | 0.002 | 0.162 | 0.138 | 0.000 | 0.001 | 0.164 | 0.069 | 0.502 | 0.005 | 0.293 |
| | R2 | 0.537 | 0.825 | 1.377 | 0.135 | 0.323 | 0.811 | 0.741 | 0.944 | 0.555 | 0.908 |
| | R3 | 0.515 | 0.819 | 1.368 | 0.126 | 0.305 | 0.801 | 0.739 | 0.940 | 0.535 | 0.784 |
| SA | | 0.003 | 0.778 | 0.839 | 0.000 | 0.000 | 0.349 | 0.357 | 0.901 | 0.009 | 0.637 |
| rBOA | | 0.015 | 0.239 | 0.254 | 0.006 | 0.009 | 0.446 | 0.049 | 1.178 | 0.019 | 1.165 |
| rGA | | 0.009 | 0.479 | 0.585 | 0.002 | 0.005 | 0.283 | 0.159 | 0.305 | 0.014 | 0.757 |

**Table A.48** Variance for performance criterion $\phi_1$ (scaled) in 25 dimensions.

| $Var(\phi_1)$ (scaled) | | sphr. | ack. | lg-ack. | whit. | rosen. | rastr. | sal. | schw. | grie. | weier. |
|---|---|---|---|---|---|---|---|---|---|---|---|
| CG | | 0.000 | 0.074 | 0.356 | 0.019 | 0.176 | 0.139 | 0.277 | 0.283 | 0.000 | 0.112 |
| CG-R | | 0.000 | 0.046 | 0.120 | 0.000 | 0.000 | 0.038 | 0.041 | 0.085 | 0.000 | ∞ |
| CMA-ES | 100 | 0.000 | 0.008 | 0.010 | 0.000 | 0.000 | 0.009 | 0.010 | 0.094 | 0.000 | ∞ |
| | 750 | 0.001 | 0.002 | 0.005 | 0.001 | 0.001 | 0.003 | 0.006 | 0.064 | 0.001 | ∞ |
| | 1250 | 0.002 | 0.001 | 0.004 | 0.002 | 0.002 | 0.002 | 0.004 | 0.048 | 0.001 | ∞ |
| | 2500 | 0.005 | 0.002 | 0.005 | 0.002 | 0.005 | 0.005 | 0.003 | 0.034 | 0.003 | ∞ |
| CMA-ES-R | 100 | 0.000 | 0.000 | 0.000 | 0.000 | 0.000 | 0.000 | 0.000 | 0.000 | 0.000 | 0.000 |
| | 750 | 0.000 | 0.000 | 0.000 | 0.000 | 0.000 | 0.000 | 0.000 | 0.000 | 0.000 | 0.000 |
| | 1250 | 0.000 | 0.000 | 0.000 | 0.000 | 0.000 | 0.000 | 0.000 | 0.000 | 0.000 | 0.000 |
| | 2500 | 0.000 | 0.000 | 0.000 | 0.000 | 0.000 | 0.000 | 0.000 | 0.000 | 0.000 | 0.000 |
| DE | .2/.9 | 0.020 | 0.010 | 0.052 | 0.005 | 0.010 | 0.028 | 0.035 | 0.035 | 0.233 | 0.061 |
| | .2/.2 | 0.002 | 0.010 | 0.021 | 0.002 | 0.002 | 0.029 | 0.033 | 0.057 | 0.038 | 0.040 |
| | .9/.2 | 0.002 | 0.010 | 0.050 | 0.000 | 0.002 | 0.034 | 0.022 | 0.043 | 0.031 | 0.038 |
| | .9/.9 | 0.053 | 0.010 | 0.086 | 0.016 | 0.037 | 0.042 | 0.055 | 0.039 | ∞ | ∞ |
| GSS | | 0.001 | 0.153 | 0.114 | 0.000 | 0.001 | 0.089 | 0.089 | 0.060 | 0.008 | 0.063 |
| GSS-R | | 0.002 | 0.073 | 0.076 | 0.000 | 0.001 | 0.073 | 0.054 | 0.076 | 0.007 | 0.062 |
| NM | | 0.051 | 0.024 | 0.404 | 0.490 | 0.222 | 0.146 | 0.145 | 0.098 | 0.207 | 0.261 |
| NM-R | | 0.045 | 0.025 | 0.330 | 0.524 | 0.149 | 0.120 | 0.063 | 0.088 | 0.196 | 0.253 |
| PSO | –0.5/2 | 0.016 | 0.014 | 0.099 | 0.001 | 0.005 | 0.091 | 0.060 | 0.123 | 0.014 | 0.000 |
| | 1/2 | 0.393 | 0.014 | 0.216 | 4.117 | 0.738 | 0.091 | 0.088 | 0.023 | 0.358 | 0.000 |
| REA-P | R1 | 0.046 | 0.022 | 0.048 | 0.094 | 0.093 | 0.031 | 0.030 | 0.026 | 0.049 | 0.035 |
| | R2 | 0.065 | 0.021 | 0.051 | 0.082 | 0.090 | 0.030 | 0.032 | 0.028 | 0.049 | 0.025 |
| REA-T | R1 | 0.000 | 0.015 | 0.021 | 0.000 | 0.000 | 0.030 | 0.009 | 0.038 | 0.000 | 0.060 |
| | R2 | 0.036 | 0.020 | 0.049 | 0.030 | 0.038 | 0.031 | 0.030 | 0.028 | 0.044 | 0.020 |
| | R3 | 0.037 | 0.022 | 0.046 | 0.027 | 0.039 | 0.032 | 0.028 | 0.027 | 0.037 | 0.048 |
| SA | | 0.000 | 0.044 | 0.058 | 0.000 | 0.000 | 0.054 | 0.038 | 0.039 | 0.000 | 0.062 |
| rBOA | | 0.005 | 0.021 | 0.020 | 0.004 | 0.003 | 0.051 | 0.005 | 0.063 | 0.005 | 0.045 |
| rGA | | 0.002 | 0.056 | 0.123 | 0.001 | 0.001 | 0.037 | 0.010 | 0.063 | 0.001 | 0.020 |

**Table A.49** Results on performance criterion $\phi_2$ (scaled) in 25 dimensions.

| $\phi_2$ (scaled) | | sphr. | ack. | lg-ack. | whit. | rosen. | rastr. | sal. | schw. | grie. | weier. |
|---|---|---|---|---|---|---|---|---|---|---|---|
| CG | | 0.000 | 1.078 | 2.705 | 0.274 | 1.202 | 0.890 | 1.449 | 1.734 | 0.001 | 1.467 |
| CG-R | | 0.000 | 1.076 | 2.777 | 0.000 | 0.000 | 0.977 | 1.211 | 1.060 | 0.002 | ∞ |
| CMA-ES | 100 | 0.093 | 0.501 | 0.733 | 0.000 | 0.024 | 0.681 | 0.361 | 1.214 | 0.079 | ∞ |
| | 750 | 0.640 | 0.847 | 1.426 | 0.237 | 0.462 | 0.902 | 0.783 | 1.201 | 0.583 | ∞ |
| | 1250 | 0.635 | 0.856 | 1.445 | 0.228 | 0.460 | 0.907 | 0.800 | 1.206 | 0.595 | ∞ |
| | 2500 | 0.519 | 0.814 | 1.341 | 0.105 | 0.313 | 0.825 | 0.726 | 1.146 | 0.582 | ∞ |
| CMA-ES-R | 100 | 0.075 | 0.485 | 0.692 | 0.000 | 0.018 | 0.670 | 0.335 | 1.185 | 0.067 | ∞ |
| | 750 | 0.530 | 0.778 | 1.290 | 0.110 | 0.326 | 0.838 | 0.703 | 1.168 | 0.486 | ∞ |
| | 1250 | 0.547 | 0.823 | 1.378 | 0.139 | 0.347 | 0.851 | 0.747 | 1.162 | 0.363 | ∞ |
| | 2500 | 0.522 | 0.810 | 1.348 | 0.106 | 0.314 | 0.828 | 0.728 | 1.146 | 0.077 | ∞ |
| DE | .2/.9 | 0.976 | 0.247 | 1.759 | 1.093 | 0.998 | 1.007 | 1.017 | 1.052 | 1.692 | 1.265 |
| | .2/.2 | 0.954 | 0.248 | 1.723 | 0.943 | 0.957 | 0.994 | 1.007 | 1.053 | 1.161 | 1.134 |
| | .9/.2 | 0.654 | 0.248 | 1.522 | 0.272 | 0.507 | 0.923 | 0.844 | 1.100 | 1.152 | 1.270 |
| | .9/.9 | 0.834 | 0.248 | 1.650 | 0.568 | 0.774 | 0.977 | 0.929 | 1.103 | ∞ | ∞ |
| GSS | | 0.865 | 1.137 | 1.612 | 0.330 | 0.522 | 0.793 | 0.760 | 0.786 | 0.849 | 0.777 |
| GSS-R | | 0.829 | 1.143 | 1.615 | 0.337 | 0.566 | 0.793 | 0.763 | 0.797 | 0.838 | 0.779 |
| NM | | 0.290 | 0.094 | 1.306 | 0.603 | 0.328 | 0.680 | 1.331 | 0.771 | 0.343 | 0.973 |
| NM-R | | 0.302 | 0.092 | 1.284 | 0.717 | 0.302 | 0.686 | 1.348 | 0.761 | 0.315 | 0.966 |
| PSO | −0.5/2 | 0.398 | 0.300 | 1.285 | 0.119 | 0.234 | 0.938 | 0.738 | 0.912 | 0.409 | 0.002 |
| | 1/2 | 1.842 | 0.298 | 2.284 | 13.307 | 3.651 | 1.321 | 1.366 | 0.666 | 1.828 | 0.000 |
| REA-P | R1 | 0.998 | 0.995 | 1.765 | 0.870 | 0.896 | 1.018 | 0.995 | 1.094 | 0.973 | 1.127 |
| | R2 | 0.905 | 0.993 | 1.772 | 0.833 | 0.900 | 1.027 | 0.993 | 1.089 | 0.975 | 1.128 |
| REA-T | R1 | 0.394 | 0.849 | 1.436 | 0.051 | 0.199 | 0.892 | 0.726 | 1.092 | 0.534 | 0.936 |
| | R2 | 0.943 | 0.984 | 1.740 | 0.936 | 0.950 | 1.020 | 0.978 | 1.100 | 0.960 | 1.117 |
| | R3 | 0.928 | 0.984 | 1.740 | 0.918 | 0.916 | 1.012 | 0.983 | 1.091 | 0.945 | 1.117 |
| SA | | 0.035 | 1.005 | 1.414 | 0.190 | 0.002 | 0.893 | 1.073 | 1.081 | 0.019 | 1.058 |
| rBOA | | 0.830 | 0.402 | 0.554 | 1.021 | 0.893 | 1.068 | 0.219 | 1.178 | 1.032 | 1.169 |
| rGA | | 0.673 | 0.874 | 1.446 | 0.541 | 0.600 | 0.813 | 0.894 | 0.836 | 0.671 | 0.976 |

**Table A.50** Variance for performance criterion $\phi_2$ (scaled) in 25 dimensions.

| $Var(\phi_2)$ (scaled) | | sphr. | ack. | lg-ack. | whit. | rosen. | rastr. | sal. | schw. | grie. | weier. |
|---|---|---|---|---|---|---|---|---|---|---|---|
| CG | | 0.000 | 0.098 | 0.551 | 0.000 | 0.000 | 0.195 | 0.298 | 0.000 | – | 0.124 |
| CG-R | | 0.000 | 0.106 | 0.519 | – | – | 0.262 | 0.394 | 0.000 | – | ∞ |
| CMA-ES | 100 | 0.022 | 0.024 | 0.057 | 0.000 | 0.009 | 0.046 | 0.044 | 0.043 | 0.020 | ∞ |
| | 750 | 0.080 | 0.037 | 0.081 | 0.107 | 0.114 | 0.061 | 0.057 | 0.041 | 0.073 | ∞ |
| | 1250 | 0.079 | 0.034 | 0.077 | 0.102 | 0.102 | 0.050 | 0.044 | 0.038 | 0.074 | ∞ |
| | 2500 | 0.064 | 0.031 | 0.073 | 0.045 | 0.066 | 0.046 | 0.047 | 0.039 | 0.072 | ∞ |
| CMA-ES-R | 100 | 0.000 | 0.000 | 0.000 | 0.000 | 0.000 | 0.000 | 0.000 | 0.000 | 0.000 | 0.000 |
| | 750 | 0.000 | 0.000 | 0.000 | 0.000 | 0.000 | 0.000 | 0.000 | 0.000 | 0.000 | 0.000 |
| | 1250 | 0.000 | 0.000 | 0.000 | 0.000 | 0.000 | 0.000 | 0.000 | 0.000 | 0.000 | 0.000 |
| | 2500 | 0.000 | 0.000 | 0.000 | 0.000 | 0.000 | 0.000 | 0.000 | 0.000 | 0.000 | 0.000 |
| DE | .2/.9 | 0.124 | 0.014 | 0.115 | 0.489 | 0.251 | 0.069 | 0.066 | 0.055 | 0.230 | 0.058 |
| | .2/.2 | 0.123 | 0.015 | 0.110 | 0.375 | 0.226 | 0.070 | 0.063 | 0.058 | 0.153 | 0.050 |
| | .9/.2 | 0.111 | 0.015 | 0.111 | 0.148 | 0.150 | 0.056 | 0.065 | 0.045 | 0.199 | 0.045 |
| | .9/.9 | 0.105 | 0.015 | 0.094 | 0.245 | 0.164 | 0.049 | 0.052 | 0.047 | ∞ | ∞ |
| GSS | | 0.154 | 0.088 | 0.150 | 0.260 | 0.210 | 0.108 | 0.104 | 0.133 | 0.165 | 0.088 |
| GSS-R | | 0.161 | 0.084 | 0.140 | 0.260 | 0.235 | 0.123 | 0.101 | 0.135 | 0.151 | 0.083 |
| NM | | 0.230 | 0.025 | 0.414 | 1.220 | 0.518 | 0.158 | 0.145 | 0.099 | 0.244 | 0.261 |
| NM-R | | 0.212 | 0.025 | 0.406 | 1.349 | 0.381 | 0.165 | 0.148 | 0.095 | 0.223 | 0.258 |
| PSO | –0.5/2 | 0.151 | 0.014 | 0.170 | 0.201 | 0.141 | 0.090 | 0.111 | 0.096 | 0.152 | 0.007 |
| | 1/2 | 0.256 | 0.014 | 0.146 | 4.109 | 0.720 | 0.080 | 0.085 | 0.135 | 0.220 | 0.000 |
| REA-P | R1 | 0.099 | 0.032 | 0.085 | 0.366 | 0.203 | 0.055 | 0.052 | 0.043 | 0.108 | 0.041 |
| | R2 | 0.098 | 0.037 | 0.077 | 0.320 | 0.185 | 0.053 | 0.056 | 0.048 | 0.104 | 0.044 |
| REA-T | R1 | 0.057 | 0.030 | 0.075 | 0.033 | 0.057 | 0.052 | 0.055 | 0.043 | 0.065 | 0.046 |
| | R2 | 0.104 | 0.033 | 0.086 | 0.352 | 0.190 | 0.057 | 0.056 | 0.043 | 0.106 | 0.042 |
| | R3 | 0.094 | 0.037 | 0.079 | 0.323 | 0.198 | 0.056 | 0.055 | 0.045 | 0.109 | 0.039 |
| SA | | 0.010 | 0.044 | 0.125 | 0.254 | 0.001 | 0.119 | 0.076 | 0.043 | 0.003 | 0.075 |
| rBOA | | 0.221 | 0.051 | 0.101 | 0.712 | 0.370 | 0.100 | 0.031 | 0.063 | 0.227 | 0.048 |
| rGA | | 0.107 | 0.042 | 0.113 | 0.272 | 0.164 | 0.066 | 0.071 | 0.055 | 0.105 | 0.040 |

**Table A.51** Results of neural network experiments using the performance criteria of Chapter 11: success probability ($\sigma_\varepsilon^N$), hitting time on success ($\hat{\psi}_\varepsilon^N$), final error ($\zeta_T$), average error ($\phi_1$), and weighted average error ($\phi_2$). The first standard deviation is provided where possible; the value for $\sigma_\varepsilon^N$ is accurate within $\pm 0.005$ with $p < 0.05$. The error threshold $\varepsilon$ was chosen separately for each task and is given in the table. For Currency Trading, raw fitness values are shown instead of errors. Neuroannealing outperforms NEAT on the Multiplexers and Concentric Spirals.

Neuroannealing

| Task | $\varepsilon <$ | $\sigma_\varepsilon^N$ | $\hat{\psi}_\varepsilon^N$ | $\zeta_T$ | $\phi_1$ | $\phi_2$ |
|---|---|---|---|---|---|---|
| Double Pole (Markov) | 1.0 | 0.845 | 7,767 ± 4,871 | 0.154 ± 0.360 | 0.203 ± 0.342 | 0.991 ± 0.071 |
| Double Pole (non-Markov) | 1.0 | 0.960 | 7,499 ± 3,157 | 0.039 ± 0.195 | 0.163 ± 0.181 | 0.998 ± 0.006 |
| Multiplexer, 1 × 2 | 0.010 | 0.964 | 11,278 ± 3,781 | 0.004 ± 0.023 | 0.030 ± 0.020 | 0.208 ± 0.045 |
| Multiplexer, 2 × 4 | 0.200 | 0.047 | 19,833 ± 10,351 | 0.247 ± 0.037 | 0.252 ± 0.025 | 0.329 ± 0.023 |
| Multiplexer, 3 × 5 | 0.250 | 0.028 | 20,566 ± 15,509 | 0.285 ± 0.013 | 0.287 ± 0.013 | 0.363 ± 0.017 |
| Multiplexer, 3 × 6 | 0.300 | 0.036 | 17,675 ± 12,449 | 0.305 ± 0.013 | 0.308 ± 0.011 | 0.385 ± 0.012 |
| Concentric Spirals | 0.300 | 0.261 | 21,687 ± 7,834 | 0.310 ± 0.021 | 0.317 ± 0.014 | 0.333 ± 0.001 |
| Currency Trading | $f > 250$ | 0.749 | 20,054 ± 11,189 | 31016.331 ± 55094.212 | 10904.020 ± 20980.836 | 0.930 ± 0.284 |

NEAT

| Task | $\varepsilon <$ | $\sigma_\varepsilon^N$ | $\hat{\psi}_\varepsilon^N$ | $\zeta_T$ | $\phi_1$ | $\phi_2$ |
|---|---|---|---|---|---|---|
| Double Pole (Markov) | 1.0 | 1.000 | 1,819 ± 2,276 | 0.000 ± 0.000 | 0.018 ± 0.044 | 0.566 ± 0.472 |
| Double Pole (non-Markov) | 1.0 | 1.000 | 4,676 ± 2,107 | 0.000 ± 0.000 | 0.012 ± 0.012 | 0.742 ± 0.415 |
| Multiplexer, 1 × 2 | 0.010 | 0.000 | 50,000 ± 0,000 | 0.166 ± 0.027 | 0.180 ± 0.012 | 0.187 ± 0.000 |
| Multiplexer, 2 × 4 | 0.200 | 0.000 | 50,000 ± 0,000 | 0.279 ± 0.001 | 0.282 ± 0.001 | 0.300 ± 0.008 |
| Multiplexer, 3 × 5 | 0.250 | 0.000 | 50,000 ± 0,000 | 0.322 ± 0.001 | 0.325 ± 0.001 | 0.340 ± 0.007 |
| Multiplexer, 3 × 6 | 0.300 | 0.000 | 50,000 ± 0,000 | 0.348 ± 0.003 | 0.351 ± 0.002 | 0.367 ± 0.005 |
| Concentric Spirals | 0.300 | 0.000 | 50,000 ± 0,000 | 0.331 ± 0.000 | 0.331 ± 0.000 | 0.332 ± 0.000 |
| Currency Trading | $f > 250$ | 0.028 | 29,425 ± 8,095 | 43.365 ± 103.898 | 15.910 ± 48.626 | 0.033 ± 0.105 |

# Appendix B
# Automated Currency Exchange Trading

This appendix describes the automated currency exchange trading task for the experiments in Chapter 16, including the relevant background, datasets, and input preprocessing.

## B.1 Currency Trading Background

The exchange rates of currencies that are freely floated present a stochastic sequence that is difficult to predict. Speculators attempt to make money by buying and selling currencies at high leverage. Most of these forex traders lose money, but a few consistently average a profit, suggesting that there is some structure to the problem that can be learned. The advantage to the domain is that there are relatively few inputs and outputs, allowing for fast prototyping and exploration. The high stochasticity of the signal is challenging, especially in light of the relevance of exogenous information, such as Central Bank announcements that cannot easily be modeled in a numeric setting.

Trading currencies bears substantial similarities to trading stocks in terms of market function and techniques. Most speculative trading takes place through brokers on exchanges that are separate from retail currency exchange markets; speculation is generally regarded as playing the role of price discovery for other markets. Speculative traders primarily trade with each other.

The focus of this study is short-term trading using Technical Analysis [89, 115, 130]. Technical Analysis is an approach to trading that is characterized by the mechanistic use of a set of statistics termed *technical indicators*. These technical indicators can be computed for any price sequence but are considered more effective or meaningful in the context of markets with high liquidity and large trading volumes, two features that are preeminently characteristic of foreign exchange (or *forex*) markets.

Price in technical trading is typically broken up into segments, the size of which depends on the desired frequency of trading. Within each period, four prices are



A. J. Lockett, *General-Purpose Optimization Through Information Maximization*,
Natural Computing Series, https://doi.org/10.1007/978-3-662-62007-6

recorded, specifically the high, low, opening and closing price. The closing val ues are used to compute technical indicators. Typical technical indicators includ various averages, Stochastics, Relative Strength Index (RSI), Bollinger Bands, an Fibonacci levels [89]. The Simple Moving Average (SMA) is an unweighted average over a fixed number of periods starting from the current period. The Exponential Moving Average (EMA) is defined by a discrete update rule, $\text{EMA}(t+1) = \alpha\text{EMA}(t) + (1-\alpha)\text{Close}(t)$, where $0 \leq \alpha \leq 1$; the EMA is said to be taken over $N$ periods where $N = \frac{2}{\alpha} - 1$. A common use of moving averages is to identify trends by examining the ratio between two moving averages computed from different periods, with the average over a shorter period on top [130]. If the ratio is greater than 1, then the price is concluded to be in an uptrend since the older data is less than newer data on average. If the ratio is less than 1, then the price is in a downtrend.

Stochastics (Stoch) and Relative Strength Index (RSI) are oscillators that vary between 0 and 100. The Stochastic is given by

$$\text{Stoch}(t) = 100\frac{\text{Close}(t) - \text{Low}(t)}{\text{High}(t) - \text{Low}(t)}$$

where the high, low and close are taken over all periods involved, e.g. a three period stochastic takes the high and low as the extrema of the prior 3 periods and the close from the last period [115]. The Stochastic encodes whether the current price (the most recent close) is high or low relative to recent history. Because the stochastic focuses on recent history, it tends to change faster than EMA ratios. It is also possible to compute a Stochastic from the EMA instead of the price; this is called the Slow Stochastic. As with moving averages, ratios of fast and slow Stochastics as well as ratios between Stochastics of different periods are used as trend change indicators. Stochastics also tend to exhibit a behavior termed *divergence* near price peaks. *Negative divergence* occurs when the Stochastic makes a peak above 80 and then makes a secondary peak at a lower value while the price makes a new high; this behavior often indicates a coming shift in trend to a bear market and can be used as a trading signal [89]. *Positive divergence* refers to the reverse situation with a trough below 20, and indicates a change to a bull market. RSI is defined differently, but tends to exhibit similar behavior.

Bollinger bands are a pair of lines placed around the price above and below at a distance from the price on either side equal to the standard deviation of the price from its average over a fixed number of preceding periods [130]. Prices can tend to reverse after penetrating the Bollinger bands on either side. Additionally, when the Bollinger bands contract due to a reduction in price variation, it generally indicates a period of volatility to follow.

A currency trader makes decisions consisting of orders to buy or sell currencies. Currencies are expressed as pairs, e.g. EUR/USD. For EUR/USD, the euro is the *base currency*, and the US dollar is the *counter currency*. An order has a type and an associated value. The order type can be BUY, SELL, SELL SHORT or BUY TO COVER. In a BUY order, the counter currency is traded for the base currency in the amount specified; once executed, the trader is said to have entered a long position. A SELL order exits all or a portion of a long positions depending on the value. A SELL SHORT order borrows the base currency in the amount specified in order to buy the counter currency, thereby entering a short position. A BUY TO COVER order exits all or part of an existing short position, repaying the borrowed money. The value of an order is expressed in *lots*, denominated in US dollars for the purpose of this research. A standard lot is US$100,000; a *mini-lot* is worth US$ 10,000. When a trade is entered where neither currency is USD, the actual currency amount purchased or sold is determined based on the current exchange rate with the US dollar.

Currency traders are allowed to purchase an amount up to 100 times the current value of the trader's account. The actual multiple of the account value currently held in long or short positions is termed *leverage*. After trades are entered, then the trader is allowed to hold the position until the account value dwindles to the *margin*, usually at 200 times the value of the trader's account. At that point, the positions are liquidated by the brokerage; this is termed a *margin call*. As long as the value of the account remains above the margin, the trader has discretion as to when the trade should end.

Orders can be executed by several means. A *market order* queues up for execution at the current exchange rate, which may fluctuate prior to execution, since trades must execute in order. A *limit order* specifies a maximum exchange rate for long trades, and will only execute if the price falls to or below the maximum; for short trades, the limit specifies a minimum rather than a maximum. Limit orders are given preference to market orders for execution. Finally, a *stop loss order*, or simply *stop* is the mirror image of a limit order; it specifies a maximum exchange rate for executing a short trade, or a minimum exchange rate for executing a long trade. Stops are used to protect against catastrophic loss on a trade. Several other order types are available that will not be used in this research.

## B.2 Currency Trading Experiments

Automated currency trading was used to verify the application of evolutionary annealing to RNNs experimentally. The task of currency trading was described in Section B.1 along with applicable technical indicators. It is an advantage of the currency trading domain that it involves relatively few inputs and outputs while still providing a difficult task where success is subject to uncertainty. These aspects make the currency domain a solid testbed for experimental verification of neuroannealing. This section describes how neural network controllers can be evaluated for the currency trading task.

### B.2.1 Experimental Setup

The task of an automated currency trader is to progressively read a sequence of technical indicators and output trading decisions. The sequence of technical indicators for this experiment consists of ten real-valued inputs derived from the exponential moving average at five, 20, and 50 periods (EMA-5, EMA-20, EMA-50), the relative strength index at 14 periods (RSI), the fast and slow stochastics at 14 and three periods respectively (FSTOCH and SSTOCH), the width of the Bollinger Bands (BB-W), the position of the closing price within the Bollinger Bands (BB-P), the absolute difference between the opening and closing price (OC), and the difference between the high and low price (HL). All of these indicators are commonly used and definitions can be readily found in any materials on the subject of Technical Analysis [89, 115, 130].

In order to train currency traders on multiple currency pairs, it is necessary to remove any reference to the absolute magnitude of the price. Three inputs are used for the ratio of price to EMA-5, EMA-20, and EMA-50, respectively. One input each is used for RSI, FSTOCH, and SSTOCH, since these indicators are already independent of price. Two more inputs are used for BB-P and the log ratio of BB-W to its exponential moving average. The final two inputs include the log ratio of OC and HL to their respective moving averages. Inputs are centered to have both positive and negative values and scaled to approximately the same order of magnitude.

For these experiment, several simplifying assumptions regarding trading are made. During each training and testing run, a trader trades a single currency pair at a time. Whenever the trader has no position in the currency, then the trader can issue BUY or SHORT decisions to enter a trade in a long or short position respectively; alternately, the trader may WAIT and do nothing. Leverage is determined by the strength of the BUY signal versus the SHORT signal or vice versa. Once a position is entered, the trader may either EXIT the position or HOLD it at each time step; the trader may also increase or decrease the leverage on the position. Positions are entered with a limit order fixed to the close of the prior period. In the simulation, all of these limit orders succeed, which is realistic if the limit order is issued substantially close in time to the closing price. Stop loss limits are not used in these simulations to simplify the problem. In a practical implementation, a large stop could be entered to prevent catastrophic loss, representing the portion of the account value to be placed at risk during the next hour. These assumptions are intended to give structure to the experiment and do not significantly restrict the generality of the task. However, these choices do exclude some trading strategies, including arbitrage trades, where a sequence of trades involving at least three currencies exploits short-term imbalances among currencies, and staged entry approaches where the position size is increased at set points if the trade proceeds as expected.

The goal of currency trading is to maximize gain during a trading run. A secondary goal is to simultaneously minimize risk, typically volatility. Thus a trader that steadily increases its account value without substantial losses during the run is preferable over a trader that increases the account value very quickly, but at the cost of substantial volatility and drawdown. In the current experiments, the objecive value of a trader was measured solely based on the final account value without taking risk into account specifically. Future implementations may explore a multi-objective setting in which gain must be maximized while minimizing risk over any period.

Training takes place in a simulated trading environment using a fixed data set described below. Since the data set consists only of hourly trading data, it is not possible to tell during simulation whether a limit or a stop would have been executed if both prices were reached in the same trading period. For the purpose of simulation, limits were always executed and stops were not used. This decision has the effect of potentially inflating gains, but as such it affects all the trading networks evenly. As mentioned above, the assumption that limits execute is realistic if the network makes decisions in real time at the close of each period.

### *B.2.2 Currency Exchange Rate Data Set*

The available training data consists of six months of hourly trading data from September 2009 to February 2010 on 18 separate currency pairs, obtained from a commercial brokerage. These include the six forex majors, EUR/USD, AUD/USD, USD/CAD, GBP/USD, USD/JPY, USD/CHF, as well as twelve other currency crosses including EUR/JPY, EUR/GBP, EUR/AUD, GBP/JPY, GBP/CHF, CHF/JPY, CAD/JPY, AUD/JPY, NZD/USD, AUD/NZD, AUD/CAD, and AUD/CHF.

The neural networks in the experiments are tested on a subset of eight of these trading sets: EUR/USD, GBP/USD, USD/CHF, USD/JPY, NZD/USD, USD/CAD, AUD/CAD, and AUD/NZD. These pairs are interesting since all but two of them include the US dollar. The further simplifying assumption is made that the trading account is denominated in the counter currency. In a practical implementation, the trading account would be denominated in a single currency, likely in US dollars.

The use of a dataset from a single time period incurs a risk that the dataset contains internal correlations that would not be reflected during other time periods. This problem is mitigated somewhat by including a variety of currencies from countries with substantially different economies, and by the inclusion of two pairs without the US dollar. Also, since trading is performed at an hourly scale, the use of six months of data (over $3,000$ hours) means that a large variety of trading situations are encountered.

There is also a risk that an automated trader trained on a particular dataset will overfit the data, learning a trading strategy that only works on this particular dataset. In some sense, this risk exists no matter what data is used. The presumption that the past is predictive of the future underlies all forms of learning. In this particular case, the use of fixed technical indicators should hide any specific price cues that are not also of general use as trading signals. Hopefully, simply memorizing the best output for each inputs should be a dangerous strategy given $27,000$ hours of trading on distinct currency pairs. It is unknown at this time whether memorizing the data is a viable strategy. Even if so, in the current context, this experiment is simply treated as an objective function to maximize, which is of value for assessing the learnin abilities of neuroannealing in any case.

# References

1. Ackley, D.H.: A connectionist machine for genetic hillclimbing. Kluwer Academic Publishers, Norwell, MA, USA (1987)
2. Ahn, C., Ramakrishna, R., Goldberg, D.: Real-coded Bayesian optimization algorithm. In: J. Lozano, P. Larrañaga, I. Inza, E. Bengoetxea (eds.) Towards a New Evolutionary Computation, *Studies in Fuzziness and Soft Computing*, vol. 192, pp. 51–73. Springer Berlin / Heidelberg (2006)
3. Akimoto, Y., Nagata, Y., Ono, I., Kobayashi, S.: Bidirectional relation between CMA evolution strategies and natural evolution strategies. Parallel Problem Solving from Nature, PPSN XI (2010)
4. Alabert, A., Berti, A., Caballero, R., Ferrante, M.: No-free-lunch theorems in the continuum. Theoretical Computer Science **600**, 98 – 106 (2015). DOI https://doi.org/10.1016/j.tcs.2015.07.029. URL http://www.sciencedirect.com/science/article/pii/S0304397515006635
5. Alden, M.: Marleda: Effective distribution estimation through Markov random fields. Ph.D. thesis, Department of Computer Sciences, the University of Texas at Austin, Austin, Texas (2007). URL http://nn.cs.utexas.edu/?alden:phd07. Also Technical Report AI07-349
6. Ali, M.M., Khompatraporn, C., Zabinsky, Z.B.: A numerical evaluation of several stochastic algorithms on selected continuous global optimization test problems. Journal of Global Optimization **31**, 635–672 (2005). URL http://dx.doi.org/10.1007/s10898-004-9972-2. 10.1007/s10898-004-9972-2
7. Aliprantis, C.C., Border, K.D.: Infinite Dimensional Analysis: A Hitchhiker's Guide, 3rd Edition. Springer, New York (2006)
8. Armijo, L.: Minimization of functions having Lipschitz continuous first partial derivatives. Pacific J. Math. **16** (1966)
9. Ashlock, D.: Taxonomic clustering of genetic algorithms using unique performance signatures (2011). Private communication
10. Audet, C., Dennis, J.: Mesh adaptive direct search algorithms for constrained optimization. SIAM J. Optim. **17** (2006)
11. Auger, A., Hansen, N.: A restart CMA evolution strategy with increasing population size. In: Evolutionary Computation, 2005. The 2005 IEEE Congress on (2005)
12. Auger, A., Teytaud, O.: Continuous lunches are free! In: Proceedings of the 9th Annual Conference on Genetic and Evolutionary Computation (GECCO-2007). ACM Press, New York (2007)
13. Auger, A., Teytaud, O.: Continuous lunches are free plus the design of optimal optimization algorithms. Algorithmica **57**(1), 121–146 (2010). DOI 10.1007/s00453-008-9244-5. URL https://doi.org/10.1007/s00453-008-9244-5


A. J. Lockett, *General-Purpose Optimization Through Information Maximization*,
Natural Computing Series, https://doi.org/10.1007/978-3-662-62007-6

14. Bagley, J.D.: The behavior of adaptive systems which employ genetic and correlation algorithms. Ph.D. thesis, University of Michigan, Ann Arbor (1967)
15. Baker, J.E.: Adaptive selection methods for genetic algorithms. In: Proceedings of an International Conference on Genetic Algorithms and their Applications (1985)
16. Baluja, S.: Population-based incremental learning: A method for integrating genetic search based function optimization and competitive learning. Tech. Rep. CMU-CS-94-163, Carnegie Mellon University (1994)
17. Baluja, S., Caruana, R.: Removing the Genetics from the Standard Genetic Algorithm. Morgan Kaufmann (1995)
18. Barabanov, N.E., Prokhorov, D.V.: Stability analysis of discrete-time recurrent neural networks. IEEE Transactions on Neural Networks **13**(2) (2002)
19. Barricelli, N.A.: Symbiogenetic evolution processes realized by artificial methods. Methodos **IX**(35–36) (1957)
20. Bellman, R.: Dynamic Programming, 1 edn. Princeton University Press, Princeton, NJ, USA (1957)
21. Berberian, S.K.: Lectures in Functional Analysis and Operator Theory. Springer-Verlag, New York, New York (1974)
22. Bersini, H., Dorigo, M., Langerman, S., Seront, G., Gambardella, L.M.: Results of the first international contest on evolutionary optimisation (1st ICEO). In: Proceedings of the IEEE International Conference on Evolutionary Computation (1996)
23. Bersini, H., Varela, F.: Hints for adaptive problem solving gleaned from immune networks. In: Parallel Problem Solving from Nature, First Workshop. Dortmund, Germany (1990)
24. Bertsimas, D., Tsitsiklis, J.: Simulated annealing. Statistical Science **8**(1) (1993)
25. Beyer, H.G.: Theory of Evolution Strategies. Springer-Verlag, Berlin, Germany (2001)
26. Billingsley, P.: Probability and Measure. John Wiley (1986)
27. Bremerman, H.J.: Optimization through Evolution and Recombination. Spartan Books, Washington, D.C (1958)
28. Brindle, A.: Genetic algorithms for function optimization. Ph.D. thesis, University of Alberta, Edmonton (1981)
29. Bryden, K.M., Ashlock, D.A., Corns, S., Willson, S.: Graph-based evolutionary algorithms. IEEE Transactions on Evolutionary Computation **10**(5) (2006)
30. Cabessa, J., Siegelman, H.T.: Evolving recurrent neural networks are super-Turing. In: Proceedings of the International Joint Conference on Neural Networks (2011)
31. Chaitin, G.J.: A theory of program size formally identical to information theory. Journal of Computing Machinery **22**(3) (1975)
32. Chung, K.L., Williams, R.: Introduction to Stochastic Integration. Birkhäuser, Boston, MA (1990)
33. Cohn, D.: Measure Theory. Birkhäuser, Boston, MA (1980)
34. Colombetti, M., Dorigo, M.: Learning to control an autonomous robot by distribution genetic algorithms. In: From Animals to Animats 2: Proceedings of the 2nd International Conference on Simulation of Adaptive Behavior. MIT Press (1992)
35. Coope, I.D., Price, C.J.: On the convergence of grid-based methods for unconstrained optimization. SIAM J. Optim. **11** (2001)
36. Cover, T.M.: Universal gambling schemes and the complexity measures of Kolmogorov and Chaitin. Tech. Rep. Rep. 12, Statistics Dept., Stanford University (1974)
37. Culberson, J.C.: On the futility of blind search: an algorithmic view of "no free lunch". Evolutionary Computation **6**(2) (1998)
38. Cybenko, G.: Approximation by superpositions of sigmoidal function. Mathematics of Control, Signals, and Systems **2** (1989)
39. Das, S., Konar, A., Chakraborty, U.: Annealed differential evolution. In: Evolutionary Computation, 2007. CEC 2007. IEEE Congress on, pp. 1926 –1933 (2007). DOI 10.1109/CEC.2007.4424709
40. Doerr, B., Winzen, C.: Towards a complexity theory of randomized search heuristics: Ranking-based black-box complexity. CoRR **abs/1102.1140** (2011)

41. Dorigo, M.: Optimization, learning and natural algorithms. Ph.D. thesis, Politecnico di Milano (1992)
42. Droste, S., Jansen, T., Wegener, I.: Optimization with randomized search heuristics – the (A)NFL theorem, realistic scenarios, and difficult functions. Theoretical Computer Science **287**, 2002 (1997)
43. Droste, S., Jansen, T., Wegener, I.: Perhaps not a free lunch but at least a free appetizer. Tech. Rep. No. CI-45/98, University of Dortmund (1998)
44. Droste, S., Jansen, T., Wegener, I.: Upper and lower bounds for randomized search heuristics in black-box optimization. Theoretical Computer Science **39** (2006)
45. Eberhart, R.C., Kennedy, J.: A new optimizer using particle swarm theory. In: Proceedings of the Sixth International Symposium on Micromachine and Human Science. Nagoya, Japan (1995)
46. English, T.: Evaluation of evolutionary and genetic optimizers: No free lunch. In: Proceedings of the Fifth Annual Conference on Evolutionary Programming (EP V), pp. 163–169 (1996)
47. English, T.: Optimization is easy and learning is hard in the typical function. In: Evolutionary Computation, 2000. Proceedings of the 2000 Congress on, vol. 2, pp. 924–931 (2000)
48. Farmer, J., Packard, N., Perelson, A.: The immune system, adaptation and machine learning. Physica D **2** (1986)
49. de Fermat, P.: Methodus ad disquirendam maximam et minimam (1638)
50. Fogel, D.B.: Evolving artificial intelligence. Ph.D. thesis, University of California at San Diego (1992)
51. Fogel, D.B.: Evolutionary Computation: Toward a New Philosophy of Machine Intelligence. IEEE Press, Hoboken, New Jersey (2006)
52. Fogel, D.B., Fogel, L.J., Porto, V.W.: Evolving neural networks. Biological Cybernetics **63** (1990)
53. Fogel, L.J.: Autonomous automata. Industrial Research **4** (1962)
54. Fraser, A.S.: Simulation of genetic systems by automatic digital computers i: Introduction. Australian Journal of Biological Science **10** (1957)
55. Fraser, A.S.: Simulation of genetic systems by automatic digital computers ii: Effects of linkage on rates of advance under selection. Australian Journal of Biological Science **10** (1957)
56. Frazier, P.I., Powell, W.B., Dayanik, S.: A knowledge gradient policy for sequential information collection. SIAM Journal on Control and Optimization **47**(5), 2410–2439 (2008)
57. Friedberg, R.M.: A learning machine: Part i. IBM Journal of Research and Development **2** (1958)
58. Genovese, C., Wasserman, L.: Rates of convergence for the Gaussian mixture sieve. Annals of Statistics **28**(4) (2000)
59. Ghosh, S., Das, S., Vasilakos, A.V., Suresh, K.: On convergence of differential evolution over a class of continuous functions with unique global optimum. IEEE Transactions on Systems, Man, and Cybernetics, Part B (Cybernetics) **42**(1), 107–124 (2012)
60. Goldberg, D.E.: Genetic Algorithms in Search, Optimization and Machine Learning. Addison-Wesley Longman Publishing Co., Inc., Boston, MA, USA (1989)
61. Goldberg, D.E.: A note on Boltzmann tournament selection for genetic algorithms and population-oriented simulated annealing. Complex Systems **4** (1995)
62. Gomez, F.: Robust non-linear control through neuroevolution. Ph.D. thesis (2003)
63. Gomez, F., Miikkulainen, R.: Incremental evolution of complex general behavior. Adaptive Behavior **5** (1997)
64. Gomez, F., Schmidhuber, J., Miikkulainen, R.: Accelerated neural evolution through cooperatively coevolved synapses. Journal of Machine Learning Research (JMLR) **9** (2008)
65. Hajek, B.: Cooling schedules for optimal annealing. Mathematics of Operation Research **13**(4) (1988)
66. Halmos, P.: Measure Theory. Springer-Verlag, New York, NY (1974)

67. Hansen, N., Ostermeier, A.: Adapting arbitrary normal mutation distributions in evolution strategies: The covariance matrix adaptation. In: Proceedings of the 1996 IEEE International Conference on Evolutionary Computation, pp. 312–317. Morgan Kaufmann (1996)
68. Hansen, N., Ostermeier, A.: Completely derandomized self-adaptation in evolution strategies. Evolutionary Computation **9**(2) (2001)
69. Harik, G.R., Lobo, F.G., Goldberg, D.E.: The compact genetic algorithm. In: Proceedings of the International Conference on Evolutionary Computation. Piscataway, NJ (1998)
70. Hastings, W.: Monte Carlo sampling methods using Markov chains and their applications. Biometrika **57**(1) (1970)
71. Haykin, S.: Neural Networks and Learning Machines, Third Edition. Prentice Hall (2008)
72. Heckerman, D., Geiger, D., Chickering, M.: Learning Bayesian networks: The combination of knowledge and statistical data. Machine Learning **20** (1995)
73. Hestenes, M., Stiefel, E.: Journal of the Bureau of National Standards **49**(6) (1952)
74. Holland, J.H.: Outline for a logical theory of adaptive systems. In: Information Processing in the Nervous System, Proceedings of the International Union of Physiological Sciences, vol. 3 (1962)
75. Holland, J.H.: Adaptation in Natural and Artificial Systems. University of Michigan Press, Ann Arbor, Michigan (1975)
76. Hooke, R., Jeeves, T.A.: Direct search solution of numerical and statistical problems. J. ACM **8** (1961)
77. Hutter, M.: The fastest and shortest algorithm for all well-defined problems. International Journal of the Foundations of Computer Science (2001)
78. Hutter, M.: Universal Artificial Intelligence: Sequential Decisions based on Algorithmic Probability. Springer, Berlin (2005)
79. Igel, C.: Neuroevolution for reinforcement learning using evolution strategies. In: Proceedings of the 2003 Congress on Evolutionary Computation (CEC 2003). IEEE Press, Piscataway, NJ (2003)
80. Igel, C.: No free lunch theorems: Limitations and perspectives of metaheuristics. In: Y. Borenstein, A. Moraglio (eds.) Theory and Principled Methods for the Design of Metaheuristics. Springer-Verlag (2014)
81. Igel, C., Toussaint, M.: A no-free-lunch theorem for non-uniform distributions of target functions. Journal of Mathematical Modelling and Algorithms **3**(4) (2004)
82. Jeong, I., Lee, J.: Adaptive simulated annealing genetic algorithm for system identification. Engineering Applications of Artificial Intelligence **9**(5), 523 – 532 (1996)
83. Jiang, P., Chen, Y.: Free lunches on the discrete Lipschitz class. Theoretical Computer Science **412**(17), 1614 – 1628 (2011). DOI http://dx.doi.org/10.1016/j.tcs.2010.12.028. URL `http://www.sciencedirect.com/science/article/pii/S030439751000719X`
84. de Jong, K.A.: An analysis of the behavior of a class of genetic adaptive systems. Ph.D. thesis, University of Michigan (1975)
85. de Jong, K.A., Spears, W.M., Gordon, D.F.: Using Markov chains to analyze GAFOs. Foundations of Genetic Algorithms 3 (1995)
86. Karatzas, I., Shreve, S.: Brownian Motion and Stochastic Calculus. Springer-Verlag, New York, NY (1991)
87. Karush, W.: Minima of functions of several variables with inequalities as side constraints. Ph.D. thesis, University of Chicago, Chicago, Illinois (1939)
88. Kennedy, J., Eberhart, R.C.: Particle swarm optimization. In: Proceedings of IEEE International Conference on Neural Networks. Piscataway, NJ (1995)
89. Kirkpatrick, C.D.: Technical Analysis: The Complete Resource for Financial Market Technicians
90. Kirkpatrick, S., Gelatt, C.D., Vecchi, M.P.: Optimization by simulated annealing. Science **220**(4598) (1983)
91. Kohl, N.: Learning in fractured problems for constructive neural network algorithms. Ph.D. thesis, University of Texas at Austin (2009)

92. Kolda, T.G., Lewis, R.M., Torczon, V.: Optimization by direct search: New perspectives on some classical and modern methods. SIAM Rev. **45** (2003)
93. Kolmogorov, A.N.: Grundbegriffe der Wahrscheinlichkeitsrechnung (1933)
94. Kolmogorov, A.N.: Foundations of the theory of Probability (English Translation). Chelsea Publishing Company (1956)
95. Kuhn, H.W., Tucker, A.W.: Nonlinear programming. In: Proceedings of 2nd Berkeley Symposium. University of California Press, Berkeley, CA (1951)
96. Kvasnicka, V., Pelikan, M., Popischal, J.: Hill climbing with learning (an abstraction of the genetic algorithm). Neural Network World **6** (1996)
97. Lagrange, J.: Méchanique Analytique (1788)
98. Lattimore, T., Hutter, M.: No Free Lunch versus Occam's Razor in Supervised Learning, *Lecture Notes in Computer Science*, vol. 7070. Springer, Berlin, Heidelberg (2013)
99. Lauritzen, S.L.: Graphical Models. Oxford University Press, New York (1996)
100. Lehman, J., Stanley, K.O.: Abandoning objectives: Evolution through the search for novelty alone. Evolutionary Computation **19**(2) (2011)
101. Lehre, P.K., Witt, C.: Black-box search by unbiased variation. Algorithmica **64**(4) (2012)
102. Lillicrap, T.P., Hunt, J.J., Pritzel, A., Heess, N., Erez, T., Tassa, Y., Silver, D., Wierstra, D.: Continuous control with deep reinforcement learning. CoRR **abs/1509.02971** (2015). URL `http://arxiv.org/abs/1509.02971`
103. Lipster, Shiryaev: Statistics of Random Processes, Second Edition. Springer-Verlag, New York (2001)
104. Lockett, A., Chen, C., Miikkulainen, R.: Evolving explicit opponent models in game playing. In: Proceedings of the Genetic and Evolutionary Computation Conference. Morgan Kaufmann, San Francisco, California (2007)
105. Lockett, A., Miikkulainen, R.: Temporal convolution machines for sequence learning. Tech. rep. (2009). URL `http://nn.cs.utexas.edu/?lockett:aitr09-04`
106. Mařík, J.: The Baire and Borel measure. Czechoslovak Math. J. **7** (1957)
107. Metropolis, N., Rosenbluth, A., Rosenbluth, M., Teller, A., Teller, E.: Equations of state calculations by fast computing machines. Journal of Chemical Physics **21**(6) (1953)
108. Mnih, V., Kavukcuoglu, K., Silver, D., Graves, A., Antonoglou, I., Wierstra, D., Riedmiller, M.A.: Playing Atari with deep reinforcement learning. CoRR **abs/1312.5602** (2013). URL `http://arxiv.org/abs/1312.5602`
109. Moriarty, D.E.: Symbiotic evolution of neural networks in sequential decision tasks. Ph.D. thesis, University of Texas at Austin (1997)
110. Mühlenbein, H., Mahnig, T.: Mathematical analysis of evolutionary algorithms. In: Essays and Surveys in Metaheuristics, Operations Research/Computer Science Interface Series, pp. 525–556. Kluwer Academic Publisher (2002)
111. Mühlenbein, H., Mahnig, T., Rodriguez, A.O.: Schemata, distributions, and graphical models in evolutionary optimization. Journal of Heuristics **5** (1999)
112. Mühlenbein, H., Paass, G.: From recombination of genes to the estimation of distributions: Binary parameters. In: H.M. Voigt (ed.) Lecture Notes in Computer Science 1141: Parallel Problem Solving from Nature – PPSN IV. Springer, Berlin (1996)
113. Mühlenbein, H., Schomisch, M., Born, J.: The parallel genetic algorithm as function optimizer. Parallel Computing (1991)
114. Munkres, J.R.: Topology. Prentice Hall, Upper Saddle River, NJ (2000)
115. Murphy, J.J.: Technical Analysis of the Financial Markets: A Comprehensive Guide to Trading Methods and Applications. New York Institute of Finance (1999)
116. Nelder, J., Mead, R.: A simplex method for function minimization. Comput. J. **7** (1965)
117. von Neumann, J., Morgenstern, O.: Theory of Games and Economic Behavior. Princeton University Press (1944)
118. Nolfi, S., Elman, J.L., Parisi, D.: Learning and evolution in neural networks. Tech. Rep. Technical Report 9019, University of California at San Diego (1990)
119. Papakonstantinou, J.: The historical development of the secant method in 1-D. In: The Annual meeting of the Mathematical Association of America. San Jose, CA (2007)

120. Pearl, J.: Probabilistic Reasoning in Intelligent Systems: Networks of Plausible Inference. Morgan Kaufmann, San Francisco, California (1988)
121. Pedersen, M.: Tuning & simplifying heuristical optimization. Ph.D. thesis, University of Southampton (2010)
122. Pelikan, M., Goldberg, D., Lobo, F.: A survey of optimization by building and using probabilistic models. Computational Optimization and Applications **21** (2002)
123. Pelikan, M., Goldberg, D.E., Cantu-Paz, E.: BOA: The Bayesian optimization algorithm. In: Proceedings of the Genetic and Evolutionary Computation Conference (1999)
124. Pelikan, M., Goldberg, D.E., Cantu-Paz, E.: Hierarchical problem solving by the Bayesian optimization algorithm. Tech. Rep. IlliGAL Report No. 2000002, University of Illinois at Urbana-Champaign (2000)
125. Pelikan, M., Mülenbein, H.: The bivariate marginal distribution algorithm. In: Advances in Soft Computing – Engineering Design and Manufacturing. London (1999)
126. Plaut, D., Nowlan, S., Hinton, G.E.: Experiments on learning by back propagation. Tech. Rep. Technical Report CMU-CS-86-126, Carnegie Mellon University, Pittsburgh, PA (1986)
127. Potter, M.A., de Jong, K.A.: Cooperative coevolution: An architecture for evolving coadapted subcomponents. Evolutionary Computation **8**(1) (2000)
128. Powell, M.: Direct search algortihms for optimization calculations. Acta Numerica (1998)
129. Powell, W.B., Ryzhov, I.O.: Optimal Learning. John Wiley and Sons (2012)
130. Pring, M.: Technical Analysis Explained : The Successful Investor's Guide to Spotting Investment Trends and Turning Points
131. Radcliffe, N., Surry, P.D.: Fundamental limitations on search algorithms: Evolutionary computing in perspective. In: Lecture Notes in Computer Science 1000, pp. 275–291. Springer-Verlag (1995)
132. Raphson, J.: Analysis aequationum universalis (1690)
133. Rechenberg, I.: Cybernetic solution path of an experimental problem. Royal Airport Establishment, Library Translation No. 1122 (1965)
134. Rolet, P., Sebag, M., Teytaud, O.: Optimal robust expensive optimization is tractable. In: Gecco 2009, p. 8 pages. ACM, Montréal, Canada (2009). URL `http://hal.inria.fr/inria-00374910`
135. Rosenberg, R.: Simulation of genetic populations with biochemical properties. Ph.D. thesis, University of Michigan, Ann Arbor (1967)
136. Rowe, J.E., Vose, M.D., Wright, A.H.: Reinterpreting no free lunch. Evolutionary Computation **17**(1) (2009)
137. Rudolph, G.: Convergence analysis of canonical genetic algorithms. IEEE Transactions on Neural Networks **5**(1) (1994)
138. Ruiz, A., Owens, D.H., Townley, S.: Existence of limit cycles in recurrent neural networks. In: Industrial Electronics, 1996. ISIE '96., Proceedings of the IEEE International Symposium on, vol. 1, pp. 104 –108 (1996). DOI 10.1109/ISIE.1996.548400
139. Rumelhart, D., Hinton, G., Williams, R.: Learning internal representations by error propagation. Computational Models Of Cognition And Perception Series (1986)
140. Santamaria, J.C., Sutton, R.S., Ram, A.: Experiments with reinforcement learning in problems with continuous state and action spaces. Adaptive Behavior **6**(2) (1998)
141. Schaul, T., Sun, Y., Wierstra, D., Gomez, F., Schmidhuber, J.: Curiosity-Driven Optimization. In: IEEE Congress on Evolutionary Computation (CEC) (2011)
142. Schmidhuber, J.: The speed prior: A new simplicity measure yielding near-optimal computable predictions. In: J. Kivinen, R.H. Sloan (eds.) Proceedings of the 15th Annual Conference on Computational Learning Theory (COLT-2002) (2002)
143. Schumacher, C., Vose, M.D., Whitley, L.D.: The no free lunch and problem description length. In: Proceedings of the Genetic and Evolutionary Computation Conference (GECCO-2001, pp. 565–570. Morgan Kaufmann (2001)
144. Schumacher, C.W.: Black box search: framework and methods. Ph.D. thesis (2000). AAI9996384
145. Schwefel, H.P.: Kybernetische Evolution als Strategie der Experimentellen Forschung in der Störmungstechnik. Ph.D. thesis, Technical University of Berlin (1965)

146. Siegelmann, H.T.: Neural networks and analog computation: beyond the Turing limit. Birkhäuser Boston Inc., Cambridge, MA, USA (1999)
147. Silver, D., Huang, A., Maddison, C.J., Guez, A., Sifre, L., van den Driessche, G., Schrittwieser, J., Antonoglou, I., Panneershelvam, V., Lanctot, M., Dieleman, S., Grewe, D., Nham, J., Kalchbrenner, N., Sutskever, I., Lillicrap, T., Leach, M., Kavukcuoglu, K., Graepel, T., Hassabis, D.: Mastering the game of Go with deep neural networks and tree search. Nature **529**(7587), 484–489 (2016)
148. Silverthorn, B., Miikkulainen, R.: Latent class models for algorithm portfolio methods. In: Proceedings of the Twenty-Fourth AAAI Conference on Artificial Intelligence (2010). URL `http://nn.cs.utexas.edu/?silverthorn:aaai2010`
149. Solomonoff, R.J.: A formal theory of inductive inference. part i. Information and Control **7** (1964)
150. Solomonoff, R.J.: Complexity-based induction systems. IEEE Transactions on Information Theory **24**(5) (1978)
151. Spendley, W., Hext, G.R., Himsworth, F.R.: Sequential application of simplex designs in optimisation and evolutionary operation. Technometrics **4** (1962)
152. Stanley, K.O., Miikkulainen, R.: Efficient neural network learning through evolving neural network topologies. In: Proceedings of the Genetic and Evolutionary Computation Conference. Morgan Kaufmann, San Francisco, California (2002)
153. Stanley, K.O., Miikkulainen, R.: Evolving neural networks through augmenting topologies. Evolutionary Computation **10**(2) (2002)
154. Stanley, K.O., Miikkulainen, R.: Efficient evolution of neural networks through complexification. Ph.D. thesis (2004)
155. Storn, R., Price, K.: Differential evolution - a simple and efficient adaptive scheme for global optimization over continuous spaces. Journal of Global Optimization **11** (1997)
156. Sutton, R.S., Barto, A.G.: Reinforcement Learning: An Introduction. MIT Press, Cambridge, MA (1998)
157. Syswerda, G.: Uniform crossover in genetic algorithms. In: Proceedings of the Third International Conference on Genetic Algorithms (1989)
158. Torczon, V.: On the convergence of pattern search algorithms. SIAM Journal on Optimization **7**(1) (1997)
159. Valsalam, V.K., Hiller, J., MacCurdy, R., Lipson, H., Miikkulainen, R.: Constructing controllers for physical multilegged robots using the ENSO neuroevolution approach. Evolutionary Intelligence **5**(1), 1–12 (2012). URL `http://nn.cs.utexas.edu/?valsalam:evolint12`
160. Vasile, M., Minisci, E., Locatelli, M.: An inflationary differential evolution algorithm for space trajectory optimization. IEEE Transactions on Evolutionary Computation **15**(2) (2011)
161. Vose, M.: The Simple Genetic Algorithm. MIT Press, Cambridge, Massachusetts (1999)
162. Vose, M.D.: Random heuristic search. Theoretical Computer Science **229**, 103–142 (1999)
163. Wallis, J.: A Treatise of Algebra both Historical and Practical (1685)
164. Watkins, C.J., Dayan, P.: Q-learning. Machine Learning **8**(3) (1992)
165. Wegener, I.: On the expected runtime and the success probability of evolutionary algorithms (2000)
166. Whitley, L.D., Garrett, D., Watson, J.P.: Quad search and hybrid genetic algorithms. In: Proceedings of the Genetics and Evolutionary Computation Conference (GECCO-2003), vol. 2724. Springer, Chicago, IL, USA (2003)
167. Wieland, A.: Evolving neural network controllers for unstable systems. In: Proceedings of the International Joint Conference on Neural Networks, vol. 2. IEEE Press, Piscataway, New Jersey (1991)
168. Wierstra, D., Förster, A., Peters, J., Schmidhuber, J.: Recurrent policy gradients. Logic Journal of IGPL **18** (2010)
169. Wierstra, D., Schaul, T., Peters, J., Schmidhuber, J.: Natural evolution strategies. In: IEEE Congress on Evolutionary Computation (CEC) (2008)
170. Wolfe, P.: Convergence conditions for ascent methods. SIAM Rev. **11** (1969)

171. Wolpert, D.H., Macready, W.G.: No free lunch theorems for optimization. IEEE Transaction on Evolutionary Computation **1**(1) (1997)
172. Wright, M.: Direct search methods: Once scorned, now respectable. In: D. Griffiths, G. Wat son (eds.) Proceedings of the 1995 Dundee Biennial Conference in Numerical Analysis Addison Wesley Longman, Harlow, UK (1996)
173. Yamauchi, B., Beer, R.: Integration reactive, sequential, and learning behavior using dynamic neural networks. In: From Animals to Animats 3: Proceedings of the 3rd International Conference on Simulation of Adaptive Behavior. MIT Press (1994)
174. Yang, R.L.: Convergence of the simulated annealing algorithm for continuous global optimization. Journal of Optimization Theory and Applications **104**(3) (2000)

# Index




A. J. Lockett, *General-Purpose Optimization Through Information Maximization*, Natural Computing Series, https://doi.org/10.1007/978-3-662-62007-6

Printed in the United States
by Baker & Taylor Publisher Services